HANDBOOK OF CONSTRUCTION MANAGEMENT

Scope, Schedule, and Cost Control

Industrial Innovation Series

Series Editor

Adedeji B. Badiru

Air Force Institute of Technology (AFIT) – Dayton, Ohio

PUBLISHED TITLES

Carbon Footprint Analysis: Concepts, Methods, Implementation, and Case Studies, *Matthew John Franchetti & Defne Apul*

Cellular Manufacturing: Mitigating Risk and Uncertainty, *John X. Wang*

Communication for Continuous Improvement Projects, *Tina Agustiady*

Computational Economic Analysis for Engineering and Industry, *Adedeji B. Badiru & Olufemi A. Omitaomu*

Conveyors: Applications, Selection, and Integration, *Patrick M. McGuire*

Culture and Trust in Technology-Driven Organizations, *Frances Alston*

Design for Profitability: Guidelines to Cost Effectively Management the Development Process of Complex Products, *Salah Ahmed Mohamed Elmoselhy*

Global Engineering: Design, Decision Making, and Communication, *Carlos Acosta, V. Jorge Leon, Charles Conrad, & Cesar O. Malave*

Global Manufacturing Technology Transfer: Africa–USA Strategies, Adaptations, and Management, *Adedeji B. Badiru*

Guide to Environment Safety and Health Management: Developing, Implementing, and Maintaining a Continuous Improvement Program, *Frances Alston & Emily J. Millikin*

Handbook of Construction Management: Scope, Schedule, and Cost Control, *Abdul Razzak Rumane*

Handbook of Emergency Response: A Human Factors and Systems Engineering Approach, *Adedeji B. Badiru & LeeAnn Racz*

Handbook of Industrial Engineering Equations, Formulas, and Calculations, *Adedeji B. Badiru & Olufemi A. Omitaomu*

Handbook of Industrial and Systems Engineering, Second Edition, *Adedeji B. Badiru*

Handbook of Military Industrial Engineering, *Adedeji B. Badiru & Marlin U. Thomas*

Industrial Control Systems: Mathematical and Statistical Models and Techniques, *Adedeji B. Badiru, Oye Ibidapo-Obe, & Babatunde J. Ayeni*

Industrial Project Management: Concepts, Tools, and Techniques, *Adedeji B. Badiru, Abidemi Badiru, & Adetokunboh Badiru*

Inventory Management: Non-Classical Views, *Mohamad Y. Jaber*

Kansei Engineering — 2-volume set

- Innovations of Kansei Engineering, *Mitsuo Nagamachi & Anitawati Mohd Lokman*
- Kansei/Affective Engineering, *Mitsuo Nagamachi*

Kansei Innovation: Practical Design Applications for Product and Service Development, *Mitsuo Nagamachi & Anitawati Mohd Lokman*

Knowledge Discovery from Sensor Data, *Auroop R. Ganguly, João Gama, Olufemi A. Omitaomu, Mohamed Medhat Gaber, & Ranga Raju Vatsavai*

Learning Curves: Theory, Models, and Applications, *Mohamad Y. Jaber*

Managing Projects as Investments: Earned Value to Business Value, *Stephen A. Devaux*

Modern Construction: Lean Project Delivery and Integrated Practices, *Lincoln Harding Forbes & Syed M. Ahmed*

PUBLISHED TITLES

Moving from Project Management to Project Leadership: A Practical Guide to Leading Groups, *R. Camper Bull*

Project Management: Systems, Principles, and Applications, *Adedeji B. Badiru*

Project Management for the Oil and Gas Industry: A World System Approach, *Adedeji B. Badiru & Samuel O. Osisanya*

Project Management for Research: A Guide for Graduate Students, *Adedeji B. Badiru, Christina Rusnock, & Vhance V. Valencia*

Project Management Simplified: A Step-by-Step Process, *Barbara Karten*

Quality Management in Construction Projects, *Abdul Razzak Rumane*

Quality Tools for Managing Construction Projects, *Abdul Razzak Rumane*

A Six Sigma Approach to Sustainability: Continual Improvement for Social Responsibility, *Holly A. Duckworth & Andrea Hoffmeier*

Social Responsibility: Failure Mode Effects and Analysis, *Holly Alison Duckworth & Rosemond Ann Moore*

Statistical Techniques for Project Control, *Adedeji B. Badiru & Tina Agustiady*

STEP Project Management: Guide for Science, Technology, and Engineering Projects, *Adedeji B. Badiru*

Sustainability: Utilizing Lean Six Sigma Techniques, *Tina Agustiady & Adedeji B. Badiru*

Systems Thinking: Coping with 21st Century Problems, *John Turner Boardman & Brian J. Sauser*

Techonomics: The Theory of Industrial Evolution, *H. Lee Martin*

Total Productive Maintenance: Strategies and Implementation Guide, *Tina Agustiady & Elizabeth A. Cudney*

Total Project Control: A Practitioner's Guide to Managing Projects as Investments, Second Edition, *Stephen A. Devaux*

Triple C Model of Project Management: Communication, Cooperation, Coordination, *Adedeji B. Badiru*

HANDBOOK OF CONSTRUCTION MANAGEMENT

Scope, Schedule, and Cost Control

edited by
Abdul Razzak Rumane

CRC Press
Taylor & Francis Group
Boca Raton London New York

CRC Press is an imprint of the
Taylor & Francis Group, an **informa** business

CRC Press
Taylor & Francis Group
6000 Broken Sound Parkway NW, Suite 300
Boca Raton, FL 33487-2742

First issued in paperback 2019

ISBN-13: 978-1-4822-2664-5 (hbk)
ISBN-13: 978-0-367-86935-9 (pbk)

Library of Congress Cataloging-in-Publication Data

Names: Rumane, Abdul Razzak, author.
Title: Handbook of construction management : scope, schedule, and cost control / editor: Abdul Razzak Rumane.
Description: Boca Raton : Taylor & Francis, CRC Press, 2017. | Series: Industrial innovation series | Includes bibliographical references and index.
Identifiers: LCCN 2016006907 | ISBN 9781482226645 (hard back)
Subjects: LCSH: Building--Superintendence--Handbooks, manuals, etc.
Classification: LCC TH438 .R76 2017 | DDC 624.068--dc23
LC record available at https://lccn.loc.gov/2016006907

Visit the Taylor & Francis Web site at
http://www.taylorandfrancis.com

and the CRC Press Web site at
http://www.crcpress.com

To

My parents

For their prayers and love

My prayers are always for my father who always encouraged me.

I wish he would have been here to see this book and give me blessings.

My prayers and love for my mother who always inspires me.

Contents

List of Figures

List of Tables

Foreword

Construction managers are frequently asked to establish management systems that conform to the latest construction practices and international standards. To do this effectively is a major challenge, especially for individuals and organizations that do not have a holistic view of the construction management process. This book is written by Dr. Abdul Rumane as a compendium of tools and a detailed methodology for completing a construction project in an efficient and cost-effective manner. This book was written to assist the modern construction manager, which includes students, professors, and practitioners, to understand the requirements of today's complex and demanding construction environment.

The *Handbook of Construction Management: Scope, Schedule, and Cost Control*, is an extension of Dr. Rumane's previous books, which laid the groundwork for the development of this construction management handbook. In this edition, the construction community is provided with management advice and concrete examples to establish and maintain quality during all aspects of the project life cycle. The interrelationship between the owners, the designers, and the contractors, as well as the need for management of scope, schedule, and budget is clearly laid out in the chapters. All practitioners in the construction business can use this book to improve both their own internal and external construction management processes and practices. Numerous figures and tables supporting the understanding of construction management are included in this book. Some of the quality tools, management techniques, and practices used by leading construction companies in the industry come from Dr. Rumane's own personal experience and well-developed understanding of the construction business. The information presented will give the reader a competitive edge when it comes to construction management processes, maintaining quality and effectively operating throughout the life cycle of a construction project.

I have always enjoyed our time together and value Dr. Rumane's professional and systemic approach to construction. You too will enjoy this journey of learning and improving your own construction management knowledge. I know this book will provide you with the tools to make your journey a rewarding one.

Raymond R. Crawford
American Society for Quality

Acknowledgments

"Share the knowledge with others" is the motto of this book.

Many thanks go to the numerous colleagues and friends who had extended their help in preparing this book by arranging reference material.

I thank all publishers and authors for permitting me to reproduce their work. I thank the reviewers—from various professional organizations—for their valuable input to improve my writing.

I thank Dr. Adedeji B. Badiru, series editor; Cindy Renee Carelli, senior acquisitions editor; Jennifer Ahringer, project coordinator CRC Press; and other staffs of CRC Press for their support and contribution to make this construction-related book a reality.

My sincere thanks to following contributors (authors and coauthors) for their contribution toward this book:

- Cliff Moser
- Edward Taylor
- Jitendra Patel
- Shirine L. Mafi
    - Marsha Huber
    - Mustafa Shraim
- Zofia Rybkowski
    - Lincoln Forbes

I thank Raymond R. Crawford, former chair, Design and Construction Division, American Society of Quality (ASQ) and director at Parsons Brinckerhoff for his support and nicely worded thought-provoking Foreword.

I extend my thanks to Dr. Ted Coleman for his good wishes and everlasting support.

I thank Eng. Adel Al Kharafi, former president of WFEO, for his good wishes.

I thank Eng. Ahmad Almershed, Eng. Ahmad Al Kandari, Dr. Hasan Al Sanad, Eng. Hashim M. El Refaai, Eng. Naeemah Al Hay, Eng. Sadoon Al Essa, Eng. Talal Qahtani, Eng. Tarek Shuaib, Eng. Yaseen Farraj, Dr. Ayed Alamri, Abdul Wahab Rumani, Mohammad Naseeruddin, Dr. Neelamani, Cdr. (Retd) A.K.Poothia, Maj. Gen. (Rtd) R.K.Sanan, and Joginder Singh IPS. (Retd) for their good wishes. I thank Dr. N.N. Murthy of Jagruti Kiran Consultants for his good wishes.

The support of Abdul Azeem, Aijaz Quraishi, Alice Ebby, Annamma Issac, Ashraf Hajwane, Asif Kadiwala, Babar Mirza, Badrinath, Bashir Ibrahim Parkar, Faseela Moidunny, Ganesan Swaminathan, Hakimuddin Challawala, Hesham Hasan, Hombali, Husain Dalvi, Imtiyaz Thakur, Joseph Panicker, Kaide Johar Manasi, Mahe Alam, Mohammed Farghal, Mohammed Ramzan, Mohammad Shaker, Naim Quraishi, Narendra Deopurkar, Shahid Kasim, Shantilal Sirsat, Sudhir Menghani, and Zahid Khan is worth mentioning here. I thank all of them for their valuable input and suggestions.

My special thanks go to H.E. Sheikh Rakan Nayef Jaber Al Sabah for his support and good wishes.

I thank members of ASQ Design and Construction Division, The Institution of Engineers (India), and Kuwait Society of Engineers for their support.

I thank my well wishers who inspired me to complete this book.

Most of the data discussed in this book are from the editor's practical and professional experience and are accurate to the best of the editor's knowledge and ability. However, in case of any discrepancies, I would appreciate it greatly if you let me know.

The contributions of my son and daughter, Ataullah and Farzeen, respectively, are worth mentioning. They helped me in the preparation of this book and were also a great source of encouragement. I thank my mother, brothers, sisters, and other family members for their everlasting support, encouragement, and good wishes and prayers.

Finally, special thanks go to my wife, Noor Jehan, for her patience, as she had to suffer a lot because of my busy schedule.

Abdul Razzak Rumane

Introduction

Construction has a history of several thousand years. The first shelters were built from stone or mud and the materials were collected from the forests to provide protection against cold, wind, rain, and snow. These buildings were constructed primarily for residential purposes, although some might have commercial utility.

In the first half of the twentieth century, the construction industry became an important sector throughout the world, employing many workers. During this period, skyscrapers, long-span dams, shells, and bridges were developed to meet new requirements and marked the continuing progress of construction techniques. The provision of services such as heating, air-conditioning, electrical lighting, water supply, and elevators to buildings became common. The twentieth century also saw the transformation of the construction and building industry into a major economic sector. During the second half of the twentieth century, the construction industry began to industrialize because of the introduction of mechanization, prefabrication, and system building. The design of building services systems changed considerably in the last 20 years of the twentieth century. It became the responsibility of the designer to follow health, safety, and environmental regulations while designing any building.

Construction projects are mainly capital investment projects. They are executed based on a predetermined set of goals and objectives. They are customized and nonrepetitive in nature. Construction projects have become more complex and technical, and the relationships and the contractual grouping of those who are involved are also more complex and contractually varied. In addition, the requirements of construction clients are increasing and, as a result, construction products (buildings) must meet various performance standards (climate, rate of deterioration, maintenance, etc.). Therefore, to achieve the adequacy of client brief, which addresses numerous complex needs of client/end user, it is necessary to evaluate the requirements in terms of manageable activities and their functional relationships and establish construction management procedures and practices. These processes and practices are implemented and followed towards all the work areas of the project to make the project successful to the satisfaction of the owner/end user and to meet needs of the owner.

A construction project involves many participants comprising the owner, designer, contractor, and many other professionals from the construction-related industries. These participants are both influenced by and depend on one another and also on "other players" involved in the construction process. Therefore, the construction projects have become more complex and technical, and extensive efforts are required to reduce the rework and costs associated with time, materials, and engineering.

There are mainly three key attributes in a construction project that the construction/project manager has to manage effectively and efficiently to achieve a successful project:

1. Scope
2. Time (schedule)
3. Cost (budget)

From the quality perspective, these three elements are known as "quality trilogy," whereas when considered from project/construction management perspective, these are known as "triple constraints."

For successful management of the project, the construction/project manager should have all the related information about construction management principles, tools, processes, techniques, and methods. A construction/project manager should also have the professional knowledge of management functions, management processes, and project phases (technical processes), and the skills and expertise to manage the project in a systematic manner at every stage of the project. Construction management is a framework for the construction/project manager to evaluate and balance these competing demands. To balance these attributes at each stage of project execution, the project phases and their subdivisions into various elements/activities/subsystems having functional relationships should be developed by taking into consideration various management functions, management processes, and interaction, and/or a combination of some or all of these activities/elements.

Construction management process is a systematic approach to manage a construction project from its inception to completion and handover to the client/end user. Construction management is an application of professional processes, skills, and effective tools and techniques to manage project planning, design, and construction from project inception through to the issuance of the completion certificate. Some of these techniques are tailored to the specific requirements that are unique to the construction projects.

The main objective of construction management is to ensure that the client/end user is satisfied with the quality of project delivery. In order to achieve project performance goals and objectives, it is required to set performance measures that define what the contractor is going to achieve under the contract. Therefore, to achieve the adequacy of client brief, which addresses the numerous complex needs of client/end user, it is necessary to evaluate the requirements in terms of activities and their functional relationships and establish construction management procedures and practices to be implemented and followed toward all the work areas of the project to make the project successful to the satisfaction of the owner/end user and to meet the owner's needs.

This book provides significant information and guidelines to construction and project management professionals (owners, designers, consultants, construction managers, project managers, supervisors, and many others from construction-related industry) involved in construction projects (mainly civil construction projects and commercial-A/E projects) and construction-related industries. It covers the importance of construction management principles, procedures, concepts, methods, and tools and their applications to various activities/components/subsystems of different phases of the life cycle of a construction project to improve construction process in order to conveniently manage the project and make the project most qualitative, competitive, and economical. It also discusses the interaction and/or combination among some of the activities/elements of management functions, management processes, and their effective implementation and applications that are essential throughout the life cycle of a project to conveniently manage it. The construction project life-cycle phases and their activities/elements/subsystems are comprehensively developed taking into consideration Henri Fayol's management function concept, which was subsequently modified by Koontz and O'Donnel and the management processes knowledge areas described in *PMBOK®* published by the Project Management Institute (PMI).

This book contains useful material and information for the students who are interested in acquiring the knowledge of construction management activities. It also provides useful information to academics about the practices followed in the construction projects.

The data discussed and derived in this book are from the editor's/author's practical and professional experience in the construction field. This book contains many tables and figures to support the editor's/author's writings and to enable the reader to easily understand the concepts of construction management. Different types of forms and transmittals that are used to plan, monitor, and control the project at different stages of the project are included for the benefit of readers.

For the sake of better understanding and convenience, this book is divided into nine chapters and each chapter is divided into a number of sections covering construction management related topics that are relevant and important to understand management concepts for construction projects.

Chapter 1 is an overview of construction projects. It presents a brief introduction of the types of construction projects, different phases of construction project life cycle, and principles of quality in construction projects.

Chapter 2 is about project delivery systems (PDS). It discusses different types of project delivery systems and the organizational relationships among various project participants and advantages and disadvantages of each of these systems. It also discusses different types of contracting systems based on pricing methods.

Chapter 3 is about construction management delivery systems. It discusses the qualifications of a construction manager and the types of construction management systems (agency CM and CM-at-risk). It also discusses the roles of a construction manager at predesign, design, construction, and postconstruction stages (testing, commissioning, and handover).

Chapter 4 is about quality tools. It gives a brief description of various types of quality tools that are in practice, mainly in the construction industry, such as classic tools of quality, management and planning tools, process analysis tools, process improvement tools, innovation and creation tools, Lean tools, cost of quality, quality function deployment, Six Sigma, and Triz. The usage of each of the tools under these categories is supplemented by tables, figures, and charts to enable the reader to easily understand their applications in construction projects.

Chapter 5 is about building information modeling (BIM) in design and construction. This chapter provides brief information about the use of BIM as a collaborative tool in construction projects to manage complex projects, and the BIM execution plan.

Chapter 6 is about construction contract documents. It gives brief information about various types of contract documents used to prepare construction documents.

Chapter 7 focuses on construction management practices and discusses in detail the management functions, management processes, and project life-cycle phases (technical processes) pertinent to the construction industry. It covers all the topics/areas and activities related to construction management that can be used by construction professionals to implement the procedures and practices in their day-to-day work to evolve a comprehensive system to conveniently manage the construction. In order to achieve "zero defect" policy during the construction phase, the designer has to develop project documents to ensure:

- Conformance to the owner's requirements.
- Compliance with the codes and standards.
- Compliance with the regulatory requirements.
- Great accuracy to avoid any disruption/stoppage/delay of work during the construction.

- Completion within the stipulated time.
- Develop project documents without errors and omissions.

This chapter elaborates applications of the principles/concepts and relevant construction-related activities of management functions, management processes, and allows these activities to interact to create comprehensive construction project life-cycle phases and its activities/subsystems/elements to achieve the successful completion of a project. It discusses five elements of management function, planning, organizing, staffing, directing, and controlling, and explains how these activities/elements of management functions can be used in construction projects. Brief information about strategic planning, operational planning, intermediate planning, and contingency planning and steps in planning with relevance to construction projection is covered in this section. Different types of organizational structures, such as simple, functional, divisional, matrix, team-based, network, and modular with sample organization charts normally applicable in construction projects, are also discussed. Staffing processes such as acquisition, roles and responsibilities, assessment, team building, training, and development are discussed.

Information about directing and controling elements of management functions is also presented. Five types of management processes, initiating, planning, executing, monitoring, and controlling, and the relevant construction-project-related knowledge based on the *PMBOK®* methodology are also discussed in this chapter. Different types of processes, tools, and techniques that are applied during the management of constructions projects are discussed along with the related construction activities to understand the construction management process to achieve the successful completion of a project. The management processes discussed in this chapter include Integration Management, Stakeholder Management, Scope Management, Schedule Management, Cost Management, Quality Management, Resource Management, Communication Management, Risk Management, Contract Management, Health, Safety, and Environment Management (HSE), Financial Management, and Claim Management. These processes are further divided into construction-related activities that are essential to manage and control construction projects in an efficient and effective manner.

This chapter also includes comprehensive information about the seven phases of construction project life cycle, conceptual design, schematic design, detail design, construction documents, bidding and tendering, construction, testing, commissioning, and handover, and also further divisions of these phases into various elements/activities/subsystems having functional relationships to conveniently manage major construction projects. The development of scope, stakeholder's roles and responsibilities, project schedule, project cost, establishing project quality requirements, managing design quality, and monitoring design progress in each of the design phases are also discussed to ensure "zero defect" policy during construction.

This chapter also lists the risks that have to be considered and managed while developing the project design. Procedures to review and verify a design to meet the owner's objectives are also discussed. Preparation of construction documents and bidding and tendering process is discussed in this chapter. This chapter elaborates various procedures and principles to be followed during the construction phase. These include mobilization, identification of project teams, identification of subcontractors, management of construction resources, communication, risks, contracts, management of execution of works, safety during construction, and inspection of executed works. It includes guidelines for contractors about preparation and submission of transmittals, construction schedule, contractor's quality control plan, and safety plan. Change management, construction schedule monitoring, cost control, quality control, and risk control during construction are also discussed in this chapter. Processes to make payment as a project progresses and cash flow are also discussed in this chapter. Reasons for claims (variations) and how to avoid them and resolve conflicts are also discussed. Different activities to be performed during testing, commissioning, and handover are also included in this chapter.

Chapter 8 is an introduction to Lean construction. This chapter presents an introduction to Lean construction, brief history of Lean construction, current challenges in the architecture, engineering, and construction (AEC) Industry, Lean construction response, Lean goals and elimination of waste, Lean project delivery system, and Lean tools and techniques.

Chapter 9 is about ISO certification in the construction industry. It covers brief information about the importance of standards and standardization bodies. It presents a case study related to the ISO implementation methodology and discusses in detail implementation of QMS, documentation, and certification. This chapter includes brief information about ISO 14001 and ISO 27001. It also presents a correlation matrix between ISO 9001:2008 and ISO 9001:2015, and the quality management system manual for the designer and contractor.

This book, I am certain, will meet the requirements of construction professionals, students, and academics and will satisfy their needs.

Abbreviations

AAMA	American Architectural Manufacturers Association
ACI	American Concrete Institute
ACMA	American Composite Manufacturers Association
AISC	American Institute of Steel Construction
ANSI	American National Standards Institute
API	American Petroleum Institute
ARI	American Refrigeration Institute
ASCE	American Society of Civil Engineers
ASHRAE	American Society of Heating, Refrigeration, and Air-Conditioning Engineers
ASQ	American Society for Quality
ASTM	American Society of Testing Materials
BMS	Building management system
BREEAM	Building research establishment environmental assessment methodology
BSI	British Standards Institute
CDM	Construction (design and management)
CEN	European Committee for Standardization
CIBSE	Chartered Institution of Building Services Engineers
CIE	International Commission on Illumination
CII	Construction Industry Institute
CMAA	Construction Management Association of America
CSC	Construction Specifications Canada
CSI	Construction Specification Institute
CTI	Cooling tower industry
DIN	Deutsches Institute fur Normung
EIA	Electronic Industry Association
EN	European norms
FIDIC	Federation International des Ingeneurs-Counceils
HQE	High Quality Environmental (Haute Qualite Environnementale)
ICE	Institute of Civil Engineers (the United Kingdom)
IEC	International Electrotechnical Commission
IEEE	Institute of Electrical and Electronics Engineers
IP	Ingress protection
ISO	International Organization for Standardization
LEED	Leadership in Energy and Environmental Design
NEC	National Electric Code
NEC	New engineering contract
NEMA	National Electrical Manufacturers Association (the United States)
NFPA	National Fire Protection Association
NWWDA	National Wood, Window and Door Association
PMBOK®	Project Management Book of Knowledge
PMI	Project Management Institute

QS Quantity surveyor
RFID Radio frequency identification
SDI Steel Door Institute
TIA Telecommunication Industry Association
UL Underwriters Laboratories

Synonyms

Consultant	Architect/engineer (A/E), designer, design professionals, consulting engineers, supervision professional
Contractor	Construction manager (agency CM), constructor, builder
Engineer	Resident project representative
Engineer's Representative	Resident engineer project manager
Owner	Client, employer
Quantity Surveyor	Cost estimator, contract attorney, cost engineer, cost and works superintendent, main contractor, general contractor

Editor

Abdul Razzak Rumane, PhD, is a registered senior consultant with The Chartered Quality Institute (UK) and a certified consultant engineer in electrical engineering. He obtained a bachelor of engineering (electrical) degree from Marathwada University (now Dr. Babasaheb Ambedkar Marathwada University), India, in 1972, and received his PhD from the Kennedy Western University (now Warren National University), Cheyenne, Wyoming, in 2005. His dissertation topic was "Quality Engineering Applications in Construction Projects." Dr. Rumane's professional career exceeds 40 years, including 10 years in manufacturing industries and over 30 years in construction projects. Presently, he is associated with SIJJEEL Co., Kuwait, as an advisor and director, construction management.

Dr. Rumane is associated with a number of professional organizations. He is a chartered quality professional-fellow of the Chartered Quality Institute (UK), fellow of the Institution of Engineers (India), and has an honorary fellowship of Chartered Management Association (Hong Kong). He is also a Senior Member of the Institute of Electrical and Electronics Engineers (United States) and of American Society for Quality; a Member of Kuwait Society of Engineers, SAVE International (The Value Society), Project Management Institute, London Diplomatic Academy, and International Diplomatic Academy; an Associate Member of American Society of Civil Engineers; and a Member board of governors of International Benevolent Research Forum.

As an accomplished engineer, Dr. Rumane has been awarded an honorary doctorate in engineering from The Yorker International University and has also been bestowed upon the following awards: World Quality Congress awarded him the Global Award for Excellence in Quality Management and Leadership, The Albert Schweitzer International Foundation honored him with a gold medal for outstanding contribution in the field of Construction Quality Management and outstanding contribution in the field of electrical engineering/consultancy in construction projects in Kuwait, European Academy of Informatization honored him with the World Order of Science–Education–Culture and a title of Cavalier. The Sovereign Order of the Knights of Justice, England, honored him with a Meritorious Service Medal. He was selected as one of the top 100 engineers in 2009 of IBC (International Biographical Centre, Cambridge, UK). He was also the honorary chairman of the Institution of Engineers (India), Kuwait Chapter, during 2005–2007 and 2013–2014. Dr. Rumane has authored the following books: *Quality Management in Construction Projects*, and *Quality Tools for Managing Construction Projects*, both published by the CRC Press, a Taylor & Francis Group Company, United States.

Contributors

Lincoln H. Forbes is a specialist in Lean project delivery and Lean management systems. He earned his PhD in industrial engineering/management at University of Miami. As a registered professional engineer, he has over 30 years of experience in quality/performance improvement as well as facilities design and construction management support. His book, *Modern Construction: Lean Project Delivery and Integrated Practices* (Forbes and Ahmed 2010, CRC Press), is an internationally recognized reference. Dr. Forbes is the principal consultant for Harding Associates, Inc., which provides performance improvement support for the design and construction environment. He is a fellow of the Institute of Industrial and System Engineers (IISE) and is the founder and past president of the IISE's Construction Division. He is a member of a number of professional organizations including the Lean Construction Institute (LCI), the American Society for Quality (ASQ), and the American Society for Healthcare Engineering (ASHE). Dr. Forbes has served in adjunct professor positions at the Florida International University, Drexel University, and the East Carolina University. His courses have included Lean project delivery, construction performance improvement, and quality management. He is a LEED accredited professional (LEED AP).

Marsha Huber is an associate professor of accounting at the Youngstown State University, earning her PhD in hospitality management at Ohio State University. She is a certified accountant and has also served as a faculty scholar at Harvard University to pioneer research in neuroaccounting. Her research interests are in accounting education, positive organizational scholarship, design thinking, and experiential sampling.

Shirine L. Mafi is a professor of management at the Otterbein University, earning her PhD at Ohio State University. Her areas of teaching include operations, performance improvement, and service management. Her innovation in teaching philantrophy-based education has won her many awards. Dr. Mafi is also a certified quality auditor since 2008. She has authored several articles on operations, philanthropy-based education, and quality of teaching.

Cliff Moser is a registered architect with over 30 years of experience in healthcare design and construction. He is the author of *Architecture 3.0: The Disruptive Design Practice Handbook*, Routledge, 2014, and his research include digital delivery, complexity, rules-based and generative design system, and collaboration with a focus on quality management systems. He is LEED accredited and his professional memberships include the American Association of Architects (AIA) and American Society for Quality (ASQ). An innovator in developing the requirements for the instruments of service for the digital practice, he has served in leadership roles at architecture firms—including Perkins+Will Global, Chicago, Illinois, and RTKL Associates, Baltimore, Maryland—and healthcare owner organizations, including Kaiser Permanente, Oakland, California, and Stanford Health Care, Stanford, California.

Jitu C. Patel is a certified professional environmental auditor (CPEA) for health and safety of the Board of Environment Auditing Certification (BEAC), United States and is an international health, safety, and environment (HSE) consultant with a BS in chemistry and an MPhil in fuel science. For 21 years, Patel has provided HSE services to Saudi Aramco, Dhahran, Saudi Arabia, while conducting research for 15 years. He also imparted technical training on fires, explosions, and HSE issues at a British chemical company. He has developed and conducted fire and safety seminars for industry operations. His work has taken him throughout the world IS internationally. He is a member of the BEAC Training & Education Board. He is also involved in establishing international American Society of Safety Engineers (ASSE) Chapters in the Middle East, Southeast Asia, and India. He is the recipient of the Howard Hiedeman Award and was honored in Chicago, Ohio, with the highest and most prestigious award of a "Fellow of the ASSE," and also with the Diversity and Practice Specialty awards. Patel is also an ASSE global ambassador.

Zofia K. Rybkowski has extensive research experience which includes integrated project delivery, productivity analysis and Lean construction, target value design, life cycle cost analysis, Lean simulations, sustainable design, and evidence-based design. She has extensive experience as a construction, architectural, and engineering researcher and consultant. She has provided consultancy services to firms in Boston, San Francisco, Tokyo, and Hong Kong. Dr. Rybkowski holds degrees from Stanford; Brown; Harvard; the Hong Kong University of Science and Technology; and the University of California, Berkeley. She earned her MArch degree in architecture at Harvard Graduate School of Design and her MS and PhD in civil and environmental engineering at University of California, Berkeley. Dr. Rybkowski is an assistant professor of construction science and teaches Lean construction to advanced construction science students at Texas A&M University, College Station, Texas. She is a fellow at the Center for Health Systems and Design and at the Institute for Applied Creativity. She is a LEED AP.

Mustafa Shraim has over 20 years of experience in the field of quality management as an engineer, manager, consultant, and trainer. He has extensive experience in quality management systems and Lean and Six Sigma methods. In addition, he coaches and mentors Green and Black Belts on process improvement projects across various industries.

Dr. Shraim obtained his PhD in industrial engineering at West Virginia University. He is a certified quality engineer and a certified Six Sigma Black Belt by American Society for Quality (ASQ). He is also a certified QMS lead auditor by International Register of Certified Auditors (IRCA). He was elected a fellow by ASQ in 2007.

Edward Taylor is the executive director of the Construction Industry Research and Policy Center (CIRPC) at University of Tennessee (UTK) in Knoxville, Tennessee. He obtained his BS degree in civil engineering from University of Tennessee; and his MBA from University of Georgia; and an MA in economics from University of Tennessee. Before coming to UTK in 2010, he spent 7 years in structural design and 10 years in the bridge construction industry. He is a registered engineer in three states and formerly served on the Tennessee Road Builders Association Board of Directors. In his current role, Taylor is responsible for CIRPC's Davis–Bacon construction wage survey activity and oversees CIRPC activities related to construction safety.

Collaborations and Affiliations

Since coming to CIRPC, Taylor

- Has participated in a collaborative effort between UTK and B&W Y-12 at Oak Ridge, Tennessee, as part of a planned $8 billion uranium processing facility. His role involved an application of Prevention through Design (PtD)—a NORA Construction Sector goal.
- Has also led a team that was awarded the grand prize in a national (United States) contest to develop a workplace safety application for young workers. The winning app, *Working Safely Is No Accident,* can be found at http://ilab.engr.utk.edu/cirpc/.

Publications and Synergistic Activities

Safety Benefits from Mandatory OSHA 10-Hour Training, Safety Science, Volume 77, August 2015, pp. 66–71.

1

Overview of Construction Projects

Abdul Razzak Rumane

CONTENTS

1.1 Construction Projects

Construction is the translation of owner's goals, and objectives into a facility built by the contractor/builder as stipulated in the contract documents, plans, and specifications on schedule and within the budget. A construction project is a custom rather than a routine, repetitive business and differs from manufacturing. Construction projects work against the defined scope, schedule, and budget to achieve the specified result.

Construction has a history of several thousand years. The first shelters were built from stone or mud and the materials collected from the forests to provide protection against cold, wind, rain, and snow. These buildings were meant primarily for residential purposes, although some may have a commercial utility.

During the New Stone Age, people introduced dried bricks, wall construction, metalworking, and irrigation. People gradually developed the skills to construct villages and cities for which considerable skills in building were acquired. This can be seen from the great civilizations in various parts of the world some 4000–5000 years ago. During the early period of Greek settlements, in about 2000 BC, the buildings were made of mud using timber frames. Later temples and theaters were built from marble. Some 1500–2000 years ago, Rome became the leading center of the world culture, which extended to construction.

Marcus Polo was a first century BC military and civil engineer, who published books in Rome. This was the world's first major publication on architecture and construction, and it dealt with building materials, the styles and design of building types, the construction process, building physics, astronomy, and building machines.

During the Medieval Age (476–1492), improvements in agriculture and artisanal productivity, and exploration and consequently the expansion of commerce took place, and in the late Middle Ages, building construction became a major construction industry. Craftsmen were given training and education in order to develop skills and to raise their social status. During those times, guilds were responsible for managing the quality.

The fifteenth century brought a *renaissance* or renewal in architecture, building, and science. Significant changes occurred during and after the seventeenth century because of the increasing transformation of construction and urban habitat.

The scientific revolution of the seventeenth and eighteenth centuries gave birth to the great Industrial Revolution of the eighteenth century. Construction followed these developments in the nineteenth century.

In the first half of the twentieth century, the construction industry became an important sector throughout the world, employing many workers. During this period, skyscrapers, long-span dams, shells, and bridges were developed to satisfy new requirements and marked the continuing progress of construction techniques. The provision of services such as heating, air conditioning, electrical lighting, mains water, and elevators to buildings became common. The twentieth century has seen the transformation of construction and building industry into a major economic sector. During the second half of the twentieth century, the construction industry began to industrialize, introducing mechanization, prefabrication, and system building. The design of building services systems changed considerably in the last 20 years of the twentieth century. It became the responsibility of the designer to follow health, safety, and environmental regulation while designing any building.

Residential and commercial, traditional A/E type of construction projects account for an estimated 25% of the annual construction volume. Building construction is labor intensive. Every construction project has some elements that are unique. No two construction or R&D projects are alike. Although it is clear that projects are usually more routine than research and development projects, some degree of customization is a characteristic of the projects.

There are several types of projects. Figure 1.1 illustrates the types of construction projects. A project is a plan or program performed by the people with assigned resources to achieve an objective within a finite duration. A project is a temporary endeavor undertaken to create a unique product or service. *Temporary* means that every project has a definite beginning and a definite end. Projects have specified objectives to be completed within certain specifications and funding limits. Projects are often critical components of the business strategy of an organization. Examples of projects include

- Developing a new product or service.
- Effecting a change in structure, staffing, or style of an organization.
- Designing a new transportation vehicle/aircraft.
- Developing or acquiring a new or modified information system.
- Running a campaign for political office.
- Implementing a new business procedure or process.
- Constructing a building or facilities.

The duration of a project is finite, projects are not ongoing efforts, and a project ceases when its declared objectives have been attained. Some of the characteristics of projects may include the following:

1. Performed by people
2. Constrained by limited resources
3. Planned, executed, and controlled

Construction projects comprised of a cross section of many different participants. These participants both influence and depend on one another in addition to *other players* involved in the construction process. Figure 1.2 illustrates the concept of traditional construction project organization.

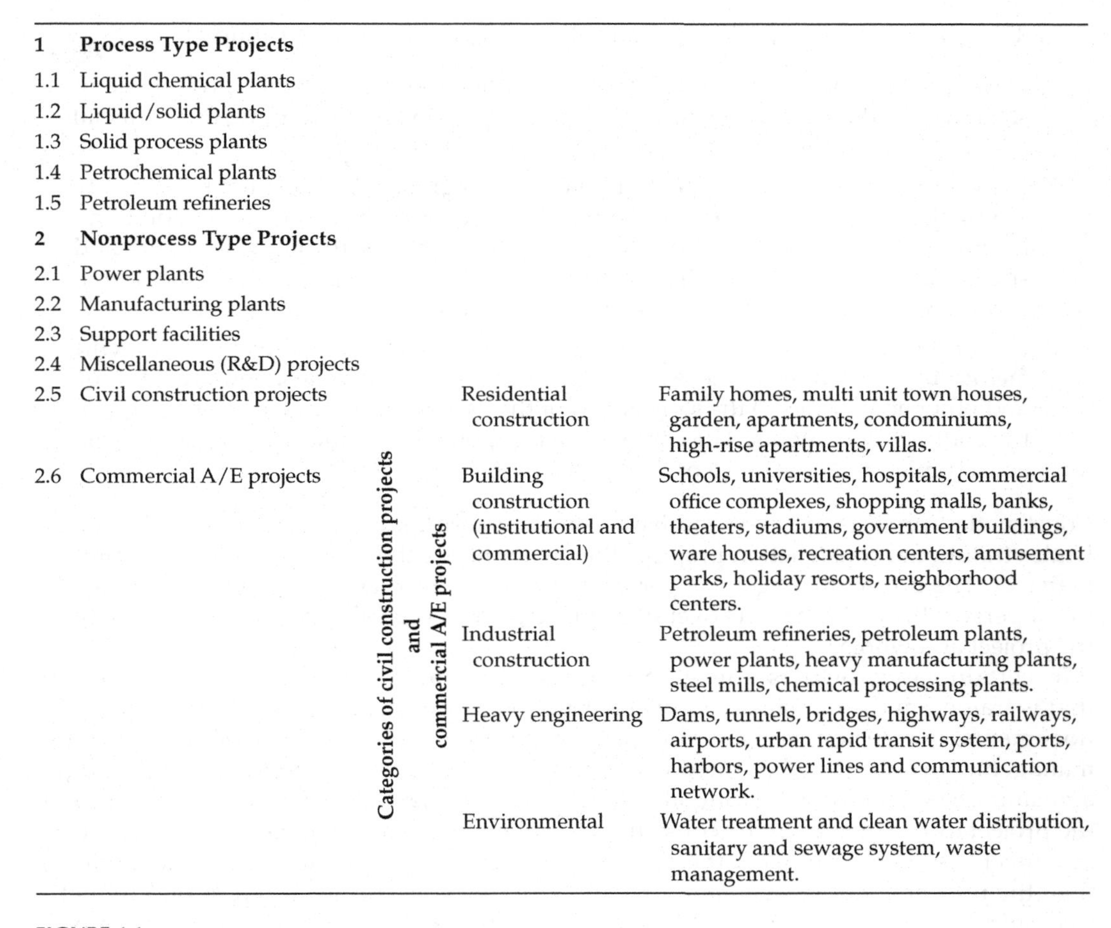

1 **Process Type Projects**
1.1 Liquid chemical plants
1.2 Liquid/solid plants
1.3 Solid process plants
1.4 Petrochemical plants
1.5 Petroleum refineries

2 **Nonprocess Type Projects**
2.1 Power plants
2.2 Manufacturing plants
2.3 Support facilities
2.4 Miscellaneous (R&D) projects
2.5 Civil construction projects
2.6 Commercial A/E projects

Categories of civil construction projects and commercial A/E projects

Category	Examples
Residential construction	Family homes, multi unit town houses, garden, apartments, condominiums, high-rise apartments, villas.
Building construction (institutional and commercial)	Schools, universities, hospitals, commercial office complexes, shopping malls, banks, theaters, stadiums, government buildings, ware houses, recreation centers, amusement parks, holiday resorts, neighborhood centers.
Industrial construction	Petroleum refineries, petroleum plants, power plants, heavy manufacturing plants, steel mills, chemical processing plants.
Heavy engineering	Dams, tunnels, bridges, highways, railways, airports, urban rapid transit system, ports, harbors, power lines and communication network.
Environmental	Water treatment and clean water distribution, sanitary and sewage system, waste management.

FIGURE 1.1
Types of construction projects. (Abdul Razzak Rumane (2013), *Quality Tools for Managing Construction Projects*, CRC Press, Boca Raton, FL. Reprinted with permission from Taylor & Francis Group.)

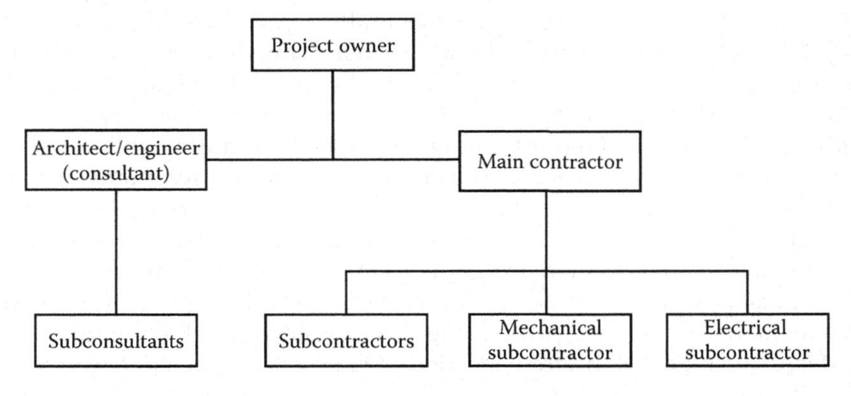

FIGURE 1.2
Traditional contracting system (design–bid–build).

Traditional construction projects involve the following three main groups:

1. *Owner*—A person or an organization that initiates and sanctions a project. He or she requests the need of the facility and is responsible for arranging the financial resources for creation of the facility.
2. *Designer* (A/E)—This consists of architects or engineers or consultants. They are the owner's appointed group accountable to convert the owner's conception and needs into the specific facility with detailed directions through drawings and specifications within the economic objectives. They are responsible for the design of the project and in certain cases supervision of the construction process.
3. *Contractor*—A construction firm engaged by the owner to complete the specific facility by providing the necessary staff, work force, materials, equipment, tool, and other accessories to the satisfaction of the owner/end user in compliance with the contract documents. The contractor is responsible for implementing the project activities and to achieve owner's objectives.

Construction projects are executed based on predetermined set of goals and objectives. Under traditional construction projects, the owner heads the team and selects the project manager. The project manager is a person/member of the owner's staff or independently hired person/firm who has the overall or principal responsibility for the management of the project as a whole.

In certain cases, owners engage a professional firm, called construction manager, that is trained in the management of construction processes, to assist in developing bid documents, overseeing, and coordinating project for the owner. The basic construction management concept is that the owner assigns a contract to a firm that is knowledgeable and capable of coordinating all aspects of the project to meet the intended use of the project by the owner. In the construction management type of project, the consultant (architect/engineer) prepares the complete design drawings and contact documents, then the project is put for a competitive bid and the contact is awarded to the competitive bidder (contractor). Owner hires a third party (construction manager) to oversee and coordinate the construction.

Construction projects are mainly capital investment projects. They are customized and nonrepetitive in nature. Construction projects have become more complex and technical, and the relationships and the contractual grouping of those who are involved are also more complex and contractually varied. The products used in construction projects are expensive, complex, immovable, and long lasting. Generally, a construction project is composed of building materials (civil), electromechanical items, finishing items, and equipment. These are normally produced by other construction-related industries/manufacturers. These industries produce products as per their own quality management practices complying with certain quality standards or against specific requirements for a particular project. The owner of the construction project or his or her representative has no direct control over these companies unless the owner or the representative or the appointed contractor commit to buy their products for use in their facility. These organizations may have their own quality management program. In manufacturing or service industries, the quality management of all in-house manufactured products is performed by manufacturer's own team, or under the control of the same organization having jurisdiction over their manufacturing plants at different locations. Quality management of vendor-supplied items/products is carried out as stipulated in the purchasing contract as per the quality control specifications of the buyer.

1.2 Construction Project Life Cycle

Systems are as pervasive as the universe in which we live. The world in which we live may be divided into the natural world and the man-made world. Systems appeared first in natural forms and subsequently with the appearance of human beings, man-made systems came into existence. Natural systems are those that came into being by natural process. Man-made systems are those in which human beings have intervened through components, attributes, or relationships.

The systems approach is a technique that represents a broad systematic approach to problem that may be interdisciplinary. It is particularly useful when problems are affected by many factors, and it entails the creation of a problem model that corresponds as closely as possible in some sense to reality. The systems approach stresses the need for the engineer to look for all the relevant factors, influences, or components of the environment that surrounds the problem. The systems approach corresponds to a comprehensive attack on a problem and to an interest in, and commitment to, formulating a problem in the widest and fullest manner that can be professionally handled.

Systems engineering and analysis when coupled with new emerging technologies reveal unexpected opportunities for bringing new improved systems and products into existence that will be more competitive in the world economy. Product competitiveness is desired by both private and public sector producers worldwide. It is the product or consumer good that must meet customer expectations.

These technologies and processes can be applied to construction projects. Systems engineering approach to construction projects helps understand the entire process of project management in order to understand and manage its activities at different levels of various phases to achieve economical and competitive results. The cost-effectiveness of the resulting technical activities can be enhanced by providing more flexibility to what they are to do before addressing what they are composed of. To ensure economic competitiveness regarding the product, engineering must become more closely associated with economics and economic facilities. This is best accomplished through the life-cycle approach to engineering. Every system is made up of components, and components can be broken down into similar components. If two hierarchical levels are involved in a given system, the lower is conveniently called a subsystem. The designation of system, subsystem, and components is relative because the system at one level in the hierarchy is the component at another level.

Most construction projects are custom oriented having a specific need and a customized design. It is always the owner's desire that his or her project should be unique and better. Further it is the owner's goal and objective that the facility is completed on time. Expected time schedule is important from both financial and acquisition of the facility by the owner/end user.

The system life cycle is fundamental to the application of systems engineering. Systems engineering approach to construction projects helps understand the entire process of project management and manage and control its activities at different levels of various phases to ensure timely completion of the project with economical use of resources to make the construction project most qualitative, competitive, and economical.

Systems engineering starts from the complexity of a large-scale problem as a whole and moves toward structural analysis and partitioning process until the questions of interest are answered. This process of decomposition is called a work breakdown structure (WBS). The WBS is a hierarchical representation of system levels. Being a family tree, the WBS consists of a number of levels, starting with the complete system at level 1 at the top and progressing downward through as many levels as necessary, to obtain elements that can be conveniently managed.

Benefits of systems engineering applications are as follows:

- Reduction in the cost of system design and development, production/construction, system operation and support, system retirement, and material disposal
- Reduction in system acquisition time
- More visibility and reduction in the risks associated with the design decision-making process

Although it is difficult to generalize project life cycle to system life cycle, considering that there are innumerable processes that make up the construction process, the technologies and processes, as applied to systems engineering, can also be applied to construction projects. The number of phases shall depend on the complexity of the project. Duration of each phase may vary from project to project. Generally, construction projects have five most common phases:

1. Conceptual design
2. Schematic design
3. Design development
4. Construction
5. Testing, commissioning, and handover

Each phase can further be subdivided on the WBS principle to reach a level of complexity where each element/activity can be treated as a single unit that can be conveniently managed. WBS represents a systematic and logical breakdown of the project phase into its components (activities). It is constructed by dividing the project into major elements with each of these being divided into subelements. This is done till a breakdown is accomplished in terms of manageable units of work for which responsibility can be defined. WBS involves envisioning the project as a hierarchy of goal, objectives, activities, subactivities, and work packages. The hierarchical decomposition of activities continues until the entire project is displayed as a network of separately identified and nonoverlapping activities. Each activity will have a single purpose, will be of a specific time duration and manageable, and will have its time and cost estimates easily derived, deliverables clearly understood, and responsibility for its completion clearly assigned. The WBS helps in

- Effective planning by dividing the work into manageable elements that can be planned, budgeted, and controlled.
- Allocating responsibility for work elements to project personnel and outside agencies.
- Developing control and information system.

WBS facilitates the planning, budgeting, scheduling, and control activities for the project manager and its team. By application of the WBS process, the construction phases are further divided into various activities. Division of these phases will improve the control and planning of the construction project at every stage before a new phase starts. The components/activities of construction project life-cycle phases, divided on the WBS principle, are as follows:

1. Conceptual design
   - Identification of need and objectives (TOR or terms of reference)
   - Identification of project team
   - Data collection
   - Identification of alternatives
   - Time schedule
   - Financial implications/resources
   - Development of concept design
2. Schematic design
   - General scope of works/basic design
   - Regulatory/authorities approval
   - Schedule
   - Budget
   - Contract terms and conditions
   - Value engineering study
3. Design development
   - Detail design of the work
   - Regulatory/authorities approval
   - Contract documents and specifications
   - Detailed plan
   - Budget
   - Estimated cash flow
   - Contract documents
   - Bidding and tender documents
4. Construction
   - Mobilization
   - Execution of works
   - Planning and scheduling
   - Management of resources/procurement
   - Monitoring and control
   - Quality
   - Inspection
5. Testing, commissioning, and handover
   - Testing
   - Commissioning
   - Regulatory/authorities approval
   - As built drawings/records
   - Technical manuals and documents
   - Training of user's personnel

TABLE 1.1

Construction Project Life Cycle

Conceptual Design	Schematic Design	Design Development	Construction	Testing, Commissioning, and Handover
Identification of objectives and goals (Terms of Reference)	General scope of work/basic design	Detail design of the work	Mobilization	Testing
Identification of project team	Regulatory approval	Regulatory/authorities approval	Execution of works	Commissioning
Data collection	Schedule	Contract documents and specifications	Planning and scheduling	Regulatory/authorities approval
Identification of alternatives	Budget	Detail plan	Management of resources/procurement	Move-in-plan
Time schedule	Contract terms and conditions	Budget	Monitoring and control	As-built drawings/records
Financial implications/resources	Value engineering study	Estimated cash flow	Quality	Technical manuals and documents
Development of concept design		Contract documents bidding and tender documents	Inspection	Training of user's personnel
				Handover of facility to owner/end user
				Substantial completion

Source: Abdul Razzak Rumane (2010), *Quality Management in Construction Projects*, CRC Press, Boca Raton, FL. Reprinted with permission from Taylor & Francis Group.

- Hand over facility to owner/end user
- Move in plan
- Substantial completion

Table 1.1 illustrates subdivided activities/components of a construction project life cycle.

These activities may not be strictly sequential; however, the breakdown allows implementation of project management functions more effectively at different stages.

1.3 Principles of Quality in Construction Projects

Quality has different meanings to different people. The definition of quality relating to manufacturing, processes, and service industries is as follows:

- Meeting the customer's need
- Customer satisfaction
- Fitness for use
- Conforming to requirements
- Degree of excellence at an acceptable price

The International Organization for Standardization (ISO) defines quality as "the totality of characteristics of an entity that bears on its ability to satisfy stated or implied needs."

However, the definition of quality for construction projects is different to that of manufacturing or services industries as the product is not repetitive, but a unique piece of work with specific requirements.

Quality in construction project is not just the quality of product and equipment used in the construction of facility, it is the total management approach to complete the facility. Quality of construction depends mainly on the control of construction, which is the primary responsibility of the contractor.

Quality in manufacturing passes through a series of processes. Material and labor are input through a series of process out of which a product is obtained. The output is monitored by inspection and testing at various stages of production. Any nonconforming product identified is repaired, reworked, or scrapped, and proper steps are taken to eliminate the roots of the problem. Statistical process control methods are used to reduce the variability and to increase the efficiency of a process. In construction projects, the scenario is not the same. If anything goes wrong, the nonconforming work is very difficult to rectify, and remedial actions are sometimes not possible.

Quality management in construction projects is different to that of manufacturing. Quality in construction projects is not just the quality of products and equipment used in the construction; it is the total management approach to complete the facility as per the scope of works to customer/owner satisfaction within the budget and to be completed within the specified schedule to meet the owner's defined purpose. The nature of the contracts between the parties plays a dominant part in the quality system required from the project, and the responsibility for achieving them must therefore be specified in the project documents. The documents include plans, specifications, schedules, and bill of quantities. Quality control in construction typically involves ensuring compliance with minimum standards of material and workmanship in order to ensure the performance of the facility according to the design. These minimum standards are contained in the specification documents. For the purpose of ensuring compliance, random samples and statistical methods are commonly used as the basis for accepting or rejecting the completed work and batches of materials. Rejection of a batch is based on nonconformance or violation of the relevant design specifications.

Quality of construction projects on the basis of the earlier discussion may be defined as follows: construction project quality is fulfillment of owner's needs as per the defined scope of works within a budget and specified schedule to satisfy owner's/user's requirements. The phenomenon of these three components can be called as *construction project trilogy* and is illustrated in Figure 1.3.

Thus quality of construction projects can be evolved as follows:

1. Properly defined scope of work.
2. Owner, project manager, design team leader, consultant, and constructor's manager are responsible for implementing the quality.
3. Continuous improvement can be achieved at different levels as follows:
   a. Owner specifies the latest needs.
   b. Designer specification includes latest quality materials, products, and equipment.
   c. Constructor uses latest construction equipment to build the facility.

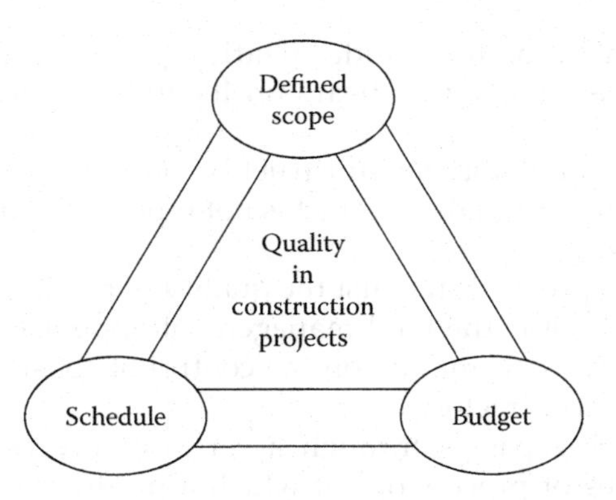

FIGURE 1.3
Construction project trilogy.

4. Establishment of performance measures
   a. *Owner*
      i. To review and ensure that the designer has prepared the contract documents that satisfy his or her needs.
      ii. To check the progress of work to ensure compliance with the contract documents.
   b. *Consultant*
      i. As a consultant designer to include the owner's requirements explicitly and clearly defined in the contact documents.
      ii. As a supervision consultant to oversee contractor's work as per the contract documents and the specified standards.
   c. Contactor—To construct the facility as specified and use the materials, products, and equipment that satisfy the specified requirements.
5. *Team approach*—Every member of the project team should know principles of the total quality management (TQM), knowing that the TQM is a collaborative effort and everybody should participate in all the functional areas to improve quality of project works. They should know that it is a collective effort by all the participants to achieve project quality.
6. Training and education consultant and contractor should have customized training plans for their management, engineers, supervisors, office staff, technicians, and labors.
7. *Establish leadership*—Organizational leadership should be established to achieve the specified quality. Encourage and help the staff and labors to understand the quality to be achieved for the project.

These definitions when applied to construction projects relate to the contract specifications or owner/end user requirements to be constructed in such a way that the construction of the facility is suitable for owner's use or it meets the owner requirements. Quality in construction is achieved through a complex interaction of many participants in the facilities development process.

The quality plan for construction projects is part of the overall project documentation consisting of

1. Well-defined specification for all the materials, products, components, and equipment to be used to construct the facility.
2. Detailed construction drawings.
3. Detailed work procedure.
4. Details of the quality standards and codes to be complied.
5. Cost of the project.
6. Manpower and other resources to be used for the project.
7. Project completion schedule.

Participation involvement of all three parties at different levels of construction phases is required to develop a quality system and application of quality tools and techniques. With the application of various quality principles, tools, and methods by all the participants at different stages of a construction project, rework can be reduced resulting in savings in the project cost and making the project qualitative and economical. This will ensure completion of construction and making the project most qualitative, competitive, and economical. Table 1.2 illustrates quality principles of construction projects.

TABLE 1.2

Principles of Quality in Construction Projects

Principle	Construction Project's Principle
Principle 1	Owner, consultant, and contractor are fully responsible for application of the quality management system to meet the defined scope of work in the contract documents
Principle 2	Consultant is responsible for providing owner's requirements explicitly and clearly defining in the contract documents
Principle 3	Method of payments (work progress, material, equipment, etc.) to be clearly defined in the contract documents. Rate analysis of bill of quantities (BOQ) or bill of materials (BOM) item to be agreed before signing of contract
Principle 4	Contract documents should include a clause to settle the dispute arising during construction stage
Principle 5	Contractor should study all the documents during tendering/bidding stage and submit his or her proposal taking into consideration all the requirements specified in the contract documents
Principle 6	Contractor shall follow an agreed upon quality assurance and quality control plan; consultant shall be responsible for overseeing the compliance with the contract documents and specified standards
Principle 7	Contractor is responsible for providing all the resources, manpower, material, equipment, and so on to build the facility as per specifications
Principle 8	Contractor shall follow the submittal procedure specified in the contract documents
Principle 9	Each member of project team should participate in all the functional areas to continuously improve the quality of the project
Principle 10	Contractor is responsible for constructing the facility as specified and use the material, products, equipment, and methods that satisfy the specified requirements
Principle 11	Contractor to build the facility as stipulated in the contract documents, plan, specifications within the budget and on schedule to meet owner's objectives
Principle 12	Contractor should perform the works as per the agreed upon construction program and handover the project as per the contracted schedule

Source: Abdul Razzak Rumane (2013), *Quality Tools for Managing Construction Projects*, CRC Press, Boca Raton, FL. Reprinted with permission from Taylor & Francis Group.

2

Project Delivery Systems

Abdul Razzak Rumane

CONTENTS

2.1 Introduction

After the establishment of project objectives, the owner/client develops the project procurement strategy by selecting a particular type of project delivery system. The type of project delivery system varies from project to project taking into consideration the objectives of the project. Project delivery method is a system to achieve satisfactory completion of a construction project from the inception to the occupancy.

A project delivery system is defined as the organizational arrangement among various participants comprising owner, designer, contractor, and many other professionals involved in the design and construction of a project/facility to translate/transform the owner's needs/goals/objectives into a finished facility/project to satisfy the owner's/end user's requirements.

A project delivery system

- Establishes the scope and responsibility for how the project is delivered to the owner.
- Includes project design and construction.
- Defines each participant's responsibility/obligations, such as scheduling, cost control, quality management, safety management, and risk management, during

various phases of the construction project life cycle (concept design, schematic design, detailed design, construction, testing, commissioning, and handover).

- Is the approach by which the project is delivered to the owner, but is separate and distinct from the contractual arrangements for financial compensation.
- Establishes procedures, actions, and sequence of events to be carried out.

2.2 Types of Project Delivery Systems

There are several types of contract delivery systems. Table 2.1 illustrates most common project delivery systems followed in construction projects.

TABLE 2.1

Categories of Project Delivery Systems

Sr. No.	Category	Classification	Sub-Classification
1	Traditional system (separated and cooperative)	Design–bid–build	Design–bid–build
		Variant of traditional system	Sequential method
			Accelerated method
2	Integrated system	Design–build	Design–build
		Design–build	Joint venture (architect and contractor)
		Variant of design–build system	Package deal
		Variant of design–build system	Turnkey method engineering, procurement, construction (EPC)
		Variant of design–build system (turnkey)	Build–operate–transfer (BOT)
			Build–own–operate–transfer (BOOT)
			Build–transfer–operate (BTO)
			Design–build–operate–maintain (DBOM)
		Variant of design–build system (funding option)	Lease–develop–operate (LDO)
			Wraparound (public–private partnership)
		Variant of design–build system	Build–own–operate (BOO)
			Buy–build–operate (BBO)
3	Management oriented system	Management contracting	Project manager (program management)
		Construction management	Agency construction manager
			Construction manager–at–risk
4	Integrated project delivery system	Integrated form of contract	

Source: Abdul Razzak Rumane (2013), *Quality Tools for Managing Construction* Projects, CRC Press, Boca Raton, FL. Reprinted with permission from Taylor & Francis Group.

2.2.1 Design–Bid–Build

In this method, the owner contracts design professional(s) to prepare detailed design and contract documents. These are used to receive competitive bids from the contractors. A design–build–bid–build contract has well defined scope of work. This method involved three steps:

1. Preparation of complete detailed design and contract documents for tendering
2. Receiving bids from prequalified contractors
3. Award of contract to the successful bidder

In this method, two separate contracts are awarded: one to the designer/consultant and one to the contractor. In this type of contract structure, design responsibility is primarily that of the architect or engineer employed by the client, and the contractor(s) is primarily responsible for construction only. In most cases, the owner contracts designer/consultant to supervise the construction process. These types of contracts are lump-sum fixed-price contracts. Any variation or change during the construction needs prior approval from the owner. Since a complete design is prepared before construction, the owner knows the cost of the project, the time of completion of the project, and the configuration of the project. The client through the architect or engineer retains the control of design during construction. This type of contracting system requires considerable time; each step must be completed before starting the next step. Table 2.2 illustrates the main aspects, advantages, and disadvantages of the design–bid–build type (traditional type) of project delivery system, and Figure 2.1 illustrates the contractual relationship.

2.2.1.1 Multiple-Prime Contracting

Multiple-prime contracting is a variation of the design–bid–build type of project delivery system. In this type of contracting system, the owner holds separate contracts with contractors of various disciplines and retains control over the project. This system can result in a lower cost to the owner because it avoids the compounded profit and overhead margins that are common to the single contract method. Since the project work is divided into a number of packages, more firms take part in bidding, which results in lower prices and also reduces the liability for delays by postponing the bidding of the follow-on work. With this type of system, *fast-tracking* of project is possible. In this type of project delivery system, the owner normally engages the construction manager (CM) or firm to manage the project. Figure 2.2 illustrates the contractual relationship in a multiple-prime contracting system.

2.2.2 Design–Build

In the design–build contract, the owner contracts a firm singularly responsible to design and build the project. In this type of contracting system, the contractor is appointed based on an outline design or design brief to understand the owner's intent of the project. The owner has to clearly define his/her needs, performance requirements, and comprehensive scope of works prior to signing off the contract. It is a must that project definition is understood by the contractor to avoid any conflict in future, as the contractor is responsible

TABLE 2.2

Design–Bid–Build

Project Delivery System	Main Aspects	Advantages	Disadvantages
Traditional contracting system (design–bid–build)	In this method the owner contracts design professional(s) to prepare detailed design and contract documents. These are used to receive competitive bids from the contractors. A design–bid–build contract has well-defined scope of work. This method involved three steps: 1. Preparation of complete detailed design and contract documents for tendering. 2. Receiving bids from pre qualified contractors. 3. Award of contract to successful bidder. In this method two separate contracts are awarded, one to the designer/ consultant and one to the contractor. In this type of contract structure, design responsibility is primarily that of the architect or engineer employed by the client and the contractor(s) is primarily responsible for construction only.	• Traditional, well-known delivery system. • Suitable for all types of client. • Client has control over design process and quality. • Project scope is defined. • Project schedule is known. • Cost certainty at the time of awarding the contract. • Client has direct contractual control over designer (consultant) and contractor. • Low risk for client. • Contractual roles and responsibilities of all parties are clearly defined and known and understood by each participant. • Well-defined relationship among all the parties. • Higher level of competition resulting in low bid cost.	• Long project life cycle. • Client has no control over selection of contractor/ subcontractor. • Client must have the resources and expertise to administer the contracts of designer and contractor. • Higher level of inspection by supervision agency. • Lack of input from constructors may result in design and constructability issues. • Project cost may be higher than estimated by the designer. • If contract documents are unclear, it raises the unexpected costs drastically. • Project likely to be delayed as it may not be possible to complete as per estimated schedule. • Exposure to risk where unreasonable time is set to complete the design phase. • No incentives to the contractor. • Not suitable for projects which are change sensitive.

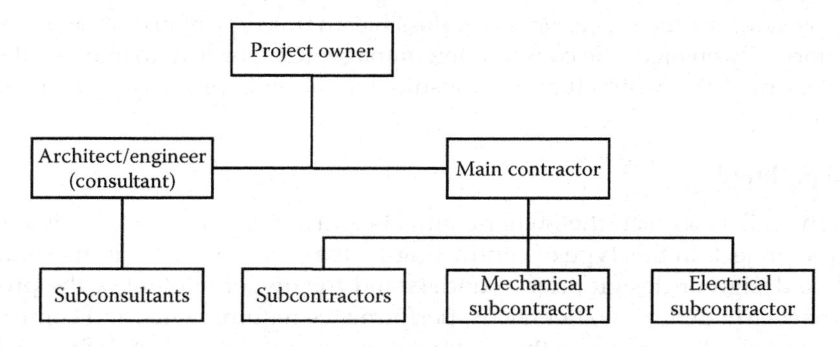

FIGURE 2.1
Design–bid–build (traditional contracting system) contractual relationship.

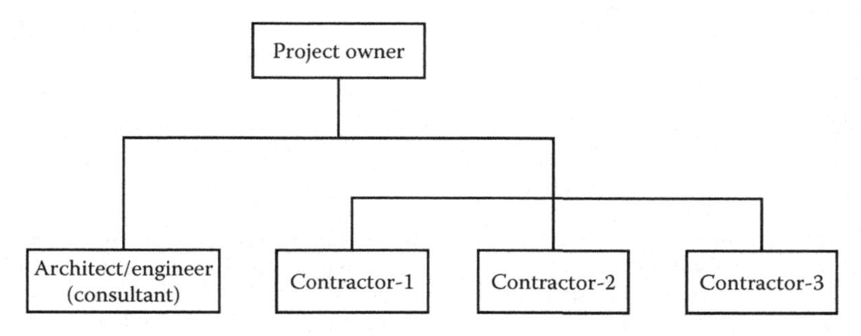

FIGURE 2.2
Multiple-prime contractor contractual relationship.

for the detailed design and construction of the project. A design–build type of contract is often used to shorten the time required to complete a project. Since the contract with the design–build firm is awarded before starting any design or construction, a cost plus contract or reimbursable arrangement is normally used instead of lump-sum, fixed-cost arrangement. This type of contract requires extensive involvement of the owner during the entire life cycle of the project. He/she has to be involved for taking decisions during the selection of design alternatives and the monitoring of costs and schedules during construction, and therefore the owner has to maintain/hire a team of qualified professionals to perform these activities. Design–build contracts are used for a relatively straightforward work, where no significant risk or change is anticipated, and when the owner is able to specify precisely what is required. Table 2.3 illustrates the main aspects, advantages, and disadvantages of the design–build type of project delivery system, and Figure 2.3 illustrates the contractual relationship.

In cases of the design–bid–build type of project delivery system, the contractor is contracted after completion of the design based on successful bidding, whereas in the design–build type of deliverable system, the contractor is contracted right from the early stage of the construction project and is responsible for the design development of the project.

Figure 2.4 illustrates a typical logic flow diagram for the design–bid–build type of construction project, and Figure 2.5 illustrates a typical logic flow diagram for the design–build type of contracting system.

2.2.3 The Turnkey Contract

As the name suggests, these are the types of contracts where, on completion, the owner turns a key in the door and everything is working to full operating standards. In this type of method, the owner employs a single firm to undertake design, procurement, construction, and commissioning of the entire works. The firm is also involved in the management of the project during the entire process of the contract. The client is responsible for the preparation of their statement requirements that become a strict responsibility of the contractor to deliver. This type of contract is used mainly for the process type of projects and is sometimes called engineering, procurement, and construction (EPC). Table 2.4 illustrates the differences between the design–build and EPC type of project delivery system.

TABLE 2.3
Design–Build

Project Delivery System	Main Aspects	Advantages	Disadvantages
Design–build	In design–build contract, owner contracts a firm (contractor) singularly responsible for design and construction of the project. In this type of contracting system, the contractor is appointed based on an outline design or design brief to understand the owner's intent of the project. The owner has to prepare comprehensive scope of work and has to clearly define his/her needs and performance requirements/specifications prior to signing of the contract. It is a must that project definition is understood by the contractor, to avoid any conflict in future, as the contractor is responsible for detailed design and construction of the project. Owner has to involve for taking decisions during the selection of design alternatives and the monitoring of costs and schedules during construction and, therefore, the owner has to maintain/hire a team of qualified professionals to perform these activities.	• Reduce overall project time because construction begins before completion of design. • Singular responsibility, contractor takes care of the design, schedule, construction services, quality, methods, and technology. • For owner/client, the risk is transferred to design–build contractor. • Project cost defined in early stage and has certainty. • Early involvement of contractor assists constructability. • Suitable for straightforward projects where significant changes or risks are not anticipated and owner is able to precisely specify the objectives/requirements. • Risk management is better than design–bid–build.	• Not suitable for complex projects. • Owner has reduced control over design quality. • Extensive involvement to ensure design deliverable meets project performance requirements. • Extensive involvement of owner during entire life cycle of project. • Real price for a contract can not be estimated by the owner/client in the beginning. • Changes by owner can be expensive and may result in heavy cost penalties to the owner. • Poor identification of owner need and wrong understanding of project brief/concept can cause main problem during the project realization. • Project quality cannot be assured if it is not monitored properly by the owner. • For contractor, more risks in this type of contract. • Not suitable for renovation projects.

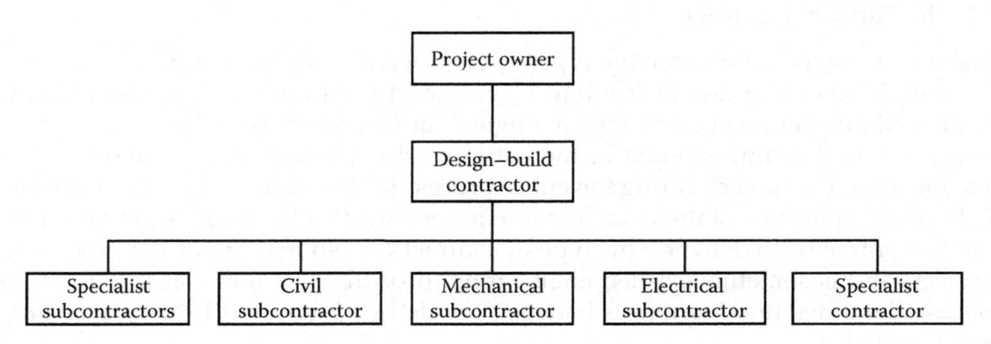

FIGURE 2.3
Design–build–delivery system.

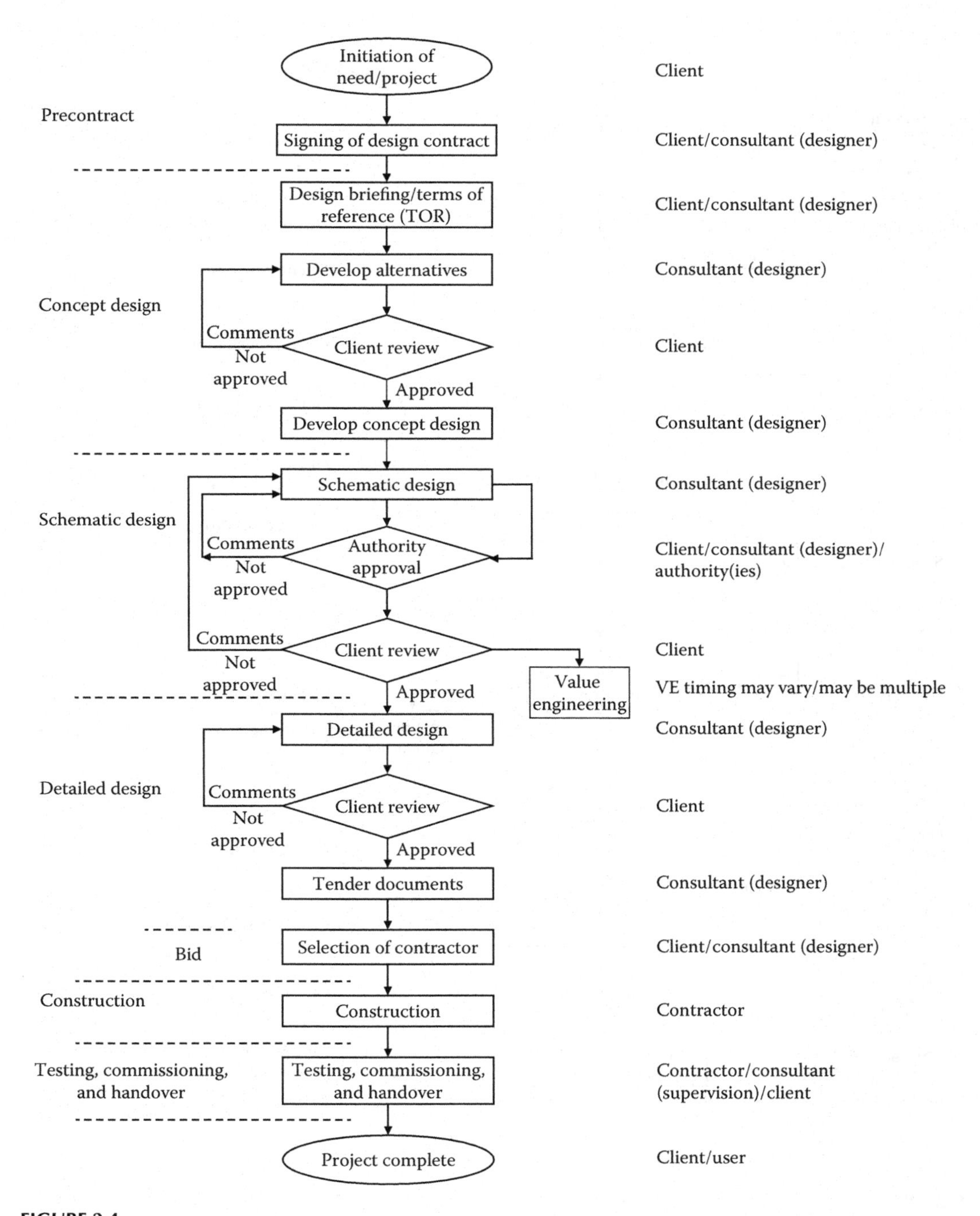

FIGURE 2.4
Logic flow diagram for a construction projects' design–bid–build system. (Abdul Razzak Rumane (2010), *Quality Management in Construction Projects*, CRC Press, Boca Raton, FL. Reprinted with permission from Taylor & Francis Group.)

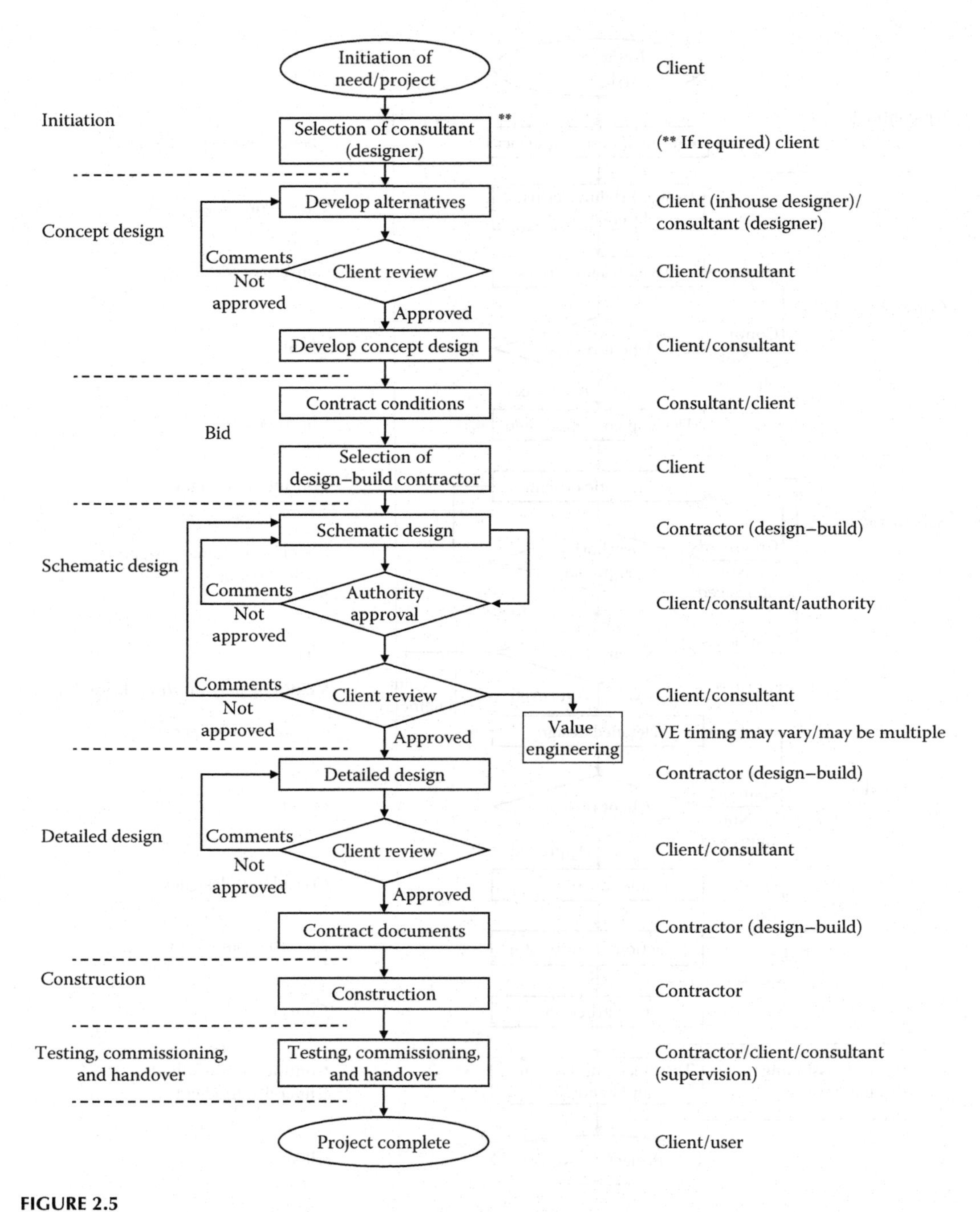

FIGURE 2.5
Logic flow diagram for a construction projects' design–build system. (Abdul Razzak Rumane (2010), *Quality Management in Construction Projects*, CRC Press, Boca Raton, FL. Reprinted with permission from Taylor & Francis Group.)

TABLE 2.4
Difference between Design–Build and EPC

Design–Build	EPC
Client/owner has an input in the outline design or design brief to understand the owner's intent of the project	Client has an input in the output (operating capacity)
Client/owner contracts a designer/consultant (in case the in-house facility is not available) to decide design outline	Engineering procurement construction (EPC) is a direct contract between the client and the owner to build the complete project to meet the agreed upon output
Detailed design is carried out by the design–build contractor (singular responsibility)	All project activities are carried out by the EPC contractor
Client/owner has to employ a team of professionals to perform supervision/management of the detail design, monitor quality, schedule, and cost	Client/owner is not involved in the detailed design process, except in the event of variation and quality procedures
Claims risk is higher	Claims risk is lower
Mainly suitable for building projects	Mainly used for projects such as power plants, process industry, and oil and gas sector

2.2.4 Build–Own–Operate–Transfer

This type of method is generally used by governments to develop public infrastructure by involving private sector to finance, design, operate, and manage the facility for a specified period and then transfer it to the same government free of charge. The terms build–own–operate–transfer (BOOT) and build–own–transfer (BOT) are used synonymously.

Examples of BOT projects include the following:

- Airports
- Bridges
- Motorways/toll roads
- Parking facilities
- Tunnels

Certain countries allow private sector to develop commercial and recreational facilities on the government land through the BOT scheme.

2.2.5 Project Manager

A project manager (PM) contract is used by the owner when the owner decides to turn over the entire project management to a professional PM. In the PM type of contract, the PM is owner's representative, and is directly responsible to the owner. The PM is responsible for planning, monitoring, and managing the project. In its broadest sense, the PM has the responsibility for all phases of the project from inception of the project till the completion and handing over of the project to the owner/end user. The PM is involved in giving advice to the owner and is responsible to appoint design professional(s), consultant, supervision firm, and select the contractor to construct the project. Table 2.5 illustrates the main aspects, advantages, and disadvantages of the PM type of project delivery system, and Figure 2.6 illustrates the contractual relationship.

TABLE 2.5
Project Manager Delivery System

Project Delivery System	Main Aspects	Advantages	Disadvantages
Project manager (PM)	A PM type delivery system is used by the owner when the owner decides to turn over entire project management to a professional PM. In PM type of contract, PM is owner's representative, and is directly responsible to the owner. The PM acts as a management consultant on behalf of the owner. The PM is responsible for planning, monitoring, and management of the project. In its broadest sense, the PM has responsibility for all the phase of the project from inception of the project till the completion and handing over of the project to the owner/end user. PM is involved in giving advice to the owner and is responsible to appoint design professional(s), consultant, supervision firm, and select the contractor/package contractor to construct the project.	• Owner/client retains full control of design. • Provide the opportunity for fast track or overlapping design and construction phases. • Reduces the owner's general management and oversight responsibilities. • Changes can be accommodated in unlit packages as long as there is no impact on time and cost. • Expert opinion and independent view toward constructability, cost, value engineering, and team member selection. • Multiprime type of delivery system is possible.	• Added project management cost to the owner. • Owner must have resources and expertise to deal with PM. • Owner cedes much of day-to-day control over the project to the PM. • PM is not at risk to the cost. • Owner continues to hold construction contracts and retains contractual liability.

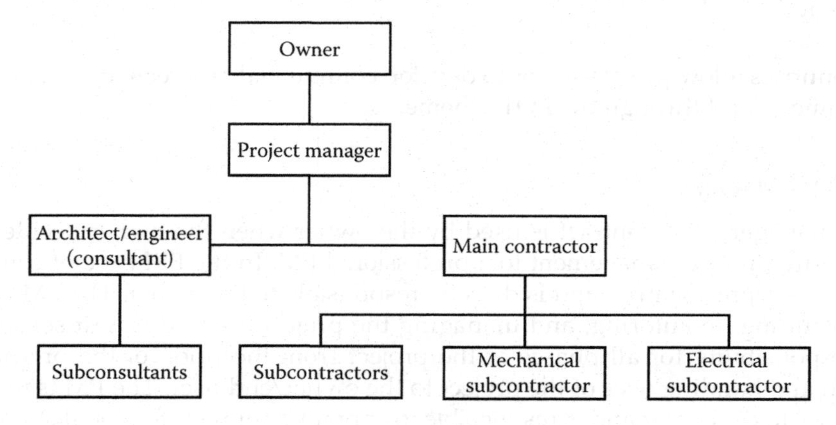

FIGURE 2.6
Project manager type delivery system contractual relationship.

2.2.6 Construction Management

In this method, the owner contracts a construction management (CM) firm to coordinate the project for the owner and provide CM services. There are two general forms of CM:

1. Agency CM
2. CM at risk (CM-at-risk)

The agency CM type is a management process type of contract system having a four-party arrangement involving the owner, designer, CM firm, and contractor. The construction manager provides advice to the owner regarding the cost, time, safety, and quality of materials/products/systems to be used on the project. The agency CM firm performs no design or construction, but assists the owner in selecting design firm(s) and contractor(s) to build the project. Agency CM could be implemented in conjunction with any type of project delivery system.

The basic concept of CM type of contract is that the firm is knowledgeable and capable of coordinating all aspects of the project to meet the intended use of the project by the owner. The agency construction manager acts as a principal agent to advise the owner/client, whereas the construction manager-at-risk is responsible for the on-site performance and actually performs some of the project works. CM-at-risk type of contract has two stages. The first stage encompasses preconstruction services and, during the second stage, the CM-at-risk is responsible for performing the construction work. The CM-at-risk project delivery system is also known as construction manager/general contractor (CM/GC).

Table 2.6 illustrates the main aspects, advantages, and disadvantages of CM-at-risk type of project delivery system; Figure 2.7 illustrates the contractual relationship for agency CM; and Figure 2.8 illustrates the contractual relationship for CM-at-risk.

2.2.7 Integrated Project Delivery

In this delivery method, the owner, designer, and contractor are contractually required to collaborate among themselves so that the risk responsibility and liability for the project delivery are collectively managed and appropriately shared. Table 2.7 illustrates the main aspects, advantages, and disadvantages of the integrated project delivery (IPD) type of project delivery system, and Figure 2.9 illustrates the contractual relationship among various parties.

In all types of project delivery system, there are mainly three participants:

1. Owner
2. Designer (A/E)
3. Contractor (known as the main contractor or general contractor)

However, if the owner engages a construction manager or PM, then it becomes a four-party contract. Table 2.8 summarizes the relationship among various project participants.

TABLE 2.6

Construction Management

Project Delivery System	Main Aspects	Advantages	Disadvantages
Construction management (CM)	In this method, owner contracts a CM firm to coordinate the project for the owner and provide CM services. There are two general types of CM type of contracts. They are 1. Agency CM 2. CM-at-risk The agency CM is a management process system, which can be implemented regardless of project delivery method. Agency CM has four party arrangement involving owner, designer, CM firm, and the contractor. CM-at-risk type of delivery system has two contracts, one between owner and designer and other one between owner and the CM-at-risk. CM-at-risk is selected on qualification basis. The construction manager, a person or a firm, provides advice to the owner regarding cost, time, safety, and about the quality of materials/ products/systems to be used on the project. The architect/engineer is responsible for design of the project. The basic concept of CM type of contract is that the firm is knowledgeable and capable of coordinating all aspects of the project to meet the intended use of the project by the owner. Agency construction manager acts as an advisor to the owner/client, whereas construction manager-at-risk is responsible for on-site performance and actually performs some of the project works while during the design phase construction manager acts as consultant to the owner and offer pre-construction services. The agency CM performs no design or construction, but assists the owner in selecting design firm(s), and contractor(s) to build the project.	• Client retains full control of project. • Agency CM helps owner providing advice during the design phase, bid evaluation, overseeing construction, managing project schedule, cost, and quality. • Suitable for large and complex projects with multiple phases and contract packages. • Since CM is an expert entity, it provides an independent view regarding constructability, cost, and value engineering. • Shorter project schedule. • Changes can be accommodated in unlit packages.	• Construction manager takes control of packages and interaction with contractors/ sub-contractors, but has no contractual role. • No cost certainty till final package is let. • Client retains all the risk. • Client has to manage the contractual agreements with each package contractor. • Client has to carry majority risk.

2.3 Types of Contracts/Pricing

Contract is a formalized means of agreement, enforceable by law, between two or more parties to perform certain works or provide certain services against agreed upon financial incentives to complete the works/services. Regardless of the type of project delivery system selected, the contractual arrangement by which the parties are compensated must also be established. In the construction projects, determining how to procure the product is as important as determining what and when. In most procurement activities, there are several options available for purchase or subcontracts. The basis of compensation type relates to the financial arrangement among the parties as to how the designer or contractor

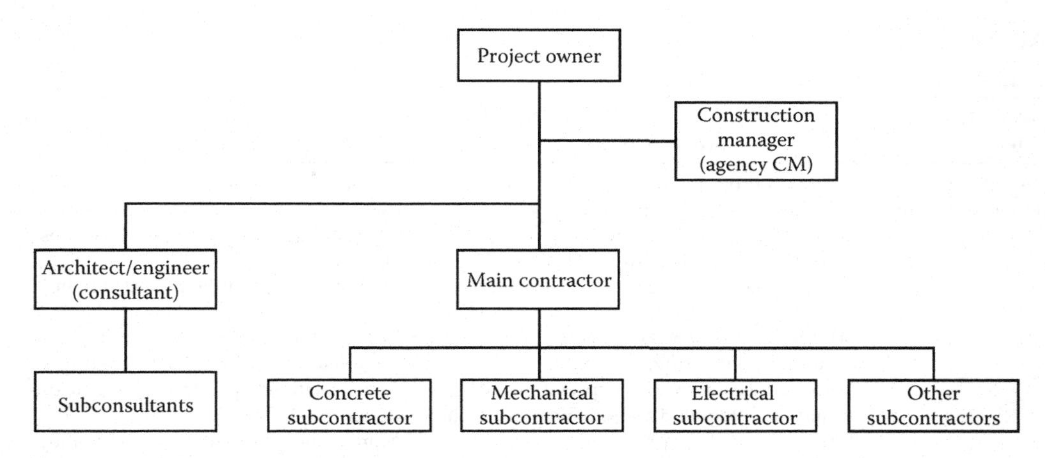

FIGURE 2.7
Agency construction management contractual relationship.

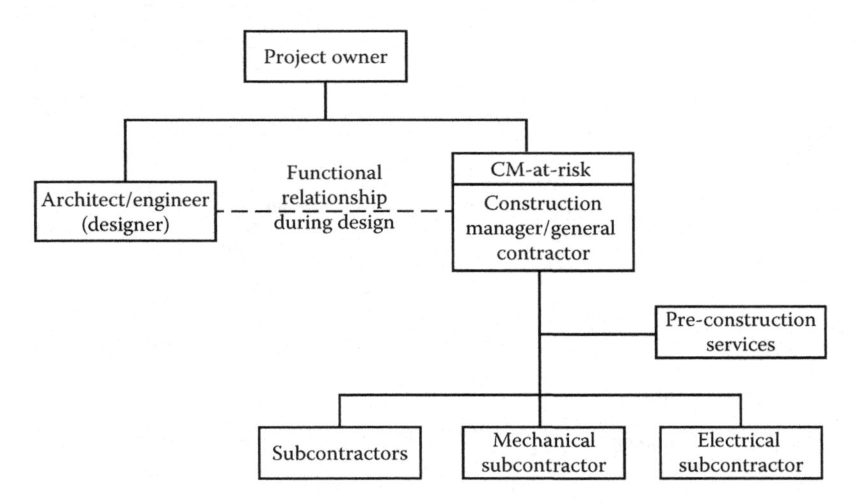

FIGURE 2.8
Construction manager at risk contractual relationship (CM-at-risk).

is to be compensated for their services. Following are the most common types of contract/compensation methods:

1. *Firm-fixed price or lump-sum contract*
   a. Firm-fixed price
   b. Fixed-price incentive fee
   c. Fixed price with economic adjustment price

   Table 2.9 illustrates advantages and disadvantages of fixed price or lump-sum price contracts.

TABLE 2.7
Integrated Project Delivery System

Project Delivery System	Main Aspects	Advantages
Integrated project delivery system	Integrated project delivery system is an alternative project delivery method distinguished by a contractual agreement at a minimum, between owner, design professional, and contractor where the risks and rewards are shared and stakeholder's success is dependent on project. It is a relational contracting, which sets performance and results expectations. It creates an integrated team, which results in optimization of the project. Value analysis throughout the design process, not at the end. It has full transparency wherein the information is shared openly and only consensus decisions are implemented.	• Integrated form of agreement between owner, architect/engineer, and contractor. • Owner's business plan is the basic design criteria. • Design and cost of the project stay in balance. • Common understanding of project parameters, objectives, and constraints. • Better relationship among project teams. • Better collaboration. • More certainty of outcome within the agreed upon target cost, schedule, and program. • Superior risk management. • Establishes a collaborative governance structure. • Incorporates Lean project delivery.

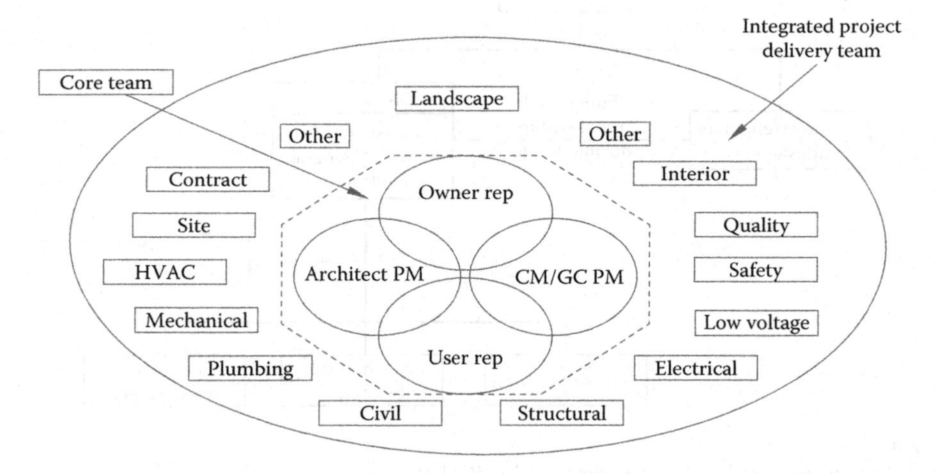

FIGURE 2.9
Integrated project delivery system contractual relationship.

2. *Unit price*
   a. Specific price for a particular task or unit of work performed by the contractor
   b. In this method, each unit of work is to be precisely defined

   Table 2.10 illustrates the advantages and disadvantages of unit price contracts.
3. *Cost reimbursement contract (cost plus)*
   a. Cost plus percentage fee
   b. Cost plus fixed fee
   c. Cost plus incentive fee
   d. Cost plus award fee

TABLE 2.8

Relationship Among the Project Participants. C—Contractual Relationship, F—Functional Relationship, and I—Integral Part of Contract

Sr. No.	Category	Classification	Sub-Classification	Owner/ Designer	Owner/ Contractor	Designer/ Contractor	Owner/ CM	CM/ Designer	CM/ Contractor
1	Traditional system	Design–bid–build	Design–bid–build	C	C	F	–	–	–
2	Integrated system	Design–build	Design–build	C	C	I	–	–	–
				Contractual relationship					
3	Management-oriented system	Construction management	Agency construction manager	C	C	–	C	F	F
			Construction manager-at-risk	C	–	C	C	F	C

TABLE 2.9

Fixed-Price/Lump-Sum Contracts

Sr. No.	Project Contracting Type	Main Aspects	Advantages	Disadvantages
1	Fixed price or lump sum	With this type of contract, the contractor agrees to perform the specified work/ services with a fixed-price (it is also called lump sum) for the specified and contracted work. Any extra work is executed only upon receipt of instruction from the owner. Fixed price contracts are generally inappropriate for work involving major uncertainties, such as work involving new technologies.	• Low financial risk to the owner. • Total cost of the project is known before construction. • Project viability is known before a commitment is made. • Suitable for projects, which can be completely designed and whose quantities are definable. • Minimal owner supervision. • Contractor will usually assign his best personnel to the work. • Maximum financial motivation of contractor. • Contractor has to solve his own problems and do so quickly. • Contractor selection by competitive bidding is fairly easy, apart from the deliberate low price.	• High financial risk to the contractor. • Owner bears the risk of poor quality from the contractor trying to maximize the profit within fixed cost. • Contractor's price may include high contingency. • Variation can be time consuming and costly. • Variations (changes) are difficult and costly. • More time taken for bidding and for developing a good design basis. • Contractor will tend to choose the cheapest and quickest solutions, making technical monitoring and strict quality control by the owner essential.

TABLE 2.10

Unit Price Contracts

Sr. No.	Project Contracting Type	Main Aspects	Advantages	Disadvantages
1	Unit price	The owner and contractor agree to structure the contract on specified unit price for the estimated quantities of the work. In unit price contract, the work to be performed is broken into various parts and a fixed price is established for each unit of work. This type of contract is well suited for repetitive or easily quantifiable tasks. A unit price contract provides benefits to both owners and contractors.	• Complete design definition is not essential for bidding process. • Suitable for competitive bidding and relatively easy contractor selection subject to sensitivity evaluation. • Typical drawings can be used for bidding. • Bidding is speedy and inexpensive, and an early start is possible. • Flexibility-depending on the contract conditions, the scope and quantity of work can be varied and easily adjustable.	• Final cost is not known at the outset. • Additional site staffs are needed to measure, control, and report on the cost and the status of the work. • Biased bidding and front end loading may not be detected.

Table 2.11 illustrates the advantages and disadvantages of cost reimbursement contracts.

4. *Remeasurement contract*
   a. Bill of quantities
   b. Schedule of rates
   c. Bill of materials

   Table 2.12 illustrates the advantages and disadvantages of reimbursement contracts.
5. *Target price contract*

   Table 2.13 illustrates the advantages and disadvantages of target price contracts.
6. *Time and material contract*

   Table 2.14 illustrates the advantages and disadvantages of time and material contracts.
7. *Guaranteed maximum price (GMP)*
   a. Cost plus fixed fee GMP contract
   b. Cost plus fixed fee GMP and bonus contract
   c. Cost plus fixed fee GMP with arrangement for sharing any cost-saving type contract

TABLE 2.11

Cost Reimbursement Contracts

Sr. No.	Project Contracting Type	Main Aspects	Advantages	Disadvantages
1	Cost reimbursement contract (cost plus)	It is a type of contract in which contractor agrees to do the work for the cost related to the project plus, an agreed upon amount of fee that covers profits and nonreimbursable overhead costs. Following are different types of cost plus contracts: 1. Cost plus % fee contract 2. Cost plus fixed fee contract 3. Cost plus incentive fee contract	• Suitable to start construction concurrently while design is in progress. • Suitable for renovation type of projects. • Suitable for projects expecting major changes. • Suitable for projects with possible introduction of latest technology. • Flexibility in dealing with changes. • An early start can be made. • Useful where site problems such as trade union actions like delays or disruptions may be encountered. • Owner can control all aspects of the work.	• Possibility of overspending by the contractor. • Difficult to predict final cost. • Project quality is likely to be affected, as the fee will be same no matter how low the costs are. • Final cost is unknown. • Difficulties in evaluating proposals, for example, L1 may not result in selecting a contractor in achieving lowest project cost. • Contractor has little incentive for early completion or cost economy. • Contractor may assign its second division personnel, make excessive use of agency personnel, or use the job as a training vehicle for new personnel. • Biased bidding of fixed fees and reimbursable rates may not be detected.

TABLE 2.12

Remeasurement Contracts

Sr. No.	Project Contracting Type	Main Aspects	Advantages	Disadvantages
1	Remeasurement contract	It is a contract in which contractor is paid as per the actual quantities of the work done. In this type of contract in which contractor agrees to do the work based on one of the following criteria: 1. Bill of quantities, or 2. Schedule of rates, or 3. Bill of material. In this type of contract, payment is linked to measure work completion.	• Suitable for competitive bidding. • Fair competition. • For contractor, low risk. • Suitable for projects where quantities of work cannot be determined in advance of construction. • Fair basis for competition.	• Require adequate breakdown and design definition of work unit. • Final cost not known with certainty until the project is complete. • Additional requirements of administrative staff to measure, control, and report.

TABLE 2.13

Target Price Contracts

Sr. No.	Project Contracting Type	Main Aspects	Advantages	Disadvantages
1	Target price contract	A target cost contract is based on the concept of top-down approach, which provides a fixed price for an agreed range of out-turn cost around the target. In this type of contract over run or under spend are shared by the owner and the contractor at a predetermined agreed upon percentages.	• Final cost is known. • Contractor may share the savings as per agreed upon percentages. • Flexibility in controlling the work. • Almost immediate start on the work, even without a scope definition. • Encourages economic and speedy completion (up to a point). • Contractor is rewarded for superior performance.	• Tight cost control is required. • Difficult to adjust major variations or cost inflation. • No opportunity to competitively bid the targets. • Difficulty in agreeing on an effective target for superior performance. • Variations are difficult and costly once the target has been established. • If the contractor fails to achieve the targets, it may attempt to prove the owner's fault.

2.3.1 Guaranteed Maximum Price

In this method, the contractor is compensated for actual costs incurred, in connection with the design and construction of the project, plus a fixed fee-all subject, however, to a ceiling above which the client is not obligated to pay. Its GMP contract price specifies a target profit (or fee), a price ceiling (but not for profit ceiling or floor), and a profit (or fee) adjustment formula. These elements are all negotiated at the outset. The GMP combines CM with

TABLE 2.14

Time and Material Contracts

Sr. No.	Project Contracting Type	Main Aspects	Advantages	Disadvantages
1	Time and material contract	A time and material contract has elements of unit price and cost plus type of contract. This type of contract is mainly used for maintenance contract and small projects.	• Owner pays the contractor based on actual cost for material and time as per pre-agreed rates. • Suitable for design services where it is difficult to determine total expected efforts in advance.	• Contractor must be accurate in estimating the price, which normally includes indirect and overhead cost.

design–build. With the GMP contract, amounts below the maximum are typically shared between the client and the contractor, whereas the contractor is responsible to absorb the cost above the maximum. Any changes that result from the specific instructions of the owner fall outside the guaranteed price. The cost plus GMP type, as it is also known, of contract is typically used

- When time pressure requires letting of the contract before design development is sufficiently advanced to allow a conventional lump-sum type of contract to be fixed.
- If this type of contract is likely to be less costly than other types.
- Where financing or other constraints preclude the use of alternatives such as the two-stage contract of CM.
- If it is impractical to obtain certain types of services with improved delivery or technical performance, or quality without the use of this type of contract.

In this method, the contractor and the owner have the knowledge that the drawings and specifications are not complete, and the contractor and the owner agree to work together to complete the drawings and specifications as provided in the contract agreement. This type of contract is weighted heavily in favor of the owner. The contractor takes on all the risk in this type of contracting system. Value engineering studies are conducted to identify design alternatives to help project maintain the budget and schedule. This type of contract needs

- Adequate cost pricing information for establishing a reasonable firm target price at the time of initial contract negotiation.
- Contractor's tendering/bidding department to have adequate information to provide the necessary data to support the negotiation of the final cost and incentive price revision.

TABLE 2.15

Cost Plus Guaranteed Maximum Price Contracts

Sr. No.	Project Contracting Type	Main Aspects	Advantages	Disadvantages
1	Cost plus guaranteed maximum price contract	With this type of contract, the owner and the contractor agree to a project cost guaranteed by the contractor as maximum. In this method, the contractor is compensated for actual costs incurred, in connection with design and construction of the project, plus a fixed fee-all subject, however, to a ceiling above which the client is not obligated to pay. This type of contract need adequate cost pricing information for establishing a reasonable firm target price at the time of initial contract negotiation.	• Maximum contract price is certain. • Construction can start at early stage.	• Contractor share the maximum risk. • High administration cost from owner side to monitor what the contractor is actually spending to get the benefit of under spending. • Contractor's tendering/bidding department should have adequate information to provide necessary data to support negotiation of final cost and incentive price revision.

- High administrative cost from the owner side to monitor what the contractor is actually spending to get the benefit of underspending.
- Evaluation of a minimum of two to three proposals for any major subcontract work.

In certain GMP contracts, the owner monitors and controls the contractor's expenses toward the project resources such as construction equipment, machinery, manpower, and staff on a monthly basis by fixing the basic price and profit percentage agreed at the initial stage.

Table 2.15 illustrates the advantages and disadvantages of the GMP type of contracts.

2.4 Selection of Project Delivery System

Each of the project delivery systems discussed earlier has advantages and disadvantages. It is a strategic decision that the owner has to take considering the suitability of an appropriate system to achieve successful completion of the project. Therefore, while selecting an appropriate project delivery system, the owner has to consider

- Size and complexity of the project.
- Type of project.
- Location of project.

- Owner's level of construction expertise (human resources available with the owner and owner's knowledge of CM practices).
- Owner's interest to exert influence/control over the design.
- Owner's interest to exert influence/control over the management of planning.
- Owner's interest to exert influence/control over the management of construction.
- Owner's interest to exert influence/control over the management of the project and the end user(s).
- Design.
- Schedule.
- Budget, funding mechanism.
- Quality.
- The risk the owner can tolerate (risk allocation).

2.4.1 Selection of Project Delivery Teams (Designer/Consultant, Contractor)

Construction projects involve three main parties:

1. Owner
2. Designer/consultant
3. Contractor

Participation involvement of all three parties at different construction phases is required to develop a successful facility/project. These parties are involved at different levels, and their relationship and responsibilities depend on the type of project delivery system and contracting system. Following are the common procurement methods for selection of project teams:

1. Low bid
    a. Selection is based solely on the price
2. Best value
    a. Total cost
        i. Selection is based on total construction cost and other factors.
    b. Fees
        i. Selection is based on weighted combination of fees and qualification.
3. Qualification-based selection (QBS)
    a. Selection is based solely on qualification

Table 2.16 illustrates the main aspects of the selection methods and selection criteria of a project delivery system team.

TABLE 2.16

Procurement Selection Types and Selection Criteria

Sr. No.	Selection Type	Description	Selection Criteria			
			D/B/B	D/B	CM-at -Risk	IPD
1	Low bid	Total construction cost including the cost of work is the sole criteria for final selection.	Yes	Yes	Not typical	Not typical
2	Best Value					
	2.1 Total cost	Both total cost and other factors are criteria in the final selection.	Yes	Yes	Not typical	Not typical
	2.2 Fees	Both fees and qualifications are factors in the final selection.	Not applicable	Yes	Yes	Yes
3	Qualification-based selection (QBS)	Total costs of the work are *not* a factor in the final selection. Qualification is the sole factor used in the final selection.	Not applicable	Yes	Yes	Yes

2.4.1.1 Selection of Designer/Consultant

Normally, design professionals (consultants) are hired on the basis of *qualifications*. The QBS can be considered as meeting one of the 14 points of Deming's principles of transformation: "End the practice of awarding business on the basis of price alone." The selection is solely based on demonstrated competence, professional qualifications, and experience for the type of services required. In QBS, the contract price is negotiated after selection of the best qualified firm. Table 2.17 lists the criteria for selection of construction project designer/consultant on the QBS basis.

Upon selection of the designer/consultant by the client, the designer is invited to submit the proposal for designing the services. Figure 2.10 illustrates the procedure for submission of design services proposal, and Figure 2.11 illustrates the overall scope of work normally performed by the designer/consultant.

2.4.1.1.1 Request for Proposal

It is a project-based process involving solution, qualifications, and price as the main criteria that define a winning proponent. It is a solicitation document requesting the proponents to submit a proposal in response to the required scope of services. The document does not specify in detail how to accomplish or perform the required services. Request for proposal (RFP) can range from a single-step process for straightforward procurement to a multistage process for complex and significant procurement. Table 2.18 illustrates the requirements to be included in the RFP document for a construction project designer/consultant.

2.4.1.2 Selection of Contractor

In construction projects, the selection process for the contractor can range from simply deciding to directly award a contract to a multistage process that involves information gathering about the contractor through the request for information (RFI), prequalification

TABLE 2.17
Qualification-Based Selection of Architect/Engineer (Consultant)

Sr. No.	Items to Be Evaluated
1	Organization's registration and license
2	Management plan and technical capability
3	Quality certification and quality management system
4	LEED or similar certification
5	Number of awards
6	Design capacity to perform the work
7	Financial strength and bonding capacity
8	Professional indemnity
9	Current load
10	Experience and past performance in a similar type of work
11	Experience and past performance of the proposed individuals in similar projects
12	Professional certification of the proposed individuals
13	Experience and past performance with the desired project delivery system
14	Design of similar value projects in the past
15	List of successfully completed projects
16	Proposed design approach in terms of 1. Performance 2. Effectiveness 3. Maintenance 4. Logistic support 5. Environment 6. Green building
17	Design team composition
18	Record of professionals in timely completion
19	Safety consideration in design
20	Litigation
21	Price schedule

Source: Abdul Razzak Rumane (2013), *Quality Tools for Managing Construction Projects*, CRC Press, Boca Raton, FL. Reprinted with permission from Taylor & Francis Group.

questionnaire (PQQ; request for qualification—RFQ), and soliciting activities. Normally, every owner maintains a list of prospective and previously qualified sellers. The existing list of potential contractors can be expanded by placing advertisement in publications such as newspapers or trade publications or professional journals. Internet search can also help generate a list of prospective sellers.

2.4.1.2.1 *Request for Information*

It is a procedure where potential contractors are provided with a general or preliminary description of the problem or need and are requested to provide information or advice about how to better define the problem or need or alternative solutions. It may be used to assist in preparing a solicitation document.

2.4.1.2.2 *RFQ/Prequalification Questionnaire*

RFQ/PQQ is a process that enables one to prequalify contractors for a particular requirement and avoid having to struggle with a large number of lengthy proposals. It is a

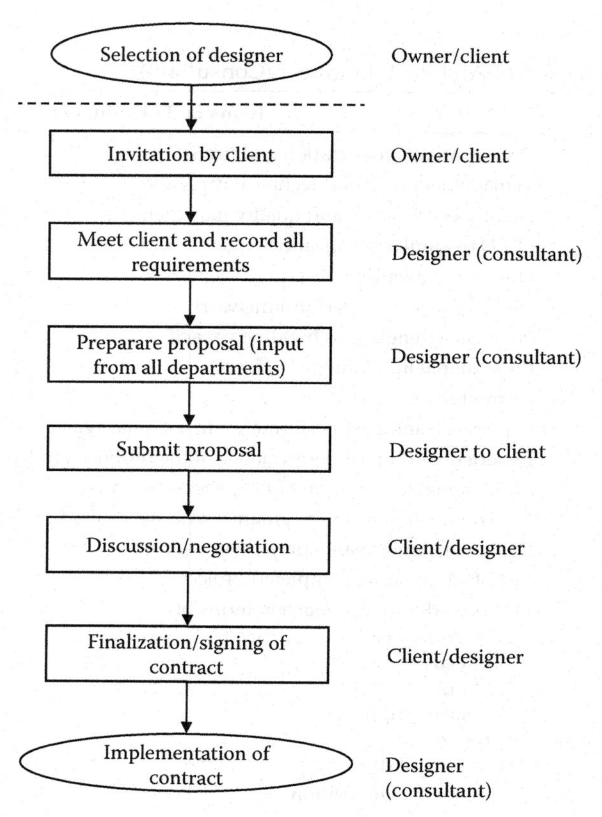

FIGURE 2.10
Contract proposal procedure for the designer. (Abdul Razzak Rumane (2013), *Quality Tools for Managing Construction Projects*, CRC Press, Boca Raton, FL. Reprinted with permission from Taylor & Francis Group.)

solicitation document requesting the proponents to submit the qualifications and special expertise in response to the required scope of services. This process is used to shortlist qualified proponents for the procurement process. Table 2.19 illustrates RFQ or the PQQ to select a contractor.

2.4.1.2.3 Request for Proposal

It is a project-based process involving solution, qualifications, and price as the main criteria that define a winning proponent. It is a solicitation document requesting the proponents to submit the proposals in response to the required scope of services. The document does not specify in detail how to accomplish or perform the required services. RFP can range from a single-step process for straightforward procurement to a multistage process for complex and significant procurement.

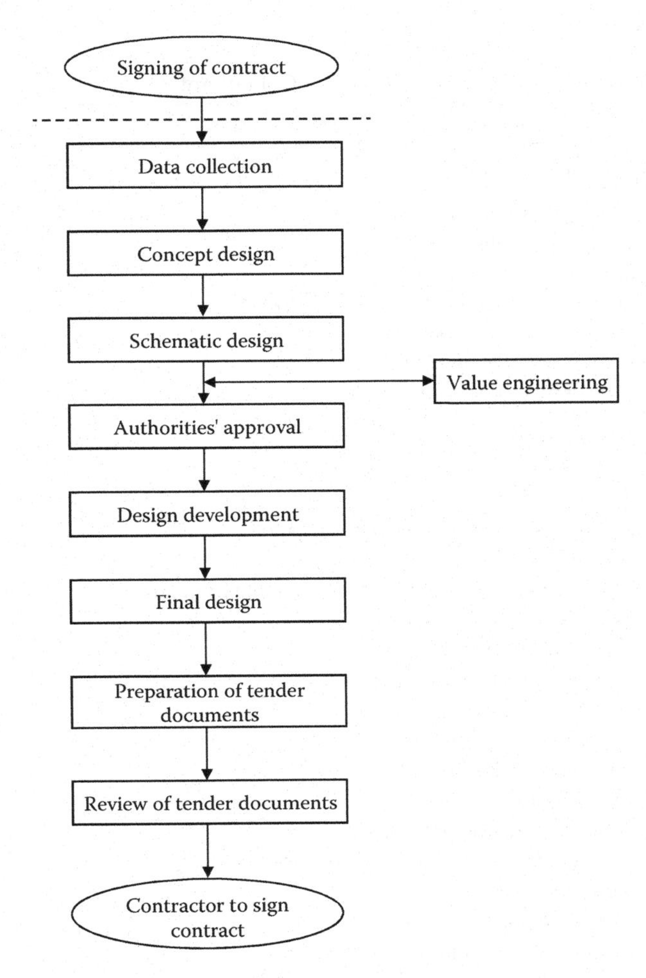

FIGURE 2.11
Overall scope of work of the designer. (Abdul Razzak Rumane (2013), *Quality Tools for Managing Construction Projects*, CRC Press, Boca Raton, FL. Reprinted with permission from Taylor & Francis Group.)

2.4.1.2.4 *Request for Quotation*

RFQ is a price-based bidding process that is used when complete documents consisting of defined project deliverables, solution, specifications, performance standards, and schedules are known. Potential bidders are provided with all the related information (documents)—except price—and are requested to submit the price, and the evaluation of the bids is done only on the basis of *price* subject to fulfilling all the required conditions. Most construction contractors are selected on the *price* basis.

TABLE 2.18

Contents of Request for Proposal for a Designer/Consultant

Sr. No.	Content
Project Details (Project Objectives)	
1	Introduction
2	Project description
3	Project delivery system
4	Designer's/consultant's scope of work
5	Preliminary project schedule
6	Preliminary cost of project
7	Type of project delivery system
Sample Questions (Information for Evaluation)	
1	Consultant name
2	Address
3	Quality management system certification
4	Organization details
5	Type of firm such as partnership or limited company
6	Is the firm listed in stock exchange?
7	List of awards, if any
8	Design production capacity
9	Current workload
10	Insurance and bonding
11	Experience and expertise
12	Project control system
13	Design submission procedure
14	Design review system
15	Design management plan
16	Design methodology
17	Submission of an alternate concept
18	Quality management during the design phase
19	Design firm's organization chart 1. Responsibility matrix 2. CVs of design team members
20	Designer's experience with green building standards or highly sustainable projects
21	Conducting value engineering
22	Authorities approval
23	Data collection during the design phase
24	Design responsibility/professional indemnity
25	Designer's relationship during construction
26	Preparation of tender documents/contract documents
27	Review of tender documents
28	Evaluation process and criteria
29	Any pending litigation
30	Price schedule

Source: Abdul Razzak Rumane (2013), *Quality Tools for Managing Construction Projects*, CRC Press, Boca Raton, FL. Reprinted with permission from Taylor & Francis Group.

TABLE 2.19

Request for Qualification (Prequalification of Contractor)

Sr. No.	Element	Description
1	General information	a. Company name b. Full address c. Registration details/business permit d. Management details e. Nature of company such as partnership, share holding f. Stock exchange listing, if any g. Affiliated/group of companies, if any h. Membership of professional trade associations, if any i. Award winning project, if any j. Quality management certification
2	Financial information	a. Yearly turnover b. Current workload c. Audited financial report d. Tax clearance details e. Bank overdraft/letter of credit capacity f. Performance bonding capacity g. Insurance limit
3	A. Organization details (general)	a. Core area of business b. How long in the same field of operation c. Quality control/assurance organization d. Grade/classification, if any
	B. Organization details (experience)	a. Number of years in the same business b. Technical capability i. Engineering ii. Shop drawing production c. List of previous contracts i. Name of project ii. Value of each contract iii. Contract period of each contract iv. Contract completion delay d. List of failed/uncompleted contracts e. Overall tender success rate f. Claims/dispute/litigation
4	Resources	a. Human resources i. Management ii. Engineering staff iii. CAD technicians iv. List of key project personnel v. Skilled labors—permanent/temporary vi. Unskilled labors—permanent/contracted b. List of equipment, machinery plant c. Human resource development plan
5	Health, safety, and environmental	a. Medical facility b. Safety record c. Environmental awareness
6	Project reference	
7	Bank reference	

Source: Abdul Razzak Rumane (2013), Quality Tools for Managing Construction Projects, CRC Press, Boca Raton, FL. Reprinted with permission from Taylor & Francis Group.

Table 2.20 illustrates the differences between RFP and RFQ. In construction projects, the contractor's bids and proposals depend on the type of delivery system and contracting methods. Table 2.21 is an example of the evaluation criterion for shortlisting/selection of contractors.

TABLE 2.20

Difference between Request for Proposal and Request for Qualification

Request for Proposal	Request for Qualification
Solicitation document requesting proposal in response to the required scope of services	Solicitation document requesting submittal of qualifications or expertise in response to the required scope of services.
Does not specify in detail how to accomplish or perform the required services	Specify specific process, critical steps, and approach to accomplish the required services.
The process is used when the services are not well defined	The process is used to shortlist qualified proponents (contractors, professionals) for use in future projects.
RFP is normally used for qualification-based selection (QBS)	Selection is based on the ability of the firm to achieve project objectives effectively and efficiently.
The document includes the statement that QBS will be used	

TABLE 2.21

Contractor Selection Criteria

Evaluation criteria	Weightage	Key points for consideration	Review result
1. General information			
1 Company information		Company's current position—a MUST information	Yes or No
2. Financial	25%		
1 Total turn over (last 5 years)	25%	Sum of the turn over for the last five years	
2 Values of current work-in-hand	25%	Project value/value of current work-in-hand	
3 Audit financial reports	10%	To confirm the ratio given in point three	
4 Financial standing	30%		
33% Assets		Current assets/current liabilities	
34% Liabilities		Total liabilities—total equity/total assets	
33% Profit/loss		Net profit before tax/total equity	
5 Bonding and insurance limit	10%		
60% a. Performace and bonding capacity		Provided or not provided	
40% b. Insurance		Provided or not provided	
3. Organization details			
3a. Business	20%		
1 Company's core area of business	30%	Degree of satisfactory answer	
2 Experience of years in business	30%	No. of years	
3 ISO certification	15%	Yes or no	
4 Registration/classification status	15%	Grade or classification	
5 Organizational chart	10%	Key staff indicated (name/title), balanced resources, departmental (specialization) diversity, lines of communication	
6 Dispute/claims		Degree of satisfactory answer	
3b. Experience	30%		
50% a. Projects' value		No. of projects with comparable value	
50% b. Projects' type (similar type and complexity)		No. of projects with similar complexity	
4. Resources	20%		
1 Personnel	60%		
30% Management		No. of managerial staff	
30% Engineers		No. of engineers and project staff	
30% Technicians		No. of CAD technicians and foreman	
10% Staff development		% turnover spent on training	
2 Technology	10%	% of turnover spent on acquiring latest construction technology	
3 Plant and equipment	30%	List of plant and equipment	
5. General	5%		
1 Bank references	30%	Provided or not provided	
2 Project references	30%	Provided or not provided	
2 Health, safety, and environment narration	40%	Degree of satisfactory answer	

Source: Abdul Razzak Rumane (2013), *Quality Tools for Managing Construction Projects*, CRC Press, Boca Raton, FL. Reprinted with permission from Taylor & Francis Group.

3

Construction Management Delivery System

Abdul Razzak Rumane

CONTENTS

3.1 Introduction

Construction is all about translating an owner's goals and objectives by the contractor to build the facility as stipulated in the contract documents, plans, and specifications within the budget and on schedule.

Construction projects are mainly capital investment projects. They are customized and nonrepetitive in nature. The technological complexity of construction projects is continuously increasing. Construction projects have become more complex and technical, and the relationships and the contractual grouping of those who are involved are also more complex and contractually varied. In addition, the requirements of construction clients are on the rise and, as a result, construction projects (buildings and other facilities) must meet various performance standards (climate, rate of deterioration, and maintenance). Therefore, to ensure the adequacy of client brief that addresses the numerous complex client/user need, it is now neccessary to evaluate the requirements in terms of activities and their interrelationship. The products used in construction projects are expensive, complex, immovable, and long lived. Generally, a construction project is composed of building materials (civil), electromechanical items, finishing items, and equipment. These are normally produced by other construction-related industries/manufacturers. These industries produce products as per their own quality management practices complying with certain quality standards or against specific requirements for a particular project. The owner of the construction project or his representative has no direct control over these companies unless he/his representative/appointed contractor commit to buy their product for use in their facility. These organizations may have their own quality management program. In manufacturing or service industries the quality management of all in-house manufactured products is performed by the manufacturer's own team or under the control of the same organization having jurisdiction over their manufacturing plants at different locations.

Construction projects comprise a cross section of many different participants. These participants are both influenced by and depend on each other in addition to *other players* involved in the construction process.

Traditional construction projects involve three main groups. These are as follows:

1. *Owner*—A person or an organization that initiates and sanctions a project. He/she requests the need of the facility and is responsible for arranging the financial resources for creation of the facility.
2. *Designer (architect/engineer)*—This consists of architects or engineers or consultant. They are appointed by the owner and are accountable to convert the owner's conception and need into specific facility (project) with detailed directions through drawings and specifications within the economic objectives. They are responsible for the design of the project and in certain cases supervision of construction process.
3. *Contractor*—A construction firm engaged by the owner to perform and complete the construction of specific facility (project) by providing the necessary staff, workforce, resources, materials, equipment, tools, and other accessories to the satisfaction of the owner/end user in compliance with the requirements of contract documents. The contractor is responsible for implementing the project activities and to achieve the owner's objectives.

Construction projects are executed based on predetermined set of goals and objectives. In order to process the construction project in an effective and efficient manner and to improve the control and planning, construction projects are divided into various phases. Traditionally, there have been five phases of a construction project life cycle that are further subdivided into various activities. These are conceptual design, preliminary design (schematic design), detail engineering (design development), construction and testing, commissioning, and handing over.

Participation involvement of all three parties at different levels of construction phases is required to ensure completion of construction and making the project most qualitative, competitive, and economical. Construction projects involve coordinated actions and input from many professionals and specialists to achieve defined objectives. There are several types of project delivery systems and contracting systems in which these parties are involved at different levels. All these contract deliverable systems follow generic life-cycle phases of construction project; however, the involvement/participation of various parties differs depending on the type of deliverable system adapted for a particular project.

Complex and major construction projects have many challenges such as delays, changes, disputes, and accidents at the site, and, therefore, the projects need to be efficiently managed from the beginning to the end to meet the intended use and owner's expectations. The main area of construction management covers planning, organizing, executing, and controlling to ensure that the project is built as per the defined scope, maintaining the completion schedule and within the agreed upon budget. The owner/client may not have necessary staff/resources in-house to manage planning, design, and construction of the project to achieve the desired results. Therefore, in such cases, the owners engage a professional firm or a person called construction manager (CM), who is trained and has expertise in the management of construction processes, to assist in developing bid documents, overseeing, monitoring, controlling, and coordinating the project for the owner. The basic construction management concept is that the owner assigns the contract to a firm or a person who is knowledgeable and capable of coordinating all the aspects of the project to meet the owner's intended use of the project. In construction management type of project delivery system, consultants (architects/engineers) prepare complete design drawings and contract documents, then the project is put up for a competitive bid and the contract is awarded to the competitive bidder (contractor). The owner hires a third party (CM) to oversee and coordinate the construction. The CM brings knowledge and experience that contribute to decisions at every stage of the project for its successful completion.

3.2 Project Management and Construction Management

As per PMI *PMBOK® Guide* (fifth edition), project management is the application of knowledge, skills, tools, and techniques to project activities to meet the project requirements. Project management completely deals with five processes (initiating, planning, executing, monitoring and control, and closing) that are divided into 13 knowledge areas spread over 47 locally grouped project management processes. Project management involves managing all types of projects from start to finish.

Construction management is a professional management practice applied effectively to the construction project from the inception to the completion of the project for the purpose of managing (planning, organizing, executing, and controlling) schedule, cost, scope, and quality. Construction management services are generally offered by registered engineering firms/professionals having the ability and expertise to manage construction projects.

Construction management is a discipline and management system specially tailored to promote the successful execution of capital and complex projects.

Construction management mainly involves construction-related management activities such as

- Planning
- Scheduling
- Monitoring and control
- Quality control/quality assurance
- Human resources
- Material and equipment
- Safety and environmental protection

TABLE 3.1

Difference between Project Manager and Construction Management Types of Project Delivery Systems

Project Manager	Construction Management
Project Manager is a conventional method of construction administration.	Construction management is a progressive, more convenient method of construction project administration.
Client/owner retains full control of design and construction.	There are two forms of construction management. 1. Agency CM 2. CM-at-risk
Client/owner retains direct control over all aspects and quality of the project.	In the agency CM type of management system, the client/owner has three contracts. One between the owner and the designer, one between the owner and the contractor, and one between the owner and the CM.
Client/owner has direct contract with the designer (consultant) and contractor, subcontractors.	In the CM-at-risk type of project delivery system, the client/owner has two contracts, one between the owner and the designer and one between the owner and the CM-at-risk/general contractor (GC).
Client/owner has direct contractual relationship with both the architect/engineer (consultant) and the contractor.	In the agency CM type of management system, the CM is not at risk for the budget, schedule, or the performance of the work.
Client/owner selects the consultant and the contractor.	In the CM-at-risk type of project delivery system, construction services contracts between the GC and the contractors of different trades.
	In the CM-at-risk type of project delivery system, the CM is responsible for performing the construction and calculates financial liability to complete the project within schedule and budget.

The CM is responsible for overseeing the performance of the contractor(s) toward construction-related activities (engineering design, construction process, testing, commissioning, and handover).

The CM and the project manager are different types of project delivery systems. Table 3.1 illustrates the basic differences between project manager and construction management types of project management systems.

3.3 Roadmap for Construction Management

The Figure 3.1 represents pictorial roadmap which covers all the activities that are discussed in Sections 3.3.1 through 3.3.5.

3.3.1 Project Initiation

Most construction projects begin with the recognition of a new facility. The owner of the facility could be an individual, a public/private sector company, or a governmental agency. The project development process begins with the project initiation and ends with the project closure and finalization of the project records. The project initiation starts with the identification of a business case and its needs.

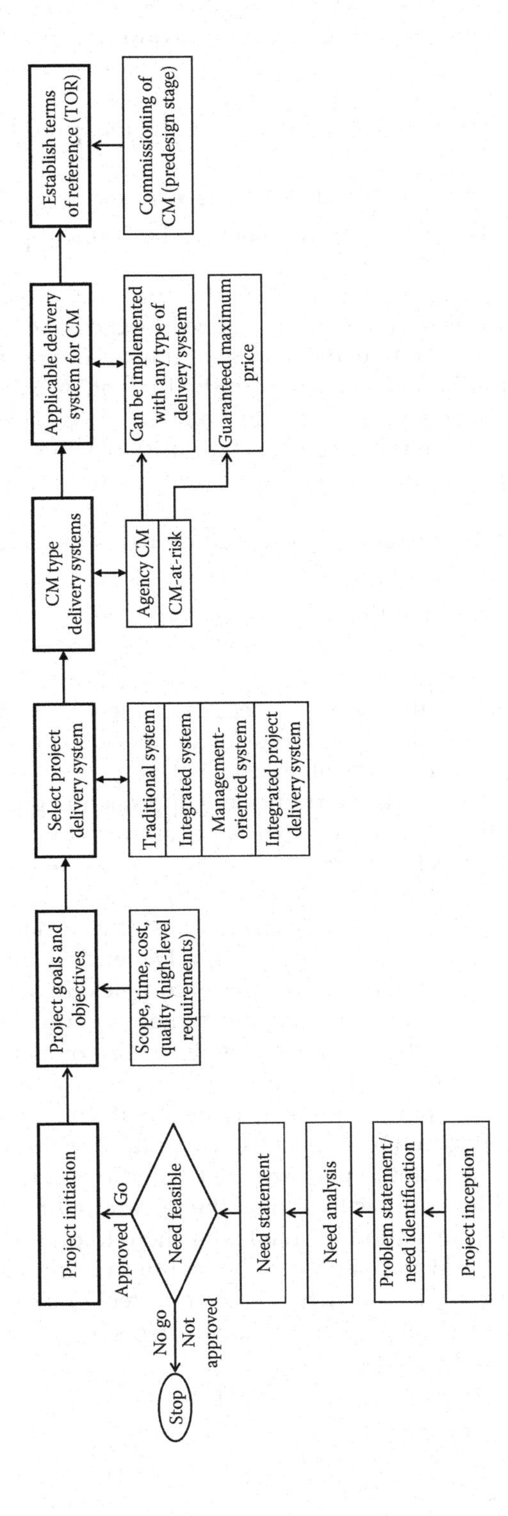

FIGURE 3.1
Construction management roadmap.

The owner creates the need of the project, which is linked to the available financial resources to develop the facility. The owner's needs are quite simple and are based on the following:

- To have the best value for money, that is, to have the maximum profit or services at a reasonable cost
- On-time completion, that is, to meet the owner's/user's schedule
- Completion within the budget, that is, to meet the investment plan for the facility

The owner's need must be well defined, indicating the minimum requirements of quality and performance, an approved main budget, and the required completion date. Sometimes, the project budget is fixed and, therefore, the quality of the building system, materials, and finishes of the project need to be balanced within the budget. A business case typically addresses the business need for the project and the value the project brings to the business (project value proposition). A value proposition is a promise of value to be delivered by the project. The following questions address the value proposition:

1. How the project solves the current problems or improves the current situation?
2. What specific benefits the project will deliver?
3. Why the project is the ideal solution for the problem?

Business need assessment is essential to ensure that the owner's business case has been properly considered before the initial project brief (need statement) is developed. Table 3.2 lists the major points to be considered for the need analysis of a construction project and Table 3.3 illustrates the need statement.

Once the owner's need is identified, the traditional approach is pursued through need analysis and then a feasibility study or an economical appraisal of the owner's needs or benefits, taking into account the relevant moral, social, environmental, and technical constraints. The feasibility study takes its starting point from the output of the project identification need. The feasibility study is conducted to assist the owner/decision-makers in deciding what will be in the best interest of the owner. Depending on the circumstances, the feasibility study may be short or lengthy, simple or complex. In any case, it is the principal requirement in project development as it gives the owner an early assessment of the viability of the project and the degree of risk involved. The owner usually performs a project feasibility study with the help of his or her own team or by engaging individuals/organizations involved in the preparation of economical and financial studies. However, the feasibility study can be conducted by a specialist consultant in this field. Table 3.4 illustrates the qualification of a consultant to perform the feasibility study.

The objective of the feasibility study is to review the technical/financial viability of the project and to give sufficient information to enable the client to proceed or abort the project. A feasibility study is undertaken to analyze the ability to complete a project successfully, taking into account various factors such as economic, technological, and scheduling. A feasibility study looks into the positive and negative effects of a project before investing the company resources, that is, time and money.

Following are the contents of a feasibility study report:

1. Purpose of the feasibility study
2. Project history (project background information)

TABLE 3.2

Major Considerations for Need Analysis of a Construction Project

Sr. No.	Points to Be Considered
1	Is the project in line with the organization's strategy/strategic plan and mandated by management in support of a specific objective?
2	Is the project a part of a mission statement of the organization?
3	Is the project a part of a vision statement of the organization?
4	Is the need mandated by a regulatory body?
5	Is the need for meeting government regulations?
6	Is the need to fulfill the deficiency/gap of such type of project(s) in the market?
7	Is the need created to meet market demand?
8	Is the need to meet the research and development requirements?
9	Is the need for technical advances?
10	Is the need generated to construct a facility/project that is innovative in nature?
11	Is the need aiming to improve the existing facility?
12	Is the need a part of mandatory investment?
13	Is the need to develop infrastructure?
14	Is the need necessary to serve the community and fulfill social responsibilities?
15	Is the need created to resolve a specific problem?
16	Is the need going to have an effect on the environment?
17	Does the need have any time frame to implement?
18	Does the need have financial constraints?
19	Does the need have major risks?
20	Is the need within the capability of the owner/client, either alone or in cooperation with other organizations?
21	Can the need be managed and implemented?
22	Is the need realistic and genuine?
23	Is the need measurable?
24	Is the need beneficial?
25	Does the need comply with environmental protection agency requirements?
26	Does the need comply with the government's health and safety regulations?

Source: Abdul Razzak Rumane (2013), *Quality Tools for Managing Construction Projects*, CRC Press, Boca Raton, FL. Reprinted with permission from Taylor & Francis Group.

3. Description of proposed project
    a. Project location
    b. Plot area
    c. Interface with adjacent/neighboring area
    d. Expected project deliverables
    e. Key performance indicators
    f. Constraints
    g. Assumptions
4. Business case
    a. Project need
    b. Stakeholders
    c. Project benefits

TABLE 3.3

Need Statement

Sr. No.	Points to Be Considered
1	Project purpose and need a. Project description
2	What is the purpose of the project? a. Project justification
3	Why is the project needed now?
4	How is the need of the project determined? a. Supporting data
5	Is it important to have the needed project?
6	Whether such facility/project is required?
7	What are the factors contributing to the need?
8	What is the impact of the need?
9	Will the need improve the existing situation and be beneficial?
10	What are the hurdles?
11	What is the time line for the project?
12	What are funding sources for the project?
13	What are the benefits of the projects?
14	What are the environmental impacts?

Source: Abdul Razzak Rumane (2013), *Quality Tools for Managing Construction Projects*, CRC Press, Boca Raton, FL. Reprinted with permission from Taylor & Francis Group.

TABLE 3.4

Consultant's Qualification for Feasibility Study

Sr. No.	Description
1	Experience in conducting feasibility study
2	Experience in conducting feasibility study in similar type and nature of projects
3	Fair and neutral with no prior opinion about what decision should be made
4	Experience in strategic and analytical analysis
5	Knowledge of analytical approach and background
6	Ability to collect large number of important and necessary data via work sessions, interviews, surveys, and other methods
7	Market knowledge
8	Ability to review and analyse market information
9	Knowledge of market trend in similar type of projects/facility
10	Multidisciplinary experienced team having proven record in the following field; a. Financial analyst b. Engineering/Technical expertise c. Policy experts d. Project scheduling
11	Experience in review of demographic and economic data

Source: Abdul Razzak Rumane (2013), *Quality Tools for Managing Construction Projects*, CRC Press, Boca Raton, FL. Reprinted with permission from Taylor & Francis Group.

   d. Financial benefits
   e. Estimated cost
   f. Estimated time
   g. Justification
5. Feasibility study details
   a. Technical
   b. Economical
   c. Financial
   d. Time scale
   e. Environmental
   f. Ecological
   g. Sustainability
   h. Political
   i. Social
6. Risk
7. Environmental impact (considerations)
8. Social impact (considerations)
9. Final recommendation

The outcome of the feasibility study helps to select a defined project that meets the stated project objectives, together with a broad plan of implementation. If the feasibility study shows that the objectives of the owner are best met through the ideas generated, then the project is moved to the next stage of the project life cycle to deliver the intended objectives.

3.3.2 Project Goals and Objectives

Project goals and objectives are prepared by taking into consideration the final recommendations/outcome of the feasibility study. Clear goals and objectives provide the project team with appropriate boundaries to make decisions about the project and ensure that the project/facility will satisfy the owner's/end user's requirements and fulfill owner's needs. Establishing properly defined goals and objectives is the most fundamental element of project planning. Therefore, the project goals and objectives must be

- Specific (Is the goal specific?).
- Measurable (Is the goal measurable?).
- Agreed upon/achievable (Is the goal achievable?).
- Realistic (Is the goal realistic or result-oriented?).
- Time (cost) limited (Does the goal have a time element?).

3.3.3 Select Project Delivery System

A *project delivery system* is defined as the organizational arrangement among various participants comprising the owner, designer, contractor, and many other professionals involved in the design and construction of a project/facility to translate/transform the owner's needs/goals/objectives into a finished facility/project to satisfy the owner's/end user's requirements. There are different types of project delivery systems followed in construction projects; however, each of the project delivery systems is a variation of the following basic types:

- Design–bid–build (traditional delivery method)
- Design–build (integrated system)
- Management oriented system
    - Project manager
    - Construction management
- Integrated project delivery system

3.3.4 Construction Management Delivery System

In the construction management type of management process, the owner retains control of the design and direction. It is a fully integrated design and construction process, thus minimizing changes, disputes, and delay in the completion of the project.

There are two forms of construction management systems:

1. Agency construction management
2. Construction management-at-risk

3.3.4.1 Agency Construction Management

Agency construction management is a management process where the CM acts as an advisor to the owner. The owner may engage the CM for the entire life cycle of the construction project or during a specific phase of the construction project. Normally, the CM is engaged as early in the project as possible to guide and assist the owner through all the phases of the project or for a specific phase. Agency CM can be used in conjunction with any project delivery system. The agency CM is always with the owner. This type of management is a fee-based service in which the CM is exclusively responsible to the owner and acts in the owner's interest. With the agency CM type of construction management system, the owner holds the contracts directly with the general contractor and also assumes the risks of delivery, including cost and schedule, quality, safety, performance, errors, and omissions. The agency CM has no financial stakes in the project and does not hold any subcontracts.

The agency CM type of management system can be used with different types of project delivery systems. Given next are figures that illustrate contractual relationship of agency CM with different types of project delivery systems and their corresponding sequential activities.

Figure 3.2 illustrates the contractual relationship of the design–bid–build–delivery system using agency CM, and Figure 3.3 illustrates the sequential activities for this type of delivery system.

Figure 3.4 illustrates the contractual relationship of the multiple-prime contractor delivery system using agency CM, and Figure 3.5 illustrates the sequential activities for this type of delivery system.

Figure 3.6 illustrates the contractual relationship of the design–build–delivery system using agency CM, and Figure 3.7 illustrates the sequential activities for this type of delivery system.

Figure 3.8 illustrates the contractual relationship of the agency CM type of management system, and Figure 3.9 illustrates the sequential activities for this type of delivery system.

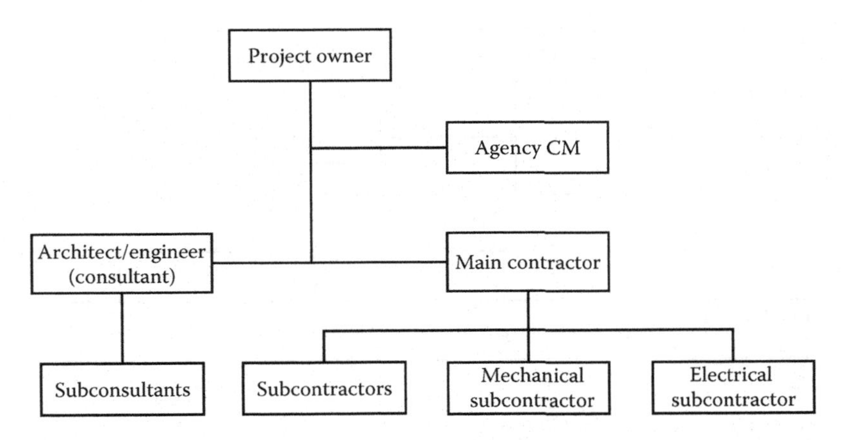

FIGURE 3.2
Design–bid–build with agency CM.

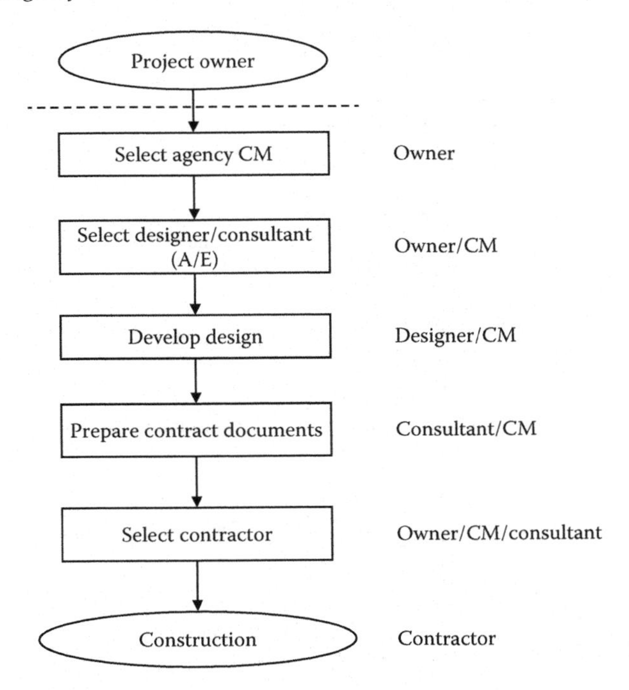

FIGURE 3.3
Sequential activities of agency CM design–bid–build–delivery system.

Figure 3.10 illustrates the contractual relationship of the CM-at-risk delivery system using agency CM, whereas Figure 3.11 illustrates the sequential activities for this type of delivery system.

3.3.4.2 Construction Management at Risk

CM-at-risk is a project delivery system. CM-at-risk is selected based on the qualification, experience, and reputation of the CM. In this system, the CM enters into a contract with the owner at an early stage and becomes a member of a collaborative project team with

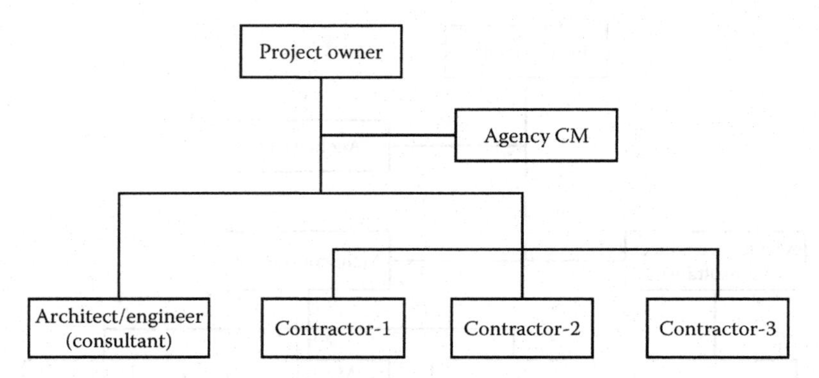

FIGURE 3.4
Multiple-prime contractor contractual relationship.

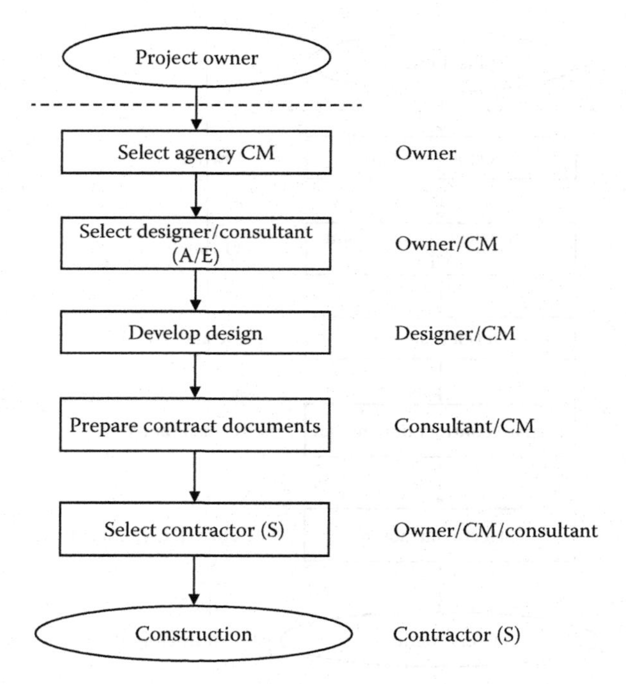

FIGURE 3.5
Sequential activities of agency CM–multiple prime contractor delivery system.

that of A/E (designer). A CM-at-risk typically contracts with the owner in two stages. Design services and construction services are contracted separately (vs. design–build, where the contracts are combined). In the first stage, the CM acts as an advisor to the owner and assists in the development of conceptual and preliminary design phase. The second phase involves development of detail design and completion of construction for a negotiated fixed or guaranteed maximum price (GMP). In a CM-at-risk arrangement, the criteria for final selection of the contactor are not based on the total construction

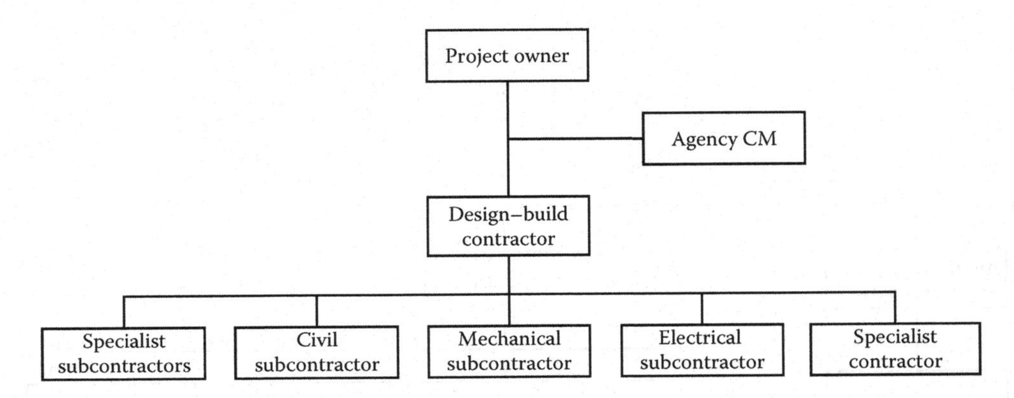

FIGURE 3.6
Design–build–delivery system with agency CM.

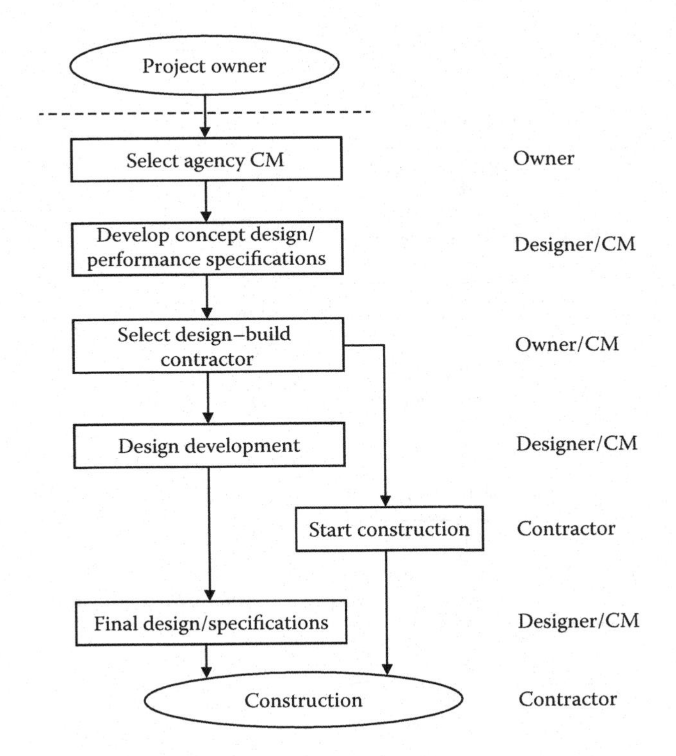

FIGURE 3.7
Sequential activities of agency CM—design–build–delivery system.

cost (vs. design–bid–build–delivery system where the total construction cost is a factor for final selection). GMP contracts method is typically used with CM-at-risk.

In the United States, certain states have promulgated statutory/regulatory guidelines to hire CM-at-risk to procure construction services. Figure 3.12 illustrates the contractual relationship of a CM-at-risk delivery system, and Figure 3.13 illustrates the sequential activities for this type of delivery system.

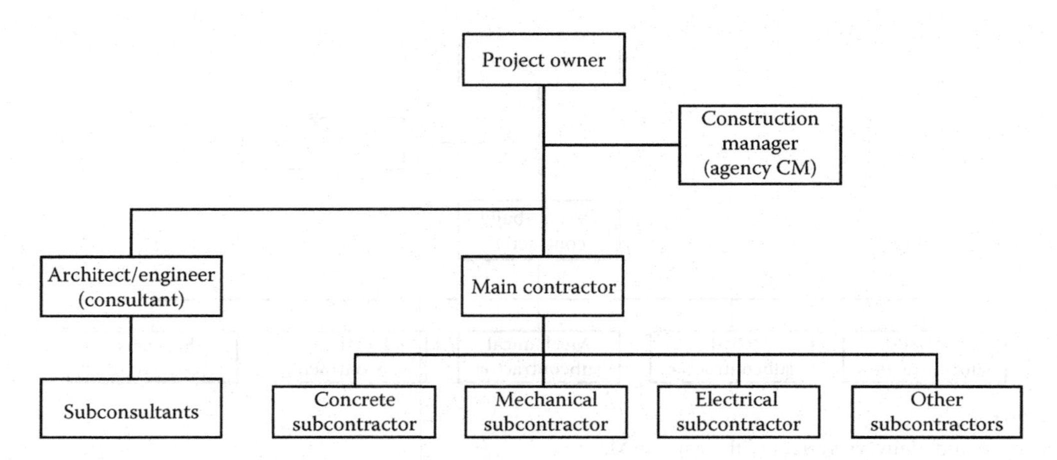

FIGURE 3.8
Construction manager type delivery system (agency CM).

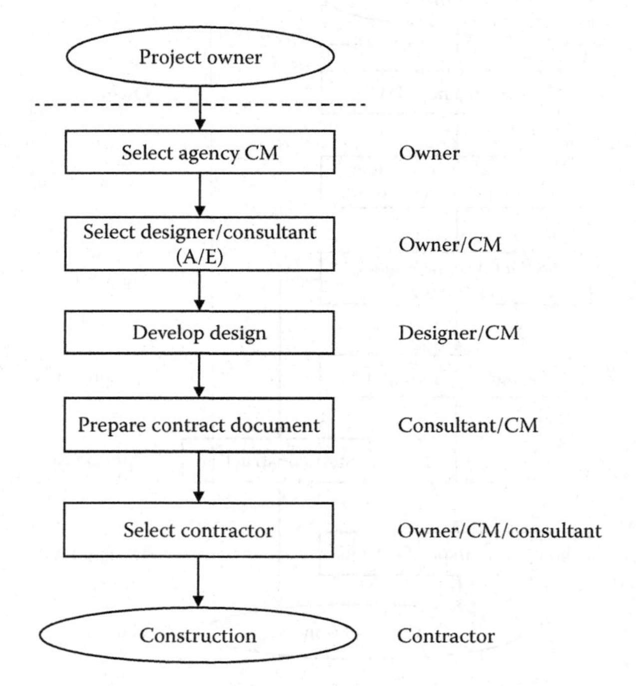

FIGURE 3.9
Sequential activities of agency CM delivery system.

3.3.5 Establish Terms of Reference

Normally, terms of reference (TOR), also known as *design brief*, are prepared by the owner/client or by the project manager on behalf of the owner describing the objectives and requirements to develop the project. In the case of construction management, the CM assists the owner/client to prepare the TOR.

A client brief (TOR) defines the objectives for the project and guides the project team to the next stage of the project. A well-prepared, accurate, and comprehensive client brief

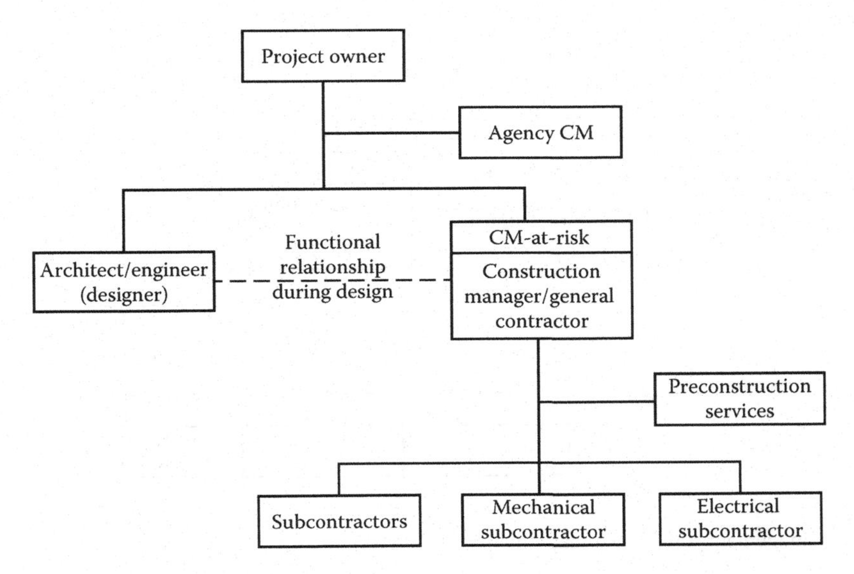

FIGURE 3.10
Construction management-at-risk (CM-at-risk) with agency CM.

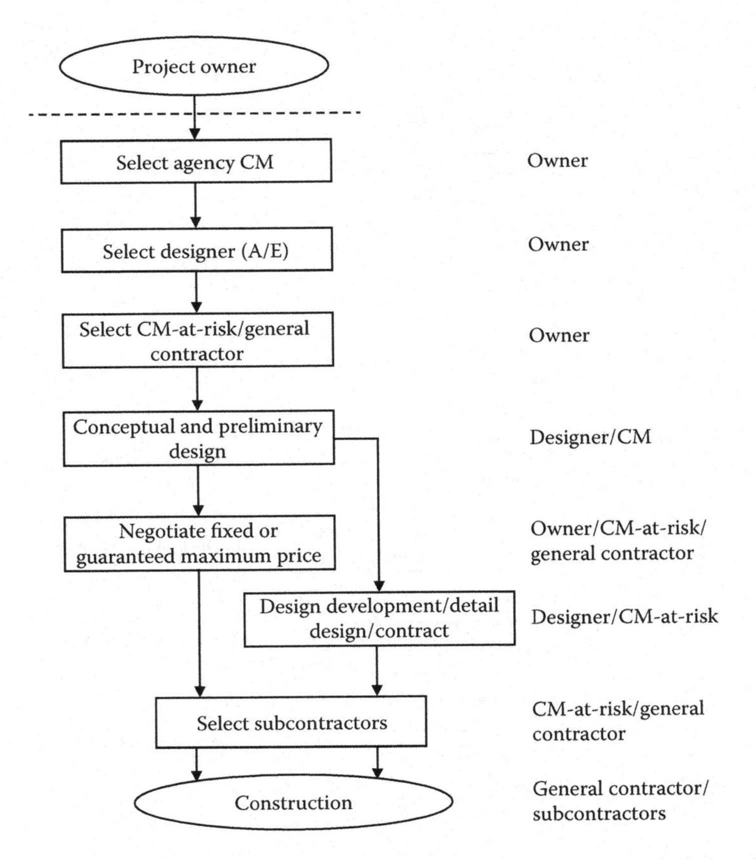

FIGURE 3.11
Sequential activities of CM-at-risk delivery system with agency CM.

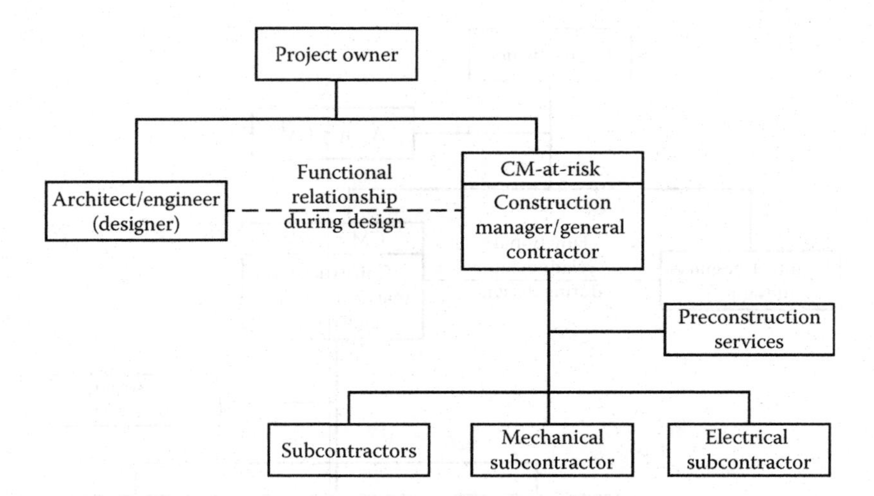

FIGURE 3.12
Construction management contractual relationship (CM-at-risk).

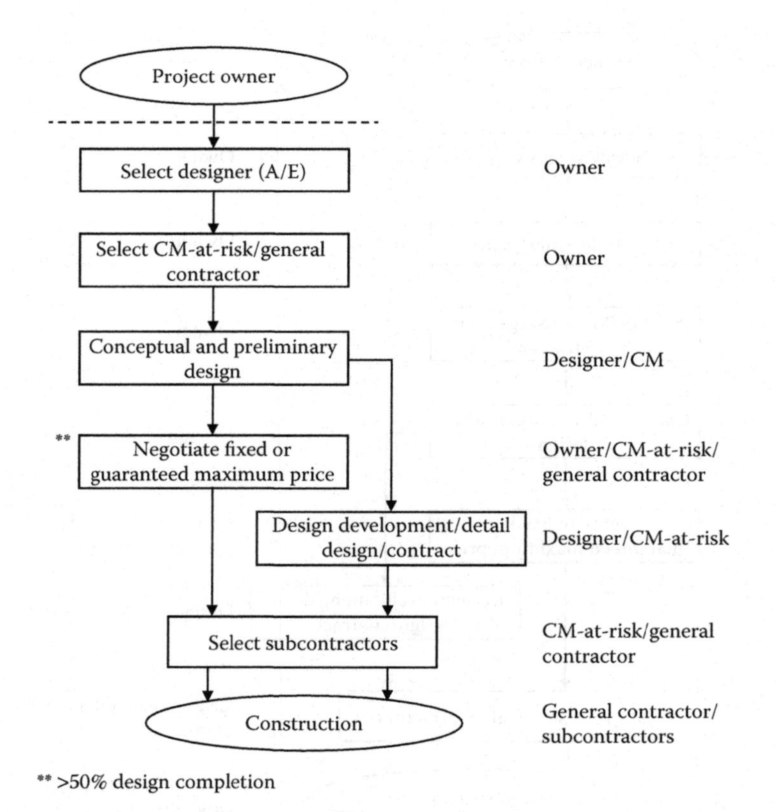

FIGURE 3.13
Sequential activities of CM-at-risk delivery system.

(TOR) is essential to achieve a qualitative and competitive project. The TOR gives the project team (designer) a clear understanding for the development of the project. Further, the TOR is used throughout the project as a reference to ensure that the established objectives are achieved. Client brief or TOR describes information such as

- The need or opportunity that has triggered the project
- Proposed location of the project
- Project/facility to be developed
- Project function and size
- Performance characteristics of the project
- Procurement strategy
- Project assumptions and constraints
- Estimated timescale
- Estimated cost
- Initial list of defined risks
- Description of approval requirements

For the development of construction projects, TOR generally details the services to be performed by the designer (consultant), which include, but are not limited to, the following:

- Predevelopment studies, collection of required data, and analyzing the same to prepare design drawings and documents for the project
- Development of alternatives
- Preparation of concept design
- Preparation of schematic design
- Preparation of detail design
- Project deliverables
- Obtaining authorities approvals
- Compliance standards, codes, and practices
- Coordinating and participation in value engineering study
- Preparation of construction schedule
- Preparation of construction budget
- Preparation of contract documents for bidding purpose
- Prequalification/selection of contractor
- Evaluation of proposals
- Recommendation of contractor to the owner/client

In cases of the construction management type of system, some of the activities mentioned earlier are performed by the CM as he or she acts as an advisor to the owner.

Following are the requirements for a building construction project, normally mentioned in TOR, to be prepared by the designer during the conceptual phase for submission to the owner:

1. Site plan
   a. Civil
   b. Services
   c. Landscaping
   d. Irrigation
2. Architectural design
3. Building and engineering systems
   a. Structural
   b. Mechanical (HVAC)
   c. Public health
   d. Fire suppression systems
   e. Electrical
   f. Low voltage systems
   g. Others
4. Cost estimates
5. Schedules

3.4 Qualification of Construction Manager

Construction management is a process wherein professional knowledge, skills, tools, and techniques are applied to project activities to meet the project requirements and successful completion of project to the satisfaction of the owner/end user. The CM should be a highly qualified person who has full knowledge of the construction management system to achieve the project objectives. Table 3.5 illustrates typical qualification requirements of a CM.

3.5 Role of CM

CMAA's *Owner's Guide to Project Delivery Methods* states that:

> In the past, most owners relied on the experience of the designer to provide a complete and responsible set of contract documents. Recently, more and more owners have found the value in utilizing the advice and expertise of those with overall process, program and construction management knowledge during the design phase.
>
> Whether provided through owner staffing or a third-party firm, the CM should be engaged as early in the project as possible to guide and assist the owner through all phases of delivering the project. The CM may also act as the owner's representative with the other members of the project team, being the point of contact for the designer, contractor, and any other specialty consultants engaged in the project by the owner.

The construction management delivery system is a project procurement/contract management process whereby the CM (firm or an individual with sound project manager skills)

TABLE 3.5

Qualifications of Construction Manager

Sr. No.	Qualification
1	Thorough knowledge and understanding of construction processes
2	In-depth knowledge of construction activities
3	Knowledge of projects delivery systems and their adoptability
4	Knowledge of contracting systems and their suitability of the project
5	Knowledge of project planning, monitoring, and control
6	Knowledge of various tools and techniques used in construction projects
7	Knowledge of construction codes and practices
8	Knowledge of applicable standards and conditions of contract
9	Knowledge of information technology
10	Knowledge and understanding of all disciplines of construction project
11	Knowledge of societal needs toward projects
12	Knowledge and skills to oversee and manage complex construction projects
13	Knowledge of quality management techniques
14	Knowledge of HSE practices
15	Knowledge of risk management
16	Excellent technical background
17	Communication skills (oral and written)
18	Strong and responsive leadership skills
19	Competency in construction management skills
20	Problem-solving skills
21	Ability to negotiate
22	Innovative and creative thinking
23	Collaborative thinking
24	Able to motivate and encourage subordinates and other team members to work as a group
25	Possess high-level university degree
26	Member of a related professional institution/society
27	Basic knowledge of design practices

undertakes to manage the work through contractors who may either be general or trade-specific contractors and oversees the performance throughout the project life cycle by systematic application of management skills and principles.

The role of the CM is to apply comprehensive management and control efforts to the project at the early project planning stages and continue until project completion. Construction management process involves the application and integration of comprehensive project controls to the design and construction process to achieve successful project delivery/completion.

The roles and responsibilities of the CM in a project may vary substantially, and can be performed under a variety of contractual terms. Regardless of the project delivery system utilized in the construction projects, the CM can play a pivotal role in all the phases of construction project life cycle.

Generally, construction projects are divided into three stages, which are as follows:

1. Preconstruction
    a. Study (predesign)
    b. Design
    c. Bidding

2. Construction
3. Postconstruction

The owner may engage the CM to oversee all the activities during all the phases of the construction project life cycle or engage the CM to perform a specific role during a specific phase of the project depending on the owner's desires and requirements for delegating responsibilities and authority. The CM is paid a fee to act as the owner's agent or advisor for a construction project. Given next are the roles played by the CM during various stages of the construction project.

3.5.1 CM Role during the Predesign Stage

During this stage, the CM acts in an advisory role, just as the agency CM, and provides the following services:

- Defining the overall performance requirements
- Defining overall project program
- Developing project's scope of work (TOR)
- Developing project procedures and standards
- Establishing a management information system
- Preparing the project schedule
- Preparing the project budget
- Identifying critical constituents of the project
- Identifying the required approvals and permits from the authorities
- Selecting the project delivery method
- Selecting the designer (consultant)

3.5.2 CM Role during Design Stage

During the design stage, the CM works collaboratively as well as independently in a design team. The CM has to provide information and recommendations that will enable the owner to make the best design decisions possible, since the owner is responsible for the contractor (S) for design errors (Spearin Doctrine).

The CM acts as an agent to the owner and provides the following services during the design stage:

- Oversees and coordinates design
- Recommends alternative solutions
- Conducts life-cycle cost analysis
- Reviews design documents
- Conducts constructability review
- Gives suggestions to improve constructability
- Provides recommendations on construction-related activities
- Coordinates with the regulatory authorities to obtain requisite permits and license

- Recommends selection of materials, building systems, and equipment
- Provides value-added engineering suggestions
- Provides value-enhancement suggestions
- Creates construction schedule
- Estimates construction cost
- Manages budget and project schedule
- Monitors and manages project risk
- Helps the designer (A/E) to develop multiple bid packages to expedite the construction process
- Reviews final drawings and specifications
- Prepares and revises project management plan

3.5.2.1 CM Role during Bidding

The role of the CM during the bidding stage is as follows:

- Prepares bid packages
- Prequalifies bidders
- Establishes bidding schedule
- Manages bidding of documents
- Conducts prebid meetings to familiarize bidders with the bidding documents
- Answers all queries related to the bid
- Addendum to bid documents
- Receives bids
- Reviews and evaluates bids
- Analyzes and compares bids
- Holds contract negotiation
- Provides recommendations to the owner to accept or reject the proposals
- Participates in contractor selection
- Conducts preaward meeting(s) with the selected contractor
- Incorporates addenda changes into contract documents
- Prepares construction contract
- Provides notice to proceed

3.5.3 CM Role during Construction Stage

A CM's roles during the construction stage vary depending on the type of contract the owner has entered with the CM. There are two forms of construction management processes:

1. Agency CM
2. CM-at-risk

3.5.3.1 Agency CM

As an agency CM, the CM is responsible for performing services related to the following activities:

- Ensure that contractor submitted performance bond
- Ensure that contractor submitted worker's insurance policy
- Selects and recommends contractor's core staff
- Selects and recommends subcontractor
- Establishes and implements procedures for processing and approval of shop drawings
- Approves construction material
- Manages contractor's request for information (RFI)
- Change order management
- Approves construction schedule
- Supervises construction
- Quality management
- Coordination of on-site, off-site inspection
- Inspection of works
- Construction contract administration
- Conducts periodic progress meetings
- Prepares minutes of meeting and distributes it as per agreed upon matrix
- Document control
- Technical correspondence between contractor
- Manages submittals
- Monitors daily progress
- Monitors contractor's performance and ensures that the work is performed as specified, as per approved shop drawings, and as per applicable codes
- Scope control
- Monitors construction scheduling
- Cost tracking and management
- Reviews, evaluates, and documents claims
- Maintains project progress record
- Evaluates payment request and recommends progress payments
- Monitors project risk
- Monitors contractor's HSE plans (health, safety, and environment)
- Coordinates work of multiple contractors
- Coordinates delivery and storage of owner-supplied materials and systems
- Tests systems
- Punch list

3.5.3.2 CM-at-Risk

Under the CM-at-risk form of project delivery system, the CM acts as a general contractor. The owner transfers the responsibility and risk to the CM-at-risk contractor for the entire construction effort, *performance risk* including subcontract administration and coordination.

3.5.4 CM Role during Testing, Commissioning, and Handover Stage

- Testing and commissioning of systems
- Review of as-built drawings
- Review of record documents and manuals
- Warranties and guarantees
- Authorities' approval for occupancy
- Coordination of hand-off procedure
- Move in plan
- Punch list
- Preparing list of lessons learned
- Substantial completion
- Archiving project documents
- Settlement of claims
- Project final account
- Project closure

4

Construction Management Tools

Abdul Razzak Rumane

CONTENTS

4.1 Introduction

Quality tools are the charts, check sheets, diagrams, graphs, techniques, and methods that are used to create an idea, engender planning, analyze the cause and process, foster evaluation, and create a wide variety of situations for continuous quality improvement. Applications of tools enhance chances of success and help maintain consistency, accuracy, increase efficiency, and process improvement.

4.2 Categorization of Tools

There are several types of tools, techniques, and methods, in practice, which are used as quality improvement tools and have variety of applications in manufacturing and process industry. However, all of these tools are not used in construction projects due to the nature of construction projects, which are customized and nonrepetitive. Some of these quality management tools that are most commonly used in the construction industry are listed under the following broader categories:

1. Classic quality tools
2. Management and planning tools
3. Process analysis tools
4. Process improvement tools
5. Innovation and creative tools
6. Lean tools
7. Cost of quality
8. Quality function deployment (QFD)
9. Six Sigma
10. Triz

A brief description of these tools is given in Sections 4.2.1 through 4.2.10.

4.2.1 Classic Quality Tools

Classic quality tools have a long history. These tools are listed in Table 4.1. All of these tools have been in use since World War II. Some of these tools date prior to 1920. The approach includes both quantitative and qualitative aspects, which taken together, focus on company-wide quality.

A brief definition of these quality tools is as follows (values shown in the figures are indicative only).

TABLE 4.1
Classic Quality Tools

Sr. No.	Name of Quality Tool	Usage
Tool 1	Cause-and-effect diagram	To identify possible cause-and-effect in processes.
Tool 2	Check sheet	To provide a record of quality. How often it occurs?
Tool 3	Control chart	A device in statistical process control to determine whether or not the process is stable.
Tool 4	Flowchart	Used for graphical representation of a process in sequential order.
Tool 5	Histogram	Graphs used to display frequency of various ranges of values of a quantity.
Tool 6	Pareto chart	Used to identify the most significant cause or problem.
Tool 7	Pie chart	Used to show classes or group of data in proportion to the whole set of data.
Tool 8	Run chart	Used to show measurement against time in a graphical manner with a reference line to show the average of the data.
Tool 9	Scatter diagram	To determine whether there is a correlation between two factors.
Tool 10	Stratification	Used to show the pattern of data collected fron different sources.

Source: Abdul Razzak Rumane (2013), *Quality Tools for Managing Construction Projects*, CRC Press, Boca Raton, FL. Reprinted with permission from Taylor & Francis Group.

4.2.1.1 Cause-and-Effect Diagram

Cause-and-effect diagram is also called as *Ishikawa diagram*, after its developer Kaoru Ishikawa, or *fishbone diagram*. It is used to identify possible causes and effect in process. It is used to explore all the potential or real causes that result in a single output. The causes are organized and displayed in graphical manner to their level of importance or details. It is a graphical display of multiple causes with a particular effect. The causes are organized and arranged mainly into four categories, which are as follows:

1. Machine
2. Manpower
3. Material
4. Method

The effect or problem being investigated is shown at the end of horizontal arrow. Potential causes are shown as labeled arrows entering the main cause arrow. Each arrow may have a number of other arrows entering, as the principal cause or factors are reduced to their subcauses. Figure 4.1 illustrates an example of cause-and-effect diagram for rejection of executed (installed) false ceiling works by the supervision engineer for not complying with contract specifications.

4.2.1.2 Check Sheet

Check sheet is a structured list prepared from the collected data to indicate how often each item occurs. It is an organized way of collecting and structuring data. The purpose of the check sheet is to collect the facts in the most efficient manner. Data is collected and ordered by adding tally or check marks against predetermined categories of items or measurements. Figure 4.2 illustrates the check sheet for checklist approval record of wire bundles.

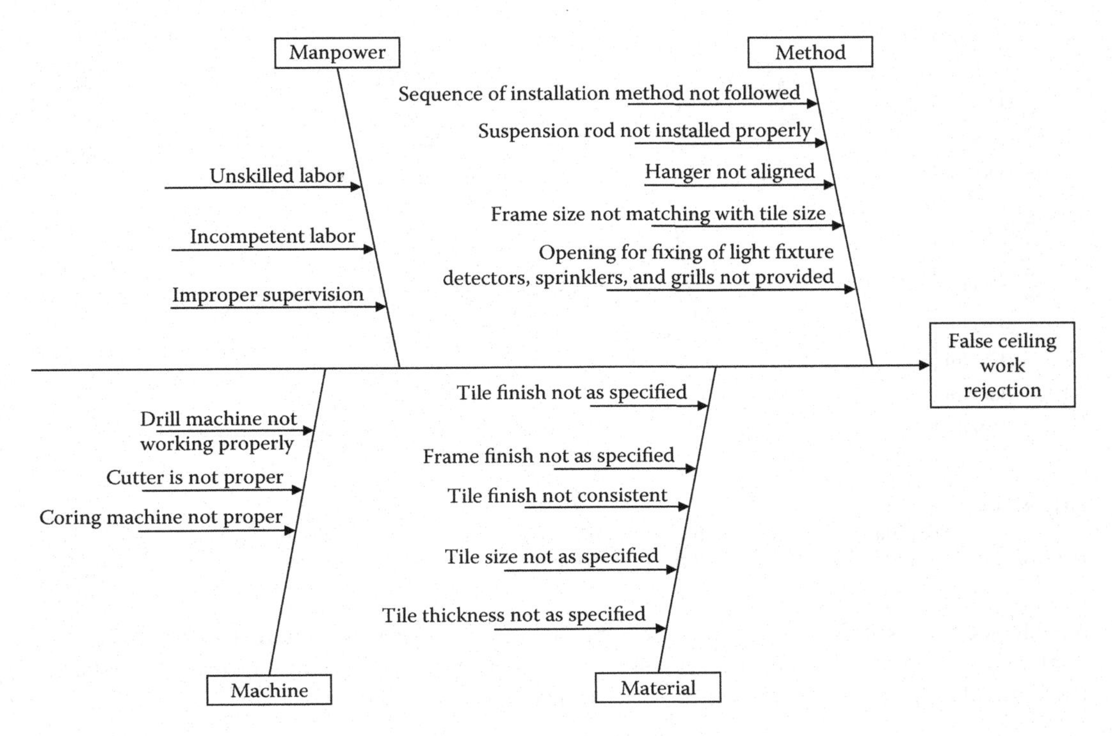

FIGURE 4.1
Cause and effect for false ceiling rejection. (Abdul Razzak Rumane (2013), *Quality Tools for Managing Construction Projects*, CRC Press, Boca Raton, FL. Reprinted with permission from Taylor & Francis Group.)

Approval record for wire bundles

	Approved	Not approved	Total	Percentage not approved
1.5 mm^2 wire	𝍸𝍸///	///	50	4
2.5 mm^2 wire	𝍸𝍸𝍸 𝍸///	///	85	6
4.0 mm^2 wire	𝍸𝍸𝍸 𝍸///	///	25	3
6.0 mm^2 wire	𝍸///	///	15	2
10.0 mm^2 wire	𝍸///	///	10	2

FIGURE 4.2
Check sheets. (Abdul Razzak Rumane (2013), *Quality Tools for Managing Construction Projects*, CRC Press, Boca Raton, FL. Reprinted with permission from Taylor & Francis Group.)

4.2.1.3 Control Chart

Control chart is the fundamental tool of statistical process control (SPC). The control chart is a graph used to analyze how a process behaves over time and to show whether the process is stable or is being affected by special cause of variation and creating an out of control condition. It is used to determine whether the process is stable or varies between the predictable limits. It can be employed to distinguish between the existence of a stable pattern of variation and the occurrence of an unstable pattern. With control charts, it is

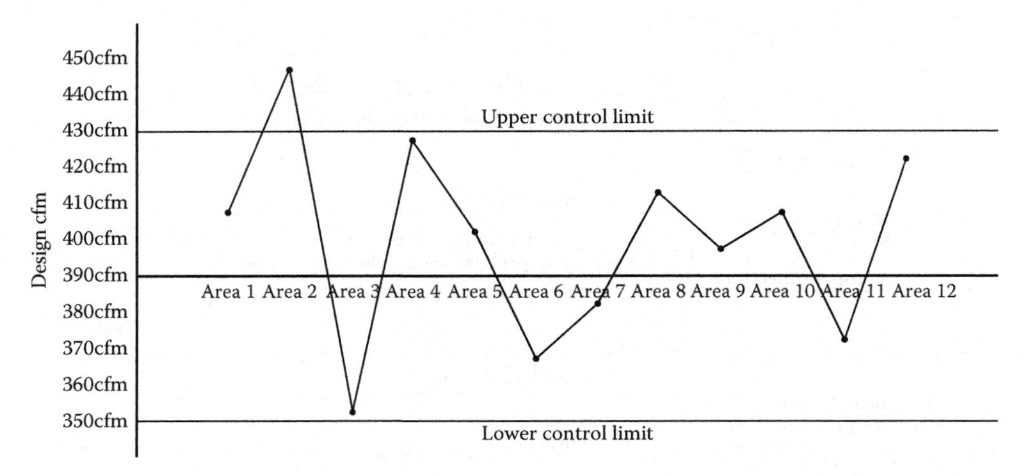

FIGURE 4.3
Control chart for air handling unit air distribution (cfm). (Abdul Razzak Rumane (2013), *Quality Tools for Managing Construction Projects*, CRC Press, Boca Raton, FL. Reprinted with permission from Taylor & Francis Group.)

easy to see both special and common cause variation in a process. There are many types of control charts. Each is designed for a specific kind of process or data. Figure 4.3 illustrates the control chart for distribution of air through air handling unit.

4.2.1.4 Flowchart

Flowchart is a pictorial tool that is used for representing a process in a sequential order. It uses graphic symbols to depict the nature and flow of the steps in a process. It helps to see whether the steps of the process are logical, uncover the problems, or miscommunications, define the boundaries of a process, and develop a common base of knowledge about a process. The flow of steps is indicated with arrows connecting the symbols. Flowcharts can be applied at all stages of the project life cycle. Figure 4.4 illustrates the flowchart for contractor's staff approval in construction projects.

4.2.1.5 Histogram

Histogram is a pictorial representation of frequency distribution of the data. It is created by grouping the data points into cell and displays how frequently different values occur in the data set. Figure 4.5 illustrates the histogram for employee reporting time.

4.2.1.6 Pareto Chart

Pareto chart is named after Vilfredo Pareto, a nineteenth-century Italian economist who postulated that a large share (80%) of wealth is owned by a small (20%) percentage of population. Pareto chart is a graph bar having a series of bars whose height reflects the frequency of occurrence. Pareto charts are used to display the Pareto principle in action and arrange data, so that the few vital factors that are causing most of the problems reveal themselves. The bars are arranged in descending order of height from left to right. Pareto charts are used to identify those factors that have greatest cumulative effect on the system, and thus less significant factors can be screened out from the process. Pareto chart can be

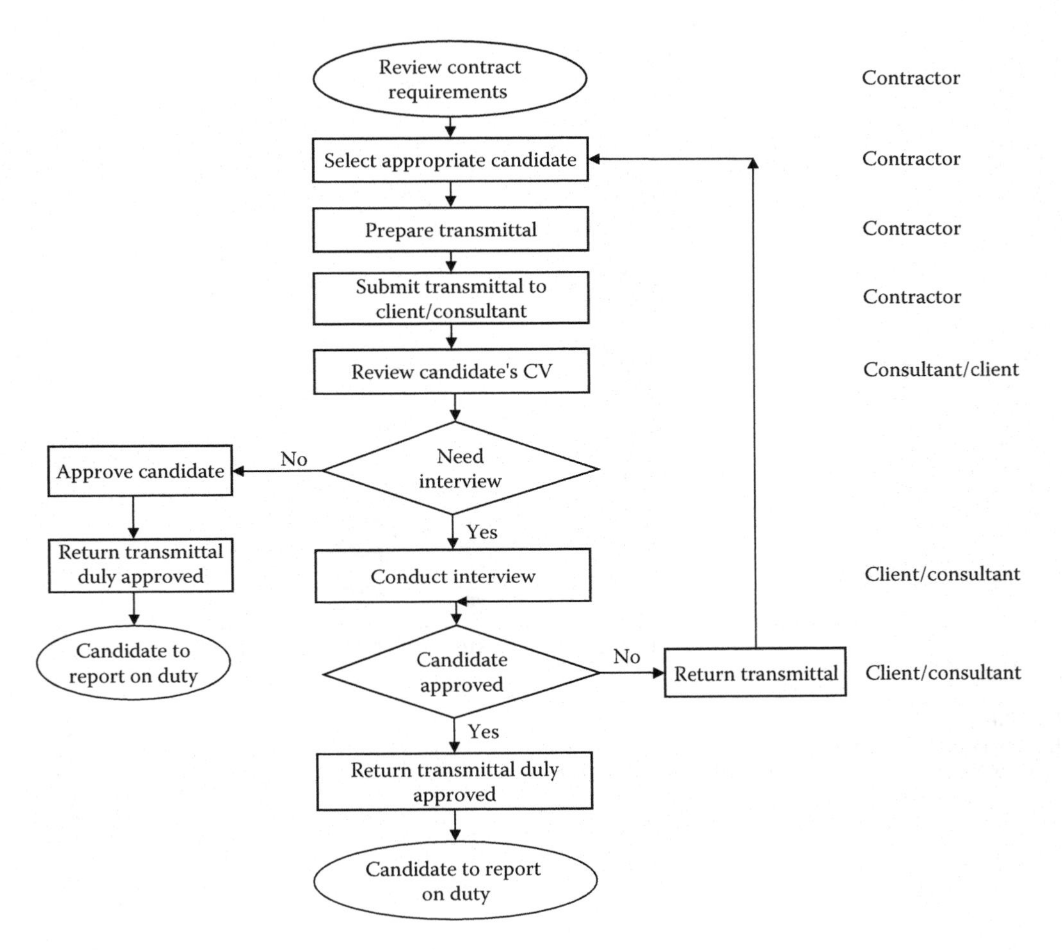

FIGURE 4.4
Flow diagram for contractor's staff approval. (Abdul Razzak Rumane (2013), *Quality Tools for Managing Construction Projects*, CRC Press, Boca Raton, FL. Reprinted with permission from Taylor & Francis Group.)

used at various stages in a quality improvement program to determine which step to take next. Figure 4.6 illustrates the Pareto chart for division cost of construction project.

4.2.1.7 Pie Chart

Pie chart is a circle divided into wedges to depict proportion of data or information in order to understand how they make up the whole. The entire pie chart represents all the data, while each slice or wedge represents a different class or group within the whole. The portions of entire circle or pie sum to 100%. Figure 4.7 illustrates the contents of contractor's site staff at construction project site.

4.2.1.8 Run Chart

Run chart is a graph plotted by showing measurement (data) against time. They are used to know the trends or changes in a process variation over time over the average and also

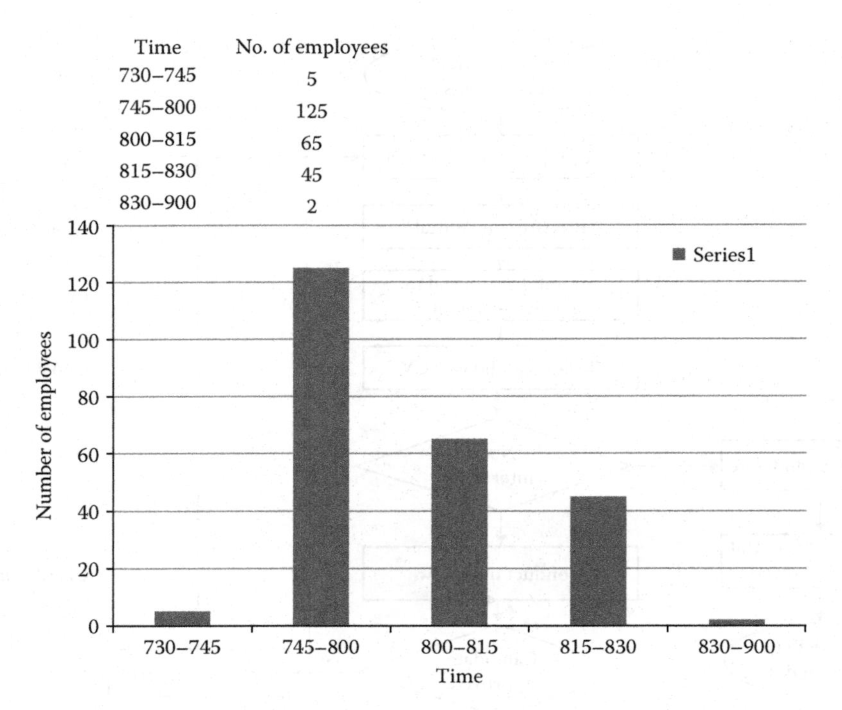

Time	No. of employees
730–745	5
745–800	125
800–815	65
815–830	45
830–900	2

FIGURE 4.5
Employee reporting histogram. (Abdul Razzak Rumane (2013), *Quality Tools for Managing Construction Projects*, CRC Press, Boca Raton, FL. Reprinted with permission from Taylor & Francis Group.)

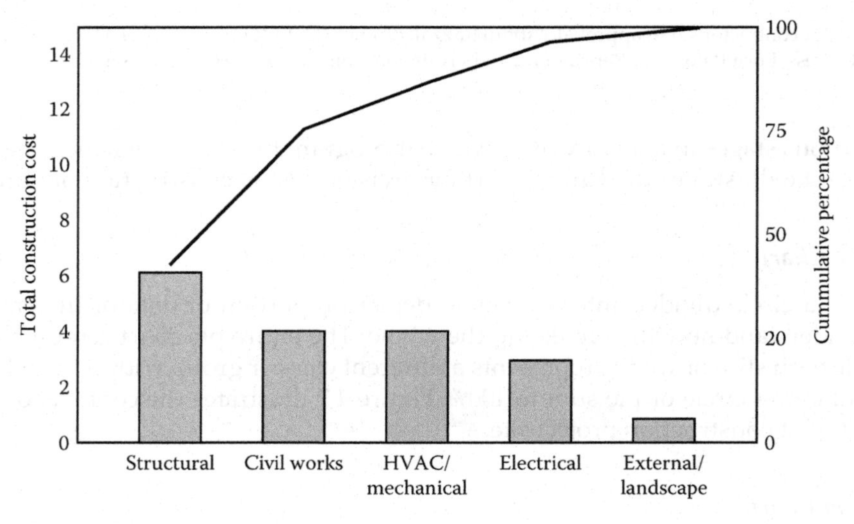

FIGURE 4.6
Pareto analysis for construction cost. (Abdul Razzak Rumane (2013), *Quality Tools for Managing Construction Projects*, CRC Press, Boca Raton, FL. Reprinted with permission from Taylor & Francis Group.)

Site staff	
Managers	2
Engineers	16
Technicians	9
Foremen	20
Skilled	230
Unskilled	570

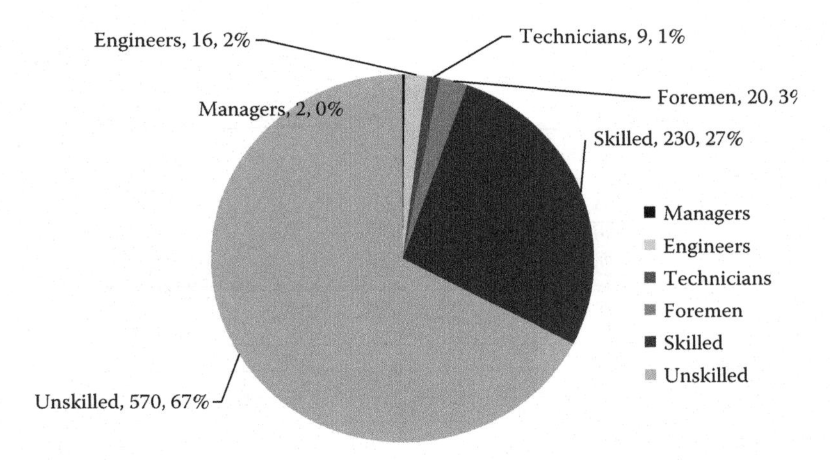

FIGURE 4.7
Pie chart for site staff. (Abdul Razzak Rumane (2013), *Quality Tools for Managing Construction Projects*, CRC Press, Boca Raton, FL. Reprinted with permission from Taylor & Francis Group.)

to determine if the pattern can be attributed to common causes of variation, or if special causes of variation were present. A run chart is also used to monitor process performance. Run charts can be used to track improvements that have been put in place, checking to determine their success. Figure 4.8 illustrates the run chart for weekly manpower of different trades of a project. It is similar to control chart but does not show control limits.

4.2.1.9 Scatter Diagram

Scatter diagram is a plot of one variable versus another. It is used to investigate the possible relationship between two variables that both relate to the same event. It helps to know how one variable changes with respect to other. It can be used to identify potential root cause of problems and to evaluate cause-and-effect relationship. Figure 4.9 illustrates the scatter diagram for beam quantity of various length.

4.2.1.10 Stratification

Stratification is a graphical representation of data collected from different data. Figure 4.10 illustrates the stratification diagram for cable drums.

4.2.2 Management and Planning Tools

Seven management tools are most popular after quality classic tools. These tools are listed in Table 4.2.

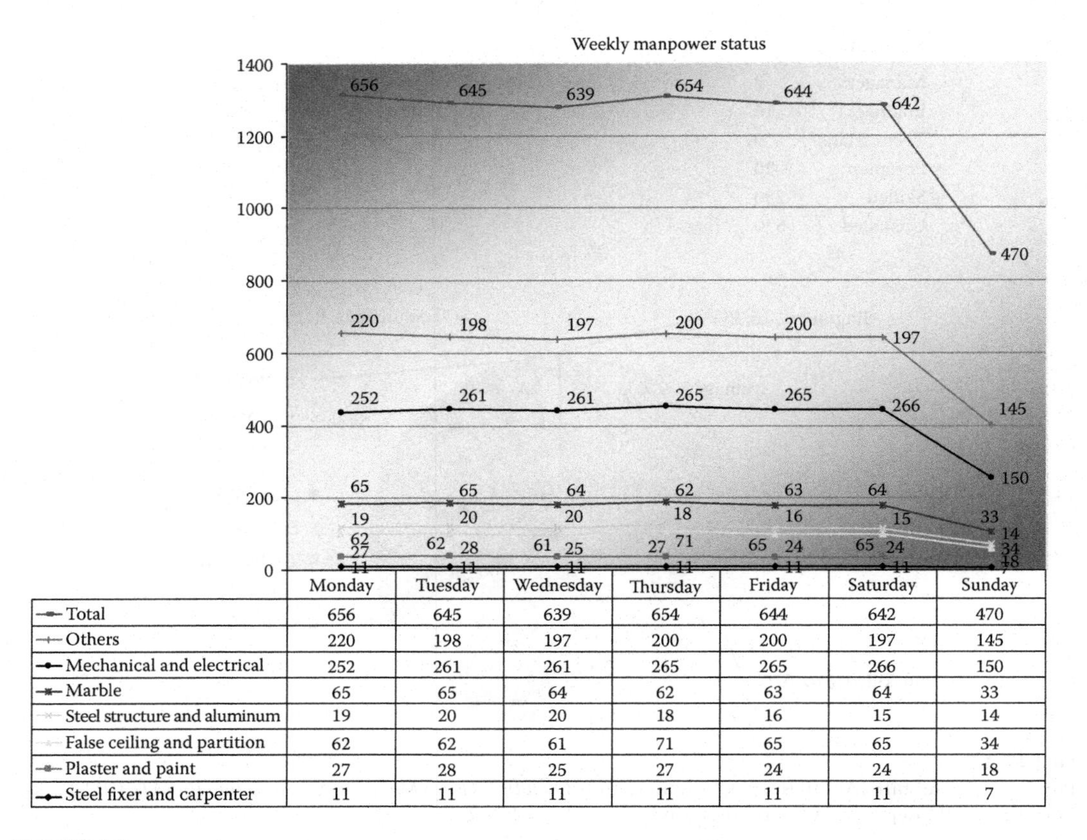

	Monday	Tuesday	Wednesday	Thursday	Friday	Saturday	Sunday
Total	656	645	639	654	644	642	470
Others	220	198	197	200	200	197	145
Mechanical and electrical	252	261	261	265	265	266	150
Marble	65	65	64	62	63	64	33
Steel structure and aluminum	19	20	20	18	16	15	14
False ceiling and partition	62	62	61	71	65	65	34
Plaster and paint	27	28	25	27	24	24	18
Steel fixer and carpenter	11	11	11	11	11	11	7

FIGURE 4.8
Run chart for manpower. (Abdul Razzak Rumane (2013), *Quality Tools for Managing Construction Projects*, CRC Press, Boca Raton, FL. Reprinted with permission from Taylor & Francis Group.)

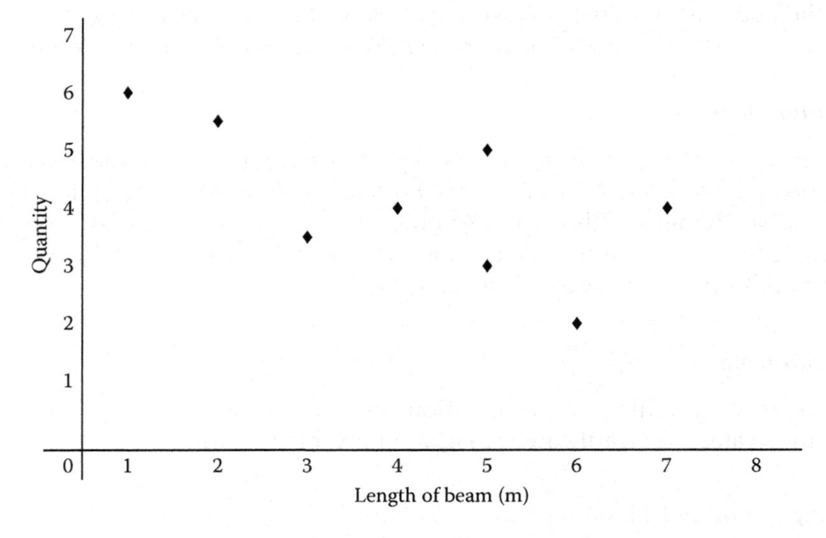

FIGURE 4.9
Scatter diagram. (Abdul Razzak Rumane (2013), *Quality Tools for Managing Construction Projects*, CRC Press, Boca Raton, FL. Reprinted with permission from Taylor & Francis Group.)

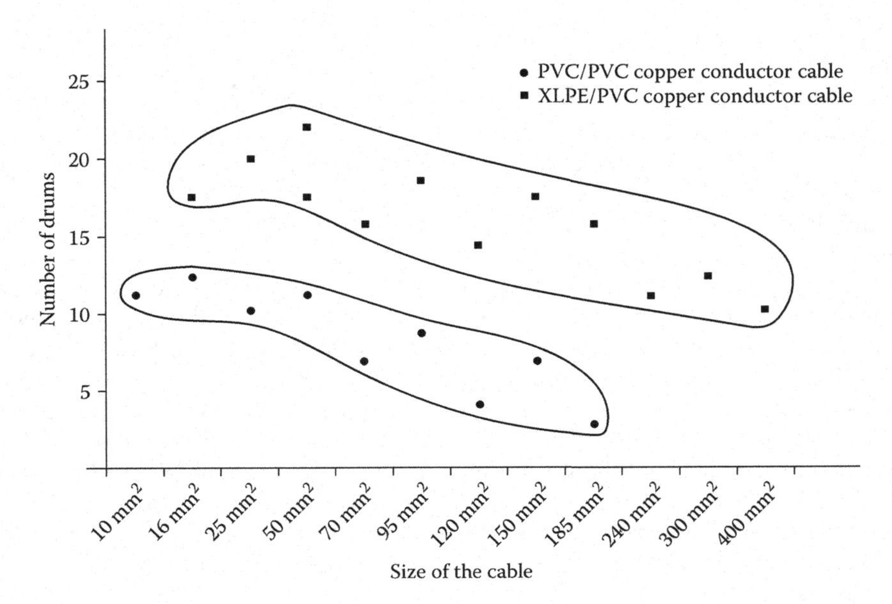

FIGURE 4.10
Stratification chart. (Abdul Razzak Rumane (2013), *Quality Tools for Managing Construction Projects*, CRC Press, Boca Raton, FL. Reprinted with permission from Taylor & Francis Group.)

TABLE 4.2
Management and Planning Tools

Sr. No.	Name of Quality Tool	Usage
Tool 1	Activity network diagram (arrow diagram)/critical path method	Used when scheduling or monitoring task is complex or lengthy and has schedule constraints.
Tool 2	Affinity diagram	Used to organize a large group of items in a smaller categories that are easier to understand and deal with.
Tool 3	Interelationship digraph (relations diagram)	Used to show logical relationship between ideas, process, cause, and effect.
Tool 4	Matrix diagram	Used to analyze the correlations between two or more groups of information.
Tool 5	Prioritization matrix	Used to choose one or two options which have important criteria from several options.
Tool 6	Process decision program chart	Used to help contingency plan.
Tool 7	Tree diagram	Used to break down or stratify ideas in progressively more detailed step.

Source: Abdul Razzak Rumane (2013), *Quality Tools for Managing Construction Projects*, CRC Press, Boca Raton, FL. Reprinted with permission from Taylor & Francis Group.

These tools are focused on managing and planning quality improvement activities. A brief definition of these quality tools is as follows (values shown in the figures are indicative only).

4.2.2.1 Activity Network Diagram

Activity network diagram (AND) is a graphical representation chart showing interrelationship among activities (task) associated with a project. An AND was developed by US Department of Defense. It was first used as a management tool for military projects. It was adapted as an education tool for business managers. In AND, each activity is represented by one and only one arrow in the network and is associated with an estimated time to perform the activity. AND analyzes the sequences of tasks necessary to complete the project. The direction of arrow specifies the order in which the events must occur. The event represents a point in time that indicates the completion of one or more activities and beginning of new ones. Figure 4.11 illustrates AND.

There are two kinds of network diagrams, the *activity-on-arrow* (A-O-A) network diagram and the *activity-on-node* (A-O-N) network diagram.

Arrow diagrams or A-O-A uses a diagramming method to represent the activities on arrows and connect them at nodes (circles) to show the dependencies. With the A-O-A method, a detailed information about each activity is placed on arrow or as footnotes at the bottom.

Figure 4.12 illustrates the arrow diagramming method for concrete foundation work.

Activities originating from a certain event cannot start until the activities terminating the same event have been completed. This is known as *precedence relations*. These relationships are drawn using precedence diagramming method (PDM). The PDM technique is also referred to as A-O-N, because it shows the activities in a node (box) with arrows showing dependencies. A-O-N network diagram has the activity information written in small boxes, which are the nodes of the diagram. Arrows connect the boxes to show the logical relationships between pairs of the activities.

In a networking diagram, all activities are related in some direct way and may be further constrained by an indirect relationship. Following are direct logical relationships or dependencies among project-related activities:

1. *Finish-to-start*—Activity A must finish before activity B can begin.
2. *Start-to-start*—Activity A must begin before activity B can begin.
3. *Start-to-finish*—Activity A must begin before activity B can finish.
4. *Finish-to-finish*—Activity A must finish before activity B can finish.

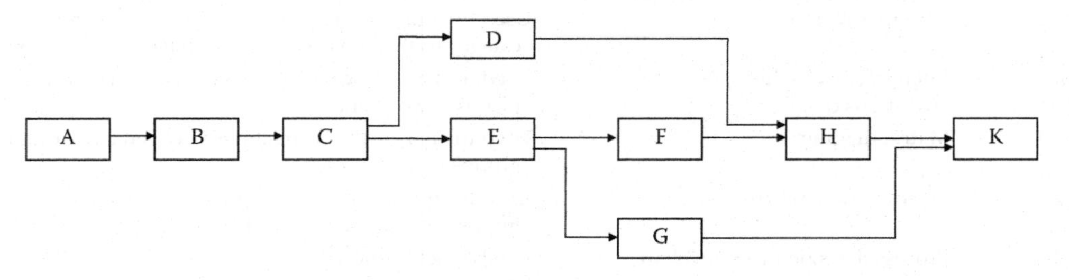

FIGURE 4.11
Activity network diagram. (Abdul Razzak Rumane (2013), *Quality Tools for Managing Construction Projects*, CRC Press, Boca Raton, FL. Reprinted with permission from Taylor & Francis Group.)

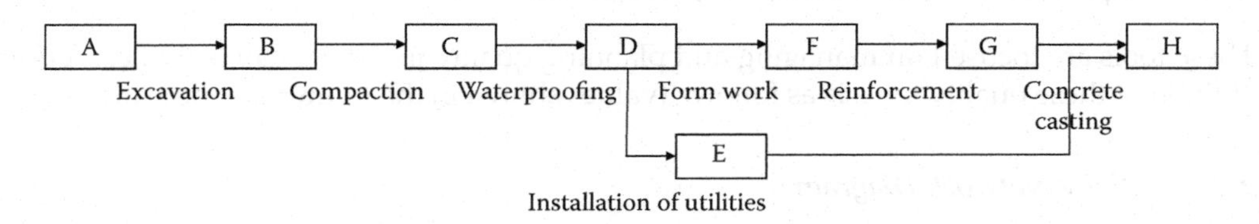

FIGURE 4.12
Arrow diagramming method for a concrete foundation. (Abdul Razzak Rumane (2013), *Quality Tools for Managing Construction Projects*, CRC Press, Boca Raton, FL. Reprinted with permission from Taylor & Francis Group.)

Apart from these, there are other dependencies, including the following:

1. Mandatory
2. Discretionary
3. External

Figure 4.13 illustrates dependency relationship diagrams and Figure 4.14 illustrates the precedence diagramming method.

AND or arrow diagram is a tool used for detailed planning, a tool for analyzing schedule during execution, and a tool for controlling complex or large-scale project. A network diagram uses nodes and arrows. Date information is added to each activity node.

Critical path method (CPM) chart is an expanded AND showing an estimated time to complete each activity and connecting these activities based on the task to be performed. Critical path is a sequence of interrelated predecessor/successor activities that determines the minimum completion time for a project. The duration of critical path is the sum of the activities' duration along the path. The activities in the critical path have the least scheduling flexibility. Any delays along the critical path would imply that additional time would

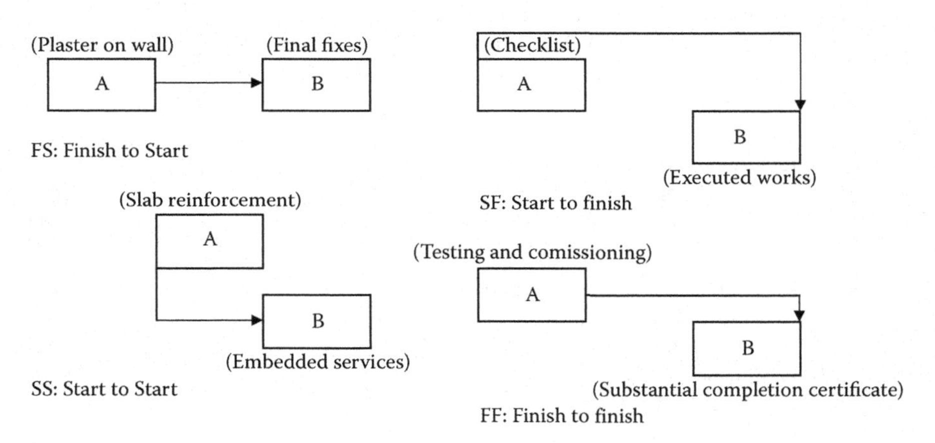

FIGURE 4.13
Dependency relationship diagram. (Abdul Razzak Rumane (2013), *Quality Tools for Managing Construction Projects*, CRC Press, Boca Raton, FL. Reprinted with permission from Taylor & Francis Group.)

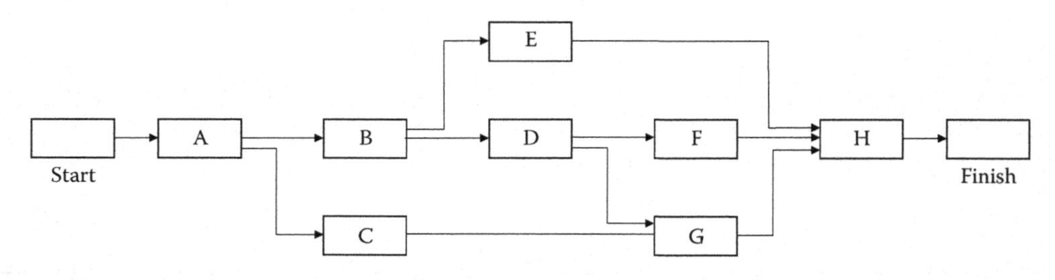

FIGURE 4.14
PDM diagramming method. (Abdul Razzak Rumane (2013), *Quality Tools for Managing Construction Projects*, CRC Press, Boca Raton, FL. Reprinted with permission from Taylor & Francis Group.)

be required to complete the project. There may be more than one critical path among all the project activities, so completion of the entire project could be affected due to delaying activity along any of the critical paths. Table 4.3 illustrates the activity relationship for a substation project and Figure 4.15 illustrates the CPM diagram for construction of a substation.

AND is also known as program evaluation and review technique (PERT). PERT is used to schedule, organize, and coordinate tasks within a project.

TABLE 4.3

Activities to Construct a Substation Building

Activity Number	Description of Activity	Duration in Days	Preceding Activity
1	Start	0	
2	Mobilization	21	1
3	Preparation of site	15	2
4	Staff approval	15	2
5	Material approval	15	2,4
6	Shop drawing approval	15	2,5
7	Procurement (structural work)	15	5
8	Procurement (pipes, ducts, sleeves)	7	5
9	Procurement (H.T Switchgear, transformer)	60	5
10	Procurement (civil , MEP, furnishing)	30	5
11	Excavation	4	3,6
12	Blinding concrete	4	5,7,11
13	Raft foundation	7	5,6,7,12
14	Utility services (embedded)	4	8,12
15	Concrete (floor)	1	6,13,14
16	Trenches	7	13
17	Embedded services/ducts in trench	2	8,13
18	Concrete (transformer area)	1	16,17
19	Walls and columns	7	15,18
20	Form work for slab	3	6,7,19
21	Reinforcement	2	7,20
22	Embedded services	1	8,20
23	Concrete (roof slab)	1	21,22
24	Masonry work	14	6,10,23
25	Installation of equipment	7	9,18,24
26	Installation of electromechanical items	14	6,10,23,24
27	Installation of ventilation system	4	10,23,24
28	Finishes	7	10,24
29	Installation of final fixes	3	10,28
30	Furnishing	2	10,28
31	Testing of equipment	2	25
32	Testing of HVAC, firefighting, electrical system	4	25,26,27
33	Handing over	1	29,30,31,32
34	End	0	

Source: Abdul Razzak Rumane (2013), *Quality Tools for Managing Construction Projects*, CRC Press, Boca Raton, FL. Reprinted with permission from Taylor & Francis Group.

Note: The duration is indicative only to understand sequencing and relationship of activities.

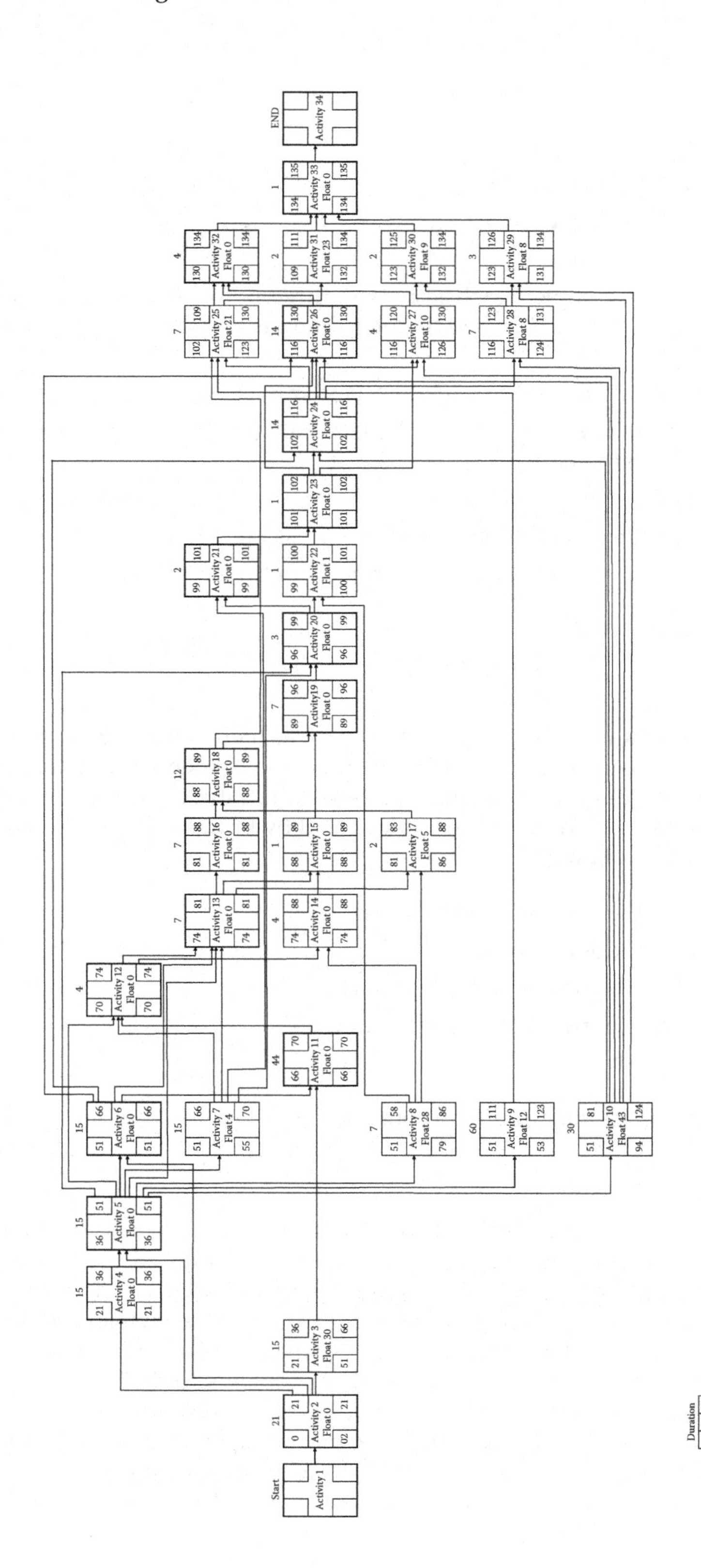

FIGURE 4.15
Critical path method. (Abdul Razzak Rumane (2013), *Quality Tools for Managing Construction Projects*, CRC Press, Boca Raton, FL. Reprinted with permission from Taylor & Francis Group.)

PERT planning involves the following steps:

1. Identify specific activity.
2. Identify milestones of each activity.
3. Determine proper sequence of each activity.
4. Construct a network diagram.
5. Estimate time required to complete each activity. A three-point estimation method using the following formula can be used to determine the approximate estimated time for each activity:

   Estimated time = (Optimistic + 4*Most likely + Pessimistic)/6
6. Compute the early start (ES), early finish (EF), late start (LS), and late finish (LF) times for each activity in the network.
7. Determine critical path.
8. Identify the critical path for possible schedule compression.
9. Evaluate the diagram for milestones and target dates in the overall project.

Figure 4.16 illustrates a Gantt chart.

4.2.2.2 Affinity Diagram

Affinity diagram is a tool that gathers a large group of ideas/items and organizes into a smaller grouping based on their natural relationships. It is the refinement of brainstorming ideas into smaller groups that can be dealt more easily and satisfy the team members. The affinity process is often used to group ideas generated by brainstorming. An affinity diagram is created as per the following steps:

1. Generate ideas and list the ideas without criticism.
2. Display the ideas in a random manner.
3. Sort the ideas and place them into multiple groups.
4. Continue until smaller groups satisfy all the members.
5. Draw the affinity diagram.

Figure 4.17 illustrates the affinity diagram for a concrete slab.

4.2.2.3 Interrelationship Digraph

Interrelationship digraph (di is for directional) is an analysis tool that allows the team members to identify logical cause-and-effect relationship between the ideas. It is drawn to show all the different relationships between factors, areas, or processes. The diagraph makes it easy to pick out the factors in a situation which are the ones which are driving many of the other symptoms or factors. Although affinity diagrams organize and arrange the ideas into groups, interrelationship digraph identify problems to define the ways in which ideas influence each other. Interrelationship digraph is used to identify the cause-and-effect relationship with the help of directional arrows among critical issues. The number of arrows coming into the node determines the outcome (key indicator), whereas the outgoing arrows determine the cause (driver) of the issue. Figure 4.18 illustrates the interrelationship digraph for causes of bridge collapse.

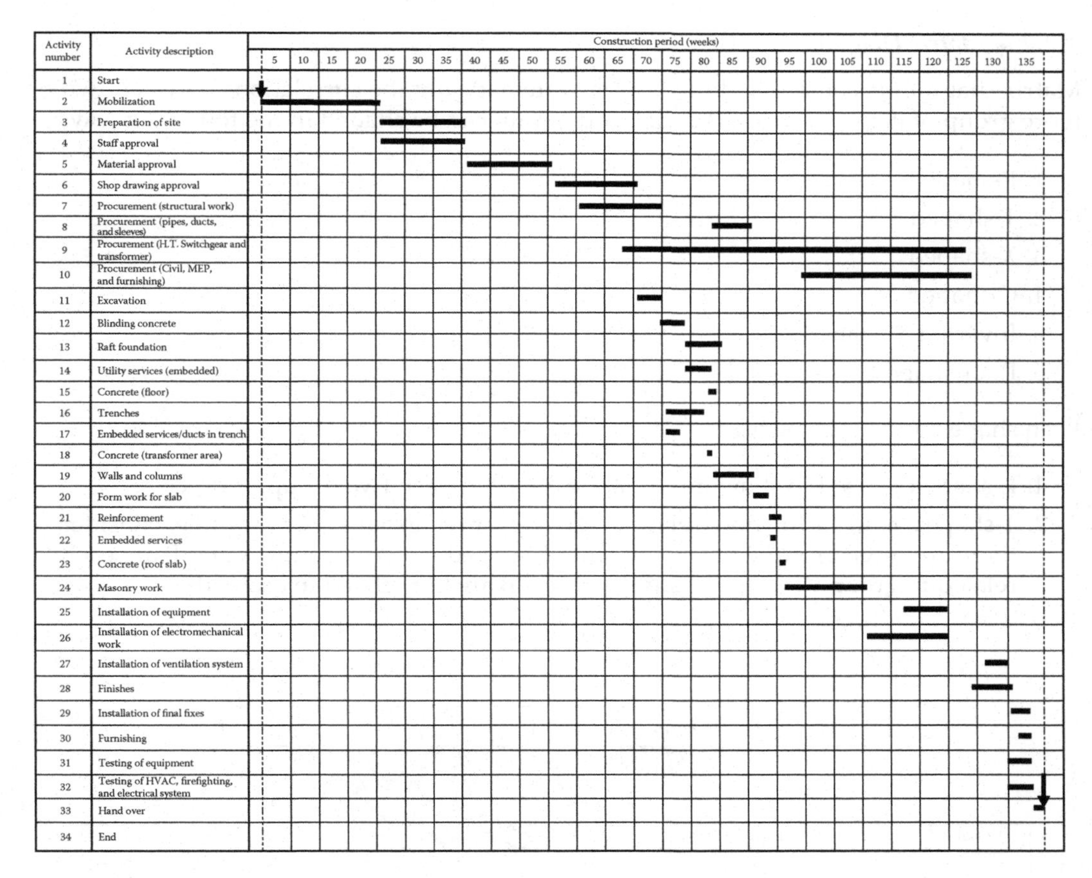

FIGURE 4.16
Gantt chart for a substation. (Abdul Razzak Rumane (2013), *Quality Tools for Managing Construction Projects*, CRC Press, Boca Raton, FL. Reprinted with permission from Taylor & Francis Group.)

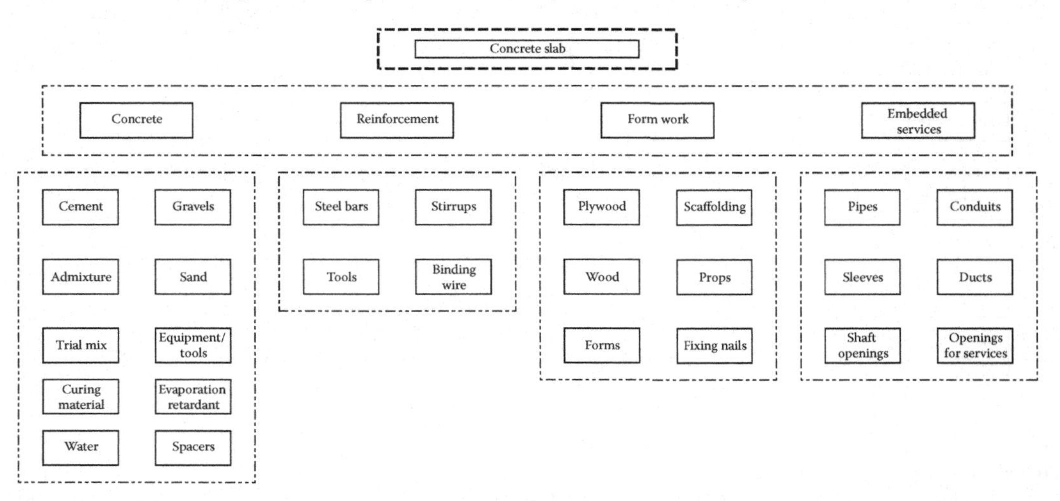

FIGURE 4.17
Affinity diagram for a concrete slab. (Abdul Razzak Rumane (2013), *Quality Tools for Managing Construction Projects*, CRC Press, Boca Raton, FL. Reprinted with permission from Taylor & Francis Group.)

4.2.2.4 Matrix Diagram

Matrix diagram is constructed to analyze systematically the correlations between two or more group of items or ideas. The matrix diagram can be shaped in the following ways:

1. L-shaped
2. T-shaped
3. X-shaped
4. C-shaped
5. Inverted Y-shaped
6. Roof-shaped

Each shape has its own purpose.

1. L-shaped is used to show interrelationships between two group or process.
2. T-shaped is used to show relation between three groups. For example, consider there are three groups A, B, and C. In a T-shaped matrix, groups A and B are each related to group C, whereas groups A and B are not related to each other.

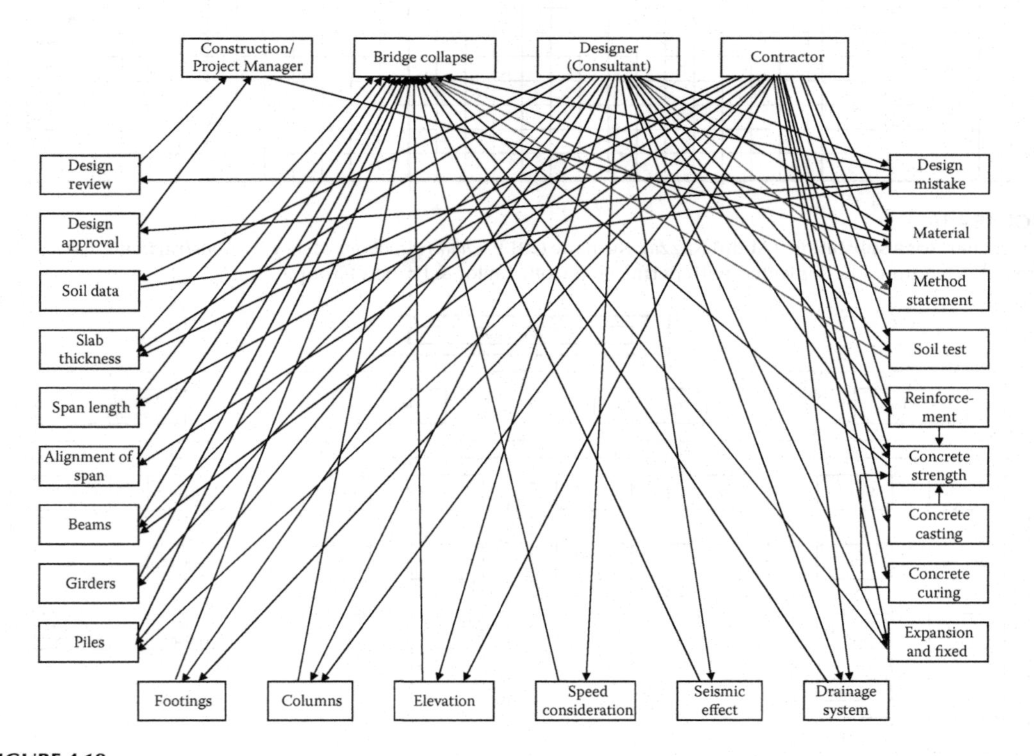

FIGURE 4.18
Interrelationship digraph. (Abdul Razzak Rumane (2013), *Quality Tools for Managing Construction Projects*, CRC Press, Boca Raton, FL. Reprinted with permission from Taylor & Francis Group.)

3. X-shaped matrix is used to show relationship among four groups. Each group is related to two other groups in a circular fashion.
4. C-shaped matrix interrelate three groups of a process or ideas in three-dimensional ways.
5. Inverted Y-shaped matrix is used to show relation between three groups. Each group is related with other groups in a circular fashion.
6. Roof-shaped matrix relates one group of items to itself. It is used with an L- or a T-shaped matrix.

Table 4.4 and Figures 4.19 and 4.20 illustrate an L-shaped matrix, a T-shaped matrix, and a roof-shaped matrix, respectively.

TABLE 4.4

L-Shaped Matrix

Customer Requirements of Distribution Boards				
		Customer		
Sr. No.	**Component Details**	**A**	**B**	**C**
1	Isolator	1	1	–
2	Molded case circuit breaker (MCCB)	–		1
3	HRC fuse	1	1	–
4	Earth leakage circuit breaker	2	1	2
5	Miniature circuit breaker (MCB)	18	12	18
6	Single bus bar	–	1	–
7	Double bus bar	2	–	2
8	Enclosure with lock	Yes	No	Yes
9	Surface mounting	Yes	No	Yes
10	Flush mounting	–	Yes	

Source: Abdul Razzak Rumane (2013), *Quality Tools for Managing Construction Projects*, CRC Press, Boca Raton, FL. Reprinted with permission from Taylor & Francis Group.

Manufacturing plant	Products				
Customer					
International				#	#
European manufacturer	≠	#	#		
Local plant	#	≠	≠		
# Large capacity # Small capacity	600 KVA Transformer	1000 KVA Transformer	1250 KVA Transformer	1600 KVA Transformer	2000 KVA Transformer
ABC company	#	≠	≠		
XYZ company	≠	#	#		
Others				#	#

FIGURE 4.19

T-shaped matrix. (Abdul Razzak Rumane (2013), *Quality Tools for Managing Construction Projects*, CRC Press, Boca Raton, FL. Reprinted with permission from Taylor & Francis Group.)

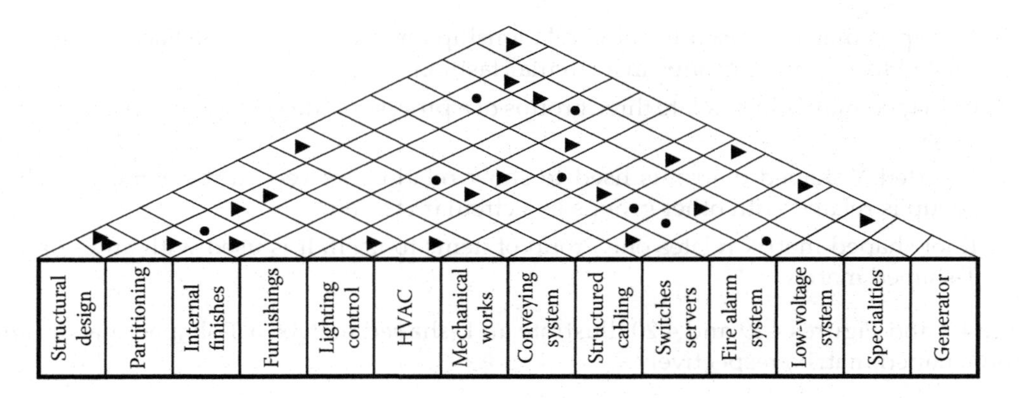

FIGURE 4.20
Roof-shaped matrix. (Abdul Razzak Rumane (2013), *Quality Tools for Managing Construction Projects*, CRC Press, Boca Raton, FL. Reprinted with permission from Taylor & Francis Group.)

4.2.2.5 Prioritization Matrix

The prioritization matrix assists in choosing between several options in order of importance and priority. It helps decision makers determine the order of importance, considering the relative merit of each of the activities or goal being considered. It focuses the attention of team members to those key issues and options that are more important for the organization or project. Figure 4.21 illustrates a prioritization matrix.

4.2.2.6 Process Decision Program

The process decision program is a technique used to help prepare contingency plans. It systematically identifies what might go wrong in a project plan or project schedule and describe specific actions to be taken to prevent the problems from occurring in the first place, and to mitigate or avoid the impact of the problems if they occur.

Figure 4.22 illustrates the process decision program for submission of contract documents.

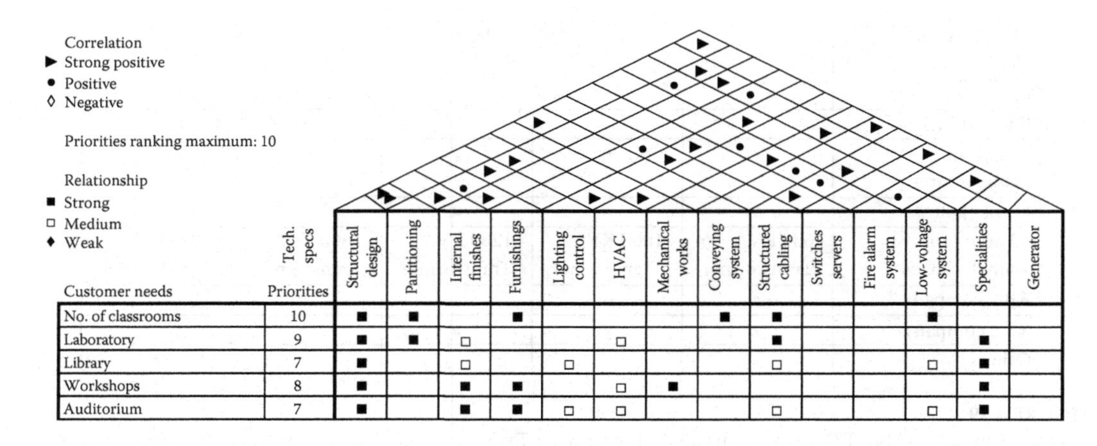

Customer needs	Tech. specs / Priorities	Structural design	Partitioning	Internal finishes	Furnishings	Lighting control	HVAC	Mechanical works	Conveying system	Structured cabling	Switches servers	Fire alarm system	Low-voltage system	Specialities	Generator
No. of classrooms	10	■	■		■				■	■			■		
Laboratory	9	■	■	□			□			■				■	
Library	7	■		□		□				□			□	■	
Workshops	8	■		■	■		□	■						■	
Auditorium	7	■		■	■	□	□			□			□	■	

FIGURE 4.21
Prioritization matrix. (Abdul Razzak Rumane (2013), *Quality Tools for Managing Construction Projects*, CRC Press, Boca Raton, FL. Reprinted with permission from Taylor & Francis Group.)

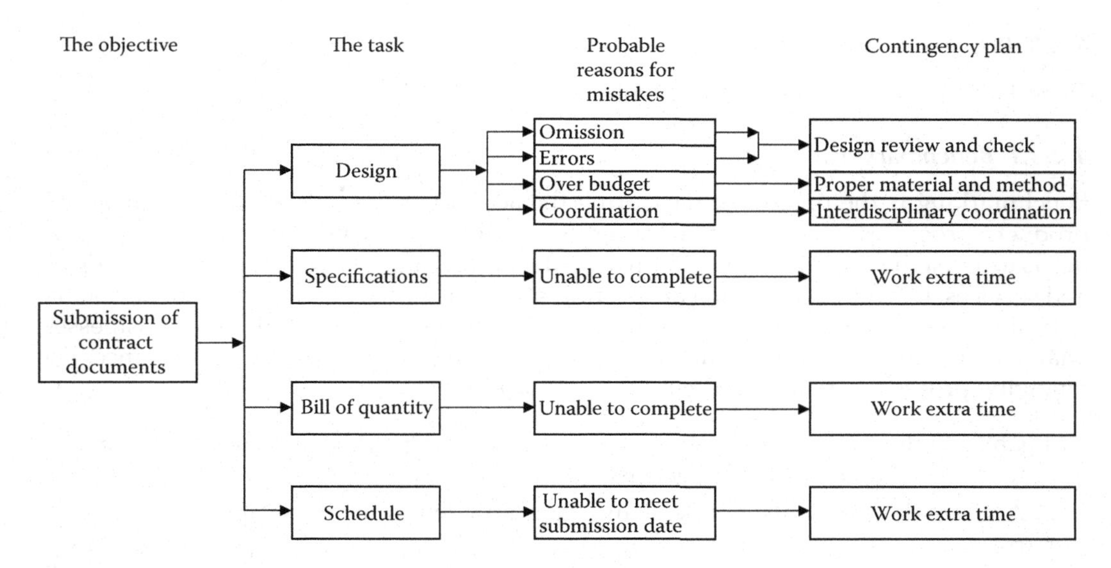

FIGURE 4.22
Process decision diagram chart. (Abdul Razzak Rumane (2013), *Quality Tools for Managing Construction Projects*, CRC Press, Boca Raton, FL. Reprinted with permission from Taylor & Francis Group.)

4.2.2.7 Tree Diagram

Tree diagrams are used to break down or stratify ideas progressively into more detailed steps. It breaks broader ideas into specific details and helps make decision easier to select the alternative. It is used to figure out all the various tasks that must be undertaken to achieve a given objective.

Figure 4.23 illustrates the tree diagram for water in the storage tank.

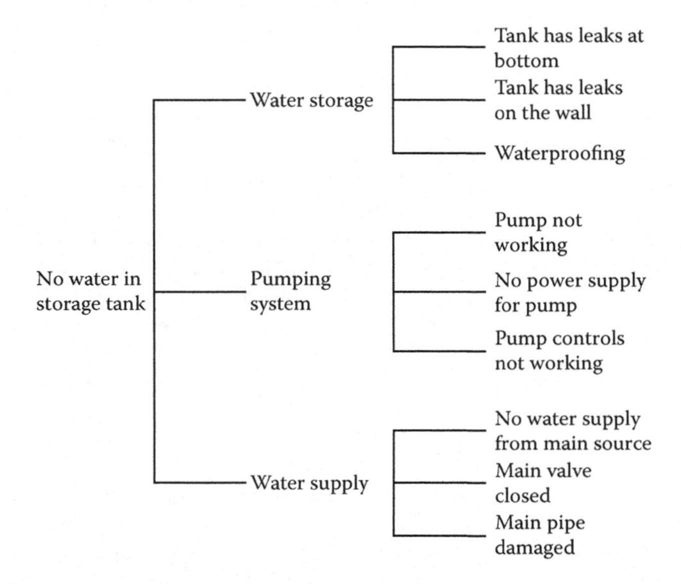

FIGURE 4.23
Tree diagram for no water in storage tank. (Abdul Razzak Rumane (2013), *Quality Tools for Managing Construction Projects*, CRC Press, Boca Raton, FL. Reprinted with permission from Taylor & Francis Group.)

4.2.3 Process Analysis Tools

Table 4.5 illustrates process analysis tools.

4.2.3.1 Benchmarking

Benchmarking is the process of measuring the actual performance of the organization's products, processes, and services, in order to compare it to the best-known industry standards to assist the organization in improving the performance of their products, processes, and services. It involves analyzing an existing situation, identifying and measuring factors critical to the success of the product or services, comparing them with other businesses, analyzing the results, and implementing an action plan to achieve better performance. The following processes are involved while benchmarking:

1. Collect internal and external data on work, process, method, product characteristics, and system selected for benchmarking.
2. Analyze data to identify performance gaps and determine cause and differences.
3. Prepare action plan to improve the process in order to meet or exceed the best practices in the industry.
4. Search for the best practices among market leaders, competitors, and non-competitors that lead to their superior performance.
5. Improve the performance by implementing these practices.

Figure 4.24 illustrates a benchmarking process.

4.2.3.2 Cause-and-Effect Diagram

It is one of the quality classic tools. It is used to analyze the cause-and-effect of defects or nonconformance and the effect on the process due to these causes.

TABLE 4.5

Process Analysis Tools

Sr. No.	Name of Quality Tool	Usage
Tool 1	Bench marking	To identify best practices in the industry and improve the process or project.
Tool 2	Cause and effect	To identify possible cause and its effect in the process.
Tool 3	Cost of quality	To identify hidden or indirect cost affecting the overall cost of product/project.
Tool 4	Critical to quality	To identify quality features or characteristics most important to the client.
Tool 5	Failure mode and effects analysis (FEMA)	To identify and classify failures according to their effect.
Tool 6	5 why analysis	Used to analyze and solve any problem where the root cause is unknown.
Tool 7	5W2H	The questions used to understand why the things happen the way they do.
Tool 8	Process mapping/ flowcharting	It is technique used for designing, analyzing, and communicating work processes.

Source: Abdul Razzak Rumane (2013), *Quality Tools for Managing Construction Projects*, CRC Press, Boca Raton, FL. Reprinted with permission from Taylor & Francis Group.

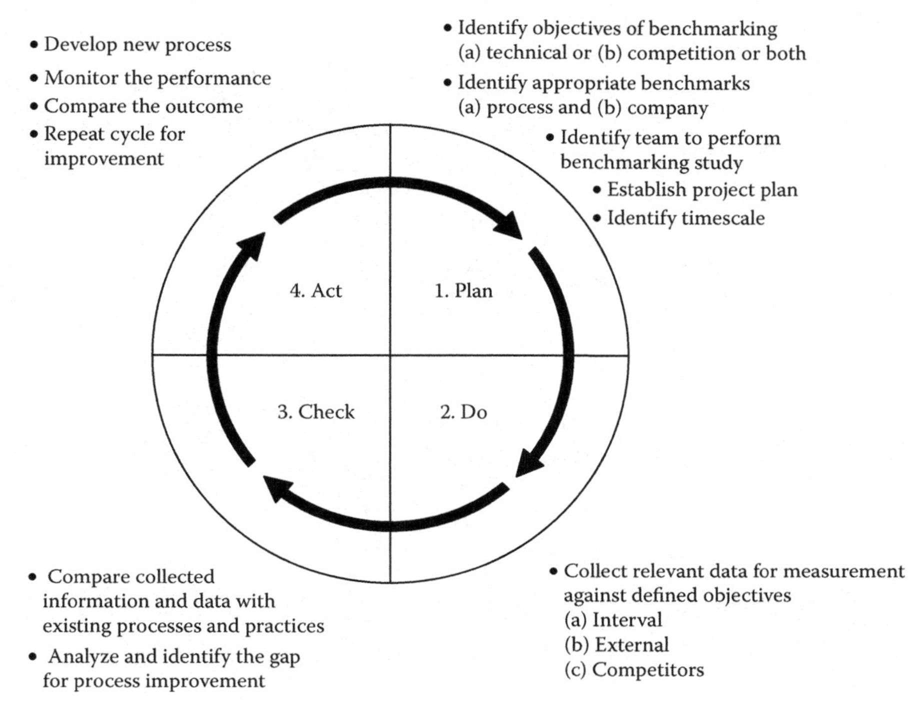

FIGURE 4.24
Benchmarking process. (Abdul Razzak Rumane (2013), *Quality Tools for Managing Construction Projects*, CRC Press, Boca Raton, FL. Reprinted with permission from Taylor & Francis Group.)

Figure 4.25 illustrates the cause-and-effect diagram for rejection of masonry work.

4.2.3.3 Cost of Quality

Cost of quality is discussed in detail under Section 4.2.7.

4.2.3.4 Critical to Quality

Critical to quality is a significant step in the design process of a product or service to identify the customer's/client's expectation to fulfill their need and requirements.

4.2.3.5 Failure Mode and Effects Analysis

Failure mode and effects analysis (FMEA) is to identify all the possible failures in the design of a product, process, service, and their effects on the product, process, and service. Its aim is to reduce risk of failure and improve the process. Figure 4.26 illustrates a FMEA process and Figure 4.27 illustrates an example form used to record FMEA readings.

4.2.3.6 Whys Analysis

It is used to analyze and solve any problem where the root cause is unknown. Table 4.6 illustrates the 5 whys analysis chart for burning of cable.

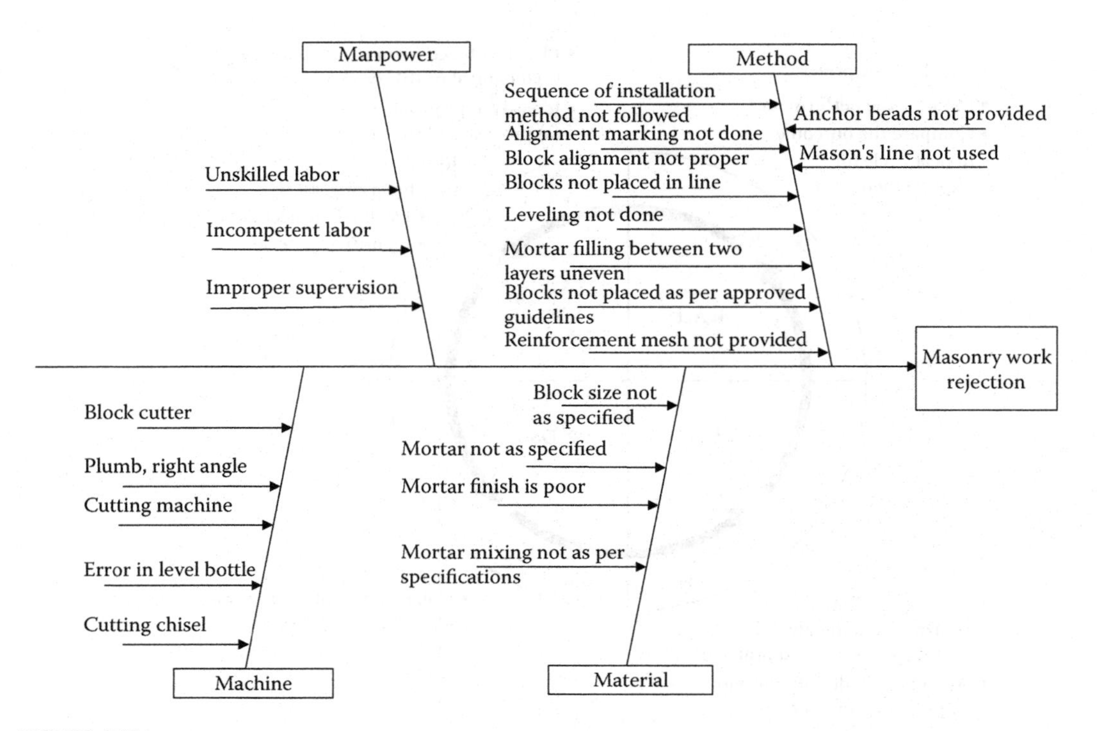

FIGURE 4.25
Cause and effect for masonry work. (Abdul Razzak Rumane (2013), *Quality Tools for Managing Construction Projects*, CRC Press, Boca Raton, FL. Reprinted with permission from Taylor & Francis Group.)

4.2.3.7 5W2H

5W2H is about asking the questions to understand about a process or problem.
The 5 Ws are as follows:

1. Why
2. What
3. When
4. Where
5. Who

And the 2 Hs are as follows:

1. How
2. How much

Table 4.7 illustrates the 5W2H for a slab collapse.

4.2.3.8 Process Mapping/Flowcharting

Process mapping/flowcharting is a graphical representation of workflow giving a clear understanding of a process or services of parallel processes. Process mapping/flowcharting

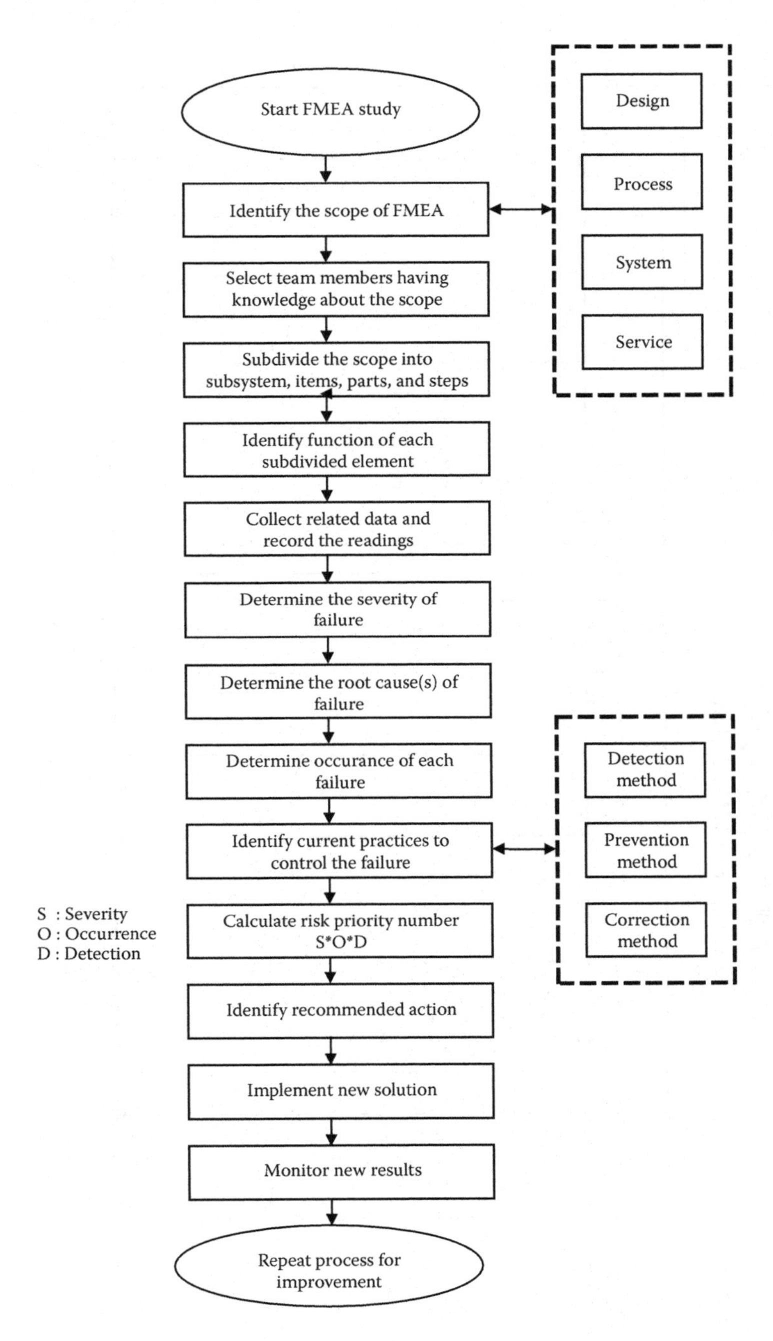

FIGURE 4.26
Failure mode and effects analysis process. (Abdul Razzak Rumane (2013), *Quality Tools for Managing Construction Projects*, CRC Press, Boca Raton, FL. Reprinted with permission from Taylor & Francis Group.)

Failure mode and effect analysis

Product name:

Drawing reference:

Revision:

EXAMPLE ANALYSIS

Team members
1.
2.
3.
4.

Operation number	Scope description	Subdivided elements	Failure mode	Effects of failure	Severity rating	Cause of failure	Occurence rating	Current practice of controls			Detection rating	Current status of product				Recommended corrective action	Action by	Action taken date	Revise status			
								Detection	Prevention	Correction		S	O	D	RPN				S	O	D	RPN
1	Emergency generator system	Operation	Generator failed to start	1. No lights 2. Life support equipment stopped functioning 3. No power supply for IT system 4. Water supply pumps stopped 5. No power supply for lift 6. Fire mode operation equipment not operate 5. HVAC system stopped	7 10 9 6 4	1. No signal from ATS 2. Automatic starting system failed 3. Low battery voltage for starter motor 4. Circuit breaker in off position 5. No diesel in day tank		1. Through BMS 2. No regular check 3. Manual 4. Manual	1. Not in practice	1. Manual						1. Regular check of starting system 2. Check starter regularly 3. Check diesel level regularly. Interface level indicator with BMS 4. Check breaker position regularly. Interface with BMS	Maintenance engineer					

Legend:
RPN: Risk priority number
S: Severity
O: Occurence
D: Detection

ATS: Automatic transfer switch
BMS: Building management system

FIGURE 4.27
FMEA recording form. (Abdul Razzak Rumane (2013), *Quality Tools for Managing Construction Projects*, CRC Press, Boca Raton, FL. Reprinted with permission from Taylor & Francis Group.)

TABLE 4.6

The 5 Whys Analysis for Cable Burning

Sr. No.	Why	Related Analyzing Question
1	Why	Why the cable burned
2	Why	Why the earth leakage relay not tripped
3	Why	Why circuit breaker not tripped
4	Why	Why poor insulation of cable was not noticed
5	Why	Why under size rating of breaker with respect to current carrying capacity of cable was not noticed

Source: Abdul Razzak Rumane (2013), *Quality Tools for Managing Construction Projects*, CRC Press, Boca Raton, FL. Reprinted with permission from Taylor & Francis Group.

TABLE 4.7

5W2H Analysis for Slab Collapse

Sr. No.	Why	Related Analyzing Question
1	Why	The slab collapse
2	What	What is the reason for collapse
3	Who	Who is responsible
4	Where	Where is the mistake
5	When	When did the slab collapse
6	How many	How many persons affected (injured or died)
7	How much	How much loss in terms of cost and time

Source: Abdul Razzak Rumane (2013), *Quality Tools for Managing Construction Projects*, CRC Press, Boca Raton, FL. Reprinted with permission from Taylor & Francis Group.

is a technique that can be employed to not only produce visual representation of the production processes but also processes related to other departments.

Figure 4.28 illustrates the mapping/flowcharting process diagram for approval of variation order.

4.2.4 Process Improvement Tools

Table 4.8 illustrates process improvement tools.

4.2.4.1 Root Cause Analysis

It is used to analyze root causes of problems. The analysis is generally performed by using the Ishikawa diagram or the cause-and-effect diagram. Figure 4.29 illustrates the root cause analysis for rejection of executed marble work.

4.2.4.2 Plan-Do-Check-Act Cycle

Plan–do–check–act (PDCA) is mainly used for continuous improvement. It consists of a four-step model for carrying changes. The PDCA cycle model can be developed as a process improvement tool to reduce cost of quality. Figure 4.30 illustrates the PDCA cycle for preparation of shop drawings.

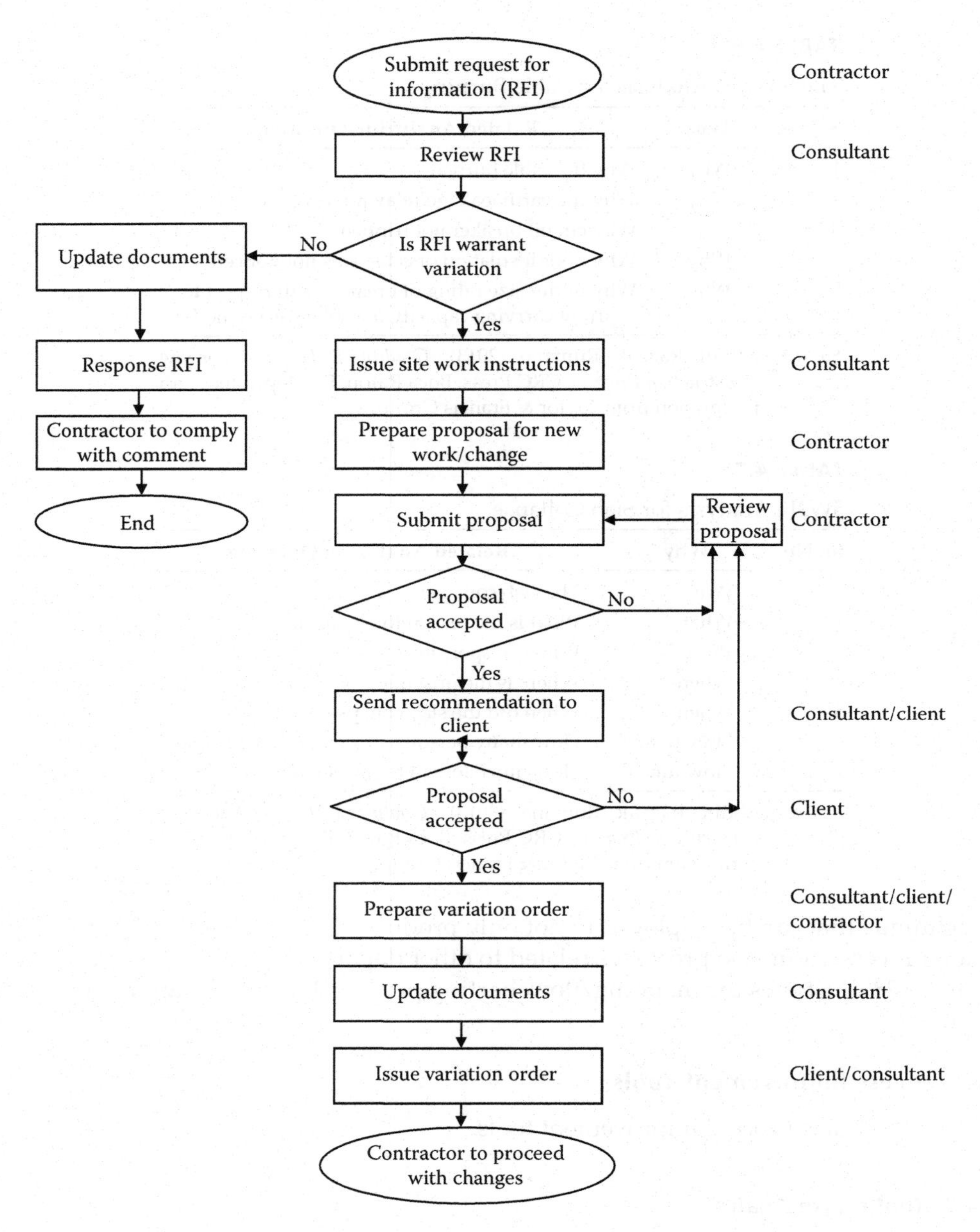

FIGURE 4.28
Process mapping/flowcharting for approval of variation order. (Abdul Razzak Rumane (2013), *Quality Tools for Managing Construction Projects*, CRC Press, Boca Raton, FL. Reprinted with permission from Taylor & Francis Group.)

4.2.4.3 Supplier–Input–Process–Output–Customer Analysis

It is used to identify supplier–input–process–output–customer (SIPOC) relationship. The purpose of the SIPOC analysis is to show the process flow by defining and documenting the suppliers, inputs, process steps, outputs, and customers. Table 4.9 illustrates the SIPOC analysis for an electrical panel.

TABLE 4.8

Process Improvement Tools

Sr. No.	Name of Quality Tool	Usage
Tool 1	Root cause analysis	To identify root causes that caused the problem to occur
Tool 2	PDCA cycle	Used to plan for improvement followed by putting into action
Tool 3	SIPOC analysis	Used to identify supplier–input–process–output–customer realationship
Tool 4	Six Sigma DMAIC	Used as analytic tool for improvement
Tool 5	Failure mode and effects analysis (FMEA)	To identify and classify failures according to effect and prevent or reduce failure
Tool 6	Statistical process control	Used to study how the process changes over a time

Source: Abdul Razzak Rumane (2013), *Quality Tools for Managing Construction Projects*, CRC Press, Boca Raton, FL. Reprinted with permission from Taylor & Francis Group.

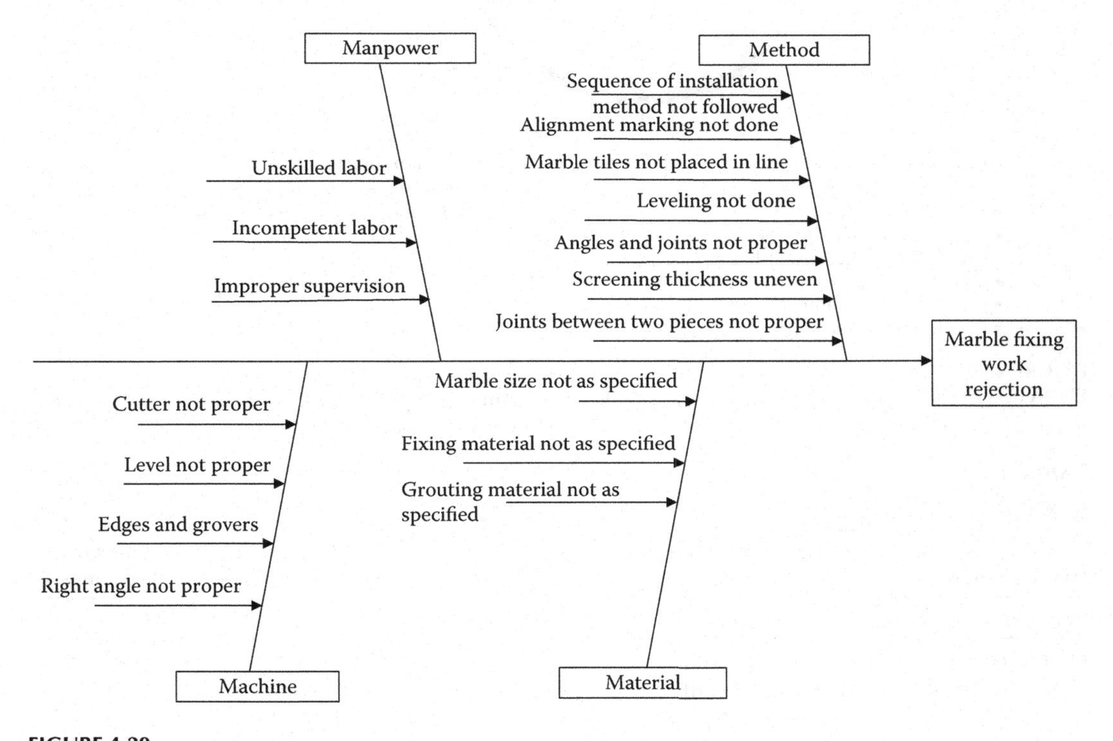

FIGURE 4.29

Root cause analysis for rejection of executed marble work. (Abdul Razzak Rumane (2013), *Quality Tools for Managing Construction Projects*, CRC Press, Boca Raton, FL. Reprinted with permission from Taylor & Francis Group.)

4.2.4.4 Six Sigma—DMAIC

Six Sigma is discussed in detail in Section 4.2.9.

4.2.4.5 FMEA

FMEA is also used as a process improvement tool. It identifies all the possible failures in the design of a product, process, service, and their effects on the product, process, and

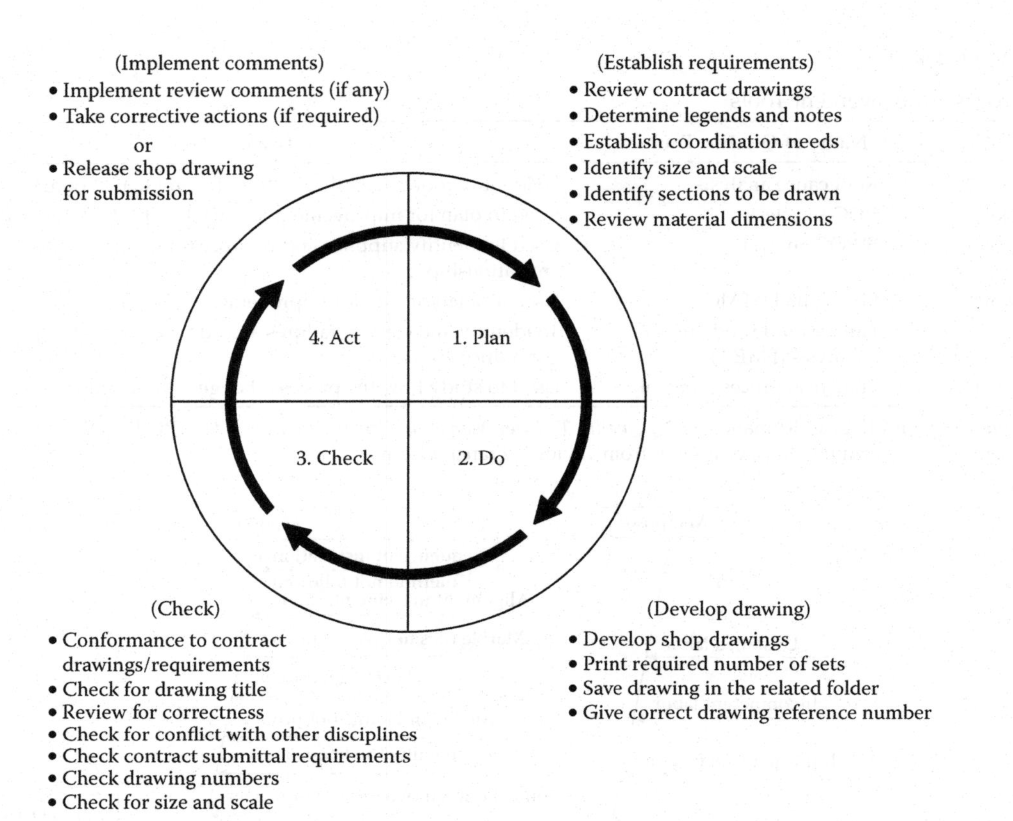
(Implement comments)
• Implement review comments (if any)
• Take corrective actions (if required)
or
• Release shop drawing for submission
(Establish requirements)
• Review contract drawings
• Determine legends and notes
• Establish coordination needs
• Identify size and scale
• Identify sections to be drawn
• Review material dimensions
4. Act
1. Plan
3. Check
2. Do
(Check)
• Conformance to contract drawings/requirements
• Check for drawing title
• Review for correctness
• Check for conflict with other disciplines
• Check contract submittal requirements
• Check drawing numbers
• Check for size and scale
(Develop drawing)
• Develop shop drawings
• Print required number of sets
• Save drawing in the related folder
• Give correct drawing reference number

FIGURE 4.30
PDCA cycle for preparation of shop drawing. (Abdul Razzak Rumane (2013), *Quality Tools for Managing Construction Projects*, CRC Press, Boca Raton, FL. Reprinted with permission from Taylor & Francis Group.)

TABLE 4.9
SIPOC Analysis for an Electrical Panel

(Who Are Suppliers)	(What the Suppliers Are Providing)	(What Is the Process)	(What Is the Output of Process)	(Who Are the Customers)
Supplier	**Inputs**	**Process**	**Outputs**	**Customer**
Electrical panel builder/assembler	Main low tension panel Main switch boards Distribution boards Starter panels Control panels	Elecrical installation work	Electrical distribution network	Power supply for project

Source: Abdul Razzak Rumane (2013), *Quality Tools for Managing Construction Projects*, CRC Press, Boca Raton, FL. Reprinted with permission from Taylor & Francis Group.

service. Its aim is to reduce risk of failure and improve the process. Refer to Figure 4.26 in Section 4.2.3.5, which illustrates the FMEA process.

4.2.4.6 Statistical Process Control

SPC is a quantitative approach based on the measurement of process control. Dr. Walter A. Shewhart developed the control charts as early as 1924. SPC charts are used for identification of common cause and special (or assignable) cause variations and assisting diagnosis of quality problems. SPC charts reveal whether a process is *in control*-stable and exhibiting only random variation or *out of control* and needing attention. Control chart is one of the key tools of SPC. It is used to monitor processes that are not in control, using measured ranges. The following are the two types of process control charts:

1. Variable charts
2. Attributes charts

Variable charts relate to variable measurement such as length, width, temperature, and weight.

Attributes charts relate to the characteristics possessed (or not possessed) by the process or the product. Figure 4.31 illustrates statistical control process charts for generator frequency.

4.2.5 Innovation and Creative Tools

Table 4.10 illustrates innovation and creative tools.

4.2.5.1 Brainstorming

Brainstorming is listing all the ideas put forth by a group in response to a given question or problem. It is a process of creating ideas by storming some objective. In 1939, a team led by advertising executive Alex Osborn coined the term *brainstorm*. According to Osborn, brainstorm means using the brain to storm a creative problem. Classical brainstorming is the most well-known and often-used technique for idea generation in a short period.

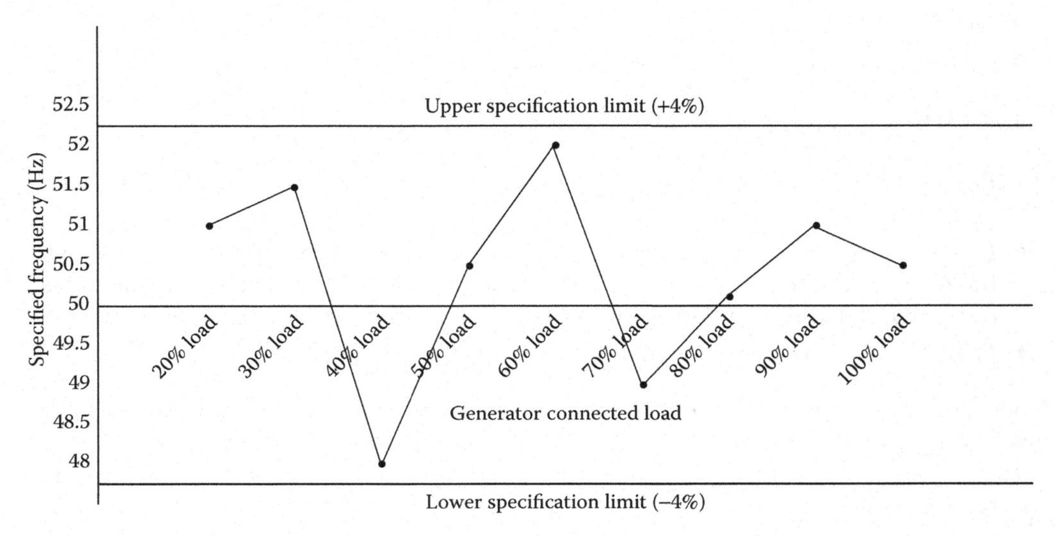

FIGURE 4.31
Statistical process control chart for generator frequency. (Abdul Razzak Rumane (2013), *Quality Tools for Managing Construction Projects*, CRC Press, Boca Raton, FL. Reprinted with permission from Taylor & Francis Group.)

TABLE 4.10

Innovation and Creative Tools

Sr. No.	Name of Quality Tool	Usage
Tool 1	Brainstorming	Used to generate multiple ideas.
Tool 2	Delphi Technique	Used to get ideas from select group of experts.
Tool 3	5W2H	The questions used to understand why the things happen the way they do.
Tool 4	Mind mapping	Used to create a visual representation of many issues that can help get more understanding of the situation.
Tool 5	Nominal group technique	Used to enhance brainstorming by ranking the most useful ideas.
Tool 6	Six Sigma DMADV	Used primarily for the invention and innovation of modified or new product, services, or process.
Tool 7	Triz	Used to provide systematic methos and tools for analysis and innovative problem solving.

Source: Abdul Razzak Rumane (2013), *Quality Tools for Managing Construction Projects*, CRC Press, Boca Raton, FL. Reprinted with permission from Taylor & Francis Group.

It is based on the fundamental principles of deferment of judgment and that quantity breeds quality. It involves the following questions:

- Does the item have any design features that are not necessary?
- Can two or more parts be combined together?
- How can we cut down the weight?
- Are these nonstandard parts that can be eliminated?

There are four rules for successful brainstorming:

1. Criticism is ruled out.
2. Freewheeling is welcomed.
3. Quantity is wanted.
4. Contribution and improvement are sought.

A classical brainstorming session has the following basic steps:

- *Preparation*: The participants are selected, and a preliminary statement of the problem is circulated.
- *Brainstorming*: A warm-up session with simple unrelated problems is conducted; the relevant problem and the four rules for brainstorming are presented and ideas are generated and recorded using checklists and other techniques if necessary.
- *Evaluation*: The ideas are evaluated relative to the problem.

Generally, a brainstorming group should consist of four to seven people, although some suggest a larger group. Figure 4.32 illustrates the brainstorming process.

4.2.5.2 Delphi Technique

Delphi technique is intended to determine a consensus among experts on a subject matter. The goal of Delphi technique is to pick brains of experts in the subject area, treating them

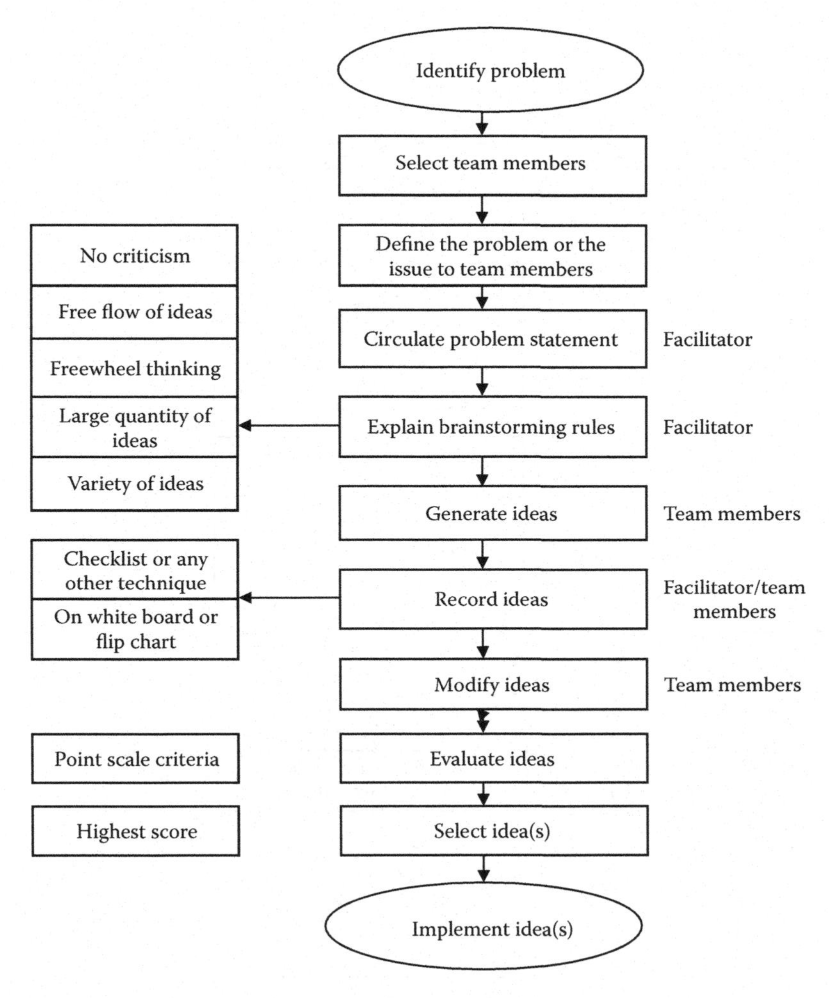

FIGURE 4.32
Brainstorming process. (Abdul Razzak Rumane (2013), *Quality Tools for Managing Construction Projects*, CRC Press, Boca Raton, FL. Reprinted with permission from Taylor & Francis Group.)

as contributors to create ideas. It is a measure and method for consensus building by using questionnaire and obtaining responses from the panel of experts in the selected subjects. Delphi technique employs multiple iterations designed to develop consensus opinion about the specific subject. The selected expert group answers questions by facilitator. The responses are summarized and further circulated for group comments to reach the consensus. The iteration/feedback process allows the team members to reassess their initial judgment and change or modify the earlier suggestions. Figure 4.33 illustrates the Delphi technique process.

4.2.5.3 5W2H

5W2H is also used as an innovation and creative tool. 5W2H is about asking the questions to understand about a process or problem.

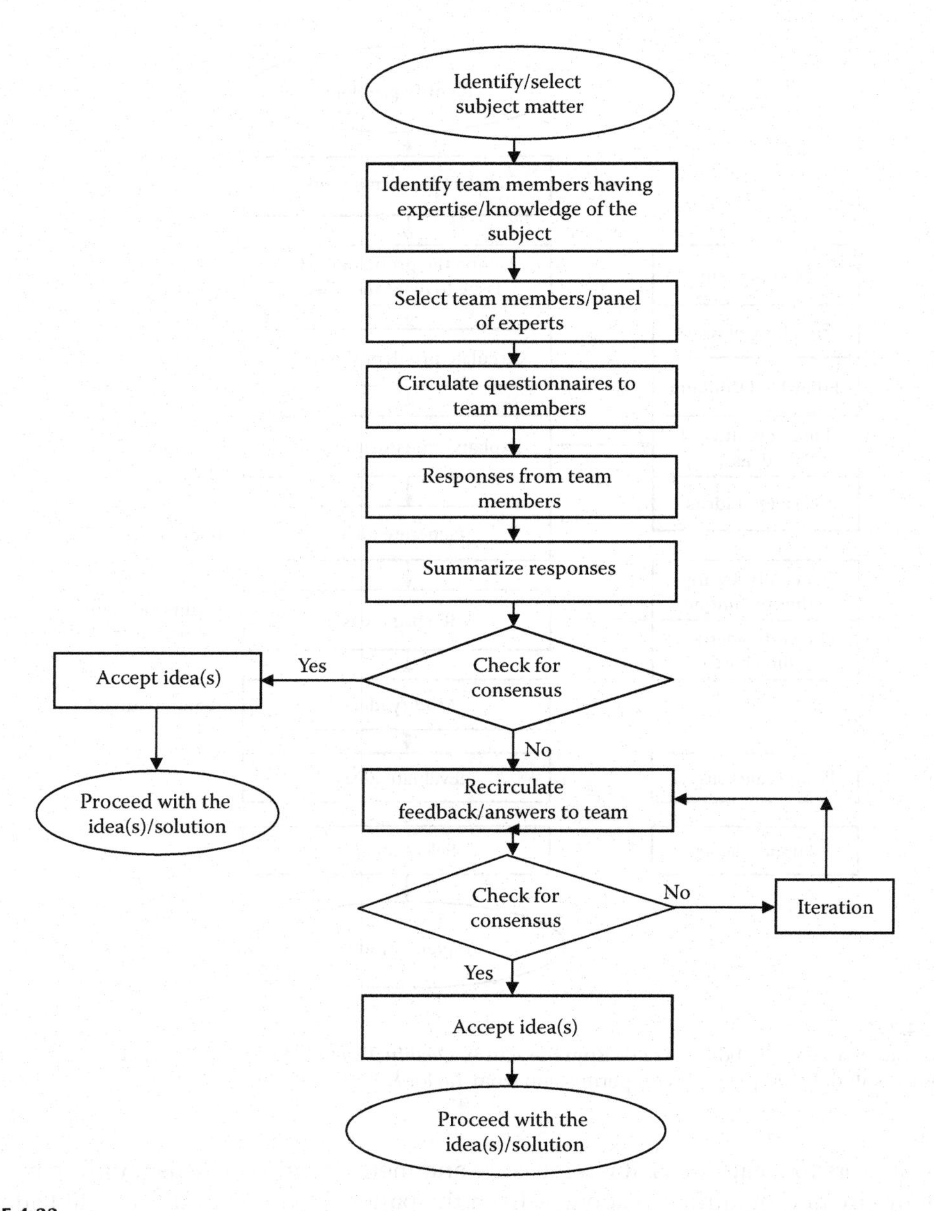

FIGURE 4.33
Delphi technique process. (Abdul Razzak Rumane (2013), *Quality Tools for Managing Construction Projects*, CRC Press, Boca Raton, FL. Reprinted with permission from Taylor & Francis Group.)

The 5 Ws are as follows:

1. Why
2. What
3. When
4. Where
5. Who

And the 2 Hs are as follows:

1. How
2. How much

Table 4.11 illustrates 5W2H for development of a new product.

4.2.5.4 Mind Mapping

Mind mapping is a graphical representation of ideas that can help get more understanding of the situation and create the solution or improve the task. Figure 4.34 illustrates mind mapping sketch to improve site safety.

TABLE 4.11

5W2H Analysis for New Product

Sr. No.	Why	Related Analyzing Question
1	Why	New product
2	What	What advantage it will have over other similar products
3	Who	Who will be the customers for this product
4	Where	Where can we market the product
5	When	When the product will be ready for sale
6	How many	How many pieces will be produced/sold per year
7	How much	How much market share we will get for this product

Source: Abdul Razzak Rumane (2013), *Quality Tools for Managing Construction Projects*, CRC Press, Boca Raton, FL. Reprinted with permission from Taylor & Francis Group.

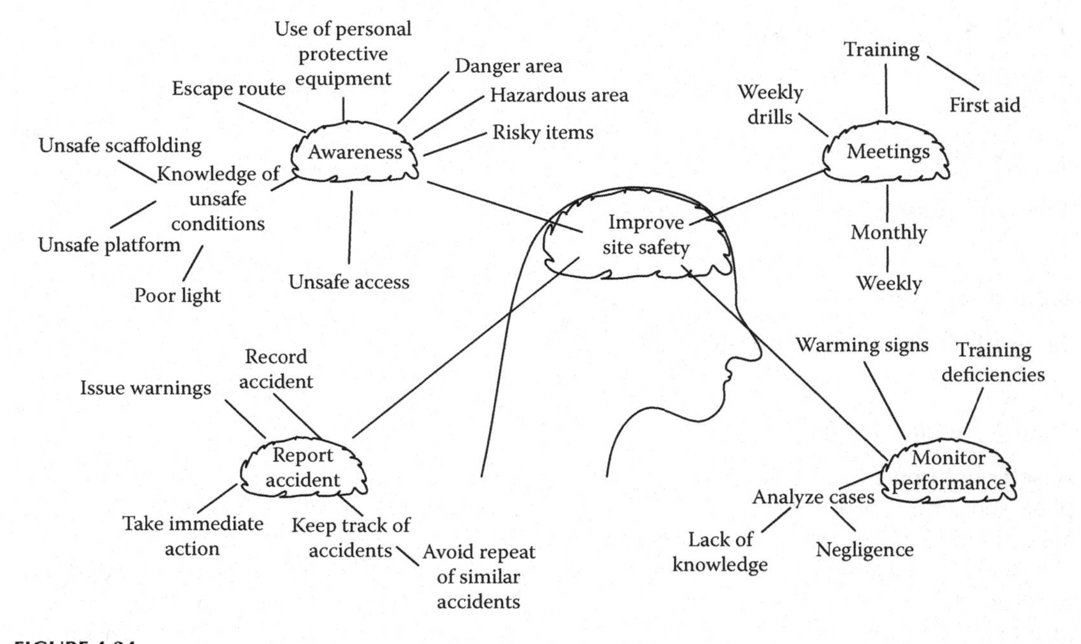

FIGURE 4.34
Mind mapping. (Abdul Razzak Rumane (2013), *Quality Tools for Managing Construction Projects*, CRC Press, Boca Raton, FL. Reprinted with permission from Taylor & Francis Group.)

4.2.5.5 Nominal Group Technique

The nominal group technique (NGT) involves a structural group meeting designed to incorporate individual ideas and judgments into a group consensus. By correctly applying the NGT, it is possible for groups of people (preferably 5 to 10) to generate alternatives or other ideas for improving the competitiveness of the firm. The technique can be used to obtain group thinking (consensus) on a wide range of topics. The technique, when properly applied, draws on the creativity of the individual participants, while reducing two undesirable effects of most group meetings:

1. The dominance of one or more participants
2. The suppression of conflict ideas

The basic format of an NGT session is as follows:

- Individual silent generation of ideas
- Individual round-robin feedback and recording the ideas
- Group's clarification of each idea
- Individual voting and ranking to prioritize ideas
- Discussion of group consensus results

The NGT session begins with an explanation of the procedure and a statement of question (s), preferably written by the facilitator.

4.2.5.6 Six Sigma—DMADV

Six Sigma is discussed in detail in Section 4.2.9.

4.2.5.7 Triz

Triz is discussed in detail in Section 4.2.10.

4.2.6 Lean Tools

Table 4.12 illustrates Lean tools.

4.2.6.1 Cellular Design

A self-contained unit dedicated to perform all the operational requirements to accomplish sequential processing. With a cellular design, individual cells can be fabricated and assembled to give same performance and saving time.

An electrical main switch board may consist of a number of cells assembled together to perform desired operations. It helps easy maneuvering and assembling at work place for proper functioning. Figure 4.35 illustrates the cellular design for an electrical panel.

TABLE 4.12
Lean Tools

Sr. No.	Name of Quality Tool	Usage
Tool 1	Cellular design	It is a self-contained unit dedicated to perform all the operational requirements to accomplish sequential processing.
Tool 2	Concurrent engineering	It is used for product cycle reduction time. It is a systematic approach of creating a product design that simultaneously considers all elements of a product life cycle.
Tool 3	5S	It is used to eliminate waste that results from improper organization of work area.
Tool 4	Just in time	It is used to reduce inventory levels, improve cash flow, and reduce space requirements for storage of material.
Tool 5	Kanban	It is used to signal that more material is required to be ordered. It is used to eliminate waste from inventory.
Tool 6	Kaizen	It is used for continually eliminating waste from manufacturing processes by combining the collective talent of company.
Tool 7	Mistake proofing	It is used to eliminate the opportunity for error by detecting the potential source of error.
Tool 8	Outsourcing	It is contracting out certain works, processes, and services to specialists in the discipline area.
Tool 9	Poka Yoke	It is used to detect the abnormality or error and fix or correct the error and take action to prevent the error.
Tool 10	Single minute exchange of die	It is used to reduce setup time for change over to new process.
Tool 11	Value stream mapping	It is used to establish flow of material or information, eliminate waste, and add value.
Tool 12	Visual management	It addresses both visual display and control and exposes waste elimination/prevention.
Tool 13	Waste reduction	It focuses on reducing waste.

Source: Abdul Razzak Rumane (2013), *Quality Tools for Managing Construction Projects*, CRC Press, Boca Raton, FL. Reprinted with permission from Taylor & Francis Group.

4.2.6.2 Concurrent Engineering

Product life cycle begins with need and extend through concept design, preliminary design, detail design, production or construction, product use, phase out, and disposal. Concurrent engineering is defined as a systematic approach to create a product design that simultaneously considers all the elements of product life cycle, thus reducing the product life-cycle time. It is used to expedite the development and launch of new product. In construction projects, construction can simultaneously start while the design is under development. Figure 4.36 illustrates concurrent engineering for a construction project life cycle.

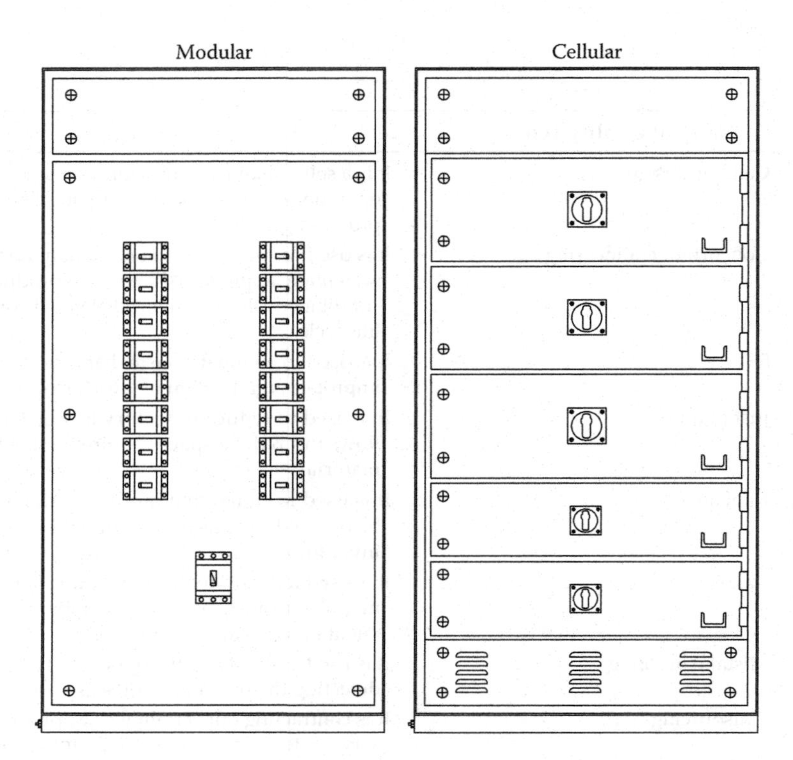

FIGURE 4.35
Cellular main switch board. (Abdul Razzak Rumane (2013), *Quality Tools for Managing Construction Projects*, CRC Press, Boca Raton, FL. Reprinted with permission from Taylor & Francis Group.)

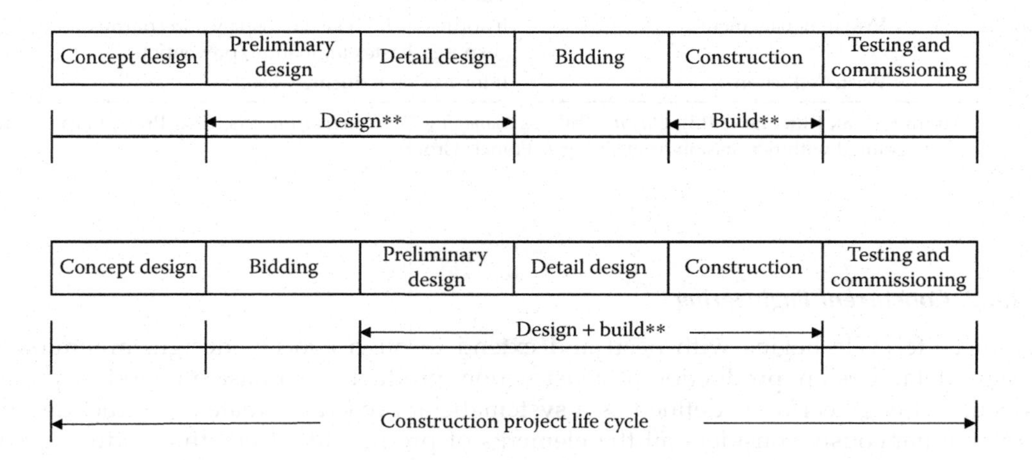

FIGURE 4.36
Concurrent engineering for construction life cycle. (Abdul Razzak Rumane (2013), *Quality Tools for Managing Construction Projects*, CRC Press, Boca Raton, FL. Reprinted with permission from Taylor & Francis Group.)

4.2.6.3 5S

5S is a systematic approach for improvement of quality and safety by organizing a workplace. It is a methodology that advocates the following:

- What should be kept?
- Where should be kept?
- How should be kept?

5S is a Japanese concept of housekeeping having reference to five Japanese words starting with letter *S*. Table 4.13 illustrates 5S for construction projects.

4.2.6.4 Just in Time

It is used to reduce inventory levels, improve cash flow, and reduce storage space requirements for material. Examples include the following:

1. Concrete block can be received at site just before the start of block work and can be stacked near the work area where masonry work is in progress.
2. Chiller can be received at site and directly placed on the chiller foundation without storing in the storage yard.

4.2.6.5 Kanban

It is used to signal more material that is required to be ordered. It is used to eliminate waste from inventory and inventory control, thus to avoid extra storage required for large inventory. In construction projects, electrical wires for circuiting can be ordered to receive at site when the wire pulling work is under progress. Similarly, concrete blocks, false ceiling tiles can be ordered and received as and when required.

4.2.6.6 Kaizen

It is used for continually improving through small changes to eliminate waste from manufacturing process by combining the collective talent of every employee of the company.

4.2.6.7 Mistake Proofing

Mistake proofing is used to eliminate the opportunity for error by detecting the potential source of error. Mistakes are generally categorized as follows:

1. Information
2. Mismanagement
3. Omission
4. Selection

Table 4.14 illustrates mistake proofing chart for eliminating design error.

TABLE 4.13
5S for Construction Projects

Sr. No.	5S	Related Action
1	Sort	• Determine what is to be kept in open and what under shed • Allocate area for each type of construction equipment and machinery • Allocate area for electrical tools • Allocate area for hand tools • Allocate area for construction material/equipment to be used/installed in the project • Allocate area for hazardous, inflammable material • Allocate area for chemicals and paints • Allocate area for spare part for maintenance
2	Set in order	• Keep/arrange equipment in such a way that their maneuvering/movement shall be easy • Vehicles to be parked in the yard in such a way that frequently used vehicles are parked near the gate • Frequently used equipment/machinery to be located near the work place • Set boundaries for different types of equipment and machinery • Identify and arrange tools for easy access • Identify and store material/equipment as per relevant division/section of contract documents • Identify and store material in accordance with their usage as per construction schedule • Determine items which need special conditions • Mark/tag the items/material • Display route map and location • Put the material in sequence as per their use • Frequently used consumables to be kept near work place • Label on the drawer with list of contents • Keep shuttering material at one place • Determine inventory level of consumable items
3	Sweeping	• Clean site on a daily basis by removing • Cut pieces of reinforced bars • Cut pieces of plywood • Left out concrete • Cut pieces of pipes • Cut pieces of cables and wires • Used welding rods • Clean equipment and vehicles • Check electrical tools after return by the technician • Attend to breakdown report
4	Standardize	• Standardize the store by allocating separate areas for material used by different divisions/sections • Standardize area for long lead items • Determine regular schedule for cleaning the work place • Make available standard tool kit/box for a group of technicians • Make every one informed of their responsibilities and related area where the things are to be placed and are available • Standardize the store for consumable items • Inform suppliers/vendors in advance the place for delivery of material
5	Sustain	• Follow the system till the end of project

Source: Abdul Razzak Rumane (2010), *Quality Management in Construction Projects*, CRC Press, Boca Raton, FL. Reprinted with permission from Taylor & Francis Group.

TABLE 4.14

Mistake Proofing for Eliminating Design Errors

Sr. No.	Items	Points to Be Considered to Avoid Mistakes
1	Information	1. Terms of reference (TORs) 2. Client's preferred requirements matrix 3. Data collection 4. Regulatory requirements 5. Codes and standards 6. Historical data 7. Organizational requirements
2	Mismanagement	1. Compare production with actual requirements 2. Interdisciplinary coordination 3. Application of different codes and standards 4. Drawing size of different trades/specialist consultants
3	Omission	1. Review and check design with TOR 2. Review and check design with client requirements 3. Review and check design with regulatory requirements 4. Review and check design with codes and standards 5. Check for all required documents
4	Selection	1. Qualified team members 2. Available material 3. Installation methods

Source: Abdul Razzak Rumane (2013), *Quality Tools for Managing Construction Projects*, CRC Press, Boca Raton, FL. Reprinted with permission from Taylor & Francis Group.

4.2.6.8 Outsourcing

It refers to contracting out certain works, processes, and services to specialist in a particular discipline. For example, in construction projects, the following is a list of some of the works that are outsourced (subcontracted):

1. Structural concrete
2. Waterproofing work
3. HVAC work
4. Fire suppression work
5. Water supply piping
6. Electrical work
7. Security system
8. Low-voltage works
9. Landscape work

4.2.6.9 Poka Yoke

Poka Yoke is a quality management concept developed by Shigeo Shingo to prevent human errors occurring in the production line. The main objective of Poka Yoke is to achieve zero defects.

4.2.6.10 Single Minute Exchange of Die

It is used to reduce set-up time for change over to new process. For example, a spare circuit breaker of similar rating can be used as immediate replacement to damaged circuit breaker in the electrical distribution board to avoid breakdown of electrical supply for long duration. Subsequently, a new circuit breaker can be fixed in place of spare breaker.

4.2.6.11 Value Stream Mapping

It is used to establish flow of material or information and eliminate waste and add value. Value stream mapping is used to identify areas for improvement. Figure 4.37 illustrates the value stream mapping diagram for an emergency power system.

4.2.6.12 Visual Management

It addresses both visual display and control, and it exposes waste elimination/prevention. Visual displays present information, whereas visual control focuses on a need to act.

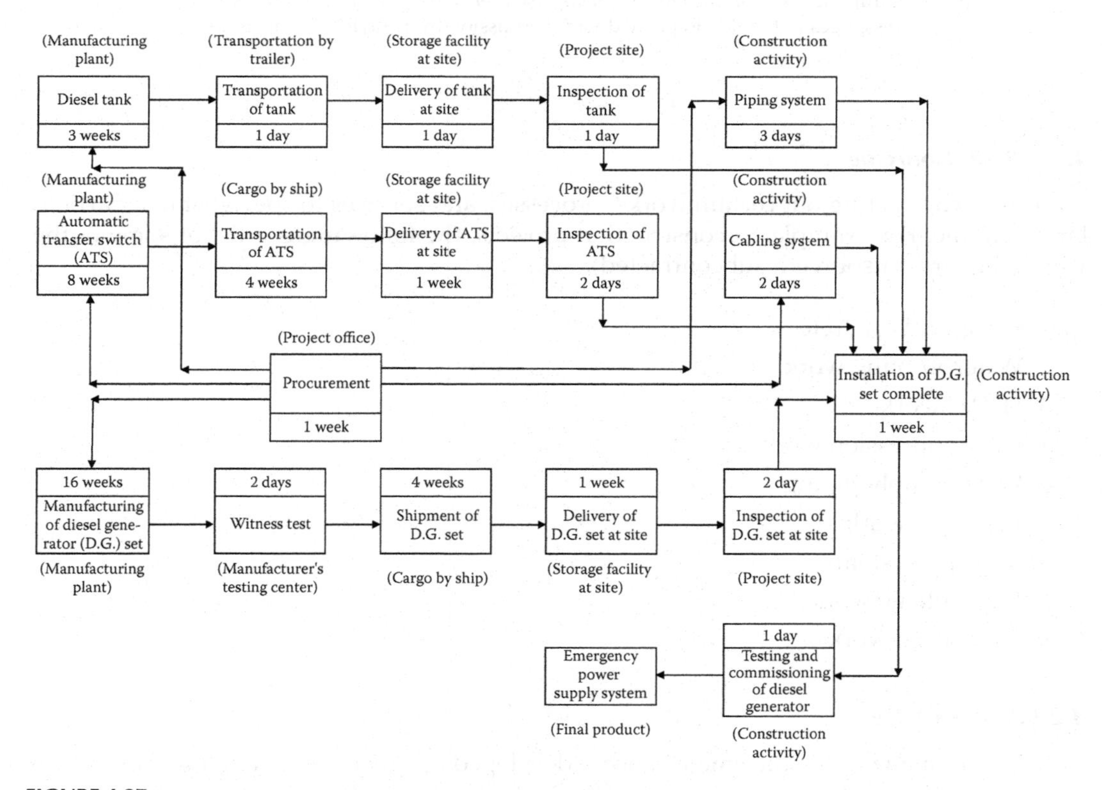

FIGURE 4.37
Value stream mapping for emergency power system. (Abdul Razzak Rumane (2013), *Quality Tools for Managing Construction Projects*, CRC Press, Boca Raton, FL. Reprinted with permission from Taylor & Francis Group.)

4.2.6.13 Waste Reduction

It focuses on reducing waste. Following are the general types of waste*:

1. Defective parts
2. Delays, waiting
3. Excess inventory
4. Misused resources
5. Overproduction
6. Processing
7. Transportation
8. Untapped resources
9. Wasted motion

4.2.7 Cost of Quality

4.2.7.1 Introduction

Quality has an impact on the costs of products and services. The cost of poor quality is the annual monetary loss of products and processes that are not achieving their quality objective. The main components of the cost of low quality are as follows:

1. Cost of conformance
2. Cost of nonconformance

Table 4.15 illustrates cost of quality elements.

4.2.7.2 Categories of Costs

Costs of poor quality are those associated with providing poor-quality products or services. These are incurred because of poor quality, costs that would not be incurred if things were done right from the time and at every time to follow thereafter in order to achieve the quality objective. There are four categories of costs:

1. *Internal failure costs*: The costs associated with defects found before the customer receives the product or service. It also consists of cost of failure to meet customer satisfaction and needs and cost of inefficient processes.
2. *External failure costs*: The cost associated with defects found after the customer receives the product or service. It also includes lost opportunity for sales revenue.
3. *Appraisal costs*: The costs incurred to determine the degree of conformance to quality requirements.
4. *Prevention costs*: The costs incurred to keep failure and appraisal costs to minimum.

---

* Abdul Razzak Rumane (2013). *Quality Tools for Managing Construction Projects*, CRC press, Boca Raton, FL. Reprinted with permission from Taylor & Francis Group.

TABLE 4.15

Cost of Quality

Cost of Compliance	Cost of Noncompliance
• Quality planning	• Scrap
• Process control planning	• Rework
• Quality training	• Corrective action
• Quality audit	• Additional material/inventory cost
• Design review	• Expedition
• Product design validation	• Customer complaints
• Work procedure	• Product recalls
• Method statement	• Warranty
• Process validation	• Maintenance service
• Field testing	• Field repairs
• Third-party inspection	• Rectification of returned material
• Receiving inspection	• Reinspection or retest
• Prevention action	• Downgrading
• In-process inspection	• Loss of business
• Outside endorsement	
• Calibration of equipment	
• Laboratory acceptance testing	

Source: Abdul Razzak Rumane (2013), *Quality Tools for Managing Construction Projects*, CRC Press, Boca Raton, FL. Reprinted with permission from Taylor & Francis Group.

These cost categories allow the use of quality cost data for a variety of purpose. Quality costs can be used for measurement of progress, for analyzing the problem, or for budgeting. By analyzing the relative size of the cost categories, the company can determine if its resources are properly allocated.

4.2.7.3 Quality Cost in Construction

Quality of construction is defined as

1. Scope of work
2. Time
3. Budget

Cost of quality refers to the total cost incurred during the entire life cycle of a construction project in preventing nonconformance to owner requirements (defined scope). There are certain hidden costs that may not affect directly the overall cost of the project; however, it may cost consultant/designer to complete design within stipulated schedule to meet owner requirements and conformance to all the regulatory codes/standards, and for contractor to construct the project within a stipulated schedule, meeting all the contract requirements. Rejection or nonapproval of executed or installed works by the supervisor due to noncompliance with specification will cause the contractor loss in terms of

- Material
- Manpower
- Time

The contractor shall have to rework or rectify the work that will need additional resources and extra time to do the work as specified. This may disturb contractor's work schedule and affect execution of other activities. The contractor has to emphasis upon *zero defect* policy, particularly for concrete works. To avoid rejection of works, the contractor has to take following measures:

1. Execution of works as per approved shop drawings using approved material
2. Following approved method of statement or manufacturers recommended method of installation
3. Conduct continuous inspection during construction/installation process
4. Employ properly trained workforce
5. Maintain good workmanship
6. Identify and correct deficiencies before submitting the checklist for inspection and approval of work
7. Coordinate requirements of other trades, for example, if any opening is required in the concrete beam for crossing of services pipe

Timely completion of project is one of the objectives to be achieved. To avoid delay in the completion schedule, proper planning and scheduling of construction activities are necessary. Since construction projects have involvement of many participants, it is essential that requirements of all the participants are fully coordinated. This will ensure execution of activities as planned resulting in timely completion of project.

Normally the construction budget is fixed at the inception of project; therefore, it is required to avoid variations during construction process, as it may take time to get approval of additional budget resulting time extension to the project. Quality costs related to construction projects can be summarized as follows*:

Internal failure costs:

- Rework
- Rectification
- Rejection of checklist
- Corrective action

External failure costs:

- Breakdown of installed system
- Repairs
- Maintenance
- Warranty

Appraisal costs:

- Design review/preparation of shop drawings
- Preparation of composite/coordination drawings
- On-site material inspection/test

* Abdul Razzak Rumane (2010). *Quality Management in Construction Projects*, CRC Press, Boca Raton, FL. Reprinted with permission from Taylor & Francis Group.

- Off-site material inspection/test
- Pre-checklist inspection

Prevention costs:

- Preventive action
- Training
- Work procedures
- Method statement
- Calibration of instruments/equipment

4.2.8 Quality Function Deployment

QFD* is a technique to translate customer requirements into technical requirements. It was developed in Japan by Dr. Yoji Akao in the 1960s to transfer the concepts of quality control from the manufacturing process into the new product development process. QFD is referred to as *voice of customer* that helps in identifying and developing customer requirements through each stage of product or service development. It is a development process that utilizes a comprehensive matrix involving project team members.

QFD involves constructing one or more matrices containing information related to others. The assembly of several matrices showing the correlation with one another is called the *house of quality*. The house of quality matrix is the most recognized form of QFD. QFD is being applied virtually in every industry and business such as aerospace, communication, software, transportation, manufacturing, services industry, and construction industry. The house of quality is made up of following major components:

1. *Whats*
2. *Hows*
3. Correlation matrix (roof)—technical requirements
4. Interrelationship matrix
5. Target value
6. Competitive evaluation

Whats is the first step in developing the house of quality. It is a structured set of needs/requirements ranked in terms of priority and the levels of importance being specified quantitatively. It is generated by using some of the following questions:

- What are the types of finishes needed for the building?
- What type of air-conditioning system is required for the building?
- What type of communication system is required for the building?
- What type of flooring material is required?
- Whether the building needs any security system?

* Abdul Razzak Rumane (2010). *Quality Management in Construction Projects*, CRC Press, Boca Raton, FL. Reprinted with permission from Taylor & Francis Group.

Hows is the second step in which project team members translate the requirements (*Whats*) into technical design characteristics (specifications) and are listed across the columns.

Correlation matrix identifies technical interaction or physical relationship among the technical specifications. Interrelationship matrix illustrates team member's perception of interrelationship between owner's requirements and technical specifications. The bottom part allows for technical comparison between possible alternatives, target values for each technical design characteristics, and performance measurement.

The right side of the house of quality is used for planning purpose. It illustrates customer perceptions observed in market survey. QFD technique can be used to translate owner's need/requirements into development of set of technical requirements during conceptual design. Figure 4.38 illustrates the house of quality for a hospital building.

4.2.9 Six Sigma

4.2.9.1 Introduction

Six Sigma is basically a process quality goal. It is a process quality technique that focuses on reducing variation in process and preventing deficiencies in product. In a process that has achieved Six Sigma capability, the variation is small compared to the specification limits.

Correlation
▶ Strong positive
● Positive
◊ Negative

Priorities ranking maximum:10

Relationship
■ Strong
□ Medium
♦ Weak

Customer needs	Tech. specs Priorities	Structural design	Partitioning	Internal finishes	Furnishings	Lighting control	HVAC	Mechanical works	Conveying system	Structured cabling	Switches servers	Fire Alarm system	Low-voltage system	Specialities	Generator
Patient bed area general	10	■	■	■	■	□	□	□	■	■	■	□	■	■	
Patient bed area special rooms	9	■	■	■	■	□	□	□	■	■	■	□		■	
Telemedicine	10	■	□	□	□	□	□	□		□	□	□	□	■	□
Surgical	8	■	□	■	■	□	□	■	□	□	□	□		■	
Internal medicine	7	■	□	■	■	□	□	□		□			□	■	
Physiotherapy	8	■	□	□					□	□	□		□	■	
Blood bank	9	■				■				■	□		■		■
Laboratory	7	■		□	□	□					■	□	□	■	□
Pharmacy	6	■									□	□	□	■	
Dental center	7	■			□	□		□			□		□	■	
OB/GYN Services	9	■			□					■	■			■	□
Out patient services	8	■			□					□	■	□	■	■	□
Administration Area	7	■	■	■	■	□	□	□	□	□	■	□	□		□
IT facility	9									■	■	□	□		■
Audio visual/paging system	8				■	□	□	□	□	■		□	■		□
CCTV/access control	9					□	■	□		■	■	□	□		■
HVAC	8					■	■	■		□		■			□
Building automation	9									■		■	□		■
ECO lighting and controls	9					■		■		□					
Fire alarm/voice evacuation	8						□			■	□	■	□		
Interior plantation	7		■	■					■						□
Parking	7	■											□		■
Staff housing	7	■		■	■	□	□	□			□	□	□		

Technologically advance hospital building

Better medical facility

Latest architectural design

Latest medical facilities

Modern ward rooms

FIGURE 4.38
House of quality for hospital building project. (Abdul Razzak Rumane (2013), *Quality Tools for Managing Construction Projects*, CRC Press, Boca Raton, FL. Reprinted with permission from Taylor & Francis Group.)

Sigma is a Greek letter "σ"—which stands for standard deviation. Standard deviation is a statistical way to describe how much variation exists in a set of data, a group of items, or a process. Standard deviation is the most useful measure of dispersion. Six Sigma means that a process to be capable at Six Sigma level, the specification limits should be at least 6σ from the average point. Therefore, the total spread between the upper specification (control) limit and lower specification (control) limit should be 12σ. With Motorola's Six Sigma program, no more than 3.4 defects per million fall outside specification limits with process shift of not more than 1.5σ from the average or mean. Six Sigma started as a defect reduction effort in manufacturing and was then applied to other business processes for the same purpose.

Six Sigma is a measurement of *goodness* using a universal measurement scale. Sigma provides a relative way to measure improvement. Universal means sigma can measure anything from coffee mug defects to missed chances to closing of a sales deal. It simply measures how many times a customer's requirements were not met (a defect), given a million opportunities. Sigma is measured in defects per million opportunities (DPMO). For example, a level of sigma can indicate how many defective coffee mugs were produced when one million were manufactured. Levels of sigma are associated with improved levels of goodness. To reach a level of Three Sigma, you can only have 66,811 defects, given a million opportunities. A level of Five Sigma only allows 233 defects. Minimizing variation is a key focus of Six Sigma. Variation leads to defects, and defects lead to unhappy customers. To keep customers satisfied, loyal, and returning, you have to eliminate the sources of variation. Whenever a product is created or a service is performed, it needs to be done the same way every time, regardless of the person involved. Only then will you truly satisfy the customer. Figure 4.39 illustrates the Six Sigma roadmap.

4.2.9.2 Six Sigma Methodology

Six Sigma is the overall business improvement methodology that focuses an organization on the following:

- Understanding and managing customer requirements
- Aligning key business process to achieve these requirements
- Utilizing rigorous data analysis to minimize variation in these processes

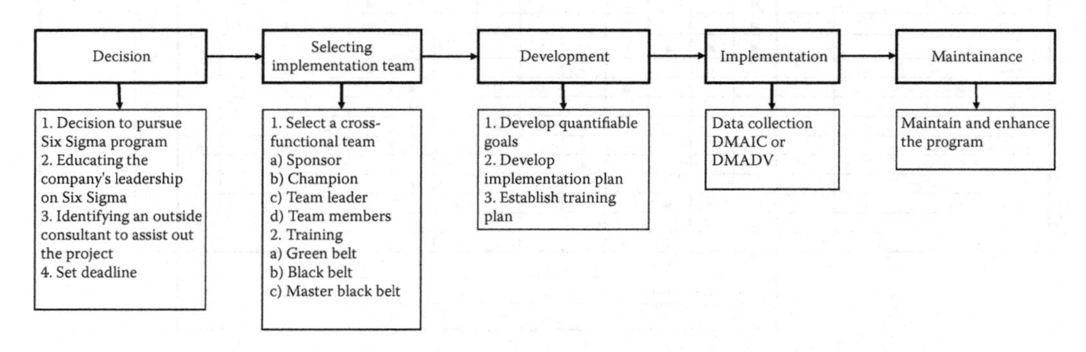

FIGURE 4.39
Six Sigma roadmap. (Abdul Razzak Rumane (2013), *Quality Tools for Managing Construction Projects*, CRC Press, Boca Raton, FL. Reprinted with permission from Taylor & Francis Group.)

- Driving rapid and sustainable improvement in business process by reducing defects, cycle time, impact to the environment, and other undesirable variations
- Timely execution

As a management system, Six Sigma is a high-performance system for executing a business strategy. It uses concept of fact and data to drive better solutions. Six Sigma is a top-down solution to help organizations:

- Align their business strategy to critical improvement efforts.
- Mobilize teams to attack high-impact projects.
- Accelerate improved business results.
- Govern efforts to ensure improvements are sustained.

The Six Sigma methodology also focuses on

- Leadership principles
- Integrated approach to improvement
- Engaged teams
- Analytic tool
- Hard-coded improvements

4.2.9.2.1 Leadership Principles

The Six Sigma methodology has four leadership principles. These are as follows:

1. Align
2. Mobilize
3. Accelerate
4. Govern

A brief description of these leadership principles are as follows:

1. *Align*: Leadership should ensure that all improvement projects are in line with the organization's strategic goals. Alignment begins with the leadership team developing a scorecard. This vital tool, the cornerstone of the Six Sigma business improvement campaign, translates strategy into tactical operating terms. The scorecard also defines metrics an organization can use to determine success. Just as a scoreboard at a sporting event tells you who is winning, the scorecard tells leadership how well the company is meeting its goals.
2. *Mobilize*: Leadership should enable teams to take action by providing clear direction, feasible scope, a definition of success, and rigorous reviews. Mobilizing sets clear boundaries, lets people go to work, and trains them as required. The key to mobilizing is focus—lack of focused action was one of the downfalls of previous business improvement efforts. True focus means the project is correctly aligned with the organization's scorecard. Mobilized teams have a valid reason for engaging in improvement efforts—they can visualize benefit for the customer.

The project has strategic importance and they know it. They know exactly what must be done and the criteria they can use to determine success.

3. *Accelerate*: Leadership should drive a project to rapid results through tight clock management, training as needed, and shorter deadlines. More than 70% of all improvement initiatives fail to achieve desired results in time to make a difference. For projects to make an impact, they must achieve results quickly, and that is what acceleration is all about. Accelerate leadership principle involves three main components:

   a. Action learning
   b. Clock management
   c. Effective planning

   Accelerate employs *action learning* methodology to quickly bridge from *learning* to *doing*. Action learning mixes traditional training with direct application. Training is received while working on a real-world project, allowing plenty of opportunity to apply new knowledge. The instructor is not simply a trainer, but a coach as well, helping work with real-world project. Action learning accelerates improvement over traditional learning methods. It helps receiving training, and also completing a worthwhile project at the same time. In addition to the four-to-six-month time frame, *accelerate* requires teams to set deadlines that are reinforced through rigorous reviews.

4. *Govern*: Leadership must visibly sponsor projects and conduct regular and rigorous reviews to make critical mid-course corrections. The fourth leadership principle is *govern*. Once leadership selects an improvement opportunity, their work is not done. They must remain ultimately responsible for the success of that project. Govern requires leaders to drive for results.

   While governing a Six Sigma project, you need the following:

   - A regular communications plan and a clear review process
   - Actively sponsor teams and their projects
   - Encourage proactive dialogue and knowledge sharing with the team and throughout the organization

4.2.9.2.2 Six Sigma Team

Team work is absolutely vital for complex Six Sigma projects. For teams to be effective, they must be engaged—involved, focused, and committed to meet their goals. Engaged teams must have a leadership support. There are four types of teams:

1. Black Belts
2. Green Belts
3. Breakthrough
4. Blitz

A brief description of these teams is as follows:

1. *Black Belt*: These teams are led by a Black Belt, and may have Green Belts and functional experts assigned to complex, high-impact process improvement projects or designing new products, services, or complex processes. Black Belts are internal

Six Sigma practitioners, skilled in the application of rigorous statistical methodologies, and they are crucial to the success of Six Sigma. Their additional training and experience provide them with the skills they need to tackle difficult problems. Black Belts have many responsibilities. They

- Function as a team leader on Black Belt projects.
- Integrate their functional discipline with statistical, project, and interpersonal skills.
- Serve as internal consultants.
- Tackle complex, high-impact improvement opportunities.
- Mentor and train Green Belts.

2. *Green Belt*: This team is led by a Green Belt and is composed of nonexperts. Green Belt teams tackle less complex, high-impact process improvement projects. These teams are often coached by Black Belts or master Black Belts. Green Belts are also essential to the success of Six Sigma. They perform many of the same functions as Black Belts, but their work requires less complex analysis. Green Belts are trained in basic problem-solving skills and the statistical tools needed to work effectively as members of process improvement teams.

   Green Belt responsibilities include the following:

   - Acting as a team leader on business improvements requiring less complex analysis
   - Adding their unique skills and experiences to the team
   - Working with the team to come up with inventive solutions
   - Performing basic statistical analysis
   - Conferring with a Black Belt as questions arise

3. *Breakthrough*: While creating simple processes, sophisticated statistical tools may not be needed. Breakthrough teams are typically used to define low-complexity, new processes.
4. *Blitz*: Blitz teams are put in place to quickly execute improvements produced by other projects. These teams can also implement digitization for efficiency using a new analytic tool set.

For typical Six Sigma project, four critical roles exist:

1. Sponsor
2. Champion
3. Team leader
4. Team member

A sponsor typically

- Remains ultimately accountable for a project's impact.
- Provides project resources.
- Reviews monthly and quarterly achievements, obstacles, and key actions.
- Supports the project champion by removing barriers as necessary.

A champion typically

- Reviews weekly achievements, obstacles, and key actions.
- Meets with the team weekly to discuss progress.
- Reacts to changes in critical performance measures as needed.
- Supports the team leader, removing barriers as necessary.
- Helps ensure project alignment.

A team leader typically

- Leads improvement projects through an assigned, disciplined methodology.
- Works with the champion to develop the team charter, review project progress, obtain necessary resources, and remove obstacles.
- Identifies and develops key milestones, timelines, and metrics for improvement projects.
- Establishes weekly, monthly, and quarterly review plans to monitor team's progress.
- Supports the work of team members as necessary.

Team members typically

- Assist the team leader.
- Follow a disciplined methodology.
- Ensure that the team charter and timeline are being met.
- Accept and execute assignments.
- Add their views, opinions, and ideas.

4.2.9.3 Analytic Tool Sets

Following are the analytic tools used in Six Sigma projects.

4.2.9.3.1 Ford Global 8D Tool

What problem needs solving? →

Who should help solve problem? →

How do we quantify symptoms? →

How do we contain it? →

What is the root cause? →

What is the permanent corrective action? →

How do we implement? →

How can we prevent this in future? →

Who should we reward? →

Ford Global 8D tool is primarily used to bring performance back to a previous level.

4.2.9.3.2 DMADV Tool Set Phases

Define → What is important?

Measure → What is needed?

Analyze → How will we fulfill?

Design → How do we build it?

Verify → How do we know it will work?

The Define, measure, analyze, design, verify (DMADV) tool is used primarily for the invention and innovation of modified or new products, services, or process. Using this toolset, Black Belts optimize performance before production begins. DMADV is proactive, solving problems before they start. This tool is also called as *design for Six Sigma* (DFSS). Table 4.16 lists fundamental objectives of DMADV.

4.2.9.3.2.1 DMADV Process

Define phase: What is important?

(Define the project goals and customer deliverables.)

Key deliverables of this phase are as follows:

- Establish the goal
- Identify the benefits
- Select project team
- Develop project plan
- Project charter

Measure phase: What is needed?

(Measure and determine customer needs and specifications.)

Key deliverable in this phase is to

- Identify specification requirements.

TABLE 4.16

Fundamental Objectives of Six Sigma DMADV Tool

DMADV	Phase	Fundamental Objective
1	*Define*—What is important?	Define the project goals and customer deliverables (internal and external)
2	*Measure*—What is needed?	Measure and determine customer needs and specifications
3	*Analyze*—How we fulfill?	Analyze process options and prioritize based on capabilities to satisfy customer requirements
4	*Design*—How we build it?	Design detailed process(es) capable of satisfying customer requirements
5	*Verify*—How do we know it will work?	Verify design performance capability

Source: Abdul Razzak Rumane (2013), *Quality Tools for Managing Construction Projects*, CRC Press, Boca Raton, FL. Reprinted with permission from Taylor & Francis Group.

Analyze phase: How we fulfill?

(Analyze process options and prioritize based on capability to satisfy customer requirements.)

Key deliverables in this phase are as follows:

- Design generation (data collection)
- Design analysis
- Risk analysis
- Design model (prioritization of data under major variables)

Design phase: How we build it?

(Design detailed process(es) capable of satisfying customer requirements.)

Key deliverables in this phase are as follows:

- Constructing a detail design
- Converting critical to quality into critical to process (CTP) elements
- Estimating the capabilities of the CTP elements in the design
- Preparing a verification plan

Verify phase: How do we know it works?

(Verify design performance capability.)

Key deliverables in this phase is to

- Design a control and transition plan.

4.2.9.3.3 DMAIC Tool

Define → What is important?

Measure → How are we doing?

Analyze → What is wrong?

Improve → What needs to be done?

Control → How do we guarantee performance?

Define, measure, analyze, improve, control (DMAIC) tool refers to a data-driven quality strategy and is used primarily for improvement of an existing product, service, or process. Table 4.17 lists fundamental objectives of DMAIC.

TABLE 4.17

Fundamental Objectives of Six Sigma DMAIC Tool

DMAIC	Phase	Fundamental Objective
1	*Define*—What is important?	Define the project goals and customer deliverables (internal and external)
2	*Measure*—How are we doing?	Measure the process to determine current performance
3	*Analyze*—What is wrong?	Analyze and determine the root cause(s) of the defects
4	*Improve*—What needs to be done?	Improve the process by permanently removing the defects
5	*Control*—How do we guarantee performance?	Control the improved process's performance to ensure sustainable results

Source: Abdul Razzak Rumane (2013), *Quality Tools for Managing Construction Projects*, CRC Press, Boca Raton, FL. Reprinted with permission from Taylor & Francis Group.

4.2.9.3.3.1 The DMAIC Process The majority of the time Black and Green Belts approach their projects with the DMAIC analytic tool set, driving process performance to never-before-seen levels.

DMAIC has following fundamental objective:

1. *Define phase*: Define the project and customer deliverables.
2. *Measure phase*: Measure the process performance and determine current performance.
3. *Analyze*: Collect, analyze, and determine the root cause(s) of variation and process performance.
4. *Improve*: Improve the process by diminishing defects with alternative remedial.
5. *Control*: Control improved process performance.

The DMAIC process contains five distinct steps that provide a disciplined approach to improving existing processes and products through the effective integration of project management, problem solving, and statistical tools. Each step has fundamental objectives and a set of key deliverables, so the team member will always know what is expected of him/her and his/her team.

DMAIC stands for the following:

- Define opportunities
- Measure performance
- Analyze opportunity
- Improve performance
- Control performance

Define opportunities (What is important?)

The objective of this phase is to

- Identify and/or validate the improvement opportunities that will achieve the organization's goals and provide the largest payoff, develop the business process, define critical customer requirements, and prepare to function as an effective project team.

Key deliverables in this phase include the following:

- Team charter
- Action plan
- Process map
- Quick win opportunities
- Critical customer requirements
- Prepared team

Measure performance (How are we doing?)

The objectives of this phase are to

- Identify critical measures that are necessary to evaluate the success or failure, meet critical customer requirements, and begin developing a methodology to effectively collect data to measure process performance.

- Understand the elements of the Six Sigma calculation and establish baseline sigma for the processes the team is analyzing.

Key deliverables in this phase include the following:

- Input, process, and output indicators
- Operational definitions
- Data collection format and plans
- Baseline performance
- Productive team atmosphere

Analyze opportunity (What is wrong?)
The objectives of this phase are to

- Stratify and analyze the opportunity to identify a specific problem and define an easily understood problem statement.
- Identify and validate the root causes and thus the problem the team is focused on.
- Determine true sources of variation and potential failure modes that lead to customer dissatisfaction.

Key deliverables in this phase include the following:

- Data analysis
- Validated root causes
- Sources of variation
- FMEA
- Problem statement
- Potential solutions

Improve performance (What needs to be done?)
The objectives of this phase are to

- Identify, evaluate, and select the right improvement solutions.
- Develop a change management approach to assist the organization in adapting to the changes introduced through solution implementation.

Key deliverables in this phase include the following:

- Solutions
- Process maps and documentation
- Pilot results
- Implementation milestones
- Improvement impacts and benefits
- Storyboard
- Change plans

Control performance (How do we guarantee performance?)
The objectives of this phase are to

- Understand the importance of planning and executing against the plan and determine the approach to be taken to ensure achievement of the targeted results.
- Understand how to disseminate lessons learned, identify replication and standardization opportunities/processes, and develop related plans.

Key deliverables in this phase include the following:

- Process control systems
- Standards and procedures
- Training
- Team evaluation
- Change implementation plans
- Potential problem analysis
- Solution results
- Success stories
- Trained associates
- Replication opportunities
- Standardization opportunities

Six Sigma methodology is not so commonly used in construction projects; however, the DMAIC tool can be applied at various stages in construction projects. These are as follows:

1. *Detailed design stage*: To enhance coordination method in order to reduce repetitive work
2. *Construction stage*:
   a. Preparation of builders' workshop drawings and composite drawings, as it needs a lot of coordination among different trades
   b. Preparation of contractor's construction schedule
   c. Execution of works

4.2.9.3.4 DMADDD Tool

Define → Where must we be leaner?

Measure → What is our baseline?

Analyze → Where can we free capacity and improve yields?

Design → How should we implement?

Digitize → How do we execute?

Draw Down → How do we eliminate parallel paths?

DMADDD Tool is primarily used to drive the cost out of a process by incorporating digitization improvements. These improvements can drive efficiency by identifying non-value-added tasks and use simple web-enabled tools to automate certain tasks and improve efficiency. In doing so, employees can be freed up to work on more value-added tasks. Table 4.18 lists fundamental objectives of DMADDD.

4.2.9.3.4.1 Impact of Six Sigma Strategy The Six Sigma strategy affects five fundamental areas of business*:

1. Process improvement
2. Product and service improvement
3. Customer satisfaction
4. Design methodology
5. Supplier improvement

4.2.10 Triz

Triz is short for `teirija rezhenijia izobretalenksh zadach` (theory of inventive problem solving), developed by the Russian scientist Genrich Altshuller. Triz provides systematic methods and tools for analysis and innovative problem solving to support decision-making process.

Continuous and effective quality improvement is critical for organization's growth, sustainability, and competitiveness. The cost of quality is associated with both chronic and sporadic problems. Engineers are required to identify, analyze the causes, and solve these problems by applying various quality improvement tools. Any of these quality tools taken individually does not allow a quality practitioner to carry out whole problem solving cycle. These tools are useful for solving a particular phase of problem, and need combination of various tools and methods to find problem solution. Triz is an approach that starts at a point where fresh thinking is needed to develop a new process or to redesign a process. It focuses on methods for developing ideas to improve a process, get something done,

TABLE 4.18

Fundamental Objectives of Six Sigma DMADDD Tool

DMADDD	Phase	Fundamental Objective
1	*Define*—Where must we be learner?	Identify potential improvements
2	*Measure*—What is our baseline?	Analog touch points
3	*Analyze*—Where can we free capacity and improve yields?	Task elimination and consolidated operations Value added/non-value added tasks Free capacity and yield
4	*Design*—How should we implement?	Future state vision Define specific projects Define draw down timing Define commercialization plans
5	*Digitize*—How do we execute?	Execute project
6	*Draw down*—How do we eliminate parallel paths?	Commercialize new process Eliminate parallel path

Source: Abdul Razzak Rumane (2013), *Quality Tools for Managing Construction Projects*, CRC Press, Boca Raton, FL. Reprinted with permission from Taylor & Francis Group.

* Abdul Razzak Rumane (2010), *Quality Management in Construction Projects*, CRC Press, Boca Raton, FL. Reprinted with permission from Taylor & Francis Group.

design a new approach, or to redesign an existing approach. Triz offers a more systematic although still universal approach to problem solving. Triz has advantages over other problem-solving approaches in terms of time efficiency and has low cost-quality improvement solution. The pillar of Triz is the realization that contradictions can be methodically resolved through the application of innovative solutions. Altshuller defined an inventive problem as one containing a contradiction. He defined contradiction as a situation where an attempt to improve one feature of system detracts from another feature.

4.2.10.1 Triz Methodology

Traditional processes for increasing creativity have a major flaw in that their usefulness decreases as the complexity of the problem increases. At times, trial-and-error method is used in every process and a number of trials increase with the complexity of the inventive problem. In 1946, Altshuller determined to improve the inventive process by developing the *science* of creativity, which led to creation of Triz. Triz was developed by Altshuller as a result of analysis of many thousands of patents. He reviewed over 200,000 patents looking for problems and their solutions. He selected 40,000 as representative of inventive solutions, the rest were direct improvements easily recognized within the system. Altshuller recognized a pattern where some fundamental problems were solved with solutions that were repeatedly used from one patent to another, although the patent subject, applications, and timings varied significantly. He categorized these patterns into five level of inventiveness. Table 4.19 summarizes his findings.

He noted that with each succeeding level, the source of the solution required broader knowledge and more solutions to consider before an ideal solution could be found.

Triz is a creative thinking process that provides a highly structured approach for generating innovative ideas and solutions for problem solving. It provides tools and methods for use in problem formulation, system analysis, failure analysis, and pattern of system evolution. Triz is contrast to techniques such as brainstorming, aims to create algorithmic approach to invent new systems, and refines old systems. Using Triz requires some training and a good deal of practice.

Triz body of knowledge contains 40 creative principles drawn from analysis of how complex problems have been solved.

- The laws of systems solution
- The algorithm of inventive problem solving

TABLE 4.19

Level of Inventives

Level	Degree of Inventiveness	Percentage of Solutions	Source of Solution
1	Obvious solution	32	Personal skill
2	Minor improvement	45	Knowledge within existing systems
3	Major improvement	18	Knowledge within the industry
4	New concept	4	Knowledge outside industry and are found in science, not in technology
5	Discovery	1	Outside the confines of scientific knowledge

Source: Abdul Razzak Rumane (2010), *Quality Management in Construction Projects*, CRC Press, Boca Raton, FL. Reprinted with permission from Taylor & Francis Group.

- Substance-field analysis
- 76 standard solutions

4.2.10.2 Application of Triz

Engineers can apply Triz for solving the following problems in construction projects:

- Nonavailability of specified material
- Regulatory changes to use certain type of material
- Failure of dewatering system
- Casting of lower grade of concrete to that of specified higher grade
- Collapse of trench during excavation
- Collapse of formwork
- Collapse of roof slab while casting in progress
- Chiller failure during peak hours in the summer
- Modifying method statement
- Quality auditor is a person who audits the construction quality

4.2.10.3 Triz Process

Altshuller has recommended four steps to invent new solution to a problem. These are as follows:

Step 1: Identify the problem
Step 2: Formulate the problem
Step 3: Search for precisely well-solved problem
Step 4: Generate multiple ideas and adapt a solution

The above referred methods are primarily used for low-level problems. To solve more difficult problems more precise tools are used. These are as follows*:

1. Algorithm for inventive problem solving
2. Separation principles
3. Substance-field analysis
4. Anticipator failure determination
5. Direct product evaluation

QFD matrix is also used to identify new functions and performance levels to achieve truly exiting level of quality by eliminating technical bottlenecks at the conceptual stage. QFD may be used to feed data into Triz, especially using the *rooftop* to help develop

* Abdul Razzak Rumane (2010). *Quality Management in Construction Projects*, CRC Press, Boca Raton, FL. Reprinted with permission from Taylor & Francis Group.

contradictions. The different schools for Triz and individual practitioners have continued to improve and add to the methodology.

4.3 Tools for Construction Project Development

Project development is a process spanning from the commencement of project initiation and complete with closeout and finalizing project records after project construction and handing over of the project. The project development process is initiated in response to an identified need. It covers a range of time framed activities extending from identification of a project need to a finished set of contract documents and to construction.

Construction project development has four major elements/stages as follows:

1. Study
2. Design
3. Bidding and tendering
4. Construction

As the project develops, more information and specifications are developed. Table 4.20 illustrates construction project development stages.

TABLE 4.20

Construction Project Development Stages

Sr. No.	Stages	Elements	Description
1	Study	Problem Statement/Need Identification	Project needs, goals, and objectives.
		Need Assessment	Identification of needs.
			Prioritization of needs.
			Leveling of needs.
			Deciding what needs to be addressed.
		Need Analysis	Perform project need analysis/study to outline the scope of issues to be considered in the planning phase.
		Need Statement	Develop project need statement.
		Feasibility Study	Technical studies, economics assessment, financial assessment, scheduling, market demand, risk, environmental and social assessment.
		Establish Project Goals and Objectives	Scope, time, cost, quality.
		Identify Alternatives	Identify alternatives based on a predetermined set of performance measures.
		Preliminary Schedule	Estimate the duration for completion of project/facility
		Preliminary Financial Implications	Preliminary budget estimates of total project cost (life cycle cost) on the basis of any known research and development requirements. This will help arrange the finances. (Funding agency.)
		Preliminary Resources	Estimate resources.

(*Continued*)

TABLE 4.20 (*Continued*)

Construction Project Development Stages

Sr. No.	Stages	Elements	Description
		Project Risk	Risk, constraints.
		Authorities Clearance	Identify issues, sustainability, impacts, and potential approvals (environmental, authorities, permits) required for subsequent design and authority approval processes.
		Select Preferred Alternative	Select preferred alternative considering technological and economical feasibility.
		Identify Project Delivery System	Establish how the participants, owner, designer (A/E), and contractor will be involved to construct the project/facility. (design–build–bid, design–build, guaranteed maximum price, CM type, PM type, BOT, turnkey, etc.)
		Identify Type of Contracting System/Pricing	Select contract pricing system such as firm fixed price or lump sum, unit price, cost reimbursement (cost plus), reimbursement, target price, time and material, guaranteed maximum price
		Develop Project Charter	Terms of reference (TOR).
		Identify project team	Select designer (A/E) firm if design–bid–build type of contract system is selected. Select other team members based on project delivery system requirements
		Project Launch	Proceed with concept design.
2	Design	Develop Concept Design	Report, drawings, models, presentation.
		Regulatory Approvals	Obtain regulatory approvals.
		Project Planning	Prepare project plan.
		Schematic Design	Preliminary design, value engineering.
		Design Development	Detailed design.
		Construction Documents	Construction contract documents.
3	Bidding & Tendering	Bid Documents	Tendering documents.
		Selection of Contractor	Advertisement for bidders.
			Pre qualification of contractors.
			Issuing tender documents/ Request for proposal.
			Receipt of bid documents.
			Tabulation of proposals.
			Analysis of proposal.
			Selection of contractor.
			Most competitive bidder (low bid, qualification based).
		Award Contract	Signing of contract.
			Bonds and Insurance.
			Notice to proceed.
4	Construction	Construction	Contractor to execute contracted works.
		Monitoring and Control	Monitor and control scope, schedule, cost, quality, risk, procurement of the project.
		Commissioning and Handover	Testing, commissioning and handover.
		Project Closeout	Close the project.

Source: Abdul Razzak Rumane (2013), *Quality Tools for Managing Construction Projects*, CRC Press, Boca Raton, FL. Reprinted with permission from Taylor & Francis Group.

4.3.1 Tools for Study Stage

Table 4.21 illustrates major elements of study stage and related tasks performed during this stage. Figure 4.40 illustrates logic flow diagram in study stage (conceptual design stage). Table 4.22 illustrates points to be considered for need assessment. Table 4.23 illustrates points to be considered for feasibility study of a construction project. Table 4.24 illustrates contents of terms of reference. Some of the tools discussed earlier can be applied during study stage. Table 4.25 is an example how 5W2H (under innovation and creative tool) can be used to analyze the project need.

TABLE 4.21
Major Elements of Study Stage

Sr. No.	Elements	Description	Related Task
1	Problem statement/ need identification	Project needs, goals, and objectives	Strategic objectives, policies, and priorities. (Please refer to Figure 4.40.)
2	Need assessment	Identification of needs Prioritization of needs Leveling of needs Deciding what needs to be addressed	Ensure that owner's business case has been properly considered. (Please refer to Table 4.22.)
3	Need analysis	Perform project need analysis/study to outline the scope of issues to be considered in the planning phase	Perform Need Analysis. (Please refer to Table 3.2.)
4	Need statement	Develop project need statement	Develop Need Statement. (Please refer to Table 3.3.)
5	Feasibility study	Technical studies, economics assessment, financial assessment, scheduling, market demand, risk, environmental and social assessment.	Perform Feasibility Study. Statement. (Please refer to Table 4.23.)
6	Establish project goals and objectives	Scope, time, cost, quality	Project initiation documents developed on SMART concept.
7	Identify alternatives	Identify alternatives based on a predetermined set of performance measures	Select conceptual alternatives.
8	Preliminary schedule	Estimate the duration for completion of project/facility	Establish project schedule.
9	Preliminary financial implications	Preliminary budget estimates of total project cost (life cycle cost) on the basis of any known research and development requirements. This will help arrange the finances. (Funding agency)	Determine project budget.
10	Preliminary resources	Estimate resources	Confirm availability of resources, manpower, material, equipment.
11	Project risk	Identify project risk, constraints	Establish risk response, mitigation plan.

(Continued)

TABLE 4.21 (*Continued*)

Major Elements of Study Stage

Sr. No.	Elements	Description	Related Task
12	Authorities clearance	Identify issues, sustainability, impacts, and potential approvals (environmental, authorities, permits) required for subsequent design and authority approval processes	Establish requirements for statutory approvals and other regulatory authorities.
13	Select preferred alternative	Assess technological and economical feasibility and compare to the preferred option/alternative to prepare business case	Discuss relative merits of various alternative schemes and evaluate the performance measures to meet owner's needs/requirements. Consider social, economical, environmental impact, safety, reliability, and functional capability.
14	Identify project delivery system	Establish how the participants, owner, designer (A/E), and contractor will be involved to construct the project/facility (design–build–bid, design–build, guaranteed maximum price, CM type, PM type, BOT, Turnkey, etc.)	Select suitable project delivery system as per strategic decision and suitability of appropriate system. (Please refer to Chapter 2.)
15	Identify type of contracting system	Select contract pricing system such as firm fixed price or lump sum, unit price, cost reimbursement (cost plus), reimbursement, target price, time and material, guaranteed maximum price	Select contracting/pricing most appropriate for the benefit of owner. (Please refer to Chapter 2.)
16	Identify project team	Select designer (A/E) firm if design–bid–build type of contract system is selected. Select other team members based on project delivery system requirements	Select project team considering the selected type of project delivery system and procurement process of the organization. (Please refer to Chapter 2.)
17	Project launch	Project charter	Prepare terms of reference (TOR). (Please refer to Table 4.24.)

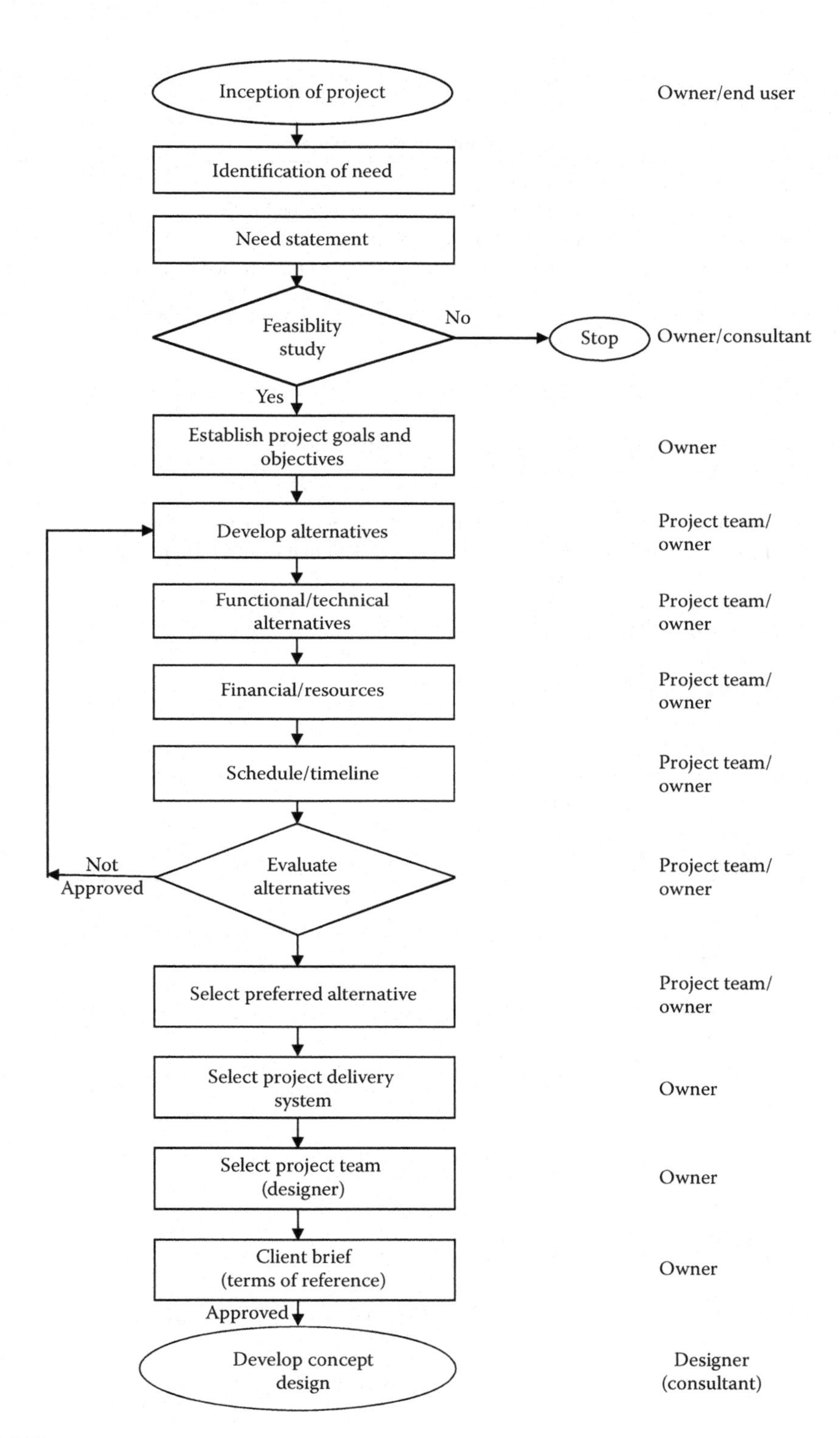

FIGURE 4.40
Logic flow of activities in the study stage.

TABLE 4.22

Need Assessment for a Construction Project

Sr. No.	Points to Be Considered
1	Ensure the need is properly defined
2	Confirm that the need outcome will benefit the owner/end user
3	Gather and analyze owner/end user requirements
4	Set priorities and establish criteria for solutions to meet the targeted demand or requirements
5	Identify and measure areas for improvement to achieve the objectives of the project

TABLE 4.23

Major Considerations for Feasibility Study of a Construction Project

Sr. No.	Points to Be Considered
1	Technical suitability of facility for intended use by the owner/end user.
2	Economical feasibility to ascertain value of benefit that results from the project exceeds the cost that results from the project.
3	Financial payback period
4	Market demand
5	Environmental impact
6	Social and cultural assessment
7	Legal and regulatory impacts
8	Political aspects
9	Resources availability
10	Scheduling of the project
11	Operational
12	Risk analysis

TABLE 4.24

Typical Contents of Terms of Reference Documents

Sr. No.	Topics
1	Project objectives
	1.1 Background
	1.2 Project information
	1.3 General requirements
	1.4 Special considerations
2	Project requirements
	2.1 Scope of work
	2.2 Work program
	2.2.1 Study phase
	2.2.2 Design phase
	2.2.3 Tender stage
	2.2.4 Construction phase
	2.3 Reports and presentations

(*Continued*)

TABLE 4.24 (*Continued*)

Typical Contents of Terms of Reference Documents

Sr. No.	Topics
	2.4 Schedule of requirements
	2.5 Drawings
	2.6 Energy conservation considerations
	2.7 Cost estimates
	2.8 Time program
	2.9 Interior finishes
	2.10 Aesthetics
	2.11 Mechanical
	2.12 HVAC
	2.13 Lighting
	2.14 Engineering systems
3	Opportunities and constraints
	3.1 Site location
	3.2 Site conditions
	3.3 Land size and access
	3.4 Climate
	3.5 Time
	3.6 Budget
4	Performance target
	4.1 Financial performance
	4.1.1 Performance bond
	4.1.2 Insurance
	4.1.3 Delay penalty
	4.2 Energy performance target
	4.2.1 Energy conservation
	4.3 Work program schedule
5	Environmental considerations
6	Design approach
	6.1 Procurement strategy
	6.2 Design parameters
	6.2.1 Architectural design
	6.2.2 Structural design
	6.2.3 Mechanical design
	6.2.4 HVAC design
	6.2.5 Electrical design
	6.2.6 Information and communication technology
	6.2.7 Conveying system
	6.2.8 Landscape
	6.2.9 External works
	6.2.10 Parking
	6.3 Sustainable architecture
	6.4 Engineering systems
	6.5 Value engineering study
	6.6 Design review by client
	6.7 Selection of products/systems
7	Specifications and contract documents
8	Project control guidelines

(*Continued*)

TABLE 4.24 (*Continued*)

Typical Contents of Terms of Reference Documents

Sr. No.	Topics
9	Submittals
	9.1 Reports
	9.2 Drawings
	9.3 Specifications
	9.4 Models
	9.5 Sample boards
	9.6 Mock up
10	Presentation
11	Project team members
	11.1 Number of project personnel
	11.2 Staff qualification
	11.3 Selection of specialists
12	Visits

Source: Abdul Razzak Rumane (2013), *Quality Tools for Managing Construction Projects*, CRC Press, Boca Raton, FL. Reprinted with permission from Taylor & Francis Group.

TABLE 4.25

5W2H Analysis for Project Need

Sr. No.	Why	Related Analyzing Question
1	Why	New project
2	What	What advantage it will have over other similar projects
3	Who	Who will be the customer for this project
4	Where	Where can we fine the market for the project
5	When	When the project will be ready
6	How many	How many such projects are in the market
7	How much	How much market share we will have by this project

4.3.2 Tools for Design Stage

Table 4.26 illustrates major elements of design stage and related tasks performed during this stage. Table 4.27 illustrates points to be considered for data collection to facilitate develop concept design. Table 4.28 illustrates points to be considered for development of concept design.

TABLE 4.26

Major Elements of Design Stage

Sr. No.	Elements	Description	Related Task
1	Develop concept design	Report, drawings, models, presentation	Data collection. (Please refer to Table 4.27.) Design development points. (Please refer to Table 4.28.)
2	Regulatory approvals	Obtain regulatory approvals	Submission of drawings and related documents to authorities and obtain their approvals

(*Continued*)

TABLE 4.26 (*Continued*)

Major Elements of Design Stage

Sr. No.	Elements	Description	Related Task
3	Project planning	Prepare project plan	Develop project plan considering (Please refer to Figure 4.41.)
4	Schematic design	Preliminary design, value engineering	Develop schematic design. (Please refer to Table 4.29.) Preliminary (outline) specifications, authority approvals, value engineering
5	Design development	Detailed design	Develop detailed design. (Please refer to Figure 4.42.)
6	Construction documents	Construction contract documents	Preparation of particular specifications, contract drawings. (Please refer to Chapter 6 for details.)

TABLE 4.27

Major Items for Data Collection during Concept Design Phase

Sr. No.	Items to Be Considered
1	Certificate of title 1. Site legalization 2. Historical records
2	Topographical survey 1. Location plan 2. Site visits 3. Site coordinates 4. Photographs
3	Geotechnical investigations
4	Field and laboratory test of soil and soil profile
5	Existing structures in/under the project site
6	Existing utilities/services passing through the project site
7	Existing roads, and structure surrounding the project site
8	Shoring and underpinning requirements with respect to adjacent area/structure
9	Requirements to protect neighboring area/facility
10	Environmental studies
11	Day-lighting requirements
12	Wind load, seismic load, dead load, and live load
13	Site access/traffic studies
14	Applicable codes, standards, and regulatory requirements
15	Usage and space program
16	Design protocol
17	Scope of work/client requirements

TABLE 4.28
Development of Concept Design

Sr. No.	Points to Be Considered
1	Project goals
2	Usage
3	Incorporate requirements from collected data
4	Technical and functional capability
5	Aesthetics
6	Constructability
7	Sustainability (environmental, social, and economical)
8	Health and Safety
9	Reliability
10	Environmental compatibility
11	Sustainability
12	Fire protection measures
13	Supportability during maintenance/maintainability
14	Cost-effective over the entire life cycle (economy)
15	Leadership in energy & environmental design (equivalent) compliance
16	Reports, drawings, and models

Figure 4.41 illustrates preliminary project plan. Table 4.29 illustrates points to be considered for development of schematic design. Figure 4.42 illustrates major activities in the detailed design phase.

Some of the tools discussed earlier can be applied during design stage. These are as follows:

- Innovative and creative tool
  - Six Sigma DMADV (please refer to Table 4.15)
- Process analysis tool
  - Cost of quality (Figure 4.43-cost of quality during design)
- Process improvement tool
  - PDCA cycle (Figure 4.44-PDCA cycle for construction projects)
- Lean tool
  - Mistake proofing (please refer to Table 4.14)

4.3.3 Tools for Bidding and Tendering Stage

Table 4.30 illustrates major activities during bidding and tendering stage, and related tasks performed during this stage.

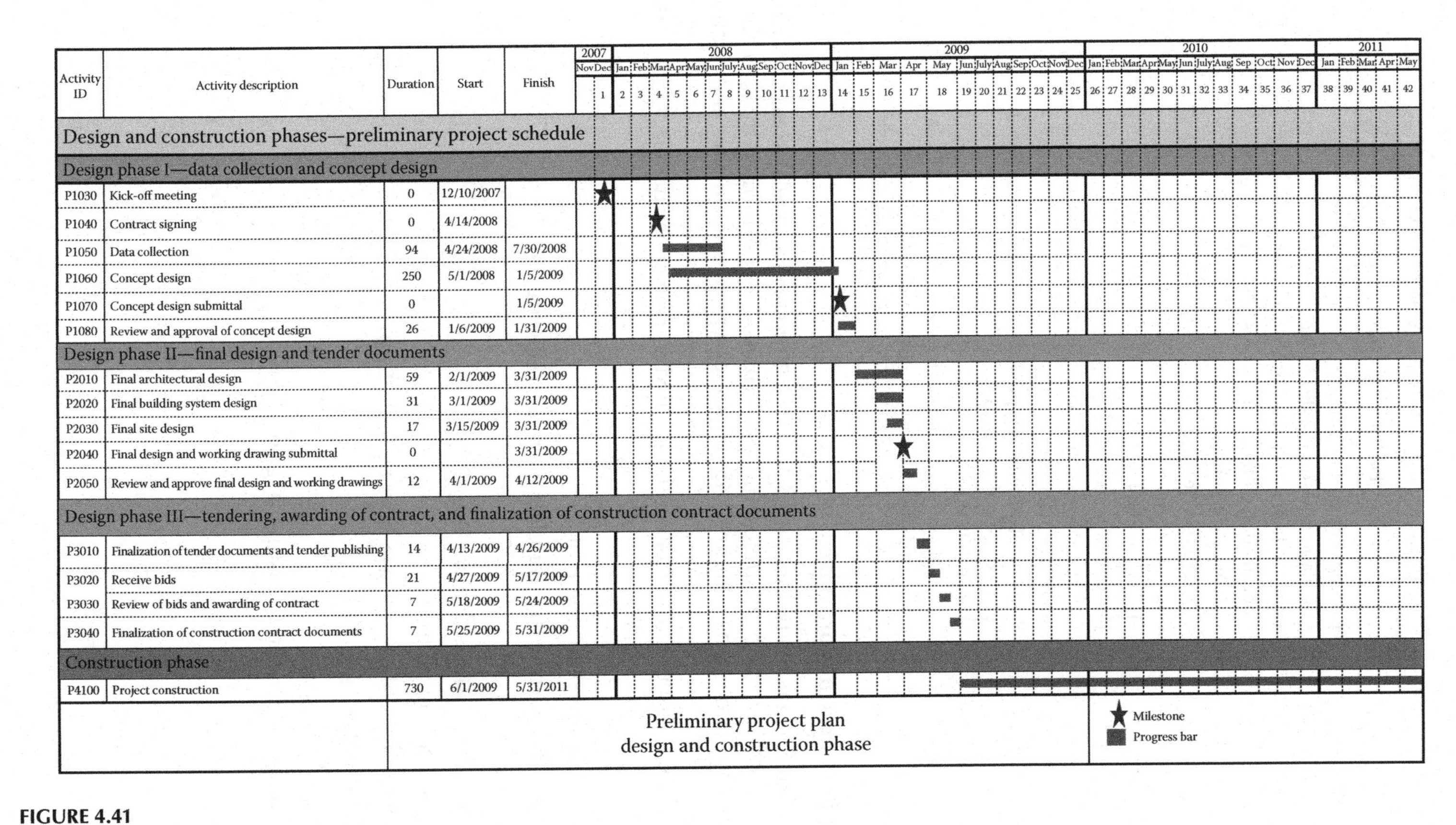

Activity ID	Activity description	Duration	Start	Finish
Design and construction phases—preliminary project schedule				
Design phase I—data collection and concept design				
P1030	Kick-off meeting	0	12/10/2007	
P1040	Contract signing	0	4/14/2008	
P1050	Data collection	94	4/24/2008	7/30/2008
P1060	Concept design	250	5/1/2008	1/5/2009
P1070	Concept design submittal	0		1/5/2009
P1080	Review and approval of concept design	26	1/6/2009	1/31/2009
Design phase II—final design and tender documents				
P2010	Final architectural design	59	2/1/2009	3/31/2009
P2020	Final building system design	31	3/1/2009	3/31/2009
P2030	Final site design	17	3/15/2009	3/31/2009
P2040	Final design and working drawing submittal	0		3/31/2009
P2050	Review and approve final design and working drawings	12	4/1/2009	4/12/2009
Design phase III—tendering, awarding of contract, and finalization of construction contract documents				
P3010	Finalization of tender documents and tender publishing	14	4/13/2009	4/26/2009
P3020	Receive bids	21	4/27/2009	5/17/2009
P3030	Review of bids and awarding of contract	7	5/18/2009	5/24/2009
P3040	Finalization of construction contract documents	7	5/25/2009	5/31/2009
Construction phase				
P4100	Project construction	730	6/1/2009	5/31/2011

FIGURE 4.41
Preliminary schedule for construction project.

TABLE 4.29

Development of a Schematic Design for a Construction Project

Sr. No.	Points to Be Considered
1	Concept design deliverables
2	Calculations to support the design
3	System schematics for electromechanical system
4	Coordination with other members of the project team
5	Authorities requirements
6	Availability of resources
7	Constructability
8	Health and safety
9	Reliability
10	Energy conservation issues
11	Environmental issues
12	Selection of systems and products that support functional goals of the entire facility
13	Sustainability
14	Requirements of all stakeholders
15	Optimized life cycle cost (value engineering)

4.3.4 Tools for Construction Stage

Table 4.31 illustrates major activities during construction stage and related tasks performed during this stage. Table 4.32 illustrates major activities to be performed by the contractor during the construction phase. Figure 4.45 illustrates project monitoring and controlling process cycle.

Some of the tools discussed earlier can be applied during construction stage. These are as follows:

- Management and planning tool
  - Networking arrow diagram (please refer to Figure 4.12)
- Process analysis tool
  - Root cause analysis (Figure 4.46)
- Process improvement tool
  - Six Sigma—DMAIC (please refer to Table 4.16)
  - PDCA cycle (Figure 4.47)
- Classic quality tool
  - Flowchart for concrete casting (Figure 4.48)

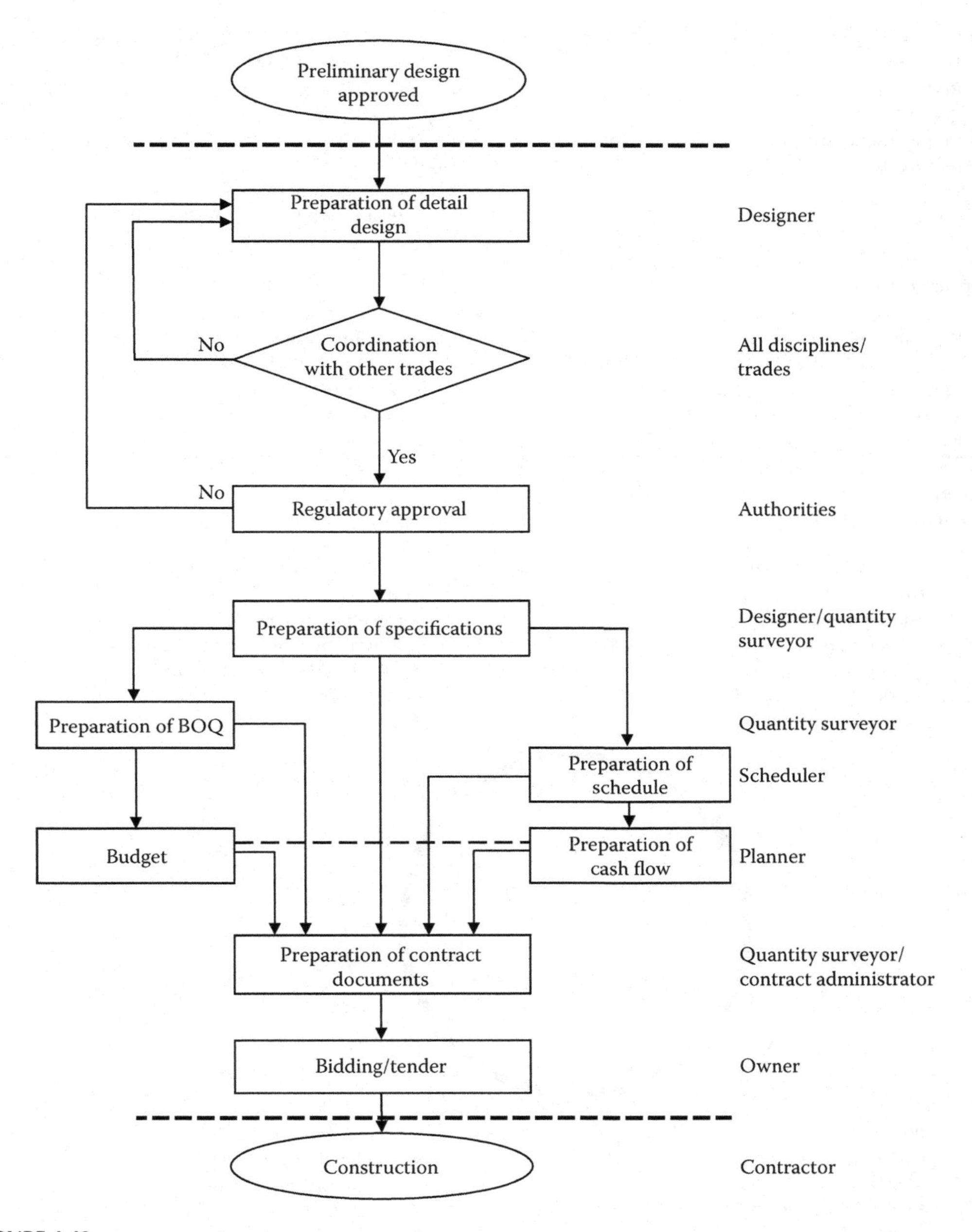

FIGURE 4.42
Major activities in the detailed design phase. (Abdul Razzak Rumane (2010), *Quality Management in Construction Projects*, CRC Press, Boca Raton, FL. Reprinted with permission from Taylor & Francis Group.)

Internal Cost	External Cost
• Redesign/Redraw to meet fully coordinated design • Rewrite specifications/documents to meet requirements of all other trades	• Incorporate design review comments by client/project manager • Incorporate specifications/documents review comments by client/project manager • Incorporate comments by regulatory authority(ies) • Resolve RFI (request for information) during construction
Appraisal Cost	**Prevention Cost**
• Review of Design Drawings • Review of Specifications • Review of Contract Documents to ensure meeting owner's needs, quality standards, constructability, and functionality • Review for regulatory requirements, codes	• Conduct technical meetings for proper coordination • Follow quality system • Meeting submission schedule • Training of project team members • Update of software used for design

FIGURE 4.43
Cost of quality during the design stage.

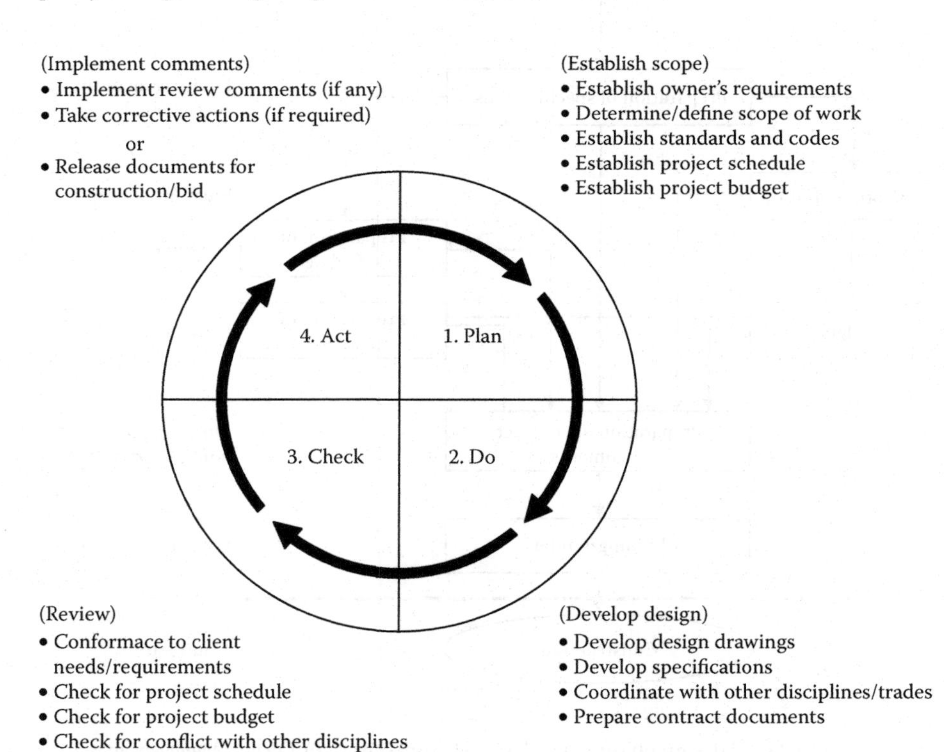
(Implement comments)
• Implement review comments (if any)
• Take corrective actions (if required)
or
• Release documents for construction/bid
(Establish scope)
• Establish owner's requirements
• Determine/define scope of work
• Establish standards and codes
• Establish project schedule
• Establish project budget
4. Act
1. Plan
3. Check
2. Do
(Review)
• Conformace to client needs/requirements
• Check for project schedule
• Check for project budget
• Check for conflict with other disciplines
• Conformance to regulatory requirements, standards, and codes
• Check for constructability
• Check for environmental compatibility
(Develop design)
• Develop design drawings
• Develop specifications
• Coordinate with other disciplines/trades
• Prepare contract documents

FIGURE 4.44
PDCA cycle for construction projects (design phases). (Abdul Razzak Rumane (2010), *Quality Management in Construction Projects*, CRC Press, Boca Raton, FL. Reprinted with permission from Taylor & Francis Group.)

TABLE 4.30

Major Elements of Bidding and Tendering Stage

Sr. No.	Elements	Description	Related Task
1	Bidding	Bid documents	Prepare Tender documents
2	Selection of contractor	Tender announcement	Advertisement for bidders
			Prequalification of contractors
3	Issue tender documents	Request for proposal	Issuing tender documents/request for proposal
4	Receipt of proposal	Receipt of bid documents	
5	Tender analysis	Tabulation of proposals	Analysis of quotation
			Low bid
			Qualification based
6	Selection of contractor		Most competitive bidder (low bid or qualification based as per the company's methodology)
7	Award of contract	Signing of contract	Performance bid
			Insurance
8	Notice to proceed		Contractor to commission

TABLE 4.31

Major Elements of the Construction Stage

Sr. No.	Elements	Description	Related Task
1	Construction	Contractor to execute contracted works	Mobilization, managing resources, work execution, QA/QC, works approval. (Please refer to Table 4.32.)
2	Monitoring and control	Monitor and control scope, schedule, cost, quality, risk, procurement of the project	Scheduling, monitoring and control. (Please refer to Figure 4.45.)
3	Commissioning and handover	Testing, commissioning, and handover	Project startup procedures.
4	Project closeout	Close the project	Project closeout documents, finalization of claims and payments.

TABLE 4.32
Major Activities by Contractor during the Construction Phase

Sr. No.	Activities to Be Performed by Contractor
1	Mobilization
2	Staff approval
3	Selection of subcontractor(s)
4	Selection of material
5	Selection of resources
6	Preparation of shop drawings
7	Execution of works
8	Project monitoring and control
9	Auditing/installation of executed works
10	Quality management
11	Approval of works
12	Health, safety, and environmental compliance

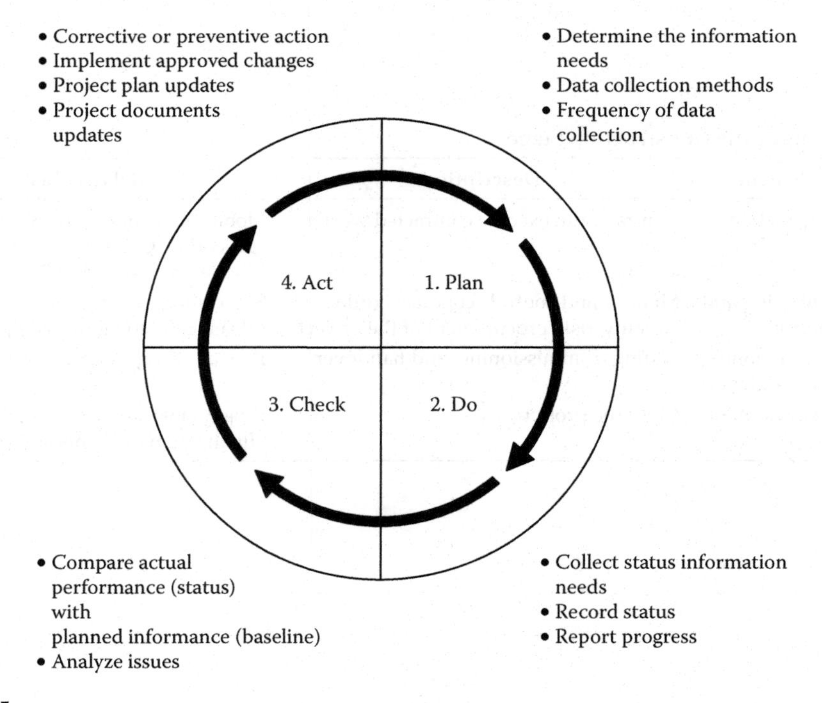

FIGURE 4.45
Project monitoring and controlling process cycle. (Abdul Razzak Rumane (2013), *Quality Tools for Managing Construction Projects*, CRC Press, Boca Raton, FL. Reprinted with permission from Taylor & Francis Group.)

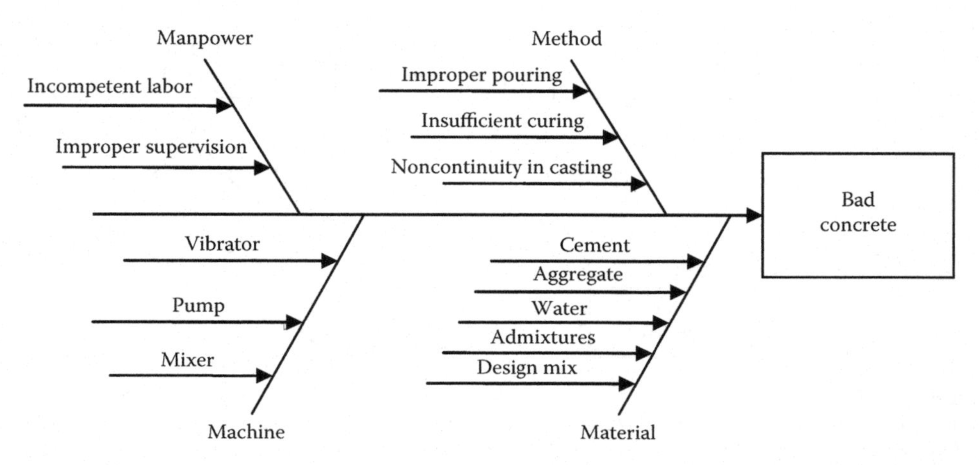

FIGURE 4.46
Root cause analysis for bad concrete. (Abdul Razzak Rumane (2010), *Quality Management in Construction Projects*, CRC Press, Boca Raton, FL. Reprinted with permission from Taylor & Francis Group.)

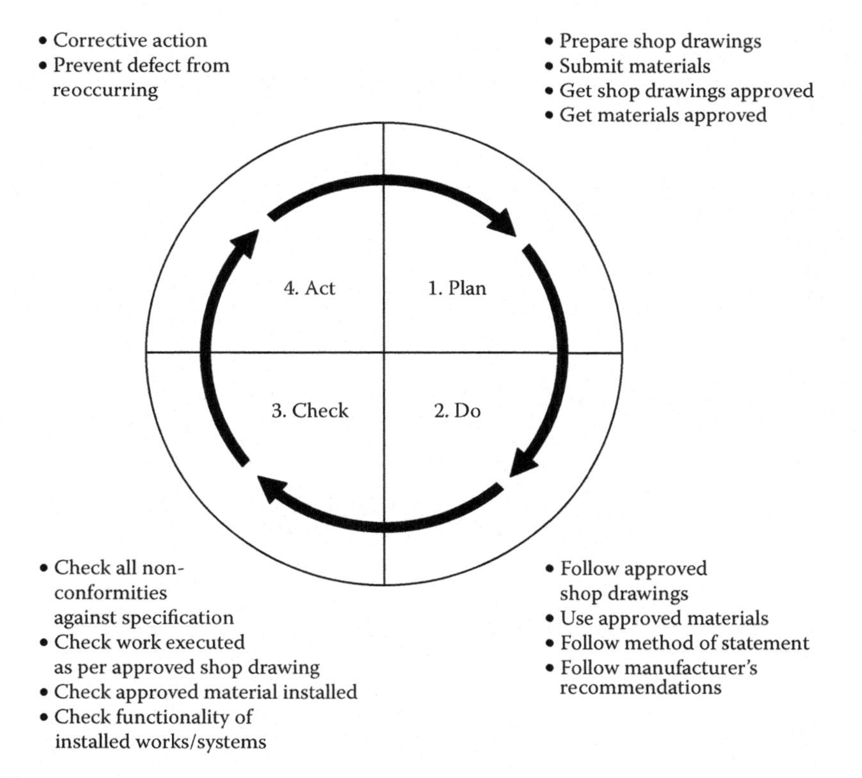

FIGURE 4.47
PDCA cycle (Deming wheel) for execution of works. (Abdul Razzak Rumane (2013), *Quality Tools for Managing Construction Projects*, CRC Press, Boca Raton, FL. Reprinted with permission from Taylor & Francis Group.)

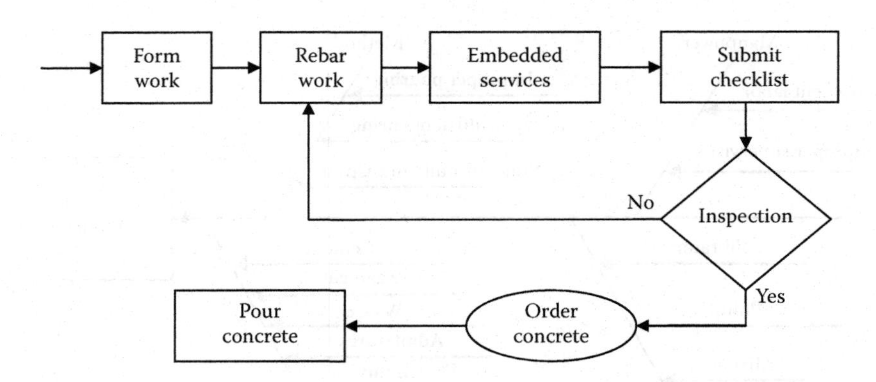

FIGURE 4.48
Flowchart for concrete casting. (Abdul Razzak Rumane (2010), *Quality Management in Construction Projects*, CRC Press, Boca Raton, FL. Reprinted with permission from Taylor & Francis Group.)

5

BIM in Design and Construction

Cliff Moser

CONTENTS

5.1 Introduction

Building information modeling (BIM) uses software and processes to digitally develop building data in a manner that is collaborative and integrative. Using BIM as a robust process and not just as a tool has transformed the design and construction industry. Different from computer-aided drafting (CAD), BIM leverages digital information about the building (including materials, furnishings, and equipment requirements) in a searchable database that is additive and scalable, enabling design and construction team members to participate in the virtual cocreation and codevelopment of a project design and operational requirements. Accurately described as virtual design and construction (VDC), BIM facilitates three-dimensional (3D) digital development including coordination, estimating, scheduling, and material selection. BIM also creates a rich data source onto which rules-based project studies, such as occupancy simulations, fire and life-safety code reviews, and energy use calculations, can be applied. The future of BIM will be leveraged in physical facility operations, where model information will be utilized in continuous facility commissioning for operating and maintaining a building.

5.2 BIM as Collaborative Tool

The advent of BIM and sharing the parametric geometry and information contained in these files during all phases of design and construction has disrupted the traditional incumbent roles within the construction industry. Currently deployed on large and complicated projects, the use of BIM has become the tool of choice for designers and builders for coordination and scheduling.

Construction quality begins with design quality, and with BIM, there is the ability to design and document a building as a single virtualized 3D object, without the need to overlay separate two-dimensional (2D) floor plans, sections, and elevation. This is a disruptive force in itself. Furthermore, the construction teams can virtually "build" a project using the myriad tools of BIM.

However, the full effect of this shift (including the parametric changeability of the model) have not been completely incorporated and promoted into the existing processes of the building designers, perhaps because it is perceived to be just another CAD tool in the arsenal of design solving techniques that support existing iterations of using a layering of plans, sections, and elevations to achieve the goals of design. Conceptually, the designer design solves through proper side to side (*X* and *Y*) requirements and adjacencies, identifying and matching requirements for area and clearances, before moving into the volume requirements of the *Z*-axis. Furthermore, designers resist the opportunity to parametrically complete the models accurately with design details, materials, and equipment, instead relying on convention (and the case law of means and methods), which identify established construction elements (materials, schedules, and detailing) as being the realm of the builder. Therefore, the design team still establishes design "intent" from which the builder "interprets" and constructs. In the past, it was 2D CAD drawings, scale solid model prototypes, as well as specifications and other design information, such as door and finish schedules.

5.2.1 History of Collaboration

Coordination in construction has existed as a contractually required instrument of service and deliverable through the building side of the project team. The process functions as the construction manager or general contractor works with the project's trade contractors to demonstrate the entire building team's understanding of the design intent of the project. In this process, all of the trade contractors, but especially the mechanical, electrical, and plumbing (MEP) trades, create coordination and installation drawings, which identify the materials needed for the work and plan the sequence of construction. In the past, these deliverables were typically 2D plans, overlaid over the design team's design drawings, which were developed to demonstrate understanding of the design documents, and to identify the necessary coordination required. For MEP systems, this became an important activity from which the ductwork was fabricated, smoke and fire dampers ordered, and hangers placed within the structural system to accommodate installation. Often, after completion and review by the design team, the coordination drawings would immediately become fabrication and installation sets, and complete systems were subsequently designed and built from the approved coordination sets. But coordination is not collaboration. As described above with AutoCodes, validating design intent can run through several new levels of "design inspection" not anticipated by the design team. Furthermore, coordination has had its own share of failures in the past.

There are numerous projects where major portions of the systems design were modified during the "builder's coordination phase." In traditional contract arrangements, the designer is responsible for design coordination. The design documents should deliver a completed and functional design as part of this phase. But in some projects, the design (especially MEP) information was out of date, incomplete and if shown, was mostly incorrect. On one project, the heating, ventilation, and air conditioning (HVAC) ductwork design on the project was flawed and could only be completed after significant "redesign" during the builder coordination phase, resulting in multimillion-dollar change orders and the initial delivery of thousands of dollars of unnecessary mechanical equipment (as a result of early order placement off of the design drawings in order to maintain an aggressive construction schedule). Project complexity and project coordination in lieu of project collaboration forced this result.

Therefore, faced with a similar project risk of verifying the design before or during construction, most owners will choose that the builder be involved during design in a formal collaboration process, rather than just construction coordination. In order to facilitate this engagement, and to mitigate risk similar to the project identified above, owners have begun pressing for the involvement of construction team members as early in the design phase of the construction project as possible. As a result, design and construction agreements have been adjusted accordingly, bringing builders in as early as possible creating design–build, design assist, gross maximum price contracts, and others, where the building team becomes involved as soon as the project design moves from the schematic design phase (and sometimes earlier). While these contract adjustments solve some design and construction problems, they are incremental and don't properly identify and address the root cause problem of complete and coordinated (and therefore, collaborative) documents and construction activities in complicated projects.

5.2.2 Collaborative Agreement

BIM enables design and construction teams, and the project owners, to collaborate beyond past traditional coordination efforts (identified more as "serial cooperation" than collaboration).

Many of the ways to remedy this was to develop collaborative agreements in the hope of formally adjusting relationships between designers, builders, and owners, at least during the design and construction phases. The AIA has addressed collaborative owner–architect–contractor agreements that in turn address digital delivery, collaboration, and shared risk and responsibility during the design and construction phases of projects, recognizing that early, preconstruction involvement by builders and their trade contractors is a net benefit to the design process, which enhances the design and minimizes risk for all parties, as well as provides strength to design goals. As a result, there are several types of integrated forms of agreement (IFOA) in use in the United States, one being the AIA documents, another one being the Consensus Documents produced by the Association of General Contractors (AGC).

5.3 BIM Maturity Model

A computational rules-based model in architectural deliverables will help redefine standard of care within the instruments of service. *Architecture 3.0* is not about hanging new inspection-based quality assurance (QA)/quality control (QC) on a building model, it is an opportunity to begin to identify liberation from the archaic confines of inspection-based QC.

Architects have continually failed in the delivery of complete and coordinated construction phase documents, embracing BIM as the next documentational improvement to which aspirational goals of coordinated success were applied. However, design teams find themselves unable to fully embrace the process quality tools within BIM. Hindered by the master–apprentice based past, there is an inability to recognize and abandon the shortcomings of inspection QC efforts.

5.3.1 Construction Phase Design Deliverables

As construction deliverables become increasingly more complicated and regulated, design teams do not have the expertise to successfully delineate the technical details for a complete and coordinated set of construction documents. However, BIM, on the other hand, has been recognized for enabling the ability to document and coordinate complex building structures, which, in the 2D past, would have been impossible.

In *Architecture 3.0: The Disruptive Design Practice Handbook*, the problems with traditional QA/QC activities are outlined within the practice of architecture. QA/QC is rooted in the legacy apprentice-based training activities. Applying quality measures to these activities and the resultant deliverables is the basis for many incumbent processes deployed not only during construction phase activities, but through the entire project life cycle for the design team.

Baseline requirements of successful project upon delivery to the owner can be specified through applied performance metrics; such as energy use, user satisfaction, and neighborhood integration. Creating and applying similar requirement to deliverables and service in support of construction phase activities is more difficult. Therefore, in order to understand the industry's problematic relationship with QA/QC and construction phase deliverables, an historical context is required.

5.3.1.1 Why Hasn't BIM Improved the Quality of Construction Phase Deliverable from the Design Team?

It's all about the separation of design and production, and the legacy effects of the master draftsman.

Instruments of service are defined by locally defined and regulated standard-of-care requirements, which skew to legacy imperatives. Furthermore, rather than conjuring up images of smiling, round spectacled professionals, specters emerge of lawyered-up old men beating down any attempted innovations.

QA/QC are activities based on legacy processes that designers bring (but don't often think about) to projects. Applying quality measures to these activities and the resultant deliverables is the basis for many incumbent processes deployed not only during construction, but through the entire project life cycle.

While the baseline requirements of a successful project upon delivery to the owner can be specified through applied performance metrics, such as energy use, user satisfaction, and neighborhood integration, creating and applying similar requirement to deliverables and services in support of construction phase activities is more difficult.

5.3.1.2 QA/QC

QA/QC began as a formalized inspection activity in the early part of the twentieth century. It was initially based on product standardization within the manufacturing industry.

QC was defined as creating measures of the internal processes for minimizing deviation in the production of readily interchangeable parts and equipment received from manufacturing suppliers. QA was the manufacturer's internal inspection of that product, prior to delivery to the customer. The need for creating specifications and standards in order to scale the mass-marketing of parts and equipment began the process of turning craftsmen into assemblers. QC created requirements and specifications for interchangeability and adaptability of parts and systems, and the requirements were applied to the deliverables and service of manufacturers and suppliers. World War II accelerated the requirements for quality, as it reduced the tolerance for deviation in order to scale manufacturing for globally deployed militaries. QA/QC requirements became even more important because millions of soldiers' lives depended on an established quality supply chain.

After the war, QA/QC continued its role within the industry, and moved into design and construction activities, as the concept of standardization and manufacturing took hold in commercial and household products. Since the design team is a key supplier of the construction industry, deliverables followed the same rules of supplier standardization and QC.

5.3.1.3 Inspection-Based Systems

Creating and enforcing internal organizational QC systems was built on an inspection-based approach. Early in the twentieth century, most architectural firms organized around a design and production approach. The design architect would create schematic level design documents which were then handed off to the production team who would develop the project through design-development and into construction-documents. The deliverables for these documents, typically plans and specifications, were internally reviewed by a QC team within the firm prior to delivery to the owner of the project. The owner would submit the construction document package to selected general contractors to obtain a price for building the project.

In this process, completed deliverables were internally measured by QC, accepted within defined tolerances and then released for use to the owner. QA measured and accepted the deliverable against defined requirements and standards. QC was enforced through internal inspection-based processes, which had naturally evolved from the past apprenticeship culture. This deliverable quality was defined by the agreements with owners and legally binding local standards of care, which included construction phase service activities by the design team. Enforcement was established through calibrated fee payment schedules. Rework was the resultant activity to correct the failure to provide deliverables within measured tolerance of those requirements.

Through established internal QC services, design firms tried to maintain services and deliverables based on applied standards with oversight. Inspection and rework requires that there be an understanding and adherence to standard of care, project requirements, and organizational standards within the firm that sets the baseline for acceptable delivery, as well as internal training and inculcation of that understanding, against a past measured standard. Therefore, in order to successfully provide deliverables and service to industry requirements, enforced through standard of care and organizational quality standards, the firm, in order to survive and scale, needed to create internal programs for training and oversight into producing quality deliverables and services.

In the past this was accomplished through an apprenticeship methodology, by pairing experienced project architects with junior staff, and then training through the iterative development of drawings and details along with continuous review and oversight. The end result was that the inexperienced drafting staff would gradually learn the requirements of the firm, industry, and profession's standards. Expertise would be inculcated through oversight and rework, similar to craftsman–apprenticeship activities in the past.

However, the advent of electronic deliverables as well as the development of a mobile and contingent workforce, introduced a huge gap in the oversight and rework process. With the advent of technology, the profession lost the role of the master draftsman; the seasoned technical member of the organization and industry, who usually without a professional architectural degree, was a master in understanding how to show the design requirements of construction phase activities.

For example, Mary Woods, in her book, *From Craft to Profession,* outlines the role of these master draftsmen, craftsmen-builders who moved from craftsman to professional architects. Their activities were shaped through interactions with the construction process. These master draftsmen organized professional societies and strived to achieve the rights to an architectural education, the receipt of appropriate compensation, and professional accreditation.

Furthermore, George Barrett Johnson, in his book, *Drafting Culture*, outlines the codification of the "draftsman's Bible" Ramsey and Sleeper's *Architectural Graphic Standards* (AGS). The authors of AGS, one a draftsman and the other an architect, created a graphic compilation of architectural details and standards that attempted to codify the shared knowledge of the integration of drafting and design. AGS delineates the boundaries of the shift from the draftsman's craft to the architect's academically based knowledge. The former "drafting culture" gave way to massive postwar changes in design and building requirements.

Today most firms no longer have the ability or staff to support an apprentice-based "architectural labor" culture within its design and production departments. Furthermore, now all firms outsource and distribute "architectural labor" tasks throughout their own supply chain internal and external design team members (including engineering and other consultants). This distribution completely outpaces the ability of inspection teams to oversee project deliverables' QC.

However, the tasks of QA, with the building contractor's reviewing the deliverables against internally inspected standards, continue today, reinforced by contracts and legal requirements. The tasks of QC are made ever more difficult through electronic, computer-generated deliverables. Oversight and inspection QC is impossible.

5.3.1.4 Process-Based Systems

Therefore, a new quality model is required in order to leverage electronic deliverables, accelerated project delivery models, globally distributed teams, and increasingly complicated building types, which are ever-more regulated and difficult to build. This model requires an integrated quality management system which goes beyond an inspection-based approach and leverages the members of the construction industry supply chain in order to deliver a more integrated and robust process-based organizational approach. Additionally, it requires agreements that differ from the legacy standards built on oversight of paper deliverables. This model is fundamentally different than the inspection-based systems of the past and requires a new approach to providing deliverables and services.

The background for this model is based on the integrated quality management systems that manufacturing has long developed and embraced in its deliverables and services. Former inspection-only QC systems in manufacturing have evolved and migrated to process-based systems. Accordingly, the workforce that serves manufacturing is cross-functionally trained in order to understand and integrate process-based systems of QC. These systems instill deliverable and service quality by incorporating quality reviews into the process as deliverables are created rather than waiting for inspection of the end results. They also incorporate quality management built on supplier-based initiatives, such as Lean, as well as individual customer-focused quality requirements, such as a balanced scorecard, and ISO 9000. Therefore, a successful quality management program is based on a blend of process-based systems, supply-chain integration, and identified deliverables and service, reinforced through oversight of processes and performance sampling of deliverables.

Creating a process-based system enables the ability for a practice to leverage tools and deliverables such as BIM and project life cycle integration, which will add value to the client services and leverage the talent within the distributed team and supply chain beyond successful construction phase activities.

Therefore, introducing parametric virtual 3D model uncomfortably binds the architect to the means and methods realm of the builder. Furthermore, since designing in 3D with a tool like BIM doesn't appear to be so radical, the volume of the space, the Z coordinates, will generate itself as a result of successful floor planning.

However, builders know that this isn't always the case. Design quality isn't inherent in the floor plans (or in the resulting elevations, sections, and details), and the design problem may not be solved as part of 2D floor planning, and scope and requirements may be missing as a result as well as important details and connections. Surely most builders can easily describe many projects where substantial amounts of required information was lacking after interpreting and representing the understanding of design intent as shown by the design team in the 2D version of the contract documents.

As a result of this quality failure, builders everywhere have embraced BIM, and most, if not all, large construction projects have large collaboration rooms where the design drawings along with building systems are coordinated virtually through electronic software sharing prior to being fabricated and installed. Furthermore, as a result of separation of construction means and method versus design intent, BIM files delivered to the construction team are often in need of revision and redesign. The builders' use of BIM in these processes began as an audit for the completeness of the design package, and soon evolved into one of the preferred ways for the builder to ensure that all the necessary design requirements of the project were included and coordinated as required by contract.

This step was required in past contracts; the builder reviewing, coordinating, and sometimes reworking the plans or model, if necessary, because this is what happened in the past with the coordination and detailing activity (CDA) phase of the project. Prior to BIM, this phase was a verification task that the construction team was required (by contract) to undertake in order to demonstrate that they understood the requirements shown in the intent of design documents. In the past, this had been accomplished by overlaying contractor-created 2D plans (which were generated by interpreting design drawings). This began with overlaid hard paper tracings, then electronic soft copies—CAD or PDFs—to ensure and demonstrate that the designed building systems (the plumbing, HVAC, electrical), as well as the architecture and structural framework were well-coordinated with each other by the contractor and the trades, and that, most importantly, that the builder understood and could communicate the designer's intent.

BIM VDC coordination has the ability to go beyond simple gross geometry and spatial testing. If a model is developed and used correctly, BIM leverages the validation of the "why" of systems in addition to the "how" of how of the geometry. However, because as designers, BIM is just a 3D geometric coordination tool, "why" modeling isn't generated because BIM is seen through the lens of gross coordination. This argument worked in the past, and builders and owners accepted this design delivery method because it appears to comply with past design instruments of service, and maintains the firm virtual "chalkline" of means and methods requirements of the past which were developed to separate design and construction activities.

5.3.1.5 Autocodes

However, this is changing. And the topic of this chapter is how to utilize BIM as a quality tool as well as design and construction management tool. The changes now occurring within the design industry as *Architecture 3.0* are resetting the design profession (and the design and construction industry as a whole) in which it intentionally disturbs this traditional arrangement, willfully encouraging all parties in a construction project to leverage the opportunity to use the "I" in BIM, and to re-envision and redefine their past roles, reaching out to all parties to use their appropriate expertise. This is an opportunity to use BIM and its associate process tools as a vehicle for true information capture and knowledge sharing throughout the entire project life cycle. Furthermore, Fully Integrated Automated Technology (Fiatech), the technology arm of the Construction Industry Institute in the United States, is spearheading a rules-based AutoCodes project (http://fiatech.org/), where, as stated on the Fiatech web page that, "in concert with the International Code Council (ICC), 14 jurisdictions, and independent software developers, (phase 1 of) the AutoCodes project demonstrated … the benefits of automated plan checking specifically for access and egress provisions for a retail structure." That phase also conducted a comparative analysis of automated code checking versus traditional 2D (paper) plan reviews.

The Regulatory Streamlining Steering Committee has already engaged in developing Phase 2, the objectives are "(1) to expand development of rule sets for other occupancy classifications and construction codes; and (2) to develop training materials to aid jurisdictions in transitioning from traditional to electronic plan review and ultimately to automated code checking" (http://www.fiatech.org/).

The many benefits of using the design BIM to check for code compliance are that the project teams have an opportunity to populate the model with smart, rule-based data that has the ability to be queried at select points during the design process. AutoCodes will work if the design team can utilize it to help build and validate their model, not as a tool to "challenge" the design intent after the project is completed.

5.4 BIM for Managing Project Complexity

Using BIM to coordinate the communication distribution on a project, decision making is often divided between the different project partners who are focused on their individual requirements and organizational needs. The fact that the details of complex projects is distributed to many separate player, whose concern goes beyond design (into fee, schedule, contract requirements, etc.), ensure that the impact of individual organization design changes is too constrained, and the sum of all of the design decisions are too voluminous and complex to be made from the top-down. Therefore, the key business drivers become local project utility maximization and the fit between these local design decisions, as well as delivering up to date information to project managers. In this model, encouraging the proper influence relationships and concession strategies becomes the primary tool available to design managers.

If these are defined inappropriately, the resulting design decisions are not timely, do not meet important project requirements, and/or miss opportunities for exploration of the design space. Plan for complexity, create a decision escalation model that enables select, high-level project decision-making to be performed by people not involved with the project. This will allow neutral power balance that will help manage the outcome. Furthermore, establish an evolving RACI (responsible, accountable, consulted, informed) diagram that monitors and controls communication and decision-making inside and outside of the project team. Identify who is responsible, accountable, consulted and informed, and identify the management and executive role for enforcing the levels of decision-making.

5.4.1 Project Complexity

Project complexity emerges as a project becomes so complex that it becomes self-adaptive. Large design and construction projects, like bridges and hospitals, are like raising children, and traditional management is ineffective. Using BIM to link emergent and adaptive system effects to achieve "change the delivery of healthcare" or to "create a new regional transportation model" rather than to build a hospital or design a bridge. Therefore, the information created inside of a project model should be rule-based to autonomously support rote project applications, such as code checking, validation of design solutions, and support the operations of the project after construction. But because BIM grew out of and still exists within the boundaries of 2D coordination, it has not been allowed or guided to evolve where it should naturally flourish and evolve.

One of the difficulties of managing complex and convoluted design and construction projects is that there are too many control points. If a team focuses on one control point, say schedule, then the budget can skyrocket. In turn, focusing on budget may force the schedule to suffer. Additionally as defined later in this section, the design and construction industry suffers from the Baumol effect; in that project manpower efficiencies are limited.

Using BIM will help the activities of complexity modeling and will enable the design and construction teams frame a baseline for success. Because strong interdependencies exist between design and construction decision-making, it is difficult to converge on a single design or schedule solution that will satisfy all of the requirements of all participants. As outlined earlier in this chapter, collaborative design approaches are limited in their deployment because they tend to rely on traditional tools which focus on familiar adversarial and litigation-primed processes. These solutions are serial, expensive, and time-consuming. Additionally, these tools are not able to incorporate new tools such as life cycle design issues (typically later life cycle issues such as environmental impact), and miss opportunities for wholesale disruption or innovation.

Complexity models rely on understanding systems thinking. Systems thinking looks at a project as a complex system that is largely uncontrollable, full of unintended consequences, and interconnected problem opportunities. Complexity modeling identifies an ecosystem that nurtures events, and highlights activities that may encourage the nurturing of negative events, which may result in destructive results. The example above of the millions of dollars of change orders due to finishing the HVAC design during construction that was a result of incremental changes made by the owner and architect during the early years of design (nearly a decade before the project moved to final design) is an example of systems thinking. The MEP engineer was a subconsultant to the architect, who was included late in all of the design changes. Blaming obfuscation doesn't enable the awareness required for creating an actionable complexity model. Clearly the architect didn't properly update the entire design team. But an awareness of complexity modeling could have helped the MEP team (and the entire project team including client, design, and construction)) ask the questions, frame the boundaries, and help prepare responses to complex and interconnected project changes. As it was, the subtle changes in ongoing operational flow modified suite layouts and egress requirements that were not properly communicated and implemented in the MEP designer's drawings.

A complexity model for a project is created by identifying the results from four project activity factors:

- Deliverables and project timeline uncertainty
- Project type uniqueness
- Project system interconnectivity
- Project size

For most projects that fall under complexity models, BIM can be used to help monitor and control a project through these different project factors. For example, regarding project deliverables and project system interconnectivity, BIM can be used to break up a project into pieces that are more manageable and connected.

Complexity research looks at the dynamics of distributed networks as well as networks in which there is no centralized controller. Within these project networks, control behavior emerges as a result of concurrent local actions around project control. Distributed networks are typically modeled on multiple nodes, each node representing a state variable with a given value. Each node in a network acts to select the value that maximizes its consistency with the influences from the other network nodes.

Communication in these distributed networks is based on

1. Team members exchanging information within project teams in order to collaborate.
2. Members exchanging information between project teams with the intent to share.
3. Members gathering information from project activities in order to aggregate, visualize, and manage a project's goals and deadlines.

5.5 BIM Execution Plan

A complexity model should be the foundation of a BIM execution plan. The tools and processes of BIM are transformational. Recognizing and leveraging this paradigm shift cannot be overstated. Construction projects are complex systems. The use of BIM in creating new quality systems for design and construction in this new realm cannot be overstressed.

The BIM execution plan (BxP) is the project roadmap that delineates the creation and stewardship of the model from design through construction into the facility operations. The BxP is created by the owner, and the plan spells out the model as an instrument of service throughout the project life cycle. A BxP template lays out the menu for the project collaborative dinner (as identified above). The BxP is created to support the contract requirements and to lay out the working relationship between all parties. Tie the BxP to the contract requirements, and outline the requirements for the joint BEP. That's the next step.

BIM tools and Lean processes facilitate collaboration, because as a process it means that multiple parties can work on the model at the same time, and at the same time the builder can start ordering materials while the owner facilitates approvals, all while using the model as a tool.

However, this new model challenges the incumbent processes of architecture.

5.5.1 Joint BIM Execution Plan

The joint JBxP (or JBxP) outlines the requirements for codelivery of the project documents. Created jointly, by the design and the building team, the JBxP should initially provide the guardrails for the quality and the coordination requirements for the project and sets the framework for complexity. The JBxP creates the rules for collaboration within the incumbent and litigious world that currently rules in a design and construction project. Both the BEP and JPEP are small but necessary steps toward an IFOA; a version of a contract document in which all complex projects will eventually formally move to. While not perfect, both the BxP and JBxP recognize that current contract conditions distill case law into formal behavior for the participants of the construction project. JBxP leverages the existing tripartite agreements of design, bid, build and the rules of owner, architect, and contractor by helping leverage the codelivery of the project documents and project construction.

Other tools like Lean processes integrated with BIM can assist in the development of project documents, where multiple team members can work on the model at the same time, while the owner facilitates approvals.

5.5.2 Defining Minimum BIM by Using the Capability Maturity Model

It is important to note that the minimum BIM (MBIM) capability maturity model (CMM) described provides a range of opportunities for BIM; MBIM is described as follows:

- Visualization or some level of improved document production may be one output from a BIM; however, neither is in and of itself BIM. MBIM has the following characteristics through the associated areas of maturity in the complete CMM:
  - *Spatial capability*—The facility need not yet be spatially located as this is a higher-level goal to be considered a MBIM.
  - *Roles or disciplines*—MBIM includes the sharing of information between disciplines and documentation of the BIM's intended uses.
  - *Data richness*—The data must be of the level of detail to support the intended use of the BIM. The level of data for a concept BIM will be different from that of a design BIM or construction BIM.
  - *Delivery method*—BIM must be implemented in a way that allows discipline information to be shared.
  - Change management or ITIL maturity assessment.
  - Business process.
  - *Information accuracy*—The BIM must be used to compute space and volume and to identify what areas have been quantified.
  - *Life cycle views*—A complete life cycle does not need to be implemented at this point. MBIMS recommends that the data should be maintained in interoperable formats that allow for future life cycle use.
  - *Graphical information*—Since all drawing output should at this point be National CAD Standard compliant, these are the requirements for a minimum BIM. This demonstrates that standards are being considered, when possible.
  - *Timeliness and response*—BIM is not yet expected as the trusted authoritative source for information about the facility for first responders.
- *Interoperability and industry foundation class (IFC) support*—The BIM must be capable of creating IFC data, exporting IFC data, importing IFC data, and operating IFC, interoperable data.

Change management, or Information Technology Infrastructure Library (ITIL), maturity assessment business process; graphical information; and spatial capability are other characteristics of MBIMs that will be required as the industry matures and requirements develop.

5.5.3 BIM: A Tool for Coordination

In support of the new contractual opportunities, BIM becomes both a tool for ongoing design and construction coordination as well as a process to facilitate and support these new design and construction activities through new contractual agreements. Furthermore, as a process, BIM is the system that facilitates owner and facility capital planning before design, as well as after construction through operation activities.

Even though the industry is beginning to adjust to the new agreements and design intent versus interpretation versus means and methods, the adoption by design professionals continues to be limited due to standard of care legal issues.

5.5.3.1 Standard of Care: A Hurdle for Collaboration?

In the United States, courts have used guidelines established by the American Institute of Architects (AIA) in outlining what the prevailing standard of care is for licensed architects. This is codified through Section 2.2 of the AIA Agreements B101 (2007, p.2), where the standard for architectural performance is described as:

> The Architect shall perform its services consistent with the professional skill and care ordinarily provided by architects practicing in the same or similar locality under the same or similar circumstances. The Architect shall perform its services as expeditiously as is consistent with such professional skill and care and the orderly progress of the Project.

Within this short open-ended paragraph, collaboration activities that may be defined as to be outside of the "same or similar locality or circumstances" could be framed negatively, and cooperation with a builder could be seen as "collaborating with the enemy" (as outlined earlier with means and methods) and outside standard of care requirements. Therefore, by enlisting design advice (or review, or assistance) from a nonprofessional designer (builder) team member, the architect risks litigation exposure from a third party's review of the architect's cooperation within the existing local professional market. This is why complicated and expensive projects have benefited first from the innovation available with BIM and collaboration, because the measure of standard of care in these "circumstances" for the design professional may include collaboration.

For example, robust builder and designer collaboration for a hospital is more acceptable than for a single family residence, because most hospital designers have already developed a standard of care based around exploring and delivering project collaboratively. Therefore, the standard of care is measured differently on a larger project than a smaller project.

On the flip side, since the standard of care metrics are ultimately measured against external, outside-of-the-profession perspectives (due to the US Court and jury system), failure to use an innovative tool like BIM could be seen as not following an established standard. In this case, the use of BIM and other collaboration tools are seen to be "reasonable," "ordinary," or "average," and as such "required." The (legal) question posed then is not "what service are you able to provide?" but, "what are other architects (in the same design agreement, and the same location, the same project type and jurisdiction or situation) doing?"

In these projects, collaboration and BIM have become the accepted innovation within the industry, posing risk for nonuse against these new incumbency standards. BIM and collaboration have become the accepted new tool and then the standard within the profession while working on these project types.

However, is this collaboration and use of BIM disruptive, or just an innovation? Reviewing new innovative systems as they are applied to mature incumbent industries is to examine a process which is described as *dematuration*. Dematuration describes the process that a "mature" industry goes through when a number of competitors adopt a host of new innovations that rejuvenate the industry, by giving it the characteristics of a younger more nimble industry. Dematuration was coined in the mid-1980s by Harvard Business School professors William Abernathy and Kim Clark. They examined the US auto industry, which was forcibly undergoing a profound operational renewal. Changes were generated by a number of factors—Japanese competition, the quality movement, and new Lean management techniques. Toyota and Honda, with their new Lean production methods, did not fully disrupt Detroit. They dematured it. Instead of collapsing, the Detroit Three slowly adopted their rivals' tools and techniques, and in turn, the entire (now global) industry advanced to higher levels of quality and customer satisfaction through the process of dematuration. BIM, a tool originally developed to support the aerospace industry (http://en.wikipedia.org/wiki/CATIA), is doing the same thing, applying incremental improvements into design and construction and bringing new innovations to existing processes in a new industry.

John Sviokla, a principal at PwC, describes the opportunities for dematuration if two or more of the following four industry activities happen simultaneously:

1. Customer's core requirements are changing.
2. The core technologies used to deliver the product and service are changing.
3. The number of large competitors interested in the same market is rising.
4. Significant regulation, deregulation, or re-regulation is beginning to occur within the industry.

The design and construction industry is continuously feeling the effects of all of these activities.

The core technologies and processes, contracts, coordination, and drawing methodologies have changed. Large outside competitors are not showing undue interest, however, new consolidated and multidisciplined competitors are making their presence known (especially in the United States with super newly merged design and construction companies like Stantec and AECOM). Through acquisitions, these large organizations now control large markets in both design and construction.

And of course, regulation and re-regulation have always been part of the industry. Dematuration is occurring within the industry, but what about disruption?

5.5.3.2 Dematuration, Disruption, and Standard of Care

Dematuration challenges incumbents through the use of new innovative tools, but is the industry experiencing disruption?

In the 1990s, Clayton M. Christensen, also from Harvard, framed the idea of disruption. These changes in business processes or deliverables disrupted industries by slowly taking away market share of the incumbents. Think music streaming versus MP3s versus CDs versus LPs, or hotels to Airbnb, or taxis to Uber.

Christensen recognized that these disruptions initially gained traction within the markets they served, usually through an "innovation" that initially began by improving a product or service offer within a particular market, and then eventually taking over that market. He further delineated the way the innovation diffused through industries, usually slowly and building acceptance and solving problems at the low ends of markets. These were areas where the incumbents (the established firms serving those affected markets) found it not as profitable (or more risky) than their main products and services.

Furthering this hypothesis, there are three levels of innovative disruption:

A level one disruption is a firm or practice making improvement to an existing product, service, or technology that it already provides. This disruption is an internal innovation, which leads to a new product or service, or a dematured existing process. In architecture, for example, the internal innovation of CAD disrupted hand drawings. This occurred by not formally changing the process of creating drawings, but improving the process of creating, developing, and managing drawings.

A level two disruption is the creation of a new innovation that surprises the market and industry. This disruption wasn't expected or planned for. It may have migrated from another industry, but its application within this new market is a surprise. More importantly, its arrival may generate doubt and scorn, as to whether it can be successful. Initially, incumbents may dismiss it out of hand. In design and construction, BIM is a level two disruption, as it completely replaces CAD as a tool, and builds new markets in the process.

A level three disruption, according to Christensen, is the required survival response to the almost total disappearance of a market. The oil shocks of the 1970s were a level three disruption for the US auto market.

For design and construction, one can see BIM and collaboration and design and construction with components of all three disruption levels. Improvements to existing processes would be the use of BIM in lieu of CAD. The use of collaborative "Big Rooms" where a project is virtually built before any material is procured or installed would be a Level Two disruption—Collocated teams were first established in the auto industry.

A level three disruption occurred with the advent of the recession of 2008 to 2012. The loss of projects and financing shocked the design and construction industry into a new awareness of the fragileness of the profession. Collaboration was forced upon architects as the only way engagement could be maintained in construction projects. Encouraged to utilize the new tools to support the collaboration became the methodology for moving forward in this new environment.

As a design professional in this new collaborative *Architecture 3.0* model, it is important to understand that an architect is not negligent because the design team efforts may have been unsuccessful, or even because errors have occurred within the reasonable delivery of services. Within the confines of a design and construction project, the architect is negligent only if they are deemed not as skillful, knowledgeable, or careful as other architects may have been in similar circumstances (California law).

Therefore, in order to mitigate this potential standard of care risk, the architect must recognize where the new standard may be and develop measures that clearly outline the parameters of standard of care within this new collaborative model. The adoptions of innovations are dematuring the profession.

5.5.3.3 Formalizing Collaboration through the PCAD Model

In enabling the dematuring or disrupting of the industry, the designer still experiences required liability issues. Therefore, it is suggested that the individual design organizations facing change, look at outlining the boundaries of this potentially new standard of care by following what is known as procedures, contract, adapt and document (PCAD) model. This model formalizes the collaborations needs of complex projects.

Procedures—New methodologies need formalized procedures. As identified in this book, all construction team members live and die according to the strength of their procedures, and therefore the subsequent delivery of quality documents and projects. Therefore, creating procedures and processes which recognize and formalize the unique use of collaboration on construction projects, such as the use of BIM execution plans will benefit all parties and help define a probable standard of care. These written and formalized procedures along with strong service agreements will help build understanding and expectations.

Contract—Formalize new collaborative internal procedures with contracts and service agreement that outline the service provided in order to mitigate any misunderstandings, and, again, define standard of care. If processes call for bringing a builder or trade-contractor onboard earlier to review and refine the design documents, then provide the boundaries for that service within formalized agreements that outline defined agreements.

In these agreements, ensure that project goals/objectives/priorities are defined and outlined

- Project assumptions and caveats
- Process and sequencing of project development
- Obligations of each party, individually and collectively, within the process and sequence

Adapt—Within the dynamic and fast changing field of technical supported coordination and collaboration, formal procedures and contracts are only the starting point. Recognize and formalize an understanding with all parties that adaption is required to meet a project's changing demands and commitments. Review and evolve the adaption of tools, processes, and agreements to support those processes regularly throughout the project life cycle.

Document—Standard of care relies on what the owner and team define as the "Standard." So document all the decisions, processes, adaptions, and handoffs in order to show a following of rigorous, if dynamic and changing, standard.

5.5.3.4 Potluck Cooperation

The current, non-BIM-enabled coordination processes in design and construction is a lot like participating in a potluck dinner; where individual participants separately (and off-site) prepare and bring dishes to the event. For example, the designer may separately prepare and bring the meat dish (design drawings) and one of the many trade contractors will separately prepare and bring a salad or drinks, or chips. The guy over there (perhaps the owner) will select the venue (the project site), and provide the chairs and tables and will force all of us to group together and jointly design and build something that matches a particular theme (Italian spaghetti night, or a highway overpass, or a medical office building). All of us participants will try to accomplish this to a forced schedule (maybe provided like a dessert by the contractor) and a forced budget (perhaps provided like the band or music by the project's financial partners).

Recognizing project complexity should help embrace true collaboration; as designers, and constructors and owners, build a single agreed-to meal, at an agreed-to time and place, for an agreed-to price, with established neutral escalation boundaries. This meal ends up with several courses to a final design (highway overpass, medical office building) planned and cooked and consumed together as a unified team. In this process, each member or group brings something additive that matches their expertise and requirement, rather than a fully cooked prebuilt activity or process.

5.6 BIM and Lean

In addition to litigious relationships, the design and construction industry has always suffered from what is described as the *Baumol effect*. Described by William J. Baumol and William G. Bowen in 1965, they described resource-stretched industries unable to improve production. Baumol and Bowen described it as a string quartet playing Mozart today using just as many resources (four people) as the seventeenth century string quartet playing Mozart. The difference today is that it costs more to enjoy that quartet, where other enjoyments cost less because of productivity. Similar to that quartet, most design and construction projects have enjoyed no increase in productivity.

Furthermore, while there has been technology productivity, it has been diminished by other factors. Without further study this may be attributed to workplace safety and environmental factors. Construction was a more dangerous and dirty industry in the past. Now work is accomplished more safely and, as a result, perhaps less quickly.

5.6.1 US Occupational Safety and Health Administration Making a Difference

For example, in four decades, US Occupational Safety and Health Administration (OSHA) and its state partners, coupled with the efforts of employers, safety and health professionals, unions and advocates, have reduced by more than 65% and occupational injury and illness rates have declined by 67%. At the same time, US employment has almost doubled.

Worker deaths in America are down on an average from about 38 worker deaths a day in 1970 to 12 a day in 2012. Worker injuries and illnesses are down from 10.9 incidents per 100 workers in 1972 to 3.4 per 100 in 2011.

5.6.2 Correlation Is Not Causation

As a high labor industry, the consequence of Baumol effect is lots of projects (and lots of jobs) when the economy is good, and no projects (and no jobs) when the economy is bad. Becoming more productive in this environment is difficult, with fluctuations in the economy managed by hiring and firing. Construction projects will always rely on the resources of unskilled labor, and the Baumol effect will limit the amount of labor that can be deployed. Using complexity models recommends decentralization control and separating projects into smaller pieces. BIM becomes the tool to identify and manage the control points, and creating models for integration through virtual building programs.

These preconstruction models comprise complete building systems including architecture, structure, and MEP components are modeled and geometrically aligned 3D space to ensure that the design components of a building are coordinated and correctly fit together prior to actual physical construction.

However, the designer is that Baumol string quartet, limited by resources. Furthermore, architects see construction as a separate and distinctly different activity, which the profession instructs to stay away from, in fear of accidently touching the third rail of means and methods. Therefore, the integration and collaboration requirements necessary to truly make design and construction a unified activity mean changing contracts, identifying and truly sharing risk while recognizing, elevating and maintaining design requirements, and focusing on total life cycle costs of the building (and not just first time design and construction costs).

Architecture 3.0 delineates the topography of this new professional landscape and industry. There will still be projects that follow the tried and tested "design–bid–build" methodologies. There will still be architects who focus and deliver design through contract document and contractors who provide services based on traditional means.

The Baumol effect along with the results of project safety (and dematuration) have created an environment of retrogression within the industry. Complexity modeling helps the project team identify that the project controls are best administered through the distributed team members. As part of its dematuration process, the construction industry has started focusing on Lean methodologies that started within the Japanese automobile industry and have since made their way into other manufacturing-based enterprises. BIM assists in enabling Lean as it provides a robust and controllable platform for transparency and collaboration. Utilized correctly, BIM can become the steward for the building during design and into construction as well as creating robust files for operations once the building is complete.

Using BIM to leverage Lean during design and into construction will facilitate single modeling, robust budgeting and scheduling, and will empower the project's team members to pull the project planning through completion. Lean coordination occurs during The Big Room "Ba" space of knowledge transfer. The project BIM files become the repository of shared project knowledge.

5.7 Future of Collaboration

The challenge with construction projects begins after the completed project is handed over to operations. The building's operations and facility managers are even more legacy-bound than designers and constructors. Even new BIM projects still have project plan storage rooms with operations and maintenance manuals lining bookshelves on the wall. While a number of vendors have attempted to capitalize on this deficiency in operations, a sustainable and scalable solution has not been created. Collaboration should not stop at the end of the construction phase of a project. Continuous commissioning is one way to continue fine-tuning and adapting a builder after construction. But once the act of building a facility is complete, should the model continue to be used, or placed on the shelf?

This is a challenge as the physical representation of the model is much more dynamic (and real) than the model. So until facility applications or building automation systems can seamlessly integrate and interconnect with the model, the opportunities will not be scalable.

6

Construction Contract Documents

Abdul Razzak Rumane

CONTENTS

6.1 Introduction

Construction is translating owner's goals and objectives, by the contractor, to build the facility as stipulated in the contract documents, plans, and specifications within the schedule and budget.

Construction projects involve the following three main participants:

1. Owner
2. Designer/consultant
3. Contractor

In certain cases, owners engage a professional, called construction manager, who is trained in the management of construction processes, to assist in developing bid documents, overseeing, and coordinating project for the owner, or use project manager type of contract when the owner decides to turn over the entire project management to a professional project manager. Participation/involvement of all three parties at different levels of construction phases vary depending upon the type of project delivery system.

Construction contract documents are written documents that constitute a set of contract documents developed to meet the owner's needs, required level of quality, schedule, and budget. Based on the type of project delivery system and type of contracting arrangements, necessary documents are prepared by establishing a framework for execution of project to satisfy owner requirement. It is essential that contract documents are clearly written in simple language that is unambiguous and convenient and is understood by all the concerned parties. These documents mainly consist of the following:

1. Bidding
    - Tendering procedures
    - Bid bond
    - Performance bond
    - Agreement forms
2. Contracting
    - General conditions
    - Particular conditions
3. Construction
    - General specifications
    - Particular specifications
    - Drawings
    - Bill of quantities

- Price analysis schedule
- Forms
- Reports
- Soil test reports
- Survey reports
- Addenda (if any)
- Special technical requirements

There are organizations/agencies producing different types of construction contract formats that are globally used as a guide to prepare construction contract documents. Following are the best-known organizations whose documents are most commonly used in the preparation of construction contract documents for a specific project:

1. The Engineers Joint Contract Documents Committee (EJCDC)—the United States
2. Federation Internationale Des Ingenieurs (FIDIC) (International Federation of Consulting Engineers)
3. MasterFormat®—The Construction Specifications Institute (CSI) and Construction Specifications Canada (CSC)
4. New Engineering Contract (NEC) or NEC Engineering and Construction Contract—the United Kingdom (Institution of Civil Engineers)

6.2 EJCDC

EJCDC is a joint venture of the American Council of Engineering Companies (ACEC), National Society of Professional Engineers/Professional Engineers in Private Practice (NSPE/PEPP), and the American Society of Civil Engineers—Construction Institute (ASCE-CI).

EJCDC has existed since 1975 to develop and update fair and objective standard documents that represent the latest and best thinking in contractual relations between all parties involved in engineering design and construction projects. EJCDC represents a major portion of the professional groups engaged in the practice of providing engineering and construction services for the constructed project.

Each EJCDC contract document is prepared by experienced engineering design and construction professionals, owners, contractors, professional liability and risk management experts, and legal counsel.

Following is the list of EJCDC contract document categories:

1. Joint venture documents
2. Environmental remediation documents
3. Construction-related documents
4. Funding agency documents
5. Procurement documents
6. Design–bid–build documents
7. Guides, commentaries, and references

8. Design–build documents
9. Owner engineer documents
10. Peer-review documents
11. Engineer subconsultant documents

Figure 6.1 illustrates table of contents of *Standard General Conditions of The Construction Contract* published by EJCDC. More information about the EJCDC contract documents is available in www.ejcdc.org.

6.3 FIDIC

Federation of Internationale Des Ingenieurs-Conseils (from French International Federation of Consulting Engineers) *FIDIC* was founded in 1913 by Belgium, France, and Switzerland. FIDIC has now about 78 member associations from all over the world.

The first edition of the *Conditions of Contract (International) for Works of Civil Engineering Construction* was published in August 1957 having been prepared on behalf of FIDIC and the Fédération Internationale des Bâtiment et des Travaux Publics (FIBTP). The form of the early FIDIC contracts followed closely the fourth edition of the *Institute of Civil Engineers (ICE) Conditions of Contract.*

FIDIC publishes international standard forms for contracts, for works, and for clients, consultant, subconsultants, joint ventures, and representatives, together with related materials such as standard prequalification forms.

FIDIC is well known for its international forms of contracts that are used worldwide by the construction industry and also endorsed by multilateral development banks (MDBs).

6.3.1 The FIDIC Suite of Contracts

In 1999, FIDIC published a revised suite of contracts. These books consist of two parts. The first part consist of general principles that should be included in the contract, whereas the second part explains how the employer can prepare the special conditions in accordance with the first part, based on the nature of the work in each construction contract.

The FIDIC Suite of Contracts include the following books:

1. *The Green Book: Short Form of Contract*
2. *The Red Book: Conditions of Contract for Construction for Building and Engineering Works Designed by the Employer* (the construction contract)
3. *The Pink Book: Harmonized Red Book (MDB Edition), Conditions of Contract for Construction for Building and Engineering Works designed by the Owner*
4. *The Yellow Book: Conditions of Contract for Plant and Design–Build*—for Electrical and Mechanical Plant and for Building and Engineering Works, Designed by the Contractor (the plant and design–build contract)
5. *The Orange Book: Conditions of Contract for Design–Build and Turnkey*
6. *The Silver Book: Conditions of Contract for EPC/Turnkey Projects* (the EPC turnkey contract)
7. *The Gold Book: Conditions of Contract for Design, Build and Operate Projects* (the design–build–operate contract)

8. *The Blue Book: Form of Contract for Dredging and Reclamation Works*
9. *The White Book: FIDIC Client/Consultant Model Services Agreement*
10. *The Conditions of Subcontract for Construction: Conditions of Subcontract for Construction for Building and Engineering Works Designed by the Employer*

This document has important legal consequences; consultation with an attorney is encouraged with respect to its use or modification. This document should be adapted to the particular circumstances of the contemplated Project and the controlling Laws and Regulations.

STANDARD GENERAL CONDITIONS
OF THE CONSTRUCTION CONTRACT

Prepared by

ENGINEERS JOINT CONTRACT DOCUMENTS COMMITTEE

and

Issued and Published Jointly by

AMERICAN COUNCIL OF ENGINEERING COMPANIES

ASSOCIATED GENERAL CONTRACTORS OF AMERICA

AMERICAN SOCIETY OF CIVIL ENGINEERS

PROFESSIONAL ENGINEERS IN PRIVATE PRACTICE
A Practice Division of the
NATIONAL SOCIETY OF PROFESSIONAL ENGINEERS

These General Conditions have been prepared for use with the Suggested Forms of Agreement Between Owner and Contractor (EJCDC C-520 or C-525, 2007 Editions). Their provisions are interrelated and a change in one may necessitate a change in the other. Comments concerning their usage are contained in the Narrative Guide to the EJCDC Construction Documents (EJCDC C-001, 2007 Edition). For guidance in the preparation of Supplementary Conditions, see Guide to the Preparation of Supplementary Conditions (EJCDC C-800, 2007 Edition).

EJCDC C-700 Standard General Conditions of the Construction Contract.
Copyright © 2007 National Society of Professional Engineers for EJCDC. All rights reserved.

7/17/07 {MW000852;1}

(a)

FIGURE 6.1
Table of contents. (Standard General Conditions of the Construction Contract (2007), Engineers Joint Contract Documents Committee. Reprinted with permission from Engineers Joint Contract Documents Committee, (EJCDC).) *(Continued)*

STANDARD GENERAL CONDITIONS OF THE CONSTRUCTION CONTRACT

SAMPLE

TABLE OF CONTENTS

EJCDC C-700 Standard General Conditions of the Construction Contract.

Page i

7/17/07 {MW000852;1}

(b)

FIGURE 6.1 (Continued)
Table of contents. (Standard General Conditions of the Construction Contract (2007), Engineers Joint Contract Documents Committee. Reprinted with permission from Engineers Joint Contract Documents Committee, (EJCDC).)

(*Continued*)

SAMPLE

EJCDC C-700 Standard General Conditions of the Construction Contract.

Page ii

7/17/07 {MW000852:1}

(c)

FIGURE 6.1 (Continued)
Table of contents. (Standard General Conditions of the Construction Contract (2007), Engineers Joint Contract Documents Committee. Reprinted with permission from Engineers Joint Contract Documents Committee, (EJCDC).)

(Continued)

SAMPLE

EJCDC C-700 Standard General Conditions of the Construction Contract.

Page iii

7/17/07 {MW000852:1}

(d)

FIGURE 6.1 (Continued)
Table of contents. (Standard General Conditions of the Construction Contract (2007), Engineers Joint Contract Documents Committee. Reprinted with permission from Engineers Joint Contract Documents Committee, (EJCDC).)
(Continued)

SAMPLE

EJCDC C-700 Standard General Conditions of the Construction Contract.

Page iv

7/17/07 {MW000852;1}

(e)

FIGURE 6.1 (Continued)

Table of contents. (Standard General Conditions of the Construction Contract (2007), Engineers Joint Contract Documents Committee. Reprinted with permission from Engineers Joint Contract Documents Committee, (EJCDC).)

6.3.1.1 The Green Book: *Short Form of Contract*

This is FIDIC's recommended form of contract for use on engineering and building works of relatively small capital value or where the construction time is short. FIDIC has provided guidance that this would probably apply to contracts with a value of less than US $500,000 or a construction time of less than six months.

However, the *Green Book* may also be suitable for simple or repetitive work. Although, typically in these types of arrangements, the contractor is to construct the works according to the employer's design, it does not matter whether the design is provided by the employer or by the contractor. There is no engineer, and payments are made in monthly intervals.

The contents of the book are as follows:

- Agreement
- General conditions
- Rules for adjudication
- Notes for guidance

The *Green Book* is a flexible document containing all the essential administrative and commercial arrangements. It is possible to easily amend and supplement the provisions of the *Green Book* with differing options incorporated via the Appendix. The *Green Book* is likely to be the most-suited fairly simple or repetitive work or work of short duration without the need for specialist subcontracts.

6.3.1.2 The Red Book: *Conditions of Contract for Construction for Building and Engineering Works Designed by the Employer (the Construction Contract)*

In 1956, the first edition of the original form of the *Red Book* was published. This was updated and amended over four editions. Then in 1999, FIDIC released a totally new set of standard forms of contract, including a brand new version of the *Red Book*, which superseded the original version. The *Red Book* is the FIDIC recommended form of contract for building or engineering works where the employer has been responsible for nearly all the design. In fact, FIDIC claim that the *Red Book* is the most widely used international construction contract where most of the works have been designed by the employer.

A key feature of the *Red Book* is that payment is made according to bills of quantities (i.e., a document in which materials, plant, and labor (and their costs are itemized), although payment can also be made on the basis of agreed lump sums for items of work.

The *Red Book* is administered by the engineer (a third party) rather than the contractor or the employer. The engineer will also be responsible for monitoring the construction work (but still keeping the employer fully informed, so that he may make variations) and certifying payments.

The contents of the book are as follows:

- General conditions
- Guidance for the preparation of the particular conditions

- Forms of tender and contract agreement
- Dispute adjudication agreement

The *Red Book* provides conditions of contract for construction works where the design is carried out by the employer. The current *Red Book* bears little resemblance to its predecessors. Earlier versions were drafted for use on civil engineering projects. The current edition drops the words *civil engineering* from the title, and this signifies a move away from the *Red Book* only being applicable to civil engineering works.

6.3.1.3 The **Pink Book:** ***MDB Harmonized Edition***

This is a variant of the *Red Book*. It is drafted for use on projects that are funded by certain MDBs, which are supranational institutions such as the World Bank, where the *Red Book* would otherwise have been applicable. Where the project is funded by an MDB but the employer is not responsible for the design, the parties should not use the *Pink Book*, but use and amend one of the other FIDIC forms of contract.

Prior to the publication of the *Red Book*, where the MDBs had originally adopted the *Red Book* for the projects they were funding, they amended the FIDIC General Conditions. As a result of negotiations between FIDIC and the MDBs, the *Pink Book* was drafted, which incorporated the amendments that were commonly inserted by the MDBs.

The contents of the book are as follows:

- General conditions
- Guidance for the preparation of the particular conditions
- Forms of tender and contract agreement
- Dispute adjudication agreement

As part of their standard bidding documents, the MDBs have for a number of years required their borrowers or aid recipients to adopt the FIDIC conditions of contract.

The FIDIC MDB edition of the *Red Book* simplifies the use of the FIDIC contract for the MDBs, their borrowers, and others involved with project procurement, such as consulting engineers, contractors, and contract lawyers.

The following MDBs have all participated in the preparation of this edition of the *Red Book*:

- African Development Bank
- Asian Development Bank
- Black Sea Trade and Development Bank
- Caribbean Development Bank
- European Bank for Reconstruction and Development
- Inter-American Development Bank
- International Bank for Reconstruction and Development (The World Bank)
- Islamic Development Bank
- Nordic Development Fund

6.3.1.4 The Yellow Book: *Conditions of Contract for Plant and Design–Build for Electrical and Mechanical Plant and for Building and Engineering Works, Designed by the Contractor (the Plant and Design–Build Contract)*

This form of contract is drafted for use on projects where the contractor carries out the majority of the design (i.e., the contractor carries out the detailed design of the project, so that it meets the outline or performance specification prepared by the employer). The *Yellow Book* is therefore traditionally used for the provision of plant and for building or engineering works on a design–build basis.

It is a lump-sum price contract with payments made according to achieved milestones on the basis of certification by the engineer (like the *Red Book*, the engineer administers the contract). The contractor is also subject to a fitness-for-purpose obligation in respect of the completed project.

The contents of the book are as follows:

- General conditions
- Guidance for the preparation of the particular conditions
- Forms of tender and contract agreement
- Dispute adjudication agreement

The *Yellow Book* provides conditions of contract for construction works where the design is carried out by the contractor. The current *Yellow Book* bears little resemblance to its predecessors. The current edition drops the words *electrical and mechanical works* from the title, and in line with the rest of the FIDIC suite, the focus is now more on the type of procurement rather than the nature of the works.

6.3.1.5 The Orange Book: *Conditions of Contract for Design–Build and Turnkey*

The contents of the book are as follows:

- General conditions
- Guidance for the preparation of the particular conditions
- Forms of tender and agreement

The *Orange book* was published in 1995 to provide a design and build option to the then-current FIDIC suite. This was the first FIDIC contract to adopt the present FIDIC style of drafting and was a template for the drafting teams when preparing the 1999 suite of contracts.

6.3.1.6 The Silver Book: *Conditions of Contract for EPC/Turnkey Projects (the EPC Turnkey Contract)*

The *Silver Book* is drafted for use on EPC (engineer, procure, and construct) projects. These are projects that require the contractor to provide a completed facility to the employer that is ready to be operated at *the turn of a key*. These contracts therefore place the overall responsibility for the design and construction of the project on the contractor.

The *Silver Book* is used where the certainty of price and completion date is important. The *Silver Book* allows the employer to have greater certainty as to a project's cost as the contractor assumes greater time and cost risks than under the *Yellow Book*.

The *Silver Book* may also be used for privately financed BOT (build, operate, transfer) projects. These are projects where the employer takes total responsibility for the design, construction, maintenance, and operation of a project and wishes to pass the responsibility of construction to the contractor. There is no engineer under the *Silver Book*, as his responsibilities are assumed by the employer. Similar to the *Yellow Book*, the contractor is also subject to a fitness-for-purpose obligation in respect of the completed project.

The contents of the book are as follows:

- General conditions
- Guidance for the preparation of the particular conditions
- Forms of tender and contract agreement
- Dispute adjudication agreement

The *Silver Book* is suitable for use on process, power, and private infrastructure projects where a contractor is to take on full responsibility for the design and execution of a project. Risks for completion within time, cost, and quality are transferred to the contractor and so the *Silver Book* is only suitable for use with experienced contractors familiar with sophisticated risk management techniques.

6.3.1.7 The **Gold Book:** *Conditions of Contract for Design, Build, and Operate Projects (the Design–Build–Operate Contract)*

As per FIDIC, the *Gold Book* is drafted to minimize the risk of rapid deterioration after the handover of a project due to poor design, workmanship, or materials. The *Gold Book* is therefore suitable where long-term operation and maintenance commitment is required along with design and build obligations. The contractor must operate and maintain the completed project on behalf of the employer for a period of typically 20 years from the date of the commissioning certificate, which is issued at completion of construction of the project. During this 20-year period, the contractor must meet certain targets and at the end of this period, the project must be returned to the employer in an agreed condition.

Throughout the 20-year period, the employer owns the plant but the contractor operates it at the contractor's own risk. However, the contractor has no responsibility for financing the project or ensuring its long-term success.

The contents of the book are as follows:

- General conditions
- Particular conditions
- Sample forms

6.3.1.8 The **Blue Book:** *Form of Contract for Dredging and Reclamation Works*

As per FIDIC, the *Blue Book* is suitable for all types of reclamation and dredging works as well as for ancillary construction works.

Typically, under such form of contract, the contractor constructs the works in accordance with the employer's design. However, this form of contract can also be adapted for contracts that include or consist entirely of contractor-designed works.

6.3.1.9 The White Book: *FIDIC Client/Consultant Model Services Agreement*

This form of contract is not for the provision of construction and engineering works, despite it also being referred to by the color of its cover. This form of contract is used to appoint consultants to provide services to the employer such as feasibility studies, design, contract administration, and project management. The *White Book* typically forms the basis of the agreement between the consultant and the employer where the construction and engineering works are being undertaken pursuant to a FIDIC contract.

6.3.1.10 The Conditions of Subcontract for Construction—Conditions of Subcontract for Construction for Building and Engineering Works Designed by the Employer

Unlike the above forms of FIDIC contract, no color is given to this form of contract. This is the FIDIC recommended form of subcontract for use with the *Red*, *Yellow*, *Silver*, and *Gold* books referred to above.

6.3.2 Scope of General Conditions of Contract and Attachments

The first part consists of 72 clauses and 160 subclauses under the title "General Administrative Conditions of Contract." In this section, fundamental principles about how to regulate a construction contract are determined.

6.3.3 Scope of Part Two "Special Administrative Conditions of Contract"

The second part explains as an example of how the employers can make regulations according to required local conditions and work in each separate construction contract, referring to the principles in the first part under the title of "Special Administrative Conditions of Contract and Provisioning Principles."

FIDIC's General Administrative Conditions of Contracts (Part One) and the Special Administrative Conditions of Contracts (Part Two) have an integrity that covers all rights and duties of parties. Therefore, a standard form is not intended for the second part, because the second part has to be prepared separately for each tender. In this context, FIDIC Manual of the Special Administrative Conditions of Contracts is published, providing options for a variety of clauses to help in this regard.

6.4 MasterFormat®

MasterFormat® is the standard for formatting construction specifications that are most often used for commercial building design and construction projects in North America. This standard is widely used for organizing specifications and other written information for commercial and institutional building projects in the United States and Canada. It lists titles and section numbers for organizing data about construction requirements, products, and activities. By standardizing such information, MasterFormat facilitates communication among architects, specifiers, contractors, and suppliers, which helps them meet building owners' requirements, timelines, and budgets. MasterFormat titles and numbers are organized into 2 groups, 5 subgroups, and 50 divisions (2014). MasterFormat® 2014 was consisting of 16 divisions. The term *49 divisions* refers to the 49 divisions of construction information, as defined by the CSI's MasterFormat®. Table 6.1 illustrates MasterFormat titles and numbers as of April 2014.

TABLE 6.1
MasterFormat® 2014

Division Numbers and Titles	
PROCUREMENT AND CONTRACTING REQUIREMENTS GROUP	
Division 00 Procurement and Contracting Requirements	
SPECIFICATIONS GROUP	
General Requirements Subgroup	
Division 01	General Requirements
Facility Construction Subgroup	
Division 02	Existing Conditions
Division 03	Concrete
Division 04	Masonry
Division 05	Metals
Division 06	Wood, Plastics, and Composites
Division 07	Thermal and Moisture Protection
Division 08	Openings
Division 09	Finishes
Division 10	Specialties
Division 11	Equipment
Division 12	Furnishings
Division 13	Special Construction
Division 14	Conveying Equipment
Division 15	Reserved
Division 16	Reserved
Division 17	Reserved
Division 18	Reserved
Division 19	Reserved
Facility Services Subgroup	
Division 20	Reserved
Division 21	Fire Suppression
Division 22	Plumbing
Division 23	Heating, Ventilation, and Air Conditioning
Division 24	Reserved
Division 25	Integrated Automation
Division 26	Electrical
Division 27	Communications
Division 28	Electronic Safety and Security
Division 29	Reserved
Site and Infrastructure Subgroup	
Division 30	Reserved
Division 31	Earthwork
Division 32	Exterior Improvements
Division 33	Utilities
Division 34	Transportation
Division 35	Waterway and Marine Construction
Division 36	Reserved
Division 37	Reserved
Division 38	Reserved
Division 39	Reserved
Process Equipment Subgroup	
Division 40	Process Interconnections
Division 41	Material Processing and Handling Equipment
Division 42	Process Heating Cooling, and Drying Equipment
Division 43	Process Gas and Liquid Handling, Purification, And Storage Equipment
Division 44	Pollution Control Equipment
Division 45	Industry-Specific Manufacturing Equipment
Division 46	Water and Waste Water Equipment
Division 47	Reserved
Division 48	Electric Power Generation
Division 49	Reserved

Source: MasterFormat® (2014), *The Construction Specifications Institute and Construction Specification,* Canada. Reprinted with permission from Construction Specification Institute.

6.5 NEC

In September 1985, the Institution of Civil Engineers (ICE) asked its Legal Affairs Committee to review contract forms used at that time. Owing to the problems and issues stated in the previous section, the committee recommended the development of a new family of contracts. The first step was to draft key principles of the contract and its functions. This work was done under the supervision of Dr. Martin Barnes, a leading construction project manager, founder and chairman of the United Kingdom's Association of Project Management and fellow member of the ICE and with the assistance of Professor John Perry, the then-head of Civil Engineering Department at the University of Birmingham.

First edition

The work of specifying, designing, and drafting the new contract was sponsored by the ICE. It was carried out by a small team of people drawn from different backgrounds and with expert legal contributions. The result was the NEC. After full consultation and trials, it was first published for general use in 1993. Since then, it has been taken up widely by the industry and its clients and has achieved considerable success.

Second edition

In July 1994, Sir Michael Latham published a report called "Constructing the team," known as the *Latham report*. In this report, he identified 13 principles that should be included in a contract to be effective in modern conditions. These principles are listed below.

- A duty of all stakeholders to act fairly
- Duties of teamwork to achieve *win–win* solutions to problems
- An interrelated package of documents
- A comprehensive language and guidance notes included
- The separation of roles, especially the roles related to the project manager and the adjudicator
- A choice of risk allocation
- Variations priced in advance
- Activity schedules or payment schedules
- Clear payment periods including interests if there are delays
- Sure trust fund for payment
- Quick adjudication of disputes
- Incentives for high level of performance
- Mobilization payments

In response to these recommendations together with feedback from early users, a second edition of the NEC was published in November 1995 and the NEC was renamed as *The Engineering and Construction Contract.*

Third edition

In July 2005, the third edition of the contract was issued. *The Engineering and Construction Contract* has been widely revised and updated. As a result, all other contracts have been updated as well, so that NEC3 with its integrated set of documents remain consistent. Two entirely new contracts were introduced. These are the *Term Service Contract* and *The Framework Contract* (NEC one stop shop, 2007b as cited in Genessay, 2007) joined in 2010 by a supply contract.

April 2013 edition (Latest)

The suite was updated and enlarged to 39 documents in April 2013, including a Professional Services Short Contract—the Association for Project Management's standard form for appointing project managers—and an enhanced set of guidance documents (https://www.neccontract.com/About-NEC/History).

6.5.1 The NEC Family of Contracts

The NEC is the only standard contract that is based on the best modern practice in project management. It stimulates forward-looking control of cost, time, and performance. Uniquely, and for the first time, it provides for effective management and control of risk. Much of the NEC is *general purpose* and can be used in all situations. The flexibility of the NEC is achieved through choosing the right contract from the NEC family of contracts and choosing the right options for clauses within the contract. The NEC family is made up of seven standard contracts. They all use the same basic set of procedures, the same management culture, and the same terminology.

NEC family of contracts includes the following (NEC3, 2005):

- NEC3 Engineering and Construction Contract (ECC, June 2005)
- NEC3 Engineering and Construction Subcontract (ECS, June 2005)
- NEC3 Professional Services Contract (PSC, June 2005)
- NEC3 Adjudicator's Contract (AC, June 2005)
- NEC3 Engineering and Construction Short Contract (ECSC, June 2005)
- NEC3 Engineering and Construction Short Subcontract (ECSS, June 2005)
- NEC3 Term Service Contract (TSC, June 2005)
- NEC3 Framework Contract (FC, June 2005)

All types of contracts are usually completed with guidance notes and flowcharts, following the recommendations of Latham report. This family is growing and other NEC forms are constantly in preparation or under consideration. Within the NEC, all contract documents share the same format, so that the user can get more rapidly familiar with its content (ECC3 Guidance Notes, 2005). Thus, each contract is divided into three main sections: nine core clauses that are identical for each contract, main and secondary options that are different for each contract.

6.5.2 The Nine Core Clauses

The nine core clauses include the following (ECC3 Guidance Notes)

1. *General*: The general core clauses deal with definitions, interpretation, ambiguities, and general introductory matters.
2. *The contractor's main responsibilities*: This section focuses on the contractor's responsibilities, the provision of works and the design, requirements concerning the equipment, and the subcontracting process.

3. *Time*: This section looks at one of the key characteristics of NEC, which is the accepted program.
4. *Testing and defects*: This part covers important issues such as testing and inspection, correction of failures, and acceptance of defects.
5. *Payment*: This section details the assessment of payments due to the contractor by the project manager.
6. *Compensation events*: This section defines compensation events and procedures to manage them.
7. *Title*: This section focuses on the employer's entitlement to plant and materials, together with the removal of equipment and materials within the site.
8. *Risks and insurance*: This section deals with the employer's and contractor's risks as well as the insurance requirements for both parties.
9. *Termination*: This section deals with reasons and procedures for termination, including the payments involved.

As far as dispute resolution is concerned, following the recommendations of the Latham report (1994), the NEC requires all disputes to be referred to an independent adjudicator prior to arbitration or litigation.

6.5.3 The Structure of ECC3

The structure of ECC is given below.

6.5.3.1 Core Documents

The ECC is published in 10 volumes (Genessay, 2007). The core documents are listed below.

- NEC3 Engineering and Construction Contract
- NEC3 Engineering and Construction Contract Guidance Notes
- NEC3 Engineering and Construction Contract Flowcharts
- NEC3 Engineering and Construction Subcontract

6.5.3.2 Main Option Clauses

The ECC has six main options based on different payment mechanisms, which offer different basic allocations of risk between the employer and the contractor and reflect modern procurement practice. All the main options can be used with boundary between design by the employer and design by the contractor to set the chosen strategy (ECC3 Guidance Notes, 2005). The six main options are listed below.

- Option A (priced contract with activity schedule)
- Option B (priced contract with bill of quantities)
- Option C (target contract with activity schedule)
- Option D (target contract with bill of quantities)

- Option E (cost-reimbursable contract)
- Option F (management contract)

6.5.3.2.1 *Option A (Priced Contract with Activity Schedule)*

An activity schedule is a list of activities prepared by the contractor, which he expects to carry out in providing the works. When it has been priced by the contractor, the lump sum for each activity is the price to be paid by the employer for that activity. The total of these prices is the contractor's price for providing the whole of the works, including for all matters that are at the contractor's risk. Option A provides for stage payments.

6.5.3.2.2 *Option B (Priced Contract with Bill of Quantities)*

A bill of quantities comprises a list of work items and quantities. It is prepared by contractor for the employer. Standard methods of measurement are published, which state the items to be included and how the quantities are to be measured and calculated.

Tenderers price the items, taking account of the information in the tender documents, including all matters that are at the contractor's risk. The employer pays for the work done on the basis of actual measurement of those items with quantities.

6.5.3.2.3 *Options C and D: Target Contracts (with Activity Schedule or Bill of Quantities)*

Target contracts are sometimes used where the extent of work to be done is not fully defined or where anticipated risks are greater. The financial risk is shared between the employer and the contractor in the following ways:

- The contractor tenders a target price in the form of the prices using either the activity schedule or a bill of quantities. The target price includes the contractor's estimate of the defined cost plus other costs, overheads, and profit to be covered by his fee.
- The contractor tenders his fee in terms of fee percentages to be applied to the defined cost.
- During the course of the contract, the contractor is paid a defined cost plus the fee. This is defined as the price for work done to date (PWDD).

The prices are adjusted for the effects of compensation events, and for inflation if option X1 is used; for option D, the prices are also adjusted as the work completed by the contractor is measured.

At the end of the contract, the contractor is paid (or pays) his share of the difference between the final total of the prices and the final PWDD according to a formula stated in the contract data. If the final PWDD is greater than the final total of the prices, the contractor pays his share of the difference. The contractor's share is paid provisionally at completion and is corrected in the final account.

6.5.3.2.4 *Option E (Cost-Reimbursable Contract)*

A cost-reimbursable contract should be used when the definition of the work to be done is inadequate even as a basis for a target price and yet an early start to construction is required. In such circumstances, the contractor cannot be expected to take cost risks other than those that entail control of his employees and other resources. He carries minimum risk and is paid a defined cost plus his tendered fee, subject only to a small number of constraints designed to motivate efficient working.

6.5.3.2.5 Option F (Management Contract)

The conditions of contract applied to management contracts are still evolving. In practice, there are several different approaches used in relation to, for example, scope of services, time of appointment, and methods of fee payment. The terms under which subcontractors are employed are also changing. The ECC management contract is based on the following framework.

The contractor's responsibilities for construction work are the same as those of a contractor working under one of the other options. However, he does only the limited amount of construction work himself as stated in the contract data. The project manager has no authority to instruct the contractor to carry out further construction work beyond that stated in the contract data. Any increase in the extent of the construction work to be carried out by the contractor must be the subject of negotiation between the employer and the contractor.

The contractor's services apply mainly to the construction phase, although he would usually be appointed before the construction starts. If substantial preconstruction services are required and the employer wishes to have the option to change the management contractor before construction starts, a separate contract should be awarded for such preconstruction services, using the NEC Professional Services Contract.

All subcontracts are direct contracts with the contractor, who acts as a management contractor. If the employer wishes to be a party to the construction subcontracts, a management contract is not appropriate. He should instead appoint a construction manager as the project manager and use the ECC with appropriate main options for the contracts with package contractors.

The contractor tenders his fee and his estimated total of the prices of the subcontracts. The subcontract prices are paid to the contractor as part of the defined cost. The contractor is responsible for supplying management services, including the management of design, if required. If the contractor wishes to be responsible for doing work other than management contracts, he must state the extent of that work in the contract data.

The contractor's fee will increase if subcontractors' prices (part of the defined cost to the contractor) increase due to compensation events. However, he will not receive separate payment for his work in dealing with compensation events and he will not receive any additional fee for work on compensation events, which does not lead to an increase in subcontractors' prices.

6.5.3.2.6 Secondary Option Clauses

Once the employer has selected the main option, he can select the available secondary options to refine the contract strategy (NEC one stop shop, 2007 cited in Genessay, 2007). The chosen main option clause and the secondary clauses have to appear on the first statement to be a part of the contract data (ECC3 Guidance Notes, 2005). The secondary options are listed below.

- Option X1—price adjustment for inflation
- Option X2—changes in the law
- Option X3—multiple currencies
- Option X4—parent company guarantee
- Option X5—sectional option
- Option X6—bonus for early completion

- Option X7—delay damages
- Option X12—partnering
- Option X13—performance bond
- Option X14—advance payment to contractor
- Option X15—limitation of contractor's liability
- Option X16—retention
- Option X17—low performance damages
- Option X18—limitation of liability
- Option X20—key performance indicator
- Option Y(UK)2—Housing Grants Act
- Option Y(UK)3—Rights of Third Parties Act
- Option Z—additional conditions

6.5.3.2.7 Dispute Resolution Options

There are two options for dispute resolution under ECC3, which are as follows:

1. *Option W1*—not HGCRA (Housing Grants, Construction and Regeneration Act, 1996) compliant
2. *Option W2*—for UK construction contracts with HGCRA compliance

6.5.4 NEC April 2013 Edition

The latest edition of the NEC3 contains 39 documents, including the new Professional Services Short Contract, which has been developed in partnership with the Association for Project Management and a series of seven new *how to* books designed to aid the use of these contracts. In addition, the latest Construction Act amendments, Project Bank Account provisions and clauses, as well as CIC BIM protocol references are included.

6.5.4.1 NEC Suite of Contracts, April 2013 Edition

1. NEC3 Engineering and Construction Contract (ECC)
2. NEC3 Engineering and Construction Contract Option A: Priced contract with activity schedule
3. NEC3 Engineering and Construction Contract Option B: Priced contract with bill of quantities
4. NEC3 Engineering and Construction Contract Option C: Target contract with activity schedule
5. NEC3 Engineering and Construction Contract Option D: Target contract with bill of quantities
6. NEC3 Engineering and Construction Contract Option E: Cost reimbursable contract
7. NEC3 Engineering and Construction Contract Option F: Management contract
8. NEC3 Engineering and Construction Subcontract (ECS)

9. NEC3 Engineering and Construction Short Contract (ECSC)
10. NEC3 Engineering and Construction Short Subcontract (ECSS)
11. NEC3 Professional Services Contract (PSC)
12. NEC3 Professional Services Short Contract (PSSC)
13. NEC3 Term Service Contract (TSC)
14. NEC3 Term Service Short Contract (TSSC)
15. NEC3 Supply Contract (SC)
16. NEC3 Supply Short Contract (SSC)
17. NEC3 Framework Contract (FC)
18. NEC3 Adjudicator's Contract (AC)

6.5.4.2 Guidance Notes and Flowcharts

1. NEC3 Engineering and Construction Contract Guidance Notes
2. NEC3 Engineering and Construction Contract Flowcharts
3. NEC3 Engineering and Construction Short Contract Guidance Notes and Flowcharts
4. NEC3 Professional Services Contract Guidance Notes and Flowcharts
5. NEC3 Term Service Contract Guidance Notes
6. NEC3 Term Service Contract Flowcharts
7. NEC3 Term Service Short Contract Guidance Notes and Flowcharts
8. NEC3 Supply Contract Guidance Notes
9. NEC3 Supply Contract Flowcharts
10. NEC3 Supply Short Contract Guidance Notes and Flowcharts
11. NEC3 Framework Contract Guidance Notes and Flowchart
12. NEC3 Adjudicator's Contract Guidance Notes and Flowcharts
13. NEC3 Procurement and Contract Strategies
14. NEC3: How to Write the ECC Works Information
15. NEC3: How to Use the ECC Communication Forms
16. NEC3: How to Write the PSC Scope
17. NEC3: How to Use the PSC Communication Forms
18. NEC3: How to Write the TSC Service Information
19. NEC3: How to Use the TSC Communication Forms
20. NEC3: How to Use BIM with NEC3 Contracts

6.5.5 Implementation of the NEC

6.5.5.1 Bodies Responsible for NEC's Implementation

This section deals with the bodies responsible for the implementation of NEC in the United Kingdom and overseas. From the literature, four major bodies are listed below (Genessay, 2007):

1. *ICE*—As a qualifying body, a network for sharing knowledge, a supplier of resources, the worldwide recognized ICE is an ideal support to the implementation of NEC that it has launched back in the 1990s. Through the NEC, the ICE has been encouraging innovation and excellence in the profession (http://www.ice.org.uk/About-ICE, 2010).
2. *Thomas Telford Limited*—Thomas Telford Limited is the business subsidiary of the ICE. It is the administrative support for the NEC through its NEC division that manages the training, talks, workshops, seminars, publications, and sales of the NEC family of contracts (http://www.thomastelford.com/, 2010).
3. *NEC Users' Group*—The NEC Users' Group is composed of more than 200 member organizations representing clients, consultants, designers, project managers, and contractors. Its aim is to stimulate members to stay actively involving themselves in the ongoing development and promotion of the ECC by sharing ideas and experiences (http://www.neccontract.com/users_group/index.asp, 2010).
4. *NEC panel*—The NEC panel has replaced the earliest NEC Working Group. It is composed of senior professionals of the construction industry who meet on a regular basis; the NEC panel is the ICE's official consultative board for the NEC (http://www.neccontract.com/news/article.asp?NEWS_ID=571, 2006).

7

Construction Management

Abdul Razzak Rumane
(Section 7.3.11 by Edward Taylor, and Jitu C. Patel)

CONTENTS

7.1 Introduction

Construction management process is a systematic approach to manage construction project from its inception to its completion and handover to the client/end user. Construction management is application of professional services, skills, and effective tools and techniques to manage project planning, design, and construction from project inception through to issuance of completion certificate. Some of these techniques are tailored to the specific requirements unique to construction projects.

The main objective of construction management is to ensure that the client/end user is satisfied with the quality of project delivery. In order to achieve project performance goals and objectives, it is required to set performance measures that define what the contractor is to achieve under the contract. Therefore, to achieve the adequacy of client brief, which addresses the numerous complex needs of client/end user, it is necessary to evaluate the requirements in terms of activities and their functional relationships and establish construction management procedures and practices to be implemented and followed toward all the work areas of the project to make the project successful to the satisfaction of the owner/end user and to meet owner needs.

Construction management involves project execution-related management works managed by a professional firm or a person called construction manager. For successful management of the project, it is essential to have professional knowledge of management functions, management processes, and project phases (technical processes). The construction project quality is the fulfillment of owner's needs per defined scope of works within a budget and specified schedule to satisfy the owner's or user's requirements. There are mainly three key attributes in construction projects, which the construction/project manager has to manage effectively and efficiently to achieve successful project. These are as follows:

1. Scope
2. Time (schedule)
3. Cost (budget)

These are known as *quality trilogy* or *triple constraints*. From the project quality perspective, the phenomenon of these three components is called the *construction project quality trilogy* and is illustrated in Figure 7.1.

This phenomenon when considered with project management/construction management perspective is known as *triple constraints*. Scope (defined scope), schedule (time), and cost (budget) are three sides of a triangle. Figure 7.2 illustrates the principle of triple constraints.

Triple constraints is a framework for the construction manager or project manager to evaluate and balance these competing demands. It became a way to track and monitor

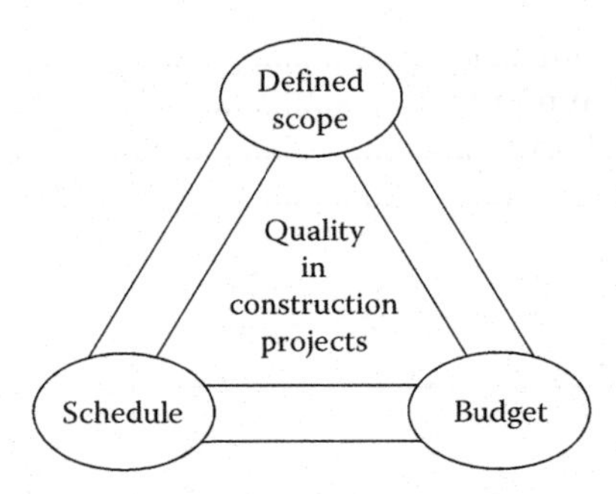

FIGURE 7.1
Construction project quality trilogy.

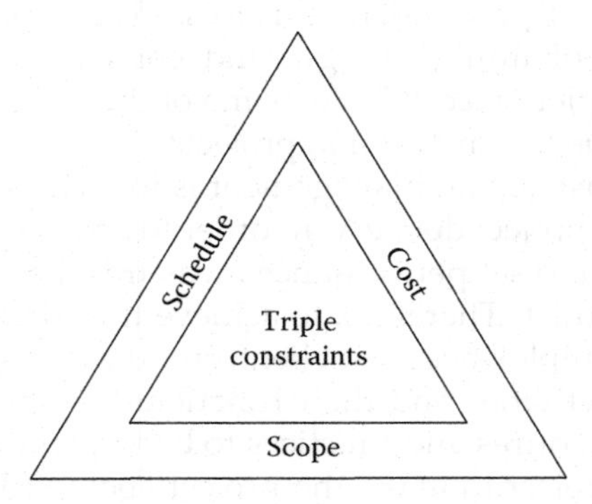

FIGURE 7.2
Triple constraints.

projects. In pictorial form, triple constraints is a triangle in which one cannot adjust or alter one side of it without any effect or altering the other side(s).

- If scope is increased, the cost will increase or the time must be extended, or both.
- It time is reduced, then cost must increase or scope must decrease, or both.
- If cost is reduced, then the scope must be decreased, or the time must increase, or both.

In order to achieve a successful project, the construction/project manager must handle these key attributes effectively and efficiently, and track the progress of the work from the inception to completion of construction and handover of the project for successful completion. These three attributes have functional relationship with many other processes, activities, and elements/subsystems of the project. To achieve a successful project to the satisfaction of owner/end user, the construction/project manager has to manage the project in a systematic manner at every stage of the project and balance these attributes in conjunction with all the other activities that may affect the successful completion of the project. This can be done by implementing, amalgamating, and coordinating some or all the activities/elements of management functions, management processes, and project life cycle phases (technical processes). Thus, construction management process can be described as implementation and interaction of following functions and processes:

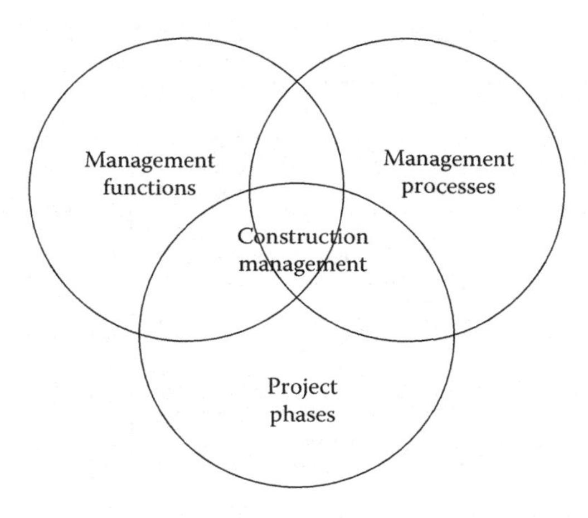

FIGURE 7.3
Construction management integration.

- Management functions
- Management processes
- Project phases (technical processes)

Figure 7.3 illustrates construction management process elements.

In practice, it is difficult to separate one element from others while executing a project. Interaction and/or combination among some or all of the elements/activities of these processes and their effective implementation and applications are essential throughout the life cycle of the project to conveniently manage the project. Figure 7.4 illustrates the integration diagram of components/activities of three major elements of the construction management process.

7.2 Management Functions

Management is a systematic way of managing processes in an efficient and effective manner toward accomplishment of organizational goals. It is a systematic application of knowledge derived from general principles, concepts, theories, and techniques, and embodied in the management functions that are variables in terms of business practices as per organizational needs and requirements. Regardless of the type and scope of processes, there must be a plan that is organized, implemented, controlled, and maintained. Henri Fayol (1841–1925), a French engineer, was one of the most influential contributors to the modern concept of management. He was the first person to identify and describe the elements or functions of management. He outlined the following four managerial functions that are the foundation for all management concepts:

1. Planning
2. Organizing
3. Leading
4. Controlling

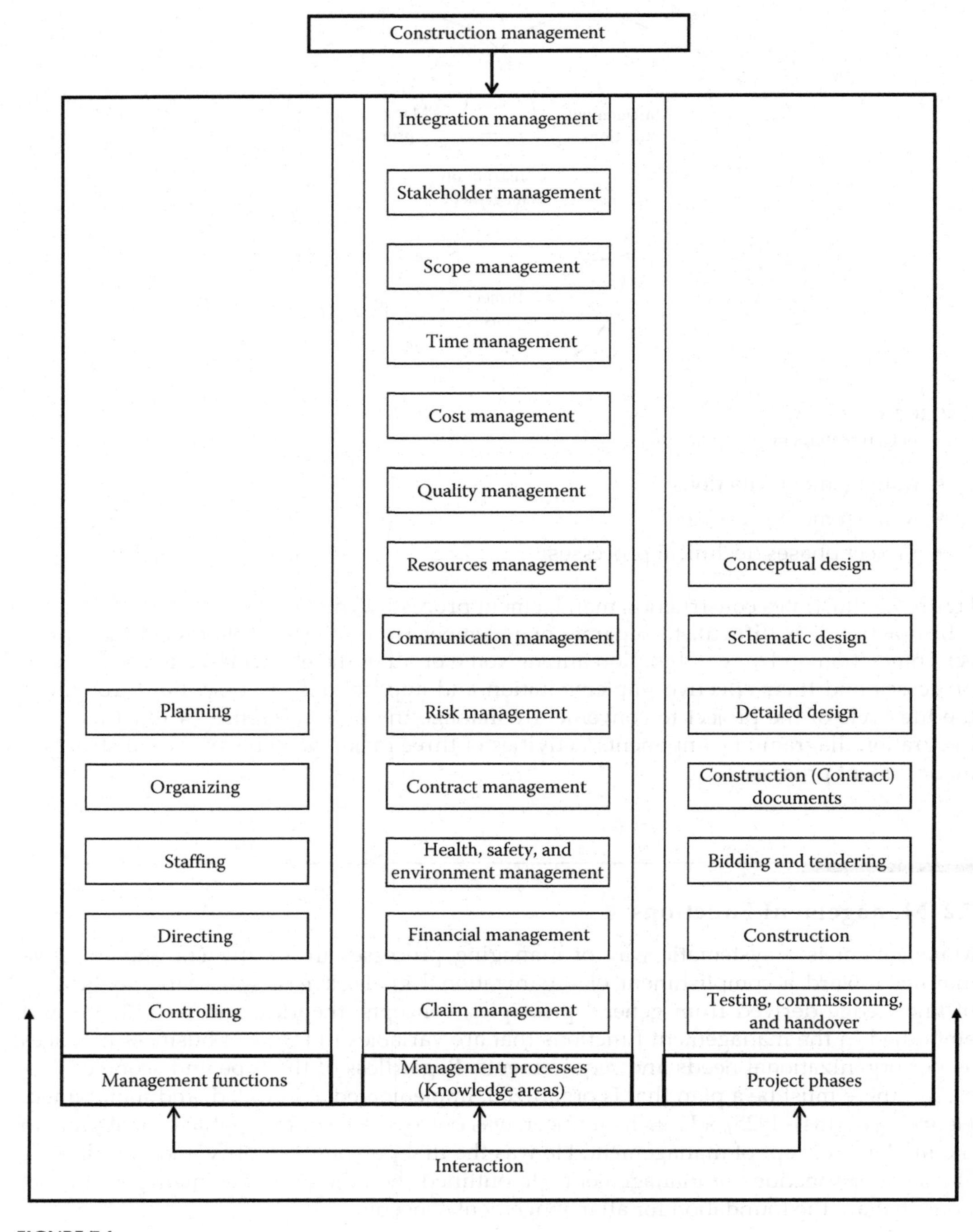

FIGURE 7.4
Construction management process elements integration diagram.

These functions are performed by managers at all the levels (top management, middle management, and first-line managers) in an organization.

Fayol's original concept of four functions was subsequently modified as follows:

1. Planning
2. Organizing
3. Commanding
4. Coordinating
5. Controlling

Luther Gulick gave a new formula to suggest the elements of management processes. These are as follows:

1. Planning
2. Organizing
3. Directing
4. Staffing
5. Coordinating
6. Reporting
7. Budgeting

These functions were modified according to the management perspective of the organizational business need. The five functions of management as explained by Koontz and O'Donnell are as follows:

1. Planning
2. Organizing
3. Staffing
4. Directing
5. Controlling

These functions or elements are briefly discussed in the following sections.

7.2.1 Planning

Planning is foreseeing future circumstances and requirements. It is deciding what is to be done in future. Planning is establishing objectives (what to do) and making long- and short-term plans (when to do). The following are the four major types of planning:

1. *Strategic planning*: Strategic planning is the process of defining and prioritizing long-term plans that include developing and analyzing the organization's mission, overall goals, general strategies, and allocation of the resources. SWOT (strength, weakness, opportunities, and threats) analysis is performed during strategic planning.
2. *Operational planning*: It is short-term planning that deals with day-to-day operations.

3. *Intermediate planning*: It is also known as mid-term planning. It is prepared based on the existing performance and is revised for the rest of the period of plan.
4. *Contingency planning*: It is developed to manage the problems that interfere with getting the work done.

7.2.1.1 Steps in Planning

The following are the steps to be considered while planning:

1. Defining the organization's mission and vision
2. Performing SWOT analysis
3. Setting goals and objectives
4. Decide which task to be done to reach these goals
5. Decide what course of action to adopt
6. Define parameters to be measured
7. Decide allocation of resources
8. Define target values
9. Update

Table 7.1 illustrates an example SWOT analysis to establish construction material testing laboratory.

TABLE 7.1

SWOT Analysis for Construction Material Testing Laboratory

Strength	Opportunities

• Advanced and modern test set up to carry out related tests and analysis
• Temperature controlled test facility
• Only laboratory in private sector offering testing and calibration facilities in
  • Soil testing
  • Aggregates
  • Concrete materials
  • Concrete blocks
  • Reinforcement material
  • Asphalt
  • Hydrocarbon
  • Calibration of instruments, meters, and gauges
  • Calibration of compression/testing machines
  • Carbon emission tests
  • Steel/metal testing
  • Weld test
  • Fabrication
  • Corrosion test
	• There is no such laboratory in local market
	• Wide range of testing facility
	• The is no local laboratory/facility to test and analyze material related to oil and gas industry products.
	• Private sector laboratories are not equipped to carry out all types of tests for the construction industry and the oil and gas sector
	• Many new construction projects in the offing
	• Many new projects in the oil sector.
	• Government sector laboratories are fully loaded with testing of materials for government projects and there is delay to get test results.
	• Lack of sophisticated calibration facility in local market.
	• Internationally accredited and recognized

Weakness	**Threat**
• Competition with government laboratories • Government projects are less likely to be referred to private sector • Products shipped to developed countries	• Delay in getting approvals • Delay in registering with local authorities • Delay in obtaining accreditation and certification • Similar laboratory set up by existing businesses

7.2.1.2 Construction Project Planning

Project planning is a logical process to ensure that the work of project is carried out:

- In an organized and structured manner
- By reducing uncertainties to minimum
- By reducing risk to minimum
- By establishing quality standards
- By achieving results within budget and scheduled time

Project planning is the heart of good project management because it provides the central communication that coordinates the work of all parties. Planning also establishes the benchmark for the project control system to track the quantity, cost, and timing of work required to successfully complete the project.

Project planning determines how the project will be accomplished. It is the process of identifying all the activities to successfully complete the project. Project planning is the key to a successful project. Table 7.2 lists basic reasons for planning the project.

Project planning is a discipline for stating how to complete a project within a certain timeframe, usually with defined stages and designated resources. Planning describes what needs to be done, when, by whom, and to what standards. The following are the steps to prepare a project plan:

1. The first step of project planning is to clearly define the problem to be solved by the project.
2. Once the problem is clearly defined, the next step is to define the project objectives or goals. Establishing properly defined objectives and goals is the most fundamental elements of project planning. Therefore, the goals/objectives must be
   - Specific
   - Measurable

TABLE 7.2

Reasons for Planning

1.	To execute work in an organized and structured manner
2.	To eliminate or reduce uncertainty
3.	To reduce risk to the minimum
4.	To reduce rework
5.	To improve the efficiency of the process
6.	To establish quality standards
7.	To provide basis for monitoring and control of project work
8.	To know duration of each activity
9.	To know cost associated with each activity
10.	To establish benchmark for tracking the quantity, cost, and timing of work required to complete the project
11.	To know responsibility and authority of people involved in the project
12.	To establish timely reporting system
13.	Integration of project activities for smooth flow of project work

Source: Abdul Razzak Rumane (2010), *Quality Management in Construction Projects*, CRC Press, Boca Raton, FL. Reprinted with permission from Taylor & Francis Group.

- Agreed upon/achievable
- Realistic
- Time (cost) limited

3. The next step in project planning is to identify project deliverables.
4. After identifying project deliverables, these are subdivided into smaller activities to enable developing schedule and cost estimates.
5. The next step is to estimate activity resources, activity duration, and develop schedule.
6. The next step is to estimate cost and develop budget based on project deliverables and schedule.
7. Supporting plans such as quality management, human resources, communication, risk management, and procurement/supply management are simultaneously developed.
8. The next step is to compare the plan for compliance with the original project objectives.
9. The plan is updated, if required, to meet original objectives/goals, keeping in mind three principal aspects of construction projects, that is, scope, time, and cost. In order to achieve a successful project, these three aspects are required to be in balance.
10. The project is launched based on this plan.

Figure 7.5 illustrates construction project planning steps.

Construction management plan is a process to define, establish, and document key management activities to manage project execution by providing baseline for scope, time, and cost to meet the project objectives. Figure 7.6 illustrates major elements in the construction project development process.

Effective project management requires planning, measuring, evaluating, forecasting, and controlling all aspects of a project quality and quality of work, cost, and schedules. The purpose of the project plan is to successfully control the project to ensure completion within the budget and schedule constraints.

Planning is a mechanism that conveys or communicates to project participants what activity to be done, and how and in what order to meet the project objectives by scheduling the same. Project planning is required to bring the project to completion on schedule, within budget, and in accordance with the owner's needs as specified in the contract. The planning process considers all the individual tasks, activities, or jobs that make up the project and must be performed. It takes into account all the resources available, such as human resources, finances, materials, plant, and equipment. It also considers the works to be executed by the subcontractors.

7.2.2 Organizing

This function relates to identification and classification of the required activities to be performed to achieve the objectives. The activities mainly include the following:

- Developing an organizational structure
- Division of labor

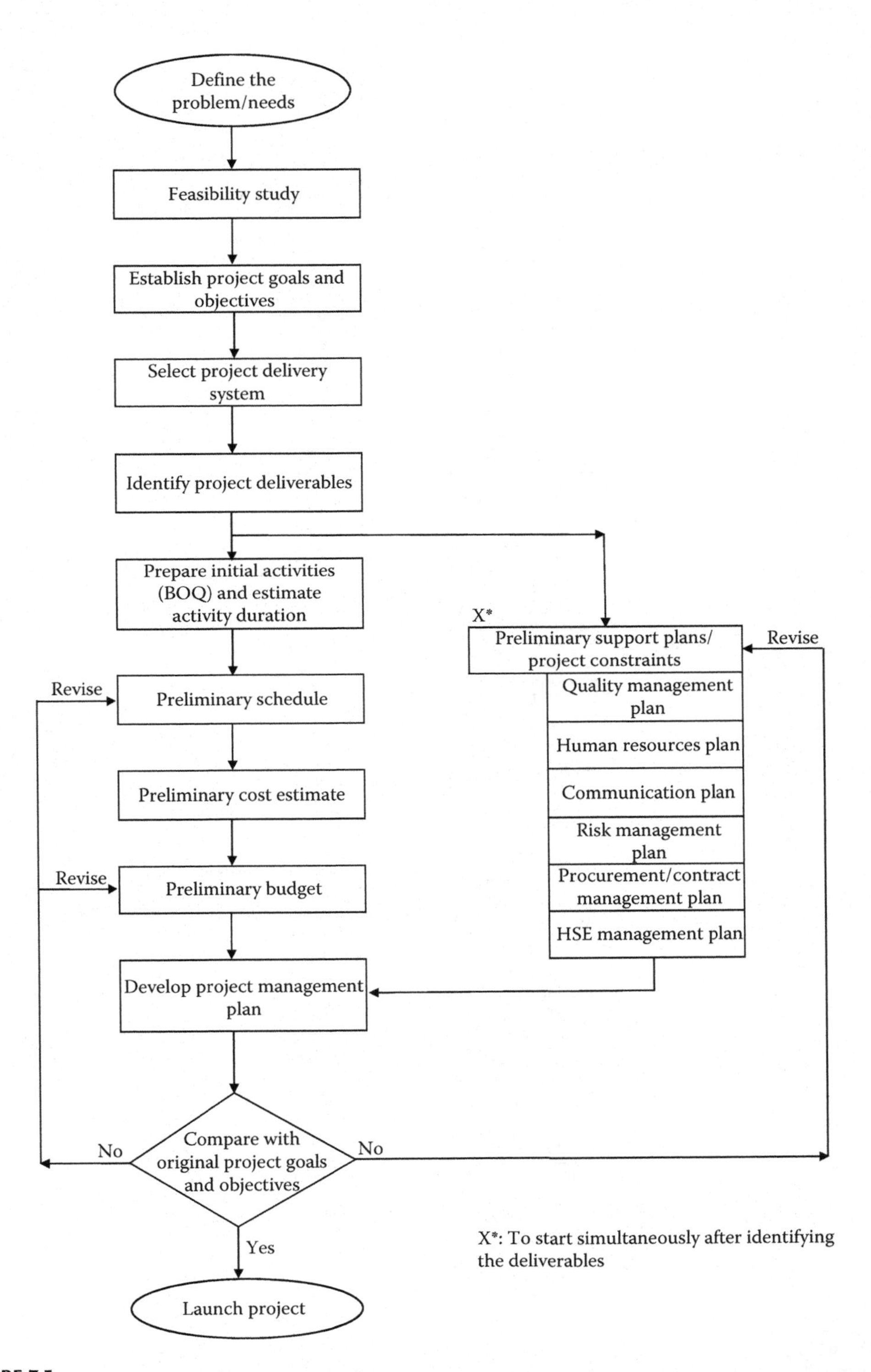

FIGURE 7.5
Construction project planning steps. (Abdul Razzak Rumane (2013), *Quality Tools for Managing Construction Projects*, CRC Press, Boca Raton, FL. Reprinted with permission from Taylor & Francis Group.)

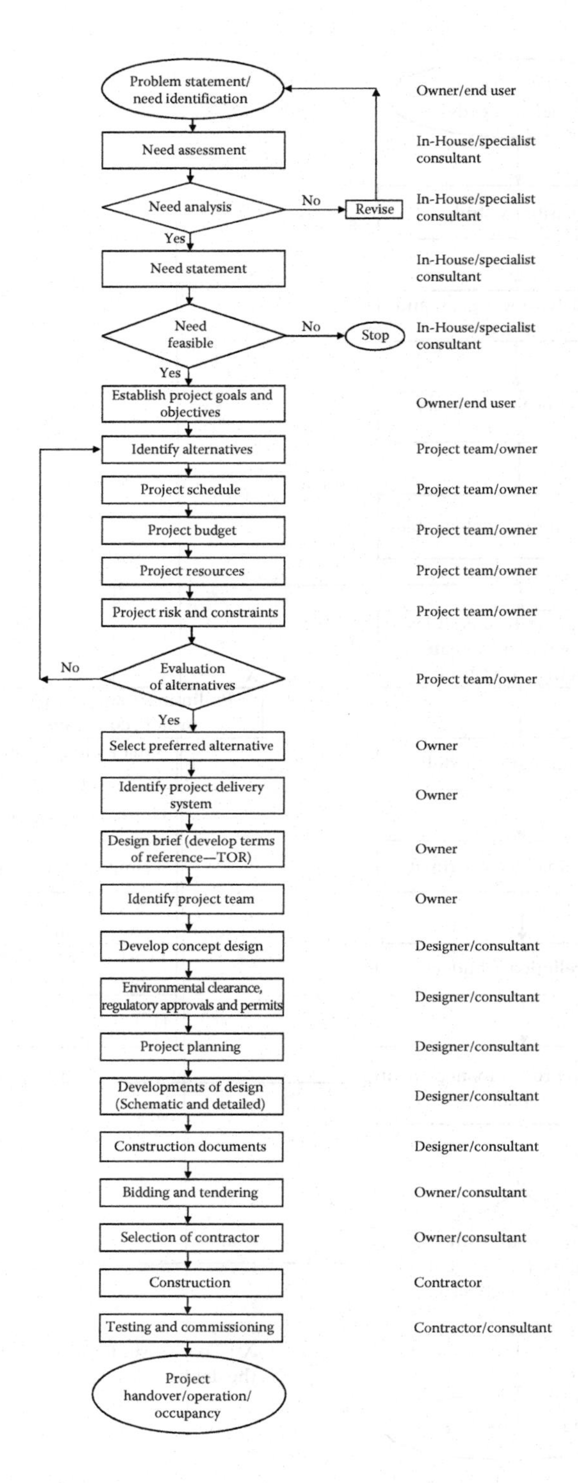

FIGURE 7.6
Major elements in construction project development process (design–bid–build system). (Abdul Razzak Rumane (2013), *Quality Tools for Managing Construction Projects*, CRC Press, Boca Raton, FL. Reprinted with permission from Taylor & Francis Group.)

- Delegation of authority
- Departmentalization
- Span of control
- Coordination

These activities may be performed by two or more people who work together in a structural way to achieve a specific goal or a set of goals. Organizing means assigning the planned tasks to various individuals or group within the organization and creating a mechanism to put plans into action. The design of organizational structure is determined mainly by the following three elements:

1. Complexity
2. Formalization
3. Centralization

These three aspects can be combined to create many different organizational designs. An organization consists of a group of people working together in a structured way to reach goals that individuals acting alone could not achieve. One of the most important factors to design organizational structure is span of control. It refers to the number of departments/managers/staff one can effectively supervise. The ideal number for span of control depends on nature of job, time required to supervise each of them, and to maintain close control.

7.2.2.1 Types of Organizational Structures

There are seven types of organizational structures and they have been briefly explained:

1. *Simple*: This is an organization with a simple structure, having the centralized authority vested with a single person. This type of structure is common in small start-up businesses. When the business expands and more employees are added, then the structure becomes complex. Figure 7.7 illustrates an example of a simple organizational structure.
2. *Functional*: In this type of organizational structure, people with similar occupational specialties are put together in a formal group. Figures 7.8 and 7.9 are examples of different types of functional organizational structure.

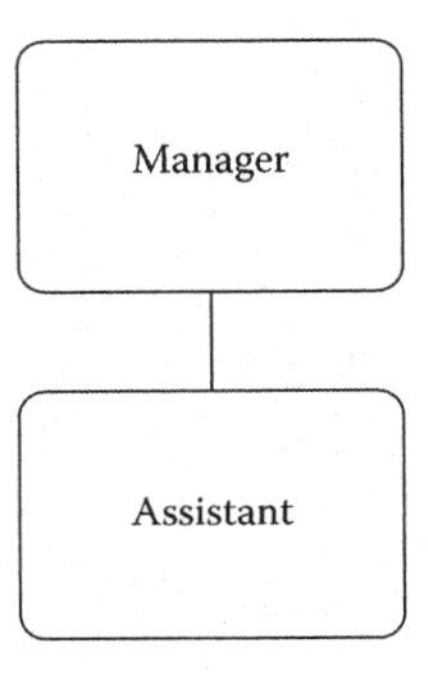

FIGURE 7.7
Simple organizational structure.

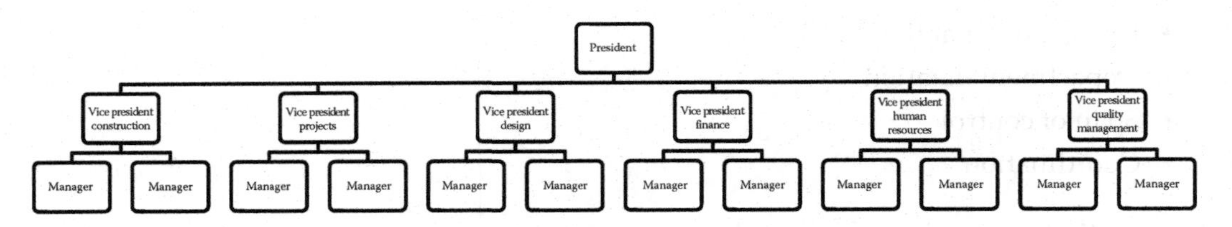

FIGURE 7.8
Functional organizational structure (departmental).

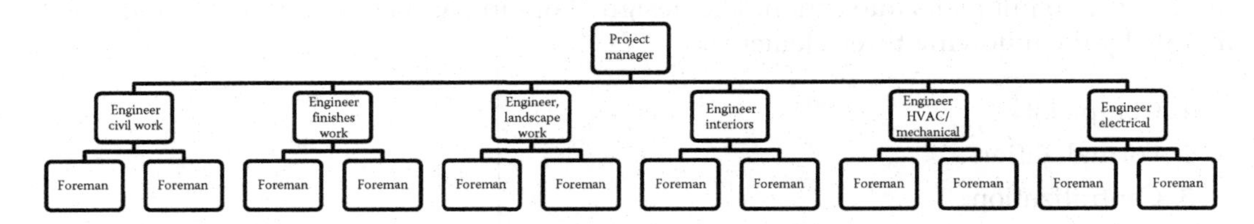

FIGURE 7.9
Functional organizational structure (engineering discipline).

3. *Divisional*: In this type of organizational structure, people with diverse specialties are put together in formal groups by similar products, services, customer, or geographical regions. Figures 7.10 through 7.13 are examples of different types of divisional organization structures.
4. *Matrix*: A matrix organization is one in which specialists from functional departments are assigned to work on one or more projects led by a project manager. It is a type of organizational structure in which there are multiple lines of authority and in certain cases, individuals report to at least two managers. This type of organizational structure brings together managers and staff from different disciplines to work toward accomplishing a common goal for the organization. It is a combination of functional and divisional organizational structures. In this type of organizational structure, the assigned staff goes back to their departments. Figure 7.14 illustrates an example of a matrix organizational structure.

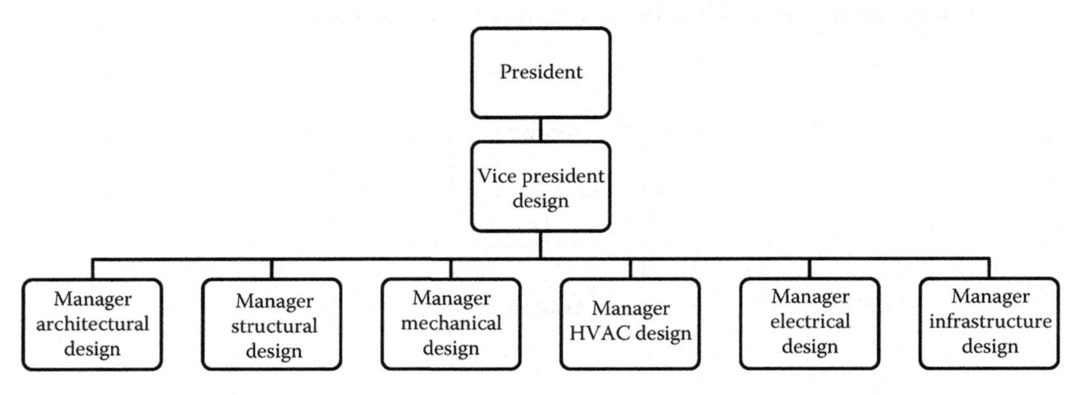

FIGURE 7.10
Divisional organizational structure.

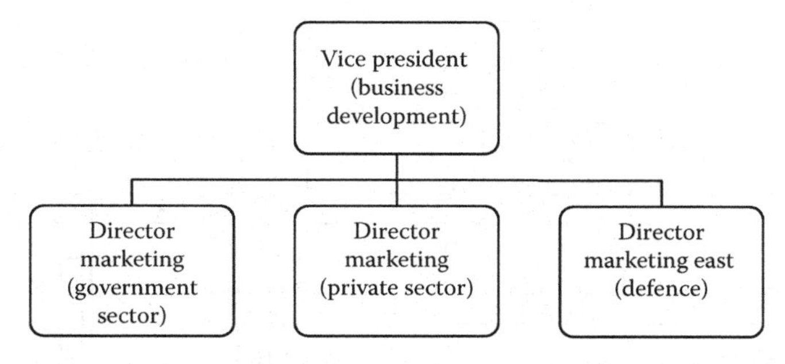

FIGURE 7.11
Divisional organizational structure (customer).

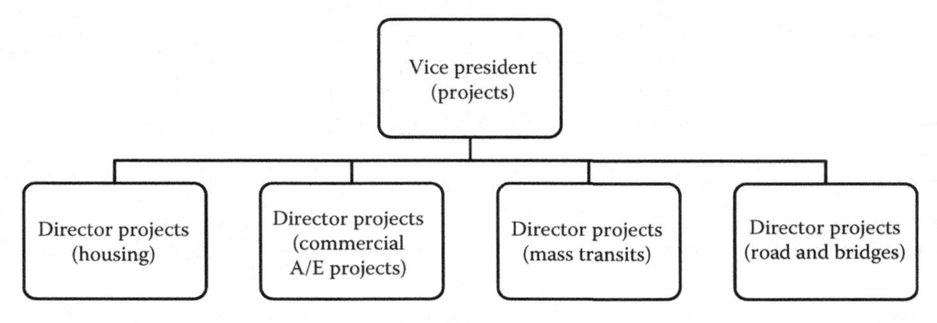

FIGURE 7.12
Divisional organizational structure (construction categories).

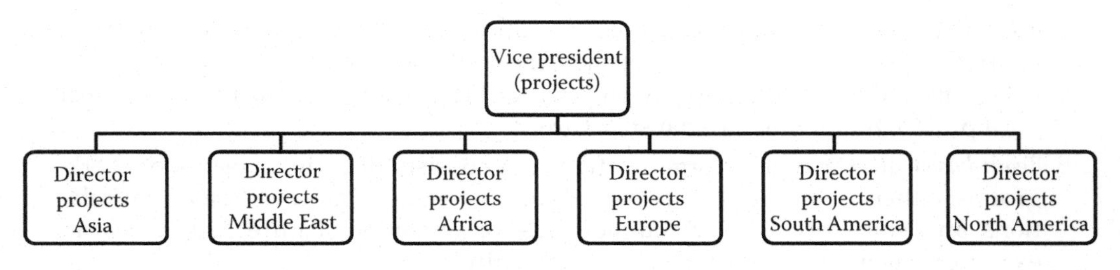

FIGURE 7.13
Divisional organizational structure (geographical).

The following are the advantages of a matrix organization:

- It has professionals who are specialists in the specific disciplines.
- It has professionals with a broader range of responsibilities and experience.
- It has access to expertise personnel.
- It facilitates use of specialized personnel and facilities, having experience in the similar projects.

The following are the disadvantages of a matrix organization:

- Since the organization has personnel attached to functional head and project head, there is likely to be confusion in command, and working relationship can become more complicated.

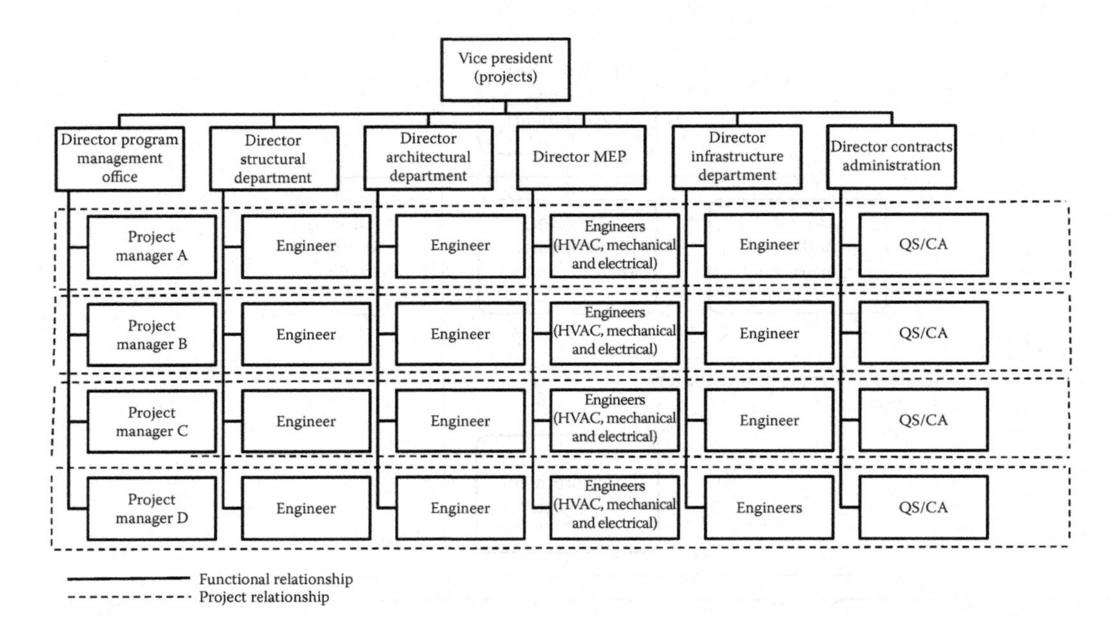

FIGURE 7.14
Matrix-type organizational structure.

- It may cause uncertainty and may lead to intense power struggle and conflicts.
- Decision process may take long time, as the personnel may have to refer to their functional head to take decision.

4.1 *Projectized organizational structure*: It is similar to a matrix organizational structure; however, when the project ends, the staff does not go back to their departments and work continuously on projects. Figure 7.15 illustrates an example of a projectized organizational structure.

5. *Team-based*: It is a type of organization that consists of activity teams or workgroups, either temporary or permanent, and is used to improve horizontal relations and solve problems throughout the organization. Figure 7.16 shows an example of team-based organizational structure.
6. *Network/boundaryless*: It is type of organization that is not defined or limited by boundaries or categories imposed by traditional organizational structures. Figure 7.17 shows an example of a network-type organizational structure.
7. *Modular*: It is a type of organization consisting of separate modules assembled together to form a modular structure. In this type of organization, each module performs a specific function, and the module can be replaced by a similar module suitable to perform the same function. Each module has its own organizational structure.

7.2.3 Staffing

Staffing is manning the organizational structure through proper and effective human resources. It involves functions related to selection and development of people to fill the roles designed into the organizational structure.

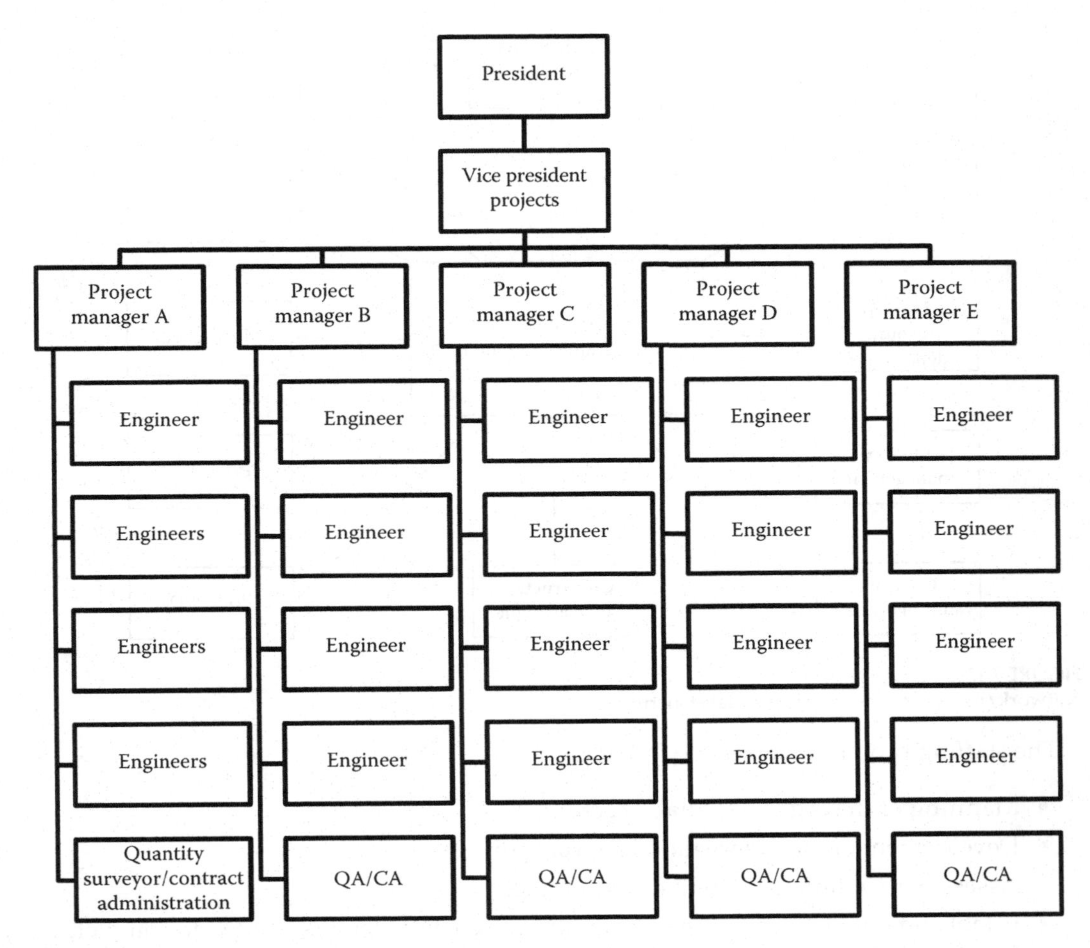

FIGURE 7.15
Functional organizational structure (projectized).

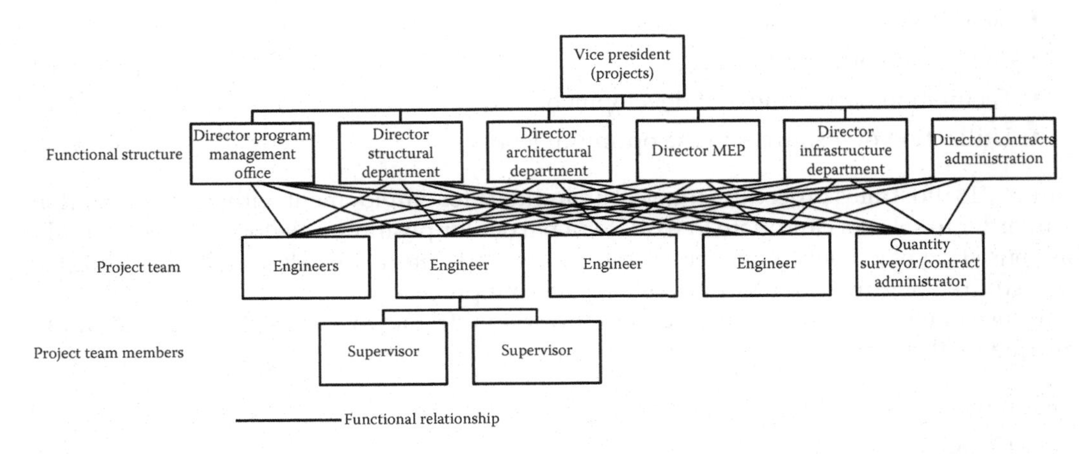

FIGURE 7.16
Team-based organizational structure.

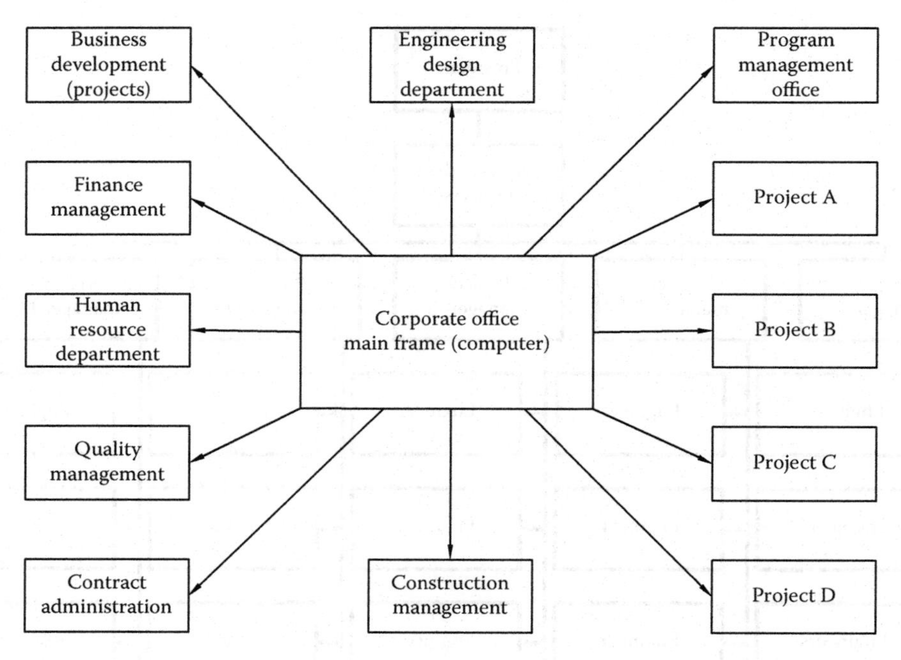

FIGURE 7.17
Network/boundaryless organizational structure.

The staffing process involves the following:

- Manning of the organizational structure
- Development of the organizational structure
- Assessment of human resources requirement
- Establishment of the roles, responsibilities, and competencies needed to complete the assigned tasks
- Establishment of the qualification, knowledge, skills required to suit the organizational requirements
- Acquisition of human resources
- Facilitation of team building
- Facilitation of training and development
- Utilization of resources to the optimum level

Figure 7.18 illustrates procedure to select a candidate for employment with the organization.

In order to select a candidate to work for a particular position, development of job analysis consisting of job definition (job requirements) and criteria for selection of the candidate are required. Figure 7.19 illustrates project staffing process.

In case of construction projects, an organizational framework is established and implemented by three major groups or parties. These are as follows:

1. Owner
2. Designer/consultant
3. Contractor

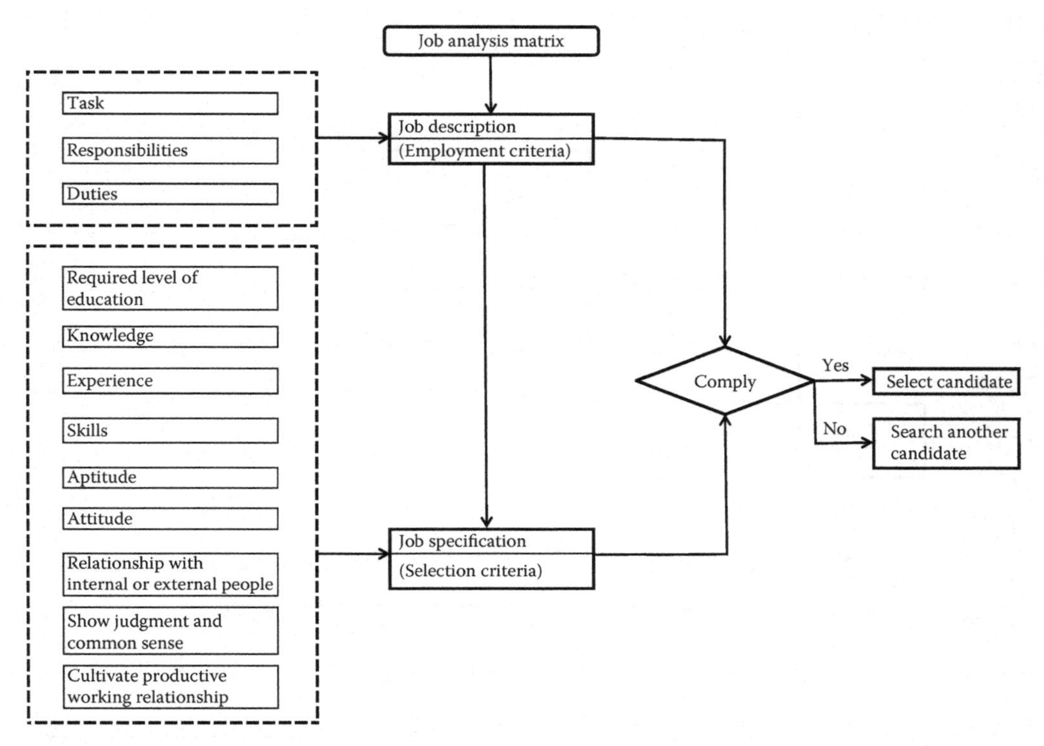

FIGURE 7.18
Candidate selection procedure.

However, because of the complex nature of major projects, the owner engages a project manager or construction manager to assist/help owner perform project management or construction management services in an efficient manner from the beginning of the project to the completion of the project. The roles and responsibilities of a project manager or construction manager may vary depending on owner's desire and requirements. Thus, the project delivery system has involvement of the following four major group or parties:

1. Owner
2. Project manager or construction manager
3. Designer/consultant
4. Contractor

The staffing requirements of these groups differ from each other. The roles and responsibilities of similar job titles differ for each of these project participants. For example, the role and responsibilities of project manager employed with owner/client differs from that of project manager working with designer (consultant) or contractor. Similarly, job requirements of engineering professionals of same discipline (e.g., structural engineer) vary with their affiliation to a particular project participant group.

The staffing (human resource) requirements of construction project-related organizations can be mainly divided into following categories:

1. Corporate team members
2. Project team members

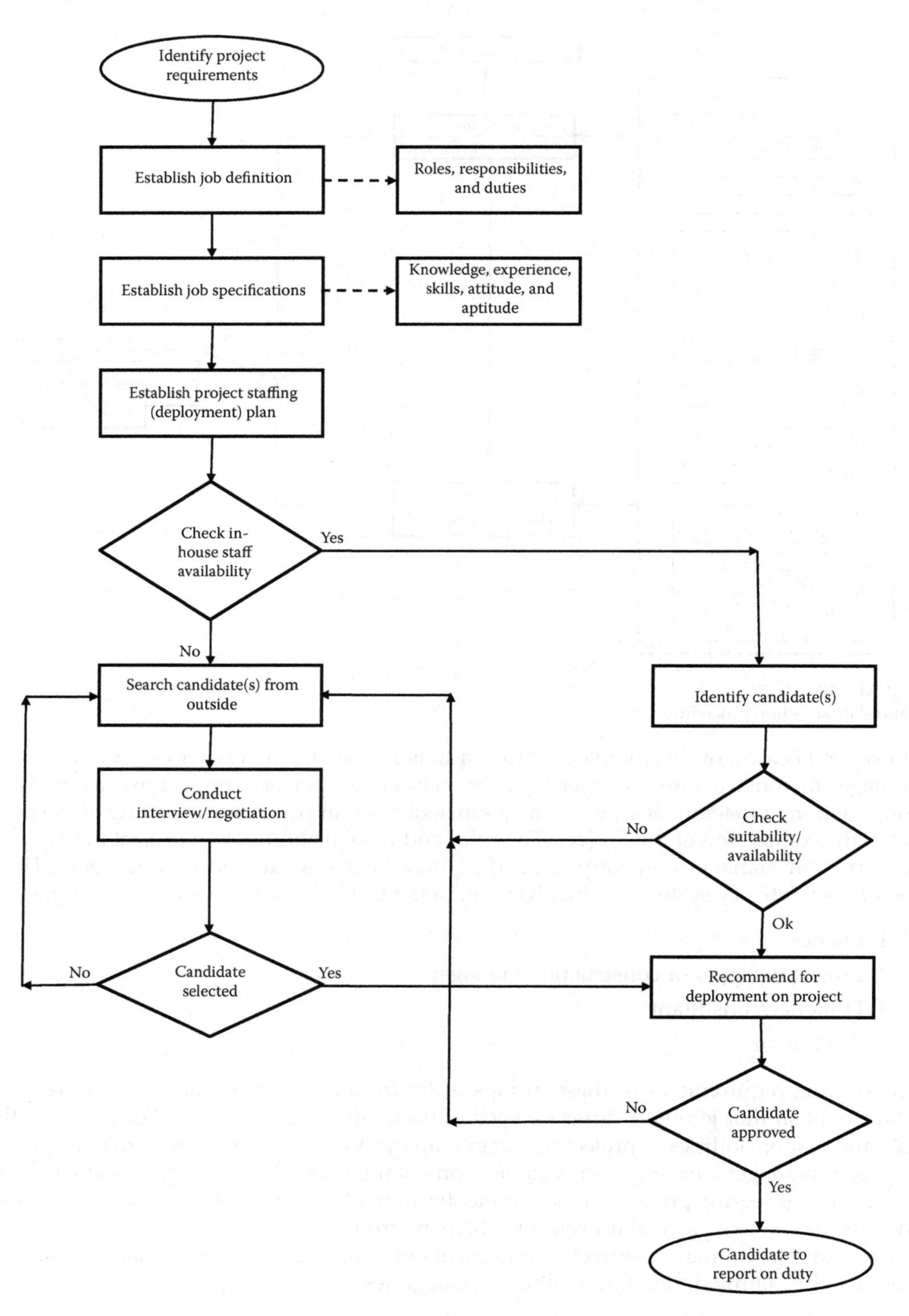

FIGURE 7.19
Project staffing process.

TABLE 7.3
Job Analysis (KESAA Requirements) for Contractor Project Manager

KESAA	Job Specifications	Description
K	Knowledge	• Thorough knowledge and understanding of construction process • In-depth knowledge of construction activities • Knowledge of construction codes and practices • Knowledge and understanding of all disciplines of construction projects • Knowledge of risk management • Knowledge of HSE
E	Experience	• Experience in execution of major projects • Experience in execution of similar projects • Experience in subcontractor management • Experience in project management • Experience in construction quality management • Excellent technical background
S	Skills	• Strong and responsive leadership skills • Communication skills (oral and written) • Information technology • Negotiation skills • Interpersonal skills • Conflict management
A	Aptitude	• Collaborative thinking • Teamwork • Societal needs toward project
A	Attitude	• Problem solving • Able to motivate and encourage subordinates and other team members

Each team member under these categories should possess job skills and competencies to perform assigned roles and responsibilities to complete the project efficiently. Table 7.3 illustrates KESAA requirements for a project manager to be employed by the contractor.

7.2.3.1 Team Development Stages

In construction projects, team members (individuals) are selected from different background and disciplines. When a group of people begins to work in a team, they go through a fairly predictable series of stages in their growth and progress. As the team matures and member relationship grow, members gradually learn to cope with emotional and group pressure they face, overcome differences, and build on each other's strength. A team should have clear understanding about what constitute the team's work and what is important.

Bruce Tuckman (1965) summarized the results of over 50 studies into a four-stage model for team development. These are as follows:

- Forming
- Storming
- Norming
- Performing

Later he added adjourning as the fifth stage.

These stages are considered as an attempt to develop a collaborative team. Table 7.4 illustrates different stages of team development.

TABLE 7.4
Team Development Stages

Sr. No.	Stages	Description
1	Forming	• Team members are new and are learning to work together • Members start feeling team experience • Most team members are positive and polite • This stage is the transition from individual to team member status • This stage is also known as orientation; members introduce themselves to each other • During this stage the team members need to identify their purpose, develop group norms, identify group processes, and define roles build relationship and trust • The team leader plays a dominant role at this stage because the roles and responsibilities of team members are not clear
2	Storming	• During this stage, individual expressions of ideas occur and there is open conflict between members • Team members have individualist thinking • Team members tend to exhibit increased conflict, confrontation, and less conformity • There is infighting, defensiveness, and competition • Team members become hostile or overzealous as a way to express their individuality and group formation • Team members resist the task • Team members resist quality improvement approaches suggested by others • Polarization of group members • Disunity, increased tension, and jealousy among members • Establishment of unrealistic and unachievable goals • Team members need to know how to resolve conflict, clarify their roles, power and structure, and build consensus through revisiting purpose • During this stage the team needs to select their desired leadership style and decision methodology • Team needs leader who is willing to identify issues and resolve conflicts • The team leader should guide the team processes toward clear goals, defined roles, acceptable team behavior, and a mutual feedback process for team communication
3	Norming	• During this stage, the team members develop sense of cohesiveness with a common sprit and goals and work habits to work together avoiding any conflict • The team members exhibit good behavior, more friendliness, cooperation, confidence, mutual trust, motivation, and positive team work • The team demonstrates an improved ability to complete the task, solve problems, and resolve conflict • Conflict among members reduces and focus is on team objective • The team establishes and maintains the habits that support group rules, values, and boundaries • The team leader continues to encourage participation and professionalism among the team members

(Continued)

TABLE 7.4 (*Continued*)

Team Development Stages

Sr. No.	Stages	Description
4	Performing	• At this stage the team becomes high-performance team • The team becomes efficient, energetic, mature, and knowledgeable about the processes and the works to be performed • The team becomes a cohesive unit • Members have insight into personal and group processes, and better understanding of each other's strength and weakness • The team becomes self directing in development of plans and strategy to meet their goals and carryout the work • The team is capable of accepting new projects and tasks and accomplishing them successfully • The team is capable of diagnosing and solving problems and making decisions • At this stage, the team is complete self-directed team and requires little, if any, management direction • During this stage every member shares the responsibility as a leader • The leader becomes facilitator aiding the team in communication processes and helping if they revert to prior stage
5	Adjourning	• At this stage the team has completed its mission or tasks • The team wraps up its work and then team dissolves or disassembled • This is a finalizing stage of team building

Source: Abdul Razzak Rumane (2013), *Quality Tools for Managing Construction Projects*, CRC Press, Boca Raton, FL. Reprinted with permission from Taylor & Francis Group.

*7.2.3.2 Training and Development**

Training and development is a set of a systematic process designed to meet learning objectives related to an organization's strategic plan. It is a subsystem of the organization. The main objective of training and development is to ensure the availability of required skilled professionals to the organization.

Training is a means of acquiring proficiency in skills or a set of skills, new knowledge, and changing attitudes to perform desired objectives in an efficient and economical manner. It is the process of identifying performance requirements and the gap between what is required and what exists. It is process that includes a variety of methods for helping the employee to attain a specific level of knowledge, skills, and abilities through professional development.

Development includes the acquisition of behavioral skills, including communication, interpersonal relations, and conflict resolution. Development may also include processes aimed at the acquisition and development by employees of knowledge, understanding, behaviors or attitude those specifically required to perform the professional duties.

Training and development helps in optimizing the utilization of human resources that further helps the employee to achieve the organizational goals as well as their individual goals. It also helps to provide opportunity and broad structure for the development of human resources' technical and behavioral skill in an organization. Training and development process cycle has mainly five steps or phases. These are as follows:

1. Needs assessment
2. Development

* Abdul Razzak Rumane (2013), *Quality Tools for Managing Construction Projects*, CRC Press, Boca Raton, FL. Reprinted with permission from Taylor & Francis Group.

3. Training implementation
4. Assessment (Evaluation)
5. Follow up

Figure 7.20 illustrates training process cycle.

Construction projects are constantly increasing in technological complexity. The competitive nature of business; comprehensive management system; accelerated changes in construction processes; and use of latest equipment, machinery, products, and methods require that the project team members and other stakeholders have technical training, in addition to formal education in their respective field. Moreover, the technological changes are happening at an ever-increasing rate, and the need for continual individual development is crucial.

There is a need for improved quality-based performance of construction projects and therefore effective management services. Training in construction project management is considered as an essential issue. Training can play an important role in improving the ability and efficiency of construction professionals, construction workers, whether the training is part of the ongoing process of professional skill development or simply about learning a specific skill. In both cases, it can improve people's skills and knowledge and can help them perform the task more effectively.

Construction projects are facing challenges of ever-changing construction technology, knowledge ideology, management techniques, project delivery system, contracting practices, and dynamic nature of site works. To meet these challenges, construction

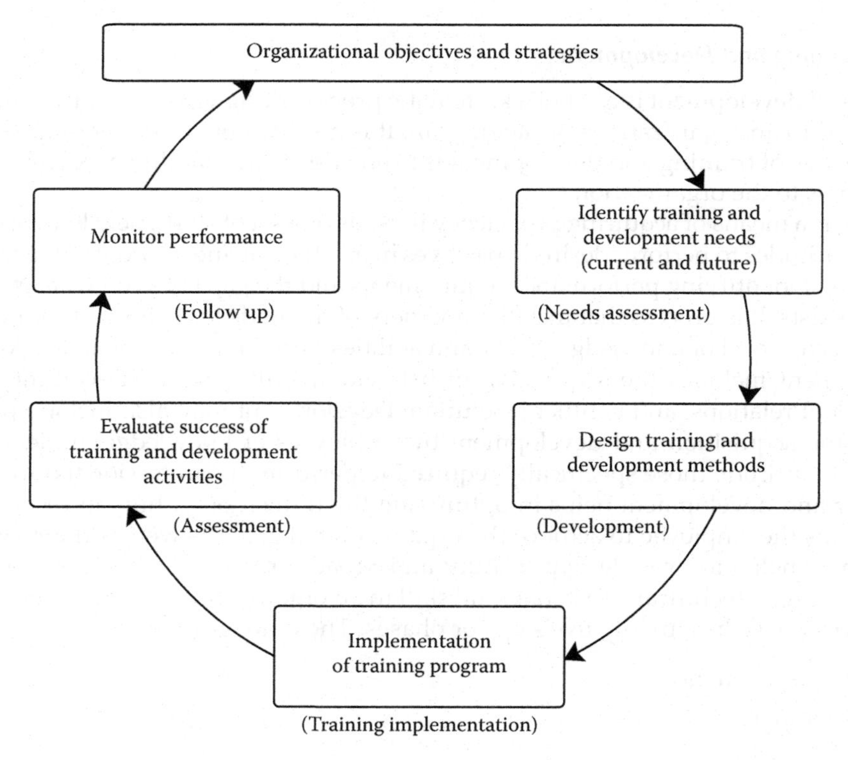

FIGURE 7.20
Training process cycle.

professionals require training on a regular basis to keep themselves abreast of these changes. Construction projects involve mainly three parties: owner, designer (consultant), and contractor. There are many other participants such as subcontractors, suppliers, and testing and auditing agencies that are directly or indirectly involved in construction projects. Professionals working with each of these parties need to have skills and working knowledge to perform the organizational requirements for competitive advantage of the organization as well as individual growth. The area of training can depend on the type of business and the field of expertise in a particular management field. In general, there is a need of training in following management areas in the construction sector:

1. Project inception/development/feasibility
2. Project cost estimation and cost management
3. Project planning and scheduling
4. Quality management
5. Contract conditions
6. Procurement and tendering procedures
7. Risk management
8. Health and safety management
9. Environmental management
10. Value management

7.2.3.2.1 Needs Assessment

Training is directed toward agreed standards and objectives. The person being trained participates with the trainer or facilitator in the training activity. Training is a means of communicating new knowledge and skills to the trainee and changing his/her attitude toward work performance. It can raise awareness and provide the opportunity to the people to explore their existing knowledge and skills. In order to be effective, training should be based on the needs of the person(s) who is (are) being trained. Training need is the gap between what somebody already knows and what he or she needs to know to do his or her job or fulfill his or her role effectively.

The first step in training and development process is to conduct needs assessment. The assessment begins with a need that can be identified in several ways, but is generally described as a gap between what is currently in place and what is needed, now and in future. In order to establish need, it is necessary to assess the current level of knowledge and skills and what level of knowledge and skills are needed to do their job or task. In general, the following are the four objectives/levels of training and development:

1. Organizational objectives (organization's goals)
2. Occupational objectives (jobs and related tasks that need to be learned)
3. Individual objectives (competencies and skills that are needed to perform the job)
4. Societal objectives

However, the first three are more commonly used in construction projects to conduct needs assessment by organizations. Societal objectives are mainly taken care by the governmental agencies. In order to establish the needs, the following are to be considered:

1. Whether the need is mandated by a regulatory body (safety, fire, handling of hazardous material from environmental perspective, health care, and food quality)
2. Whether the need is mandated by the management's strategic requirements to achieve, including:
   a. Organizational objectives
   b. Occupational objectives
   c. Individual objectives

Once the need is established, it is required to set objectives to

1. Identify
   - Who needs training (individuals who are to be trained)?
   - What training is needed?
2. Establish how the data to be collected.
3. Analyze data.
4. Summarize the findings.

7.2.3.2.1.1 Organizational Objectives The first level of training assessment is organizational objectives. The organizational analysis looks into the effectiveness of the organization's existing capability and determines where training is needed and under what conditions will it be conducted. The following are the major items to be identified and analyzed to establish organizational training objectives:

- Organizational goals
- Strategic plans and objectives
- Political impacts
- Economical impacts
- Environmental impacts
- Global market
- Future market trend
- Technological changes
- Competitor's approach to training and development
- Organizational structure
- Significance and importance of training in the organization
- Codes, standards, and regulations
- Employee willingness to participate in training

A proper assessment of needs will help the organization to prioritize the availability of resources to address the current/future requirements.

7.2.3.2.1.2 Occupational Objectives The assessment for occupational training is to determine where a change or new way of working about a job or group of jobs is needed and to establish the knowledge, skills, attitude and abilities needed to achieve optimum performance.

7.2.3.2.1.3 Individual Objectives The assessment for individual training analyzes how an individual employee is performing the job and determines which employees need training and what type of training is needed. Individual training requirements are based on annual performance (appraisal) report with the employees and their line manager. Individual training helps in enhancing the skills of the employee to improve performance on the current job, to deal with forthcoming changes, or developmental needs that will enable the individual to program their career.

All the three levels of need analysis discussed above are interrelated; therefore, the need analysis at these three levels is to be considered in conjunction with one another. The data collected from each level is critical for a thorough and effective needs assessment.

7.2.3.2.1.4 Data Collection Data collection is an important instrument and an indispensible part to complete needs assessment. The following types of data are normally collected for needs assessment:

1. Business plans and objectives
2. Interviews
3. Observations
4. Work sample
5. Questionnaires
6. Performance evaluation
7. Performance problems
8. Best practices followed by other companies/competitors
9. Analysis of operating problems (work time)
10. Survey

7.2.3.2.2 Development

Once the needs assessment is completed and training objectives are clearly identified, the development phase (design of training and development methods) is initiated. This phase/step consists of the following:

1. Selection of trainer/facilitator (internal or external)
2. Designing of course material
3. Techniques or methods to be used to facilitate training
4. Select the appropriate method (on the job or any other methods)
5. Duration of training

The type of training methods depends on the training objectives or expected learning outcomes. It depends on how the professional needs to be trained and what are the needs of the participant (trainee). While planning for the type of training, the following points are to be considered:

- Needs and abilities of the participants
- Communication method
- Regulatory requirements, if any

7.2.3.2.3 Training Implementation

The following are the most common training methods:

1. Seminars
2. Workshops
3. Lectures
4. Self-study tutorials
5. Coaching
6. On-the-job training
7. Field visits

The following are the most common materials used for training:

1. Projectors
2. Slides
3. Audio–visual aids
4. Videos
5. Computer-based instruction techniques
6. Flip charts
7. Wall charts
8. Flash cards
9. Workbooks
10. Handouts
11. User manuals
12. Case study
13. Job aids

7.2.3.2.4 Assessment

Evaluation of training is crucial for knowing the effectiveness of training program and plan for future training. Evaluation is to be done for both the trainer and the trainee. Normally, evaluation forms are distributed to all the participants at the end of training session to get their opinion and to evaluate the understanding of the subject training by the trainee and also to know their views about the trainer.

7.2.3.2.5 Follow Up

The training program is only part of the knowledge and skills acquiring process. Putting the acquired knowledge into practice is most important step. The training will be considered successful only if the positive changes have taken place and an improvement in the organizational and individual performance is recorded.

In order to know the effectiveness of training, a short session to discuss the training process and its on-the-job implementation is required. Construction projects involve owner, designer (consultant), and contractor. The roles and responsibilities of personnel vary according to their affiliation to a particular working group and also may vary from project to project. For example, an engineer working with designer must possess the knowledge and skills related to design, whereas an engineer working with contractor must know how to execute/implement the project. The knowledge and skills of the team members should be closely linked to the project requirements. Deployment of project staff is done in accordance with the specific qualification listed under contract documents. Therefore, while establishing the needs assessment for project personnel, contract requirements should be taken into consideration.

Normally, the construction professionals assigned to a particular project are, apart from client/owner representative(s), from the designer's (consultant) or contractor's regular staff, supplemented by additional hiring, if needed to fulfill contract requirements. It is assumed that these personnel are fully capable of taking up the responsibilities to perform on the assigned project. Sometimes it is required to hire local work force to comply with regulatory requirements. In such cases, it is possible that these personnel may not possess the requisite knowledge and skills to perform the required task. Therefore, special training should be arranged for these personnel to maintain the organization's reputation.

Training needs of team members assigned or about to be assigned on a construction project is different from those required for normal employees working with the organization. The assigned team members should have knowledge about the site conditions, safety, environmental rules, quality, and climatic conditions. Therefore, special training is required by establishing special need assessment, taking into consideration project-specific roles and responsibilities. Project personnel hired for specific project should be given induction training to become fully oriented in their respective job. Furthermore, they should be made fully aware of company's quality management system.

Table 7.5 lists training needs for contractor's project manager to make him or her fully conversant with project management.

TABLE 7.5

Training Needs for Contractor's Project Manager

Sr. No.	Areas of Training
1.	Human relations
2.	Construction methods
3.	Project management
4.	Communication skills
5.	Resource management
6.	Quality analysis/quality control
7.	Cost management
8.	Project planning and scheduling
9.	Regulatory requirements about labors
10.	Safety, health, and environmental requirements
11.	Basic design principles(all the trades)
12.	Human resources management
13.	Management of subcontractors
14.	Valuation of work in progress
15.	Contract conditions
16.	Building contract laws
17.	Interpersonal skills
18.	Logistical/demographic conditions
19.	Organization planning

Source: Abdul Razzak Rumane (2013), *Quality Tools for Managing Construction Projects*, CRC Press, Boca Raton, FL. Reprinted with permission from Taylor & Francis Group.

7.2.4 Directing

Directing is putting plan into action and using the resources to achieve the desired results. Directing involves the initiation of action, and it entails the following three elements that are action oriented in nature:

- Motivation
- Communication
- Leadership

Directing is also known as *leading*. Leading is the process of influencing people, so that they will contribute to the organization's goals. The main objective of leading is to improve productivity.

7.2.5 Controlling

Controlling is a process to assess and regulate works in progress and taking action to ensure that the desired results are attained. In controlling, it is required to

- Establish objectives and standards and measure actual performance.
- Compare results with established objectives and standards.
- Take necessary action for improving performance and productivity.
- Reduce and prevent unacceptable performance.

In order to have effective controlling, there must be real-time feedback.

7.3 Management Processes

A Guide to the Project Management Body of Knowledge (*PMBOK® Guide*) published by Project Management Institute describes application of the project management processes during the life cycle of projects to enhance the chances of success over a wide range of projects. *PMBOK® Guide Fifth Edition* identifies and describes the following five project management process groups required for the successful completion of any project:

1. Initiating process group
2. Planning process group
3. Executing process group
4. Monitoring and controlling process group
5. Closing process group

Figure 7.21 illustrates the overview of the project management process groups, and Table 7.6 illustrates the overview of project integration management processes.

These process groups are independent of application areas or industry focus. Project integration management is required from project initiation to the project closeout. The groups

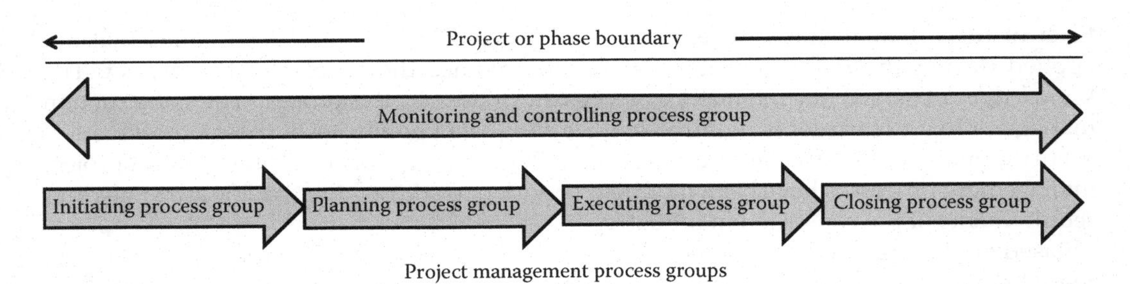

FIGURE 7.21
Overview of project management process groups.

TABLE 7.6
Project Integration Management Processes

Sr. No.	Project Management Process Group	Project Management Processes
1	Initiating process group	1.1 Develop project charter 1.2 Develop preliminary scope Statement
2	Planning process group	2.1 Develop project management plan
3	Executing process group	3.1 Direct and manage project works
4	Monitoring and controlling process group	4.1 Monitor and control project work 4.2 Perform integrated change control
5	Closing process group	5.1 Close project or phases

defined in *PMBOK® Guide Fifth Edition* consist of 47 project management processes and are further grouped into the following ten separate knowledge areas:

1. Project integration management
2. Project scope management
3. Project time management
4. Project cost management
5. Project quality management
6. Project human resource management
7. Project communication management
8. Project risk management
9. Project procurement management
10. Project stakeholder management

PMBOK® Guide Third Edition had nine knowledge areas. Additional four knowledge areas that are specific to construction industry have been added in *Construction Extension to the PMBOK® Guide Third Edition*. These are as follows:

1. Project safety management
2. Project environmental management
3. Project financial management
4. Project claim management

Each of these management areas consists of processes, tools, and techniques that are applied during the management of project to enhance the success of the project. The knowledge areas are not intended to represent phases in the project. The construction manager/project manager is required to have adequate knowledge of these processes.

Furthermore, in *Construction Extension to the PMBOK® Guide Third Edition* certain project management processes have been modified or added to address the specific attributes and requirements of the construction industry.

Based on the principles of project management processes defined in the *PMBOK® Guide*, these knowledge areas are categorized into the following management processes to manage and control various processes and activities to be performed in construction project management:

1. Integration management
2. Stakeholder management
3. Scope management
4. Schedule management
5. Cost management
6. Quality management
7. Resource management
8. Communication management
9. Risk management
10. Contract management
11. Health, safety, and environment management (HSE)
12. Financial management
13. Claim management

These processes are discussed in details referring to construction industry.

7.3.1 Integration Management

Integration management is coordination and implementation of five project management process groups (initiating, planning, executing, monitoring, and controlling) right from the time the project is conceived to closeout stage. It involves putting all the process groups. Figure 7.22 illustrates the integration management process cycle.

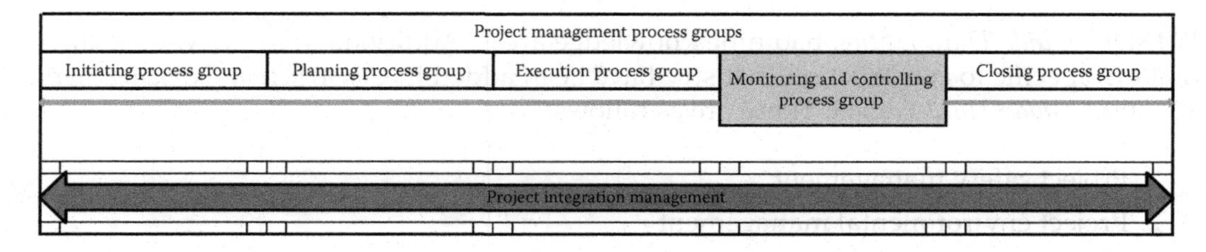

FIGURE 7.22
Project integration management.

Integration management of construction project includes all the activities performed to effectively control the final output of project production (facility), and the input of the process is owner's need for the construction project.

The management processes mentioned above have a variety of applications in construction projects. From the perspective of managing construction projects, it is difficult to generalize and evolve the exact equivalent processes of knowledge areas that are generally followed in the management of construction projects. However, each of the above-mentioned management processes can be divided into construction-related technical activity (ies) that can be subdivided into its element(s) to enable improve the efficiency and effectiveness of construction management. These management processes can be performed in any order as long as the required inputs are available. Each of the activity evolved will be single purpose and has to be managed, controlled, and completed within specific duration. This will result in improving the practices followed in construction projects and help conveniently manage the construction projects.

A construction project begins with the inception stage that results from business case, which suggests constructing a new project/facility. The owner of the facility could be an individual, a public/private company, or a government agency. Normally the need for the project is created by the owner and is linked to the financial resources available to develop the facility. Construction projects are constantly increasing in technological complexity. In addition, the requirements of construction project clients are on the increase and, as a result, construction projects must meet varied performance standards (climate, rate of deterioration, maintenance, etc.). Therefore, to ensure the adequacy of client brief, which addresses the numerous complex client/user need, it is now needed to evaluate the requirements in terms of activities and their interrelationship.

The major construction activities evolved based on the management processes are described under five project management process groups.

7.3.1.1 Initiating Process Group

Table 7.7 illustrates major construction activities relating to project initiating process group.

7.3.1.2 Planning Process Group

Table 7.8 illustrates major construction activities relating to project planning process group.

7.3.1.3 Executing Process Group

Table 7.9 illustrates major construction activities relating to project executing process group.

7.3.1.4 Monitoring and Controlling Process Group

Table 7.10 illustrates major construction activities during project monitoring and controlling process group.

7.3.1.5 Closing Process Group

Table 7.11 illustrates major construction activities relating to project closing process group.

TABLE 7.7
Major Construction Project Activities Relating to Initiating Process Group

Sr. No.	Management Processes	Activities	Elements
1	Integration management	1.1 Develop project charter	1.1.1 Project inception
			1.1.2 Problem statement/need identification
			1.1.3 Need analysis
			1.1.4 Need statement
			1.1.5 Need feasibility
			1.1.6 Project goals and objectives
			1.1.7 Project deliverables
			1.1.8 Design deliverables
		1.2 Develop preliminary scope statement	1.2.1 Project terms of reference (TOR)
			1.2.2 Contract documents
2	Stakeholder management	2.1 Identify stakeholders	2.1 Project delivery system
			2.2 Project life cycle
			2.3 Project team members
			2.4 Other parties

Note: These activities may not be strictly sequential; however, the breakdown allows implementation of project/construction management functions more effective and manageable at different stages of project life cycle phases.

7.3.2 Stakeholder Management

A stakeholder is anyone who has involvement, interest, or impact in the construction project processes in a positive or negative way. Stakeholders play a vital role in determination, formulation, and successful implementation of project processes. Stakeholders can mainly be classified as follows:

- Direct stakeholders
- Indirect stakeholders
- Positive stakeholders
- Negative stakeholders
- Legitimacy and power

Stakeholder management process consists of following activities:

1. Identify stakeholders
2. Plan stakeholder management
3. Manage stakeholder engagement
4. Control stakeholder engagement

TABLE 7.8

Major Construction Activities Relating to Planning Process Group

Sr. No.	Management Processes	Activities	Elements
1	Integration management	1.1 Project baseline plan	1.1.1 Preliminary plans
2	Stakeholder management	2.1 Responsibilities matrix	2.1.1 Owner, designer, contractor, and other stakeholders
		2.2 Stakeholders requirement (work progress)	2.2.1 Design progress
			2.2.2 Construction progress
			2.2.3 Testing, commissioning, and handover
		2.3 Change reporting	2.3.1 Updated schedule
			2.3.2 Variation report
			2.3.3 Cost variation
		2.4 Project updates	
		2.5 Status reports	2.5.1 Status logs
			2.5.2 Performance reports
			2.5.3 Issue log
		2.6 Meetings	2.6.1 Kick-off meeting
			2.6.2 Progress meetings
			2.6.3 Coordination meetings
			2.6.4 Other meetings
		2.7 Payments	2.7.1 Payment status
3	Scope management	3.1 Establish scope baseline plan	
		3.2 Collect requirements	3.2.1 Need statement
			3.2.2 Project goals and objectives
			3.2.3 Project TOR
			3.2.4 Owner's preferred requirements
		3.3 Project scope documents	3.3.1 Design development • Concept design • Schematic design • Detail design
			3.3.2 Final design
			3.3.3 Bill of quantity
			3.3.4 Project specifications
			3.3.5 Construction documents
			3.3.6 Project deliverables
		3.4 Organizational breakdown structure	3.4.1 Project delivery system
			3.4.2 Organizing
			3.4.3 Staffing
			3.4.4 Concept design
		3.5 Work breakdown structure	3.5.1 Project life cycle
			3.5.2 Work packages

(Continued)

TABLE 7.8 (*Continued*)

Major Construction Activities Relating to Planning Process Group

Sr. No.	Management Processes	Activities	Elements
4	Schedule management	4.1 Bill of quantity	4.1.1 Quantities take off
			4.1.2 Sequencing of activities
			4.1.3 Estimate activity resources
			4.1.4 Estimate duration of activity
		4.2 Identify project assumption	4.2.1 Dependencies
			4.2.2 Risks and constraints
			4.2.3 Milestones
		4.3 Develop baseline schedule	
		4.4 Develop schedule	4.4.1 Predesign stage
			4.4.2 Design development • Concept design • Schematic design • Detail design
			4.4.3 Contract documents
			4.4.4 Bidding and tendering and contract award
			4.4.5 Construction phase
			4.4.6 Testing, commissioning, and handover
		4.5 Construction schedule	4.5.1 Contractor's construction schedule
5	Cost management	5.1 Estimate cost	5.1.1 Conceptual estimate
			5.1.2 Preliminary estimate
			5.1.3 Detail estimate
			5.1.4 Definitive estimate
		5.2 Estimate budget	5.2.1 Prepare budget
		5.3 Determine project cost baseline	5.3.1 S-curve
			5.3.2 Cost loading
			5.3.3 Resource loading
		5.4 Estimate cost	5.4.1 Estimate project resources cost
			5.4.2 Estimate project material cost
			5.4.3 Estimate project equipment cost
			5.4.4 Bill of quantities
			5.4.5 BOQ price analysis
		5.5 Contracted project value	5.5.1 Progress payments
		5.6 Change order procedure	5.6.1 Change order
			5.6.2 Cost variation
6	Quality management	6.1 Project quality management plan	6.1.1 Quality codes and standards to be compiled
			6.1.2 Design criteria
			6.1.3 Design procedure
			6.1.4 Quality matrix (design stage)
			6.1.5 Well-defined specification
			6.1.6 Detailed construction drawings
			6.1.7 Quality matrix (construction phase)

(*Continued*)

TABLE 7.8 (*Continued*)

Major Construction Activities Relating to Planning Process Group

Sr. No.	Management Processes	Activities	Elements
			6.1.8 Construction process
			6.1.9 Detailed work procedures
			6.1.10 Quality matrix (Inspection, testing during execution)
			6.1.11 Defect prevention/rework
			6.1.12 Quality matrix (testing and handing over startup)
			6.1.13 Regulatory requirements
			6.1.14 Quality assurance/quality control procedures
			6.1.15 Reporting quality assurance/quality control problems
			6.1.16 Stakeholders quality requirements
7	Resource management	7.1 Project human resources	7.1.1 Construction/project manager
			7.1.2 Designer's team
			7.1.3 Supervision team
		7.2 Construction resources	7.2.1 Contractor's core team
			7.2.2 Construction material
			7.2.3 Construction equipment
			7.2.4 Construction labor
			7.2.5 Subcontractor(s)
8	Communication management	8.1 Communication plan	8.1.1 Communication matrix
		8.2 Communication methods	8.2.1 Design progress
			8.2.2 Work progress
			8.2.3 Project issues
			8.2.4 Project variations
			8.2.5 Authorities
		8.3 Submittal procedures	8.3.1 Submittal procedure
			8.3.2 Progress payments
			8.3.3 Progress reports
			8.3.4 Minutes of meetings
			8.3.5 Other meetings
		8.4 Documents	8.4.1 Design documents
			8.4.2 Contract documents
			8.4.3 Construction documents
			8.4.4 As-built documents
			8.4.5 Authority-approved documents/drawings
		8.5 Logs	8.5.1 Issue log
			8.5.2 Correspondence with stakeholders
			8.5.3 Correspondence with team members
			8.5.4 Regulatory authorities

(Continued)

TABLE 7.8 (*Continued*)

Major Construction Activities Relating to Planning Process Group

Sr. No.	Management Processes	Activities	Elements
9	Risk management	9.1 Risk identification	9.1.1 During inception
			9.1.2 During design
			9.1.3 During bidding
			9.1.4 During construction
			9.1.5 During testing and commissioning
			9.1.6 During handing over
		9.2 Managing risk	9.2.1 Risk register
			9.2.2 Risk analysis
			9.2.3 Risk response
10	Contract management	10.1 Project delivery system	10.1.1 Selection of CM
			10.1.2 Selection of designer
		10.2 Bidding and Tendering	10.2.1 Prequalification of contractors
			10.2.2 Issue tender documents
			10.2.3 Acceptance of tender
11	Health, safety, and environment	11.1 Environmental compatibility	
		11.2 Safety management plan	11.2.1 Safety consideration in design
			11.2.2 HSE plan for construction site safety
			11.2.3 Emergency evacuation plan
		11.3 Waste management plan	
12	Financial management	12.1 Financial planning	12.1.1 Payments to designer (consultant), construction/project manager, contractor
			12.1.2 Material procurement
			12.1.3 Equipment procurement
			12.1.4 Project staff salaries
			12.1.5 Bonds, insurance, guarantees
			12.1.6 Cash flow
13	Claim management	13.1 Claim identification	13.1.1 Design errors
			13.1.2 Additional works
			13.1.3 Delays in payment
		13.2 Claim quantification	13.2.1 Change order procedures Cost Time

Note: These activities may not be strictly sequential, however the breakdown allows implementation of project/construction management functions more effective and manageable at different stages of project life cycle phases.

7.3.2.1 Identify Stakeholder

Construction projects have direct involvement of following three stakeholders:

1. Owner
2. Designer
3. Contractor

TABLE 7.9

Major Construction Activities Relating to Project Execution Process Group

Sr. No.	Management Processes	Activities	Elements
1	Integration management	1.1 Design development	1.1.1 Concept design
			1.1.2 Schematic design
			1.1.3 Detail design
		1.2 Construction	1.2.1 Notice to proceed
			1.2.2 Mobilization
			1.2.3 Submittals
			1.2.4 Execution
			1.2.5 Corrective actions
			1.2.6 Project deliverables
		1.3 Implement changes	1.3.1 Approved changes
			1.3.2 Preventive actions
			1.3.3 Defect repairs
			1.3.4 Rework
			1.3.5 Update scope
			1.3.6 Update plans
			1.3.7 Update contract documents
2	Stakeholder management	2.1 Project status/performance report	2.1.1 Updated plans
		2.2 Payments	2.2.1 Progress payments
		2.3 Change requests	2.3.1 Site work instruction
			2.3.2 Change orders
			2.3.3 Schedule
			2.3.4 Materials
		2.4 Conflict resolution	
		2.5 Issue log	
3	Scope management		
4	Schedule management		
5	Cost management		
6	Quality management	6.1 Quality assurance	6.1.1 Design compliance to TOR
			6.1.2 Design coordination with all disciplines
			6.1.3 Material approval
			6.1.4 Shop drawing approval
			6.1.5 Method approval
			6.1.6 Method statement
			6.1.7 Mock up
			6.1.8 Quality audit
			6.1.9 Functional and technical compatibility

(Continued)

TABLE 7.9 (*Continued*)

Major Construction Activities Relating to Project Execution Process Group

Sr. No.	Management Processes	Activities	Elements
7	Resource management	7.1 Project staff	7.1.1 Project/construction manager staff
			7.1.2 Supervision staff
		7.2 Project manpower	7.2.1 Core staff
			7.2.2 Site staff
			7.2.3 Workforce
		7.3 Team management	7.3.1 Team behavior
			7.3.2 Conflict resolution
			7.3.3 Demobilization project workforce
		7.4 Construction resources	7.4.1 Material
			7.4.2 Equipment
			7.4.3 Subcontractor(s)
8	Communication management	8.1 Submittals	8.1.1 Shop drawings
			8.1.2 Material
			8.1.3 Change orders
			8.1.4 Payments
		8.2 Documentation	8.2.1 Status log
			8.2.2 Issue log
			8.2.3 Minutes of meetings
			8.2.4 Contract documents
			8.2.5 Specifications
			8.2.6 Payments
		8.3 Correspondence	8.3.1 Stakeholders
			8.3.2 Regulatory authorities
			8.3.3 Correspondence among team members
9	Risk management	9.1 Manage risk	9.1.1 Risk register
			9.1.2 Risk response
10	Contract management	10.1 Contract documents	10.1.1 Notice to proceed
		10.2 Selection of subcontractor(s)	
		10.3 Selection of materials, systems, and equipment	
		10.4 Execution of works	
11	Health, safety, and environment	11.1 HSE management plan	11.1.1 Site safety
			11.1.2 Preventive and mitigation measures
			11.1.2 Temporary firefighting
			11.1.3 Environmental protection
			11.1.4 Waste management
			11.1.5 Safety hazards
12	Financial management		
13	Claim management		

Note: These activities may not be strictly sequential, however the breakdown allows implementation of project/construction management functions more effective and manageable at different stages of project life cycle phases.

TABLE 7.10

Major Construction Activities Relating to Monitoring and Controlling Processes Group

Sr. No.	Management Processes	Activities	Elements
1	Integration management	1.1 Project performance	1.1.1 Design performance
			1.1.2 Construction performance
			1.1.3 Project startup
			1.1.4 Forecasted schedule
			1.1.5 Forecasted cost
			1.1.6 Issues
		1.2 Change management system	1.2.1 Design changes
			1.2.2 Design errors
			1.2.3 Change requests
			1.2.4 Scope change
			1.2.5 Variation orders
			1.2.6 Site work instruction
			1.2.7 Alternate material
			1.2.8 Specs/methods
		1.3 Change analysis	1.3.1 Review, evaluate changes
			1.3.2 Approve, delay, reject changes
			1.3.3 Corrective actions
			1.3.4 Preventive actions
		1.4 Compliance to contract documents	
2	Stakeholder management	2.1 Project performance	2.1.1 Progress reports
			2.1.2 Updates
			2.1.3 Safety report
			2.1.4 Risk report
		2.2 Project updates	2.2.1 Contract documents
		2.3 Payments	2.3.1 Payment certificate
		2.4 Change requests	2.4.1 Site work instruction
			2.4.2 Change orders
		2.5 Issue log	2.5.1 Anticipated problems
		2.6 Minutes of meetings	2.6.1 Progress meetings
			2.6.2 Other meetings
3	Scope management (contract documents)	3.1 Validate scope	3.1.1 Conformance to TOR
			3.1.2 Review of design documents
			3.1.3 Conformance to contract documents
			3.1.4 Approval of changes
			3.1.5 Authorities approval of deliverables
			3.1.6 Stakeholders approval of deliverables
			3.1.7 Quality audit
		3.2 Scope change control	3.2.1 Variation orders
			3.2.2 Change orders
		3.3 Performance measures	

(*Continued*)

TABLE 7.10 (*Continued*)

Major Construction Activities Relating to Monitoring and Controlling Processes Group

Sr. No.	Management Processes	Activities	Elements
4	Schedule management	4.1 Schedule monitoring	4.1.1 Project status
		4.2 Schedule control	4.2.1 Progress curve
		4.3 Schedule changes	4.3.1 Approved changes
		4.4 Progress monitoring	4.4.1 Planned versus actual
		4.5 Submittals monitoring	4.5.1 Subcontractors
			4.5.2 Material
			4.5.3 Shop drawings
5	Cost management	5.1 Cost control	5.1.1 Work performance
			5.1.2 S-curve
			5.1.3 Forecasted cost
		5.2 Change orders	
		5.3 Progress payment	
		5.4 Variation orders	
6	Quality management	6.1 Control quality	6.1.1 Quality metrics
			6.1.2 Quality checklist
			6.1.3 Material inspection
			6.1.4 Work inspection
			6.1.5 Rework
			6.1.6 Testing
			6.1.7 Regulatory compliance
7	Resource management	7.1 Conflict resolution	
		7.2 Performance analysis	
		7.3 Material management	
8	Communication management	8.1 Meetings	8.1.1 Progress meetings
			8.1.2 Coordination meetings
			8.1.3 Safety meetings
			8.1.4 Quality meetings
		8.2 Submittal control	8.2.1 Drawings
			8.2.2 Material
		8.3 Documents control	8.3.1 Correspondence
9	Risk management	9.1 Monitor and control risk	9.1.1 Scope change risk
			9.1.2 Schedule change risk
			9.1.3 Cost change risk
			9.1.4 Mitigate risk
			9.1.5 Risk audit
10	Contract management	10.1 Inspection	
		10.2 Checklists	
		10.3 Handling of claims and disputes	
11	Health, safety, and environment	11.1 Prevention measures	11.1.1 Accidents avoidance/mitigation
			11.1.2 Firefighting system
			11.1.3 Loss prevention measures
		11.2 Application of codes and standards	

(*Continued*)

TABLE 7.10 (*Continued*)

Major Construction Activities Relating to Monitoring and Controlling Processes Group

Sr. No.	Management Processes	Activities	Elements
12	Financial management	12.1 Financial control	12.1.1 Payments to project team members
			12.1.2 Payments to contractor(s)/ subcontractor(s)
			12.1.3 Material purchases
			12.1.4 Variation order payment
			12.1.5 Insurance and bonds
		12.2 Cash flow	
13	Claim management	13.1 Claim prevention	13.1.1 Proper design review
			13.1.2 Unambiguous contract documents language
			13.1.3 Practical schedule
			13.1.4 Qualified contractor(s)
			13.1.5 Competent project team members
			13.1.6 RFI review procedure
			13.1.7 Negotiations
			13.1.8 Appropriate project delivery system

Note: These activities may not be strictly sequential, however the breakdown allows implementation of project/construction management functions more effective and manageable at different stages of project life cycle phases.

However, there are many other stakeholders who have significant influence/impact on the outcome of construction project. It is important to identify the stakeholders who have interest and have significant influence on the outcome of the project. The stakeholders include members from within organization and people, agencies, and authorities outside the organization. A stakeholder register/log is developed using different types of classification models for stakeholder's analysis. The stakeholder register/log is maintained and updated throughout the life cycle of project. Figure 7.23 illustrates stakeholders having involvement or interest in the construction project.

7.3.2.2 Plan Stakeholder Management

In order to run a successful project, it is important to address the needs of project stakeholders effectively predicting how the project will be affected and how the stakeholders will be affected. Stakeholder management planning is a process to develop stakeholder engagement plan depending on the roles and responsibilities of the stakeholders and their needs, expectations, and influence on the project. While developing stakeholder management plan, the following factors have to be considered:

- Who are the stakeholders of the project?
- What role each stakeholder will have in the project?
- What relationship project team members will have with the stakeholders?

TABLE 7.11
Major Construction Activities Relating to Closing Process Group

Sr. No.	Management Processes	Activities	Elements
1	Integration management	1.1 Close project or phase	1.1.1 Testing and commissioning
			1.1.2 Authorities' approvals
			1.1.3 Punch list/snag list
			1.1.4 Handover of project/facility
			1.1.5 As-built drawings
			1.1.6 Technical manuals
			1.1.7 Spare parts
			1.1.8 Lesson learned
2	Resource management	2.1 Close project team	2.1.1 Demobilization
			2.1.2 New assignment
		2.2 Material and equipment	2.2.1 Excess material removal/disposal
			2.2.2 Equipment removal
3	Contract management	3.1 Close contract	3.1.1 Project acceptance/takeover
			3.1.2 Issuance of substantial completion certificate
			3.1.3 Occupancy
4	Financial management	4.1 Financial administration and records	4.1.1 Payments to all contractors, subcontractors, and other team members
			4.1.2 Bank guarantees/warranties
5	Claim management	5.1 Claim resolution	5.1.1 Settlement of claims

Note: These activities may not be strictly sequential, however the breakdown allows implementation of project/construction management functions more effective and manageable at different stages of project life cycle phases.

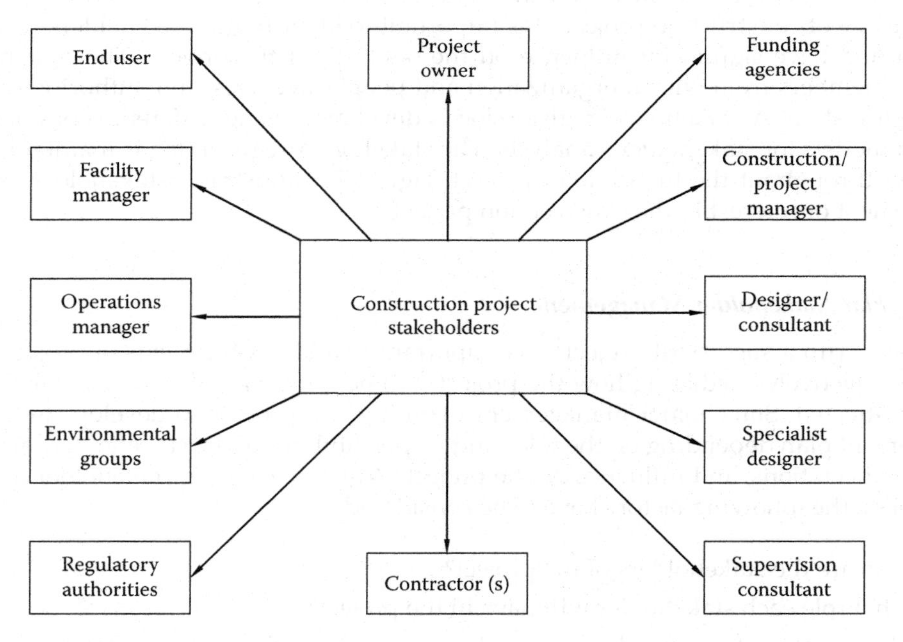

FIGURE 7.23
Construction project stakeholders.

- What opportunities do they present?
- What challenges or threats the stakeholders will have toward the project?
- What are the information or reports to be communicated to the stakeholders?
- What are the involvements of stakeholders in approvals and review of project activities?

In order to manage stakeholders' expectations in construction projects, the following construction-related activities have to be evolved:

1. Develop stakeholders' responsibility matrix
2. Develop stakeholders' requirements
3. Procedure to distribute performance report
4. Procedure to distribute project update reports
5. Stakeholders' involvement in approvals and reviews
6. Progress payment
7. Meetings

Table 7.12 is an example of stakeholders' responsibility matrix for construction project.

7.3.2.3 Manage Stakeholder Engagement

Managing stakeholder engagement is a process of involving stakeholders through communicating and working together to address the needs/expectations and issues of the stakeholders. Successful completion of construction project is dependent on meeting the expectations of stakeholders. Stakeholder engagement helps in

- Reducing risk in the project.
- Sharing experience and skills, thus mitigating the threats and uncertainties.
- Gaining stakeholders' support.
- Responding efficiently and effectively to the difficulties that may arise or to the issues that need to be resolved.
- Reducing conflict.
- Dealing with changing needs of stakeholders.
- Ensuring that the project deliverables meet stakeholder expectations.
- Successful completion of project within schedule, budget, and as per the approved scope.
- In construction project, stakeholder engagement is required to address following activities:
    - Reporting project status/performance
    - Reporting changes in scope, schedule, and budget
    - Project-related issues
    - Project payments
    - Conflicts
    - Variation orders
    - Anticipated/forecasted problems

TABLE 7.12

Stakeholders Responsibilities Matrix

LEGEND: P = Prepare/Initiate/Responsible, R = Review/Comment, B = Advise/Assist, A = Approve, E = Attend, C = Inform

Sr. No.	Activity	Owner/ Client	Construction Manager/Project Manager	Designer/ Consultant	Contractor	Supervisor	Regulatory Authority	Funding Agency	End User/ Facility Manager	Notes/ Comments
1	Project initiation	P	–	–	–	–	–	B	B	
2	Selection of construction manager	P	–	–	–	–	–	–	–	
3	Selection of designer	P	B	–	–	–	–	–	–	
4	Preparation of terms of reference (TOR)	A	P	–	–	–	–	–	–	
5	Preparation of design	A	B	P	–	–	R	–	–	
6	Value engineering	A	R	P						
7	Preparation of contract documents	A	B	P	–	–	–	–	–	
8	Project schedule	A	B	P	–	–	–	C	C	
9	Project budget	A	B	P	–	–	–	B	–	
10	Preparation of tendering documents	A	P	B	–	–	–	–	–	
11	Submission of bid	C	C	–	P	–	–	–	–	
12	Evaluation of bid	C	C	P	–	–	–	–	–	
13	Selection of contractor	A	P	B	–	–	–	C	C	
14	Approval of subcontractor	A	B	B	P	–	–	–	–	
15	Approval of contractor's staff	A	B	B	P	–	–	–	–	
16	Execution of works	C	C	R	P	R	–	–	–	
17	Supervision of works	C	C	R	P	P	–	–	–	
18	Approval of material	C	A	R	P	B	–	–	–	

(*Continued*)

TABLE 7.12 (*Continued*)

Stakeholders Responsibilities Matrix

LEGEND: P = Prepare/Initiate/Responsible, R = Review/Comment, B = Advise/Assist, A = Approve, E = Attend, C = Inform										
Sr. No.	**Activity**	**Owner/ Client**	**Construction Manager/Project Manager**	**Designer/ Consultant**	**Contractor**	**Supervisor**	**Regulatory Authority**	**Funding Agency**	**End User/ Facility Manager**	**Notes/ Comments**
19	Approval of shop drawings	C	C	A	P	B	–	–	–	
20	Construction schedule	C	A	R	P	B	–	–	–	
21	Monitoring progress	C	P	P	P	B				
22	Monitoring cost	C	P	P	B	B				
23	Payments	A	R	R	P	B	–	–	–	
24	Request for information	C	C	R	P	B				
25	Approval of change	A	B	R	P	B	–	–	–	
26	Quality plan	C	B	R	P	B	–	–	–	
27	Project quality	C	R	R	P	P				
28	Meetings	E	E	P	E	E	–	–	–	
29	Safety plan	C	B	R	P	B	–	–	–	
30	Site safety	C	C	B	P	P	–	–	–	
31	Testing and commissioning	C	C	R	P	D	–	–	C	
32	Authorities approval	C	C	B	P	B	A	–	–	
33	Snag list	C	C	R	P	P	–	–	C	
34	Substantial completion certificate	A	R	P	C	–	–	–	C	

7.3.2.4 Control Stakeholder Engagement

Controlling stakeholder engagement is a process to ensure that all the related information and reports are properly distributed to the concerned stakeholder.

In construction projects, stakeholder engagement control is done by distributing the following information and reports:

- Project status/performance
- Project status
- Project updates
- Project-related issues
- Project payments
- Project conflicts
- Change orders
- Minutes of meetings

7.3.3 Scope Management

As per *PMBOK® Guide Fifth Edition*, project scope management includes the processes required to ensure that the project includes all the work required, and only the work required, to complete the project successfully. Project scope management consists of six processes, which are as follows:

1. Plan scope management
2. Collect requirements
3. Define scope
4. Create work breakdown structure (WBS)
5. Validate scope
6. Control scope

In construction projects, scope management is the process that includes the activities to formulate and define the client's need, by establishing project objectives and goals properly addressed in order for the project to have clear direction, and controlling what is or is not involved in the project. The project scope documents explains the boundaries of the project, establishes project responsibilities for each team member, and sets up procedures for how completed works will be verified and approved. The scope describes the features and functions of the end product or the services to be provided by the project. During the project, the scope documentation helps the project team remain focused and on task. The scope statement also provides the project team with guidelines for making decisions about change requests during the project. It is essential that the scope statement should be unambiguous and clearly written to enable all the members of project team understand the project scope to achieve project objectives and goals.

Project development is a process spanning from the commencement of project initiation and ends with closeout and finalizing project records after project construction. The project development process is initiated in response to an identified need by the owner/end user. It covers a range of time-framed activities extending from identification of a project need, development of contract documents, and construction of the project.

By applying the concept of scope management processes methodology in development of construction project, the following construction-related activities can be evolved:

1. Develop scope management plan
   a. Project assumptions
   b. Constraints
   c. Key deliverables
   d. Project organization
   e. Roles and responsibilities
   f. Dependencies
   g. Milestones
   h. Project cost
   i. Quality
   j. Risks
   k. Safety regulations
   l. Environmental considerations
   m. Change control
2. Collect requirements
   a. Need statement
   b. Project goals and objectives
   c. Terms of reference (TOR)
3. Develop project scope documents
   a. Collect owner's preferred requirements
   b. Develop design
      i. Concept design
      ii. Schematic design
      iii. Detail design
   c. Project specifications
      i. General specifications
      ii. Particular specifications
   d. Contract documents
      i. General conditions
      ii. Particular conditions
      iii. Tender/bidding documents
4. WBS
   4.1 Project breakdown structure
      i. Project life cycle phases
      ii. Project scope
      iii. Bill of quantities (BOQ)
      iv. Schedule milestones

v. Cost estimates
vi. Quality management
vii. Resource management
viii. Risk management
ix. Documents

4.2 Organizational breakdown structure
4.3 WBS dictionary
4.4 Responsibility assignment matrix
4.5 Scope baseline

5. Validate scope
   a. Review of design documents
   b. Approval of design documents
   c. Review of contract documents
   d. Approval of contract documents
   e. Regulatory approvals
   f. Acceptance of project
6. Control scope
   a. Scope change control
   b. Variation orders
   c. Change orders
   d. Performance measures

7.3.3.1 Develop Scope Management Plan

Project scope management plan is a part of the overall project management plan that describes how the scope will be defined, developed, validated, and controlled. It explains how the project will be managed and how scope changes will be incorporated into the project management plan. Scope management plan establishes a structured process to ensure that the work performed by the project team is clearly within the established parameters and ensures that all project objectives are achieved.

The key benefit of this process is that it provides guidelines and direction of how the scope will be managed throughout the project. Scope management plan documents:

- Scope definition
- The scope management approach
- Roles and responsibilities of stakeholders pertaining to the project scope
- WBS (activity list)
- Organization breakdown structure
- Procedures to verify and approve the completed works
- Managing any changes in the project scope baseline
- Guidelines for making decision about change requests during execution of project and controlling project scope

7.3.3.2 Collect Requirements

Construction project development is initiated with the identification of need to develop a new facility or renovation/refurbishment of existing facility. It is essential to get a clear definition of the identified need or the problem to be solved by the new project. The owner's need must be well defined, indicating the minimum requirements of quality and performance, an approved budget, and a required completion date. The need should be based on real (perceived) requirements. The identified need is then assessed and analyzed to develop a need statement. Need assessment is conducted to determine the need. Need assessment is a systematic process for determining and addressing needs or *gaps* between current conditions and desired conditions (*want*). Need analysis is the process of identifying and evaluating need. The need statement is written based on the need analysis and is used to perform feasibility study to develop project goals and objectives and subsequently to prepare project scope documents.

The feasibility study takes its stating point from the output of project identification need. The need statement is the input to perform feasibility study. The main purpose of feasibility study is to evaluate the project need and decide whether to proceed with the project or stop. Depending on the circumstances, the feasibility study may be short or lengthy, simple or complex. In any case, it is the principle requirement in project development as it gives owner/client an early assessment of the viability of the project and the degree of risk involved.

Feasibility study can be categorized into the following functions:

- Legal
- Marketing
- Technical and engineering
- Financial and economical
- Social
- Environmental
- Risk
- Scheduling of project

The project feasibility study is usually performed by the owner through his own team or by engaging a specialist agency or individual. After completion and approval of feasibility study, it is possible to establish project goals and objectives. The goals and objectives must be

1. Specific
2. Measurable
3. Attainable/achievable
4. Realistic
5. Time (cost) limited

Once project goals and objectives are established, a comprehensive scope statement (TOR) is prepared by the owner/client or by the project manager on behalf of the owner describing in detail the project objectives and requirements to develop the project. TOR is a document that describes the intention of a project, the approach in which it will be constructed, and how it will be implemented. It can also be described as a specification of a team member's responsibilities and influence within a project. A TOR is an outline of a project, including its mission statement, its procedures and rules, and the different administrative aspects of the entire project.

7.3.3.3 Develop Project Scope Documents

Construction project scope documents are developed based on the requirements described in the TOR prepare by the owner/client or construction manager/project manager on behalf of the owner/client. The TOR gives the designer (consultant) a clear understanding for the development of the project. The designer (consultant) utilizes TOR to develop contract documents to suit the project delivery system. The contract documents for design–bid–build type of project delivery system mainly consist of the following:

1. Scope of work
2. Design drawings
3. BOQ
4. Technical specifications
5. Conditions of contract
6. Project schedule
7. Tender/bidding documents

In case of design–build type of project delivery system, tendering/bidding documents are prepared taking into consideration performance specifications for the project. Figure 7.24 shows an illustrative flowchart for the development of TOR (construction project documents), and Figure 7.25 illustrates the construction project development process.

7.3.3.4 Work Breakdown Structures

WBS is a hierarchical representation of system levels. WBS is a family tree, consisting of a number of levels, starting with the complete scope of work at level 1 at the top and progressing downward through as many levels as necessary, to obtain work elements (activities) that can be conveniently managed. WBS involves envisioning the project as a hierarchy of goal, objectives, activities, subactivities, and work packages. WBS is constructed by dividing the project into major elements, with each of these being divided into sub-elements. This is done till a breakdown is done in terms of manageable units of work for which responsibility can be defined. The hierarchical decomposition of activities continues until the entire project is displayed as a network of separately identified and non-overlapping activities. Each activity will be single purposed, of specific time duration, and manageable, its time and cost estimates easily derived, deliverables clearly understood, and responsibility for its completion clearly assigned.

In order to manage and control the project at different levels in the most effective manner, the project is broken down into a group of smaller subprojects/subsystems and then to small well-defined activities. Each element (activity) should be

- Definable
- Manageable

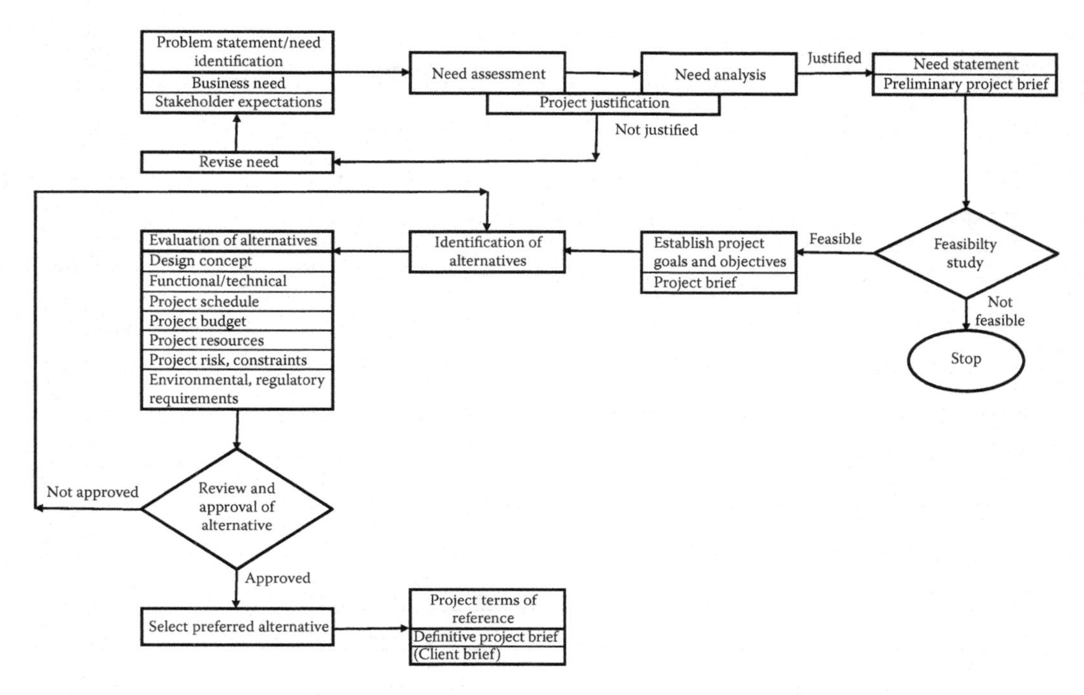

FIGURE 7.24
Flow chart for development of terms of reference.

- Measurable
- Estimable
- Independent
- Integratable
- Adaptable

Figure 7.26 illustrates an approach to development of WBS.

7.3.3.4.1 Project Breakdown Structure

Construction projects are constantly increasing in technical complexity, and the relationships and contractual groupings of those who are involved are also more complex and contractually varied. WBS approach to construction projects help understand the entire process of project management and to manage and control its activities at different levels of various phases to ensure timely completion of the project with economical use of resources to make the construction project most qualitative, competitive, and economical. WBS is a deliverable-oriented grouping of project work elements shown in graphical display and organized and subdivided the total scope of work into small and manageable components (activities).

The WBS development involves following major steps:

1. Identify final product(s) necessary to achieve total project scope.
2. Identify major deliverables.

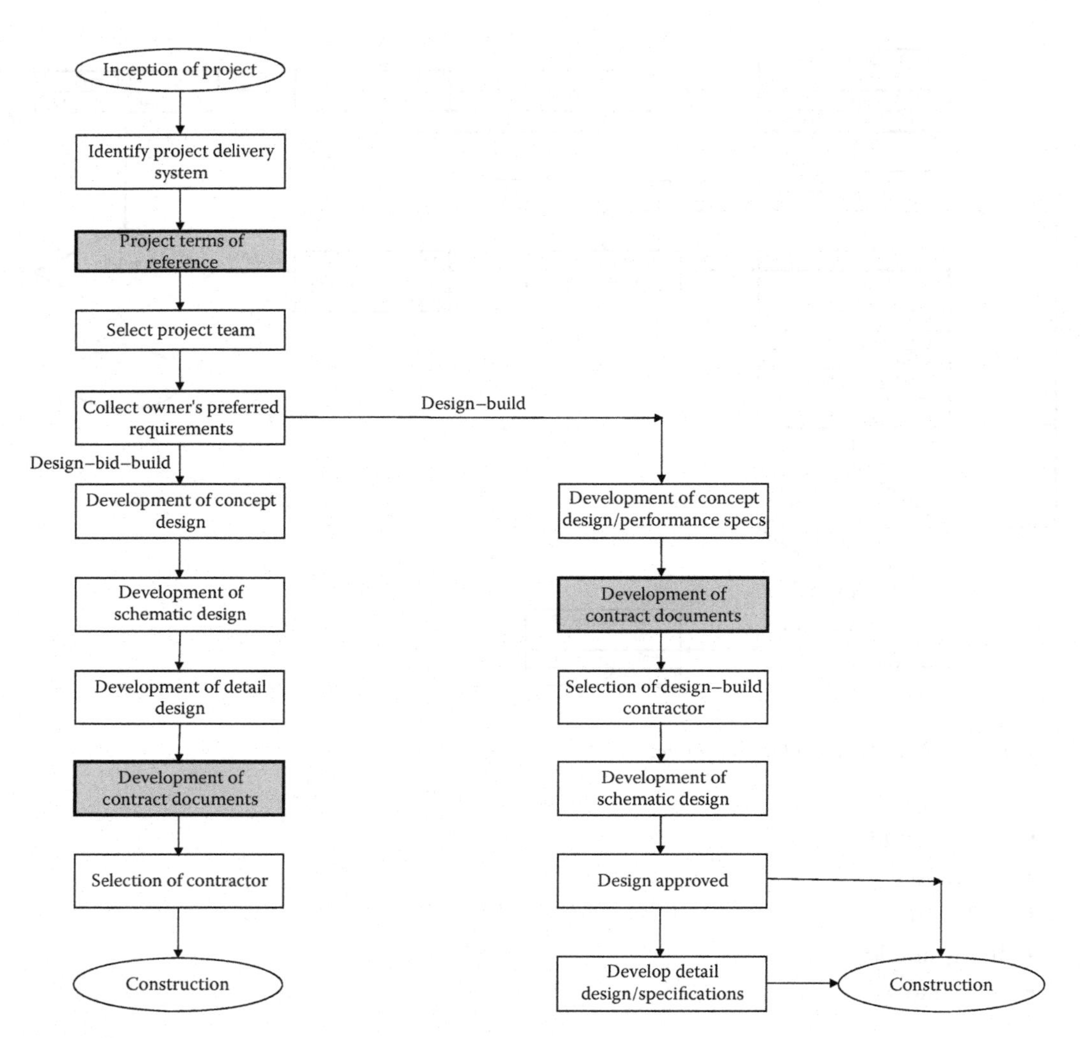

FIGURE 7.25
Development of project scope documents.

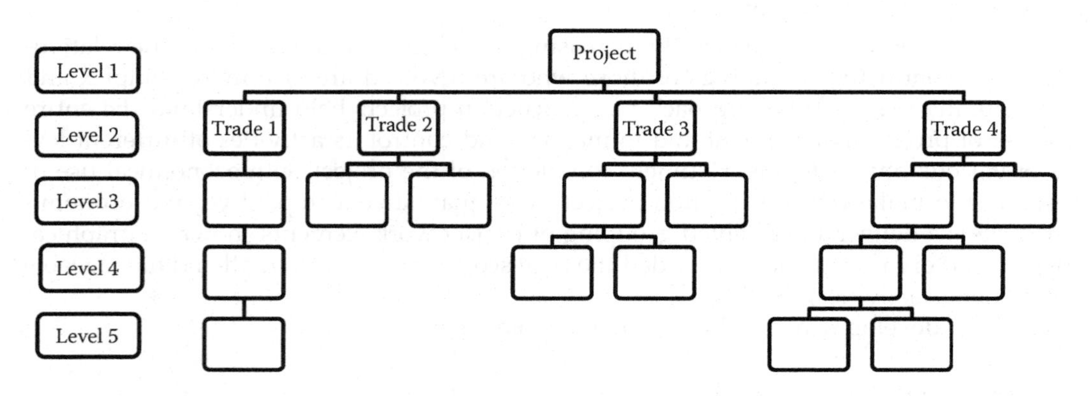

FIGURE 7.26
Approach to development of work breakdown structure.

3. Divide these major deliverables to a level of detail appropriate to meet managing and controlling requirements of the project.
4. Divide each of these deliverables into its components.

While preparing WBS, the following points need to be considered:

- Complexity of the project.
- Size of the project.
- Reporting requirements.
- Resource allocation.
- Works included in the scope of work are only included in the WBS.
- Duration of task or activities should be as per industry common practice, which is *8-80 rule*. This rule recommends that the lowest level of work should be no less than 8 hours and no more than 80 hours.
- Each level is assigned a unique identification number.

The lowest level of WBS is known as *work package*. A work package can be divided into specific activities to be performed. Figure 7.27 illustrates typical levels of WBS process in a construction project.

Traditional construction projects involve the following three main parties:

1. Owner/client
2. Designer/consultant
3. Contractor

The involvement and interaction between owner/client, designer/consultant, and contractor(s) depends on the construction project procurement strategy followed by the owner. Based on the strategy, the WBS for construction project can be developed to suit the requirements of the project development at different stages (study, design, and construction) of the project. There are typically two types of WBS:

1. Project WBS
   - It is prepared by the contractor by taking into considerations the requirements listed in contract documents. It is also known as contractor's construction schedule.
2. Project summary WBS
   - It is prepared by the designer (consultant), summarizing the entire project requirements. It is part of contract documents signed between owner/client and contractor.

There are different approaches followed for development of the WBS. These are as follows:

- Physical location—which is further divided into floor levels and zones
- Project development stages such as study, design, and construction

Figure 7.28 illustrates WBS for a project design.

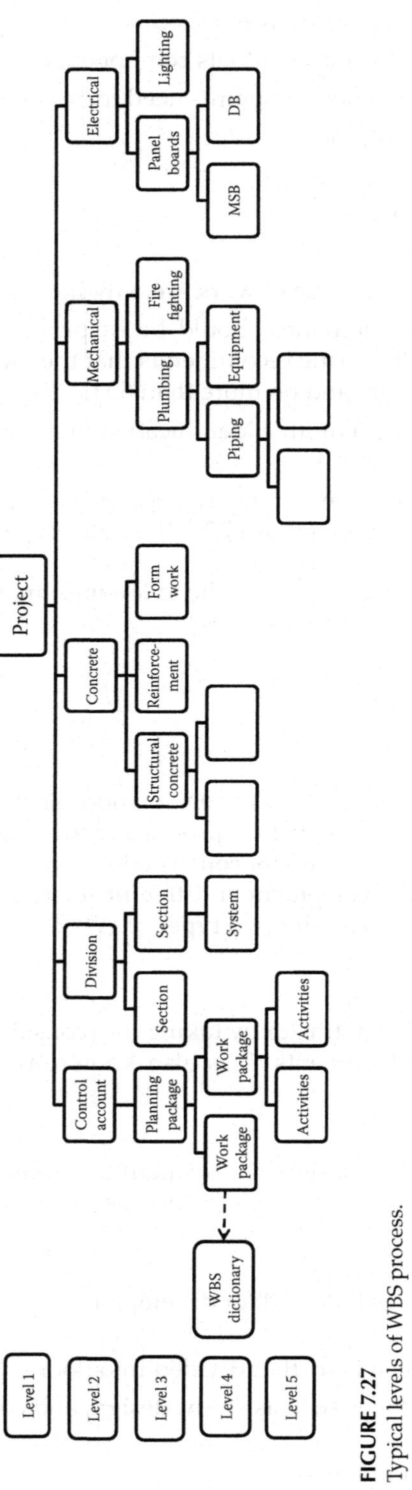

FIGURE 7.27
Typical levels of WBS process.

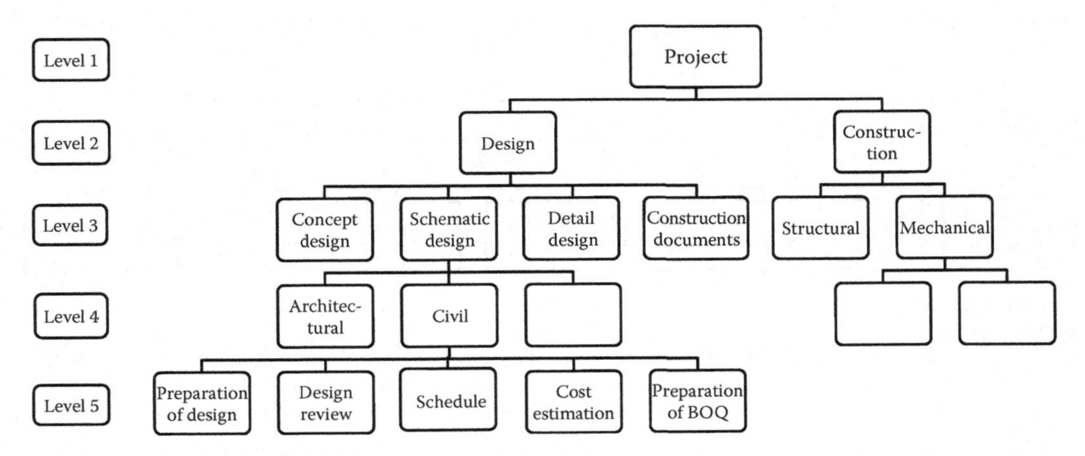

FIGURE 7.28
Typical levels of WBS process (project design).

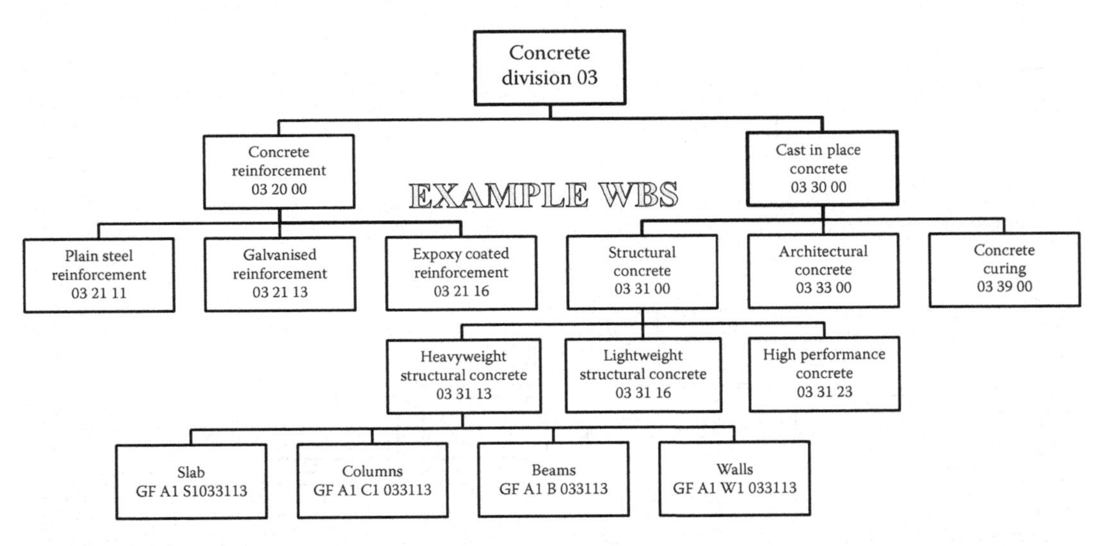

FIGURE 7.29
WBS for concrete works.

In most construction projects, Construction Specification Institute (CSI)'s MasterFormat® coding system is followed to develop WBS. Following figures are guidelines to develop WBS by taking into consideration divisions, sections, and titles from MasterFormat®. More detailed levels can be developed according to the project requirements. Figure 7.29 illustrates WBS for concrete works in the construction project.

Figure 7.30 illustrates WBS for fire suppression works in a construction project.

Figure 7.31 illustrates WBS for plumbing works in a construction project.

Figure 7.32 illustrates WBS for HVAC works in a construction project.

Figure 7.33 illustrates WBS for electrical works in a construction project.

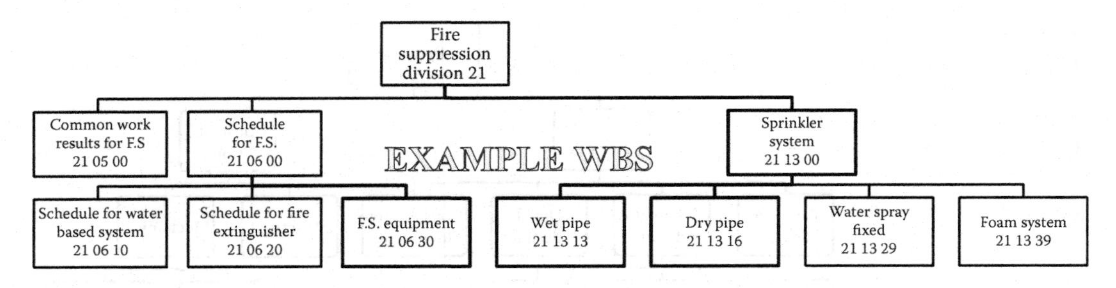

FIGURE 7.30
WBS for fire suppression works.

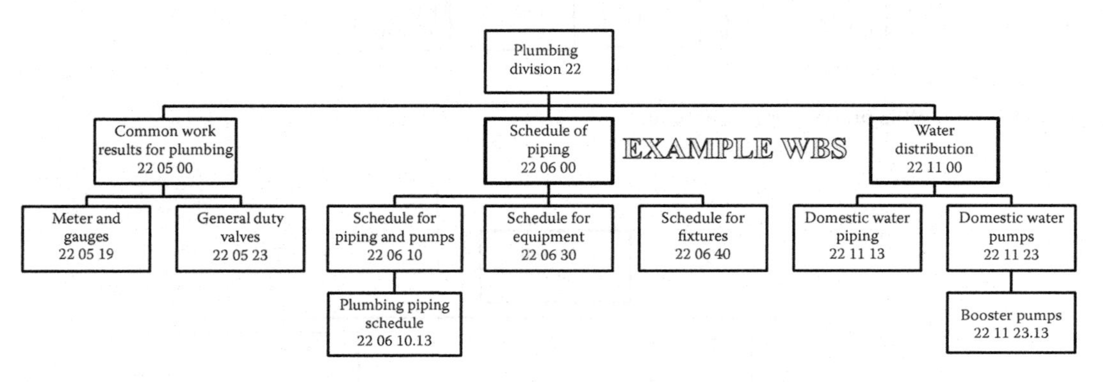

FIGURE 7.31
WBS for plumbing works.

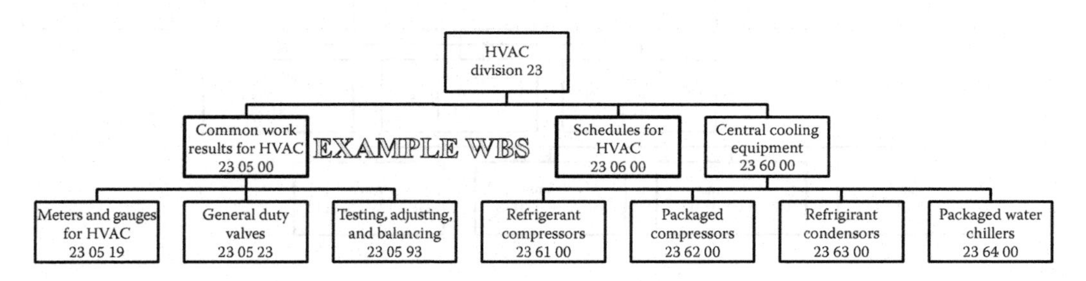

FIGURE 7.32
WBS for HVAC works.

7.3.3.4.2 Organizational Breakdown Structure

Organizational breakdown structure (OBS) is a hierarchical organizational relationship of the project teams, including subcontractors, responsible for managing the designate scope of work described within the WBS. It is used as a framework for assigning work relationship. The WBS identifies what work is to be done, whereas OBS identifies the individual that will do the work. An OBS for the project can be simple or complex depending on the size and complexity of the project.WBS with appropriate type of organization (projectized, functional, and matrix) assures that all the scope of work is accounted for, and each element of works is assigned to the level of responsibility for planning, tracking progress,

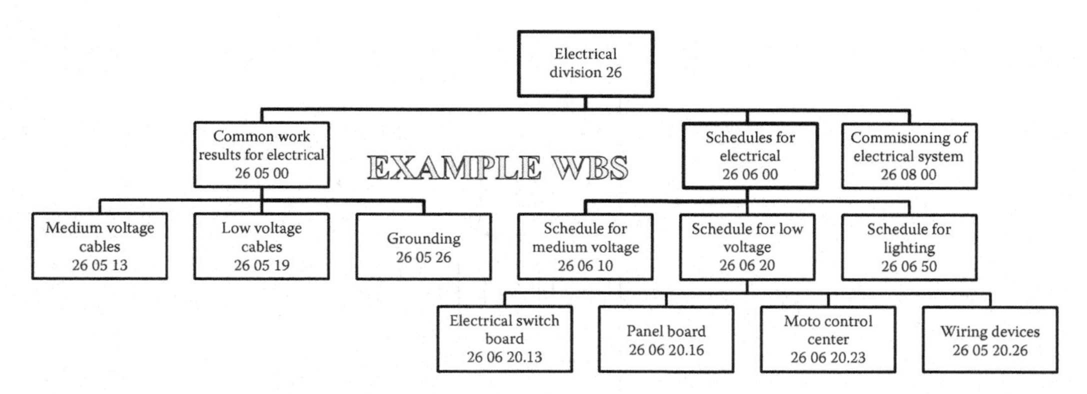

FIGURE 7.33
WBS for electrical works.

costing, and reporting. The organizational and personal relationship can be established at any of the several levels within the project and functional organization.

Figure 7.34 illustrates construction supervisor's site organization for a construction project.

Figure 7.35 illustrates contractor's site organization for a construction project.

7.3.3.4.3 WBS Dictionary

WBS dictionary is a set of companion documents to WBS, which describes WBS elements in the WBS, and includes the following:

1. Code of account identifier for each WBS elements
2. Scope description (statement of work)
3. BOQ
4. Activities
5. Schedule milestone
6. Cost estimates
7. Resources requirements
8. Quality requirements
9. Technical and other references
10. Responsible organization/person
11. Deliverables
12. Contract information

Table 7.13 illustrates an example WBS dictionary for a construction project.

7.3.3.4.4 Responsibility Assignment Matrix

Responsibility assignment matrix (RAM) depicts the intersection of WBS and OBS. It identifies the specific responsibility for specific project task and provides a realistic picture of the resources needed and can identify if the project has enough resources for its successful completion. RAM uses the WBS and OBS to link deliverables and/or activities to resources and relates the WBS element/task to the organization and the named individual who is

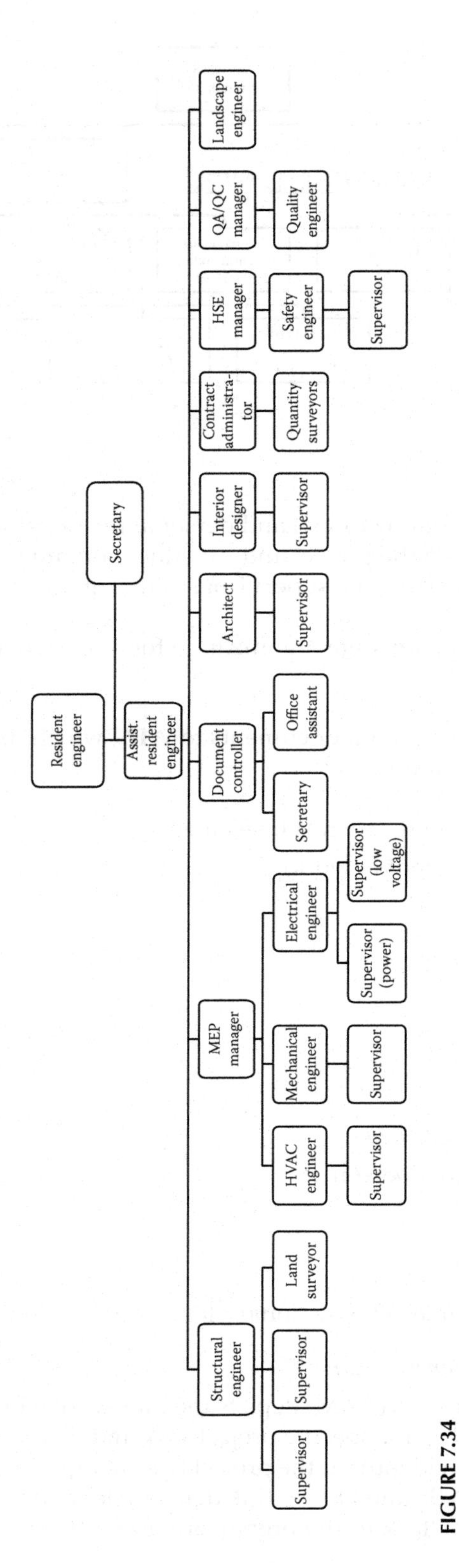

FIGURE 7.34
Consultant's (supervisor's) organizational breakdown structure.

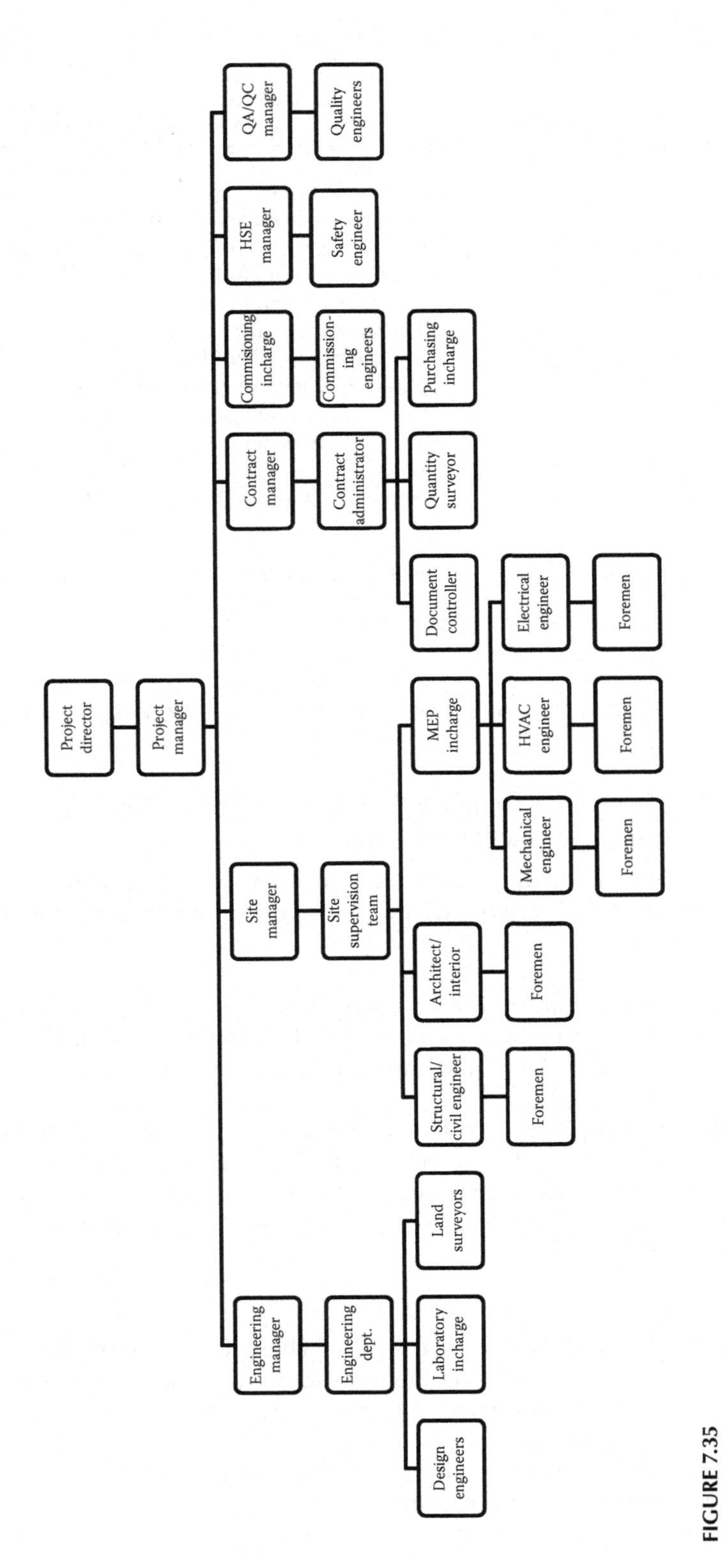

FIGURE 7.35
Contractor's organizational breakdown structure.

TABLE 7.13
WBS Dictionary

Sr. No.	WBS Element	WBS Code	WBS Element Description
1	Demolition/site preparation	024113	Clear and remove surface matters, windblown sand, heaps of soil or debris of any kind on the proposed roads, shoulders, side slopes and existing road, and shoulder, if any, all as per specifications and drawings
2	Cast in place concrete	033000	Plain concrete using sulfate resistant cement including formwork and additives
3	Concrete unit masonry assembly	042200	Block work nonload bearing including control joints, filler and sealant at wall heads, mortar, anchors, reinforcement
4	Bituminous damp proofing	071113	Two layers of torch applied 4 mm thick water proofing membrane, including laps and bituminous primer
5	Plumbing system	221400	Supply, install, commission, and handover water drainage system
6	HVAC system	230593	Testing, adjusting, and balancing of HVAC
7	Switchboards and Panelboards	262413	Supply, install, and test the following main low tension board complete with all accessories as per the requirements of the drawings and specifications.

responsible for the assigned scope of a control account. The following are the procedures to prepare RAM:

- Develop WBS and prepare a complete list of project deliverables.
- Identify team members who will be assigned for the project.
- Allocate responsibility to team members according to their field of competency, expertise, and required experience that will support the work to produce different project deliverables.

Figure 7.36 illustrates relation between WBS and OBS for construction project.

RAM can be a simple tick box (check mark) or RACI type. Table 7.14 illustrates an example of RAM for a construction project, and Table 7.15 illustrates an example of RACI matrix for a construction project.

WBS is a critical tool for organizing the work and preparing realistic schedule and cost estimates. It helps reporting, monitoring, and controlling the project. WBS supports integrating responsibilities for performing various works with various organizations and individuals by having a direct relationship between the WBS elements related to the identified individual through RAM.

7.3.3.4.5 Scope Baseline

As per PMI *PMBOK® Guide Fifth Edition* (p. 131), "The Scope Baseline is the approved version of a scope statement, work breakdown structure (WBS), and its associated WBS Dictionary, that can be changed only through formal change control procedures and is used as a basis for comparison."

For a construction project, scope baseline is technical scope baseline that describes the performance capabilities that the project must provide at the end of construction phase to meet the owner's needs. The scope base line is the contract documents developed by the

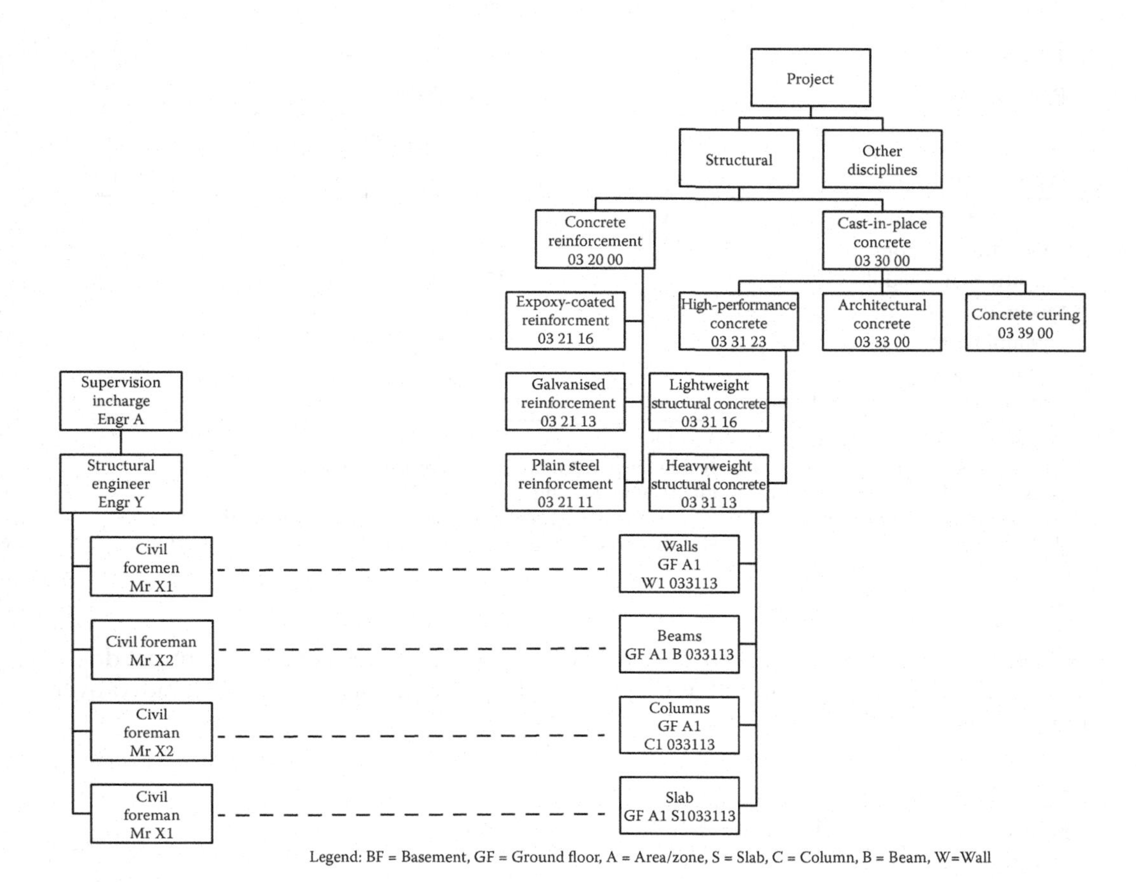

FIGURE 7.36
RAM for concrete works.

TABLE 7.14
Responsibility Assignment Matrix (RAM)

	Office, Role, or Person				
WBS Element	**Owner**	**Resident Engineer**	**Planning and Control**	**Quality Manager**	**Quantity Surveyor**
Notice to proceed	X				
Approval, permits	X				
Conduct meetings		X			
Action on request for information (RFI)		X			
Project monitoring			X		
Construction quality				X	
Review of progress (interim) payment					X

TABLE 7.15
RACI Matrix

WBS Element	Person			
	Owner	Project Manager	Consultant	Contractor
Bonds and guarantees	I	I	A	R
Subcontractor approval	A	C	C	R
Construction schedule	I	C	A	R
Meetings	C	C	R	I
Submittal logs	I	C	A	R
As-built drawings	I	I	A	R
Substantial completion certificate	A	C	R	I

R, Responsible; A, Accountable; C, Consulted; I, Informed.
R: Responsible:-The person (s) responsible for performance of the concerned activity.
A: Accountable:-The person (s) accountable for ensuring the activity is completed.
C: Consulted:-The person (s) who must be consulted prior to or during the execution of the activity.
I: Informed:-The person (s) who must be informed about the progress and outcome of the activity.

designer based on the requirements described in TOR. The scope baseline (contract documents) is developed based on following construction documents, which are handed over to the successful bidder:

- Working drawings
- Technical specifications (particular specifications)
- BOQ
- General specifications
- General conditions
- Particular condition

Contractor has to follow these documents for implementation/execution of project works.

7.3.3.5 Validate Scope

Validate scope is the process of formalizing acceptance of completed project deliverables. It is a method to ensure that

- Project design conforms to current applicable codes and standards.
- Project design is taken care of all the requirements listed under TOR.
- Assess whether value engineering analysis has been performed and the recommendations are incorporated in the project baseline.
- All the material and equipment installed comply with specifications requirements.
- All the works at site are performed per approved shop drawings by approved material.
- Regulatory approval is obtained.

- All the installed/executed works are checked, inspected at every stage to confirm that they have been installed/executed as specified, using specified and approved materials, installations method recommended by the manufacturer, to meet intended use of the project.
- Corrective actions or defect repairs completed.
- Inspection and tests are carried out to ascertain operational requirement.
- The work is documented, and changes are recorded.
- Records of inspection and tests are maintained to verify approved construction methods and materials were used.
- Outstanding defects and works (punch list) are listed and documented.
- As-built drawings, documents, and manuals are ready for handover.
- Start-up test plans are established to demonstrate that all the systems installed in the project meet required operations and safety requirements.
- Handover/takeover program is established.
- All the requirements for facility management are documented with all the information and knowledge that is required to strategically and physically manage the new facility.

In a construction project, the project elements (intermediate deliverables) need to be verified, reviewed, approved, and accepted at different stages of the project life cycle to ensure that completed project deliverables meet owner's needs and expectations (goals and objectives). It is the assessment of readiness of construction or execution and to confirm the completeness and accuracy of the project as per agreed upon scope baseline.

It is performed mainly during the following phases of the construction project:

1. Design phase
2. Construction phase
3. Testing, commissioning, and handover phase

The following are the four main steps needed to establish an assessment, review, approval, and acceptance of the project:

1. Identify the elements, items, material, system, and products to be reviewed and checked in each of the trade (architectural, structural, fire suppression, plumbing, HVAC, electrical, low-voltage systems, landscape, external, etc.).
2. Identify the frequency of checking and inspection.
3. Identify the agency/persons authorized to check and approve.
4. Identify the indications/criteria for performance monitoring and control.

Figure 7.37 illustrates scope validation process for a construction project.

7.3.3.6 Control Scope

Control scope is the process of monitoring the project scope and managing any changes to the scope baseline. It is common that despite all the efforts devoted to develop the contract documents (scope baseline), the contract documents cannot provide complete information

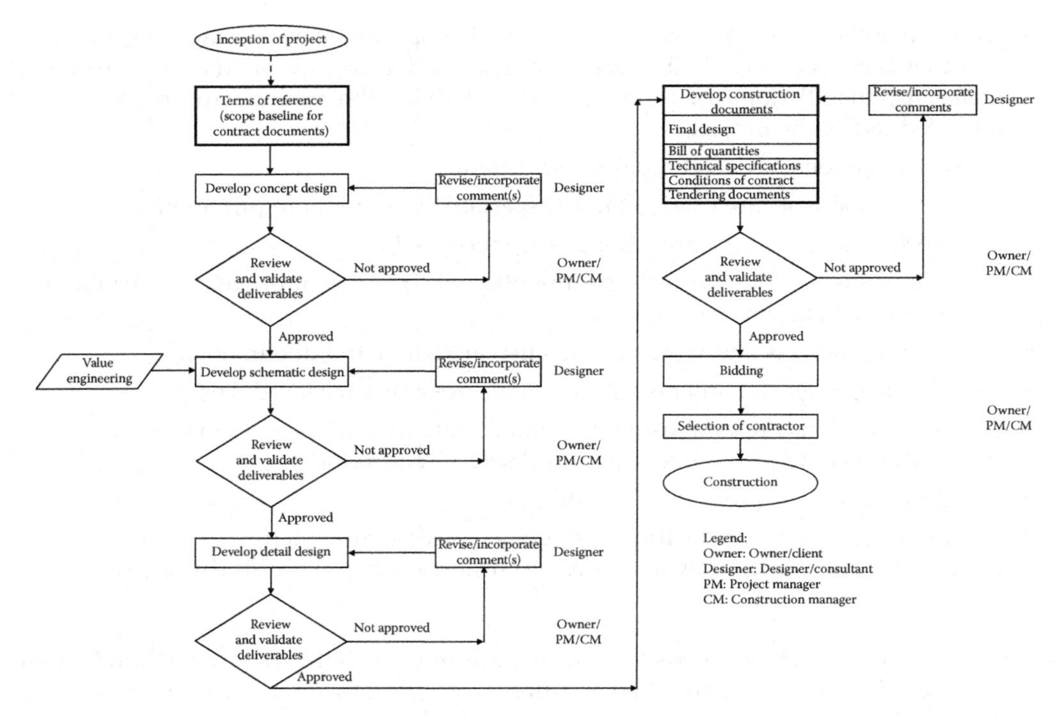

FIGURE 7.37
Scope validation process for construction project (design and bidding stage).

about every possible condition or circumstance that the construction team may encounter. Figure 7.38 illustrates the scope control process in a construction project.

During construction process, circumstances may come to light that necessitate minor or major changes to the original contract. These changes may occur due to following reasons:

1. Differences/errors in contract documents
2. Construction methodology
3. Nonavailability of specified material
4. Regulatory changes to use certain type of material
5. Technological changes/introduction of new technology
6. Value engineering process
7. Additional work instructed
8. Omission of some works

Table 7.16 lists the causes of changes in construction projects.

These changes are identified as the construction proceeds. These changes or adjustments are beneficial and help build the facility to achieve project objective. Prompt identification of such requirements helps both the owner and contractor to avoid unnecessary disruption of work and its impact on cost and time. The impacts and consequences of changes in the construction project vary according to

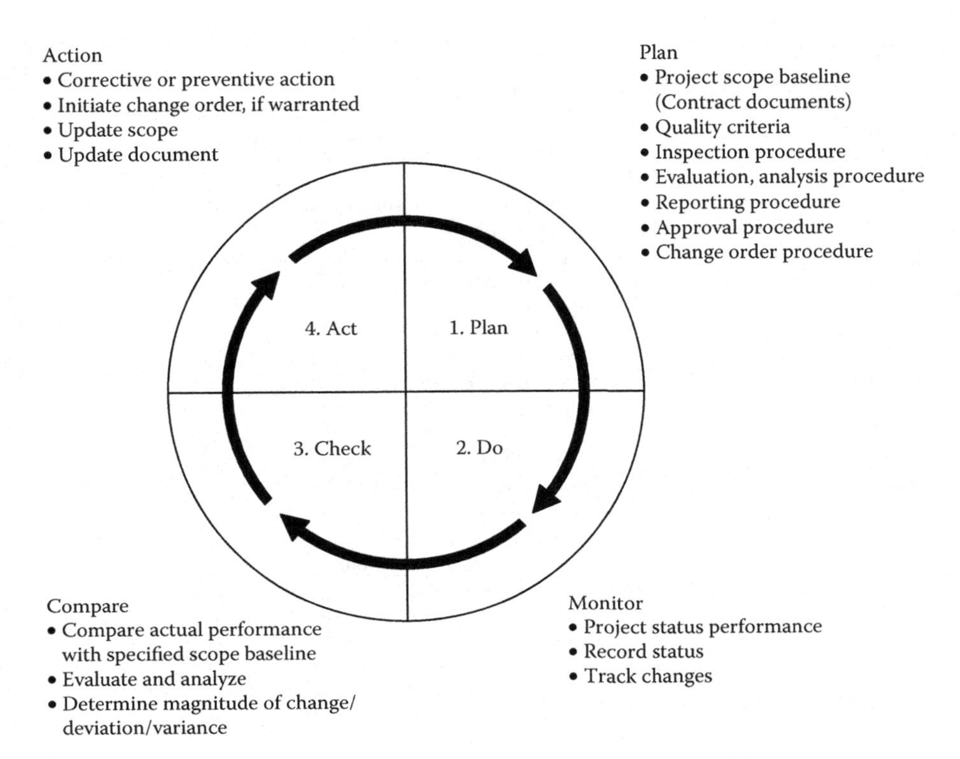
Action
• Corrective or preventive action
• Initiate change order, if warranted
• Update scope
• Update document
Plan
• Project scope baseline (Contract documents)
• Quality criteria
• Inspection procedure
• Evaluation, analysis procedure
• Reporting procedure
• Approval procedure
• Change order procedure
4. Act
1. Plan
3. Check
2. Do
Compare
• Compare actual performance with specified scope baseline
• Evaluate and analyze
• Determine magnitude of change/ deviation/variance
Monitor
• Project status performance
• Record status
• Track changes

FIGURE 7.38
Scope control process.

- Type and nature of changes.
- Time of occurrence or observance of error or omission.
- Change needed for the benefit of the project.

A critical change may have a negative impact on the project baseline(s). It is important to establish a reliable change control system to manage changes to the project scope. The changes should be managed to maximize the benefits to the project and minimize the negative impacts and effects on the project. The change control system should include the following:

- Submission procedure
- Evaluation and review procedure
- Impact on project
- Approval
- Reporting
- Baseline(s) updates
- Document update

The changes that arise during construction can be initiated either by the owner or by the contractor or even by any of the stakeholder to the construction project.

TABLE 7.16

Causes of Changes in Construction Project

Sr. No.	Causes
I	*Owner*
	I-1 Delay in making the site available on time
	I-2 Change of plans
	I-3 Financial problems/payment delays
	I-4 Change of schedule
	I-5 Addition of work
	I-6 Omission of work
	I-7 Project objectives are not well defined
	I-8 Different site conditions
	I-9 Value engineering
II	*Designer (Consultant)*
	II-1 Inadequate specifications
	—a. Design errors
	—b. Omissions
	II-2 Scope of work not well defined
	II-3 Conflict between contract documents
	II-4 Coordination among different trades and services
	II-5 Design changes/modifications
	II-6 Introduction of latest technology
III	*Contractor*
	III-1 Process/methodology
	III-2 Substitution of material
	III-3 Nonavailability of specified material
	III-4 Charges payable to outside party due to cancellation of certain items/products
	III-5 Delay in approval
	III-6 Contractor's financial difficulties
	III-7 Unavailability of manpower
	III-8 Unavailability of equipment
	III-9 Material not meeting the specifications
	III-10 Workmanship not to the mark
IV	*Miscellaneous*
	IV-1 New regulations
	IV-2 Safety considerations
	IV-3 Weather conditions
	IV-4 Unforeseen circumstances
	IV-5 Inflation
	IV-6 Fluctuation in exchange rate
	IV-7 Government policies

Source: Abdul Razzak Rumane (2013), *Quality Tools for Managing Construction Projects*, CRC Press, Boca Raton, FL. Reprinted with permission from Taylor & Francis Group.

Identification of discrepancies/errors and changes in the specified scope are common in construction projects. Prompt identification of such requirements helps both the owner and contractor to avoid unnecessary disruption of work and its impact on cost and time. Contractor uses a request for information (RFI) form to request technical information from the supervision team. These queries are normally resolved by the concerned supervision engineer. However, it is likely that the matter has to be referred to the designer as the RFI has many other considerations to be taken care, which may be beyond the capacity of supervision team member to resolve. Normally there is a defined period to respond to RFI. Such queries may result in variation to the contract documents. It is in the interest of both the owner and contractor to resolve RFI expeditiously to avoid its effect on construction schedule.

Figure 7.39 illustrates an RFI form that a contractor submits to the consultant to clarify differences/errors observed in the contract documents, change in construction methodology, change in the specified material, etc.

Figure 7.40 illustrates a process to resolve scope change (contractor initiated).

Figure 7.41 illustrates a variation order request form that a contractor submits to the owner/consultant for approval of change(s) in the contract.

Figure 7.42 illustrates a process to resolve request for variation (contractor initiated).

Figure 7.43 illustrates a site works instruction (SWI) form. It gives instruction to the contractor to proceed with the change(s). All the necessary documents are sent along with the SWI to the contractor. SWI is also used to instruct contractor for owner-initiated changes.

Similarly, if the contractor requires any modification to the specified method, then the contractor submits a request for modification to the owner/consultant. Figure 7.44 illustrates the request for modification. Usually these modifications are carried out, by the contractor, without any extra cost and time obligation toward the contract.

Figure 7.45 illustrates the process to resolve scope change (owner initiated).

It is the normal practice that, for the benefit of project, the engineer's representative assesses the cost and time related to SWI or request for change over and obtain preliminary approval from the owner and the contractor is asked to proceed with such changes. The cost and time implementation is negotiated and formalized simultaneously/later to issue the formal variation order. In all the circumstances where a change in contract is necessary, owner approval has to be obtained. Figure 7.46 illustrates the form used by the engineer's representative to obtain change order approval from owner.

Once cost and time implications are negotiated and finalized and both the owner and contractor approve the same, variation order is issued to the contractor and changes are adjusted with contract sum and schedule. Figure 7.47a illustrates variation order form issued to formalize the change order, and Figure 7.47b illustrates the attachment to variation order.

7.3.4 Schedule Management

Planning and scheduling are often used synonymously for preparing construction program because both are performed interactively. Planning is the process of identifying the activities necessary to complete the project, whereas scheduling is the process of determining the sequential order of the planned activities and the time required to complete the activity. Scheduling is the mechanical process of formalizing the planned functions and assigning the starting and completion dates to each part or activity of the work that will be carried out in such a manner that the whole work proceeds in a logical sequence and in an orderly and systematic manner. Scheduling is a time-based graphical presentation of project activities/tasks utilizing information on available resources and time constraints. Table 7.17 illustrates advantages of project planning and scheduling.

Project Name

Consultant Name

REQUEST FOR INFORMATION (TECHNICAL)

CONTRACT NO.: ______________ R.F.I. NO. : ______

CONTRACTOR.: ______________ DATE: ______

To: Resident Engineer ______________

REF:

SUBJECT:

REQUEST FOR INFORMATION (Technical)

This form is used by the contractor to request information and is normally sent to the A/E who responds on the same form.

SAMPLE FORM

CONTRACTOR: ______________

DISTRIBUTION: Employer ☐ Engineer ☐ R.E. ☐

RESPONSE BY RE.:

Signature of R. E. ______________ Date ______

RESPONSE RECEIVED:
FOR CONTRACTOR: ______________ DATE : ______

DISTRIBUTION: Employer ☐ Engineer ☐ R.E. ☐

FIGURE 7.39
Request for information.

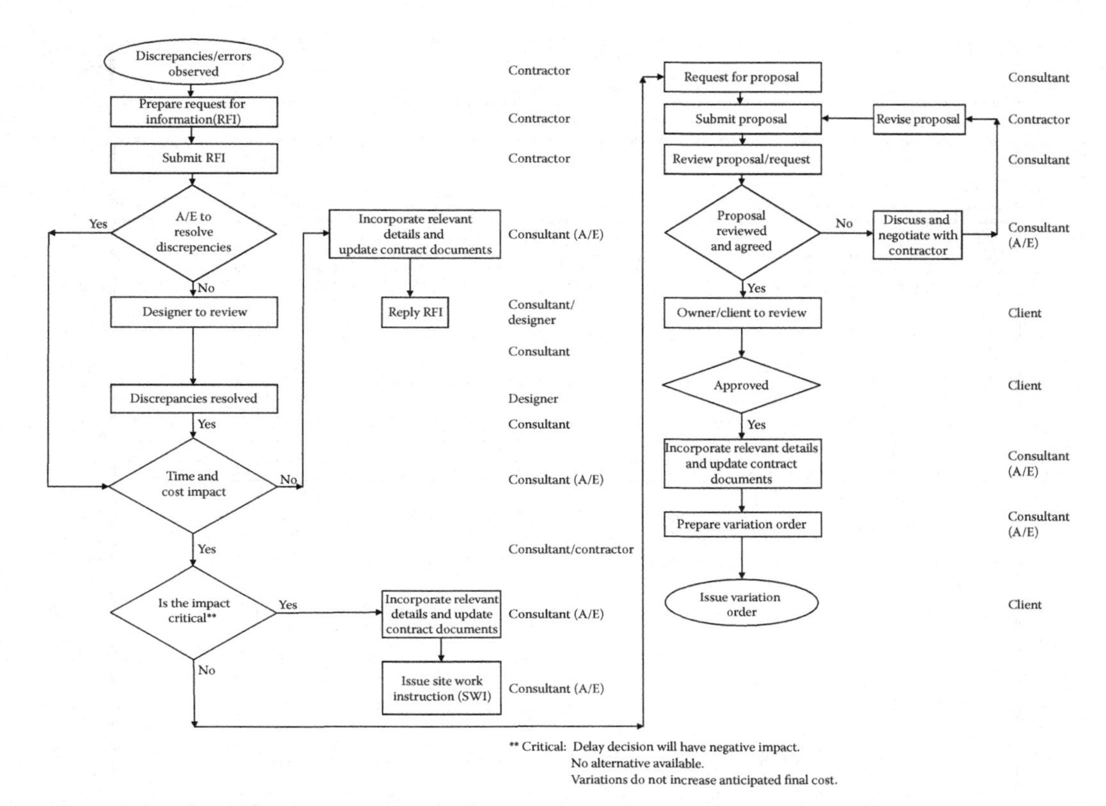

FIGURE 7.40
Process to resolve scope change (contractor initiated).

As per *PMBOK® Guide Fifth Edition,* project time management includes the processes required to manage the timely completion of the project. Project time management consists of the following seven processes:

1. Plan schedule management
2. Define activities
3. Sequence activities
4. Estimate activity resources
5. Estimate activity duration
6. Develop schedule
7. Control schedule

As per *Construction Extension PMBOK® Guide Third Edition,* there are three additional processes applicable for construction projects. These are as follows:

1. Activity weights definitions
2. Progress curves development
3. Progress monitoring

Project Name

Consultant Name

REQUEST FOR VARIATION

CONTRACT NO. : ____________ NO. : ________

CONTRACTOR: ____________ DATE : ________

TO:

SAMPLE FORM

PROPOSED VARIATION:

☐ PRODUCT ☐ METHOD OF FABRICATION ☐ METHOD OF INSTALLATION

SPECIFIED PRODUCT ____________

PROPOSED PRODUCT ____________

SPEC. SECTION # ______ PAGE # ______ ARTICLE # ______

DRG REF ______ DRG # ______ REV # ______

SPECIFIED MANUFACTURER ____________

PROPOSED MANUFACTURER ____________

BRIEF PRODUCT DESCRIPTION ____________

REASON FOR PROPOSED VARIATION

CHANGE IN DESIGN | REQUIRED BY AUTHORITIES | SITE CONDITIONS | SWI

COST AND TIME EFFECT

COST NO ☐ YES ☐ AMOUNT --------- (ADDITION)

TIME NO ☐ YES ☐ DAYS ----------

ATTACHMENTS :

1 Schedule of additions/ommissions
2 Bill Summary
3 Rate Analysis
4 Measurements

Technical and cost comparison sheets must be attached with this request, other wise it will not be reviewed. Contractor shall fill and submit two forms to the OWNER. Front sheet only shall be returned to Contractor with OWNER action.

WE (THE MAIN CONTRACTOR) CERTIFIES AND UNDERTAKES THAT:

CONTRACTOR'S REP ____________ DATE/TIME ________

RECEIVED BY A/E ____________ DATE/TIME ________

REVIEW AND ACTION BY OWNER

☐ Approved ☐ Not Approved ☐ Approved as Noted ☐ Incomplete Data Resubmit

COMMENTS

APPROVED SUBJECT TO COMPLIANCE WITH CONTRACT DOCUMENTS

Authorised Signature ________ ________ DATE/TIME: ________

THE APPROVAL OF ANY VARIATION REQUEST SHALL BE SOLELY AT THE DIRECTION OF THE OWNER AND SUCH APPROVAL SHALL IN NO WAY RELIEVE THE CONTRACTOR OF ANY OF HIS LIABILITIES AND OBLIGATIONS UNDER THE CONTRACT.

RECEIVED BY CONTRACTOR: ____________ DATE/TIME: ________

cc: OWNER ☐ EMPLOYER ☐ R.E. ☐ ☐

FIGURE 7.41
Request for variation.

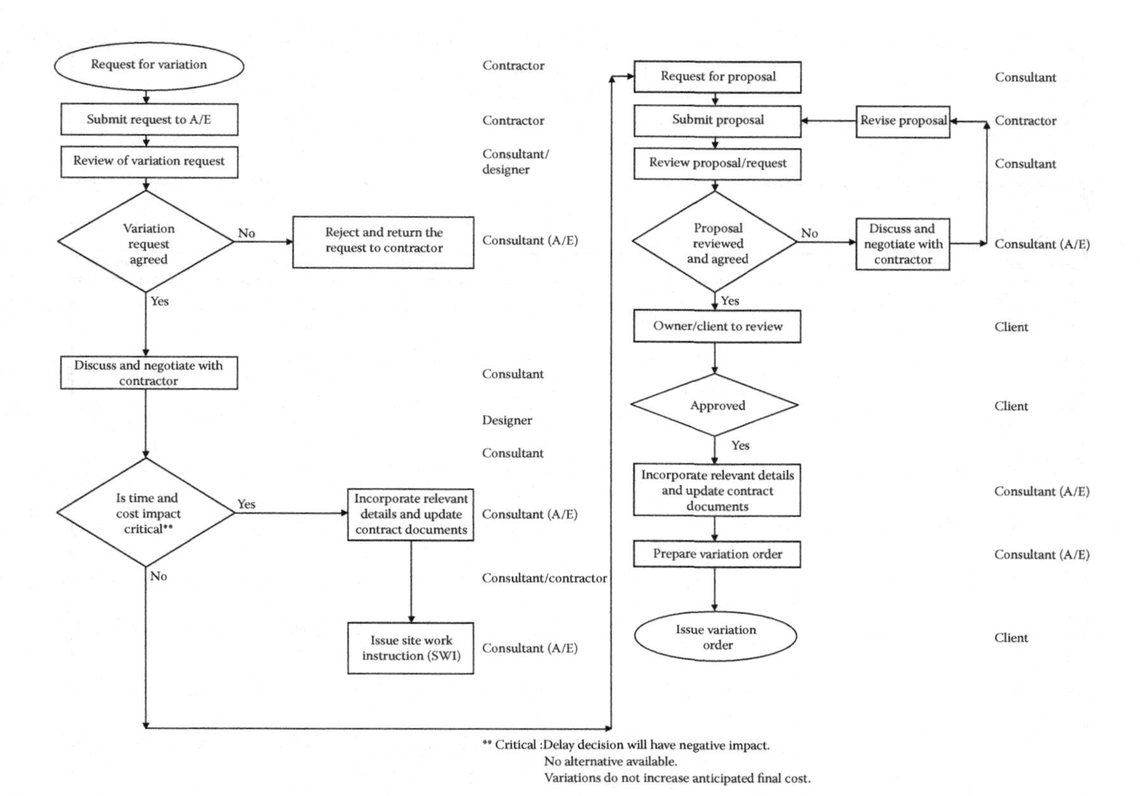

FIGURE 7.42
Process to resolve request for variation.

In construction projects, project time management methodology can be termed *schedule management* having the following construction-related activities:

1. Identify project activities/tasks
    1.1 Study stage
    1.2 Design stage
    1.3 Bidding and tendering
    1.4 Construction stage
2. Sequence project activities
    2.1 Study stage
    2.2 Design stage
    2.3 Bidding and tendering
    2.4 Construction stage
3. Develop project network diagram
    3.1 Study stage
    3.2 Design stage
    3.3 Bidding and tendering
    3.4 Construction stage

Project Name
Consultant Name

SITE WORK INSTRUCTION

CONTRACT NO. :	______	NO. :	______
CONTRACTOR :	______	DATE :	______

SUBJECT :

SITE WORK INSTRUCTION (SWI)

S.W.I. involves an anticipated change in the work. All S.W.I. must be authorized and signed by the Owner Representative OR Authorised Signatory.

A S.W.I. is an instruction to the contractor to proceed prior to the issuing of a Variation Order (V.O.). Whenever time allows, a V.O. will be issued instead of S.W.I. .

Owner Rep. Signature **Date**

THIS SITE WORKS INSTRUCTION (S.W.I.) IS A NOTICE TO PROCEED AND MAY INVOLVE CHANGE IN COST AND OR TIME. YOU ARE REQUIRED TO ADVISE THE ENGINEER WITHIN 14 DAYS OF ANY ADDITIONAL COST AND OR TIME REQUIRED TO COMPLY WITH THIS INSTRUCTION.

RECEIVED FOR CONTRACTOR: ______ **DATE:** ______

DISTRIBUTION: Owner ☐ Engineer ☐ R.E. ☐

FIGURE 7.43
Site work instruction.

Project Name
Consultant Name

REQUEST FOR MODIFICATION
(AT NO EXTRA COST & OR TIME TO THE EMPLOYER)

CONTRACTOR : ______________________ DATE : ________
CONTRACT NO : ______________________ NO : ________

TO : Owner Name **CC:** **A/E** ☐

PROPOSED MODIFICATION TO:
☐ DESIGN DRAWING ☐ METHOD OF FABRICATION ☐ METHOD OF INSTALLATION
DESIGN DRAWING NO ______________________
SECTION # ______________________
CONTRACTOR'S PROPOSED DRAWING NO. ______________________
REASON FOR MODIFICATION ______________________

COST AND TIME SAVINGS *At no extra cost and or time to the Employer*

COST	NO	☐	YES	☐	AMOUNT ------- (DEDUCTION) ________
TIME	NO	☐	YES	☐	DAYS

ATTACHMENTS :
Supplier confirmation letter

SAMPLE FORM

CONTRACTOR'S REP ______________________ DATE/TIME : ________
RECEIVED BYA/E ______________________ DATE/TIME : ________

REVIEW AND ACTION BY OWNER
☐ Approved ☐ Not Approved ☐ Approved as Noted ☐ Incomplete Data Resubmit
COMMENTS

Authorised Signatories ______________________ DATE/TIME : ________

RECEIVED BY CONTRACTOR : ______________________ DATE/TIME : ________
cc: Owner ☐ Engineer ☐ R.E. ☐ ☐

FIGURE 7.44
Request for modification.

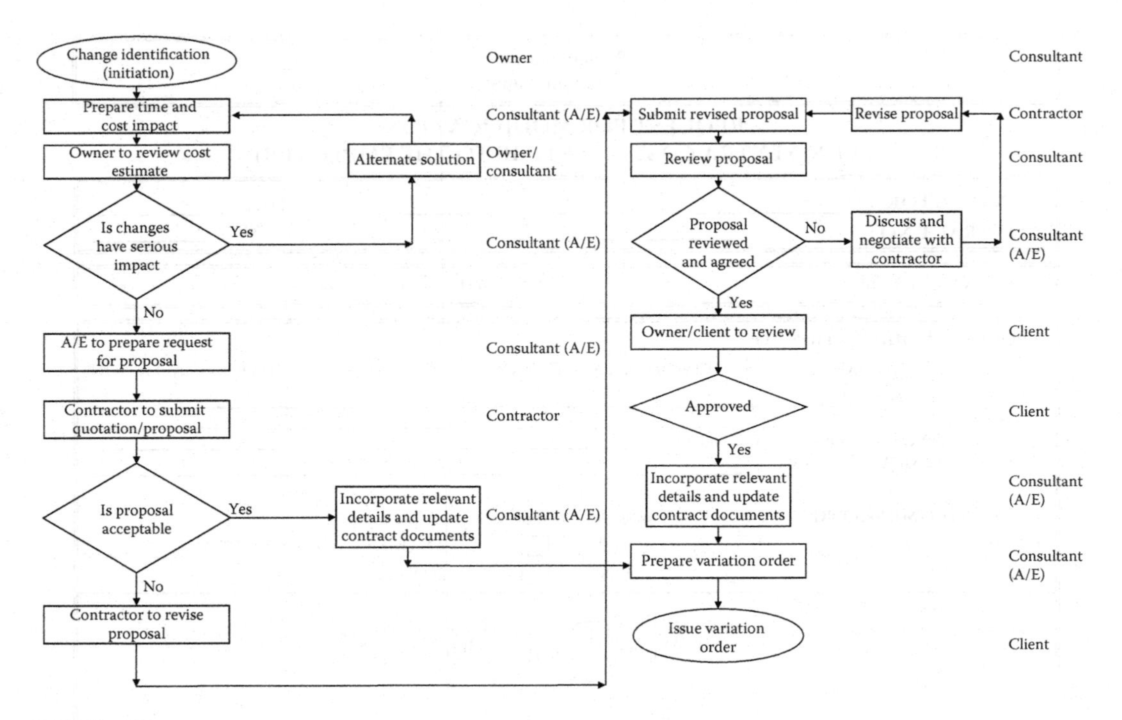

FIGURE 7.45
Process to resolve scope change (owner initiated).

4. Estimate activity resources
    4.1 Study stage
    4.2 Design stage
    4.3 Construction stage
5. Estimate activity duration
    5.1 Study stage
    5.2 Design stage
    5.3 Bidding and tendering
    5.4 Construction stage
6. Develop schedule
    6.1 Project master schedule (project life cycle)
    6.2 Design development schedule
    6.3 Construction (project) schedule (tendering documents)
    6.4 Contractor's construction schedule
        6.4.1 Detailed schedule
        6.4.2 Monthly schedule
        6.4.3 Weekly/biweekly schedule
        6.4.4 Look ahead (15 days)

Project Name
Consultant Name

VARIATION ORDER PROPOSAL (VOP)

CONTRACT NO : ______________ VOP No. ______________

CONTRACTOR : ______________ Date : ______________

The "Employer" decision is requested for approval /rejection of this Variation Order Proposal as described below. Should the VOP be approved by the "Employer" an order to proceed shall be issued to the Contractor for his further action.

INITIATED / PROPOSED BY:

☐ ENGINEER ☐ ENGINEER'S REP. ☐ CONTRACTOR ☐ OTHERS

REASONS:

☐ CHANGE IN DESIGN ☐ SWI ______________

☐ REQUIRED BY AUTHORITIES ______________

☐ REQUIRED BY SITE CONDITIONS

BRIEF DESCRIPTION & LOCATION :

SAMPLE FORM

BASIS OF V.O.EVALUATION :

PRORATA BOQ PRICES : ☐

PROPOSAL from CONTRACTOR : ☐

APPROXIMATE COST IMPACT ______________ APPROXIMATE % ______________

APPROXIMATE TIME IMPACT ______________ ANY DELAY ______________

RELATED REFERENCES :

ENGINEER'S RECOMMENDATION :

RESIDENT ENGINEER'S SIGNATURE ______________ DATE : ______________

Distribution: ☐ Employer ☐ Engineer ☐ Resident Engineer ☐

FIGURE 7.46
Variation order proposal.

Project Name
Consultant Name

VARIATION ORDER (VO)

CONTRACT NO. : ______________________ V.O. NO. : ____________

CONTRACTOR : ______________________ DATE : ____________

In accordance with clause ----- of Document ---- Conditions of Contract, you hereby ordered to undertake work and/or amend the Contract as detailed below:

DESCRIPTION OF WORK

According to Clause ---------, Extension of Time for Completion and Valuation of V.O. are as follows:

Original Contract Price: ____________ Contract Completion Date: ____________
Previous Vos: ____________ Previous Extension(s): ____________
Value This VO: ____________ Extension This VO: ____________
Revised Contract Price: ____________ Revised Completion Date: ____________

We the undersigned Contractor hereby agree to carry out the works as ordered by this Variation Order and accept in full and final settlement of all related consequential costs and time

Recommended by Resident Engineer Agreed by Contractor

Recommended by Engineer Agreed by Employer

Distribution ☐ Employer ☐ Engineer ☐ Resident Engineer ☐ Contractor
(a)

FIGURE 7.47
(a) Variation order.

(Continued)

Project Name
Consultant Name

ATTACHMENT TO VARIATION ORDER

CONTRACTOR : ____________________ CONNTRACT No. : ____________

V.O. No. : __________ Date : ____________

1. Site Works instructions incorporated into this Variation Order :

2. Previous correspondence references (attached) :

SAMPLE FORM

3. Revised Drawings and Specifications (attached) :

4. Schedule of Omissions/Additions :

5. Rate Analysis for New Items :

(b)

FIGURE 7.47 (Continued)
(b) VO attachment.

TABLE 7.17
Advantages of Project Planning and Scheduling

Sr. No.	Advantages
1	It facilitates management by objectives
2	It facilitates execute the work in an organized and structured manner
3	It eliminates or minimizes uncertainties
4	It reduces risk to the minimum
5	It helps proper coordination
6	It helps integration of project activities for smooth flow of project work
7	It help reduce rework
8	It improves the efficiency of the process and increase productivity
9	It improves communication
10	It establishes timely reporting system
11	It establishes the duration of each activity
12	It provides basis for monitoring and controlling of project work
13	It helps to establish benchmark for tracking the quantity, cost, and timing of work required to complete the project
14	It helps foresee problems at early stage
15	It helps to know the responsibility and authority of people involved in the project

7. Analyze schedule
8. Project schedule (schedule baseline)
9. Monitor and control schedule
    9.1 Schedule monitoring
    9.2 Progress status
    9.3 Progress reporting
    9.4 Forecasting
    9.5 Schedule update
    9.6 Schedule control
    9.7 Performance reporting

Figure 7.48 illustrates the schedule development process.

7.3.4.1 Identify Project Activities

Project activities are the lowest level of WBS evolved from decomposition of work package. Each project activity has a specific duration in which the activity is performed (started and completed). The project activity consumes time (duration), consumes resources, has a definable start and finish, is associated with cost, and is easy to monitor and manage. It is assignable, measurable, and quantifiable. In construction projects, the terms *project activities* and *project tasks* are used interchangeably. The definition of project activity is an important factor for identification and documentation. Many of the activities are repeated

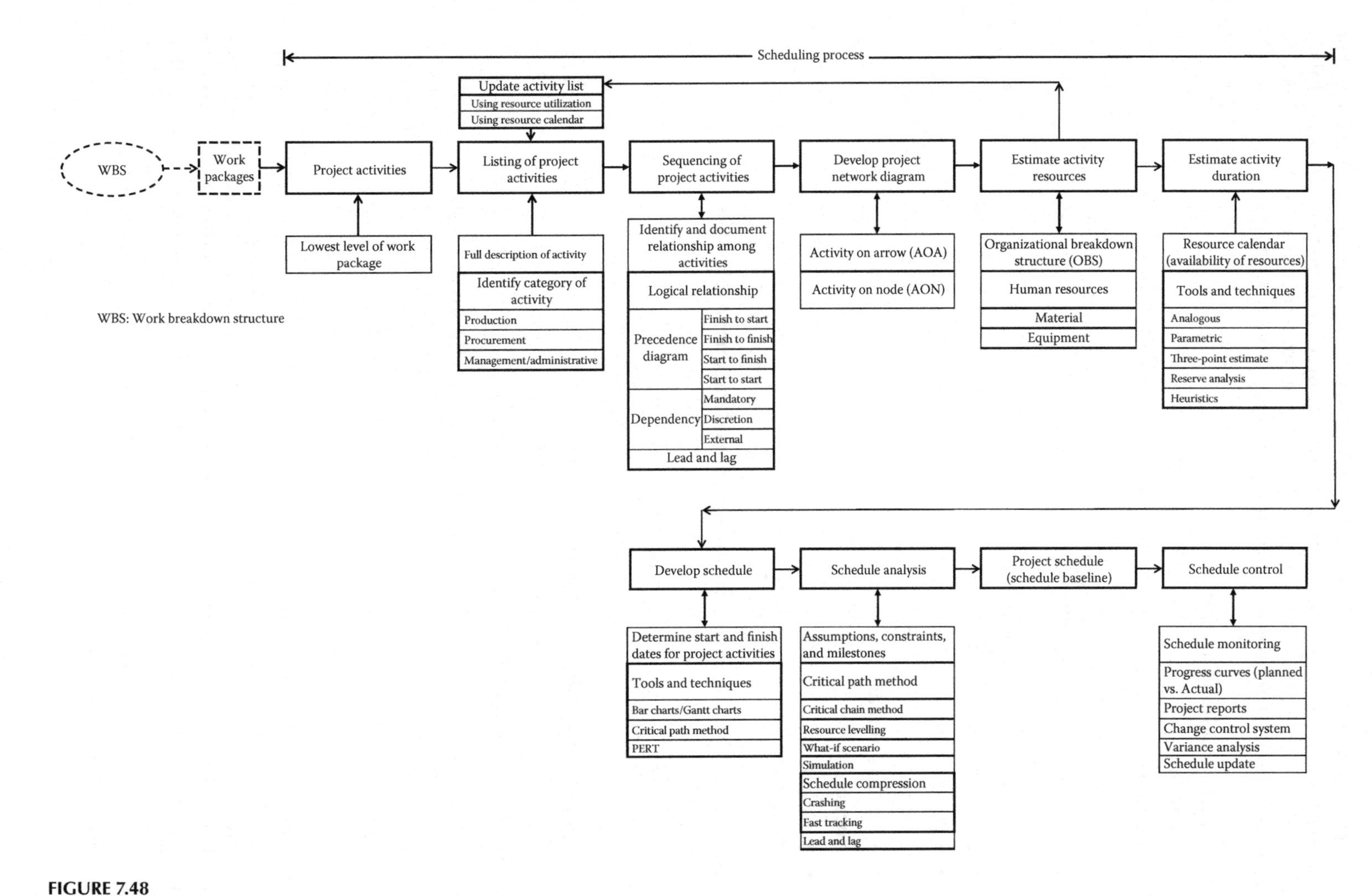

FIGURE 7.48
Schedule development process.

in the construction project, but their use and performance requirements are different. For example, pipes can be water system pipes, firefighting system pipes, and pipes for chilled water in HVAC system; therefore, proper identification and documentation of a specific activity is important. Activities are normally categorized into three types:

1. Management/administrative
2. Procurement
3. Production

Figure 7.49 illustrates an example of project activities and codes of construction project. The activity categories are as follows:

- Administration/management
  - Submittal
  - Approval
- Procurement
  - Purchase of materials
- Production
  - Installation
  - Testing

The activity code ID is developed taking into consideration the following:

- L--------General
- A--------Submittal
- W-------Women
- CSI divisions and subdivisions

7.3.4.2 List Project Activities (BOQ)

Activity list is a tabulation of activities to be performed in the project and is to be included in the project schedule. Activity list should contain the following information related to an activity:

- Activity name
- Activity identification number
- Brief description of the activity

List of activities is known as *BOQ*. It is an itemized listing of project-specific activities identified by the drawings and specifications prepared by the designer/quantity surveyor (consultant) to meet the project objectives. The quantities may be measured in number, length, area, volume, weight, and time, depending upon the scale of unit required to define the activity. BOQ is basically listed under different trade to which the activity/task belongs. BOQ is frequently used to

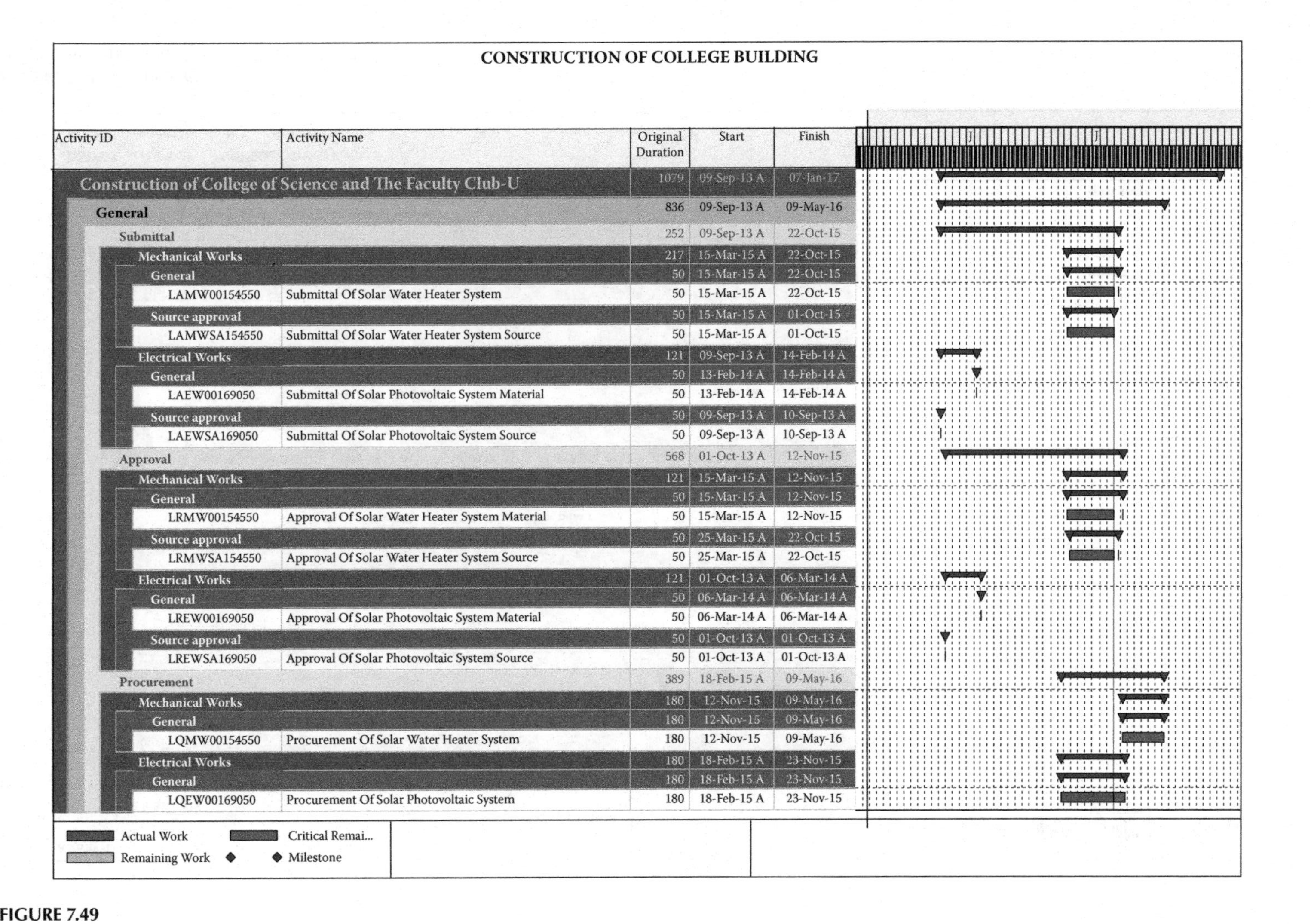

CONSTRUCTION OF COLLEGE BUILDING

Activity ID	Activity Name	Original Duration	Start	Finish
Construction of College of Science and The Faculty Club-U		1079	09-Sep-13 A	07-Jan-17
General		836	09-Sep-13 A	09-May-16
Submittal		252	09-Sep-13 A	22-Oct-15
Mechanical Works		217	15-Mar-15 A	22-Oct-15
General		50	15-Mar-15 A	22-Oct-15
LAMW00154550	Submittal Of Solar Water Heater System	50	15-Mar-15 A	22-Oct-15
Source approval		50	15-Mar-15 A	01-Oct-15
LAMWSA154550	Submittal Of Solar Water Heater System Source	50	15-Mar-15 A	01-Oct-15
Electrical Works		121	09-Sep-13 A	14-Feb-14 A
General		50	13-Feb-14 A	14-Feb-14 A
LAEW00169050	Submittal Of Solar Photovoltaic System Material	50	13-Feb-14 A	14-Feb-14 A
Source approval		50	09-Sep-13 A	10-Sep-13 A
LAEWSA169050	Submittal Of Solar Photovoltaic System Source	50	09-Sep-13 A	10-Sep-13 A
Approval		568	01-Oct-13 A	12-Nov-15
Mechanical Works		121	15-Mar-15 A	12-Nov-15
General		50	15-Mar-15 A	12-Nov-15
LRMW00154550	Approval Of Solar Water Heater System Material	50	15-Mar-15 A	12-Nov-15
Source approval		50	25-Mar-15 A	22-Oct-15
LRMWSA154550	Approval Of Solar Water Heater System Source	50	25-Mar-15 A	22-Oct-15
Electrical Works		121	01-Oct-13 A	06-Mar-14 A
General		50	06-Mar-14 A	06-Mar-14 A
LREW00169050	Approval Of Solar Photovoltaic System Material	50	06-Mar-14 A	06-Mar-14 A
Source approval		50	01-Oct-13 A	01-Oct-13 A
LREWSA169050	Approval Of Solar Photovoltaic System Source	50	01-Oct-13 A	01-Oct-13 A
Procurement		389	18-Feb-15 A	09-May-16
Mechanical Works		180	12-Nov-15	09-May-16
General		180	12-Nov-15	09-May-16
LQMW00154550	Procurement Of Solar Water Heater System	180	12-Nov-15	09-May-16
Electrical Works		180	18-Feb-15 A	23-Nov-15
General		180	18-Feb-15 A	23-Nov-15
LQEW00169050	Procurement Of Solar Photovoltaic System	180	18-Feb-15 A	23-Nov-15

FIGURE 7.49
Project activities and codes.

(Continued)

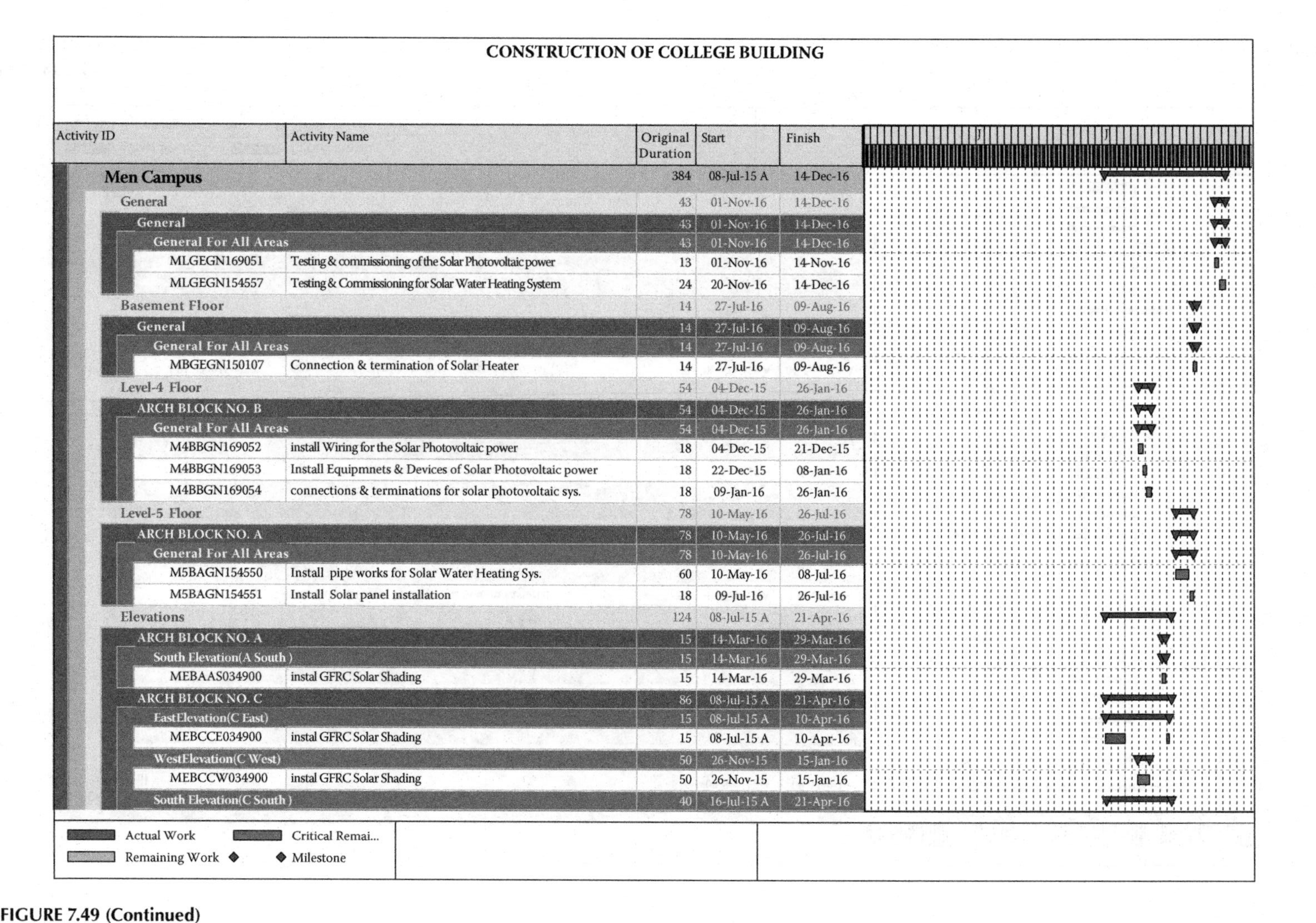

CONSTRUCTION OF COLLEGE BUILDING

Activity ID	Activity Name	Original Duration	Start	Finish
Men Campus		384	08-Jul-15 A	14-Dec-16
General		43	01-Nov-16	14-Dec-16
General		43	01-Nov-16	14-Dec-16
General For All Areas		43	01-Nov-16	14-Dec-16
MLGEGN169051	Testing & commissioning of the Solar Photovoltaic power	13	01-Nov-16	14-Nov-16
MLGEGN154557	Testing & Commissioning for Solar Water Heating System	24	20-Nov-16	14-Dec-16
Basement Floor		14	27-Jul-16	09-Aug-16
General		14	27-Jul-16	09-Aug-16
General For All Areas		14	27-Jul-16	09-Aug-16
MBGEGN150107	Connection & termination of Solar Heater	14	27-Jul-16	09-Aug-16
Level-4 Floor		54	04-Dec-15	26-Jan-16
ARCH BLOCK NO. B		54	04-Dec-15	26-Jan-16
General For All Areas		54	04-Dec-15	26-Jan-16
M4BBGN169052	install Wiring for the Solar Photovoltaic power	18	04-Dec-15	21-Dec-15
M4BBGN169053	Install Equipmnets & Devices of Solar Photovoltaic power	18	22-Dec-15	08-Jan-16
M4BBGN169054	connections & terminations for solar photovoltaic sys.	18	09-Jan-16	26-Jan-16
Level-5 Floor		78	10-May-16	26-Jul-16
ARCH BLOCK NO. A		78	10-May-16	26-Jul-16
General For All Areas		78	10-May-16	26-Jul-16
M5BAGN154550	Install pipe works for Solar Water Heating Sys.	60	10-May-16	08-Jul-16
M5BAGN154551	Install Solar panel installation	18	09-Jul-16	26-Jul-16
Elevations		124	08-Jul-15 A	21-Apr-16
ARCH BLOCK NO. A		15	14-Mar-16	29-Mar-16
South Elevation(A South)		15	14-Mar-16	29-Mar-16
MEBAAS034900	instal GFRC Solar Shading	15	14-Mar-16	29-Mar-16
ARCH BLOCK NO. C		86	08-Jul-15 A	21-Apr-16
EastElevation(C East)		15	08-Jul-15 A	10-Apr-16
MEBCCE034900	instal GFRC Solar Shading	15	08-Jul-15 A	10-Apr-16
WestElevation(C West)		50	26-Nov-15	15-Jan-16
MEBCCW034900	instal GFRC Solar Shading	50	26-Nov-15	15-Jan-16
South Elevation(C South)		40	16-Jul-15 A	21-Apr-16

FIGURE 7.49 (Continued)
Project Activities and Codes.

(Continued)

CONSTRUCTION OF COLLEGE BUILDING

Activity ID	Activity Name	Original Duration	Start	Finish
MEBCCS034900	instal GFRC Solar Shading	40	16-Jul-15 A	21-Apr-16
ARCH BLOCK NO. D		22	11-Jul-15 A	28-Jan-16
West Elevation(D West)		15	07-Jan-16	22-Jan-16
MEBDDW034900	instal GFRC Solar Shading	15	07-Jan-16	22-Jan-16
North Elevation(D North)		15	11-Jul-15 A	28-Jan-16
MEBDDN034900	instal GFRC Solar Shading	15	11-Jul-15 A	28-Jan-16
Women Campus		403	02-Dec-15	07-Jan-17
General		30	08-Dec-16	07-Jan-17
General		30	08-Dec-16	07-Jan-17
General For All Areas		30	08-Dec-16	07-Jan-17
WLGEGN154557	Testing & Commissioning for Solar Water Heating System	30	08-Dec-16	07-Jan-17
WLGEGN169051	Testing & commissioning of the Solar Photovoltaic power	13	08-Dec-16	21-Dec-16
Basement Floor		15	16-Aug-16	30-Aug-16
General		15	16-Aug-16	30-Aug-16
General For All Areas		15	16-Aug-16	30-Aug-16
WBGEGN150107	Connection & termination of Solar Heater	15	16-Aug-16	30-Aug-16
Level-4 Floor		54	02-Dec-15	24-Jan-16
ARCH BLOCK NO. H		54	02-Dec-15	24-Jan-16
General For All Areas		54	02-Dec-15	24-Jan-16
W4BHGN169052	install Wiring for the Solar Photovoltaic power	18	02-Dec-15	19-Dec-15
W4BHGN169053	Install Equipmnets & Devices of Solar Photovoltaic power	18	20-Dec-15	06-Jan-16
W4BHGN169054	connections & terminations for solar photovoltaic sys.	18	07-Jan-16	24-Jan-16
Level-6 Floor		98	10-May-16	15-Aug-16
ARCH BLOCK NO. G		98	10-May-16	15-Aug-16
General For All Areas		98	10-May-16	15-Aug-16
W6BGGN154550	Install pipe works for Solar Water Heating Sys.	80	10-May-16	28-Jul-16
W6BGGN154551	Install Solar panel installation	18	29-Jul-16	15-Aug-16

Actual Wor k
Remaining Work
Critical Remai...
Milestone

FIGURE 7.49 (Continued)
Project Activities and Codes.

- Develop total estimate of the project.
- Develop bidding documents.
- Monitor the progress of the project.
- Make progress payments.

Table 7.18 is an example BOQ for concrete in the construction project.

TABLE 7.18

Bill of Quantities (BOQ)

Owner Name
Project Name
Project Number:

ITEM	DESCRIPTION	QTY.	UNIT	RATE	AMOUNT
	DIVISION 3 - CONCRETE				
	The Contractor is referred to the				
	03300: CAST IN-PLACE CONCRETE				
	SUBSTRUCTURE				
	Plain concrete 17.5MPa using Sulphate				
A	100mm Blinding	411	m3		
	Plain concrete 17.5MPa using Ordinary				
B	Concrete filling	55	m3		
C	To kerb foundation	19	m3		
	Reinforced concrete 28MPa using sulphate				
D	Raft foundation	5,006	m3		
E	300mm Walls	51	m3		
F	400mm Ditto	66	m3		
	CARRIED TO COLLECTION				

(Continued)

TABLE 7.18 (*Continued*)

Owner Name Project Name Project Number:					
ITEM	**DESCRIPTION**	**QTY.**	**UNIT**	**RATE**	**AMOUNT**
	DIVISION 3 - CONCRETE				
	03480: PRECAST CONCRETE				
	Depressed curbs including reinforcement,				
A	250 x 180mm Barrier kerb	200	m		
B	Ditto, curve on plan	52	m		
C	350 x 150mm Flush kerb	21	m		
	Wheel stoppers, fair face finish and painting				
D	2000 x 140 x 100mm Overall size, Parking	548	no		
	03520: LIGHTWEIGHT CONCRETE				
	Lightweight concrete laid in bays of 25m2				
E	50mm Thick, laid to falls, to roof	282	m2		
F	170mm Thick, laid to falls, to pathways	6,189	m2		
	CARRIED TO COLLECTION				

(*Continued*)

TABLE 7.18 (*Continued*)

Owner Name Project Name Project Number:

ITEM	DESCRIPTION	QTY.	UNIT	RATE	AMOUNT
	DIVISION 3 - CONCRETE				
	03530: CONCRETE TOPPING / SCREED				
	Plain concrete grade 30MPa using Ordinary				
A	50 -100mm Average thickness, including	20,[illegible]	m2		
B	Ditto, ramps	1,323	m2		
C	80mm Thick, to pathway, to receive	5,748	m2		
D	Ditto, ramps	441	m2		
E	100mm Thick, to receive ceramic floor tiles	329	m2		
F	180mm Thick, to receive stone flooring	550	m2		
G	50mm Thick, to receive terrazzo floor tiles	172	m2		
	03535: SAND CEMENT SCREED				
H	20mm Thick, to roof	282	m2		
J	25mm Thick, to receive terrazzo floor tiles,	48	m2		
K	50mm Screed, to protect horizontal water	3,973	m2		
	CARRIED TO COLLECTION				

(*Continued*)

TABLE 7.18 (*Continued*)

Owner Name Project Name Project Number:		
ITEM	**DESCRIPTION**	**AMOUNT**
	DIVISION 3 - CONCRETE	
	COLLECTION	
	Total of page No. 3/1	
	Total of page No. 3/2	
	Total of page No. 3/3	
	CARRIED TO SUMMARY	

7.3.4.3 Sequence Project Activities

Once project activities/tasks have been defined and listed, it is required to identify the relationship and dependency among the project activities. The activities may have direct relationship or may be further constrained by indirect relationship. The following are direct logical relationships or dependencies among project-related activities:

- **Precedence**
  a. *Finish-to-start*: Activity A must finish before activity B can begin.
  b. *Start-to-start*: Activity A must start before activity B can start.
  c. *Finish-to-finish*: Activity A must finish before activity B can finish.
  d. *Start-to-finish*: Activity A must start before activity B can finish.
- **Dependency**
  a. *Mandatory*: It is inherent in the nature of work being performed. It is also called *hard logic*. In building construction, superstructure cannot begin unless the foundation work is complete.
  b. *Discretionary*: These are preferred or preferential logic. These are used at the discretion of project team. It is also called *soft logic*.
  c. *External*: These dependencies are outside of the project's control (e.g., approval from regulatory authority/agency).
- **Lead and lag**

  Lead may be added to start an activity before the predecessor activity is completed (a jump of the successor activity), whereas lag is inserted waiting time (time delay) between activities. Lead allows acceleration of successor activity and lag delays the start of the successor activity. For example, concrete curing time is added before start of tiling works.
- **Relationship**
  a. *Predecessor*: The relationship as to which activity must occur *before* the other
  b. *Successor*: The relationship as to which activity must occur *after* the other
  c. *Concurrent*: When one activity can occur *at the same time* as the other

Figure 7.50a diagram represents activity relationship and Figure 7.50b illustrates dependency relationship diagrams.

7.3.4.4 Develop Project Network Diagram

Network diagram is a schematic display of project activities indicating logical relationship among the activities. It shows how the work progresses from the start till the completion of the project. The following are two methods typically used to prepare the network diagram:

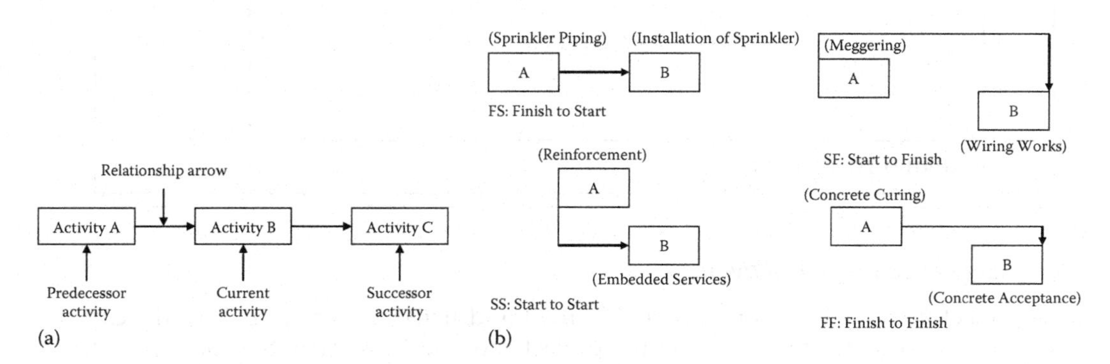

FIGURE 7.50
(a) Activity relationship and (b) dependency relationship.

Activity-on-Arrow: Arrow diagrams or activity-on-arrows (A-O-A) is a diagramming method to represent the activities on arrows and connect them at nodes (circles) to show the dependencies. With the A-O-A method, the detailed information about each activity is placed on arrow or as footnotes at the bottom.

Figure 7.51 illustrates the A-O-A diagramming method for concrete foundation work.

Activity-on-Node (A-O-N): *A-O-N* network diagram has the activity information written in small boxes that are the nodes of the diagram. Arrows connect the boxes to show the logical relationships between pairs of the activities.

Figure 7.52 illustrates the A-O-N diagramming method for concrete foundation work.

A-O-N is also referred to as the *precedence diagramming method* (PDM) technique, because it shows the activities in a node (box) with arrows showing dependencies. Figure 7.53 illustrates PDM. PDM is used to establish precedence relationship or dependencies (sequencing of activities) and also in the critical path method (CPM) for constructing the project schedule network diagram.

7.3.4.5 Estimate Activity Resources

It is the identification and description of types and quantities of resources required to perform the activity/task. Resource estimates are made by the project team, which is based on the resource productivity, related experience, and availability of resources to carry a particular activity/task. Efficient utilization of resources is critical to successful project. OBS is used to determine the related resource for a particular activity. There are different types

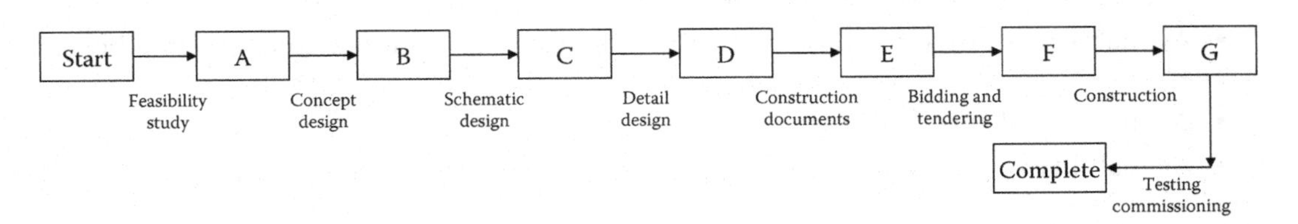

FIGURE 7.51
Arrow diagramming method for design phases.

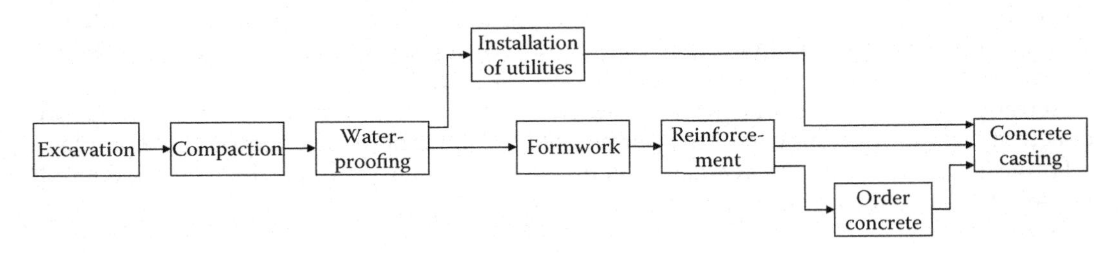

FIGURE 7.52
Activity-on-node diagram.

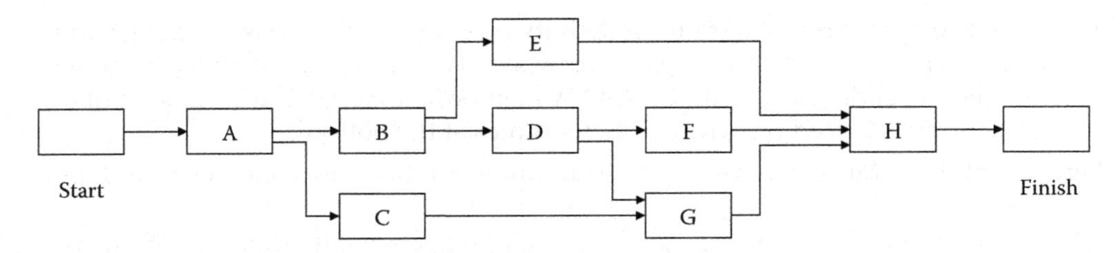

FIGURE 7.53
PDM diagramming method.

of resources required in a certain quantity to perform and complete the activity. Some of these are as follows:

- Human resources (manpower)
- Equipment
- Material

In construction projects, the availability of subcontractor (subconsultant and subcontractor for designated works) is also to be considered while estimating the resources to perform the work.

While estimating the resources, the following points should be considered:

- Availability of particular resource (resource histogram)
- Availability of exact number of resources
- Identification of the required skill (manpower) to perform the assigned activity/task
- Availability of specific type of equipment and the installation/operating crew
- Availability of space to perform certain type of activity simultaneously with other activities
- Availability of specified material
- Availability of fund to perform the activity

Figure 7.54 is illustrates resource estimation (manpower histogram) for a construction project.

7.3.4.6 Estimate Activity Duration

Once the activities/tasks are sequenced, the types and quantities of resources are identified and determined, the amount of time required to complete each activity is order to enable preparation of schedule is required to be estimated.

Activity duration is the time between the start and finish of an activity/task. Estimation of activity duration is important to approximate the amount of time required to complete all project activities with the available resources. The following points are to be considered while estimating the activity duration:

- Work methodology
- Nature of work

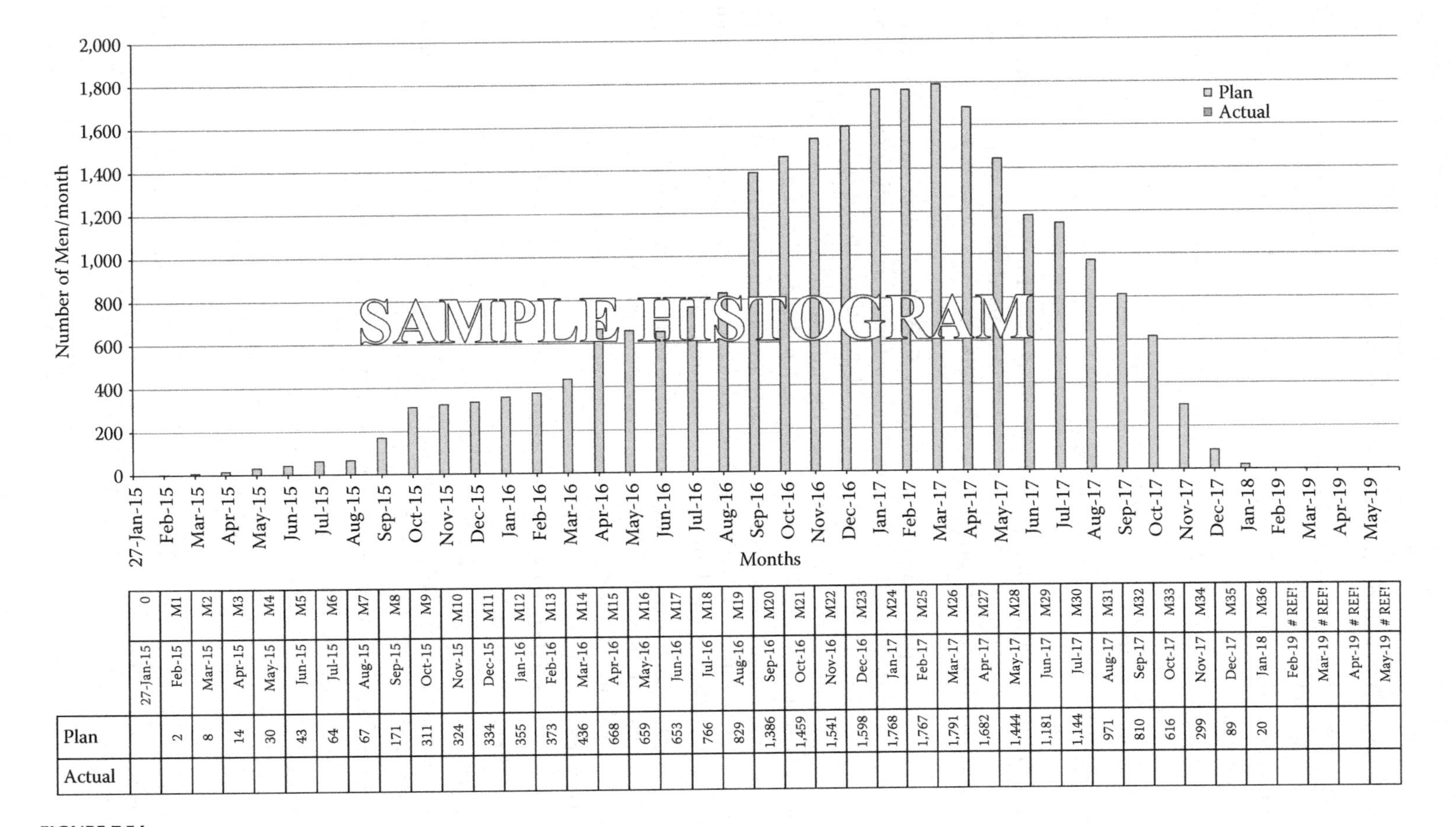

	0	M1	M2	M3	M4	M5	M6	M7	M8	M9	M10	M11	M12	M13
	27-Jan-15	Feb-15	Mar-15	Apr-15	May-15	Jun-15	Jul-15	Aug-15	Sep-15	Oct-15	Nov-15	Dec-15	Jan-16	Feb-16
Plan		2	8	14	30	43	64	67	171	311	324	334	355	373
Actual														

	M14	M15	M16	M17	M18	M19	M20	M21	M22	M23	M24	M25	M26	M27
	Mar-16	Apr-16	May-16	Jun-16	Jul-16	Aug-16	Sep-16	Oct-16	Nov-16	Dec-16	Jan-17	Feb-17	Mar-17	Apr-17
Plan	436	668	659	653	766	829	1,386	1,459	1,541	1,598	1,768	1,767	1,791	1,682
Actual														

	M28	M29	M30	M31	M32	M33	M34	M35	M36	# REF!	# REF!	# REF!	# REF!
	May-17	Jun-17	Jul-17	Aug-17	Sep-17	Oct-17	Nov-17	Dec-17	Jan-18	Feb-19	Mar-19	Apr-19	May-19
Plan	1,444	1,181	1,144	971	810	616	299	89	20				
Actual													

FIGURE 7.54
Manpower estimation.

- Resource (labor and equipment) type and availability
- Resource (labor and equipment) productivity
- Resource calendar
- Time contingency
- Quantum of work
- Quality of work
- Environmental factors
- Weather conditions
- Organizational factors
- Work restriction

There are several tools and techniques that are used to estimated activity duration. The following are the basic techniques that are most widely used by the project professionals to estimate activity estimation to prepare the schedule:

1. *Expert judgment*: Expert judgment is normally from the project members having expertise in the subject matter. It is the mix of historical information from previous projects. It also uses published database about duration estimation.
2. *Analogous estimating*: This technique uses the information from previous projects of similar nature and size. It is also known as *top-down estimation*.
3. *Parametric estimating*: This technique uses mathematical, statistical, or quantitative relationship to estimate activity duration. This method has better accuracy than that of expert judgment and analogous estimation techniques.
4. *Three-point estimating*: This technique is used to estimate the duration for an activity having higher risk and is not known very well. The following formula is used to calculate activity duration:

   Activity duration = $(P + 4M + O)/6$

   Where P is pessimistic (worst-case scenario)

   M is most likely (realistic)

   O is optimistic (best-case scenario).

   To calculate installation of block masonry work per square meter by a crew of one mason and two labors, the following assumptions can be made:

   Pessimistic—35 square meters per crew per day

   Most likely—20 square meter per crew per day

   Optimistic—15 square meter per crew per day

   Therefore, estimated duration per crew per day is as follows:

   Te = (35 + 4 * 20 + 15)/6 = 21.67 say 22 square meters

If the activity is well known with little risk, then it is possible to estimate the duration of an activity by using the available information from organization's database. This technique is known as *one-time estimate*.

7.3.4.7 Develop Schedule

Once activity resources and activity duration is estimated, it is possible to determine the start and finish dates for each activity. Thus by sequencing all the activities, the overall project schedule can be developed, showing start and finish of the project. Project schedule is calendar-based graphical representation of all the activities that need to be performed in order to achieve project scope objectives. Schedule is one of the critical elements for successful management of project. In order to develop a schedule, the following information is required:

1. Project scope statement (assumptions, milestones, and constraints)
2. Activities list
3. Network diagram
4. Resources needed
5. Duration for each activity
6. Resource calendar (availability of resources)
7. Working calendar

Various methods are available to develop project schedule. Some of the following methods/tools used to develop project schedule are as follows:

- *Gantt or bar charts*: Bar charts are graphical presentation of project schedule information by listing the project activities on the vertical axis and showing the corresponding start and finish dates on the horizontal axis as well as expected duration in a calendar format. Bar chart method was developed by Henry Gant in 1917 by listing the activities to a time scale by drawing in the bar chart format.
- *CPM*: Critical path is the longest path through a network, and hence it is the shortest project time. CPM was developed by DuPont & Remington Rand in 1958. It was developed for industrial projects (maintenance programs for chemical plant), where activity durations are generally known. CPM is a network diagramming technique use to predict total project time. CPM identifies the activities in the critical path that are likely to affect the completion of project duration as required. There could be more than one critical path if the lengths of two or more paths are same. The critical path can change as the project progresses. CPM is represented by the following types of network diagrams:
    a. A-O-A
    b. A-O-N
- *Project (program) evaluation and review technique (PERT)*: PERT is a project management tool that is used to schedule, organize, and coordinate activities/tasks within the project. It was developed by the US Navy for Polaris missile project (program). It was developed for R&D project where activity times are generally uncertain. PERT is a flowchart diagram that depicts the sequence of activities needed to complete the project and the time or cost associated with each activity. It uses three-point estimation formula (Estimated time (Te) = [To + 4Tm + Tp]/6) to calculate expected activity duration, where **To** is the optimistic estimate, **Tm** is the most likely estimate, and **Tp** is the pessimistic estimate.

CPM and PERT tools are also known as *mathematical techniques*.

Bar chart or Gantt chart and CPM (network diagramming or PDM) scheduling techniques are most commonly used techniques to develop project schedule. The following are the terminologies used to prepare network diagramming:

- *Activity block*: The activity block is a graphic display of activity information.
  Figure 7.55 is an illustrative block showing the related information about an activity.
- *Early start (ES)*: The earliest date an activity can start.
- *Early finish (EF)*: The earliest date an activity can finish.
- *Late start (LS)*: The latest date an activity can start.
- *Late finish (LF)*: The latest date an activity can finish.
- *Duration*: Number of working days or hours needed to perform the specific activity.
- *Milestone*: It is a specific point(s) having significant importance shown in the project schedule having no duration.
- *Forward pass*: Forward pass is used for calculating ES and EF of each activity.
- *Backward pass*: Backward pass is used for calculation LS and LF of each activity.
- *Float*: Float (slack) is the measure of time the work can be delayed without delaying (affecting) the project completion date.

Table 7.19 shows an illustrative example of activities for construction of guardhouse building. Figure 7.56 illustrates a schedule for construction of a guardhouse building using Gantt chart/bar chart format, whereas Figure 7.57 shows a CPM diagram for construction of the guardhouse building.

- *Advantages of CPM*:
  1. It shows logical relationship between activities and their interdependence.
  2. It is easy to understand the relationship of activities.
  3. It identifies the activities that are critical to the timely completion of the project.
  4. It shows the minimum completion time for completion of the project.
  5. It clearly demonstrates how a change in one activity impacts on other activities in the schedule.
  6. It enables easy calculation of shortest time to complete the project.

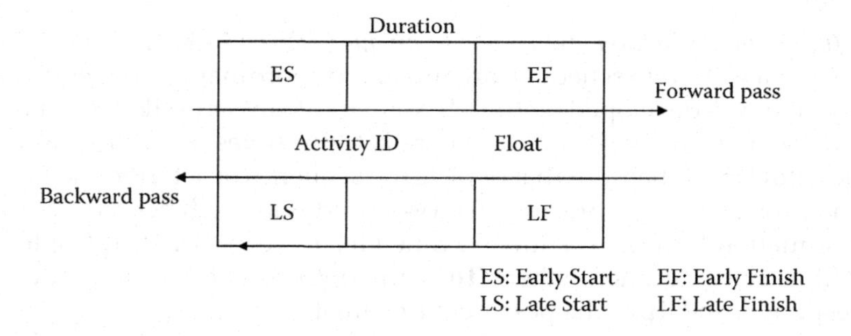

FIGURE 7.55
Activity block.

TABLE 7.19

Activities to Construct Guardhouse Building

Activity Number	Description of Activity	Duration in Days	Preceding Activity
1	Start	0	
2	Mobilization	21	1
3	Preparation of site	15	2
4	Staff approval	15	2
5	Material approval	15	2,4
6	Shop drawing approval	15	2,5
7	Procurement (structural work)	15	5
8	Procurement (pipes, ducts, and sleeves)	7	5
9	Procurement (security system equipment)	60	5
10	Procurement (Civil, MEP, Furnishing)	30	5
11	Excavation	4	3,6
12	Blinding concrete	4	5,7,11
13	Raft foundation	7	5,6,7,12
14	Utility services (embedded)	4	8,12
15	Raised flooring	3	13,14
16	Concrete (floor)	1	6,13,14,15
17	Embedded services	2	8,13
18	Walls and columns	7	16
19	Form work for slab	3	6,7,18
20	Reinforcement	2	7,19
21	Embedded services for security system	1	8,19
22	Concrete (roof slab)	1	20,21
23	Masonry work	14	6,10,22
24	Cabling for security system	3	15,17,21
25	Installation of CCTV, access control system	7	9,21,24
26	Installation of electromechanical items	14	6,10,22,23
27	Installation of ventilation system	4	10,22,23
28	Finishes	7	10,23
29	Installation of final fixes	3	10,28
30	Furnishing	2	10,28
31	Testing of security system	3	25
32	Testing of HVAC, firefighting, electrical system	4	25,26,27
33	Handing over	1	29,30,31,32
34	End	0	

Note: The duration is indicative only to understand sequencing and relationship of activities.

7. It provides accurate and detailed picture of forecasted project completion dates.
8. It is easy to apply resources and determine resources or time trade-off.
9. It shows how much float other activities have.
10. It is easy to assign lead and lag between two activities to solve a dead time problem.

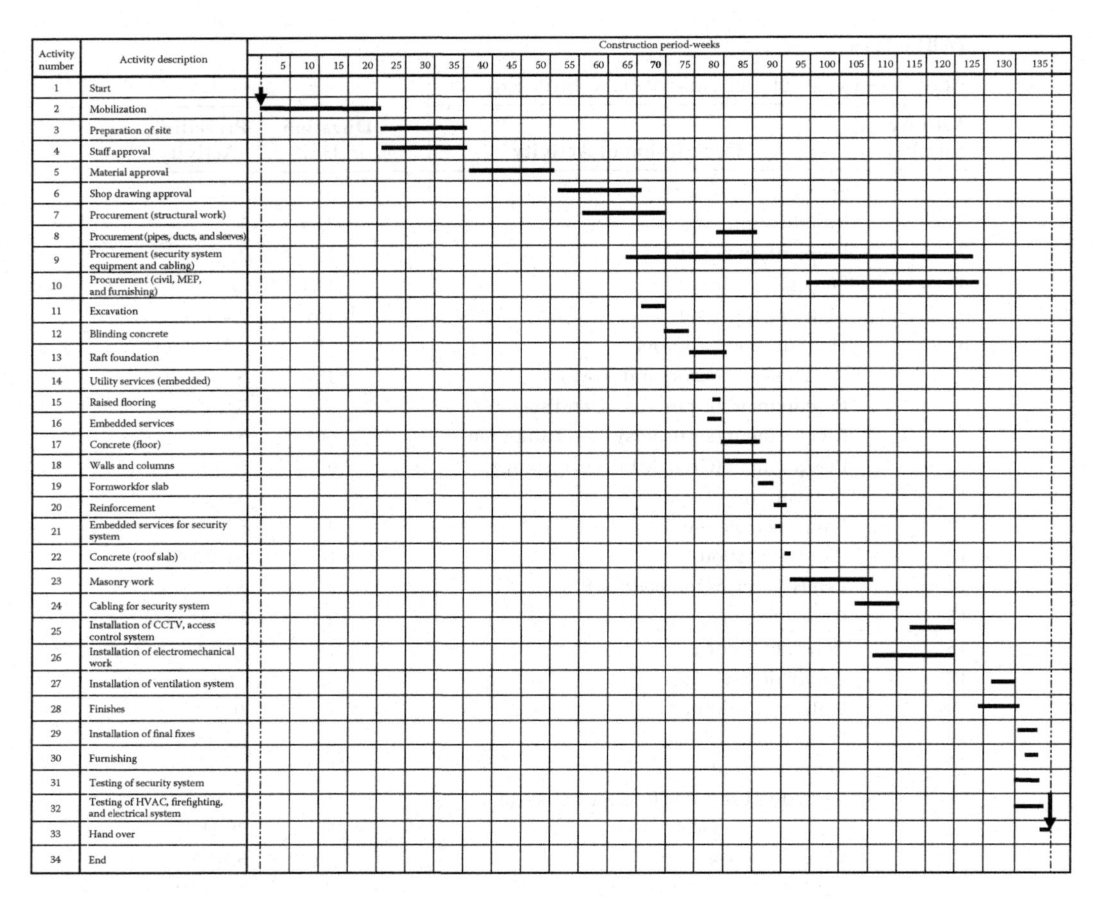

FIGURE 7.56
Gantt chart for guardhouse building.

7.3.4.8 Analyze Schedule

Once the basic schedule is developed, it needs to be reviewed and analyzed to ensure that the timing of each activity is aligned with resources and to ensure schedule accuracy with the relevant assumptions, constraints, and milestones. Scheduling is an iterative process. If the results obtained with creation of schedule matches with project requirements, the iteration must stop. Otherwise the schedule has to be reviewed and analyzed by adjusting the dependencies and resources. Schedule analysis is important to determine schedule accuracy and create project schedule (schedule baseline). The following points should be considered to develop accurate and realistic baseline schedule:

1. All the activities are properly identified and listed to meet the scope of work.
2. Project logics dependencies and relationships are properly considered while sequencing of the activities to be performed.
3. All the resources and durations are correctly estimated.
4. Sufficient resources have been identified and allocated.
5. Assumptions and constraints are sound and true.

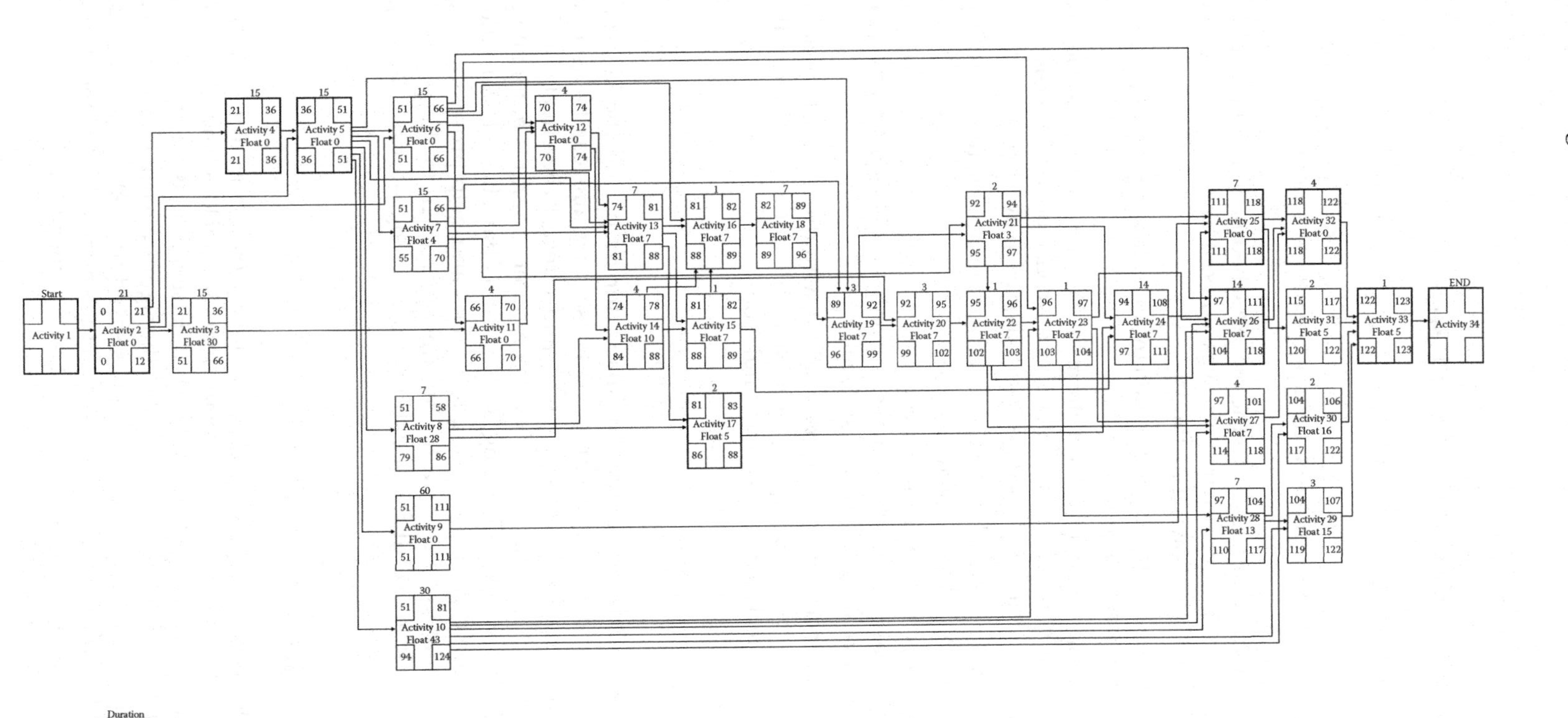

FIGURE 7.57
Critical path method (diagram for guardhouse building).

6. Schedule reserves/contingencies have been identified.
7. Realistic and feasible project duration is considered.
8. Regulatory requirements.
9. Contractual requirements.

The following tools and techniques are used to improve project schedule:

- *Schedule compression*: In order to shorten the project schedule to meet the imposed dates, schedule constraints or other objectives without changing the scope of work. This is achieved by following methods:
    - *Crashing*: To decrease total project duration by analyzing how to get maximum compression with least cost effect
    - *Fast tracking*: Compression of project duration by changing the sequences of activities to do in parallel, such as design and construction in design–build type of delivery system, or allowing some activities to overlap
- *Critical chain method*: It is a schedule analysis network technique that modifies the project schedule to account for limited resources.
- *Resource leveling*: This technique is used when shared or critically required resources are only available at certain times, are only available in limited quantities, or to keep resource usage at a constant level.
- *Simulation*: It involves calculating project durations with different sets of activity assumptions to assess the feasibility of project schedule under adverse condition. The following are the techniques used for simulation:
    - Monte Carlo analysis
    - What-if analysis

There are many computer-based programs available for preparing the network and critical path of activities for construction projects. These programs can be used to analyze the use of resources, review project progress, and forecast the effects of changes in the schedule of works or other resources. Most computer programs automate preparation and presentation of various planning tools such as bar chart, PERT, and CPM analysis. The programs are capable of storing huge data and help process and update the program quickly. It manipulates data for multiple usages from the planning and scheduling perspectives. It is useful for updating and tracking, sorting, filtering, and resource leveling.

7.3.4.9 Project Schedule (Baseline Schedule)

Baseline schedule is the project plan developed to meet the project objectives and the scope of work and accepted by the stakeholders as a benchmark for tracking and measuring the project performance and progress. An accurate baseline schedule is necessary to assess actual performance (status) of the project, determine the significance of variance, and forecast to complete the project. The baseline schedule is saved (freeze) once the schedule is approved and accepted by all the stakeholders (project team members) at the start of the project. Any change to the baseline schedule must be approved by the change control board (authorized project team members).

The baseline schedule is

- A commitment by all stakeholders (project team members) to accomplish project objectives within the agreed time.
- A representation of project assumptions, constraints, and milestones.
- A benchmark to perform the activities.
- A datum for measuring project performance and progress and for forecasting project completion date.
- Used as a framework for identification of impacts on the project.
- Saved (freeze) as original plan upon acceptance by all the stakeholders (authorized project team members).
- Updated with formal approval by change control board (authorized project team members).

Construction project development has mainly the following four stages:

1. Study
2. Design
3. Bidding and tendering
4. Construction

There are many activities/task that need to be performed during each of these stages. In order for smooth flow and performance of activities during these stages, it is necessary to prepare a schedule at every stage of construction project life cycle to perform the work in an organized and structured manner. The following are the stages/phases at which schedules are developed by various project participants:

- Pre-design phase (feasibility stage)—by owner
- Design phases—by designer (consultant)
    - Concept design
    - Preliminary design
    - Detail design
    - Construction documents preparation
- Bidding and tendering and contract award—by owner and designer
- Construction phase—by contractor
    - Construction
    - Testing, commissioning, and handover

The schedule developed at these stages (phases) serve as a baseline schedule for the specific stage and help accomplish all the key activities/tasks and to achieve overall project objectives.

7.3.4.9.1 Schedule Levels

There are different levels that the project schedule is presented. There are five levels of schedules typically developed in construction projects. The objective is to establish appropriate levels of schedule detail for planning, scheduling, monitoring, controlling, and reporting on the overall project.

Level 1: Project master schedule (executive summary)

Level 2: Summary master schedule (management summary)

Level 3: Project coordination schedule (publication schedule)

Level 4: Project working level schedule (execution schedule)

Level 5: Detail schedule

The amount of information desired at each level depends on the requirements of stakeholders and the project stage (phase). Table 7.20 illustrates schedule level deliverables.

TABLE 7.20

Schedule Levels

Schedule Level 1	
Schedule title	Project master schedule (executive summary)
Description	Level 1 schedule is a high-level schedule that reflects key milestones and high level project activities by major phase, stage or project being executed. This schedule level may represent summary activities of an execution stage, specifically engineering, procurement, construction, and start-up activities Level 1 schedules provide high-level information that assist in the decision making process (go/no go prioritization and criticality of projects) Level 1 schedule can be used to integrate multiple contractors/multiple schedules into an overall program management process Level 1 audience include, but are not limited to client, senior executives and general managers Level 1 schedule may be used in the proposal stage of a potential project/contract
Scheduling method	Graphical representation in the bar chart format
Usage in construction projects	Study stage, Predesign phase, and feasibility study
Schedule Level 2	
Schedule title	Summary master schedule (management summary)
Description	Level 2 schedules are generally prepared to communicate the integration of work throughout the life cycle of a project Level 2 schedules may reflect, at a high level, interfaces between key deliverables and project participants (contractors) required to complete the identified deliverables Level 2 schedules provide high-level information that assist in the project decision-making process (re-prioritization and criticality of project deliverables) Level 2 schedules assist in identifying project areas and deliverables that require actions and/or course correction; audiences for this type of schedule include, but are not limited to general managers, sponsors, and program or project managers
Scheduling method	Typically presented in the Gantt (bar chart) format
Usage in construction projects	Overall design schedule and overall construction schedule

(*Continued*)

TABLE 7.20 (*Continued*)

Schedule Levels

Schedule Level 3	
Schedule title	Project coordination schedule (publication schedule)
Description	Level 3 schedules are generally prepared to communicate the execution of the deliverables for each of the contracting parties Level 3 reflects the interfaces between key workgroups, disciplines, or crafts involved in the execution of the stage Level 3 schedule includes all major milestones, major elements of design, engineering, procurement, construction, testing, commissioning, and handover Level 3 schedules provide enough detail to identify critical activities Level 3 schedules assist the team in identifying activities that could potentially affect the outcome of a stage or phase of work, allowing for mitigation and course correction in short course Level 3 audiences include, but are not limited to program or project managers, CMs or owner's representatives, superintendents, and general foremen
Scheduling method	Typically presented in Gantt or CPM network format, and is generally the output of CPM scheduling software
Usage in construction projects	Detail design and detail construction
Schedule Level 4	
Schedule title	Project working level schedule (execution schedule)
Description	Level 4 schedules are prepared to communicate the production/execution of work packages at the deliverable level Level 4 schedule reflects interfaces between key elements that drive completion of activities Level 4 schedules usually provide enough detail to plan and coordinate contractor or multi-discipline/craft activities Level 4 schedule displays the activities to be accomplished by identifying all the required resources Level 4 audiences include but are not limited to project managers, superintendents, and general foremen
Scheduling method	Typically presented in Gantt or CPM network format
Usage in construction projects	Monthly schedule and look ahead schedules
Schedule Level 5	
Schedule title	Detail schedule
Description	Level 5 schedules are prepared to communicate task requirements for completing activities identified in a detailed schedule Level 5 schedules are usually considered working schedules that reflect hourly, daily, or weekly work requirements Level 5 schedules are used to plan and schedule utilization of resources (labor, equipment, and materials) in hourly, daily, or weekly units for each task Level 5 audiences include but are not limited to superintendents, general foremen and foremen
Scheduling method	Typically presented in an activity listing format without time scaled graphical representation of work to accomplish
Usage in construction projects	Daily work schedule and daily progress report

Source: AACE International Recommended Practice No. 37R-06 (2010). *Schedule Levels of details: As Applied in Engineering Procurement and Construction*. Reprinted with permission from AACE International.

In order to improve the understanding and the communication among stakeholders involved with preparing, evaluating, and using project schedule, AACE International has published the guideline to classify schedules into five classes and five levels. Figure 7.58 illustrates schedule classifications versus schedule levels that schedule can be developed and/or presented.

Table 7.21 illustrates generic schedule classification matrix, and Table 7.22 illustrates characteristics of schedule classifications.

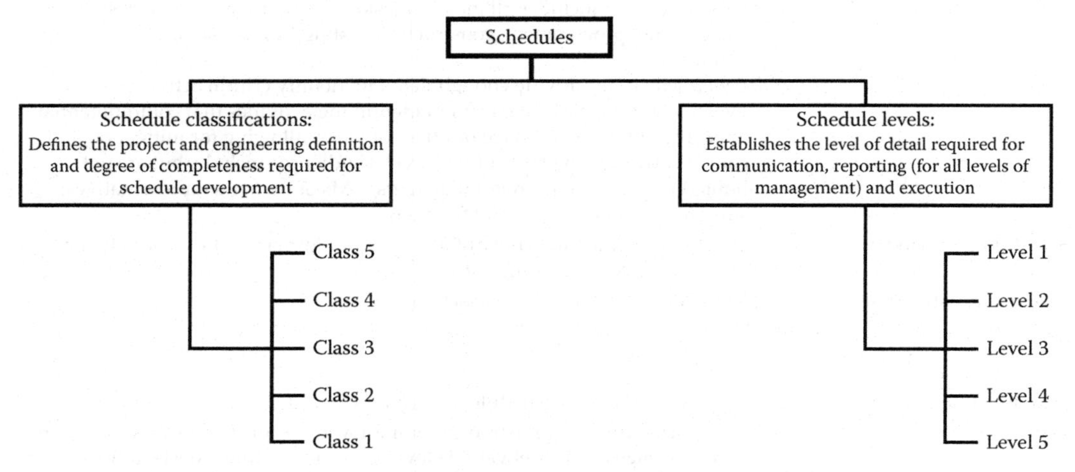

FIGURE 7.58
Schedule: Classifications versus levels. (AACE International Recommended Practice No. 27R-03 (2010). *Schedule Classification System*. Reprinted with permission from AACE International.)

TABLE 7.21
Generic Schedule Classification Matrix

	Primary Characteristic	Secondary Characteristic	
Schedule Class	**Degree of Project Definition (Expressed as % of Complete Definition)**	**End Usage**	**Scheduling Methods Used**
Class 5	0% to 2%	Concept screening	Top-down planning using high-level milestones and key project events
Class 4	1% to 15%	Feasibility study	Top-down planning using high level milestones and key project events, semi-detailed
Class 3	10% to 40%	Budget, authorization, or control	*Package* top-down planning using key events, semi-detailed
Class 2	30% to 70%	Control or bid/tender	Bottom-up planning, detailed
Class 1	70% to 100%	Bid/Tender	Bottom-up planning, detailed

Source: AACE International Recommended Practice No. 27R-03 (2010). *Schedule Classification System*. Reprinted with permission from AACE International.

TABLE 7.22

Characteristics of Schedule Classification

Class 5 Schedule Characteristics	
Description	Class 5 schedules are generally prepared based on very limited information, and subsequently have very wide accuracy range. The Class 5 schedule is considered a preliminary document, usually presented in either Gantt (bar chart) or table form. The Class 5 schedule should have, as a minimum, a single summary per stage with major project milestones identified.
Degree of project definition required	0% to 2%
End usage	Class 5 schedules are prepared for any number of strategic business planning purposes, such as but not limited to: market studies, assessment of initial viability, evaluation of alternative schemes, project screening, project location studies, evaluation of resource need and budgeting, long range capital planning, etc.
Scheduling methods used	Gantt, Bar chart, and milestone/activity table Top-down planning using high-level milestones and key project events
Class 4 Schedule Characteristics	
Description	Class 4 schedules are generally prepared based on limited information and subsequently have fairly wide accuracy ranges. They are typically used for project screening, determination of *do-ability*, concept evaluation, and to support preliminary budget approval. The Class 4 schedule is usually presented in either Gantt (bar chart) or table form. The Class 4 schedule should define the high level deliverables for each specific stage going forward (since the previous stage has passed). This document should also provide an understanding regarding the timing of key events, such as independent project reviews, committee approvals, as well as determining the timing of funding approvals. A high level WBS may be established at this time.
Degree of project definition required	1% to 15%
End usage	Class 4 schedules are prepared for a number of purposes, such as but not limited to: detailed strategic planning, business development, project screening at more developed stages, alternate scheme analysis, confirmation of "do-ability", economic and/or technical feasibility, and to support preliminary budget approval or approval to proceed to next stage. It is recommended that the Class 4 schedule be reconciled to the Class 5 schedule to reflect the changes or variations identified as a result of more project definition and design. This will provide an understanding of the changes from one schedule to the next.
Scheduling methods used	Gantt, bar chart, milestone/activity table, top-down planning using high-level milestones and key project events, semi-detailed.
Class 3 Schedule Characteristics	
Description	Class 3 schedules are generally prepared to form the basis of execution for budget authorization, appropriation, and/or funding. As such, they typically form the initial control schedule against which all actual dates and resources will be monitored. The Class 3 schedule should be a resource loaded, logic-driven schedule developed using the precedence diagramming method (PDM). The schedule should be developed using relationships that support the overall true representation of the execution of the project (with respect to start to start and finish to finish relationships with lags). The amount of detail should define, as a minimum, the work package (WP) level (or similar deliverable) per process type/unit and any intermediate key steps necessary to determine the execution path. (The WP rolls up into the predefined WBS.) In some circumstances, where there is a high degree of parallel activities, the critical nature of the project, or the extreme complexity and/or size of the project, it may be warranted to provide further detail of the schedule to assist in the control of the project.

(Continued)

TABLE 7.22 (*Continued*)

Characteristics of Schedule Classification

Degree of project definition required	10% to 40%
End usage	Class 3 schedules are typically prepared to support full project funding requests, and become the first of the project phase *control schedules* against which all start and completion dates and resources will be monitored for variations to the schedule. They are used as the project schedule until replaced by more detailed schedules. It is recommended that the Class 3 schedule be reconciled to the Class 4 schedule to reflect the changes or variations identified as a result of more project definition and design.
Scheduling methods used	PDM, PERT, and Gantt/bar charts *Package* top-down planning using key events, semi-detailed
Class 2 Schedule Characteristics	
Description	Class 2 schedules are generally prepared to form a detailed control baseline against which all project work is monitored in terms of task starts and completions and progress control. The Class 2 schedule is a detailed resource loaded, logic-driven schedule that should be developed using the critical path method (CPM) process. The amount of detail should define as a minimum, the required deliverables per contract per work package (WP). The schedule should further define any additional steps necessary to determine the critical path of the project necessary for the appropriate degree of control.
Degree of project definition required	30% to 70%
End usage	Class 2 schedules are typically prepared as the detailed control baseline against which all actual start and completion dates and resources will now be monitored for variations to the schedule, and form a part of the change/variation control program. It is recommended that the Class 2 schedule be reconciled to the Class 3 schedule to reflect the changes or variations identified as a result of more project definition and design.
Scheduling methods used	Gantt/bar charts, PDM, and PERT Bottom-up planning, detailed
Class 1 Schedule Characteristics	
Description	Class 1 schedules are generally prepared for discrete parts or sections of the total project rather than generating this amount of detail for the entire project. The updated schedule is often referred to as the current control schedule and becomes the new baseline for the cost/schedule control of the project. The Class 1 schedule may be a detailed, resource loaded, logic-driven schedule and is considered a "production schedule" used for establishing daily or weekly work requirements.
Degree of project definition required	70% to 100%
End usage	Class 1 schedules are typically prepared to form the current control schedule to be used as the final control baseline against which all actual start and completion dates and resources will now be monitored for variations to the schedule, and form a part of the change/variation control program. They may be used to evaluate bid-schedule checking, to support vendor/contractor negotiations, or claim evaluations and dispute resolution. It is recommended that the Class 1 schedule be reconciled to the Class 2 schedule to reflect the changes or variations identified as a result of more project definition and design.
Scheduling methods used	Gantt/bar charts, PDM, and PERT Bottom-up planning, detailed

Source: AACE International Recommended Practice No. 27R-03 (2010). *Schedule Classification System.* Reprinted with permission from AACE International.

7.3.4.10 Monitor and Control Schedule

Monitor and control schedule is the process to determine the current status of the schedule, identify the influencing factors that causes the schedule changes, determine that the schedule has changed, and manage the changes in the approved project schedule baseline by updating and taking appropriate actions, if necessary, to minimize deviation from the approved schedule.

Monitoring is collecting, recording, and reporting information concerning project performance. Monitoring involves measurement of current status of the project accomplishment and performance.

Current schedule (As-built schedule) is reflection of current situation of all activities/tasks, milestones, sequencing, resources, duration, constraints, and project update. Control process is established for managing the current schedule. Controlling is using the actual data collected through monitoring and comparing the same to the planned performance to bring actual performance to planned performance by correcting the variances or implementing approved changes. Analysis of variance between the baseline and current schedule dates and duration provides necessary information for management and stakeholders' decision.

Monitoring in construction projects is normally done by compiling status of various activities in the form of progress reports. These are prepared by the contractor, supervision team (consultant), and construction/project management team. The objectives of project monitoring and control are as follows:

1. Report the necessary information in details and in appropriate form that can be interpreted by management and other concerned personnel to provide them with the information about how the resources are being used to achieve project objectives.
2. Provide an organized and efficient means of measuring, collecting, verifying, and quantifying data reflecting the progress and status of execution of project activities, with respect to schedule, cost, resources, procurement, and quality.
3. Provide an organized, efficient, and accurate means of converting the data from the execution process into information.
4. Identify and isolate the most important and critical information about the project activities to enable decision-making personnel to take corrective action for the benefit of the project.
5. Forecast and predict the future progress of activities to be performed.

The following information is required to prepare current (as-built) schedule and compare with the baseline schedule:

- Percentage of completion of each activity based on approved checklist
- Actual start/finish dates for the completed activities
- Activities scheduled to start but not yet started
- Activities scheduled to complete but under progress
- Remaining duration to complete each activity
- Percentage of completed activities

- Percentage of partially completed activities
- Percentage of activities not yet started
- Material and equipment yet to be received
- Available resources
- Regulatory approvals
- Milestones not yet reached
- Logic and duration revision to keep the schedule unchanged
- Problems and issues
- Risks
- Change orders

After analyzing the current status with the actual, the schedule performance report is prepared consisting of the following information:

- *Project status*: Where the schedule stands at current situation.
- *Project progress*: Planned versus actual.
- *Forecasting*: Prediction of future status and progress trend.

Figure 7.59 illustrates schedule monitoring and controlling process.

7.3.5 Cost Management

Cost management is the process involving planning, cost estimating, budgeting, and cost controlling to ensure that the project is successfully completed within approved budget. Construction projects are mainly capital investment projects. They are customized and nonrepetitive in nature. Cost management in construction project is planning and

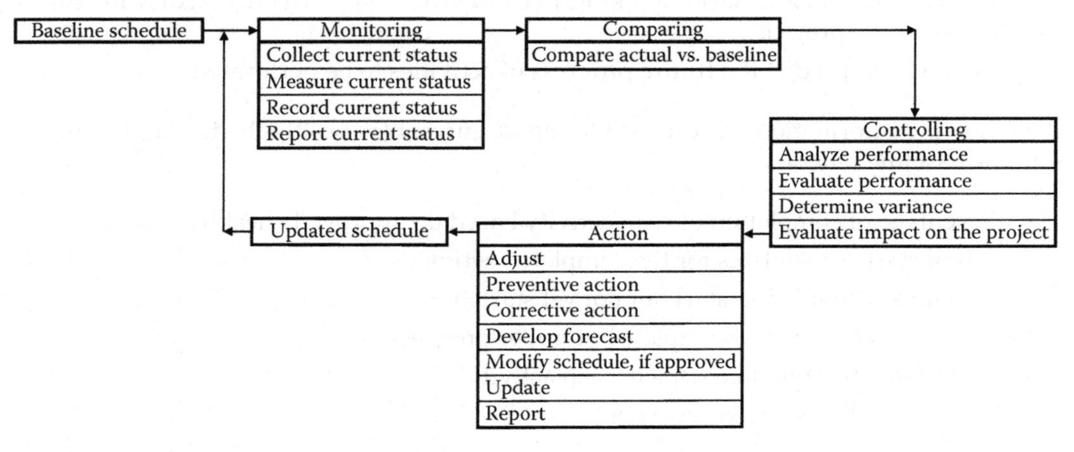

FIGURE 7.59
Schedule monitoring and controlling process.

managing the cost of facility throughout the project life cycle. The capital cost for a construction project includes the following expenses:

1. Land acquisition
   - This cost depends on the location and market price of the land. The owner may purchase the land on ownership or may lease the land to construct the facility.
2. Project construction
   - Cost expended by the owner for
     - Project study stage
     - Project design development (concept, schematic, detail, construction documents, and bidding and tendering)
     - Construction supervision (construction management and project management)
     - Site preparation
     - Construction of project (material, equipment, systems, furnishing, and human resources)
     - Inspection and testing (May be included as part of costs for different stages/phases)

Apart from the above expenses, such as license fee, permits, regulatory taxes, insurance, and project-related owner's overhead cost are incurred by the owner. In addition to above costs, operation and maintenance costs also required till the functional life of the facility.

Total cost related to construction is also known as *hard costs*. Hard costs are those that are incurred mainly by the decisions of the designer (consultant) engaged in preparation of design for the construction project.

The construction project initiation starts with identification of need and a business case. The owner is interested to have the best facility that meets the owner's schedule and is within the budget to meet investment plan and objectives. Cost management in the construction project include following processes:

1. Estimate cost
2. Prepare project budget
3. Control cost

7.3.5.1 Estimate Cost

It is important to estimate the total cost involved in the construction project in order to evolve the project budget and manage cost variance of the project. Cost estimation is a prediction of most likely total cost of the identified need and business case. Construction project estimates are based on identifying, quantifying, and estimating the cost of all the resources (material, equipment, systems, furnishing, and people) required to complete all the activities within the required schedule and specifications. Cost estimates in construction projects vary as it progresses through feasibility to the bidding and tendering stage. Construction project costs are generally estimated at following stages:

- Project feasibility study stage
- Client brief or TOR
- Concept design
- Schematic design
- Detail design
- Contract document
- Contract sum agreed during bidding and tendering stage (considered as cost baseline [contract cost] and used to manage cost variance in the project)

7.3.5.1.1 Cost Estimation Tools

The following are the basic tools and techniques generally used for cost estimation in construction projects:

- *Analogous or top-down estimates*

  Analogous estimating method relates to using the actual cost of a previous, similar nature of project as the basis for estimating the cost of current project. Analogous estimation is used when limited amount of project details is available.

  Top-down estimating approach recognizes all the activities, systems, and equipment required to complete the project and are included in the total cost of the project.
- *Bottom-up estimates*

  Bottom-up estimates involves in estimating cost of all the individual activities, material, equipment, resources, and aggregate all the costs to get total project cost.

  Bottom-up estimates assures that all known activities, components, and systems are accounted to get total cost.
- *Parametric estimate*

  It uses statistical relationship between the historical data and other variables to estimate the total cost of the project.

For example, in construction of buildings, the following factors are used:

- Buildings (offices, factories, and mix use)—square meter of floor area
- Schools—number of students
- Hospitals—number of patents
- Hotels—number of bed/rooms
- Theater/cinema—number of seats
- Parking facility—per parking bay

In construction projects, the cost estimates vary as the project design progresses. At the inception of project, the cost estimate is based on rough order of magnitude. When detail design is available, the cost estimates become definitive.

The cost estimated by the designer (consultant) is based on assumptions and historical data available from experience on similar projects. The contractor's estimation is more realistic and is based on the actual costs the contractor will incur to execute the project. Table 7.23 illustrates cost estimation levels for construction projects.

TABLE 7.23

Cost Estimation Levels for Construction Projects

Project Stage/Phase	Tools/Methodology	Accuracy	Purpose
Inception	Analogous	−50% to +100%	Project initiation (rough order of magnitude)
Feasibility	Analogous	−25% to +75%	Justification to proceed (screening estimate)
Concept design	Parametric	−10% to +25%	Budgetary (conceptual estimate)
Preliminary design	Elemental parametric	−10% to +25%	Budgetary (preliminary estimate)
Detail design	Elemental parametric/ detailed costing	−5% to +10%	Detailed estimate
Bidding and tendering	Detailed costing	−5% to +5%	Bid estimate/definitive estimate
Construction	Detailed costing	Project cost baseline (contracted value)	Contract cost (control estimate)

Contract awarded (negotiated) total project cost is considered as most accurate cost, as it is derived from the detailed information provided in the bidding and tendering documents. In the case of a design–bid–build type of project delivery system, detail BOQ is worked out by the designer (consultant) and is included in the bidding and tendering documents. The BOQ is used to prepare the estimate.

In order to estimate the accurate cost of the project, the contractor has to consider the following points:

- Location of the project
- Site conditions
- Project specifications
- Codes and standard to be followed
- Project schedule
- Risks and constraints present in the project
- Inflation
- Project delivery system
- Taxes
- Currency exchange rate, if applicable
- Contingency
- Cost toward subcontracted items

In the case of a design–build type of project delivery system, the project cost is developed based on the performance specifications and conceptual BOQ. Figure 7.60 is an example of analysis of estimation of prices form to work out activity unit price and total project cost estimate.

7.3.5.2 Establish Budget

Establish budget is the process of aggregating in a time-phased manner the estimated costs to perform all the known activities to establish an authorize baseline.

Project name Analysis of estimation of prices													
Sr. No.	BOQ reference	Brief description	Unit	Material cost	Labor cost	Equipment cost	Sub total -1 activity cost estimate	Overhead and profit (% of e)	Sub total-2 (e + f)	Main contractor's overhead {% of (g)}	Sub total-3 (g + h)	Main contractor's profit (mark up) {% of (j)}	Total (j + k) unit price
			(a)	(b)	(c)	(d)	(e) = (b + c + d)	(f)	(g)	(h)	(j)	(k)	(l)

FIGURE 7.60
Analysis of prices.

In a construction project, budget is established at various stages as the project life cycle progresses. It helps project owner control, adjust the project cost, and achieve the intended project objectives within the investment limit. The degree of accuracy of establishing project budget varies at different stages of the project.

At early stage of project, that is, the feasibility stage, the budget is developed on specific project information using the order of magnitude estimation tool. The budget is further refined in the succeeding phases as more information and details are available. The budget prepared by the designer (consultant) at the bidding and tendering stage is of definitive nature, as at this stage, the project details are clearly defined and detailed drawings, BOQ, technical specifications, and project schedule are available for cost estimation. The project budget is the approved funding amount required to complete the project within the approved schedule. The project budget is based on the total anticipated cost to complete the project and also includes contingency and management reserves. Figure 7.61 illustrates the process of establishing construction budget (hard cost).

7.3.5.2.1 Cost Baseline

Construction project's cost baseline is developed based on the construction budget. Cost baseline is a time-phased budget that is used as a basis against which the overall cost performance of the project is measured, monitored, and controlled. The cost baseline is usually shown in an S-curve graph. The S-curve predicts the amount that will be spent over the established project schedule (time). The S-curve is used to measure project performance

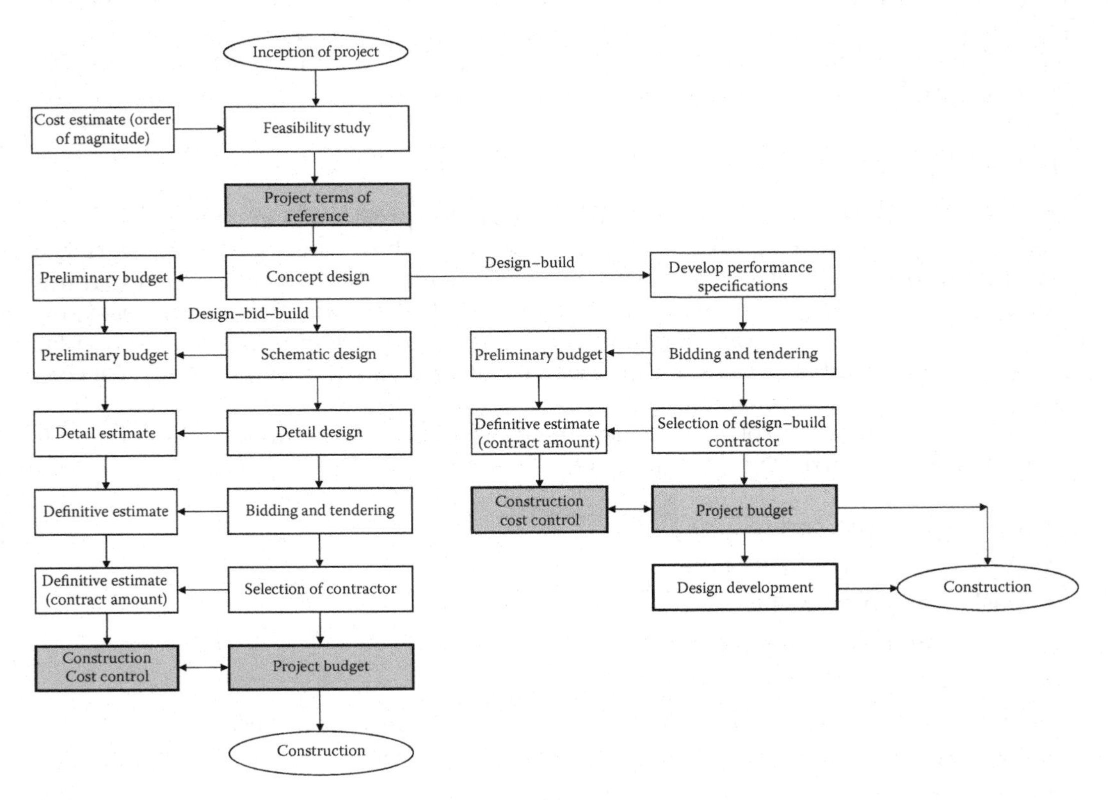

FIGURE 7.61
Process of establishing construction budget.

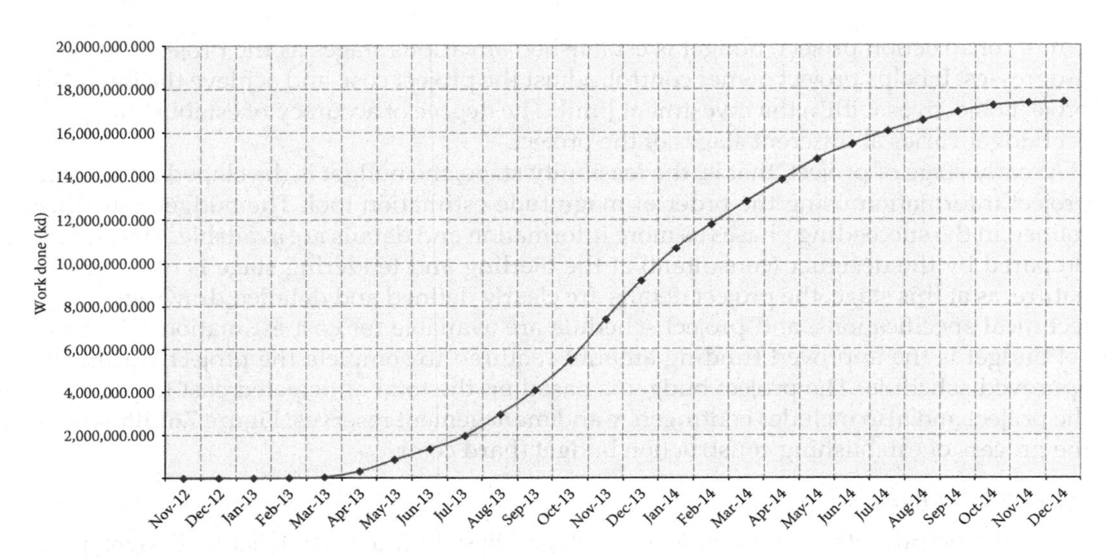

FIGURE 7.62
Project S-curve (budgeted).

and predict the expenses over project duration. The cost baseline helps owner/client to know the project funding requirements. Funding requirements, total and periodic, are derived from the cost baseline. Figure 7.62 illustrates example S-curve (budgeted).

Any changes to the baseline need approval of the Change Control Board. Figure 7.63 illustrates baseline change request form.

7.3.5.3 Control Cost

Cost control is the process of monitoring the status of the project to update the project cost and managing changes to baseline. This process provides the means to recognize variance from the approved plan, evaluate possible alternative, and take corrective action to minimize the risk. In order to have successful cost control, it is essential to have the necessary information and data to take appropriate action. If the necessary information and regular updates are not available or if the action is inefficiently executed, then the risk to cost control on a project is raised considerably.

The purpose of cost control is to manage the project delivery within the approved budget. Regular cost reporting will facilitate the following:

- Establish project cost to date
- Anticipate final budget of the project
- Cash flow requirements
- Understanding potential risk to the project
- Cost control process focused on
- Identifying the factors that influence the changes to the cost baseline
- Determining whether the cost baseline has changed or not
- Ensuring that the changes are beneficial for the project
- Establishing the cost control structure and policy

Project Name

Consultant Name

BASELINE CHANGE REQUEST

CONTRACT NO. : ______________________________ BCR. NO. : _____________

CONTRACTOR : ______________________________ DATE : _____________

BCR ORIGINATOR ______________________________

In accordance with clause ----- of Document ---- Conditions of Contract, We hereby request for Baseline Change as follows;

SCOPE ☐ SCHEDULE ☐ COST ☐

Reason for Change:

Change Impact:

Following Supporting Documents Attached:

1) Cost Estimate

2) Schedule

3) Other

APPRIVALS: **YES** _______ **NO** ________

Signature:

Resident Engineer	**PM/CM**	**Employer**

Date:

Distribution ☐ Employer ☐ Engineer ☐ Resident Engineer ☐ Contractor

FIGURE 7.63
Baseline change request.

- Managing the actual changes when and as they occur
- Monitoring cost performance to detect cost variance from the actual budget
- Recording all appropriate changes
- Informing/reporting concerned stakeholders about the approved changes
- Preventing unauthorized changes to the cost baseline
- Working to bring cost overruns within acceptable limits

Construction project development has mainly the following four stages:

1. Study
2. Design
3. Bidding and tendering
4. Construction

It is essential to perform cost controlling process at these stages in order to ensure that total project cost does not increase the investment plan set by the owner and to achieve project value proposition.

In construction projects, the designer plays an important role to influence the final cost of project to ensure the owner that there will not be budget overrun. The designer has to not only control the project cost but also to ensure that design has

- Minimum errors
- Minimum omissions
- Properly written specifications
- Realistic construction schedule
- All the requirements to achieve owner's need and objectives

In the case of the design–bid–build type of project delivery system, cost control during study stage, design stage, and bidding and tendering stage is carried out by the owner/designer. The contractor is responsible to control the cost during the construction phase.

In the case of the design–build type of project delivery system, the contractor carries out most of cost control. During the design phase, if the designer (consultant) finds that the approved budget is inadequate, then they should seek revision/adjustment to the budget from any contingency amount that the owner might have.

During the construction phase, it is the contractor who is interested to complete the construction within the contracted amount, and the construction manager/project manager controls the cost to ensure that there is no overrun and changes to the cost baseline.

The contractor has to mainly control the following costs:

- Material
- Labor
- Resource (labor) productivity
- Equipment/plant productivity
- Rework
- Delays
- Subcontractor's work
- Currency exchange

The construction manager/project manager, on behalf of the owner, should

- Compare and monitor project progress
- Identify and control approved variations

- Identify cash flow forecasts
- Compare current budget forecast for remaining works

7.3.5.3.1 Earned Value Management

Earned value management (EVM) is a methodology used to measure and evaluate project performance against cost, schedule, and scope baseline. It compares the amount of planned work with what is actually accomplished to determine whether the project is progressing as planned or not. Earned value (EV) analysis is used to

- Measure progress of the project budget, schedule, and scope to know how much percentage of
    - Budget is spent
    - Time has elapsed
    - Work is done
- Forecast its completion date and cost
- Provide budget and schedule variances

The following are the basic terminologies used in EVM:

1. BCWS → Budgeted cost of work scheduled or planned value (PV)

   It is the planned cost of the total amount of work scheduled to be performed by the milestone date.
2. BCWP → Budgeted cost of work performed or EV

   It is the actual cost incurred to accomplish the work that has been done to date.
3. ACWP → Actual cost of work performed or actual cost

   It is the planned cost to complete the work that has been done.

These three key values are used in various combinations to determine cost and schedule performance and provide an estimated cost of the project at its completion. Table 7.24 illustrates the terms used in EVM and their interpretation, whereas Table 7.25 illustrates the formulas used in EVM and their interpretations. Figure 7.64 illustrates diagram for earned value method.

7.3.6 Quality Management

Quality management is an organization-wide approach to understand customer needs and deliver the solutions to fulfill and satisfy the customer. Quality management is management and implementation of quality system to achieve customer satisfaction at the lowest overall cost to the organization, while continuing to improve the process. Quality system is a framework for quality management. It embraces the organizational structure, policies, procedures, and processes needed to implement quality management system.

Quality management in construction projects is different to that of manufacturing. Quality in construction projects is not only the quality of products and equipment used in the construction, but it is also the total management approach to complete the facility as per the scope of works to customer/owner satisfaction within the budget and to be

TABLE 7.24

Earned Value Management Terms

Sr. No.	Term	Description	Interpretation
1	PV (BCWS)	Planned value (budgeted cost of work scheduled)	Planned cost of total amount of work scheduled to be performed by the milestone date
2	EV(BCWP)	Earned value (budgeted cost of work performed)	Planned/budgeted cost to complete the work that has been done
3	AC(ACWP)	Actual cost (actual cost of work performed)	Actual cost incurred to accomplish the work that has been done to date
4	BAC	Budget at completion	Estimated total cost of project when completed
5	EAC	Estimate at completion	Expected total cost of project when completed
6	ETC	Estimate to completion	Expected additional cost needed to complete the project
7	VAC	Variance at completion	Amount over budget OR under budget we expect at the end of the project

TABLE 7.25

Parameters of Earned Value Management

Sr. No.	Name	Formula	Interpretation
1	Cost variance (CV)	BCWP–ACWP (EV–AC)	A comparison of the budgeted cost of work performed with actual cost *Positive* result means under budget *Negative* result means over budget
2	Schedule variance	BCWP–BCWS (EV–PV)	A comparison of amount of work performed during a given period of time to what was scheduled to be performed *Positive* result means ahead of schedule *Negative* result means behind schedule
3	Cost performance index (CPI)	EV/AC BCWP/ACWP	Greater than 1 means work is being produced for less than planned Less than 1 means the work is costing more than planned
4	Schedule performance index (SPI)	EV/PV BCWP/BCWS	Greater than 1 means project is ahead of schedule Less than 1 means project has accomplished less than planned and behind schedule
5	Estimate at completion (EAC)	AC + ETC	Expected total cost of project when completed
6	Estimate to complete (ETC)	EAC–AC	Expected additional cost needed to complete the project
7	Variance at Completion (VAC)	BAC–EAC	How much over budget *or* under budget we expect at the end of the project

completed within specified schedule to meet owner's defined purpose. Quality management in construction addresses both the management of project and the product of the project and all the components of the product. It also involves incorporation of changes or improvements, if needed. Construction project quality is fulfillment of owner's needs as per the defined scope of works within a budget and specified schedule to satisfy owner's/user's requirements.

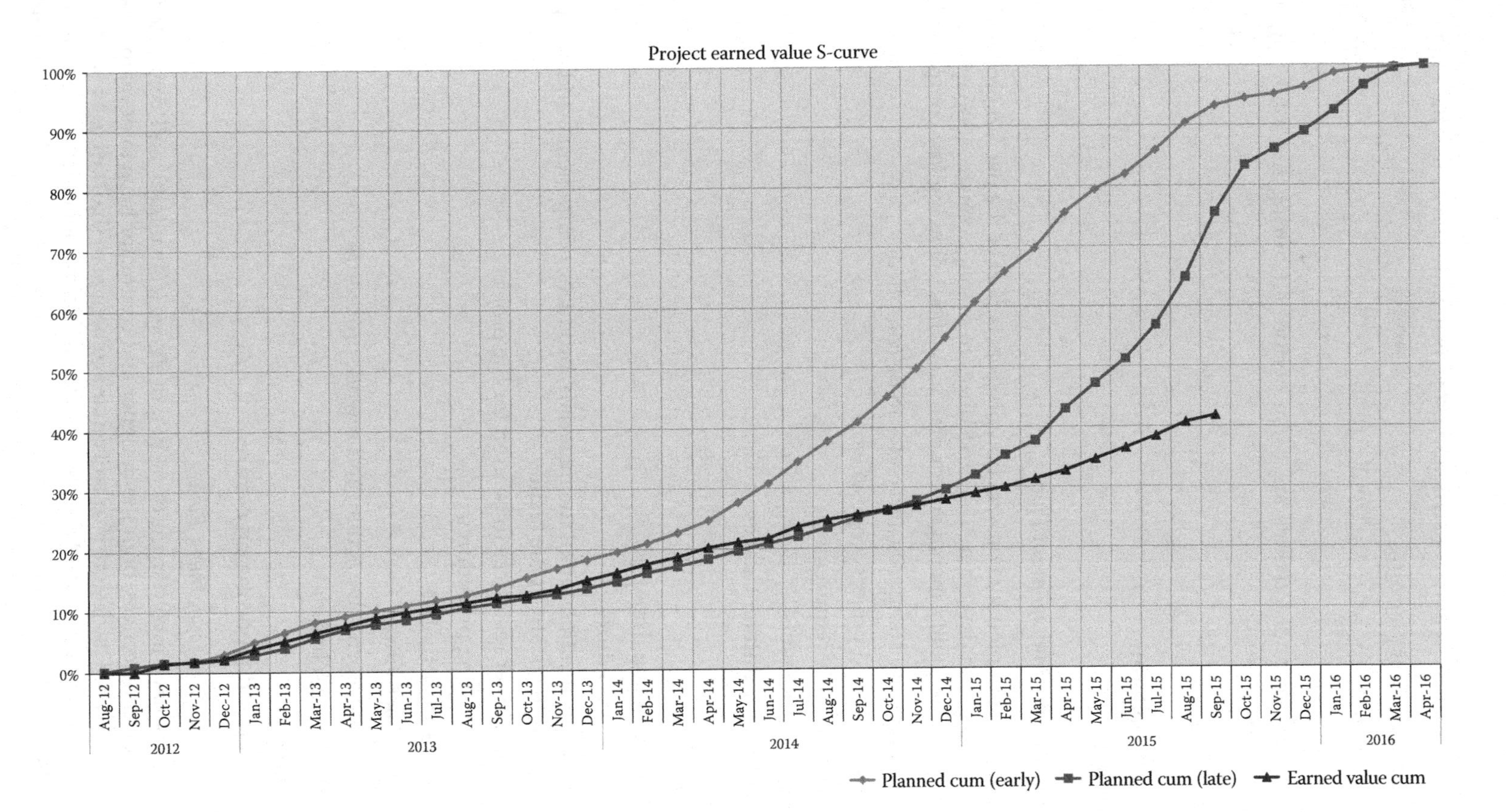

FIGURE 7.64
Earned value S-curve.

Quality management system in construction projects mainly consists of the following:

- Quality management planning
- Quality assurance
- Quality control

Each of these processes and activities are to be performed during the following main stages of construction project:

1. Study
2. Design
3. Bidding and tendering
4. Construction

7.3.6.1 Develop Quality Management Plan

The quality management plan for construction projects is part of the overall project documentation, addressing and describing the procedures to manage construction quality and project deliverable. The quality management plan identifies following key components:

- Details of the quality standards and codes to be complied
- Project objectives and project scope of work
- Stakeholders' quality requirements
- Regulatory requirements
- Quality matrix for different stages
- Design criteria
- Design procedures
- Detailed construction drawings
- Detailed work procedure
- Well-defined specification for all the materials, products, components, and equipment to be used to construct the facility
- Manpower and other resources to be used for the project
- Inspection and testing procedures
- Quality assurance activities
- Quality control activities
- Defect prevention, corrective action, and rework procedure
- Project completion schedule
- Cost of the project
- Documentation and reporting procedure

7.3.6.1.1 Designer's Quality Management Plan

Construction projects have involvement of owner, designer (consultant), and contractor. In order to achieve project objectives, both the designer and contractor have to develop project quality management plan. The designer's quality management plan shall be based on owner's project objectives, whereas the contractor's plan shall take into consideration requirements of contract documents. Figure 7.65 illustrates project quality management plan for design stage.

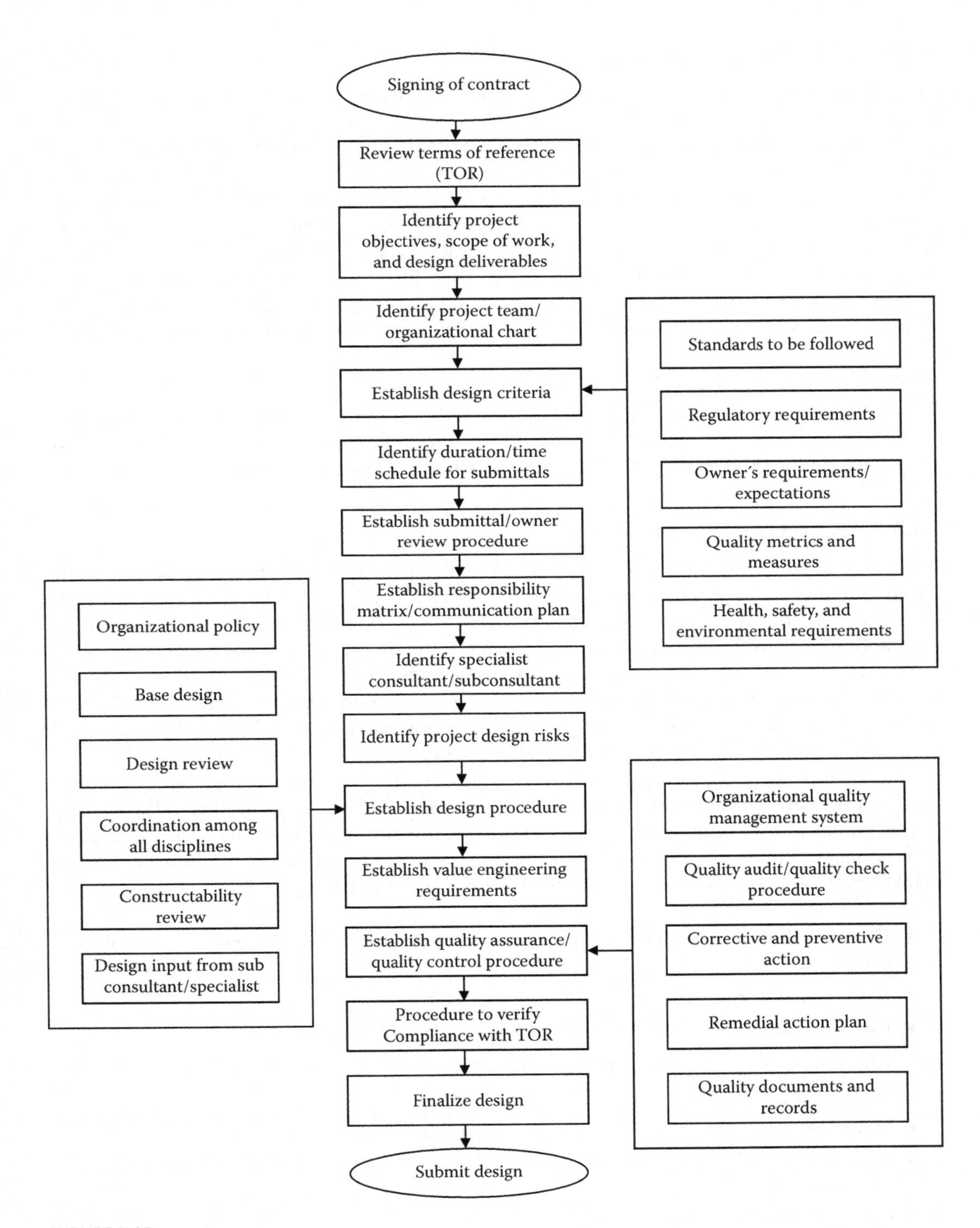

FIGURE 7.65
Project quality management plan for design stage. (Abdul Razzak Rumane (2013), *Quality Tools for Managing Construction Projects*, CRC Press, Boca Raton, FL. Reprinted with permission from Taylor & Francis Group.)

7.3.6.1.2 Contractor's Quality Control Plan

During the construction stage, the contractor prepares contractor's quality control plan (CQCP) based on project specific requirements. The CQCP is the contractor's everyday tool to insure meeting the performance standards specified in the contract documents. Its contents are drawn from the company's quality system, the contract, and related documents. It is a framework for the contractor's process for achieving quality construction. It is a document setting out the specific quality activities and resources pertaining to a particular contract or project. It is the documentation of contractor's process for delivering the level of construction quality required by the contract. A quality plan is virtually a quality manual tailor-made for the project and is based on contract requirements. Figure 7.66 illustrates the process for development of CQCP and Table 7.26 illustrates the table of contents of CQCP.

7.3.6.1.3 Quality Matrix

In the case of construction projects, an organizational framework is established and implemented mainly by three parties: owner, designer/consultant, and contractor. Table 7.27 illustrates the responsibilities matrix for quality control-related personnel.

7.3.6.2 Perform Quality Assurance

Quality assurance in construction projects covers all activities performed by design team, contractor, and quality controller/auditor (supervision staff) to meet owner's objectives as specified and to ensure and guarantee that the project/facility is fully functional to the satisfaction of owner/end user. Auditing is part of the quality assurance function.

Quality assurance is an activity for providing evidence to establish confidence among all concerned, that quality-related activities are being performed effectively. All these planned or systematic actions are necessary to provide adequate confidence that a product or service will satisfy given requirements for quality.

Quality assurance covers all activities from design, development, production/construction, installation, servicing to documentation, and also includes regulations of the quality of raw materials; assemblies, products, and components, services related to production; and management, production, and inspection processes. The following are major activities to be performed for quality assurance of the construction project:

- Confirm that owners needs and requirements are included in the scope of works (TOR)
- Review and confirm design compliance to TOR
- Executed works comply with the specified standards and codes
- Conformance to regulatory requirements
- Works executed as per approved shop drawings
- Installation of approved material, equipment on the project
- Method of installation as per approved method statement or manufacturer's recommendation
- Coordination among all the trades
- Continuous inspection during construction/installation process
- Identify and correct the deficiencies
- Timely submission and review of transmittals

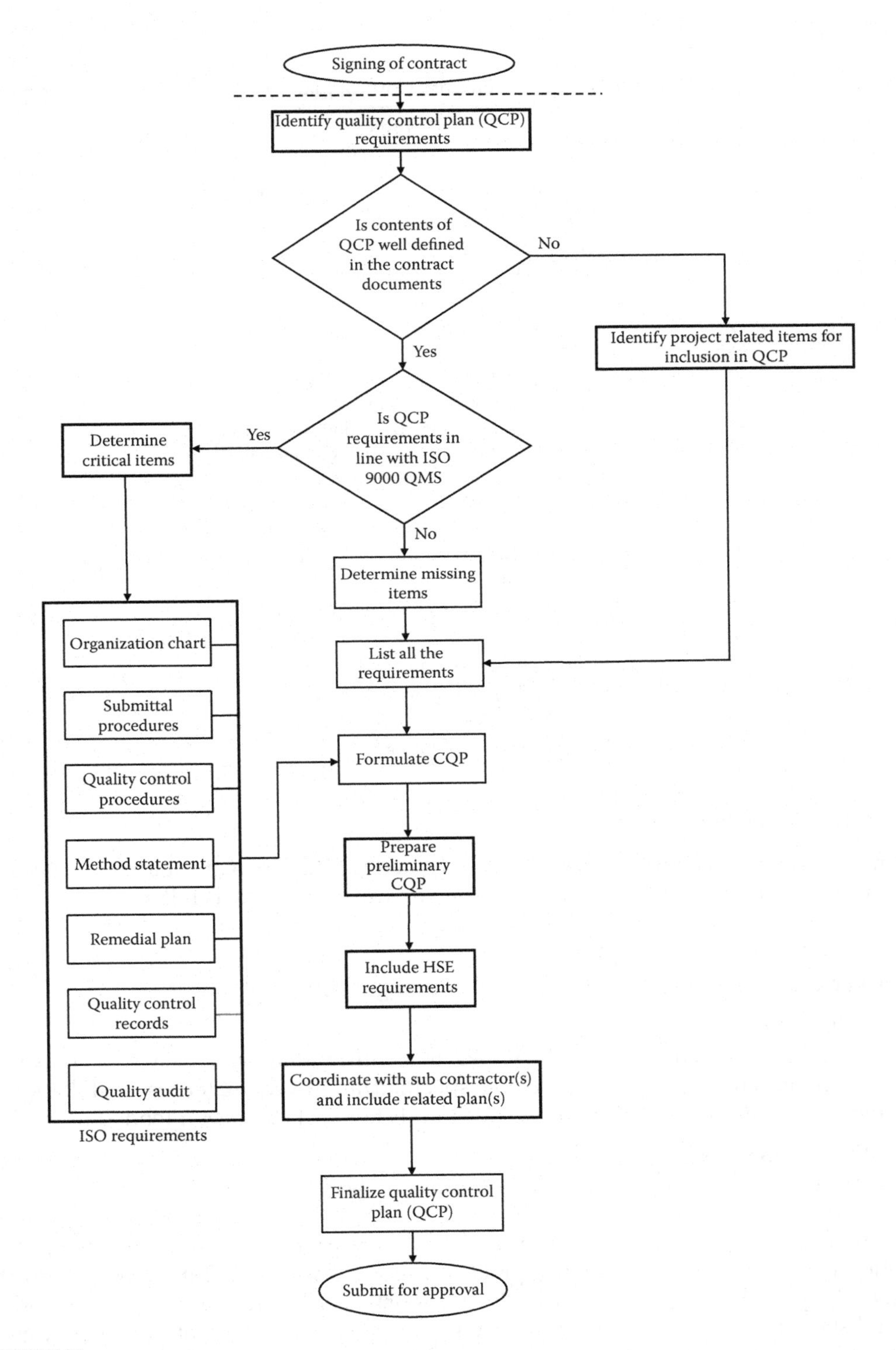

FIGURE 7.66
Logic flow diagram for development of contractor's quality control plan.

TABLE 7.26
Contents of Contractor's Quality Control Plan

Section	Topic
1	Introduction
2	Description of project
3	Quality control organization
4	Qualification of QC staff
5	Responsibilities of QC personnel
6	Procedure for submittals
7	Quality control procedure
7.1	Procurement
7.2	Inspection of site activities (checklists)
7.3	Inspection and testing procedure for systems
7.4	Off-site manufacturing, inspection, and testing
7.5	Procedure for laboratory testing of material
7.6	Inspection of material received at site
7.7	Protection of works
8	Method statement for various installation activities
9	Project-specific procedures
10	Quality control records
11	Company's quality manual and procedure
12	Periodical testing
13	Quality updating program
14	Quality auditing program
15	Testing, commissioning, and handover
16	Health, safety, and environment

7.3.6.3 Control Quality

Quality control in construction projects is performed at every stage through use of various control charts, diagrams, checklists, and so on and can be defined as follows:

- Checking and review of project design
- Checking and review of bidding and tendering documents
- Analysis of contractor's bids
- Checking of executed/installed works to confirm that works have been performed/executed as specified, using specified/approved materials, installation methods and specified references, codes, and standards to meet intended use
- Controlling budget
- Planning, monitoring, and controlling project schedule

The construction project quality control process is a part of contract documents that provide details about specific quality practices, resources, and activities relevant to the project. The purpose of quality control during construction is to ensure that the work is accomplished in accordance with the requirements specified in the contract. Inspection of construction works is carried out throughout the construction period either by the construction supervision team (consultant) or by the appointed inspection agency. Quality is an important aspect of construction project. The quality of construction project must

TABLE 7.27

Responsibilities for Site Quality Control

		Linear Responsibility Chart							
		Owner	Consultant	Contractor					
Sr. No.	Description	Owner/Project Manager Construction Manager	Consultant/Designer	Contractor Manager	Quality Incharge	Quality Engineers	Site Engineers	Safety Officer	Head Office
1	Specify quality standards	□	■						
2	Prepare quality control plan			□	■	□			□
3	Control distribution of plans and specifications			□	■	□			□
4	Submittals			■	□		■		
5	Prepare procurement documents			□			■		■
6	Prepare construction method procedures			□	□	□	■		
7	Inspect work in progress		■	■		□	■		
8	Accept work in progress		■						
9	Stop work in progress	■	□						
10	Inspect materials upon receipt		■	□	■		■		
11	Monitor and evaluate quality of works		■	□	■	■	■		
12	Maintain quality records			□	■				
13	Determine disposition of nonconforming Items	□	□	■					
14	Investigate failures	□	□	■	■	□	■		
15	Site safety			□				■	
16	Testing and commissioning	□	■	□			■		
17	Acceptance of completed works	■	□	□					

Source: Abdul Razzak Rumane (2010), *Quality Management in Construction Projects*, CRC Press, Boca Raton, FL. Reprinted with permission from Taylor & Francis Group.
■ Primary Responsibility
□ Advise/Assist

meet the requirements specified in the contact documents. Normally contractor provides onsite inspection and testing facilities at construction site. On a construction site, inspection and testing is carried out at three stages during the construction period to ensure quality compliance.

1. During construction process, it is carried out with a checklist request submitted by the contractor for testing of ongoing works before proceeding to next step.
2. Receipt of subcontractor or purchased material or services is performed by a material inspection request submitted by the contractor to the consultant upon receipt of material.
3. Inspection and testing is also conducted before final delivery or commissioning and handover.

7.3.7 Resource Management

Resource management in construction is mainly related to management of the following processes:

1. Human resources (project teams)
    - Project owner team (project manager and construction manager)
    - Designer (consultant)
2. Construction resources (contractor)
    - Manpower
    - Equipment
    - Material

Construction projects are of nonrepetitive nature and have definite beginning and definite end. The important factor in construction project is to complete the facility as per the defined scope within the specified schedule and budget. In order to meet the project quality, the construction manager firm, designer firm, and contractor have to ensure that right human resources and right type of materials and equipment are available at the right time. Since construction projects are of temporary nature, every time a new project starts, the human resource configuration changes, depending on the company policies and practices to engage human resources for a particular project.

Similarly, it has to be ensured that specified materials and equipment are available to meet the installation/execution schedule.

7.3.7.1 Manage Human Resource

Construction project human resource management process includes the following:

1. Plan human resources (project teams)
2. Acquire project team
3. Develop project team
4. Manage project team
5. Release/demobilize project team

7.3.7.1.1 Plan Human Resources (Project Teams)

Every project needs people to complete the project. Traditional construction projects involve the following three main groups:

- Owner
- Designer (consultant)
- Contractor (s)

Depending on the type of project delivery system, each of these groups needs human resources to manage the project. Human resource planning is executed in the initial stage for the overall project requirement and is performed iteratively and interactively during each of the project stage with other aspects of project such as scope, schedule, and cost of the project and adjusted as per project requirements.

For example, the construction/project management firms plan their human resources as per the scope of work and project schedule. The designer (consultant) plans their resources depending upon the size of the project and design deliverables. The contractor is responsible for all type of human resources to complete the project.

Human resource planning process includes organizational planning, taking into consideration the requirements of the project and the stakeholders involved, detailing the project roles, responsibilities, required skills, and reporting relationships to the appropriate people or group of people. Human resource planning includes the following:

- Documenting staffing requirements (roles, responsibilities, and skills)
- Project organization chart
- Staff deployment charts
- Team acquiring process
- Team development and training needs
- Release/demobilization plan

7.3.7.1.2 Acquire Project Team

Acquiring project team is process of obtaining project team members for completing the project. There are mainly three project teams associated with three groups involved in a construction project. These are as follows:

1. Construction management or project management firm (depending on the type of project delivery system selected by the owner)
2. Designer (consultant)
3. Contractor's core staff

Acquiring project team is a process of getting right people having right knowledge, skills, and experience that is required to work on the project and to perform the assignment within a given time frame. A project team is a group of professionals committed to achieving a common goal to complete the project successfully. Acquiring project team is a continuous process that is performed throughout the project as and when a particular category of team member is required as per the project deployment chart. For example, a resident engineer for a project is acquired at the beginning of the project and is released after completion of the project. Other team members are acquired as per the agreed upon deployment chart depending on the project needs and project organizational structure. Acquiring project team includes the following:

- Negotiation
- Internal or externally hired member
- Deployment chart

While engaging a team member, it is necessary to consider following criteria:

- Required level of experience
- OBS
- Project-specific requirements/needs
- Professional qualification
- Member's experience in a similar type of project
- Project knowledge
- Technical skills and capabilities in relevant areas
- Management skills
- Collaborative skills
- Communication capabilities
- Availability of the candidate to meet work schedule
- Professional membership and training certification
- Cost
- Staff approval requirements by the project owner/client
- Regulatory constraints

Figure 7.67 illustrates the project team acquiring process.

7.3.7.1.3 Develop Project Team

Project team is a combination of professionals acquired to do project-specific activities to complete the project successfully. Each of the project team members is committed to achieve a common goal to complete the project as per defined scope within specified time and budget. Most construction projects have projectized or matrix-type organizational structure. (Types of organizational structures have already been discussed under Section 7.2.2.1) In either case, the team members are composed of different background and disciplines.

The following are the five stages of team development:

1. Forming
2. Storming
3. Norming
4. Performing
5. Adjourning

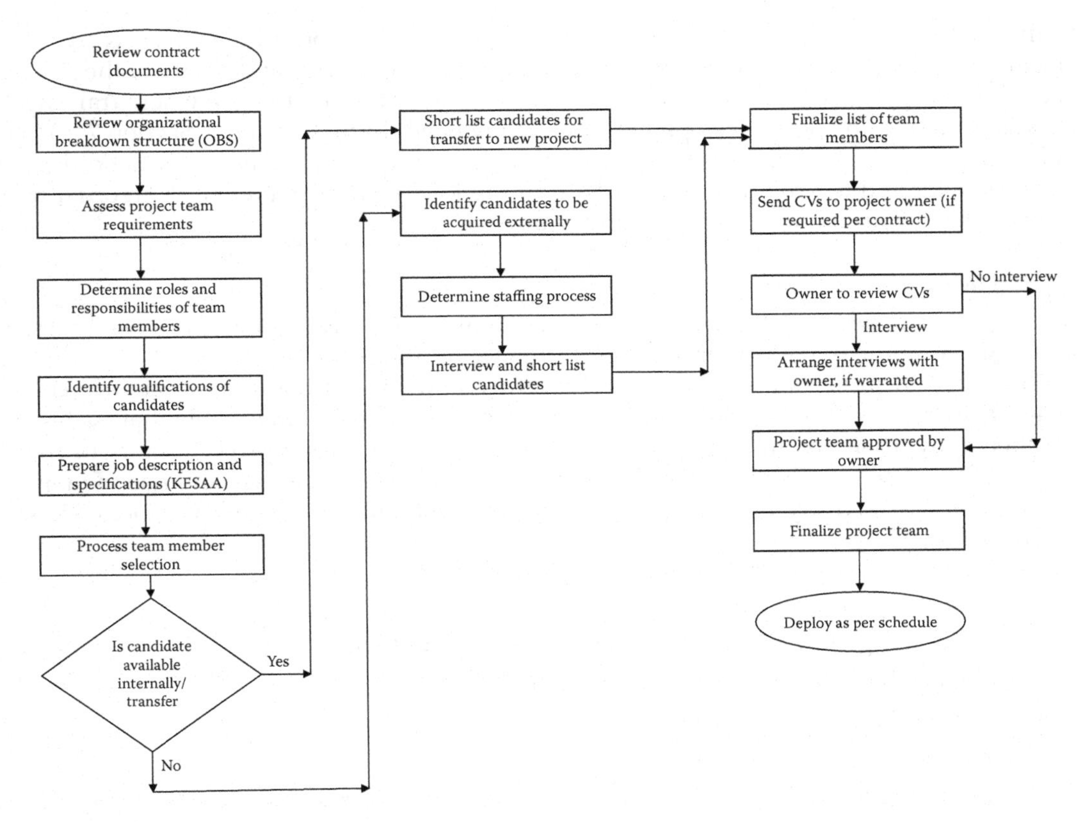

FIGURE 7.67
Project team acquisition process.

These stages have already been discussed under Section 7.2.3.1.

In construction projects, there are mainly three project teams. Once the team members are assigned and a team is formed, a meeting is to be arranged where every member is introduced and is given a clear understanding about

- Project goals
- Project overall mission
- Roles and responsibilities of each team member
- Coordination process
- Working in collaborative climate
- Communication method
- Member's relationship with each other
- Commitment of each member toward the project

Normally, project team members are selected on the basis of having required competency in the specific field for which they are assigned and responsible to perform efficiently. However, it is necessary to improve the knowledge and skills of team members in order to increase their ability and competency to complete project work to meet or exceed the specified quality of the project. In order to enhance the competency and ability of the project member, appropriate training can be arranged by ascertaining the gap where training is required. Training and development has already been discussed under Section 7.2.3.2. Performance evaluation process helps determine the training requirements for the project team member and to provide the required training identified by gap analysis in order to improve technical competencies of project team member.

7.3.7.1.4 Manage Project Team

It is a process to keep track team member performance, provide feedback, resolve issues, and manage changes to ensure project performance optimization.

Construction projects are of temporary nature and project team members are collected from different background and disciplines; therefore, it is inevitable that issues/conflicts may arise among the members. There are mainly three project teams in the construction. Each project manager has to resolve the issue as it arises and has to manage the team effectively by maintaining cohesion among all the team members. This can be achieved by

- Establishing ground rules
- Coordinating with all team members to understand their issues
- Creating shared vision among team members
- Tracking the team's performance
- Training, recognition, and rewards
- Conducting meetings, exchanging relevant information, and resolving issues
- Problem identification and providing quick solution
- Conflict management

7.3.7.1.4.1 Manage Conflict There exist several types of conflict. Each conflict can assume a different intensity at different stages of the project. The causes of disagreement vary in different phases of the project, as different members are involved at different phases of the project. The following are typical sources of conflict that may arise during the project:

- Priorities
- Project schedule
- Resources
- Cost
- Scope change
- Technical opinions
- Personality conflict
- Communication problem
- Lack of coordination by team members
- Administrative procedures

The project manager has to resolve conflict by searching for an alternate solution. The following methods are normally used to resolve conflicts:

1. *Withdrawing/avoiding*: It means both parties retreat from the conflict issue.
2. *Smoothing/accommodating*: It is emphasizing upon friendly relationship and agreement rather than differences of opinions.
3. *Collaborating*: In this method, parties try to incorporate multiple viewpoints in order to lead to consensus.
4. *Confronting/problem solving*: It is a fact-based approach where both parties solve their problems by focusing on the issues, looking at alternatives, and selecting the best alternative.
5. *Compromising*: Compromising is finding solution that brings some degree of satisfaction to both parties.
6. *Forcing*: Forcing is use of authority and power in resolving the conflict by exerting one's viewpoint over another at the expense of another party.

Figure 7.68 illustrates the conflict management flowchart.

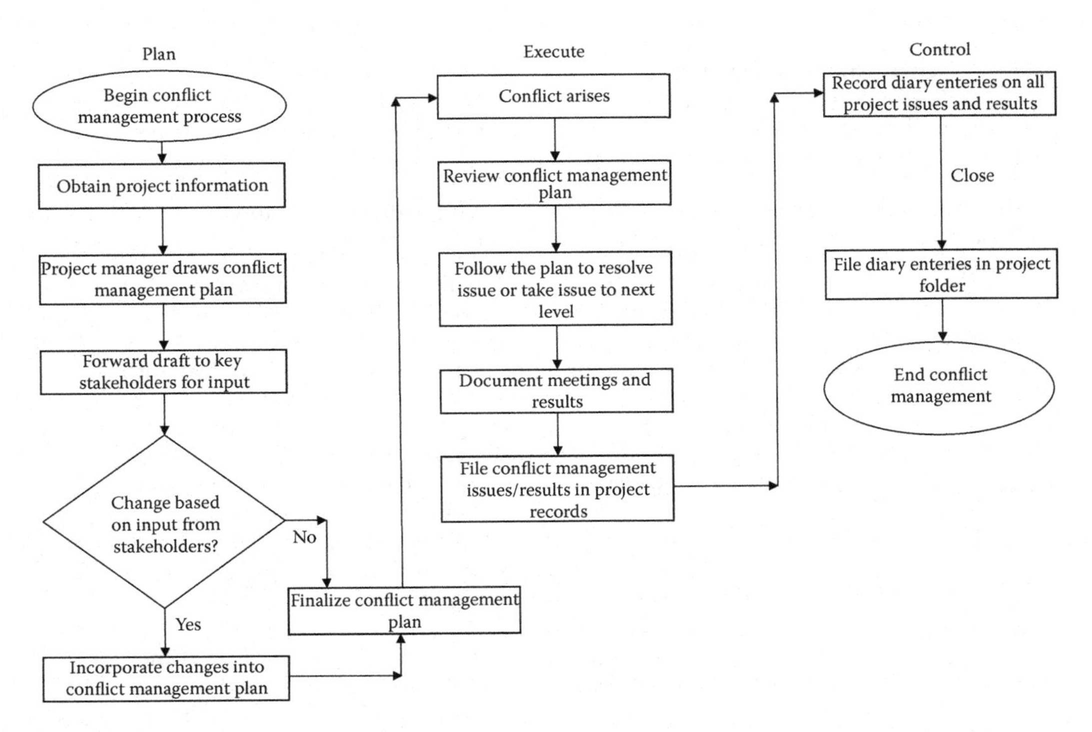

FIGURE 7.68
Conflict management flowchart. (*Project Management Handbook*, 2nd Edition (2007), Reprinted with permission from California Department of Transportation, Office of Project Management Process Improvement, Sacramento, CA.)

7.3.7.1.4.2 Dealing with Conflicts in Project Teams

The below article is by Michael Stanleigh [1].

Team conflict is challenging for project leaders but it is not necessarily bad. Conflict can lead to new ideas and approaches and facilitate the surfacing of important issues between team if it is managed well.

According to various research studies on team conflict, the major source of conflict among project teams are project *goals not agreed upon, disagreement of the project priorities and conflicting work schedules*. This is no surprise since most organizations today run multiple projects and employees often find themselves serving on variety of project teams. To add fuel to the fire, employees may report to a variety of project managers while reporting directly to functional managers. This sets the stage for further conflict opportunities due to communication and information flow. After all, when reporting relationships are complex it becomes more difficult to share information.

Personality and interpersonal issues may also draw conflict, particularly in high technology environments, where cross-functional, self-directed teams with technical background must rely on work of others to get their own work done.

So what is the learning from these research studies? The lesson here is clear. It's very important for cross-functional team members to receive training in communication and interpersonal skills.

Here are some recommendations for project leaders and project team members:

1. Hold more frequent meetings and status review sessions to increase *communication* between functions and reduce misinterpretations of project goals and priorities
2. Give team members soft skill training in human relations and facilitate more active team-building efforts.
3. Foster an environment of mutual respect. No method of managing conflict will work without mutual respect and a willingness to disagree and resolve disagreements

If conflict erupts in your team, try taking the following four steps to calm things down; Listen, Acknowledge, Respond, and Resolve Remaining Differences.

Listen: Clear your mind from distraction and really try to concentrate on what the person is trying to communicate-listen to both the words and no-verbal cues from gesture and body language. Gestures are often more important than words. After all, when resolving disagreements, you often have to deal with feelings first.

Acknowledge: Acknowledgement does not mean agreement. You can acknowledge what someone is saying without necessarily agreeing with them. Everyone's opinion and feelings are valid, even if different from yours. For example, statements like, "I understand you're angry," "Let's explore your suggestion further" or "If I understand you, you're saying that you disagree?" These are all ways of acknowledging their communication and their point of view. While you may not agree with what you are saying, it shows that they are being heard.

Respond: Now you have acknowledged to them that you have heard what they have to say, it's your turn to respond. If you don't agree with what the person is saying, be sure that your feedback is constructive and offer alternative suggestion. A good way to pose your response is to speak from personal experience. For example, "In my experience I have found that approach is ineffective because, "Be prepared to explain your position and remain open to being challenged or questioned about it.

Resolve Remaining Differences: If you have listened carefully to people around you, you have probably figured out what's causing the disagreement. Once you have defined the real problem, you should be able to break it down into manageable parts. This will help you to generate alternative solutions and then select the alternative which everyone can agree.

Remember: for individuals to work effectively in a team there has to be a certain level of trust among the group. Each team member needs to feel "safe" about sharing their ideas. While others may not be in full agreement with another person's ideas, it's important for the team leader to foster an appreciation for the different views presented and explore the best alternative in a respectful manner with honest intent.

Most importantly, when conflict does occur in a team, leaders can gain ground by not avoiding it. By managing it well, acknowledging it and tolerating it, a team leader can actually use conflict as a tool to generate revitalized team engagement and innovation.

If the conflict is too great and the project leader is unable to get the team past it, then hiring a *team coach* is a good idea.

7.3.7.1.5 Release/Demobilize Project Team

Construction projects are a temporary endeavor. Every time a project is initiated, project team members are acquired to complete the project. The organizational structure depends on the strategic policy of the organization. Once the project is completed, the project team members are released and sent back to their original functional department or engaged in other projects or terminated from the work.

7.3.7.2 Construction Resources Management

In most construction project, the contractor is responsible for managing the resources. It includes the following:

1. Manpower
2. Material
3. Equipment
4. Subcontractor(s)

7.3.7.2.1 Manage Manpower

It is the contractor's responsibility to arrange necessary manpower to execute the project. This includes the following:

- Contractor's site staff to supervise the construction (core staff)
- Field workers to execute the works

Contract documents normally specify a list of minimum number of core staff to be available at site during construction period. Absence of any of these staff may result in penalty to be imposed on contractor by the owner. The contractor has to consider the qualification and related experience mentioned in the contract documents while acquiring the team members to work on the project.

7.3.7.2.2 Manage Equipment

The contractor has to provide the required equipment to execute the construction works. Normally, contract documents specify that the minimum number of equipment is to be available at site during the construction process to ensure smooth operation of all the construction activities. The typical equipment list is as follows:

- Tower crane
- Mobile crane
- Normal mixture
- Concrete mixing plant
- Dump trucks
- Compressor
- Vibrators
- Water pumps
- Compactors
- Concrete pumps
- Trucks
- Concrete trucks
- Diesel generator set(s)

Figure 7.69 illustrates an example equipment status listing quantities of different types of equipment that are available at site during the construction period for a major building construction project.

The contractor has to maintain the equipment in good conditions to ensure its efficient utilization.

Sr. No.	Description	Months																												
		Jan/15	Feb/15	Mar/15	Apr/15	#####	Jun/15	Jul/15	Aug/15	Sep/15	Oct/15	Nov/15	Dec/15	Jan/16	Feb/16	Mar/16	Apr/16	#####	Jun/16	Jul/16	Aug/16	Sep/16	Oct/16	Nov/16	Dec/16	Jan/17	Feb/17	Mar/17	Apr/17	#####
1	Tower crane						1	1	2	2	2	2	2	2	2	2	2	2	2	2	2	2								
2	Mobile crane								1	1	2	2	2	2	2	2	2	2	2	2	2	2								
3	Tractor (Loader)				1	1	1	1	1	1	1	1	1	1	1	1	1	1	1	1	1	1	1	1	1					
4	Trailer							1	1	1	1	1	1	1	1	1	1	1	1	1	1	1	1	1						
5	Excavator				1	1	1	1	1	1	1	1	1																	
6	Truck				2	2	2	2	2	2	2	2	2	1	1	1	1	1	1	1	1	1	1	1	1					
7	Tipper truck				2	2	2	2	2	2	2	2	2																	
8	Generator					1	1	2	2	2	2	2	2	2	2	2	2	2	2	2	2	2	2	2	2	1	1	1	1	1
9	Half lorry							1	1	1	1	1	1	1	1	1	1	1	1	1	1	1	1	1	1	1	1	1		
10	Pickup							1	1	1	1	1	1	1	1	1	1	1	1	1	1	1	1	1	1	1	1	1		
11	Air compressor							2	2	2	2	2	2	2	2	2	2	2	2	2	2	2								
12	Welding machine							2	2	2	2	2	2	2	2	2	2	2	2	2	2	2	2	2	2					
13	Bar bending machine							2	2	2	2	2	2	2	2	2	2	2	2	2	2	2								
14	Bar cutting machine							2	2	2	2	2	2	2	2	2	2	2	2	2	2	2								
15	Vibrator							2	2	2	2	2	2	2	2	2	2	2	2	2	2	2								
16	Compactor						1	1	1	1	1	1	1	1	1	1	1	1	1	1	1	1								
17	Wood, marble, brick cutter									2	2	2	2	2	2	2	2	2	2	2	2	2	2	2	2	2	2			
18	Total station				1	1	1	1	1	1	1	2	1	1	1	1	1	2	1	1	1	1	1	2	1	1	1			
19	Dumpy level				2	2	2	2	2	2	2	2	2	2	2	2	2	2	2	2	2	2	2	2	2	2	2			
20	Bobcat										2	2	2	2	2	2	2	2	2	2	2	2	2	2	2	2	2			
21	Water tanker						1	1	1	1	1	1	1	1	1	1	1	1	1	1	1	1	1	1	1	1	1			
22	Boom truck										1	1	1	1	1	1	1	1	1	1	1	1	1	1	1					
23	Forklift										1	1	1	1	1	1	1	1	1	1	1	1	1	1	1	1	1			
24	Bus coaster							2	2	2	2	2	2	2	2	2	2	2	2	2	2	2	2	2	2	2	2			
25	Mini bus							2	2	2	2	2	2	2	2	2	2	2	2	2	2	2	2	2	2	2	2			
	Total equipment				9	10	13	31	33	35	40	41	40	36	36	36	36	37	36	36	36	36	23	24	22	16	16	3	1	1

FIGURE 7.69
Equipment status.

7.3.7.2.3 *Manage Material*

In most construction projects, the contractor is responsible for procurement of material, equipment, and systems to be installed on the project. The contractors have their own procurement strategies. While submitting the bid, the contractor obtains the quotations from various suppliers/subcontractors. The contractor has to consider the following, as a minimum while finalizing the material procurement:

- Contractual commitment
- Specification compliance
- Statutory obligations
- Time
- Cost
- Performance

Figure 7.70 illustrates the material management process for construction projects, and Figure 7.71 illustrates the contractor's procurement log (Report E-2) normally called *Log E-1*.

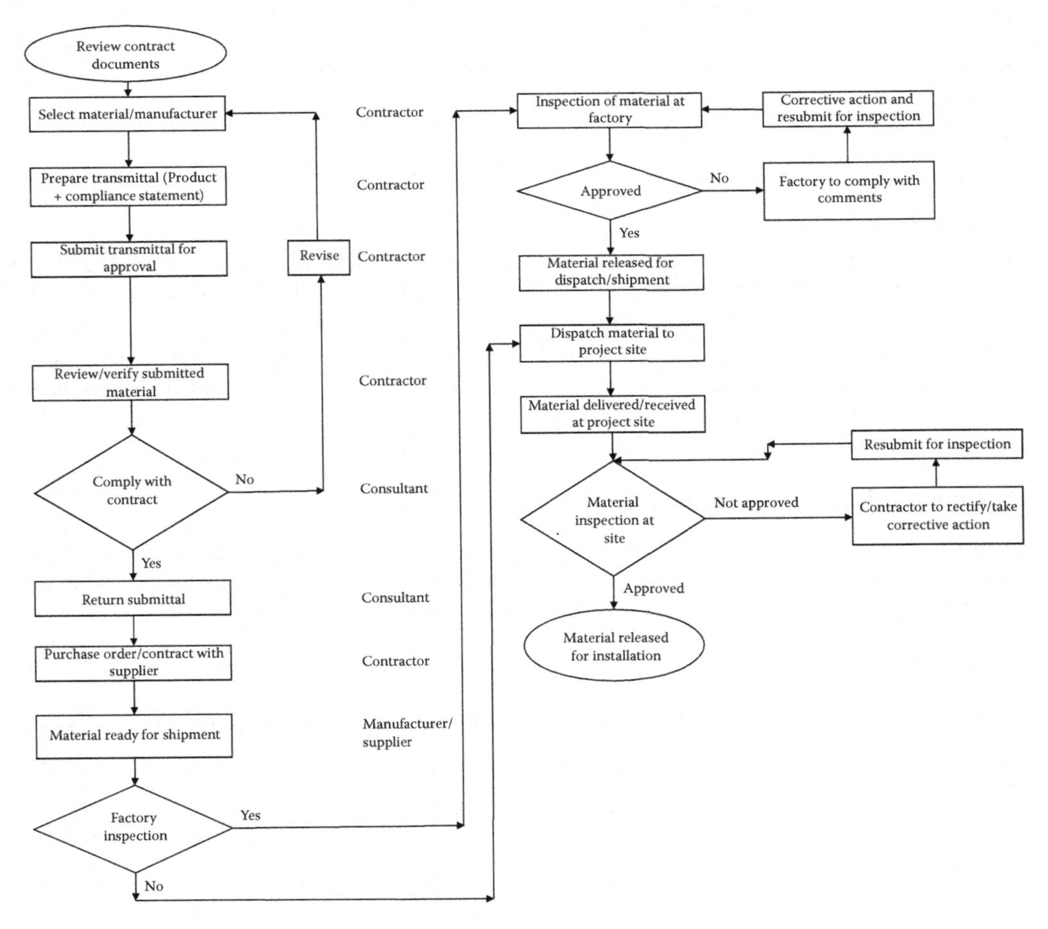

FIGURE 7.70
Material management process for construction project.

Project Name
Consultant Name

CONTRACTOR'S PROCUREMENT LOG (REPORT E-2)

CONTRACT NO. : ______________________ NO. : ______
CONTRACTOR : ______________________ DATE : ______

Activity No.	Description	B.O.Q./ Specification. No.	Estimated Quantity	Ordered Quantity	Reqd. Order Date	Date Pur. Ord. Issued	Purchase Order No.	Supplier	Method Of Shipping	Shipping Date to Kuwait	Required On Job Date	E.D.A. to Site	A.D.A. to Site	Lead Time	Remarks

FIGURE 7.71
Contractor's procurement log.

7.3.7.2.4 Manage Subcontractor

In construction projects, the main contractor (general contractor) is responsible for managing subcontractors' work. Much of the works required to be carried out on major projects are performed by subcontractors, that is, specialist contractors. The main contractor has to manage their works carefully by planning, scheduling, and coordinating to complete the project successfully. Areas of subcontracting are generally listed in the *Particular Conditions* of the contract document. The main contractor is responsible to manage subcontractors' work.

7.3.8 Communication Management

Communication is the process by which information is transmitted from a sender to a receiver via a channel/medium. The receiver encodes the message and gives feedback to the sender. Figure 7.72 illustrates a communication model.

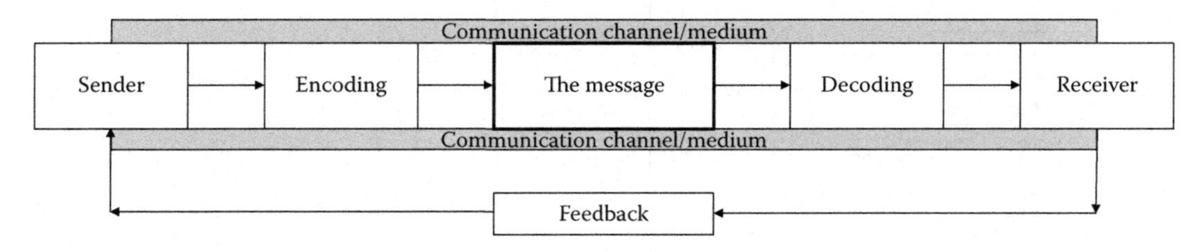

FIGURE 7.72
Communication model.

The standard communication methods widely used are either written or oral. Written communication can be letters, notices, email, or messaging (electronic media). Oral communication can be through meeting and telephonic conversation. Apart from these two mechanisms, nonverbal communication is another method used to assess communication within organization. Communication is either formal or informal.

Formal communication is planned and delivered as part of the standard operating policies and procedures of the organization. Informal communication is not mandated or otherwise required, but occurs as part of people functioning collectively.

Communication can be internal and external. Direction of communication can be top to bottom (vertical, i.e., downward) communication or bottom to top (horizontal, i.e., upward).

As per *PMBOK® Guide Fifth Edition*, project communication management includes the processes that are required to ensure timely and appropriate planning, collection, creation, storage, retrieval, management, control, monitoring, and ultimate disposition of project information. Project communication management includes following processes:

1. Plan communication
2. Manage communication
3. Control communication

Construction project has involvement of many stakeholders. The project team must provide timely and accurate information to identified stakeholders that will receive communications. Effective communication is one of the most important factors contributing to the success of project. Project communication is the responsibility of everyone on the project team. Communication management process in construction project consists of the following activities:

1. Develop a communication plan
2. Manage communication
3. Control documents

7.3.8.1 Develop a Communication Plan

A communication plan helps project team members to identify internal and external stakeholders involved in the project. Figure 7.73 illustrates the communication plan development process.

A comprehensive communication plan for construction project can be developed by analyzing the questions listed in Table 7.28.

7.3.8.1.1 Establish a Communication Matrix

For smooth flow of communication in construction project, a proper communication matrix among all the stakeholders needs to be established at the start of each stage of the project. The communication matrix is used as a guideline indicating what information to communicate, which team member initiates, who will receive and take appropriate action, when to communicate, and the method of communication. Table 7.29 illustrates an example guideline to prepare the communication matrix for site administration during the construction phase.

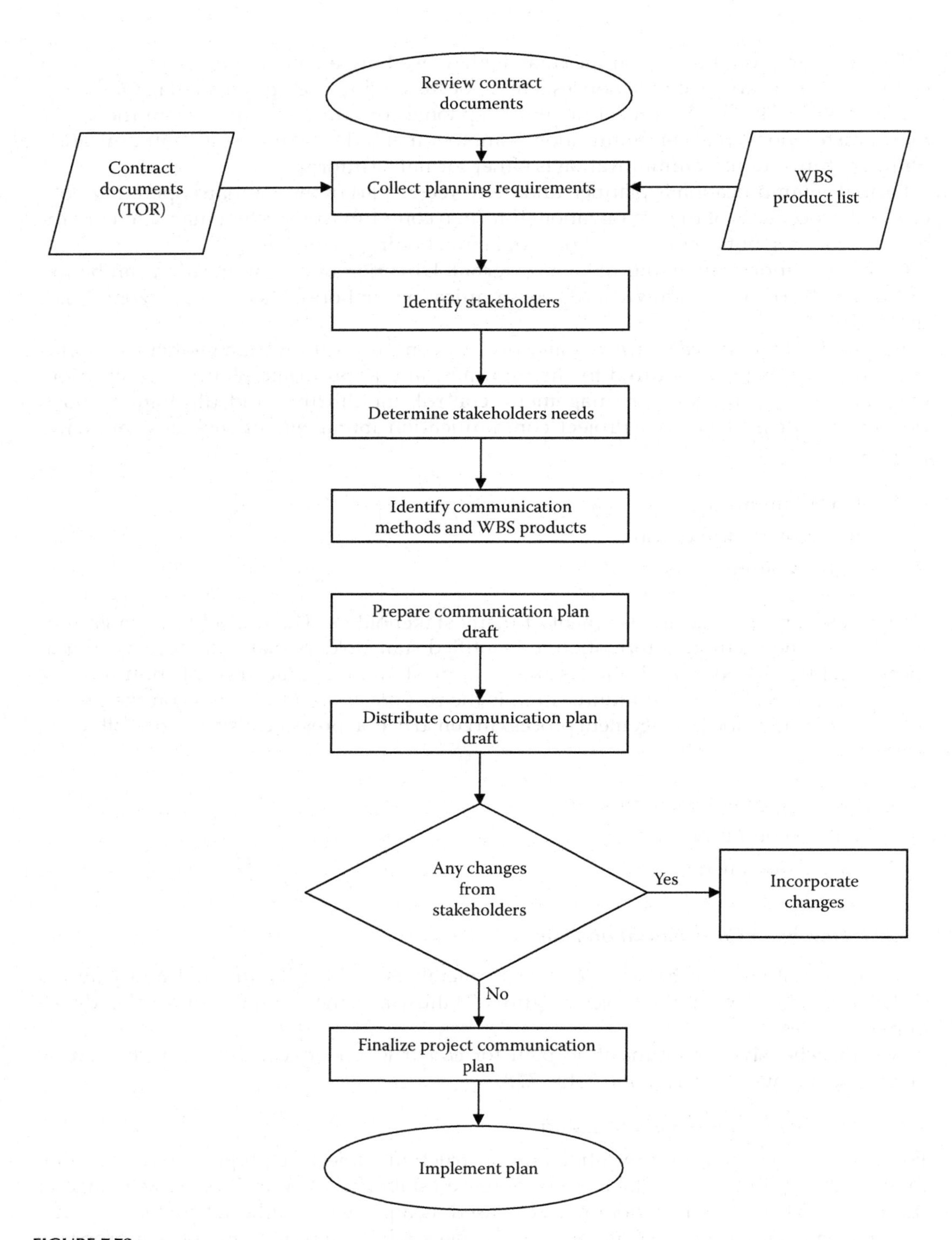

FIGURE 7.73
Communication plan development process. (*Project Management Handbook*, 2nd Edition (2007), Reprinted with permission from California Department of Transportation, Office of Project Management Process Improvement, Sacramento, CA.)

TABLE 7.28

Analysis for Communication Matrix

Sr. No.	W+H	Related Analyzing Question
1	What	What is the purpose of communication?
2	What	What type of information needs to be communicated?
3	Who	Who will initiate (send) the communication?
4	Who	Who are the stakeholders to receive the information?
5	When	When the information to be sent (frequency)?
6	What	What method of communication to be used?
7	How many	How many copies to be distributed?
8	How much	How much time to wait to receive the feedback?
9	How	How to archive the documents?

TABLE 7.29

SAMPLE FORM

Communication Matrix

Project Name								
Name of Construction/Project Manager:					**Name of Consultant:**			
Contractor Name:					**Project Number:**			
Sr. No.	Type of Document	Originator	Receiver (s)	Purpose	Frequency	Method	Responsible Person for Action	Comments

7.3.8.1.2 Determine Communication Methods

The method of communication in construction projects depends on the project stage and needs of the stakeholders. There are number of methods for determining requirements of stakeholders; however, it is imperative that they are completely understood in order to effectively manage their interest, expectations, and influence and ensure a successful project. The following are the common methods used in construction projects:

- Letters and other hard copies
- Specific type of transmittal forms
- Meeting

- Emails
- Telephone calls
- Conferencing (voicemail)
- Web-based communication

7.3.8.1.3 Establish Submittal Procedure

In construction projects, there are various types of documents that are to be sent to different stakeholders. A proper correspondence and reporting method is important to distribute this information.

Construction projects have involvement many stakeholders. A large number of documents move forward and backward between these stakeholders for information or action. During the design stage, the communication is mainly between the owner (project/construction manager) and designer (consultant). However, once the contractor is selected, depending on type of project delivery system—design–build contractor or construction contractor or construction manager (CM) at risk—then the contractor is actively involved in the project communication system. Apart from these three parties, some other stakeholders who have interests and expectations, and who influence the project are also sent the copies of information/documents for their information or action.

During the design stage, the following are the major documents the designer sends to the owner:

- Design drawings
- Reports
- Minutes of meetings
- Review comments
- Contract specifications
- Bidding and tendering documents

These documents are normally transmitted by the designer using the transmittal form. Figure 7.74 illustrates a sample transmittal form.

During the construction phase, many documents are sent back and forth between the owner, construction supervisor (architect/engineer [A/E] and consultant), and contractor. The originator of these documents is mentioned in the communication matrix. The following are the types of documents exchanged among the owner, supervisor, and contractor:

- Administrative
- Contract-related
- Engineering submittals
- Project monitoring and control
- Quality

Table 7.30 illustrates a list of forms normally used during the construction phase to communicate different types of project related documents.

The detail procedure for submitting materials/products/systems, samples and shop drawings is specified under section *SUBMITTAL* of contract specifications. The contractor has to submit the same to owner/consultant for review their review and approval.

PROJECT NAME			
PROJECT/CONSTRUCTION MANAGER			
DESIGNER (CONSULTANT NAME)			
TRANSMITTAL FORM			
To : ______		Date : ______	
Attn : ______		Ref. No. : ______	
Phase No. : ______		Copies to : ______	
Drawings ☐	Specifications ☐	Report ☐	Minutes ☐
No.	**Copies**	**Name of Document**	**Description**
Purpose of issuing :			
Sent by : ______	Signed : ______	Date : ______	
Rec'd by : ______	Signed : ______	Date : ______	

FIGURE 7.74
Transmittal form.

TABLE 7.30
List of Project Control Documents

Sr. No.	Document Name
I	**Administrative**
	I-1 Material Entry Permit
	I-2 Material Removal Permit
	I-3 Vehicular Entry Permit
	I-4 Site Entry Permit
	I-5 Visitor Entry Permit
	I-6 Municipality Permit
	I-7 Request for Overtime
	I-8 Theft and Damage Report
	I-9 Performance Bonds
	I-10 Advance Payment Guarantee
	I-11 Insurance
	I-12 Accident Report
	I-13 Sample Tag
II	**Contracts-Related**
	II-1 Notice to Proceed
	II-2 Job Site Instruction
	II-3 Site Works Instruction
	II-4 Attachment to Site Works
	II-5 Request for Staff Approval
	II-6 Request for Subcontractor Approval
	II-7 Variation Order
	II-8 Attachment to Variation Order
	II-9 Material Delivered at Site
	II-10 Baseline Change Request Form
	II-11 Extension of Time
	II-12 Suspension of Work
	II-13 Attendees
	II-14 Minutes of Meeting
	II-15 Transmittal for Minutes of Meeting
	II-16 Submittal Form
III	**Engineering Submittal**
	III-1 Master Schedule
	III-2 Cost Loaded Schedule
	III-3 Material Approval
	III-4 Specification Comparison Statement
	III-5 Product Data
	III-6 Product Sample
	III-7 Workshop Drawings
	III-8 Builders Drawings
	III-9 Composite Drawings
	III-10 Method Statement
	III-11 Request for Information
	III-12 Request for Modification
	III-13 Variation Order (Proposal)
	III-14 Request for Alternative or Substitution
IV	**PCS Reporting Forms**
	IV-1 Contractor's Submittal Status Log E-1
	IV-2 Contractor's Procurement Log E-2
	IV-3 Contractor's Shop Drawing Status Log

(*Continued*)

TABLE 7.30 (*Continued*)

List of Project Control Documents

Sr. No.	Document Name
	IV-4 Daily Progress Report
	IV-5 Weekly Progress Report
	IV-6 Look Ahead Schedule
	IV-7 Monthly Progress Report
	IV-8 Progress Photographs
	IV-9 Daily Checklist Status
	IV-10 Progress Payment Request
	IV-11 Payment Certificate
	IV-12 Submittal Schedule
	IV-13 Schedule Update Report
V	**Management Plans**
	V-1 Quality Control Plan
	V-2 Safety Plan
	V-3 Environmental Protection Plan
VI	**Quality Control Forms**
	VI-1 Checklist (Request for Inspection)
	VI-2 Checklist for Form Work
	VI-3 Notice for Daily Concrete Casting
	VI-4 Checklist for Concrete Casting
	VI-5 Quality Control of Concreting
	VI-6 Report on Concrete Casting
	VI-7 Notice for Testing at Lab
	VI-8 Concrete Quality Control Form
	VI-9 Checklist for Mechanical Work
	VI-10 Checklist for Electrical Work
	VI-11 Checklist for Finishing
	VI-12 Checklist for External Work
	VI-13 Checklist for Landscape
	VI-14 Remedial Note
	VI-15 Nonconformance/Compliance Report
	VI-16 Material Inspection Report
	VI-17 Safety Violation Notice
	VI-18 Notice of Commencement of New Activity
	VI-19 Removal of Rejected Material
	VI-20 Testing and Commissioning
VII	**Closeout Forms**
	VII-1 As-Built Drawings
	VII-2 Substantial Completion Certificate
	VII-3 Handing Over Certificate
	VII-4 Taking Over Certificate
	VII-5 Manuals
	VII-6 Handing Over of Spare Parts
	VII-7 Defect Liability Certificate

Source: Abdul Razzak Rumane (2013), *Quality Tools for Managing Construction Projects*, CRC Press, Boca Raton, FL. Reprinted with permission from Taylor & Francis Group.

The Contract Documents under section 013300 Submittal Procedures (CSI—Format General Requirements) specifies administrative and procedural requirements for submission of submittals and other documents. Contractor has to comply with the contractual requirements for submittal requirements.

Submittal process, in construction projects, is essential to ensure that the contractor's understanding of product specifications, contract drawings, and installation method matches with the designer's intent of product usage and installation method. The submittal process provides the owner the assurance that the contractor is complying with the design concept and the installed material will function as required by the contract documents. Submittals are documents that are presented by the contractor for approval, review, decision, or consideration.

Generally these submittals fall into the following three categories:

1. Approval submittals
2. Review submittals
3. Information submittals

Prior to the start of execution/installation of work, the contractor has to submit specified material/product/system, shop drawings to the A/E/consultant and construction/project manager, as per project specification requirements, for approval, review, or information. The contractor while preparing the submittal for shop drawing has to consider the following:

1. Review contract specification
2. Review contract drawings
3. Determine and verify field/site measurements
4. Installation information about the material to be used
5. Installation details relating to the axis or grid of the project
6. Dimensions of the product and equipment to be installed
7. Roughing in requirements
8. Coordination with other trade (disciplines) requirements
9. Clearly marking the changes and deviations to the contract drawings

The consultant reviews the submittal to verify that the proposed product/sample/shop drawing comply with the contract specifications and returned the transmittal to the contractor, mentioning one of the following actions on the transmittal:

- **A**—Approved
- **B**—Approved as Noted
- **C**—Revise and Resubmit or Not Approved
- **D**—For Information or More Information Required

In case of deviation to that of specified items, the contractor has to submit a schedule of such deviation(s) listing all the points that do not conform to the specifications.

Figure 7.75 illustrates a site transmittal form for material approval, Figure 7.76 illustrates a specification comparison form, and Figure 7.77 illustrates a site transmittal form for shop drawing approval.

Project Name
Consultant Name
SITE TRANSMITTAL
Request for Material Approval

CONTRACT No. :

TRANSMITTAL NO. : ____________ REV. : ______

CONTRACTOR :

TO :

WE REQUEST APPROVAL OF THE FOLLOWING MATERIALS/GOODS/PRODUCTS/EQUIPMENT

ITEM NO.	DWG., SPEC. OR BOQ. REF	DESCRIPTION	SUBMITTAL CODE *	ACTION CODE **
		SAMPLE FORM		

DETAILS OF INFORMATION, LITERATURE, CATALOG CUTS, AND THE LIKE ATTACHED ARE:

SAMPLES:
Enclosed ☐ Submitted under separate cover ☐ Not applicable ☐

N.B: We certify that above items have been reviewed in detail & and are correct & in strict performance with the Contract Drawings & Specification except as otherwise stated.

CONTRACTOR'S REP. : ______________________ DATE : ____________

RECEIVED BY CONSULTANT : ______________________ DATE : ____________

cc: Owner Rep.

Resident Engineer to enter ACTION CODE and REMARKS

R.E.'s REMARKS :

Initials ____________ Date ______

Corrections or comments made relative to submittals during this review do not relieve Contractor from compliance with the requirements of the Drawings and Specifications. This check is only for review of general conformance with the design concept of the project and general compliance with the information given in the Contract Documents. Contractor is responsible for confirming and correlaing all quantities and dimensions, selecting fabrication process and techniques of construction; coordinating his work with that of other trades, and performing his work in a safe and satisfactory manner.

Resident Engineer : ______________________ DATE : ____________

Received by Contractor : ______________________ DATE : ____________

cc: Owner Rep.

*** SUBMITTAL CODE:**	**** ACTION CODE:**	
1: Submitted for Approval	A: Approved	C: Not Approved

FIGURE 7.75
Site transmittal for material.

Project Name
Consultant Name

SPECIFICATION COMPARISON STATEMENT (SCS)

Contractor: ______________________ Date : ____________

Contract No. ______________________ A/S No.: ____________

Submittal No. : ____________ Revision: ______ Transmittal Ref. : ____________

Submittal Title : ____________ Specification Ref : ____________

S No.	PECIFICATION REQUIREMENTS	CONTRACTOR'S PROPOSAL	REMARKS

SAMPLE FORM

FIGURE 7.76
Specification comparison statement.

7.3.8.2 Manage Communication

Contract document specifies the number of original (paper print) and copies to be transmitted to A/E (consultant) for review and approval. Figure 7.78a illustrates an example of the submittal process (paper based) and Figure 7.78b illustrates an example of the submittal process (electronic).

Figure 7.79 illustrates a sample submittal transmittal form.

With the advent of the electronic submittal transmittal system, electronic documents such as portable document format (PDF) and building information model (BIM) are used for submittal purposes. Contractor, A/E (consultant), or owner may use internet-based project management software that has transmission, tracking, and email features; however, utilization of electronic documentation system should be specified in the contract documents.

7.3.8.2.1 Manage Submittals

There are different types of logs used in construction projects to monitor the submission of material, shop drawings, and other submittals. Figure 7.80 illustrates a contractor's submittal status log (report E-1) normally called *Log E-1*, and Figure 7.81 illustrates a contractor's shop drawing status log.

7.3.8.2.2 Conduct Meetings

Project meeting refers to a face-to-face communication method to exchange information among team members and stakeholders.

In construction projects, there are many types of meetings held during all the stages/phases of the project. Each meeting has its own purpose and structure. The meetings are used to distribute information, discuss issues, make proposals, suggest solutions, share

Project Name
Consultant Name
SITE TRANSMITTAL
Request for Shop Drawings Approval

CONTRACT No. : TRANSMITTAL NO. : ____________ REV. : _____

CONTRACTOR :

TO :

WE REQUEST APPROVAL OF THE FOLLOWING ENCLOSED DRAWINGS

ITEM NO.	DWG., SPEC. OR BOQ. REF	DRAWING TITLE	DWG. NOS.	Rev.	SUBMITTAL CODE *	ACTION CODE **

N.B: We certify that above items have been reviewed in detail & and are correct & in strict performance with the Contract Drawings & Specification except as otherwise stated.

CONTRACTOR'S REP. : ______________________ DATE : ____________

RECEIVED BY CONSULTANT : ______________________ DATE : ____________

cc: Owner Rep.

Resident Engineer to enter ACTION CODE and REMARKS

R.E.'s REMARKS :

____

____

____

____

____

______________________________ Initials__________ Date__________

Corrections or comments made relative to submittals during this review do not relieve Contractor from compliance with the requirements of the Drawings and Specifications. This check is only for review of general conformance with the design concept of the project and general compliance with the information given in the Contract Documents. Contractor is responsible for confirming and correlating all quantities and dimensions, selecting fabrication process and techniques of construction; coordinating his work with that of other trades, and performing his work in a safe and. satisfactory manner.

Resident Engineer : ______________________ DATE : ____________

Received by Contractor : ______________________ DATE : ____________

cc: Owner Rep.

cc: Project Manager	** **ACTION CODE:**	
1: Submitted for Approval	A: Approved	C: Not Approved
2: Submitted for yourInformation	B: Approved as Noted	D: For Information
3:		

FIGURE 7.77
Site transmittal for workshop drawings.

information, and contribute ideas for improvement and successful completion of project. The following types of meetings are normally held in construction projects:

1. Kick-off meeting
2. Planning meeting
3. Pretender meeting
4. Preconstruction meeting
5. Coordination meeting
6. Progress meeting
7. Change control meeting
8. Quality meeting
9. Safety meeting

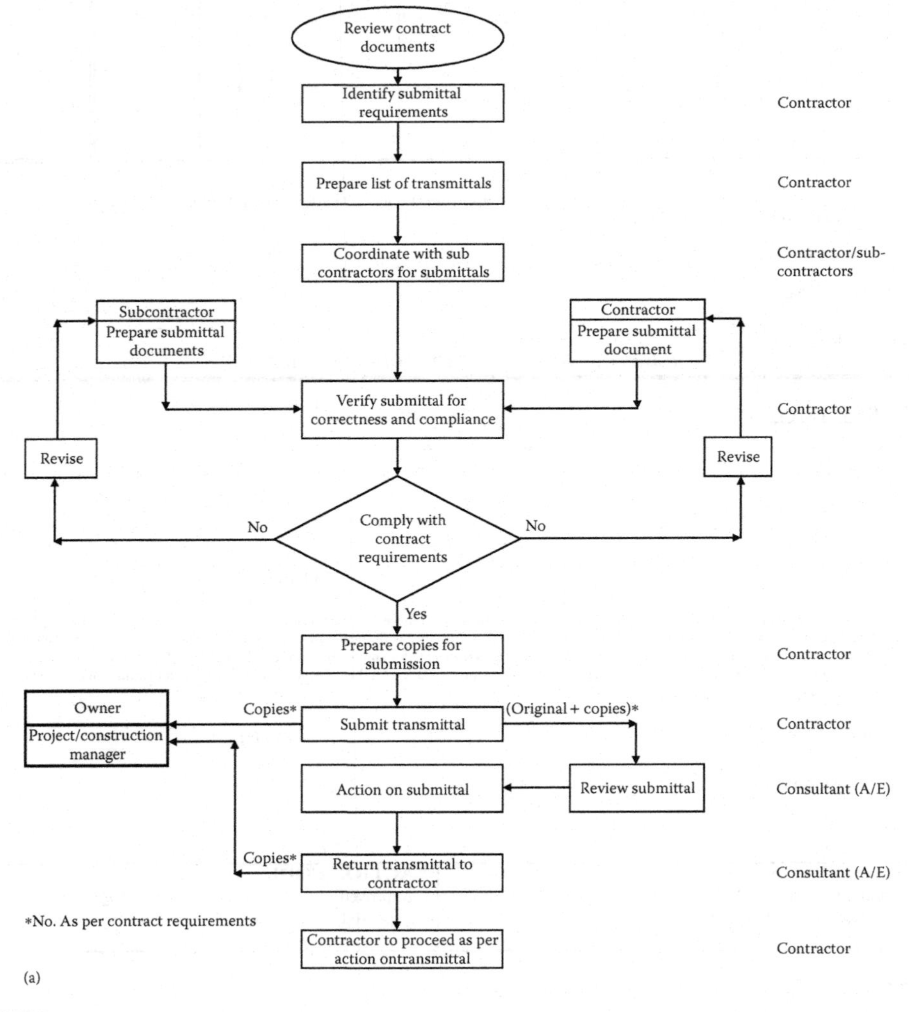

FIGURE 7.78
(a) Submittal process (paper-based).

(Continued)

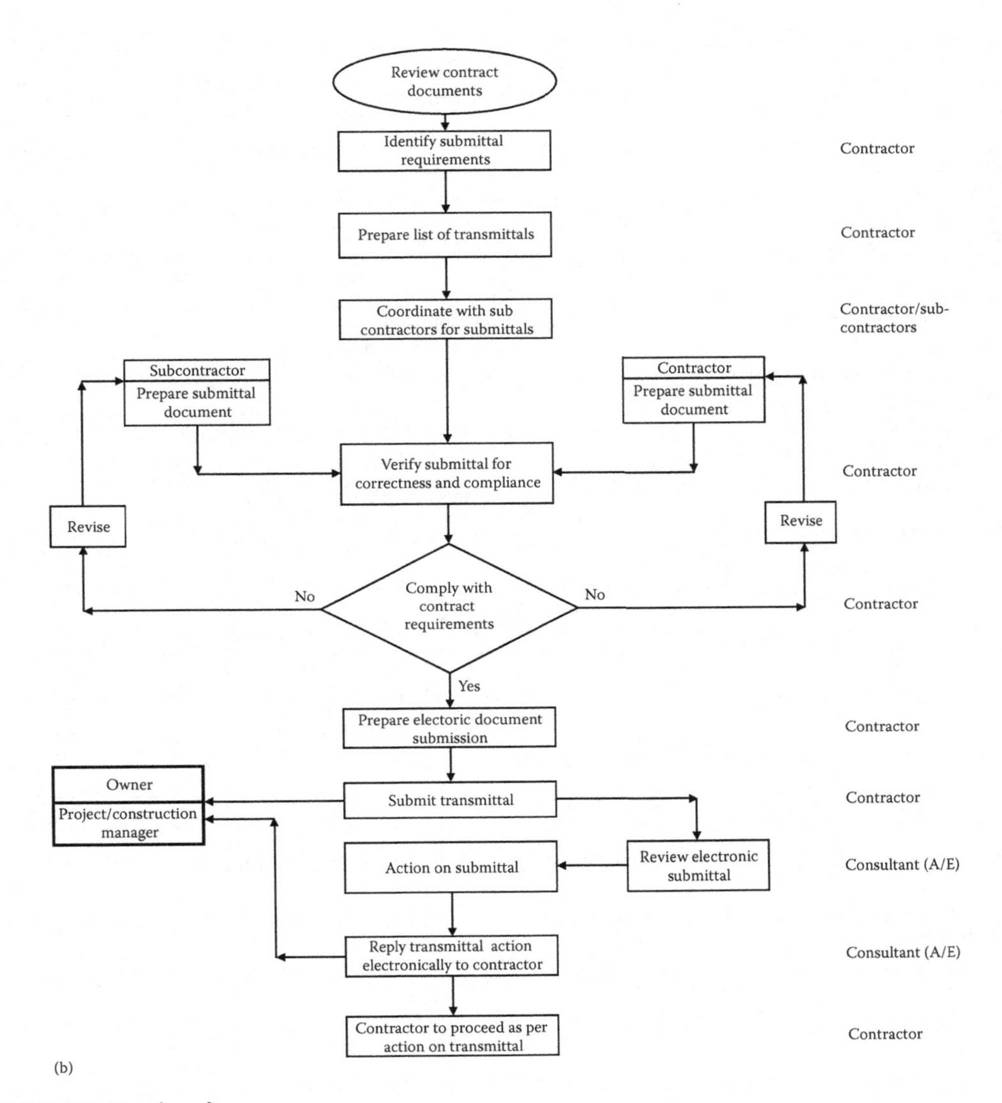

FIGURE 7.78 (Continued)
(b) Submittal process (electronic).

The frequency of meetings depends on the project stage/phase. The frequency of meetings in construction phase normally specified in the contract documents. To conduct a project meeting, it is advisable to prepare and circulate the agenda well in advance. In the case of progress meeting during the construction phase, the resident engineer prepares the agenda and circulates it to the participants who are expected to attend the meeting. The contractor informs the resident engineer in advance about the points the contractor would like to discuss. These points are also included in the agenda. Figure 7.82 illustrates a typical agenda format for the meeting. Figure 7.83 is a sample form to list the meeting attendees. Figure 7.84 illustrates a format for preparation of minutes of the meeting, and Figure 7.85 is a sample transmittal form for distribution of minutes of the meeting.

Project Name					
Consultant Name					
SUBMITTAL TRANSMITTAL FORM					
Contractor Name :					
Contract No. :					
To.	**Resident Engineer**				
Transmittal No.:			**Date :**		

Submittal Type:		**Action Requested:**	
DG	Shop Drawings	1	For Approval
SK	Sketches	2	For Review and Comment
PR	Material/Product/System	3	For Information
MD	Manufacturer's Data	4	For Construction
SM	Sample	5	For Incorporation Within the Design
MM	Minutes of Meeting	6	For Costing
RP	Reports	7	For Tendering
LG	Logs	SAMPLE FORM	
OT	Others (please specify)		

We are sending herewith the following:

ENCLOSURES

Item	**Qty**	**Ref. No.**	**Description**	**Type**	**Action**

Comments

____

____

____

Issued by:	**Received by:**
Signature:	Signature:
Date:	Date:

FIGURE 7.79
Transmittal form.

Project Name
Consultant Name

CONTRACTOR'S SUBMITTAL STATUS LOG (REPORT E-1)

CONTRACT NO. : ____________ NO. : ________
CONTRACTOR : ____________ DATE : ________

Document Specification or BOQ Page	Submittal Sequence No.	Rev	Description	Type of Submittal								Action Code A= Approved as submitted B= Approved as noted C = Not Approved D = For Information						Notes
				Shop Drawing	Samples	Guarantee	Mf's Dats	Mt's Dats	Certificates	Test	Otr							
				SD	SM	GT	MF	MT	CT	TT	OT	Planned Sub. date	Submitted By Contrac.	Returned By A/E	Action	Approval Needed By	Required On Job	

FIGURE 7.80
Contractor's submittal status log.

Project Name
Consultant Name

CONTRACTOR'S SHOP DRAWINGS STATUS LOG

CONTRACT NO. : ______________________ NO. : ________

CONTRACTOR : ______________________ DATE : ________

Document Specification or BOQ Page	Item No.	Submittal Sequence No.						Drawing No.	Description	Submittal Type	Action Code: A = Approved as submitted; B = Approved as noted; BR = Approved as noted, Resubmit; C = Not Approved; D = For Information						Remarks
		Div	Sec	Typ	Sr.	Rev	Ser			SD	Planned Sub. date	Submitted By Contrac.	Returned By A/E	Action	Approval Needed By	Required On Job	

SAMPLE FORM

FIGURE 7.81
Contractor's shop drawing submittal log.

PROJECT NAME

Contract Number :			
Type of Meeting :		Date of Meeting :	
Place of Meeting :		Time of Meeting :	
Owner :			
Project/Construction Manager			
A/E (Consultant)			

AGENDA

1. Points to be discussed :

1.1 ....

1.2 ....

1.3 ....

1.4 ....

....

....

2. Any other Issues :

Signed by : .... Position : ....

Date : ....

FIGURE 7.82
Agenda format for meeting.

7.3.8.2.3 Manage Documents

In construction projects, all the related documents are sent and received (exchanged) by using the transmittal form. These forms are generally issued to the contractor along with other contract documents. Project team members are required to follow the procedures specified in the contract.

For any communication with external stakeholders such as regulatory bodies, the company letterhead is used.

PROJECT NAME

Construction/Project Manager Name

CONSULTANT(A/E) NAME

ATTENDEES

Meeting Type : ____________ Meeting No. : ____________

Location : ____________

Project Phase : ____________ Date : ____________

Time : ____________

S No.	NAME	POSITION	COMPANY	SIGNATURE
1				
2				
3				
4				
5				
6				
7				
8				
9				
10				
11				
12				
13				
14				
15				

FIGURE 7.83
Meeting attendees.

7.3.8.3 Control Documents

Contract conditions specify the time allowed to process the transmittal and other contract-related communication. Contractor, consultant, and construction/project manager maintain logs for all incoming and outgoing documents. Follow up is also done among internal project team members to expedite the required action to be taken against the transmittal. Table 7.31 illustrates the logs maintained by the consultant during site supervision.

PROJECT NAME

Minutes of Meeting

Contract NO. :	
Owner Name :	
Project/Construction Manager :	
Contractor Name :	

Meeting Type :		Minutes Number :	
		Date :	
Meeting Location :		Time :	

SAMPLE FORM

Attendees

Number	**Name**	**Position Number**	**Company**

ITEM	DESCRIPTION OF DISCUSSION	STATUS	PRIORITY	ACTION			
				By	Due	Started	Completed

Distribution	Owner	PM/CM	Engineer's Representative	Contractor	Other
Original	☐	☐	☐	☐	☐
Copies	☐	☐	☐	☐	☐

FIGURE 7.84
Minutes of meeting format.

PROJECT NAME

Transmittal for Minutes of Meeting

Transmittal Number :		Date :	
Contract Number :		Contractor :	
Owner :			
Project/Construction Manager	SAMPLE FORM		
A/E (Consultant)			

Meeting Type :		Minutes Number :	
		Date :	
Meeting :		Time :	

ATTENDEES			
NO	NAME	POSITION	COMPANY

Your Use	☐	For Review and Comment	☐
Your Approval	☐	As Requested	☐

Engineer's Representative

Name	Signature	Date

The attached minutes constitutes our understanding of the points discussed and conclusions reached. All participants are requested to review these minutes and inform Engineer's Representative of their comments, if any, latest in 72 hours of the date of receiving these minutes.

FIGURE 7.85
Transmittal for minutes of meeting.

7.3.9 Risk Management

Risk management is the process of identifying, assessing, and prioritizing different kinds of risks, planning risk mitigation, implementing mitigation plan, and controlling the risks. It is a process of thinking systematically about the possible risks, problems, or disasters before they happen and setting up the procedure that will avoid the risk, minimize the impact, or cope with the impact. The objectives of project risk management are to increase the probability and impacts of positive events and decrease the probability and impacts

TABLE 7.31

List of Logs

Section	Log
1	Incoming and outgoing letters (owner, contractor)
2	Staff approval
3	Subcontractor approval
4	Transmittal for shop drawing
5	Material source approval
6	Transmittal for material
7	Transmittal for sample
8	Request for alternative
9	Request for information
10	Request for modification
11	Job site instruction
12	Variation order
13	Request for substitution
14	Noncompliance report
15	Remedial note
16	Payment request/certificate
17	Daily report
18	Checklist
19	Daily checklist status
20	Checklist for concrete casting
21	Material delivered at site
22	Material inspection report
23	Concrete test report
24	Weekly progress report
25	Monthly progress report
26	Photographs
27	Notice of meeting
28	Minutes of meeting
29	Safety violation report
30	Accident report
31	Request for proposal
32	Construction schedule
33	Safety management plan
34	Quality control plan
35	Authority-approved drawings
36	Correspondence with authorities
37	Issue log
38	Interoffice memo
39	Video recording

of events adverse to the project objectives. Risk is the probability that the occurrence of an event may turn into undesirable outcome (loss, disaster). It is virtually anything that threatens or limits the ability of an organization to achieve its objectives. It can be unexpected and unpredictable events that have the potential to damage the functioning of organization in terms of money, or in worst scenario, it may cause the closure of the business.

7.3.9.1 Develop Risk Management Plan

Risk management plan identifies how risk associated with the project will be identified, analyzed, managed, and controlled. Risk management is an integral part of project management as the risk is likely to occur at any stage of the project. Therefore, the risk has to be continually monitored and response actions have to be taken immediately. The risk management plan outlines how risk activities will be performed, recorded, and monitored throughout the project life cycle. It is intended to maximize the positive impact for the benefit of the project and decrease/minimize or eliminate the impact of events adverse to the project. The risk management must commence early in project development stage (study stage) and proceed as the project evolves and more information about the project in available. The project plan should

- Define risk management strategy/approach.
- Define project objectives and goals related to risk management.
- Identify risk owner and team members.
- Define risk decisions.
- Detail about risk resources.
- Include risk management processes, including:
  - Methods of risk identification
  - Methods of risk assessment
  - Level of risk
  - Response to risk
  - Management of risk
  - Control of risk
- Integrate risk management activities into project scope, schedule, cost, and quality.
- Document and record risks.
- Communicate procedure for risk reporting.
- Update of risk management plan.

7.3.9.2 Risk Management Process

Risk management process is designed to reduce or eliminate the risk of a certain kind of events happening (occurring) or having an impact on the project. The risk process consists of following steps:

1. Identify the potential sources of risk on the project.
2. Analyze their impact on the project.
   a. Qualitative
   b. Quantitative
3. Select those with a significant impact on the project.
   a. Prioritization

4. Determine how the impact of risk can be reduced.
   a. Avoidance
   b. Transfer
   c. Reduction
   d. Retention (acceptance)
5. Select best alternative.
6. Develop and implement mitigation plan.
7. Monitor and control the risks by implementing risk response plan, tracking identified risks, identifying new risks, and evaluating the risk impact.

Figure 7.86 illustrates risk management cycle (process).

7.3.9.2.1 Identify Risk

Risk identification involves determining the source and type of risk that may affect the project. The following tools and techniques are used to identify risks:

- Benchmarking
- Brainstorming
- Delphi technique
- Interviews
- Past database, historical data from similar projects
- Questionnaires
- Risk breakdown structure
- Workshops

The identified risks are classified as follows:

1. Internal
2. External

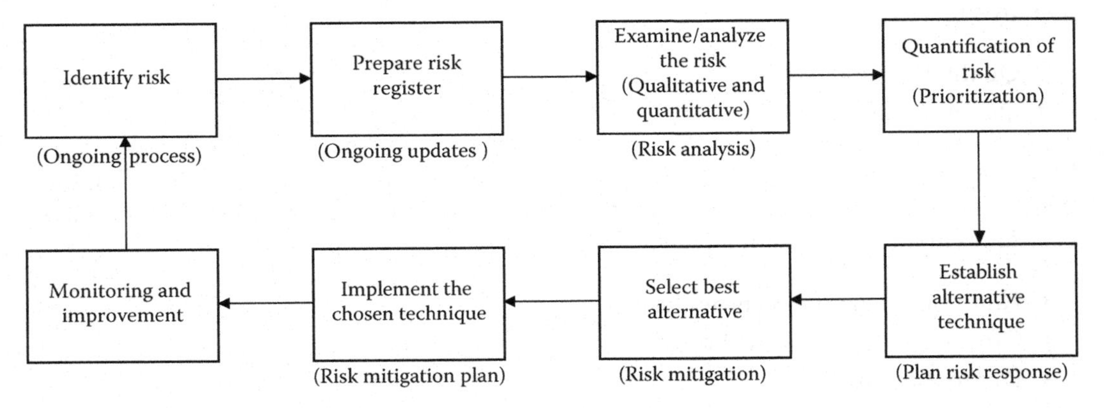

FIGURE 7.86
Risk management cycle. (Abdul Razzak Rumane (2013), *Quality Tools for Managing Construction Projects*, CRC Press, Boca Raton, FL. Reprinted with permission from Taylor & Francis Group.)

These are further divided into the following main categories:

- Management
- Project (contract)
- Technical
- Construction
- Physical
- Logistic
- Health, safety, and environmental
- Statutory/regulatory
- Financial
- Commercial
- Economical
- Political
- Legal
- Natural

Each identified risk is documented in a risk register. Please refer Figure 4.24, for Benchmarking, Figure 4.32 for Brainstorming, and Figure 4.33 for Delphi technique.

7.3.9.2.1.1 Risk Register Risk register is a document recording details of all the identified risks at the beginning of the project and during the life cycle of the project in a format that consist of comprehensive lists of significant risks along with the actions and cost estimated with the identified risks. Risk register is updated every time a new risk is identified or relevant actions are taken. Table 7.32 illustrates an example of a risk register.

7.3.9.2.2 Analyze Risk

Risk analysis is the process to analyze the listed risks. The following are the two methods of analyzing risks:

1. Qualitative analysis
2. Quantitative analysis

7.3.9.2.2.1 Qualitative Analysis Qualitative analysis is a process to assess the probability of occurrence (likelihood) of the risk and its impact (consequence). The following tools and techniques are used for qualitative analysis:

- Failure mode and effects analysis
- Group discussion (workshop)
- Pareto diagram
- Probability and impact assessment
- Probability levels
- Risk categorization

TABLE 7.32

Risk Register

Project Name Risk Register																
Sr. No.	Risk Identification Number (Risk ID)	Description of Risk	Owner of Risk	Estimated Likelyhood of Risk	Impact	Estimated Severity	Prioritization	List of Activities Influenced	Leading Indicators for Risk	Risk Mitigation Plan	Risk Mitigation Plan on Leading Indicator	Timeline for Mitigation Action	Tracking of Leading Indicators	Date of Review/ Update	Forecating Risk Happenings	Action to be Taken in Future

7.3.9.2.2.2 Quantitative Analysis Quantitative analysis is a process to quantify the probability of risk and its impact based on numerical estimation. The following tools and techniques are used for quantitative analysis:

- Event tree analysis
- Probability analysis
- Sensitivity analysis
- Simulation techniques (Monte Carlo simulation)

Table 7.33 illustrates risk levels normally assumed for a probability impact matrix. The percentage of probability of occurrence shown in the figure is indicative. The organization can determine the probability level as per the nature of business.

7.3.9.2.2.3 Prioritization It is the process of prioritizing a list of quantified risks. The results of risk assessment are used to prioritize risks to establish very high to very low ranking. Prioritization of risks depends on the following factors:

1. Probability (occurrence)
2. Impact (consequences)
3. Urgency
4. Proximity
5. Manageability
6. Controllability
7. Responsiveness
8. Variability
9. Ownership ambiguity

The prioritization list helps the project manager to plan actions and assign the resources to mitigate the realization of high-value probability. Table 7.34 illustrates an example of risk assessment for material deliver-loading and unloading of material.

TABLE 7.33

Risk Probability Levels

Sr. No.	Value	Definition	Meaning	Percentage Probability of Occurrence
1	Level 5	Very high (frequent)	• Almost certain that the risk will occur • Frequency of occurrence is very high	40–80
2	Level 4	High (likely)	• It is likely to happen • Frequency of occurrence is less	20–40
3	Level 3	Moderate (occasional)	• Its occurrence is occasional	10–20
4	Level 2	Low (unlikely)	• It is unlikely to happen	5–10
5	Level 1	Very low (rare)	• The probability to occur is rare	0–5

TABLE 7.34

RISK ASSESSMENT

Risk Assessment — **OWNER NAME**

PROJECT NAME

ACTIVITY = MATERIAL DELIVERY-LOADING AND UNLOADING OF MATERIALS

SL. NO.	Basic Job Activity/ Hazard	Hazard	Initial Risk Rating		Risk Factor	CONTROL MEASURES IN PLACE / MITIGATION ACTIONS	Final Risk Rating		Residual Risk Factor	Risk Acceptable
			Probability x Severity				Probability x Severity			
1	Uncontrolled Movement of the load.	Cause injury to the person. Damage to the property	4	2	8	a)Ensure that the personnel are not positioned in corners where emergency movement is restricted b)Beware of pinch points. c)Wear PPE d)Do not place any part of the body below the hoisted load e)Use tag lines to guide the load. f)Manual guiding should be avoided. Preplan the lift g)Adopt proper signaling method and ensure proper communication and co-ordination h)Adopt proper rigging method based on packing dimensions. i)Provide padding for the materials at sharp corners and delicate portions. j)Ensure proper guiding of the load to avoid jerks and swayingof the load k)Ensure the access is clear for the transportation of the load	2	1	2	**Low**
2	Failure of Slings while lifting the materials.	May cause serious injury to the personal. Damage to the property	4	3	12	a)Use tested slings b)Do not exceed SWL c)Ensure the slings used are suitable for the purpose intended. d)Ensure padding wherever the slings come in contact with sharp corners. e)Adopt proper rigging method. f)Examine the conditions of the sling prior to use	3	1	3	**Low**

Example Risk Assessment

RPN-Risk Priority Number	
8----16	High Risk
4---6	Medium Risk
1----3	Low Risk

7.3.9.2.3 Plan Risk Response

Plan risk response is a process that determines the action (if any) to be taken in order to address the identified and assessed risks that are listed under risk register on prioritization basis. Risk response process is used for developing options and actions to enhance opportunities and reduce the threats to the identified risk activities in the project.

For each identified risk, a response must be identified. The risk owner and project team have to select the risk response for each of the identified risk. The probability of the risk event occurring and the impacts (threats) is the basis for evaluating the degree to which the response action is to be evolved.

Generally, risk response strategies for impact (consequences) on the project fall into one of the following categories:

1. Avoidance
2. Transfer
3. Mitigation (reduction)
4. Acceptance (retention)

7.3.9.2.3.1 Avoidance Avoidance is changing the project scope, objectives, or plan to eliminate the risk or to protect the project objectives from the impact (threat).

7.3.9.2.3.2 Transfer Transfer is transferring the risk to someone else who will be responsible to manage the risk. Transferring the threat does eliminate the threat, it still exists; however, it is owned and managed by other party.

7.3.9.2.3.3 Mitigation Mitigation is reduction in the probability and/or impact to an acceptable threshold. It is done by taking a series of control actions.

7.3.9.2.3.4 Acceptance It is acceptance of consequences after response actions, understanding the risk impact, should it occur.

7.3.9.2.4 Reduce Risk

It is identifying various steps to reduce the probability and/or impact of the risk. Taking early steps to reduce the probability of risk is more effective and less costly than repairing the damage after the occurrence of the risk.

7.3.9.2.5 Monitor and Control Risk

It is a systematic process of tracking identified risks, monitoring residual risks, identifying new risks, executing risk response plan, and evaluating the effectiveness of implementation of actions against established levels of risk in the area of scope, time, cost, and quality throughout the project life cycle. It involves timely implementation of risk response to identified risk to ensure the best outcome for a risk to a project.

7.3.9.3 Risks in Construction Projects

Construction projects have many varying risks. Risk management throughout the project life cycle is important and essential to prevent unwanted consequences and effects on the project. Construction projects have involvement of many stakeholders such as project owners, developers, design firms (consultants), contractors, banks, and financial institutions funding the project who are affected by the risk. Each of these parties has involvement with certain portion of the overall construction project risk; however, the owner has a greater share of risks as the owner is involved from the inception until completion of project and beyond. The owner must take initiatives to develop risk consciousness and awareness among all the parties, emphasizing upon the importance of explicit consideration of risk at each stage of the project as the owner is ultimately responsible for overall project cost. Traditionally

1. Owner/client is responsible for the investment/finance risk.
2. Designer (consultant) is responsible for design risk.
3. Contractors and subcontractors are responsible for construction risk.

Construction projects are characterized as very complex projects, where uncertainty comes from various sources. Construction projects involve a cross section of many different participants. They have varying project expectations. Both of them influence and depend on each other in addition to involving *other players* in the construction process. The relationships and the contractual groupings of those who are involved are also more complex and contractually varied. Construction projects often require a large amount of materials and physical tools to move or modify these materials. Most items used in construction projects are normally produced by other construction-related industries/manufacturers. Therefore, risk in construction projects is multifaceted. Construction projects inherently contain a high degree of risk in their projection of cost and time as each is unique. No construction project is without any risk. Risk management in construction projects is mainly focused on delivering the project with

1. What was originally accepted (as per defined *scope*)
2. Agreed upon time (as per *schedule* without any delay)
3. Agreed upon budget (no overruns to accepted *cost*)

Risk management is an ongoing process. In order to reduce the overall risk in construction projects, the risk assessment (identification, analysis, and evaluation) process must start as early as possible to maximize project benefits. There are a number of risks that can be identified at each stage of the project. Early risk identification can lead to better estimation of the cost in the project budget, whether through contingencies, contractual, or insurance. Risk identification is the most important function in construction projects.

Risk factors in construction projects can be categorized into a number of ways according to level of details or selected viewpoints. These are categorized based on various risks factors and source of risk. Contractor has to identify related risks affecting construction, analyze these risks, evaluate the effects on the contract, and evolve the strategy to counter these risks, before bidding for a construction contract. Construction project risks mainly relate to the following:

- Scope and change management
- Schedule/time management
- Budget/cost management
- Quality management
- Resources and manpower management
- Communication management
- Procurement/contract management
- Health, safety, and environmental management

Table 7.35 illustrates typical categories of risks in construction projects.

TABLE 7.35
Typical Categories of Risks in Construction Projects

Sr. No.	Category	Types
1	Management	Selection of project delivery system
		Selection of project/construction manager
		Selection of designer
		Selection of contractor
2	Contract (project)	Scope/design changes
		Schedule
		Cost
		Conflict resolution
		Delay in changer order negotiations
3	Statutory	Statutory/regulatory delay
4	Technical	Incomplete design
		Incomplete scope of work
		Design changes
		Design mistakes
		Errors and omissions in contract documents
		Incomplete specifications
		Ambiguity in contract documents
		Inconsistency in contract documents
		Inappropriate schedule/plan
		Inappropriate construction method
		Conflict with different trades
		Improper coordination with regulatory authorities
		Inadequate site investigation data
5	Technology	New technology
6	Construction	Delay in mobilization
		Delay in transfer of site
		Different site conditions to the information provided
		Changes in scope of work
		Resource (labor) low productivity
		Equipment/plant productivity
		Insufficient skilled workforce
		Union and labor unrest
		Failure/delay of machinery and equipment
		Quality of material
		Failure/delay of material delivery
		Delay in approval of submittals
		Extensive subcontracting
		Subcontractor's subcontractor
		Failure of project team members to perform as expected
		Information flow breaks

(*Continued*)

TABLE 7.35 (*Continued*)
Typical Categories of Risks in Construction Projects

Sr. No.	Category	Types
7	Physical	Damage to equipment
		Structure collapse
		Damage to stored material
		Leakage of hazardous material
		Theft at site
		Fire at site
8	Logistic	Resources availability
		Spare parts availability
		Consistent fuel supply
		Transportation facility
		Access to worksite
		Unfamiliarity with local conditions
9	Health, safety, and environment	Injuries
		Health and safety rules
		Environmental protection rules
		Pollution rules
		Disposal of waste
10	Financial	Inflation
		Recession
		Fluctuations in exchange rate
		Availability of foreign exchange (certain countries)
		Availability of funds
		Delays in payment
		Local taxes
11	Economical	Variation of construction material price
		Sanctions
12	Commercial	Import restrictions
		Custom duties
13	Legal	Permits and licenses
		Professional liability
		Litigation
14	Political	Change in laws and regulations
		Constraints on employment of expatriate workforce
		Use of local agent and firms
		Civil unrest
		War
15	Natural	Flood
		Earthquake
		Cyclone
		Sandstorm
		Landslide
		Heavy rains
		High humidity
		Fire

Source: Abdul Razzak Rumane (2013), *Quality Tools for Managing Construction Projects*, CRC Press, Boca Raton, FL. Reprinted with permission from Taylor & Francis Group.

The risk in construction projects starts from the inception of the project and is traced until completion of the project and beyond. In order to achieve a successful project, the risk management processes are to be performed during each of the following main stages of the construction project:

1. Study
2. Design
3. Bidding and tendering
4. Construction

Construction projects involve many participants. Each of the participants has their own risks at each stage/phase of project life cycle. Therefore, it is necessary to

1. Identify and record the risks that affect the project at each stage.
2. Assess the likelihood of each risk occurring in the future and the impact on the project should this happen.
3. Plan cost-effective management actions with clearly identified responsibilities, to avoid, eliminate, or reduce any significant risk identified.
4. Monitor and report the status of these risks and the effectiveness of planned risk management action.

Risk identification is the most important function in the construction project. There are various risk factors in each of the project stage affecting the construction projects. Table 7.36 illustrates major risk factors that may affect owner at different stages of the project, whereas Table 7.37 lists major risk variables to be considered while conducting feasibility study of a project.

TABLE 7.36

Major Risk Factors Affecting Owner

Sr. No.	Risk Factor	
1	**Study**	
	1.1	Incomplete/improper feasibility study
	1.2	Long procedure for authority approval and permits
	1.3	Delay in raising the funds to proceed with the work
	1.4	Change in law
	1.5	Late internal approval process from the owner team
	1.6	Selection of project delivery system
	1.7	Errors in estimated schedule
	1.8	Errors in estimated cost
2	**Design**	
	2.1	Incompetent consultant
	2.2	Incomplete design scope
	2.3	Inadequate or ambiguous specification
	2.4	Delays in authority approval
	2.5	Delay in completion of design
3	**Tendering**	
	3.1	Inadequate tendering price if estimate is wrong

(*Continued*)

TABLE 7.36 (*Continued*)

Major Risk Factors Affecting Owner

Sr. No.	Risk Factor
4	**Construction**
	4.1 Contractor has lack of knowledge and experience on similar type of construction
	4.2 Delay in site transfer
	4.3 Impractical planning and scheduling
	4.4 Omissions in design and specifications
	4.5 Inflation
	4.6 Fluctuation in exchange rate
	4.7 Increase in material cost
	4.8 Coordination problems
	4.9 Ineffectiveness and lack of supervision by consultant
	4.10 Delays in approval of transmittals
	4.11 Contracting system
	4.12 Lack of coordination between project participants
	4.13 Unpredicted contingencies, Force majeure
	4.14 Political disturbances

Source: Abdul Razzak Rumane (2013), *Quality Tools for Managing Construction Projects*, CRC Press, Boca Raton, FL. Reprinted with permission from Taylor & Francis Group.

TABLE 7.37

Risk Variables for Feasibility Study

Sr. No.	Risk Variables
1	Marketing aspect
2	Investment risk in next 10 years
3	Financial aspect (payback)
4	Economic aspect
5	Technical and technological risk
6	Environmental and spatial plan aspect
7	Regulations and policy aspect
8	Political
9	Social and cultural

Source: Abdul Razzak Rumane (2013), *Quality Tools for Managing Construction Projects*, CRC Press, Boca Raton, FL. Reprinted with permission from Taylor & Francis Group.

Table 7.38 illustrates major risks affecting the designer (consultant) and Table 7.39 illustrates major risks affecting the contractor.

7.3.10 Contract Management

Project procurement/contract management in construction projects is an organizational method, process, and procedure to obtain the required construction products. It includes the process to acquire construction facility complete with all the related product/material, equipment, and services from outside contractors/companies to the satisfaction of the owner/client/end user.

TABLE 7.38

Major Risk Factors Affecting Designer (Consultant)

Sr. No.	Risk Factor
1	**Concept Design**
	1.1 Lack of owner input
	1.2 Incomplete requirements and specifications
	1.3 Lack of information or information transfer with wrongly estimated objectives of the proposal
	1.4 Project objectives not defined clearly
	1.5 Changing requirements and specification
	1.6 Selected alternatives not suitable for final solution
	1.7 Environmental considerations
2	**Detail Design**
	2.1 Project performance specifications
	2.2 Technical specifications
	2.3 Follow up of codes and standards
	2.4 Difficulties in dealing specifications and standards concerning existing conditions and clients requirements
	2.5 Lack of knowledge about the technical conditions
	2.6 Incomplete design
	2.7 Prediction of possible changes in design during construction phase
	2.8 Environmental considerations
	2.9 Regulatory requirements
	2.10 Constructability
	2.11 Impractical project schedule
	2.12 Incorrect cost estimation
	2.13 Design completion schedule

Source: Abdul Razzak Rumane (2013), *Quality Tools for Managing Construction Projects,* CRC Press, Boca Raton, FL. Reprinted with permission from Taylor & Francis Group.

TABLE 7.39

Major Risk Factors Affecting Contractor

Sr. No.	Risk Factor
1	**Bidding and Tendering**
	1.1 Low bid
	1.2 Poor definition of scope of work
	1.3 Overall understanding of project
	1.4 Errors in resource estimation
	1.5 Errors in resource productivity
	1.6 Errors in resource availability
	1.7 Errors in material price
	1.8 Improper schedule
	1.9 Quality standards
	1.10 Exchange rate
	1.11 Unenforceable conditions or contract clauses

(*Continued*)

TABLE 7.39 (*Continued*)

Major Risk Factors Affecting Contractor

Sr. No.		Risk Factor
2	**Construction**	
	2.1	Delay in transfer of site
	2.2	Different site conditions to the information provided
	2.3	Delay in mobilization
	2.4	Changes in scope of work
	2.5	Resource (labor) low productivity
	2.6	Equipment/plant productivity
	2.7	Insufficient skilled workforce
	2.8	Union and labor unrest
	2.9	Failure/delay of machinery and equipment
	2.10	Quality of material
	2.11	Failure/delay of material delivery
	2.12	Delay in approval of submittals
	2.13	Incompetent subcontractor
	2.14	Failure of project team members to perform as expected
	2.15	Improper communication system
	2.16	Conflict with different trades
	2.17	Inappropriate construction method
	2.18	Damage to equipment
	2.19	Structure Collapse
	2.20	Damage to stored material
	2.21	Theft at site
	2.22	Fire at site
	2.23	Resources availability
	2.24	Spare parts availability
	2.25	Consistent fuel supply
	2.26	Transportation facility
	2.27	Access to worksite
	2.28	Injuries
	2.29	Health and safety rules
	2.30	Environmental protection rules
	2.31	Disposal of waste
	2.32	Inflation
	2.33	Delays in payment
	2.34	Variation in construction material price
	2.35	Delay in permit and licenses
	2.36	Change in laws and regulations
	2.37	Constraints in employment of expatriate workforce
	2.38	New technology
	2.39	Force majeure

Conventional notions of the procurement/purchasing cycle, which is normally applied in batch production, mass production, or in merchandising, are less appropriate to the realm of construction projects. The procurement of construction projects also involves commissioning professional services and creating a specific solution. The process is complex, involving the interaction of owner/client, project/construction manager, designer/consultant, contractor (s), suppliers, and various regulatory bodies. Generally, a construction project comprise building materials (civil), electromechanical items, finishing items, and equipment. Construction involves installation and integration of various types of materials/products, equipment, systems, or other components to complete the project/

facility to ensure that the facility is fully functional to the satisfaction of the owner/end user. Contract management involves

- Identification of
  - The in-house services available.
  - The services to be procured from outside agencies/organizations.
  - Method of procurement (direct contract and competitive bidding).
  - The quantity to procure.
  - Selection of a supplier/contractor.
  - Appropriate price, terms, and conditions.
- Signing of contract
- Timely delivery
- Receiving the right type of material/system
- Timely execution of work
- Inspection of work to maintain quality of the project
- Completion of project within agreed upon schedule
- Completion of project within agreed upon budget
- Documenting reports and plans

7.3.10.1 Plan Procurement Management Process

In construction projects, the involvement of outside companies/parties starts at the early stage of the project development process. The owner/client has to decide which work to be procured, constructed by others. Every organization has their procurement system to procure services, contracts, product from others. Figure 7.87 illustrates the stages at which outside agency (contractor) is selected as per the procurement strategy for a particular type of project delivery system. At each of these stages, the procurement management process (bidding and tendering process) takes place. Figure 7.88 illustrates the procurement management process life cycle to select the contractor (outside agency).

In addition, the owner has to determine the type of contract/pricing methods to select the contractor (outside agency). We have already discussed the same under Sections 2.3 and 2.4 in Chapter 2.

7.3.10.2 Select Project Delivery System

Construction project delivery system is an organizational relationship of three elemental parties namely the owner, designer (consultant), and contractor, and the roles and responsibilities of the parties and the general sequence of activities to be performed to deliver the project. The roles and responsibilities of these parties vary considerably under different project delivery systems. The following are the main categories of project delivery system:

1. Traditional system
2. Integrated system
3. Management-oriented system
4. Integrated project delivery system

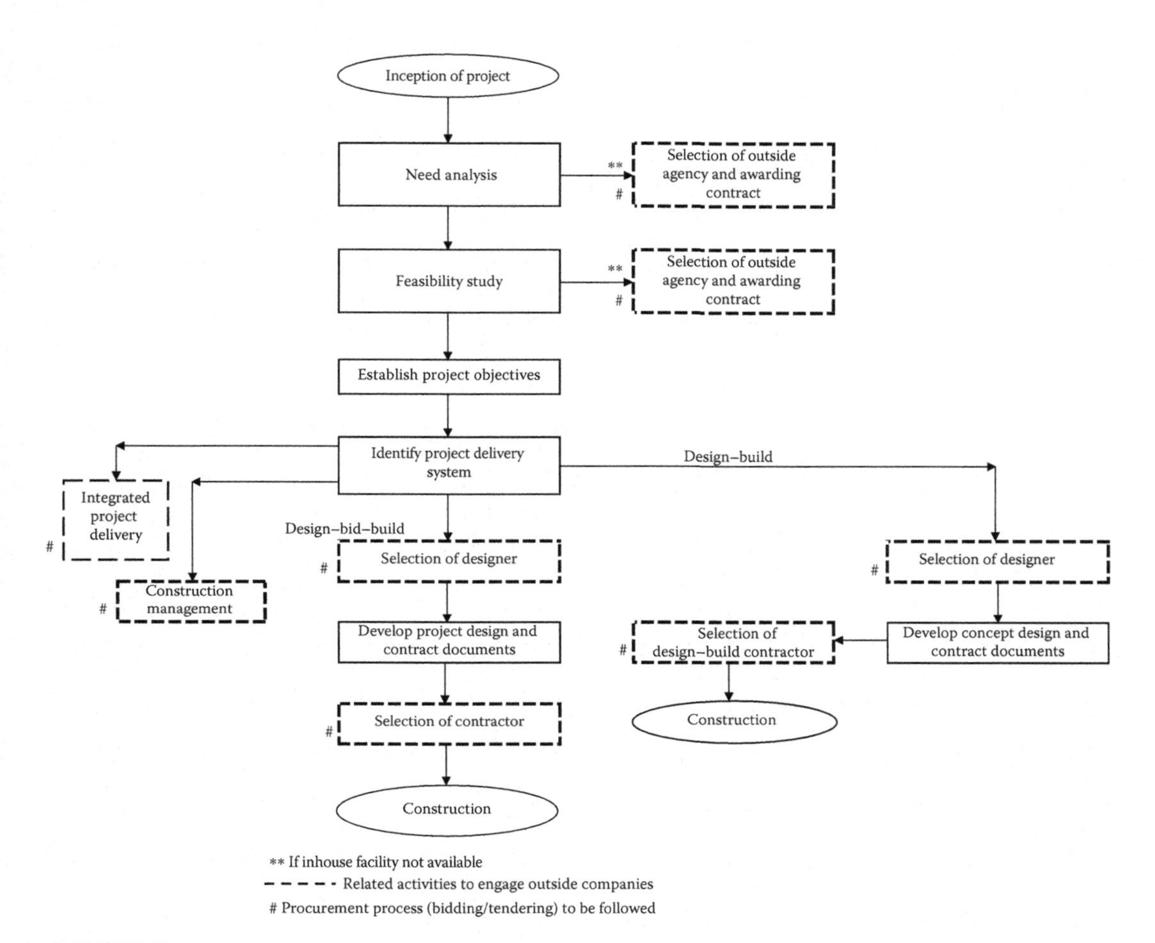

FIGURE 7.87
Procurement management processes stages for construction projects. (Abdul Razzak Rumane (2013), *Quality Tools for Managing Construction Projects*, CRC Press, Boca Raton, FL. Reprinted with permission from Taylor & Francis Group.)

Each category is further classified and subclassified into various project delivery system (see Table 2.1). Section 2.2 details various types of project delivery systems, advantages, and disadvantages of each of these systems.

The selection of project delivery system mainly depends on the project size, complexity of the project, innovation, uncertainty, urgency, and the degree of involvement of the owner. Section 2.4 details the selection of project delivery system.

7.3.10.3 Select Project Team (Contractor)

In construction projects, the owner engages outside agencies such as the construction/project manager, designer, consultant, and contractor to perform the following works:

1. Feasibility studies, if in-house facility is not available or is inadequate
2. Project management and construction management
3. Project design
4. Construction

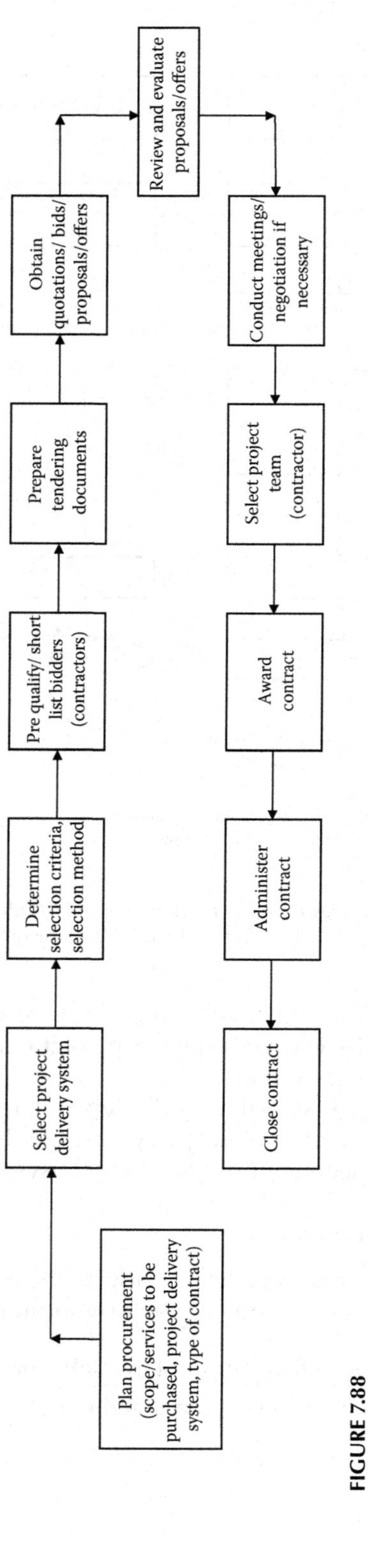

FIGURE 7.88
Contract management process.

The selection of project team (contractor) is mainly done as follows:

1. Screening of qualified contractors (prequalification of contractor)
2. Selecting contractor using contracting method such as
   a. Competitive bidding
   b. Competitive negotiations
   c. Direct negotiation
3. Awarding contract

While engaging an outside agency (contractor), the following selection criteria are to be considered as a minimum to prequalify the agency (contractor):

- Available skill level
- Relevant/past performance on similar type of work
- Reputation about their works
- Number of projects (works) successfully completed
- Technical competence
- Knowledge about the type of projects (works) for which they are likely to be engaged
- Available resources
- Commitment to creating best value
- Commitment to containing sustainability
- Rapport/behavior
- Communication

The above information is gathered through a request for information, request for prequalification or prequalification questionnaires (PQQ) from the prospective agency, consultant, and contractor.

After short listing of contractors, bid documents are distributed to submit the proposal. The following are the most common contracting methods for selection of project team:

1. Low bid
2. Best value
   a. Total cost
   b. Fees
3. Qualification-based selection

7.3.10.3.1 Perform Short Listing (Prequalification) Process

The first stage in selection of project team (contractor) is by short listing the prospective contractors. The information is gathered through

- Request for information
- Request for qualification
- Prequalification questionnaires

7.3.10.3.1.1 Prequalification of Feasibility Study Consultant Table 3.4 discussed earlier under Section 3.3.1 illustrates the required qualifications to select a consultant to conduct feasibility study. Normally the consultants are selected on quality based selection method.

7.3.10.3.1.2 Prequalification of Construction Manager Construction manager (CM) is a professional firm or individual having expertise in the management of construction processes. There are two types of construction manager (construction management systems). These are as follows:

1. Agency CM
2. CM at-risk

While selecting the construction manager, the owner has to ensure that CM has professional expertise in managing the following main areas of the project:

- Scope
- Schedule
- Cost
- Monitoring and control
- Quality
- Safety

The owner has to consider regulatory requirements (licensing) while selecting the CM. Table 7.40 lists PQQ to select construction manager.

TABLE 7.40

PQQs for Selecting Construction Manager

Sr. No.	Question	Answer
1	Name of the organization and address	
2	Organization's registration and license number	
3	ISO certification	
4	Total experience (years)	
	4.1 Agency construction manager	
	4.2 CM at-risk	
5	Size of project (maximum amount of single project)	
	5.1 Agency construction manager	
	5.2 CM at-risk	
6	List similar type (type to be mentioned) of projects completed	
	6.1 Agency construction manager	
	6.2 CM at-risk	
7	Type of services provided as agency CM/CM for above-mentioned projects	
	7.1 During project study	
	7.2 During design	
	7.3 During bidding and tendering	
	7.4 During construction	
	7.5 During startup	

(Continued)

TABLE 7.40 (*Continued*)

PQQs for Selecting Construction Manager

Sr. No.	Question	Answer
8	Total experience in green building construction	
9	Total management experience	
	9.1 Project scope	
	9.2 Project planning and scheduling	
	9.3 Project costs	
	9.4 Project quality	
	9.5 Technical and financial risk	
	9.6 HSE	
	9.7 Stakeholder management	
	9.8 Project monitoring and control	
	9.9 Conflict management	
	9.10 Negotiations	
	9.11 Information technology	
10	Experience in conducting value engineering	
11	Resources	
	11.1 Management	
	11.2 Engineering	
	11.3 Technician	
	11.4 Construction equipment	
12	Total turnover for the past 5 years	
13	Submit audited financial reports for the past 3 years	
14	Experience in training of owner's personnel	
15	Knowledge about regulatory procedures	
16	Litigation (dispute and claims) on earlier projects	

7.3.10.3.1.3 Prequalification of Designer (A/E) Designer (A/E) consists of architects, engineers, or consultant. They are the owner's appointed entity accountable to convert owner's conception and need into specific facility with detailed directions through drawings and specifications within the economic objectives. They are responsible for the design of the project and in certain cases supervision of construction process. Table 7.41 lists PQQ to select the designer (A/E).

TABLE 7.41

PQQs for Selecting Designer (A/E)

Sr. No.	Question	Answer
1	Name of the organization and address	
2	Organization's registration and license number	
3	ISO certification	
4	LEED or similar certification	
5	Total experience (years) in designing following type of projects	
	5.1 Residential	
	5.2 Commercial (mix use)	
	5.3 Institutional (governmental)	
	5.4 Industrial	
	5.5 Infrastructure	
	5.6 Design–build (specify type)	

(*Continued*)

TABLE 7.41 (*Continued*)

PQQs for Selecting Designer (A/E)

Sr. No.	Question	Answer
6	Size of project (maximum amount of single project)	
	6.1 Residential	
	6.2 Commercial (mix use)	
	6.3 Institutional (governmental)	
	6.4 Industrial	
	6.5 Infrastructure	
	6.6 Design–build (specify type)	
7	List successfully completed projects	
	7.1 Residential	
	7.2 Commercial (mix use)	
	7.3 Institutional (governmental)	
	7.4 Industrial	
	7.5 Infrastructure	
	7.6 Design–build	
8	List similar type (type to be mentioned) of projects completed	
	8.1 Project name and contracted amount	
	8.2 Project name and contracted amount	
	8.3 Project name and contracted amount	
	8.4 Project name and contracted amount	
	8.5 Project name and contracted amount	
9	Total experience in green building design	
10	Joint venture with any international organization	
11	Resources	
	11.1 Management	
	11.2 Engineering	
	11.3 Technical	
	11.4 Design equipment	
	11.5 Latest software	
12	Design production capacity	
13	Design standards	
14	Present work load	
15	Experience in value engineering (list projects)	
16	Financial capability (turnover for the past 5 years)	
17	Financial audited report for the past 3 years	
18	Insurance and bonding capacity	
19	Organization details	
	19.1 Responsibility matrix	
	19.2 CVs of design team members	
20	Design review system (quality management during design)	
21	Experience in preparation of contract documents	
22	Knowledge about regulatory procedures and requirements	
23	Experience in training of owner's personnel	
24	List of professional awards	
25	Litigation (dispute and claims) on earlier projects	

7.3.10.3.1.4 Prequalification of Contractor (Design–Build) Table 7.42 lists PQQ to select contractor (design–build).

TABLE 7.42

PQQs for Selecting Design–Build Contractor

Sr. No.	Question	Answer
1	Name of the organization and address	
2	Organization's registration and license number	
3	ISO certification	
4	Registration/classification status of the organization	
5	Joint venture with any international contractor	
6	Total turnover for the past 5 years	
7	Audited Financial Report for the past 3 years	
8	Insurance and Bonding Capacity	
9	Total experience (years) as design–build contractor	
10	Total experience (years) as contractor	
Design–Build Contracts information		
11	Total experience (years) in construction of following type of projects	
	11.1 Residential	
	11.2 Commercial (mix use)	
	11.3 Institutional (governmental)	
	11.4 Industrial	
	11.5 Infrastructure	
12	Size of project (maximum amount of single project)	
	12.1 Residential	
	12.2 Commercial (mix use)	
	12.3 Institutional (governmental)	
	12.4 Industrial	
	12.5 Infrastructure	
13	List successfully completed projects	
	13.1 Residential	
	13.2 Commercial (mix use)	
	13.3 Institutional (governmental)	
	13.4 Industrial	
	13.5 Infrastructure	
14	List similar type (type to be mentioned) of projects completed	
	14.1 Project name and contracted amount	
	14.2 Project name and contracted amount	
	14.3 Project name and contracted amount	
	14.4 Project name and contracted amount	
	14.5 Project name and contracted amount	
15	Resources	
	15.1 Management	
	15.2 Engineering	
	15.3 Technical	

(*Continued*)

TABLE 7.42 (*Continued*)

PQQs for Selecting Design-Build Contractor

Sr. No.	Question	Answer
	15.4 Foreman/supervisor	
	15.5 Skilled manpower	
	15.6 Unskilled manpower	
	15.7 Plant and equipment	
16	Quality management policy	
17	Health, safety, and environment policy	
	17.1 Number of accidents during the past 3 years	
	17.2 Number of fires at site	
18	Current projects	
19	Staff development policy	
20	List of delayed projects	
21	List of failed contract	
Designer's Information		
22	Total years of experience in design–build type of projects	
23	Size of project (maximum value)	
24	List similar type of successfully completed projects	
25	List successfully design–build projects	
26	Resources	
	26.1 Architect	
	26.2 Structural engineer	
	26.3 Civil engineer	
	26.4 HVAC engineer	
	26.5 Mechanical engineer	
	26.6 Electrical engineer	
	26.7 Low-voltage engineer	
	26.8 Interior designer	
	26.9 Landscape engineer	
	26.10 CAD technicians	
	26.11 Quantity surveyor	
	26.12 Equipment	
	26.13 Design software	
27	LEED or similar certification	
28	Total experience in green building design	
29	Design philosophy/methodology	
30	Quality management system	
31	HSE consideration in design	
32	List of professional awards	
33	Litigation (dispute and claims) on earlier projects	

7.3.10.3.1.5 Prequalification of Contractor (Design–Bid–Build) Table 7.43 lists PQQ to select contractor.

TABLE 7.43

PQQs for Selecting Contractor

Sr. No.	Question	Answer
1	Name of the organization and address	
2	Organization's registration and license number	
3	ISO certification	
4	Registration/classification status of the organization	
5	Joint venture with any international contractor	
6	Total turnover for the past 5 years	
7	Audited financial report for the past 3 years	
8	Insurance and bonding capacity	
9	Total experience (years) in the construction of following type of projects	
	9.1 Residential	
	9.2 Commercial (mix use)	
	9.3 Institutional (governmental)	
	9.4 Industrial	
	9.5 Infrastructure	
10	Size of project (maximum amount of single project)	
	10.1 Residential	
	10.2 Commercial (mix use)	
	10.3 Institutional (governmental)	
	10.4 Industrial	
	10.5 Infrastructure	
11	List successfully completed projects	
	11.1 Residential	
	11.2 Commercial (mix use)	
	11.3 Institutional (governmental)	
	11.4 Industrial	
	11.5 Infrastructure	
12	List similar type (type to be mentioned) of projects completed	
	12.1 Project name and contracted amount	
	12.2 Project name and contracted amount	
	12.3 Project name and contracted amount	
	12.4 Project name and contracted amount	
	12.5 Project name and contracted amount	
13	List of subcontractors	
14	Resources	
	14.1 Management	
	14.2 Engineering	
	14.3 Technical	
	14.4 Foreman/supervisor	
	14.5 Skilled manpower	
	14.6 Unskilled manpower	
	14.7 Plant and equipment	
15	Current projects	
16	Quality management policy	

(*Continued*)

TABLE 7.43 (*Continued*)

PQQs for Selecting Contractor

Sr. No.	Question	Answer
17	Health, safety, and environment policy	
	17.1 Number of accidents during the past 3 years	
	17.2 Number of fires at site	
18	Staff development policy	
19	List of delayed projects	
20	List of failed contract	
21	List of professional awards	
22	Litigation (dispute and claims) on earlier projects	

7.3.10.3.1.6 Evaluation of Prequalification Documents Upon receiving the information relating to the qualification data, the documents are evaluated as per the selection criteria determined earlier by the project owner. Table 7.44 is evaluation criteria to qualify the designer to participate in the tender.

TABLE 7.44

Designer's (A/E) Selection Criteria

Sr. No.	Evaluation Criteria	Weightage		Notes
1	General information			
	a. Company information			
2	Business		10%	
	a. LEED or similar certification	5%		
	b. ISO certification	5%		
3	Financial		20%	
	a. Turnover	5%		
	b. Financial standing	5%		
	c. Insurance and bonding limit	10%		
4	Experience		30%	
	a. Design experience	10%		
	b. Similar type of projects	10%		
	c. Current projects	10%		
5	Design capability		10%	
	a. Design approach	5%		
	b. Design capacity	5%		
6	Resources		20%	
	a. Design team qualification	10%		
	b. Design team composition	5%		
	c. Professional certification	5%		
7	Design quality		5%	
8	Safety consideration in design		5%	

Note: The weightage mentioned are indicative only. The percentage can be determined as per the owner's strategy.

7.3.10.4 Manage Construction Documents

The construction documents reviewed and approved by the owner are organized by the consultant for bidding purpose. Necessary information is inserted on these documents prior to release for distribution to quailed bidders.

7.3.10.5 Conduct Bidding and Tendering

Regardless of the type of the project delivery system, the contract arrangement between the owner and the contractor has to be established. The following are the most common types of contract/compensation methods that are used in construction projects:

1. Firm fixed price or lump-sum contract
2. Cost reimbursement contract
3. Remeasurement contract
4. Target price contract
5. Time and cost contract
6. Guaranteed maximum price (GMP) contract

Tendering and bidding documents are prepared based on the strategy to procure the contract. Bid/tendering documents are distributed to short-listed contractors. Figure 7.89 illustrates bidding process to select the contractor as per the procurement method in accordance with the contractor selection strategy for selection of contractor.

7.3.10.5.1 Evaluate Bids

The submitted bids are reviewed and evaluated by the contractor selection committee members for full compliance to the tender requirements. The main purpose of bid evaluation is to determine that the bid responses are complete in all respects in accordance with the evaluation and selection methodology specified in the tender documents.

Most government, public sector projects follow low bid selection method. There are three international bidding procedures that may be selected by the project owner to suit the nature of project procurement. These are as follows:

1. *Single stage—one envelope*: In this procedure, the bidders submit bids in one envelope containing both technical and financial proposals. The bids are evaluated by a selection committee that sends its recommendation to the owner. Following the review and concurrence by the owner, the contract is awarded to the lowest bidder.
2. *Single stage—two envelopes*: In this procedure, the bidders submit two envelopes, one containing technical proposal and the other the financial proposal. Initially, technical proposals are evaluated without referring to the price. Bidders whose proposal does not conform to the requirements may be rejected/not accepted. Following the technical proposal evaluation, the financial proposals of technically responsive bidders are reviewed. Upon review and concurrence by the owner, the contract is awarded to the lowest bidder.
3. *Two stages*: In this procedure, during the first stage, the bidders submit their technical offers on the basis of operating and performance requirements, but without price.

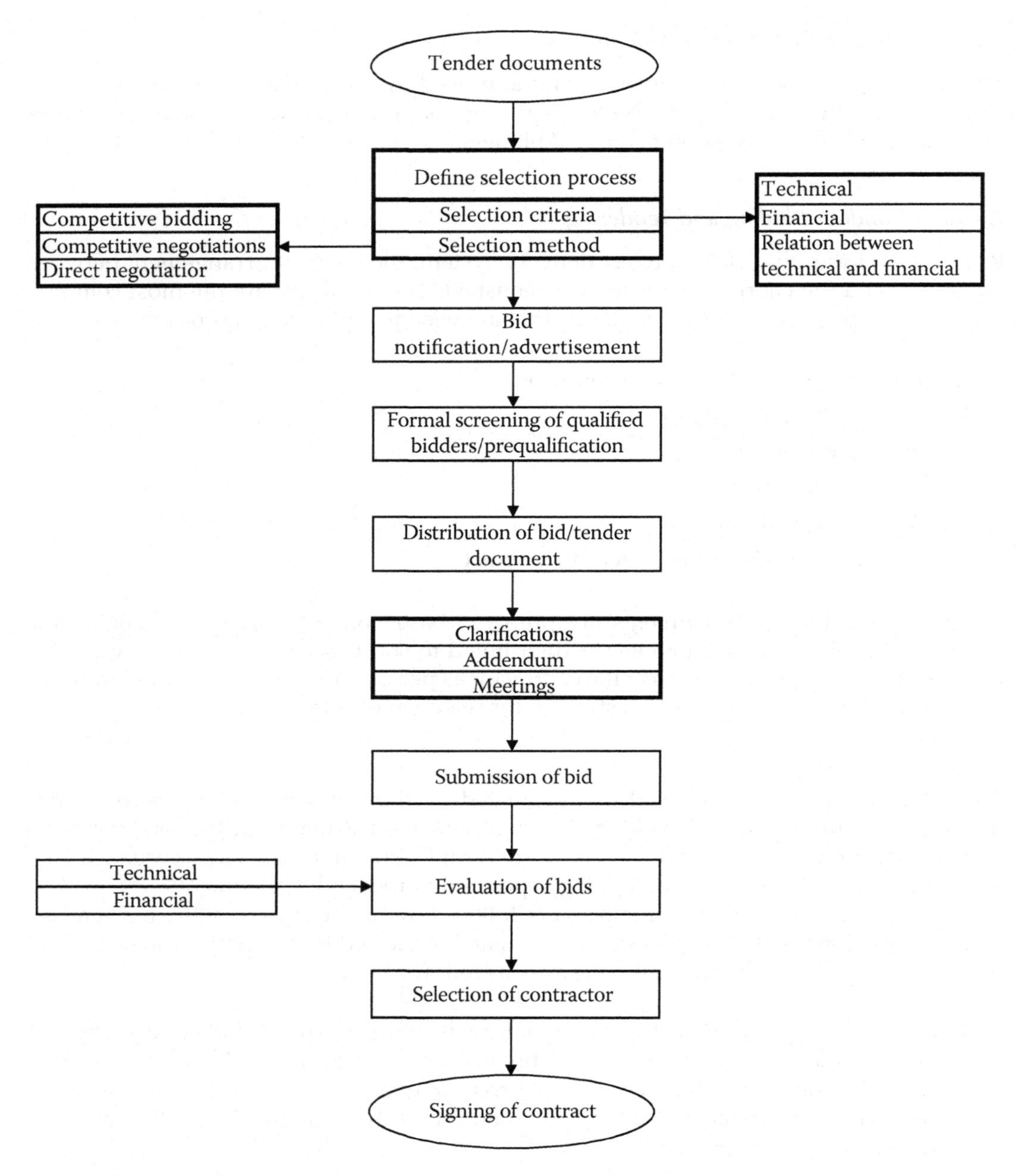

FIGURE 7.89
Bidding tendering (procurement process).

> The technical offers are evaluated by the selection committee. Any deviations to the specified performance requirements are discussed with the bidders who are allowed to revise or adjust the technical offer and resubmit the same.

During the second stage, the bidders, whose technical offers are accepted, are invited to submit final technical proposal and financial proposal. Both the proposals are reviewed by the selection committee and following review and concurrence by the owner, the contract is awarded to the lowest bidder. This procedure is mainly applicable for turnkey projects and complex plants. Table 7.45 lists the items to be reviewed prior to evaluation of bid documents.

7.3.10.5.2 Select Contractor

After the bids have been reviewed and evaluated by the consultant, they are sent to the owner for the approval and further action. In certain cases, the bids are reviewed by tender-approving agency. The comments raised by the agency are to be taken care before finalization of bid. In most cases, the bidder with lowest bid amount wins the contract. In case the bid amount is more than the approved project budget, the owner has to update the budget amount or negotiate with the contractor to meet owner's expectations. Upon approval from the relevant agency, the owner awards the contract to one bidder.

7.3.10.6 Administer Contract

Contract administration is the process of formal governance of contract and changes to the contract document. It is concerned with managing contractual relationship between various participants to successfully complete the facility to meet owner's objectives. It includes tasks such as

TABLE 7.45

Checklist for Bid Evaluation

Sr. No.	Description	Yes	No	Notes
A: Documents				
A.1	Bid submitted before closing time on the date specified in the bid documents			
A.2	Bidders identification is verified			
A.3	Bid is properly signed by the authorized person			
A.4	Bid bond is included			
A.5	Required certificates are included			
A.6	Bidders confirmation to the validity period of bid			
A.7	Confirmation to abide by the specified project schedule			
A.8	Bid documents have no reservation or conditions (limitation or liability)			
A.9	Preliminary method statement			
A.10	List of equipment and machinery			
A.11	List of proposed core staff as listed in the tender documents			
A.12	Complete responsiveness to the commercial terms and conditions			
A.13	All the required information is provided (completeness of information)			
A.14	All the supporting documents required to determine technical responsiveness is submitted			
B: Financial				
B.1	All the items are priced			
B.2	Bid amount clearly mentioned			
B.3	Prices of provision items			

- Administration of project requirements and project team members
- Communication and management reporting
- Execution of contract
- Monitoring contract performance (scope, cost, schedule, quality, and risk)
- Inspection and quality
- Variation order process
- Making changes to the contract documents by taking corrective action as needed
- Payment procedures

It is required that the contract administration procedure is clearly defined for the success of the contract and that the parties to the contract understand who does what, when, and how. The following are some typical procedures that should be in place to manage contract management activities:

1. Contract document maintenance and variation
2. Performance review system
3. Resource management and planning
4. Management reporting
5. Change control procedure
6. Variation order procedure
7. Payment procedure

Table 7.46 lists the contents of a contract management plan.

7.3.10.7 Close Contract

It is a process of completing the contract, verification of completed activities, acceptance of all the deliverables, takeover of project, and issuance of substantial completion certificate. Normally, the contract documents stipulate all the requirements to be completed to close the contract. Table 7.47 illustrates the lists of all the activities required to be recorded to close the contract.

7.3.11 Health, Safety, and Environment Management

Safety is a cornerstone of a successful project. Accidents cause needless loss of human and physical capital. Logically, these losses can only be detrimental to the project schedule and cost. Safety, long neglected, is increasingly being recognized for its value as contractors and owners strive to avoid the increasing costs of injuries and fatalities. A recent study [2] determined that the direct costs, indirect costs, and quality of life losses average $5.2 million and (in 2010 dollars) for a single fatality. Another source computes the average direct and indirect costs of a strain or sprain injury at nearly $60,000 [3]. Numerous researchers have demonstrated the business case for safety through benefit–cost analyses. In one such analysis, the US Occupational Safety and Health Administration (OSHA)'s Office of Regulatory Analysis has stated "… our evidence suggests that companies that implement effective safety and health can expect reductions of 20% or greater in their injury and illness rates and a return of $4 to $6 for every $1 invested … " [4].

TABLE 7.46

Contents of Contract Administration Plan

Sr. No.	Topics
1	Contract summary, deliverables, and scope of work
2	Type of contract
3	Contract schedule
4	Contract cost
5	Project team members with roles and responsibilities
6	Core staff approval procedure
7	Contract communication matrix/management reporting
8	Coordination process
9	Liaison with regulatory authorities
10	Material/product/system review/approval process
11	Shop drawing review/approval process
12	Project monitoring and control process
13	Contract change control process a. Scope b. Material c. Method d. Schedule e. Cost
14	Review of variation/change requests
15	Project holdup areas
16	Quality of performance
17	Inspection and acceptance criteria
18	Risk identification and management
19	Progress payment process
20	Claims, disputes, conflict, and litigation resolution
21	Contract documents and records
22	Postcontract liabilities
23	Contract closeout and independent audit

Source: Abdul Razzak Rumane (2013), *Quality Tools for Managing Construction Projects*, CRC Press, Boca Raton, FL. Reprinted with permission from Taylor & Francis Group.

The trend toward improved safety seems gradual but recognizable. As in other fields, technology advances continually provide better materials, methods, and equipment. Increasingly prefabricated materials and modular construction reduce worker exposure to more hazardous construction activities found with older conventional methods and materials. In recent years, researchers have found that incorporating safety in the design phase has a huge potential to impact exposures to hazardous situations. A concept called *prevention thru design* (PtD) is well established in Europe and gaining a foothold in the United States. With PtD, safety is *designed* in a project. For instance, steel columns can receive shop-drilled holes for the easy attachment of safety lanyards during steel erection. This small and insignificant planning effort can save a life. Innovative concepts such as PtD require strong advocates to demonstrate their safety value and make the business case to the industry.

TABLE 7.47

Contract Closeout Checklist

Sr. No.	Description	Yes/No
Project Execution		
1	Contracted works completed	
2	Site work instructions completed	
3	Job site instructions completed	
4	Remedial notes completed	
5	Noncompliance reports completed	
6	All services connected	
7	All the contracted works inspected and approved	
8	Testing and commissioning carried out and approved	
9	Any snags	
10	Is the project fully functional	
11	All other deliverable completed	
12	Spare parts delivered	
13	Is the waste material disposed	
14	Whether safety measures for use of hazardous material established	
15	Whether the project is safe for use/occupation	
16	Whether all the deliverable accepted	
Project Documentation		
17	Authorities approval obtained	
18	Record drawings submitted	
19	Record documents submitted	
20	As-built drawings submitted	
21	Technical manuals submitted	
22	Operation and maintenance manuals submitted	
23	Equipment/material warrantees/guarantees submitted	
24	Test results/test certificates submitted	
Training		
25	Training to owner/end user's personnel imparted	
Negotiated Settlements		
26	Whether all the claims and dispute negotiated and settled	
Payments		
27	All payments to subcontractors/specialist suppliers released	
28	Bank guarantees received	
29	Final payment released to main contractor	
Handing over/Taking over		
30	Project handed over/taken over	
31	Operation/maintenance team taken over	
32	Excess project material handed over/taken over	
33	Facility manager in action	

Source: Abdul Razzak Rumane (2010), *Quality Management in Construction Projects*, CRC Press, Boca Raton, FL. Reprinted with permission from Taylor & Francis Group.

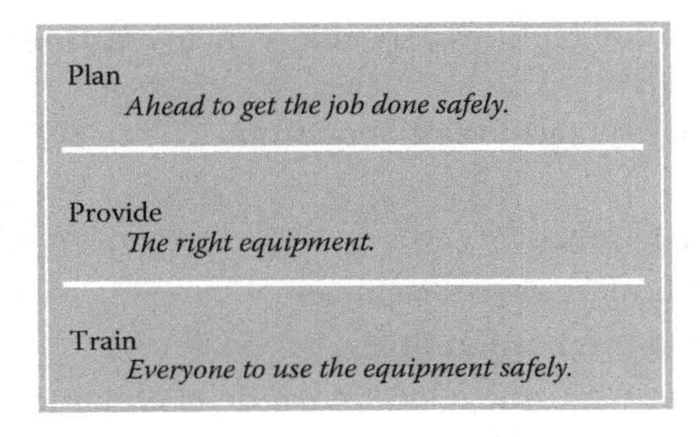

FIGURE 7.90
Safety framework.

Though safety gains have been made, the situation remains critical. The US annual construction fatality rate hovers around 10 per 100,000. In contrast, the US fatality rate for texting driver incidents is roughly 0.5 per 100,000 [5]. It is well established that texting and driving is risky. Sadly, construction workers incur risks 20 times greater solely as a result of their chosen occupation. How can construction be made safer? The process is detailed but the solution can be simply summarized as in Figure 7.90 [6].

7.3.11.1 Safety Management Framework

Strive to incorporate safety into the project as early as possible. A safety conscious contractor can provide valuable input in the early stages of design. As seen in Figure 7.91, early processes offer the most potential to affect safety.

Prior to construction, the contractor needs to compile a coordinated, comprehensive site plan. Although the details of a comprehensive safety management plan are beyond the scope of this text, we can certainly define its critical elements using our *plan*, *provide*, and *train* framework.

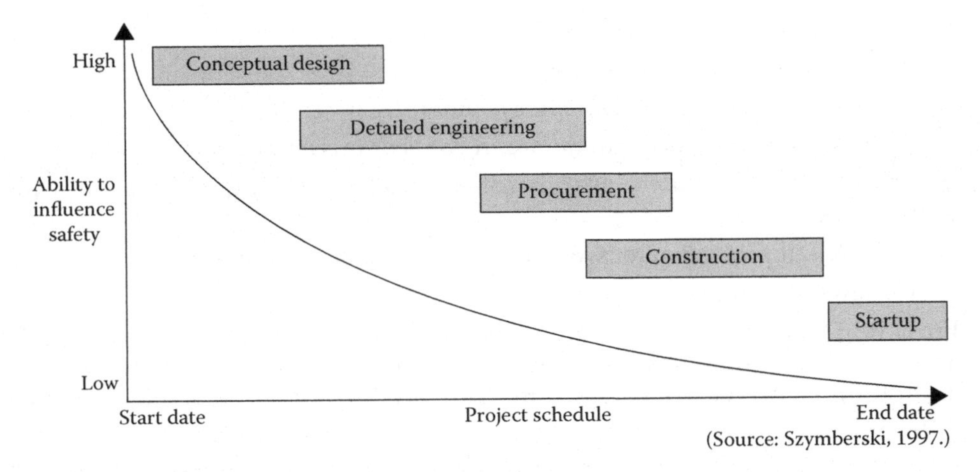

FIGURE 7.91
The ability to influence safety.

Plan: Each site is unique and requires a specific plan. In creating a plan, here are some important items:

- Secure emergency contact information from
  - Hospitals
  - Fire and rescue squad
  - First-aid medical treatment facilities
- Locate buried and overhead utilities
  - Notify utilities requiring relocation or de-energizing
  - Coordinate the project schedule with relocations
- Provide safe access and egress from the site for
  - Typical material and equipment deliveries
  - Oversize loads
- Other miscellaneous
  - Conduct regular safety inspections and site training
  - Plan for traffic control
  - Plan for emergency response scenarios

The above contains only critical items. A compete and comprehensive safety plan should incorporate sufficient detail to address all hazards, including those that are site specific.

Provide: Workers have the right to expect that their employer will provide them with the proper equipment, tools, and training to complete all tasks safely. At a minimum, employers should:

- Provide well-maintained equipment (e.g., cranes and backhoes)
  - Require periodic safety inspections
  - Keep maintenance logs
  - Furnish operating manuals and load charts
  - Utilize certified crane operators
- Require safety assurance checks
  - Inspect chains, slings, and hoists daily
  - Replace hand tools with damaged cords or missing guards
- Provide the basic safety gear: hardhats, safety glasses, gloves, and hearing protection
- Provide fall protection for workers at heights of 6 feet or more
- Designate a company safety officer

Providing workers with well-maintained equipment not only enhances safety and improves productivity but also boosts morale. Poorly functioning tools and equipment create hazards, diminish productivity, and damage worker morale.

Train: Even the best plan and top-notch equipment require well-trained employees. In developed countries, construction training is commonly available through unions, trade associations, individual company programs, and even the Internet.

In 2012, approximately 350,000 workers [7] in the US construction workforce received a standardized OSHA 10-hour training course in construction safety. Some localities require such training for all workers on public projects. This is an excellent example of informed owners realizing that increased safety pays long-term benefits by reducing costs. More extensive training courses appropriate for supervisors are also widely available.

Typical basic construction safety training might include the following:

- Four most common fatal hazards (i.e., falls, electrocutions, struck by, and caught in between)
- Material handling
- Hazard communication
- Lockout/tag out procedures
- Vehicular safety
- Personal protective equipment

Employee turnover, job reassignments, and changing technology necessitate that training be ongoing and frequently updated.

7.3.11.2 Major Causes of Site Accidents

The results of site accidents range from minor injuries such as cuts and bruises to major ones such as broken limbs and paralysis. The worst accidents may result in fatalities. Thankfully, injury severity and the frequency of that injury tend to be inversely related. This discussion focuses primarily on the causes of fatalities but in reality the causes of accidents resulting in minor and major injuries are often remarkably similar. There is an element of randomness in injury outcomes. A few inches may be the only difference between a near miss and a fatality.

Almost 60% of construction fatalities are caused by four primary means. These hazards, mentioned previously, are known as the *fatal four* [8]. They consist of falls, electrocutions, struck by incidents, and caught in between incidents. Preventing these is the core of any construction safety program.

Falls alone constitute one-third of all construction fatalities in the United States [8]. Categories include falls from roofs, scaffolding, ladders, equipment, and through openings. Though falls occur in a multitude of ways, and they are all preventable. A national campaign in the United States sponsored by a consortium of construction stakeholders and safety advocates [9] emphasized the ability to prevent falls using the theme outlined in the safety management framework from a previous section. Properly trained workers with fall protection systems in place should never become fall fatalities. Common fall protection systems include guardrails, safety nets, and body harnesses. A typical body harness is shown in Figure 7.92.

Electrocutions, another leading fatality cause, are responsible for about 10% of construction deaths. Primary causes of electrocution are as follows:

- Contact with overhead power lines
- Contact with energized sources
- Improper use of extension and flexible cords

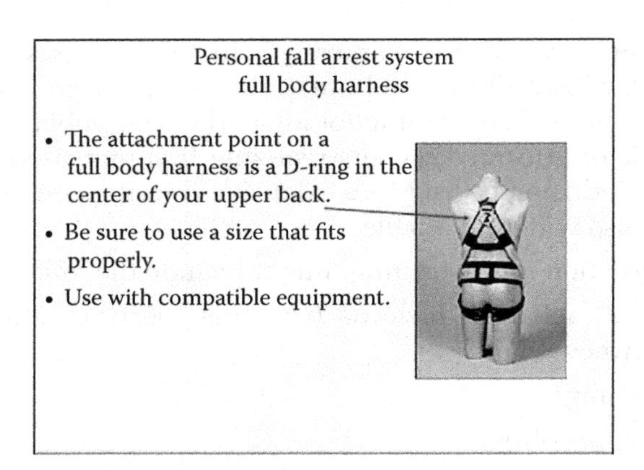

Personal fall arrest system
full body harness
• The attachment point on a full body harness is a D-ring in the center of your upper back.
• Be sure to use a size that fits properly.
• Use with compatible equipment.

FIGURE 7.92
Personal fall arrest system.

Safety countermeasures are as follows:

- Maintain a safe distance from overhead lines.*
- Use ground fault circuit interrupters.
- Inspect portable tools and extension cords for damage.

When impact alone creates the injury, this is considered as a struck by event. Workers can be struck by flying, falling, swinging, or rolling objects. These events may be minimized by staying clear of lifted loads, maintaining awareness and respect for the swing radius of cranes, backhoes, and so on, and providing equipment operators protection from falling debris.

Struck by and caught in between events are somewhat similar. When crushing between two objects or being pulled into or caught in running machinery occurs, it is considered as caught in between. Some common causes of this type are structural collapse, trench cave-in, or becoming trapped between equipment and a stationary object. A safe worksite will have bracing for unstable structures as well as sloping, benching, or shielding for trenching operations. Workers should be trained to maintain awareness of moving equipment, stay clear when possible, and avoid pinch points.

Thus far, the focus has been on accidents that cause fatalities and serious injuries. Any safety discussion should also include common causes of less serious injuries. Most construction work is physically challenging and often performed in harsh environments. Strains, sprains, bumps, bruises, and cuts are common. The inevitable bumps and scrapes can be minimized with personal protective equipment (PPE). Using hardhats, gloves, safety glasses, and hearing protection must become second nature. To avoid strains, use two people to lift heavy loads and avoid awkward working positions.

Nature's elements can create a variety of problems. Rain, wind, cold, and heat each present special challenges. Heat illness is a major concern in summer months. When temperatures are high, workers should take special care to keep hydrated, work in the shade when possible, and complete higher effort tasks in the cooler morning hours.

* Ten feet is considered to be the minimum safe clearance for power lines up to 50 kV. For a 200 kV line, the minimum clearance doubles to 20 feet.

Less obvious and often overlooked, a cluttered and dirty worksite is one common and easily preventable cause of accidents. Scattered tools, materials, and construction debris invite slips, trips, and falls. Loose nails create cuts and punctures. Site cleanliness and orderliness cannot be overemphasized. There is no doubt that a strong correlation exists between cleanliness and safety.

Another less obvious cause of accidents is the inability to recognize potential hazards on the jobsite. Young workers are particularly vulnerable because they often lack experience. Immigrant workers comprise another vulnerable group. Poor language skills make recognition training and awareness more difficult. Because the lack of ability to recognize potential hazards is frequently found in both groups, they may require specialized training, to build recognition and teach avoidance. Mentoring and close supervision is also recommended to reduce their risk of injury.

7.3.11.3 Safety in Construction Projects and Process Industries

Loss prevention philosophy makes process industries, construction projects, and other industries believe that protection of its resources, including employees and physical assets, against human distress and financial loss resulting from accidental occurrences can be controlled. Disasters have taught us for an improved loss control through a professional management system. This section sets out approaches to loss prevention that may be adopted by any industry and not limited to the process industry. It discusses the major elements and methodology of the overall program. These elements include policy declaration, employee training, permit to work procedures, safety inspections, predictive maintenance, safety meetings, safety talks, safety suggestions, loss prevention compliance reviews, executive management safety reviews, safe operations committee, contractor safety program, nonoperating personnel safety orientation, emergency drills, disaster control plan, and employee incentive schemes.

Fundamental concepts and methodology of the program is to identify all loss exposures, evaluate the risk of each exposure, plan how to handle each risk, and manage according to plan. People are the first source of losses. These could be managers, engineers, or workers who plan, design, build, operate, and maintain the plants. In addition, general public who may be subjected to the hazards! Second, equipment—whether fixed plant, machines, tools, protective gear, or vehicles. Third, materials such as process substances, supplies, and products that have physical and chemical hazards affecting people, equipment, and environment. Fourth is surrounding that includes buildings, surfaces and subsurfaces, atmosphere, lighting, noise, radiation, hot or cold weather, and social or economic conditions, which can affect safe performance of people, equipment, and materials.

Human sufferings and economic aspects of costs lost and work delays or property damage varies from minor to major. As we all know, a loss is the effect of an undesired event that could result in injury, property damage, and production upset. Root causes of accidents in the process industry are 30% due to design failure, operational error, equipment failure, and 20% attributed by maintenance and inspection deficiencies, 45% by inadequate supervision and training, and 5% due to natural phenomena and external influences.

7.3.11.3.1 Policy Declaration

It is important that the highest authority of the corporation/company and his management considers no phase of operation or administration as being of greater importance that accident prevention. It should be the policy of management to provide and maintain safe, healthful working conditions to follow operating practices that will safeguard employees, resulting in safe working conditions and efficient operations. A policy declaration to this effect should be developed and displayed for the knowledge of employees and that its implementation is ensured by the management.

7.3.11.3.2 Training

Safety training should be conducted for all employees that include the operations, maintenance, and engineering, and other administration personnel.

New employees assigned to work including clerical staff are given an introduction to general activities of the industry with emphasis on loss prevention. As a part of this training program, new employees should receive training on first aid/CPR, firefighting, usage of personal protective equipment and hazardous chemical handling, and work permit procedures. Over and above the normal work-related training, regular employees should also be given refresher safety training on a set frequent scheduled basis. The employees' supervisor usually determines the training requirements.

7.3.11.3.3 Work Permit Procedures

Management should have the certification program for the employees to issue and receive work permits for a good control and safe execution of tasks to be performed by other than normal operating personnel within the defined hazardous areas. Professionals from the loss prevention/safety department should assist work permit procedure certification. Validity of this certification should be decided by the management based on the nature of activities, number of employees, and other relevant criteria such as employee's knowledge, ability, attitude, and behavior. During this certification period, employees should undergo a refresher program to update and upgrade themselves on work permit procedures that may be organized by in-house operations/maintenance training groups.

Generally, process industries have work permits for hot and cold works and a permit for entry into confined space. Other work permits include precautionary permit for potential for accidental release of hydrocarbons, steam condensate, or any other toxic or reactive injurious materials during the course of any activity, excavation permits, permit to work on higher elevation or under water, and so on. The functioning of the work permit system and its strict adherence to the procedural requirements should be monitored by performing field surveys by dedicated management representative that should be backed by spot check reviews by loss prevention/safety professionals.

7.3.11.3.4 Safety Inspections

Over and above all daily, weekly, and monthly regular work-level inspections, the process industry should make a commitment that all facilities, such as operating plants, maintenance shops, offices and other work places be inspected as spot checks by the management at least once every quarter to identify and eliminate hazards and to provide a safe working condition.

A typical operating plant safety inspection team should consist of members from operations, maintenance, process engineering, loss prevention/safety, and fire protection. An operations' superintendent should lead the team from an unaffected division along with the process and maintenance engineers of the division to be surveyed and a representative each from loss prevention and fire protection. The team should review procedures and related software and conduct walk through inspection of the operating plants to not only verify satisfactory completion of previous items but also to make a note of other additional observations. The findings should be discussed at a subsequent critique meeting and should be consolidated in a report. The operations management should follow up on the corrective actions taken on the findings.

Similar quarterly safety inspections should also be conducted on maintenance facilities such as machine shops, welding shops, sheet metal shops, relief valve shops, instrumentation shops, offices, material warehouses, and so on.

7.3.11.3.5 Other Predictive Maintenance Inspections

Some other inspections, certifications, and predictive maintenance programs that have proved beneficial in process industries are as follows:

7.3.11.3.5.1 On-Stream Inspection Program This records thickness measurements for all pressure vessels, equipment, and selected process line circuits. The on-stream inspection program involves various survey points in the plants on a preplanned schedule.

7.3.11.3.5.2 Relief Valve Inspection Program This program should normally monitor relief valves through a relief valve coordinator or a specialist from the inspection group. The program is to forecast valves to be removed, to revise test intervals, and to evaluate specific problems.

7.3.11.3.5.3 Rotating Equipment Monitoring Program Rotating equipment should be monitored on a biweekly/monthly/quarterly cycle based on the critical nature of the equipment by dedicated maintenance specialists using state-of-the-art sophisticated vibration analysis equipment.

7.3.11.3.5.4 Crane/Heavy Equipment Inspection All cranes (mobile and fixed) used in the process industry should be inspected and certified by a crane inspection specialist. The crane/heavy equipment operators and riggers should also be certified for a particular crane/load during lifting operations.

7.3.11.3.5.5 Safety Meetings Safety meetings and talks are considered vital in communicating loss prevention/safety topics, motivating employees and acknowledging contributions in the loss prevention/safety program.

7.3.11.3.5.6 Operations Safety Meetings Every operations unit should organize a safety meeting at least once in two months. These meetings should be chaired by the plant foreman and attended by plant operation supervisors, lead operators, and a representative from loss prevention/safety, if available. Various issues related to operations safety, observations during quarterly plant safety inspections and near misses/incidents having lesson learning potential should be discussed during these meetings.

7.3.11.3.5.7 Maintenance Safety Meetings The maintenance department should organize safety meetings more frequently, preferably at least once two weeks. Owing to the nature of their work, often the maintenance employees are exposed to unsafe conditions or commit unsafe acts. These meetings, usually chaired by the maintenance foreman, are organized based on crafts and held at the work places. During these meetings, a loss prevention/safety representative should also deliver a safety talk on specialized topics. Safety films, as visual training aid, should be shown during these meetings.

7.3.11.3.5.8 Safety Talks Frequent—weekly/monthly/quarterly—safety talks should be conducted by the industry line organizations. The appropriate management should recognize the best safety talk. Discussions on related operating instructions as well as recently occurred on or off the job incidents/accidents involving losses should be included as topics of safety talks. The inmate discussion concept is the best and effective way to educate and update the employees with relevant information to help increase and improve their safety awareness both at work and at home.

7.3.11.3.5.9 Safety Suggestion Program A safety suggestion program should be introduced that encourages employee involvement in loss prevention. Employees whose suggestions are accepted by the management should be recognized for their contributions. Implementation of accepted safety suggestions should receive a greater prioritized attention for early benefits. Process industry history shows that the implementation of many of these suggestions have resulted in significant improvement in plant and personal safety.

7.3.11.3.5.10 Loss Prevention Compliance Reviews Loss prevention compliance reviews of the facility should be performed on a selective basis to monitor adherence with the industry/company loss prevention/safety policy. These independent reviews provide objective feedback on the effectiveness of the loss prevention/safety effort and should be performed by specialists such as operations/maintenance/engineering representatives from the proponent organization, other expertise from electrical/instrumentation, rotating equipment, on-stream inspector, crane/heavy equipment inspector, loss and fire prevention and industrial hygiene/health/environmental. Guidelines for performing reviews should be tailored to the size and type of review undertaken. Management should follow the team's recommendations for timely corrective actions taken.

7.3.11.3.5.11 Executive Management Safety Reviews It is a good industry practice to conduct an overall safety review of the facility by corporate executive management at least once a year. Participants should review the status of implementation to all items reported during the previous tour. They should also review new constructions and modifications to equipment in respect of loss prevention factors. These executives should allow time for presentations on items requiring capital budgetary approval, which could improve process safety and personnel safety.

7.3.11.3.5.12 Safe Operations Committee The safe operations committee should be chartered to evaluate the company organization loss prevention program and recommend ways and means for improvement. The management level company/organization representative should chair the committee. Generally, members of the committee should include managers/superintendents of operations, maintenance, and engineering, representatives from training, loss prevention, and fire protection. Key issues that should be addressed in such committees are analysis of deficiencies, effectiveness of programs, compliance with corporate general and engineering standards, review and follow up of corrective actions taken on safety problems identified by safety inspection or other reviews, examine levels of priority given to various safety problems, and discuss employee involvement in various on or off the job activities. Reviewing of the incidents/accidents/near-misses should be given a greater importance for lesson learning potentials and make recommendations.

7.3.11.3.5.13 Contractor Safety Program Special care and attention should be given to activities performed by the contractors. Safe execution of any tasks within the process facility is the sole responsibility of the proponent organization management regardless of contractor involvement. The proponents must ensure that the contractors have a good workable safety program that is well communicated and is fully implemented. The program should assure that contractors meet their responsibilities to protect personnel, equipment, and plant facilities. Contractual obligations are reviewed at precontract meetings and during regular site inspections while work is in progress. At the end, an evaluation report should be prepared for future reference to determine whether the contractor is worthy of being considered again.

7.3.11.3.5.14 Contractor Safety Orientation Program It is important that company management recognizes the hazards and the needs of nonoperating personnel who are not familiar with the process plant facilities and the procedures to follow in case of emergency. They should organize a safety program for all such employees, especially contractor employees working inside their facility. During the program, a safety awareness presentation should be made on various hazards in the facility, work permit procedural requirements, safety policy, and general functional procedures for their operations and emergency/evacuation procedures.

7.3.11.3.5.15 Fire/Emergency Drills In industry, announced and unannounced fire/emergency drills have been found of great value to tune employees to face any real in-plant emergencies. A written scenario should be prepared for each drill that sufficiently describes the role of operating personnel, location, and magnitude of fire/emergency, equipment affected, source of a release, and direction of flow. A team consisting of representatives from operations, maintenance and process engineering, fire protection, and loss prevention/safety should review these drills by making observations. A meeting should be conducted after such drills to evaluate the opportunities for improvement.

7.3.11.3.5.16 Disaster Control Plan Every process industry/organization should develop a disaster control plan that should define the procedures for obtaining assistance from inside or outside the company support organizations during disasters and detail the functions of essential personnel assigned to control the disaster. At least two drills per annum, one of which preferably unannounced, are recommended to evaluate the effectiveness of the plan and readiness of the people and corrective actions are taken on any deficiencies noted.

7.3.11.3.5.17 Incentive Programs Companies/organizations should develop good safety promotional incentive programs such as safety competitions of employees and also for the contracting firms who may be involved in major project construction and/or maintenance/repair jobs. Employee's individual and team performance should be recognized by the company to encourage and motivate employees to think and execute safety at all times and reduce losses that could occur due to injury or property damage or operations/production upsets.

7.3.11.4 Safety Hazards during Summer

During summer, people suffer from fatigue, exhaustion, heat-related illnesses, tornadoes, lightning storms, stinging insects, allergies, wildfires, poisonous plants, poisonous animals, and kids drowning are reported each week in hot areas. Vehicle safety is also the most important as a killer. Do not ever leave children unattended inside a vehicle, not even for a split second! Temperature inside a car reaches well over 120° F in less than one hour.

In addition, do not keep perfumes, cologne, hair sprays, or aerosol cans of any kind inside a car to avoid car explosions! Remove them quickly if you have them now!

As the temperature rises, the body stress also rises. There are two critical actions can help to battle the heat. Acclimatization to the heat, that is, get accustomed and used to heat and consume lot of water. Human body is a good regulator of heat. It reacts to heat by circulating blood and raising skin temperature. Excess heat is then released by sweating. Sweating maintains a stable body temperature if the humidity level is low enough to permit evaporation and also if the body fluids and salts you lose are adequately replaced. Many factors can cause the body to unbalance its ability to handle heat, for example, age, weight, fitness, medical condition, and diet.

When the heat is combined with other stresses such as physical work, loss of body fluids, fatigue, or some medical conditions, it may lead to heat-related illnesses, disability, and even to death. Therefore, caution to everyone, including workers and children, that this can happen to anybody—even to young and fit people.

Heat stress is a serious hazard. When body temperature rises even a few degrees above normal, that is, 98.6° F, we can experience tiredness, irritability, inattention, muscle cramps, weakness, disoriented, and become dangerously ill. Heat stress reduces our work capacity and efficiency. People who are overweight, physically unfit, suffer from heart conditions, drink too much alcohol, and are not used to summer temperatures, are at a greater risk and should seek medical advice. First aid for heat stress, is nutritious food, and, gets acclimatized with heat and drink a lot of water.

Heat rash: When people are constantly exposed to hot and humid air, heat rash can substantially reduce the ability to sweat and the subsequent body tolerance to heat. First aid for heat rash is to clean the affected areas thoroughly and dry them completely. Calamine or other soothing lotion may help relieve the discomfort.

Heat cramps is the final warning for heat stress. This may occur after prolonged exposure to heat. They are the painful intermittent spasms of the abdomen and other muscles. First aid for heat cramps varies. The best care is rest; move the victim to a cool environment. Give him or her plenty of water. Do not give pops, sparkling water, or alcohol.

Heat exhaustion may result from physical exertion in hot environments. Heat exhaustion develops when a person fails to replace body fluids and salt that is lost through sweating. You experience extreme weakness, fatigue, nausea, or a headache as heat exhaustion progresses. First aid for heat exhaustion is to rest in a shade or cool place, drink plenty of water, loosen clothing to allow for the body to cool, and use cool wet rags to aid cooling.

Heat stroke is a serious medical condition that urgently requires medical attention in which sweating stops, making the skin hot and dry. Body temperature is very high (105° F and rising). Symptoms of heat stroke are mental confusion, delirium, that is, disordered speech, chills, dizziness, strong fast pulse, loss of consciousness, convulsions or coma, a body temperature of 105° F or higher, and hot, dry skin that may be red or bluish. Remember, first aid for heat stroke is a medical emergency because a brain damage and death are possible. Until medical help arrives, move the victim to a cool place. Call the Emergency telephone number for help such as 999, 110, or 911. You must use extreme caution when soaking clothing or applying water to a victim. Shock may occur if done too quickly with very cold water. Use a fan or ice packs. Douse or sprinkle the body continuously with a cool liquid and summon medical help.

Other Sun-Related Health Problems

Exposure to UV radiation can lead to most common skin cancers. Getting one or two sunburn blisters before the age of 18, doubles the risk for developing melanoma. This is important for children. Prolonged exposure to sunlight can lead to eye problems later in life, such as cataracts. Another one is a *burning* of the eye surface, called *snow blindness,* from sunlight, which may lead to complications later in life.

Repeated exposure to the sun can cause premature aging effects. Sun-induced skin damage causes wrinkles, easy bruising, and brown spots on the skin. For all the ladies here, to maintain your pretty and young look and also for men their handsome appearance, avoid being exposed to sun. It is not too late!

Furthermore, sunburns can alter the distribution and function of disease-fighting white blood cells for up to 24 hours after exposure to the sun. Repeated overexposure to UV radiation can cause more damage to the body's immune system. Mild sunburns can directly suppress the immune functions of human skin where the sunburn occurred, even in people with dark or brown skin. This means that it applies to all of us as Indian origin.

The sting of the summer: With increased temperatures, insects become very active. Insects are nuisance and can cause many health-related problems. Common stinging insects are bees, wasps, hornets, and yellow jackets, fire ants, and so on. Over 2 million people are allergic to stinging insects. An allergic reaction to an insect sting can occur within minutes, or even hours after the sting. Stinging insects are especially attracted to sweet fragrances of perfumes, colognes, hair sprays, picnic food, open soda or beer cans and garbage areas. In order to lessen chances of being stung, one should avoid these attractants.

Symptoms of an allergic reaction are itching and swelling in areas other than the sting site, tightness in the chest and difficulty in breathing, hoarse voice, dizziness and unconsciousness, or cardiac arrest. The treatment to an allergic reaction is the use of epinephrine and other treatments. Epinephrine can be self-injected if directions are followed or administered by a doctor. Intravenous fluids, oxygen, and other treatments may also be needed. It is very important to call for medical assistance immediately, even if a person says, "I am okay," after administering epinephrine.

To remove the stinger, scrape a credit card or other object. *Do not* pinch and pull out the stinger, this will inject more venom. If breathing difficulties develop, *dial* Emergency telephone, for example, 999, 110, or 911 for prompt medical care. Wash the stung area with soap and water. If the sting is on a finger or hand, remove jewelry in case of swelling. Apply a cold compress.

Snakes, scorpions, disease-carrying mosquitoes, desert flies, and ticks are also prevalent during the summer months.

Ticks can carry a wide variety of diseases because they contract the diseases from the host they attach to. Other than medical help, there are over 150 *natural repellents* to protect from ticks. The most common ones are citronella, eucalyptus, lemon leaves, peppermint, lavender, cedar oil, canola, rosemary, pennyroyal, and Cajuput. Generally, these are considered safe to use in low dosage but their effectiveness is limited to 30 minutes.

How can we have fun in summer? Like anything else, moderation is best. Avoid those beehives and hornet nests. Keep wastes, beverages, and food in enclosed containers. To prevent tick and mosquito bites, wear protective clothing to prevent tick and mosquito bites. Wear proper sunscreens. Wear sunglasses to protect your eyes from UV light. If you are sensitive to sunburns, avoid being in the sun at the peak time. Consume a lot of water to stay hydrated. Cool down in air-cooled rooms or near fans. Wear light-colored, natural fiber clothing to help your body to repel heat absorption and cool easier. Avoid strenuous activities.

Control the heat at its source using insulating and reflective barriers on walls and windows of your home and keep blinds or curtains drawn. Keep doors and windows closed. Switch off lights behind you when done; also, switch off your car engines radiating heat.

Lower the shower temperatures. Reduce physical demands of work using mechanical assistance; for example, instead of lifting manually, use a wheeled trolley. Increase the frequency and length of rest breaks if you are doing hard work. Schedule jobs to cooler times: either early mornings or late evenings. Drink one cup of water every 20 minutes or so. Salt your food well, particularly while acclimatizing to a hot job. People, who may be on low-salt diet due to blood pressure problems, should consult their doctor. To avoid shocks due to sudden exposure to high temperature heat, acclimatize the body with slow exposure to heat. Recognize symptoms of heat stress by using a *buddy* system.

7.3.11.5 Loss Prevention during Construction

Various controls are used to prevent losses during construction activity. The standard hierarchy of controls is shown in Figure 7.93.

Eliminate the hazard when possible. This can be extremely simple. For example, modular components assembled at ground level can eliminate exposure to falls. Other hazards associated with placing conventional reinforcing steel and cast-in-place concrete can be avoided with the use of precast components manufactured in a controlled environment. No other means of control will be as effective as elimination.

Elimination is not always feasible. The next options are substitution and engineering controls. For example, a contractor might substitute scaffolding for ladders to reduce the danger of falls. After elimination and substitution, seek engineering controls to manage the hazard at its source. Isolation and guarding such as limiting access to hazardous areas and removing defective equipment from service are examples.

Next in the hierarchy are administrative controls such as a written safety and health plan that do not eliminate hazards but build the ability to recognize and avoid them. Although less effective than the previous controls, this aspect of loss prevention should not be neglected. To complement the written plan, employees should receive initial and refresher training in fall protection, electrical safety, lockout/tag out procedures, proper use of PPE, and so on. Small firms often utilize a professional safety consultant to develop and implement training. Larger firms might invest in a full-time safety officer for training and troubleshooting. General knowledge training is vital for administrative controls to be effective.

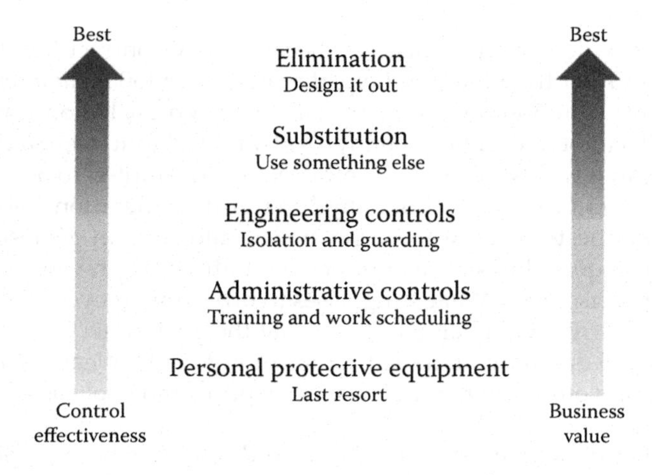

FIGURE 7.93
Hierarchy of control.

In addition to general knowledge training, employees should receive site-specific training because each site has its own peculiarities and special concerns. Site conditions are constantly changing. These changing conditions and other areas of special emphasis should be addressed in weekly jobsite safety meetings sometimes referred to as *toolbox talks.*

When other more effective controls are infeasible, the last line of defense is personal protective equipment or PPE. Types of PPE include hardhats, safety glasses, gloves, hearing protection, and respirators. These items should be furnished by the employer with strict attention to compliant utilization by all. PPE will go far in eliminating many common injuries.

7.3.11.6 OSHA

The OSHA, a division of the US Department of Labor, was formed in 1970 to assure healthful and safe working conditions. It has been greatly successful in reducing workplace accidents. However, the construction industry remains among the most dangerous industries with disproportionate injury rates. Construction occupations incur approximately 15% of all workplace fatalities, with only about 8% of the total workforce.

OSHA's construction industry regulations are found in 29 CFR Part 1926. Not surprisingly, there are major sections for fall protection (subpart M) and electrical (subpart K) since falls and electrocutions annually rank in the top three causes of construction fatalities. Because of their less specific nature, the other fatal four areas (caught in between and struck by) are addressed within multiple provisions in the OSHA 1926 standard.

To enforce its regulations, OSHA conducts jobsite investigations using compliance safety and health officer (CSHO) staff members. These investigations may be random, complaint-driven, or targeted to a special initiative such as fall protection. Like most enforcement agencies, OSHA's resources limit its ability to enforce. Only a very small percentage of active projects undergo an investigation. Given its scarce resources, OSHA continually faces the question of how to optimize its efforts for maximum worker safety.

Knowing that prevention trumps enforcement and punishment, OSHA actively promotes and certifies construction training for supervisors and workers. Although prevention is preferred and the most effective use of effort, the issuance of citations and levying punishment are frequently necessary. OSHA has the power to issue citations with penalties up to $70,000 for violations discovered during investigations or after accidents with fatalities or serious injuries.

7.3.11.7 Conclusion

Safety is always the best option. Each construction fatality is someone's son or daughter. Although economic loss can be compensated, a human being can never be replaced. Construction firms have a paramount obligation to return each worker to their family safely each and every day. Increasingly, owners are beginning to assume this obligation as well. It is now common for owners to require that contractors meet safety benchmarks for bidding prequalification, thus eliminating unsafe contractors from the bidding process.

There is no question that construction is a hazardous occupation but often safety is compromised by too much emphasis on speed of construction. This short-sighted approach neglects losses in productivity from the disruptive effects of accidents on completion schedules. Construction's macho culture, another compromising factor, creates problems by undervaluing safe work practices. It is disheartening to know that fall fatalities occur when fall protection is available but unused.

To achieve safety improvements, the industry must assume its moral obligation to workers, put aside its macho attitudes, and embrace the cultural change that can produce safety. Workers often understand how to work safely but fail to be compliant. Changing culture is a slow process and requires commitment from management, strong supervisory leadership, and positive role models. Change will be incremental and hard fought because habits are deeply embedded. Safety is no accident. *Plan. Provide. Train.*

Application of loss prevention measures discuss above can produce favorable results in the field of accident prevention in construction and process industries. The trend of this decade is toward improved safety awareness and performance in all categories of functions. Healthy action plans have proved successful to meet the above challenges and the results are encouraging.

7.3.12 Financial Management

Construction projects are mainly capital investment projects that involve three main parties:

1. Owner
2. Designer
3. Contractor

The owner is responsible for financing the project to procure the services of the designer (A/E), contractor, and other parties, and has to compensate for their services for which the owner has contracted with these parties to construct the facility/project. Other parties may include project manager, construction manager, specialists, material/equipment suppliers, and many other players as per the procurement strategy adopted by the owner. The relationship between owner and these parties is buyer–seller relationship and is bound by the contract signed between owner and these parties. The capital funding required for a construction project mainly include the following:

1. Land acquisition cost
   - There is not much to manage the finances for land acquisition cost, which are normally known as per market conditions.
2. Cost related to license fee, permits, regulatory taxes, insurance, and project-related owner's overhead cost
   - The cost toward each of these items is fixed and known and does not have major effect or variations to be managed and controlled.
3. Construction project cost
   - The overall cost expended during the life cycle of the construction project consists of
   - Designer (A/E) fee
   - Construction supervisor (consultant, project manager, and construction manager)
   - Construction cost, which normally include supply, installation/execution, testing and commissioning, and maintenance for a period of one or two years (depends on the contract)

The construction cost is generally 80%–85% of the construction project cost (construction phase). Balance is the fee for designer and supervisor.

The construction project cannot proceed unless adequate funds are arranged by the owner. These funds are arranged by either public financing or private financing.

Finance management is a process bringing together project planning, scheduling, budget, procurement, accounting, disbursement, control, auditing, reporting, and physical performance (progress) of the project, with the aim of managing project resources and achieving owner's objectives. Finance management is an important aspect of project/construction management functions. Project management has great impact on both the owner and contractor. Timely and relevant financial information provides a basis for better decisions, thus speeding the progress of the project.

7.3.12.1 Develop Financial Management Plan

Financial plan is

- Estimating the total funds required to complete the project.
- Identifying the source of funds (financial means to cover the project requirement).
- Preparing project cash flow (usage of funds).
- Ensuring that the needed finance is available at right time.
- Identifying the alternate process of securing the funds to mitigate risk or contingencies.

Most construction projects begin with recognition of new facility or refurbishment/repair of existing facility. In either case, the owner has to arrange the funds to develop/refurbish the facility. Once the need is identified, and feasibility study is conducted, project cost is estimated based on rough order of magnitude. This estimate helps the owner for initial commitment, in principle, for funding the project and managing the financial resources to build the facility. As the project phases advances to next stage, the level of accuracy is refined. Depending on the type of the project delivery system and contract methods, more accurate and a definitive estimate is evolved. At every stage, the funding requirements for construction is reviewed with the earlier stage and adjusted, if needed. The contracted amount is considered as project cost baseline for both the owner and contractor. However, both the owner and contractor have to consider reserves while estimating the financial resources. These reserves, contingency and management, are included in a cost estimate to mitigate cost risk by allowing for future situations that are difficult to predict.

Figure 7.94 illustrates the process to develop financial management plan, and Table 7.48 lists the contents of the owner's financial plan.

7.3.12.2 Control Finance

In construction projects, fund requirements for the project and the availability of funds are reviewed, compared, managed, and controlled at each stage of the project life cycle. Generally, the initial cost estimates are analogous (rough order of magnitude). These estimates are further refined as project development stages progresses and greater understanding of the project scope, schedule, quality, and resources are evolved. The contract price agreed between the owner and contractor is the cost that the owner has to pay to the contractor for the services rendered to construct/build the project/facility as specified in the contract and to complete as per agreed upon schedule and quality.

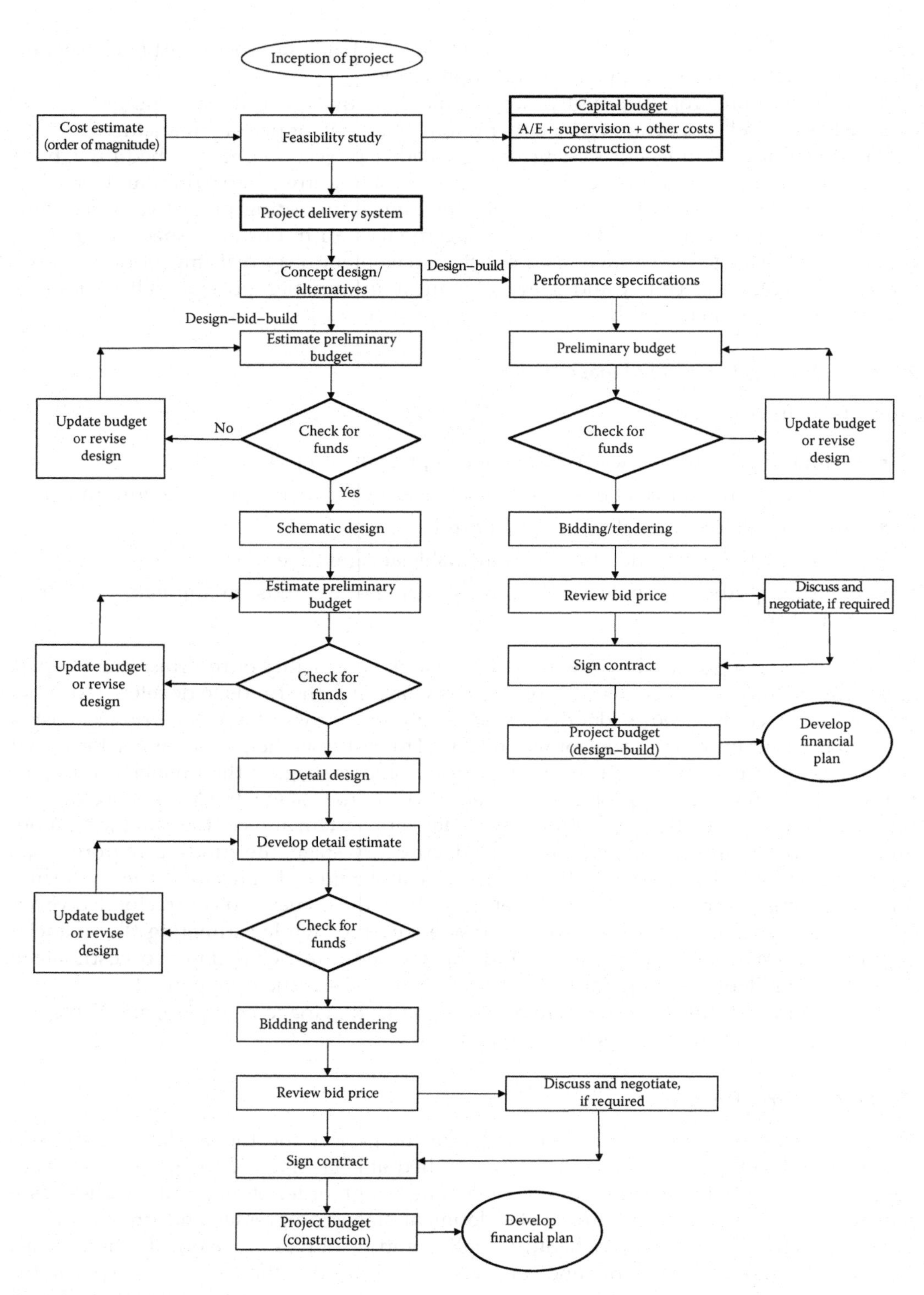

FIGURE 7.94
Process for establishing financial plan.

TABLE 7.48
Contents of Contractor's Financial Plan

Section	Topic
1	Description of project
2	Finance management organization
3	Estimated cost of project
4	Project schedule
5	Type of contracting method
6	Source of funds
7	Payment (receivable) schedule breakdown
	7.1 Advance payment
	7.2 Monthly progress payment
8	Accounts payable schedule breakdown
	8.1 Insurance and bonds
	8.2 Staff salaries
	8.3 Labor payment
	8.4 Material purchases
	8.5 Equipment purchases
	8.6 Equipment rent
	8.7 Subcontractor payment
	8.8 Consumables
	8.9 Taxes
9	Retention
10	Cash flow
11	Project risk and response strategy
12	Contingency plan to secure funds from alternate source
13	Internal control
	13.1 Accounting system
	13.2 Accounts auditing
	13.3 Track changes in budget
14	Records management
	14.1 Project expenses
	14.2 Finance closure report

The bid price (or negotiated price) is the contracted amount that is paid by the owner to the contractor as per the terms of contract. The contractor's expenses and cash flow is based on the contracted amount. The bidding price includes the following:

- Direct cost
  - Material
  - Labor
  - Equipment
- Indirect cost
  - Overhead cost
- Risks and Profit

The contractor has to plan, manage, and control all the construction activities within the contracted amount.

The owner, however, adds contingency and management reserve to the base cost estimate to mitigate the risks. The major components of a project budget for the owner are as follows:

- Base cost estimate (definitive cost at the end of detail design or bid amount).
- *Contingency*: It is the amount set aside to allow for responding to identified risks. Sometimes, it is known as known–unknown.
- *Management reserve*: It is the amount set aside to allow for future situations that are unpredictable or could not have been foreseen. This includes changes to the scope of work or unidentified risks.

Monitoring and control of project cost is essential for both parties. Normally, the S-curve is used as a reference to track the progress of the project over time. It also helps monitoring cash flow and determines the slippage in the project schedule and cost baseline. Contractor uses the cost-loaded schedule for determining the contract payments during the project against the approved progress of works. Contractor can plan the project expenses against the expected payments and has to control the finances toward

- Labor salaries
- Staff salaries
- Material purchases
- Equipment purchases
- Taxes and other regulatory expenses
- Retention

The owner has to arrange funds to pay progress payments on regular basis. Funds required to paying the changes to the scope of work or approved claims can be managed from the reserves. The contractor's monthly expenses are different to the actual progress payment expected from the owner. During the initial stage of project execution, the contractor's expenses are much higher than the progress payment. To meet these expenses, the contractor has to arrange funding from other sources. In certain projects, the contractor receives advance payment from the owner. However, it is necessary that the contractor plan and control the finances for smooth and uninterrupted execution of the project. Figure 7.95 is an example of expense (accumulated cost) and payment (accumulated net sale) schedule for a project. Plotting of cost versus sale helps the contractor control project finances.

The schedule is updated regularly to reflect ongoing changes in the progress of work and forecasted payments. Normally, it takes considerable time between the expenditure on resources and payment to be received from the owner for the work contractor has done and approved for payment by the owner. The contractor has to secure funds to cover the cash flow difference.

During execution of the construction project, the owner has to consider for payments toward unexpected changes such as

- More work done (progress of work) than planned
- Claims
- Failure to adjust extra payments
- Default from funding agency

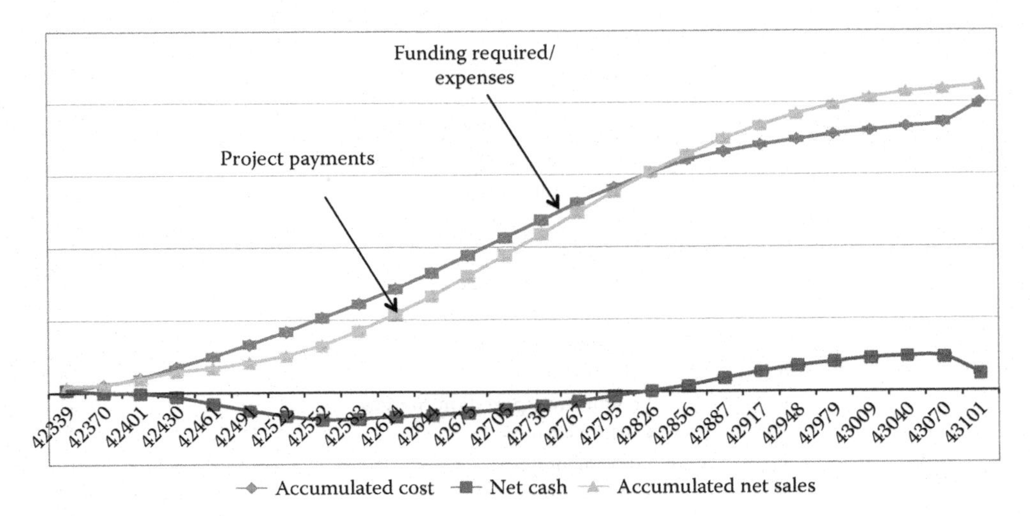

FIGURE 7.95
Contractor's cash flow.

Similarly, the contractor has to consider following:

- Slow progress of work resulting less payments than expected/planned
- Productivity is less than estimated
- Early delivery of material
- Delay in payment by the owner
- Resources to do extra (additional) work required by the owner

7.3.12.3 Administer and Record Finance

Project accounting system is required to manage the project needs and expenses and provide the financial information to all the interested stakeholders. Table 7.49 lists different logs maintained by the finance department.

7.3.13 Claim Management

Claim management is the process to mitigate the effects of the claims that occur during the construction process and resolve them quickly and effectively. If the claims are not managed effectively, it can lead to disputes ending in litigation. Even under most ideal circumstances, contract documents cannot provide full information; therefore, claims do occur in the construction projects. In construction projects, a claim is defined as seeking adjustment or consideration by one party against the other party with respect to

- Extension of time
- Scope
- Method
- Payment

TABLE 7.49
Logs by Finance Department

Section	Log
1	Incoming and outgoing correspondence
2	Progress (interim) payment
3	Subcontractor payment
4	Material purchase
5	Equipment purchase
6	Procurement (general)
7	Procurement (consumables)
8	Letter of credit
9	Freight/transportation charges
10	Custom clearance
11	Equipment rent
12	Vehicle rent
13	Insurance, bonds, and guarantees
14	Regulatory and license fee
15	Staff salaries
16	Labor salaries
17	Office rent
18	Camp rent
19	Cash in hand

The following are the three main types of claims:

1. *Contractual claims*: Claims that fall within the specific clause of the contract.
2. *Extra-contractual claims*: Claims that result from the breach of contract.
3. *Ex-gratia claims*: Claims that the contractor believes the rights on moral ground.

Table 7.50 lists major causes of claims in construction projects.

The claims in construction projects can be attributed to

1. Owner
2. Contractor

Figure 7.96 illustrates the claim management process.

7.3.13.1 Identify Claims

Claims are a common part of almost every construction project. With utmost care taken to coordinate all the activities and related contract documents, claims in construction projects may arise under any form of construction contracts. Table 7.51 illustrates the effects of claim on the construction project.

Most construction contracts impose time deadlines for submitting claims. Normally, the claims to be made within 30 days after the claim arose and also to be recorded in the monthly progress report, which the contractor submits to the owner. Claim identification

TABLE 7.50
Major Causes of Construction Claims

Sr. No.	Causes
I	*Owner Responsible*
	I-1 Delay in issuance of notice to proceed
	I-2 Delay in making the site available on time
	I-3 Different site conditions
	I-4 Project objectives are not well defined
	I-5 Inadequate specifications
	a. Design errors
	b. Omissions
	I-6 Scope of work not well defined
	I-7 Conflict between contract documents
	I-8 Change/modification of design
	I-9 Change in schedule
	I-10 Addition of work
	I-11 Omission of work
	I-12 Delay in approval of subcontractor
	I-13 Delay in approval of materials
	I-14 Delay in approval of shop drawings
	I-15 Delay in response to contractor's queries
	I-16 Delay in payment to contractor
	I-17 Lack of coordination among different contractor directly under the control of owner
	I-18 Interference and change by the owner
	I-19 Delay in owner-supplied material
	I-20 Acceleration
II	*Contractor Responsible*
	II-1 Delay to meet milestone dates
	II-2 Noncompliance with specifications
	II-3 Changes in specified process/methodology
	II-4 Substitution of material
	II-5 Noncompliance to regulatory requirements
	II-6 Charges payable to outside party due to cancellation of certain items/products
	II-7 Material not meeting the specifications
	II-8 Workmanship not to the mark
	II-9 Suspension of work
	II-10 Termination of work
III	*Miscellaneous*
	III-1 New regulations
	III-2 Weather conditions
	III-3 Unforeseen circumstances

Source: Abdul Razzak Rumane (2013), *Quality Tools for Managing Construction Projects*, CRC Press, Boca Raton, FL. Reprinted with permission from Taylor & Francis Group.

involves proper interpretation of contract requirements and gathering complete information to substantiate the claim. The identified claim needs to be submitted with all the supportive documents for justification.

7.3.13.2 Prevent Claims

Table 7.52 illustrates preventive actions to mitigate effects of claims on construction projects.

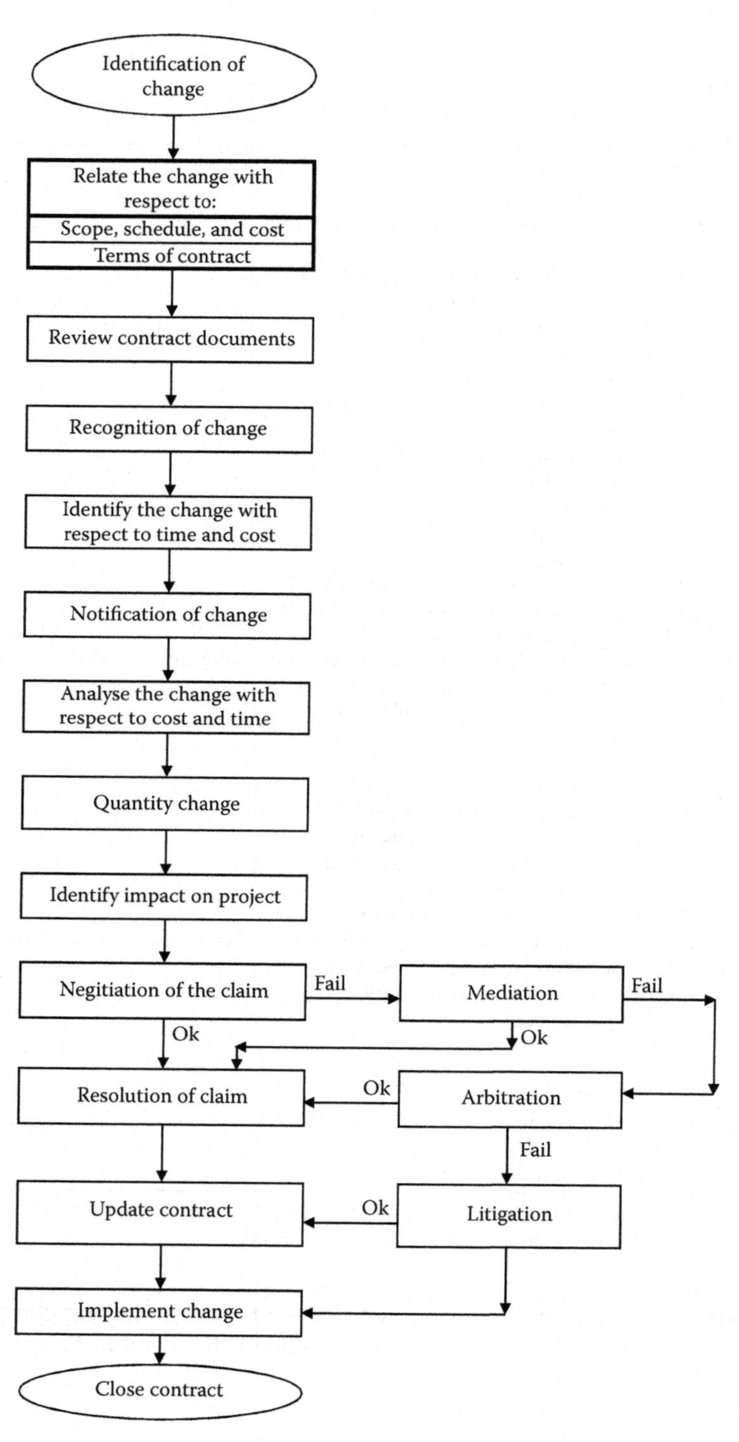

FIGURE 7.96
Claim management process.

TABLE 7.51

Effects of Claims on Construction Projects

Sr. No.	Causes		Effects on Project
I	*Owner Responsible*		
	I-1	Delay in issuance of notice to proceed	• Delay in completion of project
	I-2	Delay in making the site available on time	• Delay in completion of project
	I-3	Different site conditions	• Delay in completion of project • Change in project cost
	I-4	Project objectives not well defined	• Delay in completion of project • Change in project cost
	I-5	Inadequate specifications design errors, omissions	• Delay in completion of project • Change in project cost
	I-6	Scope of work not well defined	• Delay in completion of project • Changes in project cost
	I-7	Conflict between contract documents	• Changes in schedule
	I-8	Change/modification of design	• Change in project cost
	I-9	Change in schedule	• Change in project cost
	I-10	Addition of work	• Change in project cost • Change in completion of project
	I-11	Omission of work	• Change in project cost
	I-12	Delay in approval of subcontractor	• Delay in completion of project
	I-13	Delay in approval of materials	• Delay in completion of project
	I-14	Delay in approval of shop drawings	• Delay in completion of project
	I-15	Delay in response to contractor's queries	• Delay in completion of project
	I-16	Delay in payment to contractor	• Delay in completion of project
	I-17	Lack of coordination among different contractor directly under the control of owner	• Changes in schedule
	I-18	Interference and change by the owner	• Delay in completion of project. • Change in project cost
	I-19	Delay in owner-supplied material	• Delay in completion of project
	I-20	Acceleration	• Change in project cost
II	*Contractor Responsible*		
	II-1	Delay to meet milestone dates	• Delay in completion of project
	II-2	Noncompliance with specifications	• Changes in project cost
	II-3	Changes in specified Process/methodology	• Changes in project cost • Changes in project schedule
	II-4	Substitution of material	• Changes in project cost
	II-5	Noncompliance to regulatory requirements	• Changes in project cost
	II-6	Charges payable to outside party due to cancellation of certain items/products	• Changes in project cost
	II-7	Material not meeting the specifications	• Changes in project cost
	II-8	Workmanship not to the mark	• Changes in project cost
	II-9	Suspension of work	• Changes in project cost
	II-10	Termination of work	• Changes in project cost

(*Continued*)

TABLE 7.51 (*Continued*)

Effects of Claims on Construction Projects

Sr. No.		Causes	Effects on Project
III	*Miscellaneous*		
	III-1	New regulations	• Delay in project completion • Possible changes in cost
	III-2	Weather conditions	• Delay in project completion
	III-3	Unforeseen circumstances	• Delay in project completion

Source: Abdul Razzak Rumane (2013), *Quality Tools for Managing Construction Projects*, CRC Press, Boca Raton, FL. Reprinted with permission from Taylor & Francis Group.

TABLE 7.52

Preventive Actions to Mitigate Effects of Claims on Construction Projects

Sr. No.		Causes	Preventive Actions
I	*Owner Responsible*		
	I-1	Delay in issuance of notice to proceed	• Complete all the required documentation, permits before signing of contract
	I-2	Delay in making the site available on time	• Obtain legal documents and title deeds in time
	I-3	Different site conditions	• Conduct predesign study/survey through specialist consultant/contractor • Collect historical data prior to start of design • Proper site investigations and if any existing utility services under the site • Ensure that the contractor is aware of geographic/geological conditions of the area.
	I-4	Project objectives not well defined	• Business case to be properly defined • Designer to collect all required data and clarification from the owner
	I-5	Inadequate specifications, design errors, omissions	• Review of design drawings by the designer • Review of specifications by the designer • Designer to ensure design drawings are reviewed and coordinated • Designer to ensure clear and complete Bill of Quantities • Specifications and documents are properly prepared as per standard format • Quality Management system during design phase • Allow reasonable time to the designer to prepare complete and clear drawings and specifications
	I-6	Scope of work not well defined	• Construction documents are accurate and all the items/activities are properly defined • Review all the documents • Prepare clear and unambiguous documents
	I-7	Conflict between contract documents	• Review documents and design drawing and coordinate with all the trades
	I-8	Change/modification of design	• Resolve issue without major effects on schedule

(*Continued*)

TABLE 7.52 (*Continued*)

Preventive Actions to Mitigate Effects of Claims on Construction Projects

Sr. No.		Causes	Preventive Actions
	I-9	Improper schedule	• Designer to check schedule for all activities and precedence • Check for constraints and consistency
	I-10	Change in schedule	• Keep to contracted schedule for completion of project
	I-11	Addition of work	• Review scope of work to ensure all the requirements are included • Establish proper mechanism to process changes
	I-12	Omission of work	• Negotiate and resolve without any impact on the project
	I-13	Delay in approval of subcontractor	• Take action within specified time to response the transmittals • Ensure that review period for transmittals is appropriate
	I-14	Delay in approval of materials	• Take action within specified time to response the transmittals • Ensure that review period for transmittals is appropriate
	I-15	Delay in approval of shop drawings	• Take action within specified time to response the transmittals • Ensure that review period for transmittals is appropriate
	I-16	Delay in response to the contractor's queries	• Take action within specified time to response the transmittals • Ensure that review period for transmittals is appropriate
	I-17	Delay in payment to the contractor	• Arrange funds to pay the progress payments as per specified time in the contract documents
	I-18	Lack of coordination among different contractor directly under the control of owner	• Division of works and packages are properly coordinated and clearly identified for each contractor • Develop cooperative and problem solving attitudes for successful completion of project
	I-19	Interference and change by the owner	• Avoid interference and define clearly the roles and responsibilities of each part. • Follow change order procedure specified in the contract documents
	I-20	Delay in owner-supplied material	• Owner to ensure timely delivery of material suitable for installation
	I-21	Acceleration	• Study contractual consequences before issuance of instruction to accelerate or fast track the activities
II	*Contractor Responsible*		
	II-1	Delay to meet milestone dates	• Check the milestone activities and perform as included in the schedule • Identify constraints to complete milestone activities • Establish strategy to deal with tight schedule
	II-2	Noncompliance with specifications	• Ensure compliance to the specifications • Establish proper quality management system
	II-3	Bill of quantities not matching with design drawings	• Contractor to check quantities while bidding
	II-4	Changes in specified process/methodology	• Follow specified method • Request for change if recommended by the manufacturer • Submit and get approval on method of statement for installation of works
	II-5	Substitution of material	• Follow relevant sections of contract specifications
	II-6	Noncompliance to regulatory requirements	• Study and keep track of regulatory requirement for execution of works

(*Continued*)

TABLE 7.52 (*Continued*)

Preventive Actions to Mitigate Effects of Claims on Construction Projects

Sr. No.		Causes	Preventive Actions
	II-7	Charges payable to outside party due to cancellation of certain items/products	Settle the matter amicably
	II-8	Material not meeting the specifications	Submit request for substitution and follow the contract requirements for substitution
	II-9	Workmanship not to the mark	Use skilled manpower
	II-10	Suspension of work	Follow contract conditions
	II-11	Termination of work	Follow contract conditions
III	*Miscellaneous*		
	III-1	New regulations	Follow change order procedures for such conditions
	III-2	Weather conditions	Follow change order procedures for such conditions
	III-3	Unforeseen circumstances	Follow change order procedures for such conditions

Source: Abdul Razzak Rumane (2013), *Quality Tools for Managing Construction Projects*, CRC Press, Boca Raton, FL. Reprinted with permission from Taylor & Francis Group.

7.3.13.3 Resolve Claims

All the claims are to be resolved and contract to be closed. The method of resolution depends on the type of claim, size of the claim, severity of the claim, and effects and consequences of the claim on the project. The submitted claims are to be resolved in a justifiable manner. The following are the methods used to resolve the claim:

- Negotiation
- Mediation
- Arbitration
- Litigation

In all these cases, a comprehensive analysis is necessary to come to an amicable solution. Table 7.53 lists the documents required for analysis of construction claims.

7.4 Construction Project Phases

Most construction projects are custom oriented having specific need and customized design. Construction projects are constantly increasing in technological complexity and the relationship and the contractual grouping of those who are involved are also complex and contractually varied. There are innumerable processes that make up the construction project. It is always the owner's desire that his/her project should be unique. Furthermore, it is the owner's goal and objective that the facility is completed on time and within agreed upon budget. Systems engineering approach to construction projects help understand the

TABLE 7.53

Documents Required for Analysis of Construction Claims

Sr. No.	Name of Document
1	Contract documents
	1.1 Tender documents (related to claim, if required)
	1.2 Conditions of contract (general conditions and specific conditions)
	1.3 Particular specifications
2	Drawings
	2.1 Contract drawings
	2.2 Approved shop drawings
3	Construction schedule
	3.1 Contract schedule
	3.2 Approved construction schedule
4	Reports
	4.1 Daily report
	4.2 Monthly report
	4.3 Test reports
	4.4 Noncompliance report
	4.5 Material delivered at site
5	Minutes of Meetings
	5.1 Progress meetings
	5.2 Coordination meetings
	5.3 Safety meetings
	5.4 Quality meetings
6	Submittal Logs
	6.1 Material, sample
	6.2 Shop drawings
	6.3 Request for information
	6.4 Request for substitution (alternative)
7	Job site instruction
8	Variation order
9	Site progress records
	9.1 Photographs
	9.2 Videos
10	Payment request
11	Checklists
12	Correspondence with regulatory authorities

entire process of project management and to manage and control its activities at different levels of various phases to ensure timely completion of the project with economical use of resources to make the construction project most qualitative, competitive, and economical.

Although it is difficult to generalize the number of phases of project life cycle, the number of phases depends on considering the innumerable processes that make up the construction process, the technologies, processes, and the complexity of construction projects. The duration of each phase may vary from project to project.

Generally, construction projects have the following five most common phases:

1. Conceptual design
2. Schematic design

3. Detail design
4. Construction
5. Testing, commissioning, and handover

As per *Construction Extension PMBOK® Guide Third Edition* (p. 14), "Most Construction projects can be viewed in Five Phases, although they are sometimes shortened to four, each one of these phases can be treated like a project itself, with all of the process groups operating as they do for the overall project. These phases are concept, planning (and development), detail design, construction, and start-up and turnover."

However, in major projects, preparation of construction (contract) documents, its review and approval, and the process of bidding and tendering take considerable time. Thus, for a major construction project, it is ideal to divide the construction project life cycle into the following seven phases:

1. Conceptual design
2. Schematic design
3. Detail design
4. Construction (contract) document
5. Bidding and tendering
6. Construction
7. Testing, commissioning, and handover

The methodology applied in this book is based on seven phases. These phases are listed in Table 7.54.

Each of these phases are treated as a project itself with all the five process groups operating, as they do for the overall project, and each phase is composed of activities, elements having functional relationship to achieve a common objective for useful purpose.

Figure 7.97 illustrates the construction project life cycle for design–bid–build type of project delivery system, whereas Figure 7.98 illustrates construction project life cycle for design–build type of project delivery system.

The design–bid–build type of project delivery system has a sequential relationship, whereas in the case of design–build type, the construction phase commences while the design development process is still under progress. The construction phase overlaps the design phases.

TABLE 7.54

Phases in Construction Project Life Cycle

Phase	*Construction-Extension-PMBOK® Guide-Third Edition* Methodology	Phase	Most Common Phases for Major Projects (Methodology in This Book)
1	Concept	1	Concept design
2	Planning (and development)	2	Schematic design
3	Detail design	3	Design development
		4	Construction documents
		5	Bidding and tendering
4	Construction	6	Construction
5	Startup and turnover	7	Testing, commissioning, and handover

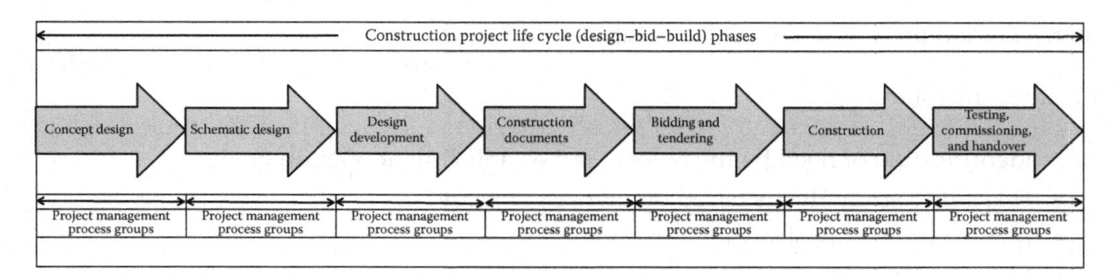

FIGURE 7.97
Construction project life cycle (design–bid–build) phases.

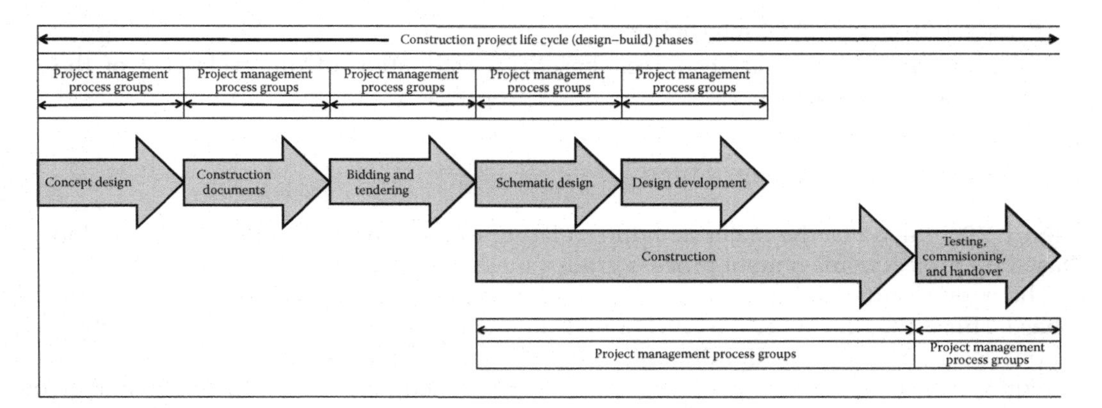

FIGURE 7.98
Construction project life cycle (design–build) phases.

In agency construction management type of management system, the construction manager is normally commissioned by the project owner once project objectives and goals are established by the owner/client. The construction manager takes up the responsibility to act as an advisor to the client and manage the project over the entire life cycle of the project, or specific phase/stage of the project.

7.4.1 Conceptual Design Phase

Conceptual design is the first phase of construction project life cycle. This phase/stage is concerned with preparation of concept design. Conceptual design is often viewed as most critical to achieving outstanding project performance. Most important decisions about planning, organization, and type of contract take place during this phase. This phase is also called as follows:

- Programming phase
- Definition phase
- Pre-design phase

The conceptual design phase commences once the need is recognized. In this phase, the idea is conceived and given initial assessment. During the conceptual phase, project goals and objective are established, preliminary project plan is prepared, alternatives are analyzed and preferred alternative is selected, the environment is examined, forecasts are prepared, and cost and time objectives of the project are performed. Conceptual phase includes the following:

- Identification of need by the owner, and establishment of main goals
- Feasibility study that is based on owner's objectives
- Identification of project delivery system
- Identification of project team by selecting other members and allocation of responsibilities
- Identification of alternatives
- Selection of preferred alternative
- Financial implications, resources, based on estimation of life cycle cost of the favorable alternative
- Time schedule
- Development of concept design

Figure 7.99 illustrates the major activities relating to the concept design phase developed based on project management process groups methodology.

The most significant impacts in the quality of the project begin during the conceptual phase. This is the time when specifications, statement of work, contractual agreements, and initial design are developed. Initial planning has the greatest impact on a project because it requires the commitment of processes, resources schedules, and budgets. A small error that is allowed to stay in the plan is magnified several times through subsequent documents that are second or third in the hierarchy. Figure 7.100 illustrates a logic flow diagram for the conceptual design phase.

7.4.1.1 Development of Project Charter

The project charter defines the objectives, scope, deliverables, and overall approach for the new project. In construction projects, the project charter is the document prepared by the owner/client or by the project manager on behalf of the owner describing the project objectives and requirements to deliver the project. Project charter is also known as follows:

- Client brief
- Definitive project brief
- TOR

TOR is developed based on the need statement, which is further analyzed for feasibility of the project. The outcome of feasibility report is used to prepare project brief (project goals and Objectives). Alternatives are identified and the preferred alternative is selected after evaluating/analyzing them. Thereafter, project delivery system is selected based on the complexity of project and the strategy to be followed by the owner. TOR is developed to proceed with design and construction of the project.

7.4.1.1.1 Identification of Need

Most construction projects begin with recognition of new facility. The need of the project is created by the owner. The owner of the facility could be an individual, a public/private

Conceptual design phase

Management processes	Project management process groups				
	Initiating process	Planning process	Execution process	Monitoring and controlling process	Closing process
Integration management	Problem statement/need statement	Preliminary project management plan	Development of alternatives	Monitor concept design progress	Concept design deliverables
	Feasibility study		Development of concept design	Manage owner's need and changes	
	Project goals and objectives		Implement changes	Review of concept design	
	Project terms of reference		Concept drawings, reports, models		
Stakeholder management	Project delivery system	Responsibility matrix	Design performance report	Design progress	
	Project team members	Establish stakeholders requirements		Design status	
	Regulatory authorities				
Scope management		Identification of alternatives		Evaluation and approval of alternative	
		Concept design scope		Aesthetics, constructability	
		Data collection		Sustainability,economy, and environmental compatibility	
		Owner's requirements		Authorities approval	
		Design deliverables		Compliance to owner's need	
				Stakeholder's approval of concept design	
Schedule management		Preliminary schedule		Concept design submission schedule	
				Approval of project schedule	
Cost management		Conceptual estimate		Control project cost	
Quality management		Quality codes and standards, regulatory requirements	Design compliance to codes, standards, and regulatory requirements	Conformance to technical and functional capability and energy efficiency,	
Resource management		Assign project design team	Manage team members from different disciplines	Performance of team members	Assign new phase/project
		Roles and responsibilities of team members			
Communication management		Design progress information	Liaison and coordination with all parties	Design status information	
			Coordination meetings		
Risk management		Management of design risk		Control design risk	
Contract management		Project delivery system	Design to comply contract type/pricing	Administer project delivery system requirements	Demobilize team members
		Contracting system			
		Selection of consultant for feasibility study			
		Selection of design team (consultant)		Check for contracting system	
HSE management		Safetyconsideration in design	HSE consideration is design	Check for regulatory requirements	
Financial management		Designer/consultant payment		Cost effectiveness over the project life cycle	Designer/consultant payment
Claim management		Management of owner's need changes		Control changes/claims	Settle claims by designer

FIGURE 7.99
Major activities relating to conceptual design processes. *Note*: These activities may not be strictly sequential; however, the breakdown allows implementation of project management function more effective and easily manageable at different stages of the project phase.

sector company, or a governmental agency. The need should be based on real (perceived) requirements or deficiency. The owner's need must be well defined by describing the minimum requirements of quality, performance, project completion date, and approved budget for the project. Tables 3.2 and 3.3 list major points to be considered for need analysis of a construction project and need statement, respectively.

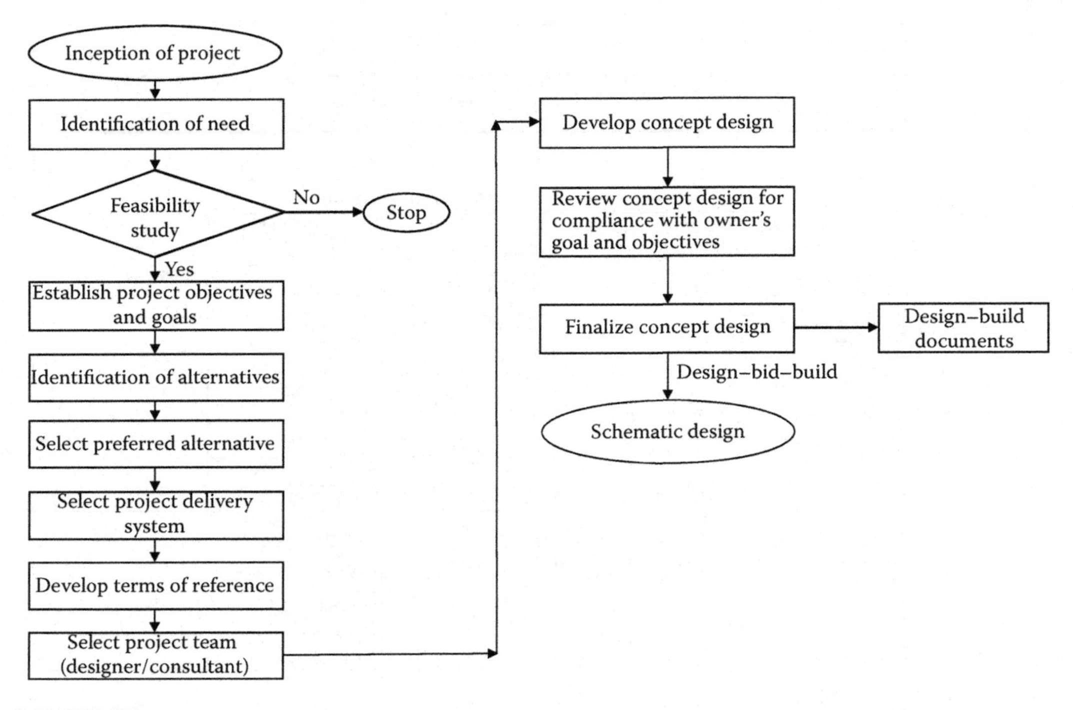

FIGURE 7.100
Logic flow process for conceptual design phase.

7.4.1.1.2 Feasibility Study

Feasibility study is defined as an evaluation or analysis of the potential impact of the identified need of the proposed project. Feasibility study takes its starting point from the output of project identification need. Once the owner's need is identified, the traditional approach is pursued through a feasibility study or an economical appraisal of owner needs for benefits. Feasibility study is performed to more clearly define the viability and form of the project that will produce the best or most profitable results. The feasibility study assists decision makers (investors/owners/clients) in determining whether to implement the project or not. The feasibility study may be conducted in-house if capability exists. However, the services of a specialist or an architect/engineer (A/E) are usually commissioned by the owner/client to perform such study. Since the feasibility study stage is very crucial stage, in which all kinds of professionals and specialists are required to bring many kinds of knowledge and experience into broad-ranging evaluation of feasibility, it is required to engage a firm having expertise in the related fields. The feasibility study establishes the broad objectives for the project and so exerts an influence throughout subsequent stages. The successful completion of the feasibility study marks the first of several transition milestones and is therefore the most important to determine whether to implement a particular project or program or not. The feasibility study decides the possible design approaches that can be pursued to meet the need.

After completion of the feasibility study, it is possible to define the project objectives. The project objective definition will usually include the following information:

1. Project scope and project deliverables
2. Preliminary project schedule

3. Preliminary project budget
4. Specific quality criteria the deliverables must meet
5. Type of contract to be employed
6. Design requirements
7. Regulatory requirements
8. Potential project risks
9. Environmental considerations
10. Logistic requirements

Completion of feasibility study is the starting point for the project by the owner/client, if the project/objective is approved. Thus, the project goals and objectives are established.

7.4.1.1.3 Identification of Alternatives

Once the owner/client defines the project objectives, a project team (in-house or outside agency) is selected to start development of alternatives. In certain cases, the owner assigns the designer (consultant) to develop conceptual alternatives, evaluation of conceptual alternatives and selection of preferred alternative in consultation with the owner and is included in the TOR.

The project goals and objectives serve as a guide for development of alternatives. The team develops several alternative schemes and solutions. Each alternative is based on the predetermined set of performance measures to meet the owner's requirements. In case of construction projects, it is mainly the extensive review of development options that are discussed between the owner and the team members. The team provides engineering advice to the owner to enable him or her assess its feasibility and relative merits of various alternative schemes to meet his or her requirements.

7.4.1.1.4 Analyze Alternatives

Quantitative comparison and evaluation of identified alternatives is carried out by considering the advantages and disadvantages of each item systematically. Social, economic, and environmental impacts; functional capability; safety; and reliability should be considered while development of alternatives. Each alternative is compared by considering the advantages and disadvantages of each systematically to meet a predetermined set of performance measures and owner's requirements. The team makes a brief presentation to the owner, and the project is selected based on preferred conceptual alternatives.

The following elements are considered to analyze and evaluate each of the identified alternatives:

1. Suitability to the purpose and objectives
2. Performance parameters
3. Economy
4. Cost efficiency
5. Life cycle costing
6. Sustainable (environmental, social, and economical)
7. Environmental impact
8. Environment preferred material and products

9. Physical properties, thermal comfort, insulation, and fire resistance
10. Utilization of space
11. Accessibility
12. Ventilation
13. Indoor air quality
14. Water efficient
15. Energy consumption and energy saving measures, use of renewable energy, alternate energy
16. Daylighting
17. High-performance lighting
18. Green building concept
19. Aesthetic
20. Safety and security
21. Statutory/regulatory requirements
22. Codes and standards
23. Any other critical issues

7.4.1.1.5 Select Preferred Alternative

Based on the analysis of identified alternatives, the preferred alternative that satisfies the project goals and objectives is selected.

With approval of the preferred alternative by the owner, the project proceeds toward the next stage of the project development process.

7.4.1.1.6 Develop TOR

A TOR is a written document stating what will be done by the designer (consultant) to develop the project/facility. It is issued to the designer (consultant) by the owner to develop project design and construction documents. The TOR generally requires the designer (consultant) to perform the following:

- Development of concept design
- Preparation of detailed design and contract documents for tendering purpose
- Preparation of preliminary design, budget, and schedule and obtaining approvals from authorities

Table 4.9 illustrates an example of contents of TOR documents for a building construction project.

7.4.1.2 Develop Preliminary Project Management Plan

The designer prepares the preliminary project management plan. It should document key management and oversight tasks and will be updated throughout the project. The plan should include the definition of the owner's goals and objectives, technical requirements, schedules, budget, resources, quality plan and standards, risk plan, and financial plan.

7.4.1.3 Identify Project Stakeholder

The following stakeholders have direct involvement in the project during conceptual design:

- Owner
- Consultant
- Designer
- Regulatory authorities
- Project manager/agency construction manager (as applicable)

7.4.1.3.1 Determine Project Delivery System

During this phase, the owner has to decide the procurement method and contract strategy. The selection of project delivery system mainly depends on the project size, complexity of the project, innovation, uncertainty, urgency, and the degree of involvement of the owner. Section 2.4 details the selection of project delivery system.

7.4.1.3.2 Select Design Team

The owner is the first member of the project team. The relationship and responsibilities with other team members depend upon the type of deliverable system the owner prefers.

For design–bid–build type of contract system, the first thing the owner has to do is to select design professionals/consultants. Generally, the owner selects designer/consultant on the basis of qualifications (qualifications based system) and prefers to use the one he or she has used before and with which he or she has had satisfactory results. Figure 7.101 illustrates the logic flow diagram for selection of a designer (A/E).

Upon selection of the designer (A/E), an agreement is made between the owner and designer (A/E). The following are typical contents of the contract between the owner and designer (A/E):

1. Project definition
2. Project schedule
3. Scope of work
4. Design deliverables
5. Owner responsibilities
6. Fees for the services
7. Variation order
8. Penalty for delay
9. Liability toward design errors
10. Insurance
11. Arbitration/dispute resolutions
12. Taxes
13. Appointment of subconsultant
14. Selection of team members

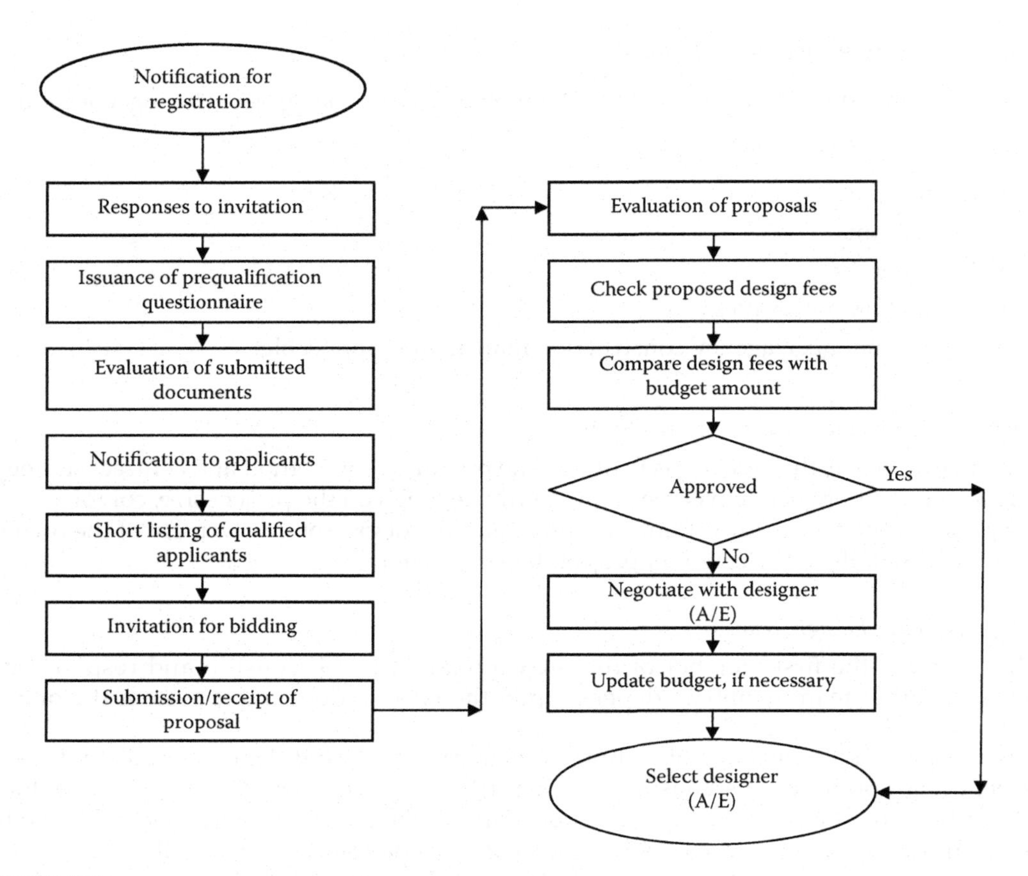

FIGURE 7.101
Logic flow diagram for selection of designer (A/E).

15. Duties (responsibilities)
16. Compliance to authorities' requirements
17. Suspension of contract
18. Termination
19. Glossary

7.4.1.3.3 Establish Stakeholder's Requirement

A distribution matrix listing key project activities and stakeholders who are involved to advice, monitor, review, and/or approve various activities, documents, or deliverables is prepared. All related documents are distributed to these stakeholders for appropriate action.

7.4.1.3.4 Develop Responsibility Matrix

Upon signing of the contract with the client to design the project and offer other services, the designer (consultant) assigns the project manager to execute the contract and is responsible to manage the development of design and contract documents to meet the client need and objectives. The project manager coordinates with other departments and

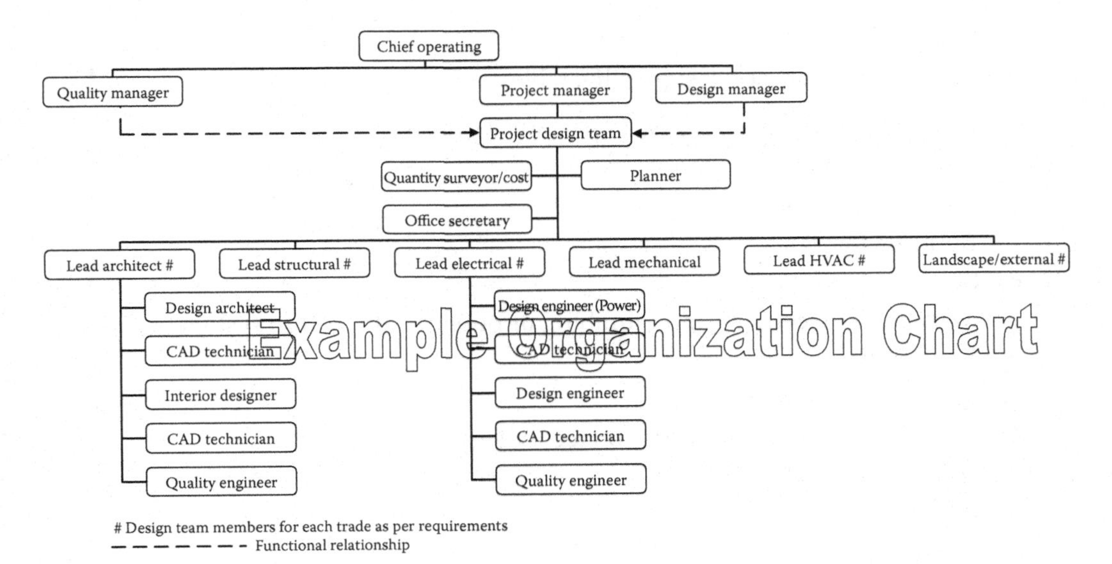

FIGURE 7.102
Project design team organizational chart. (Abdul Razzak Rumane (2013), *Quality Tools for Managing Construction Projects*, CRC Press, Boca Raton, FL. Reprinted with permission from Taylor & Francis Group.)

acquires design team members to develop project design. A project team leader along with respective design engineer(s), quality engineer, and AutoCAD technician from each trade is assigned to work for the project. The project team is briefed by the project manager about the project objectives and the roles and responsibilities and authorities of each team member. Quality manager also joins the project team to ensure compliance with organization's quality management system. Figure 7.102 illustrates the project design organization structure and Table 7.55 illustrates the responsibilities of various participants during conceptual phase of the construction project life cycle.

7.4.1.4 Develop Concept Design Scope

The TOR is a guideline to prepare concept design deliverables.

TABLE 7.55
Responsibilities of Various Participants (Design–Bid–Build Type of Contracts) during the Conceptual Design Phase

	Responsibilities		
Phase	**Owner**	**Designer**	**Regulatory**
Conceptual design	• Identification of need • Selection of alternative • Selection of team members • Approval of time schedule • Approval of budget • TOR	• Feasibility • Development of alternatives • Schedule • Cost estimates • Development of concept design	• Approval of project submittals

7.4.1.4.1 Identify Concept Design Deliverables

The following are the concept design deliverables.

1. Concept design report (narrative/descriptive report)
   a. Space program
   b. Building exterior
   c. Building interior
   d. Structural system
   e. Mechanical, electrical, and plumbing or public health (MEP) system
   f. Fire safety
   g. Conveyance system
   h. Information and communication technology
   i. Landscape
   j. HSE issues
   k. Traffic plan (if applicable)
2. Drawings
   a. Overall site plan
   b. Floor plan
   c. Elevations
   d. Sketches
   e. Sections (indicative to illustrate overall concept)
3. Data collection, studies reports
4. Existing site conditions
5. Concept schedule of material and finishes
6. Lighting/daylight studies
7. Leadership in Energy and Environmental Design (LEED) requirements
8. Facility management requirements
9. Preliminary project schedule
10. Preliminary cost estimate
11. Regulatory approvals
12. Models
13. Evaluation criteria of the alternatives (if part of TOR)

7.4.1.5 Develop Concept Design

Selected preferred alternative is the base for development of concept design. While developing the concept design, the designer must consider following:

1. Project goals
2. Usage
3. Technical and functional capability
4. Aesthetics

5. Constructability
6. Sustainability (environmental, social, and economical)
7. Health and safety
8. Reliability
9. Environmental compatibility
10. Fire protection measures
11. Provision for facility management requirements
12. Supportability during maintenance/maintainability
13. Cost-effective over the entire life cycle (economy)

It is the designer's responsibility to pay greater attention to improving environment and of achieving sustainable development. Numerous UN meetings (such as the first UN conference on human development held in Stockholm in 1972; the 1992 Earth Summit in Rio-de-Janeiro; 2002 Earth Summit in Johannesburg; 2005 World Summit) and Brundtland Commission on Environment and Development in 1987, emphasizes upon *sustainability*, whether it be a sustainable environment, sustainable economic development, sustainable agricultural and rural development, and so on. Accordingly, the designer has to address environmental and social issues and comply with the local environmental protection codes. A number of tools and rating systems have been created by LEED (United States), BREEAM (United Kingdom), and HQE (France) in order to assess and compare the environmental performance of the buildings. These initiatives have a great impact on how the buildings are designed, constructed, and maintained. Therefore, while development of building projects, the following need to be considered:

1. Accretion with the natural environment by using natural resources such as sun light, solar energy, and ventilation configuration
2. Energy conservation by energy efficient measures to diminish energy consumption (energy efficient construction)
3. Environmental protection to reduce environmental impact
4. Use material having harmony with environment
5. Aesthetic harmony between a structure and its surrounding nature and built environment
6. Good air quality
7. Comfortable temperature
8. Comfortable lighting
9. Comfortable sound
10. Clean water
11. Less water consumption
12. Integration with social and cultural environments

During the design stage, the designer must work jointly with the owner to develop details on the owner's need and objectives and give due consideration to each part of requirements. The owner on his part should ensure that the project objectives are

- Specific
- Measurable

- Agreed upon by all the team members
- Realistic
- Possible to complete within definite time
- Within the budget

7.4.1.5.1 Collect Data/Information

The purpose of data collection is to gather all the relevant information on existing conditions, both on project site and surrounding area that will impact the planning and design of the project. The data related to the following major elements is required to be collected by the designer:

1. Certificate of title
   a. Site legalization
   b. Historical records
2. Topographical survey
   a. Location plan
   b. Site visits
   c. Site coordinates
   d. Photographs
3. Geotechnical investigations
4. Field and laboratory test of soil and soil profile
5. Existing structures in/under the project site
6. Existing utilities/services passing through the project site
7. Existing roads, structure surrounding the project site
8. Shoring and underpinning requirements with respect to adjacent area/structure
9. Requirements to protect neighboring area/facility
10. Environmental studies
11. Daylighting requirements
12. Wind load, seismic load, dead load, and live load
13. Site access/traffic studies
14. Applicable codes, standards, and regulatory requirements
15. Usage and space program
16. Design protocol
17. Scope of work/client requirements

7.4.1.5.2 Collect Owner's Requirements

Based on the scope of work/requirement mentioned in the TOR, a detailed list of requirements is prepared by the designer (consultant). The project design is developed taking into consideration the owner's requirements and all other design criteria.

Table 7.56 is a checklist for the owner's preferred requirements related to architectural works.

Table 7.57 is a checklist for the owner's preferred requirements related to structural works.

TABLE 7.56

Checklist for Owner Requirements (Architectural)

Sr. No.	Description	Yes	No	Notes
1	Building Type			
	a) Commercial			
	b) Residential			
	c) Public			
	d) Other Type (specify)			
2	Multistory			
3	Basement			
	Floor number:			
	a) Mezzanine			
	b) Ground			
	c) Typical number (...)			
4	Type of Façade System			
	a. Curtain wall - Glass			
	b. Fair face-Concrete			
	c. Painting-Plaster			
	d. Stone			
	e. Ceramic			
	f. Aluminum composite panels			
	g. Precast panels			
	h. Brick			
	i. Glass fiber reinforced gypsum			
	j. Louvers			
	k. Façade cleaning system			
5	Type of roof			
	a. Space frame			
	b. Brick			
	c. Sky light			
	d. Louvers			
	e. Canopies			
6	Type of Partition			
	a. Block work			
	b. Concrete			
	c. Gypsum board			
	d. Demountable partition			
	e. Operable partition			
	f. Glass partition			
	g. Wood partition			
	h. Sandwich panels			
7	Type of Doors			
	a. Wooden			
	b. Steel			
	c. Aluminum			
	d. Glazing			
	e. Fire-rated			
	f. Rolling shutter			
	g. Sliding			
8	Type of Windows			
	a. Wooden			
	b. Aluminium			
	c. Metal			

(*Continued*)

TABLE 7.56 (*Continued*)

Checklist for Owner Requirements (Architectural)

Sr. No.	Description	Yes	No	Notes
9	Wall Finishes			
	a. Wood cladding			
	b. Painting plaster			
	c. Stone			
	d. Ceramic			
	e. Curtain glass			
	f. Mirror			
	g. Wallpaper			
	h. Polished plaster			
10	Type of ceiling			
	a. Acoustic			
	b. Wooden			
	c. Metallic			
	d. Plain gypsum panels			
	e. Curve light			
	f. Metal tiles			
	g. Gypsum tiles			
	h. Perforated tiles			
	i. Glass fiber reinforced gypsum tiles			
	j. Spider glass			
11	Floor finishes			
	Stone			
	Carpet			
	Wood parquet			
	Ceramic			
	Epoxy			
	Terrazzo			
	Rubber			
	Raised floor			
12	Toilet			
	Wash basin			
	Solid surface			
	Mirror			
	WC			
	Accessories			
13	Bathroom			
	Wash basin			
	Mirror			
	Solid surface			
	WC			
	Shower			
	Accessories			
14	Kitchen			
	Rough carpentry			
	Solid surface			
	Sink type			
	Accessories			
15	Staircase			
16	Elevator			

(*Continued*)

TABLE 7.56 (*Continued*)

Checklist for Owner Requirements (Architectural)

Sr. No.	Description	Yes	No	Notes
17	Entrance			
	Front			
	Rear			
	Service			
18	Terraces			
	Indoor			
	Outdoor			
19	Landscape garden			
20	Parking			
	Car shade			
	Open area			
21	Swimming pool			
	Indoor			
	Outdoor			
22	Other services			

Source: Abdul Razzak Rumane. (2013). *Quality Tools for Managing Construction Projects*. Reprinted with permission of Taylor & Francis Group.

TABLE 7.57

Checklist for Owner's Preferred Requirements (Structural)

Sr. No.	Description	Yes	No	Notes
1	Building Type			
	a) Commercial			
	b) Residential			
	c) Other type (specify)			
2	Multistory			
3	Basement			
4	Type of Construction			
	a) Reinforced cast in situ			
	b) Reinforced precast			
	c) Steel structure			
5	Type of roof			
	a) Concrete			
	b) Steel			
	c) Other type (specify)			
6	Type of basement			
	a) Heated and cooled			
	b) Unconditioned			
7	Type of partition walls			
	a) Block			
	b) Concrete			
	c) Other type			

Source: Abdul Razzak Rumane (2013). *Quality Tools for Managing Construction Projects*. CRC Press, Boca Raton, FL. Reprinted with permission from Taylor & Francis Group.

Table 7.58 is a checklist for the owner's preferred requirements related to mechanical works.

Table 7.59 is a checklist for the owner's preferred requirements related to heating ventilation and cooling (HVAC) works.

Table 7.60 is a checklist for the owner's preferred requirements related to electrical works.

7.4.1.5.3 Prepare Concept Design

The designer can use techniques such as quality function deployment to translate the owner's need into technical specifications. Figure 7.103 illustrates house of quality for college building project based on certain specific requirements by the customer.

While preparing concept design, the designer has to consider various available options/systems to ensure project economy, performance, and operation. The concept design should be suitable for further development.

7.4.1.5.3.1 Concept Design Drawings Table 7.61 lists the elements that comprise the concept design drawings of various trades (disciplines).

TABLE 7.58

Checklist for Owner' Preferred Requirements (Mechanical)

Sr. No.	Description	Yes	No	Notes
1	Equipment Type-Pumps			
	a) Normal			
	b) Standby unit			
	c) Preferential country of origin			
2	Water Storage Tank			
3	Water System Type			
	a) Cold water only			
	b) Cold/Hot water			
4	Type of Pipe Water Supply			
	a) Metallic			
	b) Nonmetallic			
5	Type of Pipe Drainage			
	a) Metallic			
	b) Non Metallic			
6	Boiler System			
	a) Central			
	b) Individual unit			
	c) Preferential type of boiler			
7	Preferential fittings and accessories			
8	Water pumps on Emergency			
9	Fuel storage type			
10	Interface with BMS			

Source: Abdul Razzak Rumane (2013). *Quality Tools for Managing Construction Projects*, CRC Press, Boca Raton, FL. Reprinted with permission from Taylor & Francis Group.

TABLE 7.59

Checklist for Owner's Preferred Requirements (HVAC)

Owner's Preferred Requirements (Mechanical)				
Sr. No.	**Description**	**Yes**	**No**	**Notes**
1	Equipment Type			
	a) Normal			
	b) Special			
	c) Preferential country of origin			
2	Standby Unit			
3	System Type			
	a) Cooling only			
	b) Cooling/Heating			
4	BMS			
	a) Manual operation			
	b) Automatic			
	c) Addressable			
	d) Web-based			
5	Power for Equipment			
	a) Normal			
	b) Emergency			
6	Thermostat Type			
	a) Analogue			
	b) Digital			

Source: Abdul Razzak Rumane (2013). *Quality Tools for Managing Construction Projects.* Reprinted with permission of Taylor & Francis Group.

TABLE 7.60

Checklist for Owner's Preferred Requirements (Electrical)

Sr. No.	Description	Yes	No	Notes
1	Lighting Type			
	a) Normal			
	b) LED			
	c) Optic Fiber			
2	Lighting Switching			
	a) Normal Switching			
	b) Central Switching			
	c) Smart Building			
	d) Dimming			
3	Power			
	a) Normal			
	b) Emergency			
	c) Through UPS			
4	Diesel Generator Set			

(Continued)

TABLE 7.60 (*Continued*)

Checklist for Owner's Preferred Requirements (Electrical)

Sr. No.	Description	Yes	No	Notes
5	UPS			
6	Capacitor Bank			
7	Fire Alarm System			
	a) Analogue			
	b) Addressable			
8	Communication System			
	a) Analogue			
	b) Digital			
	c) IP Telephony			
9	IT System			
	a) Passive Network			
	b) Active Components			
10	Security System			
	a) Analogue Cameras			
	b) Digital Cameras			
	c) IP Cameras			
	d) Analogue System			
	e) Digital System			
	f) IP Integration			
	g) Guard Tour System			
	h) X-Ray Machine			
	i) Metal Detectors			
	j) Other Requirements			
11	Access Control System			
12	Parking Control System			
13	Road Blocker			
14	Public Address System			
	a) Analogue			
	b) Digital			
	c) IP			
15	Audio Visual System			
16	Conference System			
17	Satellite Antenna System			
	a) Digital			
	b) IP			
18	Central Clock System			
19	System Integration			
20	Provision for Integration with Alternate Energy			
21	Any Other System			

Source: Abdul Razzak Rumane (2013). *Quality Tools for Managing Construction Projects*, CRC Press, Boca Raton, FL. Reprinted with permission from Taylor & Francis Group.

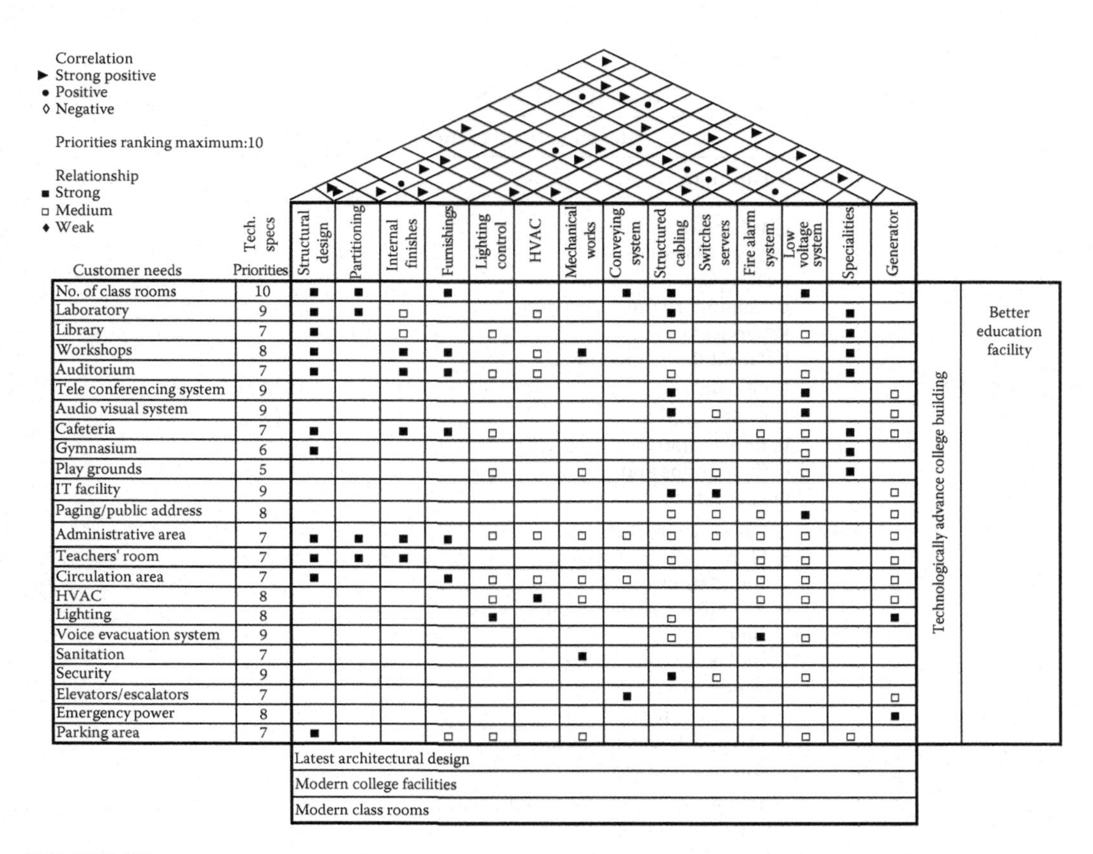

FIGURE 7.103
House of quality for college building project. (Abdul Razzak Rumane (2013), *Quality Tools for Managing Construction Projects*, CRC Press, Boca Raton, FL. Reprinted with permission from Taylor & Francis Group.)

TABLE 7.61
Elements to Be Included in Concept Design Drawings

Sr. No.	Elements to Be Included in Drawing
1	Architectural
1.1	Overall site plans
	a. Existing site plan b. Location of building, roads, parking, access, and landscape c. Project boundary limits d. Site utilities e. Water supply, drainage, and storm water lines f. Zoning g. Reference grids and axis h. Demolition plan, if any
1.2	Floor plans
	a. Floor plans of all floors b. Structural grids c. Vertical circulation elements d. Vertical shafts

(Continued)

TABLE 7.61 (*Continued*)

Elements to Be Included in Concept Design Drawings

Sr. No.	Elements to Be Included in Drawing
	e. Partitions f. Doors g. Windows h. Floor elevations i. Designation of rooms j. Preliminary finish schedule k. Services closets l. Raised floor, if required
1.3	Roof plans
	a. Roof layout b. Roof material c. Roof drains and slopes
2	Structural
2.1	a. Building structure b. Floor grade and system c. Foundation system d. Tentative size of columns, beams e. Stairs f. Roof and general sections
3	Elevator
3.1	a. Traffic studies b. Elevator/escalator location c. Equipment room
4	Plumbing and fire suppression
4.1	a. Sprinkler layout plan b. Piping layout plan c. Water system layout plan d. Water storage tank location e. Development of preliminary system schematics f. Location of mechanical room
5	HVAC
5.1	a. Ducting layout plan b. Piping layout plan c. Development of preliminary system schematics d. Calculations to allow preliminary plant selections e. Establishment of primary building services distribution routes f. Establishment of preliminary plant location and space requirements g. Determine heating and cooling requirements based on heat dissipation of equipment, lighting loads, type of wall, roof, and glass h. Estimation of HVAC electrical load i. Development of BMS schematics showing interface j. Location of plant room, chillers, and cooling towers
6	Electrical
6.1	a. Lighting layout plan b. Power layout plan c. System design schematic without any sizing of cables and breakers d. Substation layout and location e. Total connected load f. Location of electrical rooms and closets g. Location of MLTPs, MSBs, SMBs, EMSB, SEMBs, DBs, and EDBs

(*Continued*)

TABLE 7.61 (*Continued*)

Elements to Be Included in Concept Design Drawings

Sr. No.	Elements to Be Included in Drawing
	h. Location of starter panels and MCC panels i. Location of generator and UPS j. Raceway routes k. Riser requirements l. Information and communication technology —i. Information technology (computer network) —ii. IP telephone system (telephone network) —iii. Smart building system m. Loss prevention systems —i. Fire alarm system —ii. Access control security system layout —iii. Intrusion system n. Public address and audio–visual system layout o. Schematics for fire alarm and other loss prevention systems p. Schematics for ICT System, and other low voltage systems q. Location of low voltage equipment
7	Landscape
7.1	a. Green area layout b. Selection of plants c. Irrigation system
8	External
8.1	a. Street/road layout b. Street lighting c. Bridges (if any) d. Security system e. Location of electrical panels (feeder pillars) f. Pedestrian walkways g. Existing plans
9	Narrative description

7.4.1.5.3.2 Concept Design Report Concept design reports should be concise and should cover all the related information about concept design considerations. Reports are prepared for each of the trades mentioned in the TOR. The following is an example table of contents of a concept design report for architectural works and civil/structural works for building project:

A. Table of Contents for Architectural Works

1. Introduction
   1.1 Project goals and objectives
2. Owner's schedule requirements
3. Project directory
4. Architecture
   4.1 Applicable codes and standards
   4.2 International codes
   4.3 Local codes

  - 4.4 Building height
  - 4.5 Projections
  - 4.6 Fire safety requirements
  - 4.7 Usage of building
  - 4.8 Stairways
5. Project requirements
  - 5.1 Basement
  - 5.2 Ground floor
  - 5.3 Other floors
6. Existing site conditions
  - 6.1 Location maps
  - 6.2 Project site and surrounding area survey reports
7. Data collection
  - 7.1 Site analysis
  - 7.2 Traffic study
  - 7.3 Existing site views
8. Design options
9. Building structure
  - 9.1 Design loads
  - 9.2 Wind loads
  - 9.3 Deflection criteria
10. Design program software
11. Geographical investigation
12. Shoring
13. Dewatering
14. Drawings

B. Table of Contents for Structural/Civil Works

1. Introduction
  - a. Project goals and objectives
  - b. Existing site conditions
2. Codes and standards
3. Regulatory compliance
4. Design criteria
5. Design systems
  - a. Design system alternatives
  - b. Key construction elements
  - c. Environmental risk assessment
  - d. Geotechnical investigation and foundation design
  - e. Analysis model

f. Material
g. Durability
h. Fire safety
i. Design loads
j. Lateral loads
k. Structural movement criteria
l. System suitability for structure
m. Shoring and dewatering
n. Retaining wall and lateral stability system
o. Substructure
p. Design program/software

6. Testing systems
7. Risk assessment

C. Table of Contents for Mechanical Works

1. Introduction
2. Codes and standards
3. Regulatory compliance
4. Incoming services routes and regulatory requirements
5. Fire suppression work
    a. Design criteria
    b. Sprinkler system
    c. Smoke ventilation system
    d. Fire fighting pumps
    e. Piping system
    f. Fire suppression system for diesel generator room
    g. Fire suppression for low-voltage equipment room
6. Mechanical (plumbing) work
    a. Design criteria
    b. Water supply system
        i. Cold water
        ii. Hot water
    c. Water distribution network
    d. Water storage system
    e. Plumbing, fittings, and fixtures
    f. Plant room (pumps and heaters) location
    g. Storm water drainage system
    h. Sewage system
    i. Irrigation water system
    j. Piping for mechanical works

D. Table of Contents for HVAC Works

1. Introduction
2. Codes and standards
3. Regulatory compliance
4. HVAC system design criteria
5. Environmental conditions
6. HVAC equipment selection
   a. Chiller
   b. Cooling system
   c. AHU
   d. Energy recovery equipment
   e. Direct digital control system
   f. Pumps
   g. Fans and ventilation
   h. Ducting
   i. Duct insulation
   j. Piping
   k. Risers
   l. Piping insulation
   m. Acoustic and vibration
7. Location of plant room
8. Building management system (BMS)

E. Table of Contents for Electrical Works

1. Introduction
2. Codes and standards
3. Regulatory compliance
4. Design criteria
   a. Safety and protection
5. Electrical substation
6. Cables and wires
7. Main electrical supply
   a. Main distribution system
   b. Secondary distribution system
   c. Distribution equipment (panels, boards, and switchgear)
   d. Circuits
8. Internal and external lighting
   a. Light fixtures
9. Power receptacles
10. Emergency power system

11. Earthing (grounding) system
12. Lightning protection system
13. Alternate energy (solar) system
14. Fire alarm and voice evacuation system
15. Audio–visual system
16. Security system (CCTV and access control)
17. Communication system

7.4.1.6 Prepare Preliminary Schedule

The duration of a construction project is finite. It has a definite beginning and a definite end; therefore, during the conceptual phase, the expected time schedule for the completion of the project/facility is worked out. Expected time schedule is important from both financial and acquisition of the facility by the owner/end user. It is the owner's goal and objective that the facility is completed in time for occupancy/usage. During this phase, very limited information and details about project activities are available and have very wide variance in the schedule range. The preliminary project schedule is developed using top-down planning using key events. It is also known as Class 4 schedule or Schedule Level 2 (see Table 7.20 and Table 7.21).

Figure 7.104 illustrates typical project schedule (life cycle) for a construction project. The example schedule is construction of three branches for a bank.

7.4.1.7 Estimate Conceptual Cost

A cost estimate during concept design is required by the owner to know how much the capital cost of construction is. This is required by the owner to enable arrange the financial resources. Conceptual cost is also known as *budgetary cost*. Parametric cost estimation methodology is used to estimate conceptual cost (see Table 7.23). The designer has to ensure that the conceptual cost does not exceed cost estimated during the feasibility stage. The accuracy and validity of conceptual cost estimation is related to the level of information available while developing concept design. The designer has to properly estimate the resources required for the successful completion of project. Table 7.62 illustrates quality check for cost estimation.

7.4.1.8 Establish Quality Requirements

During this phase, the designer has to plan and establish quality criteria for the project. This includes mainly the following:

- Owner's requirements
- Quality standards and codes to be complied
- Regulatory requirements
- Conformance to owner's requirements
- Conformance to requirements listed under TOR
- Design review procedure
- Drawings review procedure
- Document review procedure
- Quality management during all the phases of project life cycle

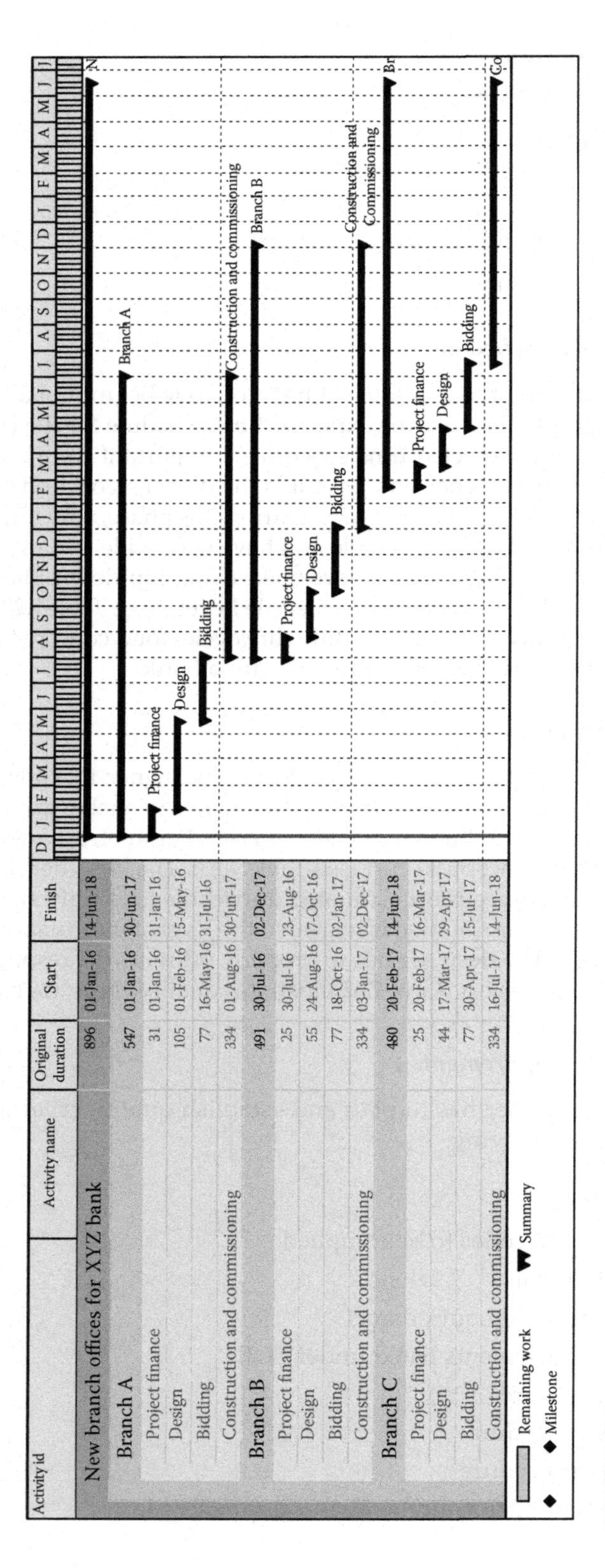

FIGURE 7.104
Typical schedule.

TABLE 7.62
Quality Check for Cost Estimate during Concept Design

Sr. No.	Points to Be Checked	Yes/No
1	Check for use of historical data	
2	Check if estimate factors used to adjust historical data	
3	Check the estimate is updated with revision or update of concept	
4	Check if the updates/revisions are chronologically listed and cost estimate updated	
5	Whether scope of work is descriptive/narrative, well enough for estimation purpose	
6	Whether estimate is updated taking into consideration feedback from each trade	
7	Is the estimate based on area schedule provided by the architect	
8	Is the cost estimate clearly identify the quantities and associated work	
9	Whether all the assumption are as per current market data	
10	Is cost estimates from all the relevant trades included in the final sum	
11	Is the estimate include all the requirements of specialist consultants	
12	Whether the total estimate is reviewed and verified	

Source: Abdul Razzak Rumane (2013), *Quality Tools for Managing Construction Projects*, CRC Press, Boca Raton, FL. Reprinted with permission from Taylor & Francis Group.

7.4.1.9 Estimate Resources

The designer has to estimate the resources required to complete the project. This includes the estimation of manpower required during the construction phase, testing, commissioning, and handover phase. The designer has to also estimate the manpower required during design and tendering stages of the construction project.

7.4.1.10 Manage Risk

The designer has to identify the risks that will affect the successful completion of project. The following are typical risks that normally occur during the conceptual design phase:

- Lack of input from the owner about the project goals and objectives.
- Project objectives are not clearly defined.
- Feasibility study is not properly performed.
- Alternative selection is not suitable for further development.
- The related project data and information collected is incomplete.
- The related project data and information collected is likely to be incorrect and wrongly estimated.
- Environmental consideration.
- Regulatory requirements.
- Errors is estimating the project schedule.
- Errors in cost estimation.

The designer has to take into account the above-mentioned risk factors while developing the concept design. Furthermore, the designer has to consider the following risks while planning the duration for completion of the conceptual phase:

- Impractical conceptual design preparation schedule
- Delay to obtain authorities' approval
- Delay in environmental approval
- Delay in data collection
- Delay in deciding project delivery system

7.4.1.11 Monitor Work Progress

The designer has to prepare a work schedule in order to prepare and submit the design to the owner for review and approval. The concept design progress has to be monitored to meet the submission schedule. Regular meetings to monitor the progress and coordination among different trades are required to ensure timely submission of concept design package.

7.4.1.12 Review Concept Design

Table 7.63 illustrates major points for analysis of concept design.

TABLE 7.63
Analysis of Concept Design

Sr. No.	Description	Yes	No	Notes
A: Concept Design				
A.1	Does the design support owner's project goals and objectives?			
A.2	Does the design meets all the elements specified in TOR?			
A.3	Does the design meet all the performance requirements?			
A.4	Whether constructability has been taken care?			
A.5	Whether technical and functional capability considered?			
A.6	Whether design confirm with fire and egress requirements?			
A.7	Whether health and safety requirements in the design are considered?			
A.8	Whether environmental constraints considered?			
A.9	Whether design risks been identified, analyzed, and responses planned for mitigation?			
A.10	Does energy conservation is considered?			
A.11	Does sustainability considered in design?			
A.12	Whether cost-effectiveness over the entire project life cycle is considered?			
A.13	Whether the design meets LEED requirements?			
A.14	Whether accessibility is considered?			
A.15	Does the design meet ease of maintenance?			
A.16	Does the design have provision for inclusion of facility management requirements?			
A.17	Whether all reasonable alternative options/systems are considered for design?			
A.18	Whether all reasonable alternative options are considered for project economy?			
A.19	Whether all the regulatory/statutory requirements taken care?			
A.20	Does the design support proceeding to next design development stage?			

(Continued)

TABLE 7.63 (*Continued*)

Analysis of Concept Design

Sr. No.	Description	Yes	No	Notes
B: Financial				
B.1	Is project cost properly estimated?			
C: Schedule				
C.1	Is project schedule practically achievable?			
D: Reports				
D.1	Whether the reports are complete and include adequate information about the project?			
D.2	Whether the reports are prepared for all the trades mentioned in TOR?			
D.3	Whether the report is properly formatted and has table of contents for each report?			
E: Drawings, Sketches				
D.1	Whether drawings, sketches for all trades prepared as per TOR?			
F: Models				
F.1	Whether the models meet the design objectives?			
G: Submittals				
G.1	Whether numbers of sets prepared as per TOR?			

7.4.1.13 Finalize Concept Design

Final designs are prepared incorporating the comments, if any, found during analysis and review of the drawings and documents for submission to the owner/client.

7.4.1.14 Submit Concept Design Package

Normally, the following are submitted to the owners for their review and approval in order to proceed with development of schematic drawings:

1. Outline specifications
2. Concept design drawings
3. Concept design reports
4. Project schedule
5. Cost estimate
6. Model

7.4.2 Schematic Design Phase

Schematic design is mainly a refinement of the elements in the conceptual design phase. Schematic design is also known as *preliminary design*. Design-intent documents quantify functional performance expectations and parameters for each system to be commissioned. It is traditionally labeled as 30% design. Schematic design adequately describes information about all proposed project elements in sufficient details for obtaining regulatory approvals, necessary permits, and authorization. At this phase, the project is planned to the level where sufficient details are available for initial cost and schedule. This phase also include the initial preparation of all documents necessary to build the facility/construction project. Figure 7.105 illustrates major activities relating to the schematic design phase developed based on project management process groups methodology.

7.4.2.1 Develop Schematic Design Requirements

The central activity of preliminary design is the architect's design concept of the owner's objective that can help making detail engineering and design for the required facility. Preliminary design is a subjective process transforming ideas and information into plans, drawings, and specifications of the facility to be built. Component/equipment configurations, material specifications, and functional performance are decided during this

Schematic design phase

Management processes	Project management process groups				
	Initiating process	Planning process	Execution process	Monitoring and controlling process	Closing process
Integration management	Concept design deliverables	Preliminary project management plan	Develop general layout of project (site plans)	Building code requirements	Schematic design deliverables
	Concept design comments		Architectural plans	Concept design comments	
	Tor requirements		Structural scheme plans	Existing conditions	
	Authorities' requirements		Electromechanical services	Design calculations	
			Landscape and infrastructure	Compliance to regulatory/ authorities' requirements	
			Develop schematic design	Constructability	
Stakeholder management	Identify project team	Stakeholders requirements	Owner's need	Stakeholders requirements	
	Identify design team				
Scope management		Site conditions		Authorities approval	
		Energy conservation requirements		Compliance to owner's needs	
		Technical and functional capability		Stakeholders approval	
		Outline specifications			
		System schematics			
		Value engineering			
		Design deliverables			
Schedule management		Project schedule (CPM/bar chart)	Preliminary schedule	Project schedule	
Cost management		Cost of activities	Preliminary cost	Project cost estimate	
		Cost of resources			
		Preliminary estimate			
Quality management		Design criteria	Design coordination with all disciplines	Compliance to codes, standards, and authorities	
		Codes and standards			
		Authorities requirements			
Resource management		Assign project design team	Manage team members of difference discipline	Performance of team members	Assign new phase/project
		Estimate resources			
Communication management		Design progress information	Liaison and coordination with all parties	Design status information	
			Coordination meetings		
Risk management		Management of design risk		Control design risk	
Contract management		Contract terms and conditions	Preliminary contract documents	Check for contracting system	
HSE management		Safety considerations in design	Environmental requirements	Life safety in design requirements	
Financial management		Designer/consultant payment			Designer/consultant payment
Claim management		Design change payments		Control changes	Settle claim by designer

FIGURE 7.105
Major activities relating to schematic design processes. *Note*: These activities may not be strictly sequential; however, the breakdown allows implementation of project management function more effective and easily manageable at different stages of the project phase.

stage. Design is a complex process. Before design is started, scope must adequately define deliverables, that is, what will be provided. These deliverables are design drawings, contract specifications type of contracts, construction inspection record drawings, and reimbursable expenses.

For development of the preliminary design phase, the designer has to

- Make investigations of site conditions.
- Collect and analyze the required data.
- Analyze building codes requirements.
- Analyze energy conservation requirements.
- Study fire and other regulatory codes and requirements.

Based on the evaluation and analysis of the required data, the designer can proceed with development of preliminary design by considering the following major points:

1. Concept design deliverables
2. Calculations to support the design
3. System schematics for electromechanical system
4. Coordination with other members of the project team
5. Authorities requirements
6. Availability of resources
7. Constructability
8. Health and safety
9. Reliability
10. Energy conservation issues
11. Environmental issues
12. Selection of systems and products that support functional goals of the entire facility
13. Sustainability
14. Requirements of all stakeholders
15. Optimized life cycle cost (value engineering)

7.4.2.1.1 Identify Schematic Design Requirements

In order to identify requirements to develop schematic design, the designer has to gather comments made by the owner/project manager on the submitted concept design, collect TOR requirements, regulatory requirements, and other related data to ensure that the developed design is error free and with minimum omissions.

7.4.2.1.1.1 Gather Concept Design Comments The objective of the schematic design is to refine and develop a clearly defined design based on the client's requirements. While developing the schematic design, the designer reviews the submitted concept design drawings, reports, and takes into consideration the comments, if any, made on the concept design. The designer can discuss in detail with the owner/project manager and incorporate all their requirements to develop schematic design.

7.4.2.1.1.2 Identify TOR Requirements Normally, the TOR lists the requirements guidelines to develop the schematic design. It mainly consists of the following:

1. Complete schematic drawings for the selected/approved concept design
2. Schematic design report
3. Preliminary models

The TOR details that during schematic design phase, the following items to be developed taking into consideration the approved concept design and comments:

A. Drawings
 A.1 Site plan
 A.2 Architectural design
 A.3 Interior design
 A.5 Structural design
 A.6 Mechanical design
  A.6.1 Public health
  A.6.2 Fire suppression
 A.7 HVAC design
 A.8 Electrical design
 A.9 Smart building systems
 A.10 Special systems
 A.11 Landscape design
 A.12 External works
B. Outline specifications
C. Preliminary contract documents
D. Project schedule (preliminary)
E. Cost estimate (preliminary)
F. Authorities approvals
G. Value engineering report
H. Preliminary model

7.4.2.1.1.3 Collect Owner's Requirements The owner's requirements discussed under Section 7.4.1.5.2 are further developed in details in order to consider in the schematic design.

7.4.2.1.1.4 Collect Regulatory Requirements During this phase, the schematic design drawings are submitted to the regulatory bodies for their review and approval for compliance with the regulations, codes, and licensing procedures. Any comments on the drawings are incorporated on the drawings and are resubmitted, if required. Normally, for building projects, the following schematic drawings are submitted for approval by the concerned authorities:

1. Architectural drawings approval by municipality
2. Fire department
   a. Escape route (stairs)
   b. Fire alarm system
   c. Fire suppression (sprinkler) system
   d. Smoke ventilation system
   e. Inflammable material and fuel system
   f. Conveyance system (elevator)
   g. Exit and emergency lighting system
3. Public works for utilities connections such as water supply
4. Public works for storm water, sanitary, and drainage system
5. HVAC system for compliance with energy conservation
6. Electricity agency for approval of total connected load (electrical)
7. Electrical substation location
8. Fuel point for building

For infrastructure and road works, the following approvals are required:

1. Environmental impact
2. Sanitary, storm water, and drainage system discharge
3. Landscape
4. Traffic signage and markings
5. Right of way
6. Street lighting
7. Substation for street lights

The requirements differ depending upon the particular country's applicable rules and regulations.

7.4.2.1.1.5 Collect LEED Requirements LEED is an internationally recognized green building certification system, providing verification by third party that the building is designed and constructed to the standards and requirements outlined by the LEED rating system. The design team and contractor have to integrate all the required features in their design to ensure that building is more durable, healthy, and more energy efficient.

In order to construct a building that merits LEED certification, the designer has to consider the following:

- Optimize site selection
- Orientation of building to make maximum advantages of sunlight
- Water efficiency
- Use of energy as efficiently as possible
- Use of water as efficiently as possible
- Indoor environmental quality

    - Indoor air quality
    - Ventilation
    - Outdoor air flow monitoring
- Maximum use of alternate (renewable) energy
- Minimize waste water
- Sustainable material selection
- Regional priority
- Innovation in design

7.4.2.1.2 Establish Schematic Design Requirements

The schematic design requirements are developed considering all the items discussed under Section 7.4.2.1.1. These are mainly as follows:

1. Gathering of comments on concept design deliverables
2. Collection of owner's preferred requirements
3. Collection of regulatory requirements
4. TOR requirements
5. Provision for facility management system/equipment

7.4.2.2 Develop Preliminary Project Management Plan

Preliminary project management plan is an update based on additional information collected during the schematic design phase. Typical contents of a project management plan are as follows:

1. Project description
2. Project objectives
3. Project organization
    a. Organizational chart
    b. Responsibility matrix
    c. Project directory
4. Project (scope) deliverables
5. Project schedule
6. Project budget
7. Project quality plan
8. Project resources
9. Risk management
10. Communication matrix
11. Contract management
12. HSE management
13. Project finance management
14. Claim settlement

7.4.2.3 Identify Project Stakeholder

During the schematic design phase, the following stakeholders have direct involvement in the project:

- Owner
- Consultant
- Designer
- Regulatory authorities
- Project manager/agency construction manager (as applicable)

7.4.2.3.1 *Identify Project Team Members*

During this phase, most project team members such as the owner's representative, project manager (design) and other design personnel are selected and identified.

7.4.2.3.2 *Identify Design Team Members*

Design team members are selected based on the organizational structure and suitable skills required to perform the job. Normally, the design team consists of the following:

1. Project manager
2. Design managers (one for each trade)
3. Quality manager
4. Team leader (principal engineer)—each trade
5. Team members (engineers and CAD technicians for each disciplines)
6. Quantity surveyor (cost engineer)

Figure 7.106 illustrates an example structural/civil engineering design team organizational chart.

7.4.2.3.3 *Establish Stakeholder's Requirement*

A distribution matrix based on the interest and involvement of each stakeholder is prepared and the appropriate documents are sent for their action/information as per the agreed upon requirements.

7.4.2.3.4 *Develop Responsibility Matrix*

Table 7.64 illustrates responsibilities of various participants during the schematic design phase.

7.4.2.4 Develop Schematic Design Scope

The schematic design scope is developed taking into considerations the requirements established under Section 7.4.2.1.2. These are as follows:

- TOR requirements
- Owner's preferred requirements
- Approved concept design documents

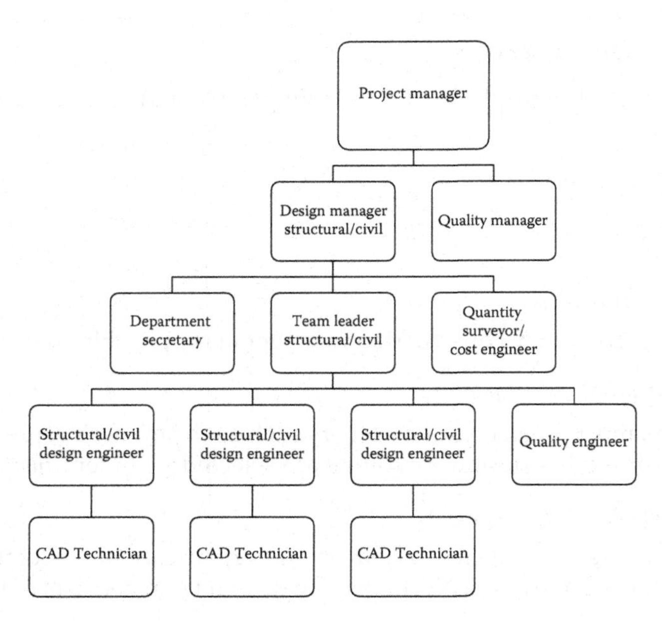

FIGURE 7.106
Structural/civil design team organizational chart.

TABLE 7.64
Responsibilities of Various Participants (Design–Bid–Build Type of Contracts) during Schematic Design Phase

	Responsibilities		
Phase	**Owner**	**Designer**	**Regulatory Authorities**
Schematic Design	• Approval of preliminary (schematic) design	• Develop general layout/scope of facility/project • Regulatory approval • Schedule • Budget • Contract terms and conditions • Value engineering	• Approval of project submittals

- Approved preliminary project schedule
- Approved preliminary cost estimate
- Regulatory requirements
- Codes and standards

The purpose of development of the schematic design is to provide sufficient information to identify the work to be performed and to allow the detail design to proceed without significant changes that may affect the project schedule and budget. The scope of work during schematic design phase mainly comprise

- Preparation of schematic drawings
- Outline specifications
- Preliminary schedule
- Preliminary cost estimate
- Authorities approvals
- Narrative reports

7.4.2.4.1 Identify Schematic Design Deliverables

Based on the scope of the schematic design, the following are the deliverables to be developed:

1. Design drawings
   a. Architectural
   b. Structural
   c. Conveying system
   d. Mechanical (public health and fire suppression)
   e. HVAC
   f. Electrical
   g. Landscape
   h. External
2. Outline specifications
3. Narrative reports
4. Preliminary schedule
5. Preliminary cost estimate
6. Regulatory approvals
7. Value engineering
8. Models

Table 7.65 lists the schematic design deliverables to be developed during the schematic design phase.

7.4.2.5 Develop Schematic Design

The primary goal of this phase is to develop a clearly defined design based on client requirements. Figure 7.107 illustrates a logic flow process for the schematic design phase.

In order to develop a schematic design, the designer has to collect related data/information and perform site investigations.

7.4.2.5.1 Collect Data/Information

The following data are to be collected to develop a schematic design for a building project:

- Needs of the owner
- Building/project usage
- Space program

TABLE 7.65

Schematic Design Deliverables

Sr. No.	Deliverables
1	General
1.1	a. Preliminary/outline specifications
	b. Zoning
	c. Permits and regulatory approvals
	d. Energy code requirements
	e. Construction methodology narration
	f. Descriptive report of environmental, health, and safety requirements
	g. Estimate construction period (preliminary schedule)
	h. Estimated cost
	i. Value engineering suggestions and resolutions
	j. Life safety requirements
	k. Sketches/perspective i. Interior ii. Exterior
	l. Graphic presentation
2	Preliminary design drawings
2.1	Architectural
2.1.1	Overall site plans a. Existing site plans b. Location of building, roads, parking, access, and landscape c. Project boundary limits d. Site utilities e. Water supply, drainage, and storm water lines f. Zoning g. Reference grids and axis
2.1.2	Floor plans a. Floor plans of all floors b. Structural grids c. Vertical circulation elements d. Vertical shafts e. Partitions of various types f. Doors g. Windows h. Floor elevations i. Designation of rooms j. Preliminary finish schedule k. Door schedule l. Hardware schedule m. Ceiling plans n. Services closets o. Raised floor, if required
2.1.3	Roof plans a. Roof layout b. Roof material c. Roof drains and slopes

(Continued)

TABLE 7.65 (*Continued*)

Schematic Design Deliverables

Sr. No.	Deliverables
2.1.4	Elevations a. Wall cladding b. Curtain wall c. Stones cladding d. Exterior insulation and finishing system e. Sections at various locations
2.2	Structural a. Building structure b. Floor grade and system c. Foundation system d. Tentative size of columns and beams e. Stairs f. Elevations and sections through various axis g. Roof
2.3	Elevator a. Traffic studies b. Elevator/escalator location c. Equipment room
2.4	Plumbing and fire suppression a. Sprinkler layout plan b. Piping layout plan c. Water system layout plan d. Water storage tank location e. Development of preliminary system schematics f. Location of mechanical room
2.5	HVAC a. Ducting layout plan b. Piping layout plan c. Development of preliminary system schematics d. Calculations to allow preliminary plant selections e. Establishment of primary building services distribution routes f. Establishment of preliminary plant location and space requirements g. Determine heating and cooling requirements based on heat dissipation of equipment, lighting loads, type of wall, roof, glass.. etc. h. Estimation of HVAC electrical load i. Development of BMS schematics showing interface j. Location of plant room, chillers, and cooling towers
2.6	Electrical a. Lighting layout plan b. Power layout plan c. System design schematic without any sizing of cables and breakers d. Substation layout and location e. Total connected load f. Location of electrical rooms and closets g. Location of MLTPs, MSBs, SMBs, EMSB, SEMBs, DBs, and EDBs h. Location of starter panels and MCC panels i. Location of generator and UPS j. Raceway routes k. Riser requirements

(*Continued*)

TABLE 7.65 (*Continued*)

Schematic Design Deliverables

Sr. No.	Deliverables
	l. Information and communication technology
	i. Information technology (computer network)
	ii. IP telephone system (telephone network)
	iii. Smart building system
	m. Loss prevention systems
	i. Fire alarm system
	ii. Access control security system layout
	iii. Intrusion system
	n. Public address, audio–visual system layout
	o. Schematics for F.A and other loss prevention systems
	p. Schematics for ICT system, and other low-voltage systems
	q. Location of LV equipment
2.7	Landscape
	a. Green area layout
	b. Selection of plants
	c. Irrigation system
2.8	External
	a. Street/road layout
	b. Street lighting
	c. Bridges (if any)
	d. Security system
	e. Location of electrical panels (feeder pillars)
	f. Pedestrian walkways

Source: Abdul Razzak Rumane (2013), *Quality Tools for Managing Construction Projects*, CRC Press, Boca Raton, FL. Reprinted with permission from Taylor & Francis Group.

- Technical and functional capability requirements
- Zoning requirements
- Aesthetics requirements
- Fire protection requirements
- Indoor air quality
- Lighting/daylighting requirements
- Conveying system traffic analysis
- Health and safety features
- Environmental compatibility requirements
- Energy conservation requirements
- Sustainability requirements
- Facility management requirements
- Regulatory/authority requirements (permits)
- Codes and standards to be followed
- Social responsibility requirements
- Project constraints

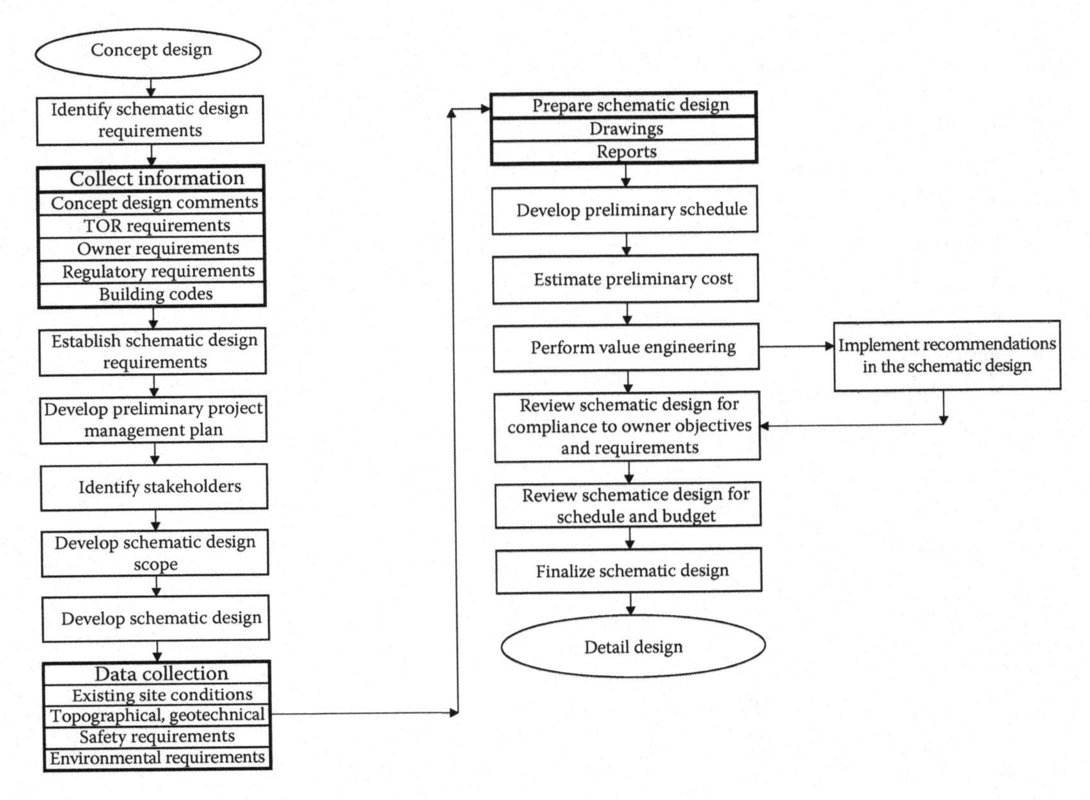

FIGURE 7.107
Logic flow process for the schematic design phase.

- Ease for constructability
- Number of drawings to be produced
- Milestone for development of each phase of design
- Disabled (special needs) access requirements

7.4.2.5.2 Perform Site Investigations

- Soil profile and laboratory test of soil
- Topography of the project site
- Hydrological information
- Wind load, seismic load, dead load, and live load
- Existing services passing through the project site
- Shoring and underpinning requirements with respect to adjacent area/structure

7.4.2.5.3 Prepare Schematic Design

During this phase, several alternative schemes are reviewed, and one scheme that meets the owner objectives is selected.

7.4.2.5.3.1 Schematic Design Drawings The following schematic drawings for building project are generally prepared during this phase:

- Architectural
  - Overall site plans
  - Floor plans
  - Roof plans
  - Sections
  - Elevations
- Structural
  - Building structure
  - Foundation system
  - Stairs
  - Roof
- Conveying system
  - Elevator location
  - Machine room
- MEP
  - General arrangement of each system
  - Single-line riser diagram
  - Electrical schematic diagram
  - Layout of equipment rooms
  - Layout of electrical substation
  - Electrical rooms
  - Vertical shafts
- Security system
- External works
- Landscape plans

The designer can use the BIM tool for designing the project.

- *BIM*: It is an innovative process of generating digital database for collaboration and managing building data during its life cycle and preserves the information for reuse and additional industry-specific applications. BIM is autodesk's strategy for the application of information technology to the building industry. It helps better visualization and clash detection. It is an excellent tool to develop project staging plans, study phasing, and coordination issues during construction project life cycle, preparation of as-built, and also during maintenance of the project. See Chapter 5 for more details about use of BIM.

7.4.2.5.3.2 Schematic Design Report The schematic reports mainly identify

- Systems
- Materials

- Finishes
- Design features describing the selected option
- Construction methodology
- Project risk

Outline Specifications

Outline specifications indicating project-specific features of major equipment, systems, and material are prepared during this phase. Normally, specifications are prepared per MasterFormat contract documents produced jointly by the CSI and Construction Specifications Canada (CSC), which are widely accepted as a standard practice for preparation of contract documents. Different types of construction contract documents used by the construction industry are discussed in Chapter 6.

Preliminary contract documents: Based on the type of contracting arrangements, the owner would like to handle the project, and necessary documents are prepared by establishing a framework for execution of project. Generally, FIDIC (Federation International des Ingénieurs—Counseils, the) model conditions for international civil engineering contracts are used as a guide to prepare these contract documents. Preliminary documents are prepared in line with model contract documents.

7.4.2.6 Perform Value Engineering

Value engineering (VE) studies can be conducted at various phases of the construction project life cycle; however, the studies conducted in the early stage of project tend to provide greatest benefit. In most projects, VE studies are performed during the schematic phase of the project. At this stage, the design professionals have considerable flexibility to implement the recommendations made by the VE team, without significant impacts to the project schedule or design budget. In certain countries, for a project over US $5 million, VE study must be conducted as part of the schematic design process. Normally, VE study is performed by specialist VE consultant. The number of team members depend on client's/owner's requirement to perform VE study. It is advisable that SAVE International registered Certified Value Specialist is assigned to lead this study. Figure 7.108 illustrates VE process activities.

7.4.2.7 Prepare Preliminary Schedule

The project schedule is developed using top-down planning using key events. It is also known as Class 3 schedule or Schedule Level 2 (see Table 7.20 and Table 7.21). Figure 7.109 illustrates a typical preliminary schedule for a construction project. The example schedule is construction of three branches for a bank.

7.4.2.8 Estimate Preliminary Cost

The cost estimate during this phase is based on elemental parametric methodology. It is also known as *budgetary* (preliminary estimate) (see Table 7.23).

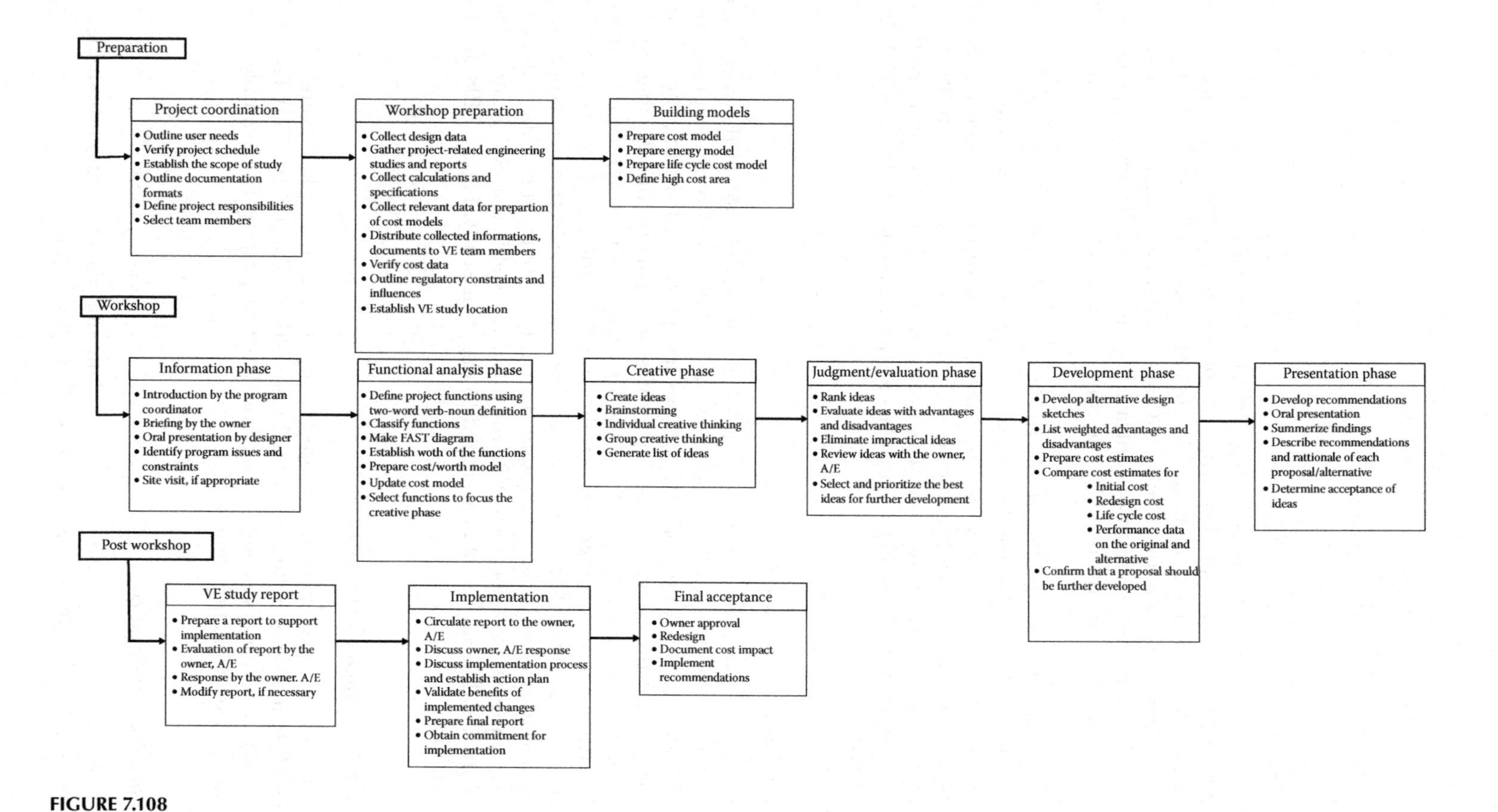

FIGURE 7.108
VE study process activities. (Abdul Razzak Rumane (2010), *Quality Management in Construction Projects*, CRC Press, Boca Raton, FL. Reprinted with permission from Taylor & Francis Group.)

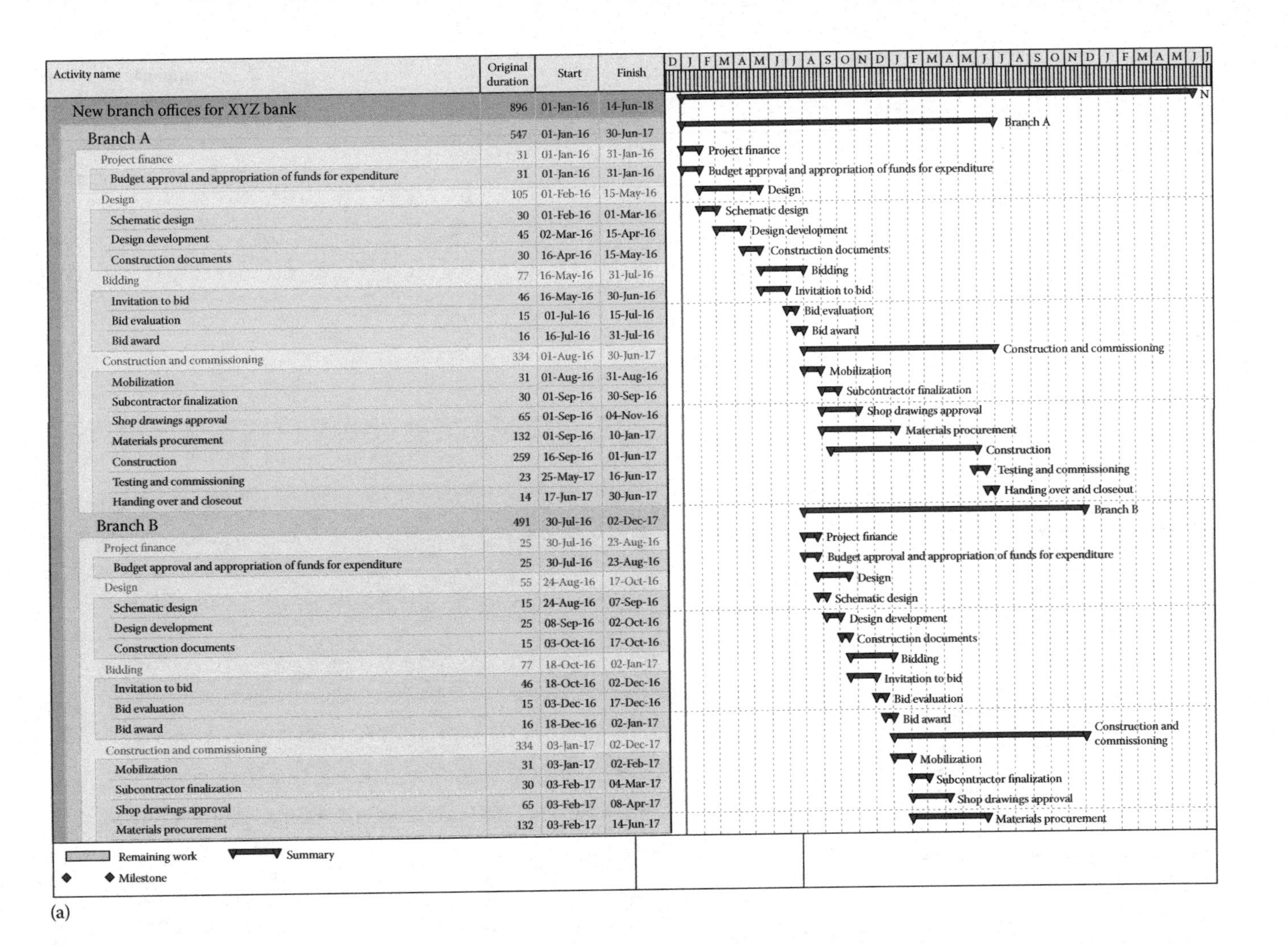

Activity name	Original duration	Start	Finish
New branch offices for XYZ bank	896	01-Jan-16	14-Jun-18
Branch A	547	01-Jan-16	30-Jun-17
Project finance	31	01-Jan-16	31-Jan-16
Budget approval and appropriation of funds for expenditure	31	01-Jan-16	31-Jan-16
Design	105	01-Feb-16	15-May-16
Schematic design	30	01-Feb-16	01-Mar-16
Design development	45	02-Mar-16	15-Apr-16
Construction documents	30	16-Apr-16	15-May-16
Bidding	77	16-May-16	31-Jul-16
Invitation to bid	46	16-May-16	30-Jun-16
Bid evaluation	15	01-Jul-16	15-Jul-16
Bid award	16	16-Jul-16	31-Jul-16
Construction and commissioning	334	01-Aug-16	30-Jun-17
Mobilization	31	01-Aug-16	31-Aug-16
Subcontractor finalization	30	01-Sep-16	30-Sep-16
Shop drawings approval	65	01-Sep-16	04-Nov-16
Materials procurement	132	01-Sep-16	10-Jan-17
Construction	259	16-Sep-16	01-Jun-17
Testing and commissioning	23	25-May-17	16-Jun-17
Handing over and closeout	14	17-Jun-17	30-Jun-17
Branch B	491	30-Jul-16	02-Dec-17
Project finance	25	30-Jul-16	23-Aug-16
Budget approval and appropriation of funds for expenditure	25	30-Jul-16	23-Aug-16
Design	55	24-Aug-16	17-Oct-16
Schematic design	15	24-Aug-16	07-Sep-16
Design development	25	08-Sep-16	02-Oct-16
Construction documents	15	03-Oct-16	17-Oct-16
Bidding	77	18-Oct-16	02-Jan-17
Invitation to bid	46	18-Oct-16	02-Dec-16
Bid evaluation	15	03-Dec-16	17-Dec-16
Bid award	16	18-Dec-16	02-Jan-17
Construction and commissioning	334	03-Jan-17	02-Dec-17
Mobilization	31	03-Jan-17	02-Feb-17
Subcontractor finalization	30	03-Feb-17	04-Mar-17
Shop drawings approval	65	03-Feb-17	08-Apr-17
Materials procurement	132	03-Feb-17	14-Jun-17

FIGURE 7.109
Typical preliminary schedule.

(Continued)

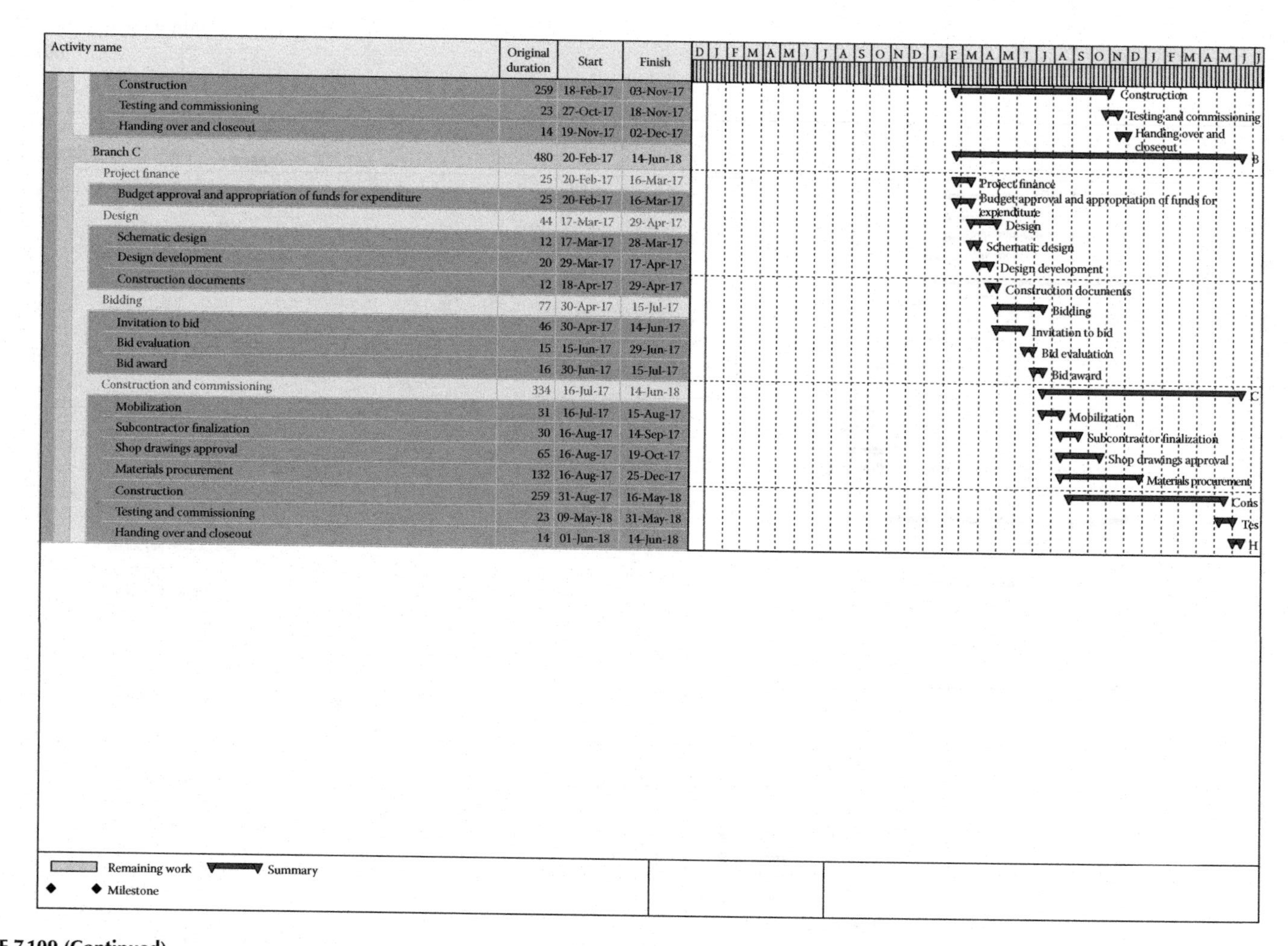

Activity name	Original duration	Start	Finish
Construction	259	18-Feb-17	03-Nov-17
Testing and commissioning	23	27-Oct-17	18-Nov-17
Handing over and closeout	14	19-Nov-17	02-Dec-17
Branch C	480	20-Feb-17	14-Jun-18
Project finance	25	20-Feb-17	16-Mar-17
Budget approval and appropriation of funds for expenditure	25	20-Feb-17	16-Mar-17
Design	44	17-Mar-17	29-Apr-17
Schematic design	12	17-Mar-17	28-Mar-17
Design development	20	29-Mar-17	17-Apr-17
Construction documents	12	18-Apr-17	29-Apr-17
Bidding	77	30-Apr-17	15-Jul-17
Invitation to bid	46	30-Apr-17	14-Jun-17
Bid evaluation	15	15-Jun-17	29-Jun-17
Bid award	16	30-Jun-17	15-Jul-17
Construction and commissioning	334	16-Jul-17	14-Jun-18
Mobilization	31	16-Jul-17	15-Aug-17
Subcontractor finalization	30	16-Aug-17	14-Sep-17
Shop drawings approval	65	16-Aug-17	19-Oct-17
Materials procurement	132	16-Aug-17	25-Dec-17
Construction	259	31-Aug-17	16-May-18
Testing and commissioning	23	09-May-18	31-May-18
Handing over and closeout	14	01-Jun-18	14-Jun-18

FIGURE 7.109 (Continued)
Typical preliminary schedule.

7.4.2.9 Manage Design Quality

In order to minimize design errors and design omissions, and reduce rework during schematic design, the designer has to plan quality (planning of design work), perform quality assurance, and control quality for preparing schematic design. This will mainly consist of the following:

1. *Plan quality*:
    - Establish owner's requirements
    - Determine the number of drawings to be produced
    - Establish the scope of work
    - Identify quality standards and codes to be complied
    - Establish design criteria
    - Identify regulatory requirements
    - Identify requirements listed under TOR
    - Establish quality organization with responsibility matrix
    - Develop design (drawings and documents) review procedure
    - Establish submittal plan
    - Establish design review procedure
2. *Quality assurance*:
    - Collect data
    - Investigate site conditions
    - Prepare preliminary drawings
    - Prepare outline specifications
    - Ensure functional and technical compatibility
    - Coordinate with all disciplines
    - Select material to meet owner objectives
3. *Control quality*:
    - Check design drawings
    - Check specifications/contract documents
    - Check for regulatory compliance
    - Check preliminary schedule
    - Check cost of project (preliminary cost)

The BIM tool can be used to control quality of the project (see Chapter 5 for more information).

7.4.2.10 Estimate Resources

The designer has to estimate the resources required to complete the project. At this stage of the project, more detail about the activities and works to be performed during the construction and testing, commissioning phase is available, the designer has to update the earlier estimated resources and prepare manpower histogram. In addition, the designer can estimate the total number of design team members to develop design and construction documents.

7.4.2.11 Manage Risks

The following are typical risks that normally occur during the schematic design phase:

- The designer not taking into considerations concept design deliverables and review comments while preparing the schematic design.
- Regulatory authorities' requirements are not taken into consideration.
- Incomplete scope of work for preparing schematic design.
- The related project data and information collected are incomplete.
- The related project data and information collected are likely to be incorrect and wrongly estimated.
- Site investigations for existing conditions are not carried out.
- Fire and safety considerations are not taken into account.
- Environmental considerations are not taken into account.
- Incomplete design.
- Prediction of possible changes in design during the construction phase.
- Inadequate and ambiguous specifications.
- Wrong selection of materials and systems.
- Undersized HVAC equipment selection.
- Incorrect water supply requirements.
- Estimated total electrical load is much lower than expected actual consumption.
- Errors in calculating traffic study for conveying system.
- Errors in estimating the project schedule.
- Errors in cost estimation.
- The number of drawings is not as per TOR requirements.

The designer has to take into account the above-mentioned risk factors while developing the schematic design. Furthermore, the designer has to consider the following risk while planning the duration for completion of the schematic phase.

- Impractical schematic design preparation schedule
- Delay to obtain authorities' approval
- Delay in site investigations
- Delay in data collection

7.4.2.12 Monitor Work Progress

The designer has to prepare work schedule to prepare schematic design package in order to submit the same to the owner for review and approval. The schematic design progress has to be monitored to meet the submission schedule. Regular meetings to monitor the progress and coordination among different trades and regulatory authorities are required to ensure timely submission of schematic design package.

7.4.2.13 Review Schematic Design

Table 7.66 illustrates the major points for analysis of the schematic design.

TABLE 7.66

Analysis of Schematic Design

Sr. No.	Description	Yes	No	Notes
A: Schematic Design (General)				
A.1	Does the design support owner's project goals and objectives?			
A.2	Does the design meets all the elements specified in TOR?			
A.3	Whether comments on concept design taken care while preparing schematic design?			
A.4	Whether regulatory approvals obtained?			
A.5	Does the design meet all the performance requirements?			
A.6	Whether constructability has been taken care?			
A.7	Whether technical and functional capability considered?			
A.8	Whether health and safety requirements in the design are considered?			
A.9	Whether design confirm with fire and egress requirements			
A.10	Whether design risks been identified, analyzed, and responses planned for mitigation?			
A.11	Whether environmental constraints considered?			
A.12	Does energy conservation is considered?			
A.13	Does sustainability considered in design?			
A.14	Whether cost-effectiveness over the entire project life cycle is considered?			
A.15	Whether the design meets LEED requirements?			
A.16	Whether accessibility is considered?			
A.17	Does the design meet ease of maintenance?			
A.18	Does the design have provision for inclusion of facility management requirements?			
A.19	Does the design support proceeding to next design development stage?			
A.20	Whether all the drawings are numbered?			
B: Architectural				
B.1	Drawings coordinated with other discipline?			
B.2	Whether grid system is established?			
B.3	Whether zoning is taken care?			
B.4	Whether overall site plans showing all the major areas?			
B.5	Whether floor plans showing overall dimensions?			
B.6	Whether roof plans are shown?			
B.7	Whether preliminary elevations are shown?			

(*Continued*)

TABLE 7.66 (*Continued*)

Analysis of Schematic Design

Sr. No.	Description	Yes	No	Notes
B.8	Whether all the rooms are numbered?			
B.9	Whether entrances, stairways, lobbies, and corridors identified?			
B.10	Whether services rooms, equipment rooms, plant rooms identified?			
B.11	Whether finishes schedule/requirements prepared?			
C: Structural				
C.1	Whether preliminary foundation plan shown?			
C.2	Whether structural systems identified?			
C.3	Whether preliminary building structure prepared?			
C.4	Whether preliminary framing plans for all floors and roof?			
C.5	Whether relevant codes regarding seismic zone, wind speed considered for structural load calculations?			
C.6	Whether slab loading for equipment considered?			
D: Elevator				
D.1	Whether traffic analysis performed?			
D.2	Whether elevator locations shown?			
D.3	Whether equipment room location shown?			
E: Mechanical				
E.1	Whether preliminary plans for toilets/rest rooms and pantry are shown?			
E.2	Whether main water supply, sanitary, and storm water system are shown?			
E.3	Whether riser diagram/single line diagram for plumbing system is shown?			
E.4	Whether plumbing fixtures have been identified?			
E.5	Whether fire protection system comply with regulatory requirements?			
E.6	Whether single line diagram for fire protection system (sprinkler) has been prepared?			
E.7	Whether equipment room (plant room) location is shown?			
E.8	Whether special fire suppression system for electrical substation and generator room considered?			
F: HVAC				
F.1	Whether single line diagram is prepared for all related systems?			
F.2	Whether shaft locations and approximate pipe sizes, and duct size are shown?			

(*Continued*)

TABLE 7.66 (*Continued*)

Analysis of Schematic Design

Sr. No.	Description	Yes	No	Notes
F.3	Whether location of plant room has been considered?			
F.4	Whether approximate load for chiller and pump sizes is considered on available data?			
F.5	Whether gross HVAC zoning and typical individual space zoning is considered while preparing the design?			
G: Electrical				
G.1	Whether preliminary lighting plans prepared?			
G.2	Whether preliminary power layout plans prepared?			
G.3	Whether substation location is shown?			
G.4	Whether electrical room located?			
G.5	Whether cable tray and other raceways route shown?			
G.6	Whether riser shaft for cables, bus ducts considered?			
G.7	Whether system schematic diagram prepared?			
G.8	Whether fire alarm system complies with regulatory requirements?			
G.9	Whether preliminary plans for all low voltage systems (communication, public address, audio–visual, access control, and security system) prepared?			
G.10	Whether emergency diesel generator is considered?			
H: Landscape				
H.1	Whether landscape plans are prepared?			
H.2	Whether plants selected?			
H.3	Whether irrigation system layout prepared?			
I: External				
I.1	Whether site layout plans showing roads, walkways, and parking areas prepared?			
I.2	Whether street lighting plans prepared?			
I.3	Whether project site boundaries are properly marked and demarketed?			
J: Financial				
J.1	Is project cost properly estimated?			
K: Schedule				
K.1	Is the project schedule practically achievable?			
L: Value Engineering				
L.1	Whether value engineering study performed and recommendations taken care?			

(*Continued*)

TABLE 7.66 (*Continued*)
Analysis of Schematic Design

Sr. No.	Description	Yes	No	Notes
M: Reports				
M.1	Whether the narrative description complete and include adequate information about the project?			
M.2	Whether outline specifications include all the works?			
M.3	Whether preliminary contract documents have taken care all the TOR requirements?			
N: Drawings, Sketches				
N.1	Whether drawings, sketches for all trades prepared as per TOR?			
O: Models				
O.1	Whether the models meet the design objectives?			
P: Submittals				
P.1	Whether numbers of sets prepared as per TOR?			

7.4.2.14 Finalize Schematic Design

The final schematic design is prepared incorporating the comments, if any, found during analysis and review of the drawings and documents for submission to the owner/client.

7.4.2.15 Submit Schematic Design

Normally, the following items are submitted to the owner for his or her review and approval in order to proceed with the development of detail design:

1. Schematic design drawings
2. Outline specifications
3. Preliminary contract documents
4. Preliminary schedule
5. Budgetary cost estimate
6. Schematic design reports
7. Model

7.4.3 Design Development Phase

Design development is the third phase of the construction project life cycle. It follows the preliminary design phase and takes into consideration the configuration and the allocated baseline derived during the preliminary phase. Design development phase is also known as *detail design/detailed engineering*. The client-approved schematic (preliminary) design is the base for the preparation of design development or development of detail design. All the comments, suggestions on the schematic design from client, and regulatory bodies are reviewed and resolved to ensure that changes will not detract from meeting the project design goals/objectives. Detail design involve the process of successively breaking down, analyzing, and designing the structure and its components, so that it complies with the recognized codes and standards of safety and performance, while rendering the design in the form of drawings and specifications that will tell the contractors exactly how to build the facility to meet the owner's need. Figure 7.110 illustrates the major activities relating to design development phase developed based on project management process groups methodology.

7.4.3.1 Develop Design Development Requirements

Design development is enhancement of the work carried out during the schematic design phase. During this phase, a comprehensive design of works with detailed work breakdown structure of design, drawings, specifications, and contract documents are prepared. The

Design development phase

Management processes	Process management groups				
	Initiating process	Planning process	Execution process	Monitoring and controlling process	Closing process
Integration management	Schematic design deliverables	Project management plan	Data collection/site investigations	Design calculations	Detail design drawings
	Comments on schematic design		Detail design drawings	Interdisciplinary Coordination	
	TOR requirements		Bill of quantities	Compliance to TOR	
	Authorities' requirements		Model		
Stakeholder management	Identify design team	Identify stakeholders		Design progress	
		Stakeholders requirements			
		Stakeholders matrix			
Scope management		Owner's needs, project goals and objectives	Project specifications	Authorities approval	
		Design development		Compliance to owner's need	
		Design documents		Stakeholders approval	
		Design deliverables			
		Bill of quantity			
Schedule management		Activity duration		Project schedule	
		Precedence diagram			
		Construction schedule			
Cost management		Price analysis		Project budget	
		Bill of quantities			
		Resources			
		Detail estimate			
Quality management		Codes and standards	Design coordination with all disciplines	Design compliance to owner's goals and	
		Regulatory requirements	Assure design quality	Coordination with all disciplines	
		Design crieteria		Control design quality	
		Well-defined specifications			
		Plan design quality			
Resource management		Estimate project resources	Manage team members From all discipline	Performance of team Project members	
Communication management		Communication matrix	Liaison with all disciplines	Design status information	
			Coordination meetings		
Risk management		Identification of risk during bidding, construction, testing, and commissioning		Design risk control	
Contract management		Bidding and tendering Documents		Check for contracting System	Contract documents
HSE management		Safety in design	Safety requirements		
		Environmental compatibility	Environmental requirements	HSE compliance in design	
Financial management		Designer/consultant payment		Payment to designer/consultant	
Claim management		Design change payment		Control changes	Settle designer's claim

FIGURE 7.110
Major activities in relation to design development processes. *Note*: These activities may not be strictly sequential; however, the breakdown allows implementation of project management function more effective and easily manageable at different stages of the project phase.

design development phase is the realm of design professionals, including architects, interior designers, landscape architects, and several other disciplines such as civil, mechanical, electrical, and other engineering professionals as needed.

During this phase detailed plans, sections, and elevations are drawn to scales, principle dimensions are noted, and design calculations are checked to conform the accuracy of design and its compliance to the codes and standards.

7.4.3.1.1 Identify Design Development Requirements

In order to identify requirements to develop the detail design, the designer has to gather comments made by the owner/project manager on the submitted schematic design, collect TOR requirements, regulatory requirements, and other related data to ensure that the developed design is accurate to the possible extent, free of errors, and with minimum omissions. Detail design activities are similar, although more in-depth than the design activities in the preliminary design stage. The size, shape, levels, performance characteristics, technical details, and requirements of all the individual components are established and integrated into the design. Design engineers of different trades have to take into consideration all these as a minimum while preparing the scope of works. The range of design work is determined by the nature of the construction project.

For developing a detail design, the designer has to consider following points:

1. Review of comments on the preliminary design by the client (project manager)
2. Review of comments on the preliminary design by the regulatory authorities.
3. Preparation of detail design for all the works
4. Interdisciplinary coordination to resolve the conflict
5. Obtain regulatory approval
6. Prepare project schedule
7. Prepare project budget
8. Prepare BOQ
9. Preparing specifications

7.4.3.1.1.1 Gather Schematic Design Comments The objective of detail design is the enhancement of the schematic design and to develop a clearly defined design based on the client's requirements. While developing the detail design, the designer has to review the approved schematic design drawings, reports, and take into consideration the comments, if any, made on the schematic design.

7.4.3.1.1.2 Identify TOR Requirements Normally, the TOR lists the requirements guidelines to develop the detail design. It mainly consists of the following:

1. Detail design drawings for the approved schematic design
2. BOQ
3. Project specifications
4. Contract documents
5. Project schedule
6. Definitive estimate
7. Format, scales, and size of reports and drawings

The TOR lists the following items to be developed during the detail design, taking into consideration the approved schematic design and comments:

A. Drawings

A.1 Project title

A.2 Drawing index, legends, and symbols

A.3 Existing site condition

A.4 Site plans

A.5 Architectural design

A.6 Interior design

A.7 Structural design

A.8 Mechanical design

A.8.1 Public health

A.8.2 Fire suppression

A.9 HVAC design

A.10 Electrical design

A.11 Smart building systems

A.12 Special systems

A.13 Landscape design

A.14 External works

B. Project specifications

C. Contract documents

D. Project schedule

E. Definitive cost estimate

F. Authorities approvals

7.4.3.1.1.3 Collect Owner's Requirements Any additional requirements/changes to those considered during development of schematic design are considered while developing the detail design. This may include additional systems or changes in the preferred requirements.

7.4.3.1.1.4 Collect Regulatory Requirements Government agency regulatory requirements have considerable impact on precontract planning. Some agencies require that the design drawings are submitted for their preliminary review and approval to ensure that the designs are compatible with local codes and regulations. These include the following:

- Submission of drawings to electrical authorities showing the anticipated electrical load required for the facility
- Fire alarm and fire fighting system drawings
- Water supply and drainage system
- HVAC drawings to ensure compliance with energy conservation codes
- Technical details of conveying system

7.4.3.1.2 Establish Detail Design Requirements

Detail design requirements are developed considering all the items discussed under Section 7.4.3.1.1.

7.4.3.2 Develop Project Management Plan

Preliminary project management plan developed during the schematic phase is updated based on additional information collected during the detail design phase. The contents of the project management plan are same as that of schematic design but additional information is added. A typical TOC of a project management plan is as given below:

1. Project description
2. Project objectives
3. Project organization
   3.1 Organizational chart
   3.2 Responsibility matrix
   3.3 Project directory
4. Project (scope) deliverables
5. Project schedule
6. Project budget
7. Project quality plan
8. Project resources
9. Risk management
10. Communication matrix
11. Contract management
12. HSE management
13. Project finance management
14. Claim settlement

7.4.3.3 Identify Project Stakeholder

The following stakeholders have direct involvement in the project during the detail design phase:

- Owner
- Consultant
- Designer
- Regulatory authorities
- Project manager/agency construction manager (as applicable)

7.4.3.3.1 Identify Project Team Members

Generally, the project team members selected to develop schematic design continues during the detail design phase. Additional design personnel are included to meet the work load to develop detail design.

7.4.3.3.2 Identify Design Team Members

Figure 7.111 illustrates the design management team and their major responsibilities.

Each of the managers has many other team members. These members are selected based on the organizational structure and suitable skills required to perform the job. These include the following:

1. Team leader (principal engineer)—each trade
2. Team members (engineers and CAD technicians for each discipline)
3. Quantity surveyor (cost engineer)
4. Owner's representative
5. End user

7.4.3.3.3 Establish Stakeholders' Requirement

A distribution matrix based on the interest and involvement of each stakeholder is prepared, and the appropriate documents are sent for their action/information as per the agreed upon requirements.

7.4.3.3.4 Develop Responsibility Matrix

Table 7.67 illustrates responsibilities of various participants during the detail design phase.

7.4.3.4 Develop Design Development Scope

Design development scope is developed taking into consideration the requirements established under Section 7.4.3.1.1, which are as follows:

- Approved schematic design documents
- TOR requirements

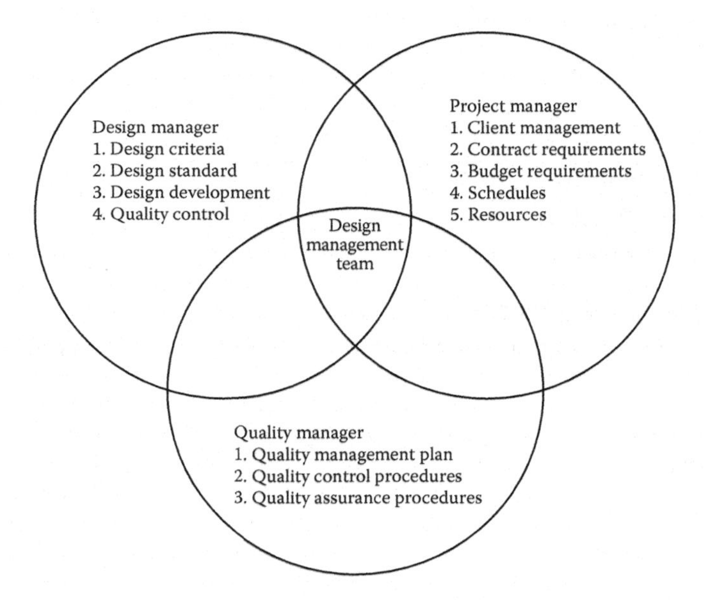

FIGURE 7.111
Design management team. (Reprinted with permission from Rumane, A.R., *Quality Tools for Managing Construction Projects*, CRC Press, Boca Raton, FL, 2013.)

TABLE 7.67

Responsibilities of Various Participants (Design–Bid–Build Type of Contracts) during Design Development Phase

	Responsibilities		
Phase	**Owner**	**Designer**	**Regulatory Authorities**
Design development	• Approval of design • Approval of time schedule • Approval of budget	• Development of detail design • Submission for Authorities approval • Detail plans • Schedule • Budget • BOQ • Verification of design	• Review and approval of project submittals

- Owner's preferred requirements
- Regulatory requirements
- Approved preliminary project schedule
- Approved preliminary cost estimate
- Codes and standards

The purpose for the development of the detail design is to provide sufficient information and detail to ensure that the work to be performed by the contractor is properly identified and appropriately addressed taking necessary measure to mitigate errors and omissions in the design. The scope of work during the detail design phase mainly comprise

- Design development drawings
- Technical specifications
- BOQ
- Project schedule
- Detail cost estimate
- Authorities approvals
- Design development report

7.4.3.4.1 Identify Design Deliverables

Based on the scope for design development, the following are the major deliverables to be developed during this phase:

1. Design drawings
   a. Site plans
   b. Architectural
   c. Interior
   d. Structural
   e. Conveying system
   f. Mechanical (public health and fire suppression)

   g. HVAC
   h. Electrical
   i. Low-voltage system
   j. Landscape
   k. External
2. Specifications
3. Project schedule
4. Definitive cost estimate
5. Regulatory approvals

Table 7.68 lists the design development deliverables to be developed during the design development phase.

7.4.3.5 Develop Detail Design

Figure 7.112 illustrates stages to develop detail design.

The designer has to collect data/information, perform site investigations, and verify the information is same as collected earlier.

7.4.3.5.1 Collect Data/Information

The designer has to collect all the missing data to ensure that detail design is developed without any errors and omissions.

7.4.3.5.2 Perform Site Investigations

The designer has to revisit the site to ensure that there are not many changes to the earlier performed investigations.

7.4.3.5.3 Prepare Detail Design

The following detail drawings for building project are generally prepared during this phase:

- Architectural
    - Site plans
    - Floor plans
    - Roof plans
    - Partitions
    - Sections
    - Elevations
    - Reflected ceiling plan
    - Finishes schedule
    - Door schedule
    - Typical window details
    - Furnishings
- Structural
    - Building structure
    - Foundation system
    - Footings

TABLE 7.68

Design Development Deliverables

Sr. No.	Deliverables
1	General
1.1	l. Project specifications
	m. Project schedule
	n. Project estimate
	o. Bill of quantities
	p. Calculations (all trades)
	q. Site investigations
	r. Design report
	s. Model
2	Detail design drawings
2.1	Architectural
2.1.1	Overall site plans
2.1.2	Floor plans
2.1.3	Roof plans
2.1.4	Elevations
2.2	Structural
2.3	Elevator
2.4	Plumbing and fire suppression
2.5	HVAC
2.6	Electrical
2.7	Landscape
2.8	External

    - Stairs
    - Roof
    - Sections
    - Structural floor plans
- Conveying system
    - Elevator location
    - Machine room
- MEP
    - General arrangement of each system
    - Single line riser diagram
    - Electrical schematic diagram
    - Layout of equipment rooms
    - Layout of electrical substation
    - Electrical rooms
    - Vertical shafts
- Security system
- External works
- Landscape plans

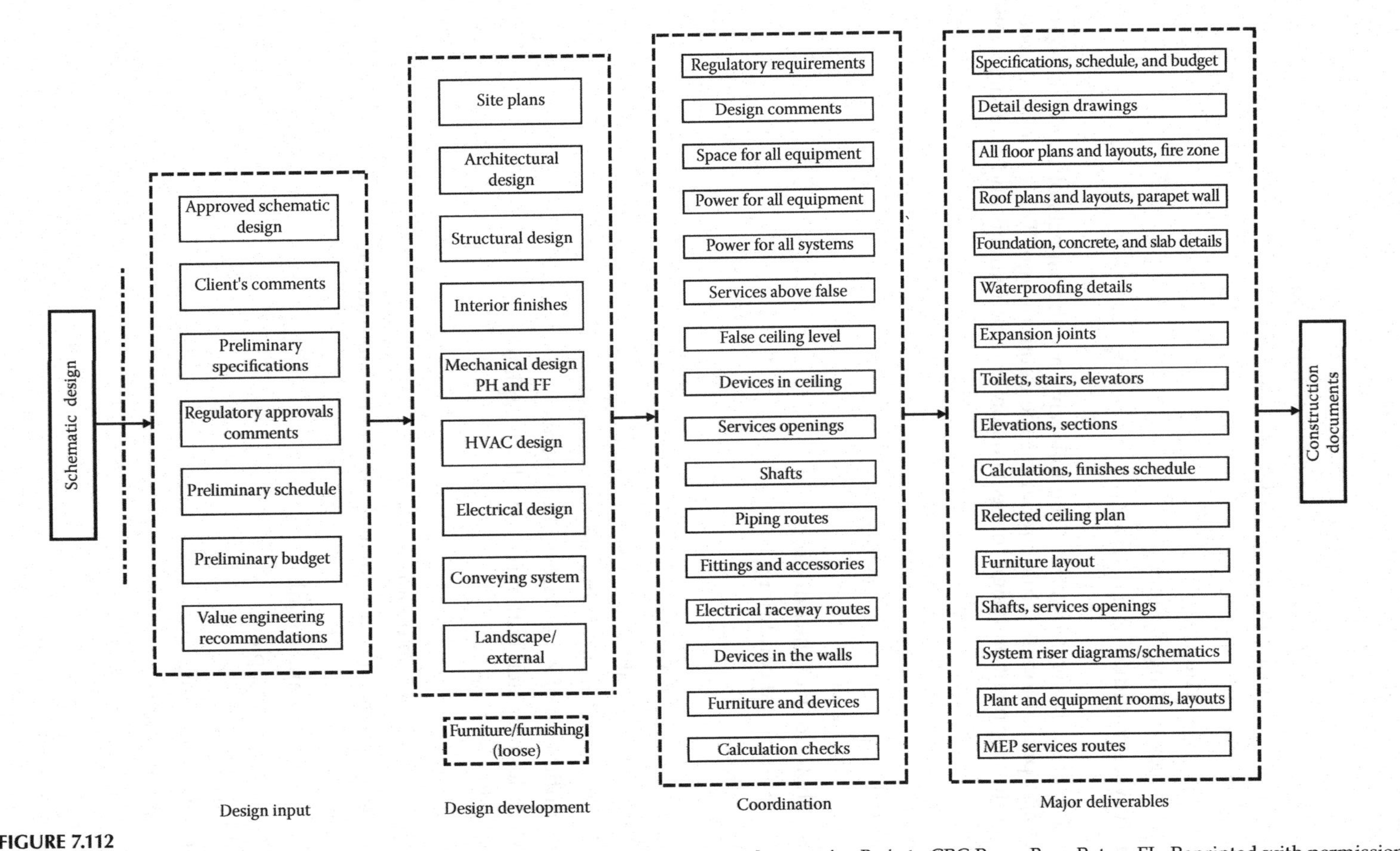

FIGURE 7.112
Design development stages. (Abdul Razzak Rumane (2013), *Quality Tools for Managing Construction Projects*, CRC Press, Boca Raton, FL. Reprinted with permission from Taylor & Francis Group.)

7.4.3.5.3.1 Design Calculations The designer has to submit the following calculations along with the detail design:

- Calculation of all structural elements (structural works)
- Calculation of lighting (Isolux calculations)
- Short circuit calculations for cables
- Lightening protection system
- Earthing (grounding) system
- Sound system (audio–visual system)

7.4.3.5.3.2 Detail Design of the Works The following are the aspects of work to be considered by design professionals while preparing the detail design. These can be considered as a base for development of design to meet customer requirements and shall help achieve the qualitative project.

Architectural design

- Intent/use of building/facility
- Property limits
- Aesthetic look of the building
- Environmental conditions
- Elevations
- Plans
- Axis, grids, and levels
- Room size to suit the occupancy and purpose
- Zoning as per usage/authorities requirements
- Identification of zones, areas, and rooms
- Modules to match with structural layout/plan
- Number of floors
- Ventilation
- Thermal insulation details
- Stairs, elevators (horizontal and vertical transportation)
- Fire exits
- Ceiling height and details
- Reflected ceiling plan
- Internal finishes
- Internal cladding
- Partition details
- Masonry details
- Joinery details
- Schedule of doors and windows

- Utility services
- Toilet details
- Required electromechanical services
- External finishes
- External cladding
- Glazing details
- Finishes schedule
- Door schedule
- Windows schedule
- Hardware schedule
- Special equipment
- Fabrication of items, such as space frame, steel construction, and retaining wall, having special importance for appearance/finishes
- Special material/product to be considered, if any
- Any new material/product to be introduced
- Conveying system core details
- Maintenance access for equipment/services requirements
- Ramp details
- Hard and soft landscape
- Parking areas
- Provision for future expansion (if required)

Concrete structure

- Property limits/surrounding areas
- Type of foundation
- Energy efficient foundation
- Design of foundation based on field and laboratory test of soil investigation, which gives following information:
    i. Subsurface profiles, subsurface conditions, and subsurface drainage
    ii. Allowable bearing pressure and immediate and long-term settlement of footing
    iii. Coefficient of sliding on foundation soil
    iv. Degree of difficulty for excavation
    v. Required depth of stripping and wasting
    vi. Methods of protecting below grade concrete members against impact of soil and ground water (water and moisture problems, termite control, and radon where appropriate)
    vii. Geotechnical design parameters such as angle of shear resistant, cohesion, soil density, modulus of deformation, modulus of sub-grade reaction, and predominant soil type
    viii. Design loads such as dead load, live load, wind load, and seismic load

- Footings
- Grade and type of concrete
- Size of bars for reinforcement and the characteristic strength of bars
- Clear cover for reinforcement for
    i. Raft foundation
    ii. Underground structure
    iii. Exposed to weather structure such as columns, beams, slabs, walls, and joists
    iv. Not exposed to weather columns, beams, slabs, walls, and joists
- Reinforcement bar schedule and stirrup spacing
- Expansion joints
- Concrete tanks (water)
- Insulation
- Services requirements (shafts and pits)
- Shafts and pits for conveying system
- Location of columns in coordination with architectural requirements
- Number of floors
- Height of each floor
- Beam size and height of beam
- Openings for services
- Substructure
    i. Columns
    ii. Retaining walls
    iii. Walls
    iv. Stairs
    v. Beams
    vi. Slab
- Superstructure
    vii. Columns
    viii. Stairs
    ix. Walls
    x. Beams
    xi. Slabs
- Consideration of water proofing requirements for roof slab against water leakage
- Deflection that may cause fatigue of structural elements, crack or failure of fixtures, fittings or partitions, or discomfort of occupants
- Movement and forces due to temperature
- Equipment vibration criteria
- Load sensors to measure deflection
- Reinforcement bar schedule and stirrup spacing

- Building services to fit in the building
- Environmental compatibility
- Parapet wall
- Excavation
- Dewatering
- Shoring
- Backfilling

Elevator works

- Type of elevator
- Loading capacity
- Speed
- Number of stops
- Travel height
- Cabin, cabin accessories, cabin finishes, and car operating system
- Door, door finishes, and door system
- Safety features
- Drive, size, and type of motor
- Floor indicators and call button
- Control system
- Cab overhead dimensions
- Pit depth
- Hoist way
- Machine room
- Operating system

Fire suppression system: The fire protection system is mainly to provide protection against fire to life and property. The system is designed taking into consideration local fire code and National Fire Protection Association (NFPA) standards. The system includes the following:

- Sprinkler system for fire suppression in all the areas of the building
- Hydrants (landing valve) for professional fighting
- Hose reel for public use throughout the building
- Gaseous fire protection system for communication rooms
- Fire protection system for diesel generator room
- Size of fire pumps and controls
- Pump size (gpm)-based on applicable codes and standards
- Fittings and accessories for pressure drop in the system
- Friction loss in the most remote area
- Size of elbow or tees fitted on the upstream of fire pump
- Space requirements for pumps and other equipment (pump room)

- Source of water supply for fire pumps
- Water storage facility
- Interface with other related systems

Plumbing works

- Maximum working pressure to have adequate pressure and flow of water supply
- Location of pressure relief valves
- Maximum design velocity
- Maximum probable demand
- Demand weight of fixture in fixture units public use
- Friction loss calculation
- Pipe slope
- Maximum hot water temperature at fixture outlet
- Water heater outlet hot temperature
- Providing isolating valves to ensure that the system shall be easy for maintenance
- Hot water system
- Central water storage capacities
- Location of storage tank
- Size of pumps and controls
- Grounding of equipment and metal piping
- Space requirements for pump and other equipment (pump room)
- Schematic diagram for water distribution system

Drainage system: While designing the drainage system, schedule of foul drainage demand units and frequency factors for following items to be considered for sizing of piping system, no. of manholes, capacity of sump pit, and size of sump pump and following:

- Wash basins
- Showers
- Urinals
- Water closets
- Kitchen sinks
- Location of vents and traps
- Location of cleanouts
- Other equipment such as dishwashers and washing machines

HVAC works

- Environmental conditions
  - Outdoor design conditions
  - Indoor design conditions
- Air-conditioning calculations
  - Cooling load calculations

  - Heating load calculations
  - Space temperature and humidity at required set point
  - Occupancy load
  - Lighting load
- Room pressurizing and leakage calculations
- Energy consumption calculations
- Air-conditioning calculations for IT equipment room(s) based on heat emission of equipment
- Air distribution system calculations
- Smokes extract ventilation calculations
- Exhaust ventilation calculations
- Ductwork sizing calculations
- Selection of the ductwork components such as balancing dampers, constant volume boxes, variable air volume boxes, attenuators, grilles and diffusers, fire dampers, and pressure relief dampers
- Pipe work sizing calculations
- Selection of the inline pipe work components, for example, valves, strainers, air vents, commissioning sets, flexible connections, and sensors
- Selection of boilers, pressurization units, and air-conditioning calculations
- Pipe work and duct work insulation selection
- Details of grilles and diffusers and control valves
- Selection of the duct work systems plant and equipment, for example, air handling units, fan coil units, filters, coils, fans, humidifiers, and duct heaters
- Selection of chillers and cooling towers
- Selection of pumps
- Selection of fans
- Equipment system calculations
- Space requirements for chillers, cooling towers, pumps, and other equipment (plant room)
- Mechanical room location and access
- Preparation of the plan and section layouts and plant room drawings
- Electrical load calculations
- Comparison of electrical consumption with electrical conservation code
- Preparation of equipment schedules
- HVAC-related electrical works
- Control details
- Starter panels, motor control center (MCC) panels, and schematic diagram of MCC
- Selection of program equipment
- Preparation of point schedule for building management system
- Schematic diagram for BMS

Electrical system

- Lighting calculations for different areas based on illumination level recommended by CIE/CEN/CIBSE and Isolux diagrams
- Selection of light fittings and type of lamps
- Selection of control gear for light fixture
- Environmental consideration for selection of light fixture and control gear
- *Exit/emergency* lighting system
- Circuiting references, normal as well as emergency
- Sizing of conduits
- Power for wiring devices
- Power supply for equipment (HVAC, PH&FF, conveying system, and others)
- Sizing of cable tray
- Sizing of cable trunking
- Selection (type and size) of wires and cable
- Voltage drop calculations for wires and cables
- Selection of upstream and downstream breakers
- Derating factor
- Sensitive of breakers (degree of protection)
- Selection of isolators
- IP ratings (degree of ingress protection) of panels, boards, and isolators
- Schedule of distribution boards, switch boards, and main low tension boards
- Cable entry details
- Location of distribution boards, switch boards, and low tension panels
- Short circuit calculations
- Sizing of diesel generator set for emergency power supply
- Sizing of automatic transfer switch
- Generator room layout
- Sizing of capacitor bank
- Provision for solar system integration
- Schematic diagrams
- Sizing of transformers
- Substation layout
- Calculations for grounding (earthing) system
- Grounding system layout
- Calculations for lightning protection system
- Lightning protection system layout

Fire alarm system: Fire alarm system is designed by taking into consideration the local fire code and NFPA standards. The system includes the following:

- Conduiting and raceways

- Type of system—analog/digital/addressable
- Type of detectors based on the area and spacing between the detectors and the walls
- Break glass/pull station
- Type of horns/bells
- Voice evacuation system, if required
- Type of wires and cables
- Mimic panel, if required
- Repeater panel, if required
- Main control panel
- Location of panels
- Interface with other systems such as HVAC, elevator
- Riser diagram

Information and communication technology

- Structured cabling considering type and size of cable, copper, and fiber optic
- Racks
- Wiring accessories/devices
- Access/distribution switches
- Internet switches
- Core switch
- Access gateway
- Router
- Network management system
- Servers
- Telephone handsets

Public address system

- Conduiting and raceways
- Type of system—analog/digital/IP-based
- Types of wires and cables
- Types of speakers
- Distribution of speakers
- Required noise level in different areas
- Calculations for sound pressure level
- Zoning of system, if required
- Size and type of premixer
- Size and type of amplifier
- Microphones
- Paging system
- Message recorder/player
- Interface with other systems

Audio–visual system

- Conduiting and raceways
- Type of system—analog/digital/IP-based
- Types of wires and cables
- Racks
- Type, size, and brightness of projectors
- Type and size of speakers and sound pressure level
- Type and size of screens
- Microphones
- Cameras (visualizers)
- CD/DVD players and recorders
- Control processors
- Video switch matrix
- Mounting details of equipment

Security system/CCTV

- Type of system—digital/IP-based
- Conduiting and raceways
- Wires and cabling network
- Level of security required
- Type and size of cameras
- Types of monitors/screens
- Video/event recording
- Video servers
- Database server
- System software
- Schematic diagram
- System console

Security system/access control

- Conduiting and raceways
- Wires and cabling network
- Proximity RFID reader
- Fingerprint and proximity combine reader
- Magnetic lock
- Release button
- Door contact
- RFID card
- Reader control panel
- Multiplexers

- Monitors
- Server
- Workstation
- Metal detector
- Road blocker
- Parking control

Landscape: As a landscape architect, the following points are to be considered while designing the landscape system:

- Property boundaries
- Size and shape of the plot
- Shape and type of dwelling
- Integration with surrounding areas
- Orientation to the sun and wind
- Climatic/environmental conditions
- Ecological constraints (soil, vegetation, etc.)
- Location of pedestrian paths and walkways
- Pavement
- Garage and driveway
- Vehicular circulation
- Location of sidewalk
- Play areas and other social/community requirements
- Outdoor seating
- Location of services, positions of the both under and above utilities and their levels
- Location of existing plants, rocks, or other features
- Site clearance requirements
- Foundation for paving including front drive
- Top soiling or top soil replacement
- Soil for planting
- Planting of trees, shrubs, and ground covers
- Grass area
- Sowing grass or turfing
- Lighting poles/bollard
- Special features, if required
- Signage, if required
- Surveillance, if required
- Installation of irrigation system
- Marking out the borders
- Storage for landscape maintenance material

External works (infrastructure and road): External works are part of the contract requirements of a project that involves construction of service road and other infrastructure facilities to be connected to the building and also to take care of existing services passing through the project boundary line. The designer has to consider following while designing external works.

- Grading material
- Asphalt paving for road or street
- Pavement
- Pavement marking
- Precast concrete curbs
- Curbstones
- External lighting
- Cable routes
- Piping routes for water, drainage, storm water system
- Trenches or tunnels
- Bollards
- Manholes and hand holes
- Traffic marking
- Traffic signals
- Boundary wall/retaining wall, if required

Bridges: Designer shall use relevant authorities' design manual and standards and consider following points while designing the bridges

- Soil stability
- Alignment with road width, property lines
- Speed
- Intersections/interchanges
- Number of lanes, width
- Right of way lines
- Exits, approaches, and access
- Elevation datum
- Superelevation
- Clearance with respect to rail road, roadway, and navigation (if applicable)
- High and low level of water (if applicable)
- Utilities passing through the bridge length
- Slopes
- Number and length of span
- Live loads, bearing capacity
- Water load, wind load, earthquake effect (seismic effect)
- Bridge rails, protecting screening, guardrails, and barriers

- Shoulder width
- Footings, columns, and piles
- Abutment
- Beams
- Substructure
- Super structure, deck slab
- Girders
- Slab thickness
- Reinforcement
- Supporting components, deck hanger, and tied arch
- Expansion and fixed joints
- Retaining walls, crash wall
- Drainage
- Lighting
- Aesthetic
- Sidewalk, pedestrian, and bike facilities
- Signage, signals
- Durability
- Sustainability

Highways: Designer shall use relevant authorities' design manual and standards and consider following points while designing the highways.

- Type of highway
- Soil stability
- Speed
- Number of lanes, width
- Shoulder width
- Gradation
- Subgrade level
- Sub-base level
- Types of asphalt courses
- Pavement details
- Type of pavement and thickness
- Right of way lines
- Exits, approaches, access, and ramp
- Superelevation
- Slopes, curvature, and turning
- Median, barriers, and curb
- Sidewalks and driveways

- Pedestrian accommodation
- Bridge roadway width
- Drainage
- Gutters
- Special conditions such as snow and rains
- Lighting
- Substation to feed street lights
- Signage, markings
- Signals
- Trees and plants
- Parking area
- Rest area
- Toll station, if any
- Speed detectors, if applicable
- Durability
- Sustainability

Furnishings/furniture (loose): In building construction projects, loose furnishings/furniture is tendered as a special package and is normally not part of main contract. In order to express all the features of the furnishing/furniture products in the specification the descriptive features of the product is not enough. In order to give enough information and understanding of the product, the product specifications are accompanied with the pictorial view/cut out sheet/photo of the product and the furniture layout.

Solar system: The designer should consider the following points while developing solar power system:

- Avoid shading from trees and buildings (especially during peak sunlight hours).
- Check the proposed plan for the proposed site to ensure that future, neighboring construction will not cast shade on the array.
- Determine where solar array can be placed (roof, carport, facade, curtain wall, boundaries, double skin or elsewhere).
- Keep the south-facing section obstruction-free, if possible. If the roof is sloped, the south-facing section will optimize the system performance.
- Minimize rooftop equipment to maximize available open area for solar collector placement.
- Ensure that the design of type of roof will have adequate space to install solar system at later stage to optimize cost of installing solar system at later stage.
- Ensure that the roof is capable of carrying the load of the solar equipment.
- Analyze wind loads on rooftop solar equipment in order to ensure that the roof structure is sufficient.
- Add additional safety equipment for solar equipment access and installation.

7.4.3.5.3.3 Develop Documents The designer/consultant is responsible to prepare detail specifications and contract documents that meets the owner's needs, and specifies the required level of quality, schedule, and budget.

Prepare specifications: Specifications of work quality are an important feature of construction project design. Specifications of required quality and components represent part of the contract documents and are detailed under various sections of particular specifications. Generally, the contract documents include all the details as well as references to generally accepted quality standards published by international standards organizations. Proper specifications and contract documentations are extremely important as these are used by the contractor as a measure of quality compliance during construction process.

Particular specifications consist of many sections related to specific topic. Detailed requirements are written in these sections to enable contractor understand the product or system to be installed in the construction project. The designer has to interact with project team members and owner while preparing the contract documents.

Generalized writing of these sections is as follows:

Section No.

Title

Part 1- General

1.01- General reference/related sections

1.02- Description of work

1.03- Related work specified elsewhere in other sections

1.04- Submittals

1.05- Delivery, handling, and storage

1.06- Spare parts

1.07- Warranties

In addition to above, a reference is made for items such as preparation of mock up, quality control plan, and any other specific requirement related to the product or system specified herein.

Part 2- Product

2.01- Materials

2.02- Manufacturer's qualification/list of recommended manufacturers

Part 3- Execution

3.01- Installation

3.02- Site quality control

Prepare contract documents: Preparation of detailed documents and specifications as per MasterFormat is one of the activities performed during this phase of the construction project. The contract documents must specify the scope of works, location, quality, and duration for completion of the facility. As regards the technical specifications of the construction project, MasterFormat specifications are included in the contract documents. Normally, construction documents are prepared as per MasterFormat contract documents produced jointly by the CSI and CSC are widely accepted as standard practice for preparation of contract document.

MasterFormat is a master list of section titles and numbers for organizing information about construction requirements, products, and activities into a standard sequence. MasterFormat is a uniform system for organizing information in project manuals, for organizing cost data, for filling product information and other technical data, for identifying drawing objects, and for presenting construction market data. MasterFormat 2014 edition consists of 48 divisions (49 is reserved). For more information, see Chapter 6.

7.4.3.6 Prepare Schedule

The project schedule is developed using bottom-up planning details using key events. It is also known as Class 1 schedule or Schedule Level 3 (see Table 7.20 and Table 7.21).

Figure 7.113 illustrates typical time schedule for a construction project. The example schedule is construction of a branch for a bank.

7.4.3.7 Estimate Cost

The cost estimate during this phase is based on elemental parametric methodology. It is also known as *detailed costing* (detailed estimate) (see Table 7.23).

7.4.3.8 Manage Design Quality

In order to reduce errors and omissions, it is necessary to review and check the design for quality assurance by the quality control personnel from the project team through itemized review checklists to ensure that design drawings fully meet owner's objectives/goals. It is also required to review the design with the owner to ensure a mutual understanding of the build process. The designer has to ensure that the installation/execution specification details are comprehensively and correctly described and also the installation quality requirements for systems are specified in details.

The designer has to plan quality (planning of design work), perform quality assurance, and control quality for preparing detail design. This will mainly consist of the following:

1. *Plan quality*:
    - Review comments on schematic design
    - Determine number of drawings to be produced
    - Establish scope of work for preparation of detail design
    - Identify requirements listed under TOR
    - Identify quality standards and codes to be complied
    - Establish design criteria
    - Identify regulatory requirements
    - Identify environmental requirements
    - Establish quality organization with responsibility matrix
    - Develop design (drawings and documents) review procedure
    - Establish submittal plan
    - Establish design review procedure
2. *Quality assurance*:
    - Collect data
    - Investigate site conditions

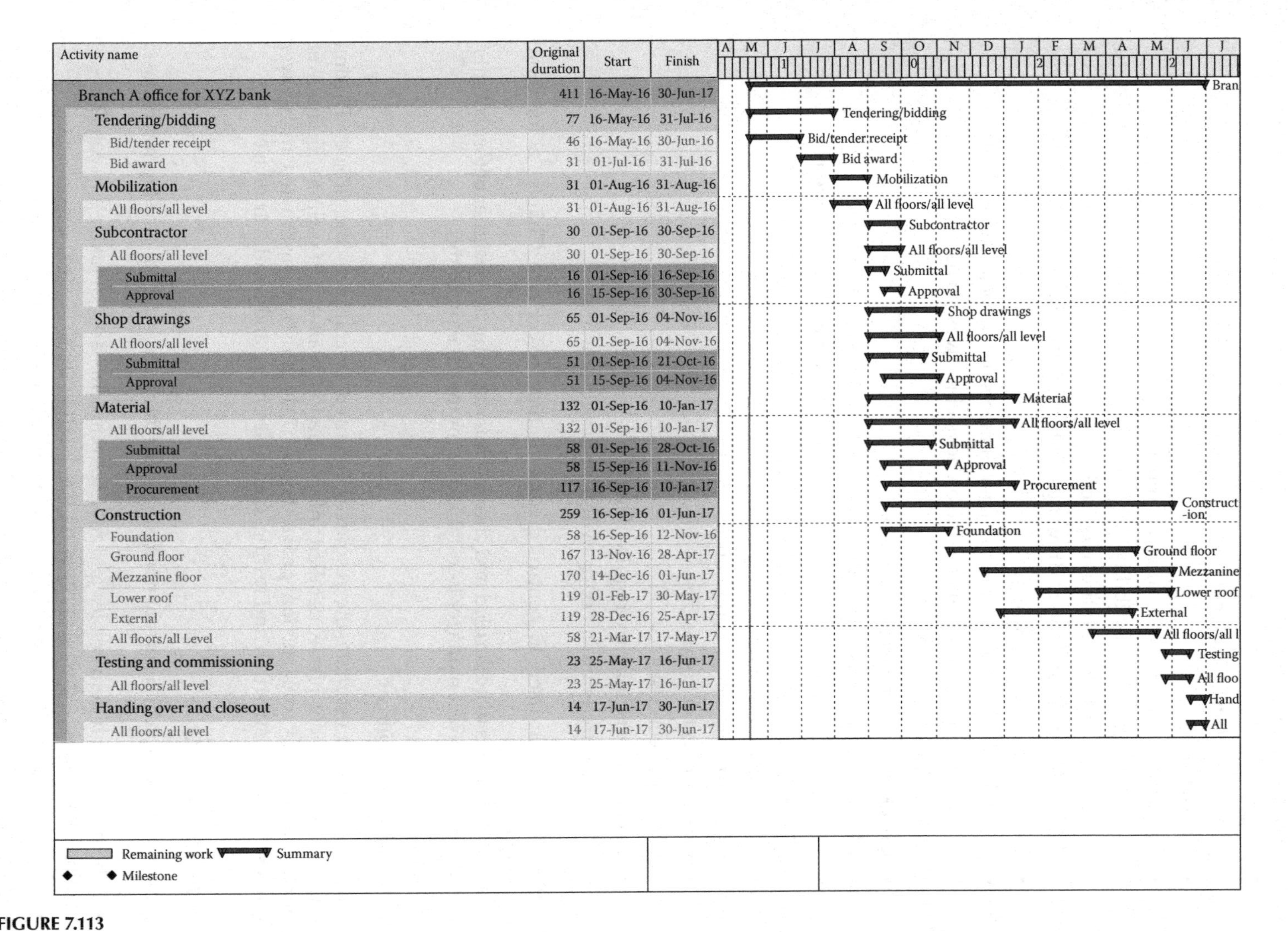

Activity name	Original duration	Start	Finish
Branch A office for XYZ bank	411	16-May-16	30-Jun-17
Tendering/bidding	77	16-May-16	31-Jul-16
Bid/tender receipt	46	16-May-16	30-Jun-16
Bid award	31	01-Jul-16	31-Jul-16
Mobilization	31	01-Aug-16	31-Aug-16
All floors/all level	31	01-Aug-16	31-Aug-16
Subcontractor	30	01-Sep-16	30-Sep-16
All floors/all level	30	01-Sep-16	30-Sep-16
Submittal	16	01-Sep-16	16-Sep-16
Approval	16	15-Sep-16	30-Sep-16
Shop drawings	65	01-Sep-16	04-Nov-16
All floors/all level	65	01-Sep-16	04-Nov-16
Submittal	51	01-Sep-16	21-Oct-16
Approval	51	15-Sep-16	04-Nov-16
Material	132	01-Sep-16	10-Jan-17
All floors/all level	132	01-Sep-16	10-Jan-17
Submittal	58	01-Sep-16	28-Oct-16
Approval	58	15-Sep-16	11-Nov-16
Procurement	117	16-Sep-16	10-Jan-17
Construction	259	16-Sep-16	01-Jun-17
Foundation	58	16-Sep-16	12-Nov-16
Ground floor	167	13-Nov-16	28-Apr-17
Mezzanine floor	170	14-Dec-16	01-Jun-17
Lower roof	119	01-Feb-17	30-May-17
External	119	28-Dec-16	25-Apr-17
All floors/all Level	58	21-Mar-17	17-May-17
Testing and commissioning	23	25-May-17	16-Jun-17
All floors/all level	23	25-May-17	16-Jun-17
Handing over and closeout	14	17-Jun-17	30-Jun-17
All floors/all level	14	17-Jun-17	30-Jun-17

FIGURE 7.113
Project schedule.

- Prepare design drawings
- Prepare detailed specifications
- Prepare contract documents
- Prepare BOQ
- Ensure functional and technical compatibility
- Ensure design constructibility
- Ensure meeting of operational objectives
- Ensure full coordination of drawings with all disciplines
- Ensure cost-effective design
- Ensure selected/recommended material meets owner objectives
- Ensure that design fully meets the owner's objectives/goals

3. *Control quality*:
    - Check quality of design drawings
    - Check accuracy and correctness of design
    - Verify BOQ for correctness as per design drawings and specifications
    - Check specifications
    - Check contract documents
    - Check for regulatory compliance
    - Check project schedule
    - Check project cost
    - Check interdisciplinary requirements
    - Check required number of drawings prepared drawing

The BIM tool can be used to control quality of the project (see Chapter 5 for more information).

Table 7.69 lists items to be verified and checked internally by the designer before submission to the owner/project manager (see Table 4.14 for mistake proofing).

7.4.3.9 Estimate Resources

The designer has to estimate the resources required to complete the project. During the design development phase, detailed information is available to estimate manpower resources during the construction phase. Figure 7.114 illustrates estimated manpower resources during construction and testing, commissioning, and handover phase.

7.4.3.10 Manage Risks

The following are typical risks that normally occur during the design development phase:

- Schematic design deliverables and review comments are not taken into consideration by the designer while preparing the detail design.
- Regulatory authorities' requirements are not taken into consideration.
- Detail design scope of work is not properly established and is incomplete.

TABLE 7.69
Checklist for Design Drawings

Sr. No.	Items to Be Checked
1	Whether design meets owner requirements and complete scope of work (TOR)
2	Whether designs were prepared using authenticated and approved software
3	Whether design calculation sheets are included in the set of documents
4	Whether design is fully coordinated for conflict between different trades
5	Whether design has taken into consideration relevant collected data requirements
6	Whether reviewer's comments responded
7	Whether regulatory approval obtained and comments, if any, incorporated and all review comments responded
8	Whether design has environmental compatibility
9	Whether energy efficiency measures are considered
10	Whether design constructability is considered
11	Whether design matches with property limits
12	Whether legends matches with layout
13	Whether design drawings are properly numbered
14	Whether design drawings have owner logo and designer logo as per standard format
15	Whether the design format of different trades have uniformity
16	Whether project name and contract reference is shown on the drawing

Source: Abdul Razzak Rumane (2013), *Quality Tools for Managing Construction Projects*, CRC Press, Boca Raton, FL. Reprinted with permission from Taylor & Francis Group.

- The related project data and information collected is incomplete.
- The related project data and information collected are likely to be incorrect and wrongly estimated.
- Site investigations for existing conditions are not verified
- Fire and safety considerations recommended by the authorities not incorporated in the design.
- Environmental considerations are not taken into account.
- Incomplete design drawings and related information.
- Inappropriate construction method.
- Conflict with different trades.
- Interdisciplinary coordination not performed.
- Wrong selection of materials and systems.
- Undersize HVAC equipment selection.
- Incorrect water supply requirements.
- Estimated total electrical load is much lower than expected actual consumption.
- Traffic study for conveying system is not verified, taking into consideration final load.
- Prediction of possible changes in design during the construction phase.
- Inadequate and ambiguous specifications.

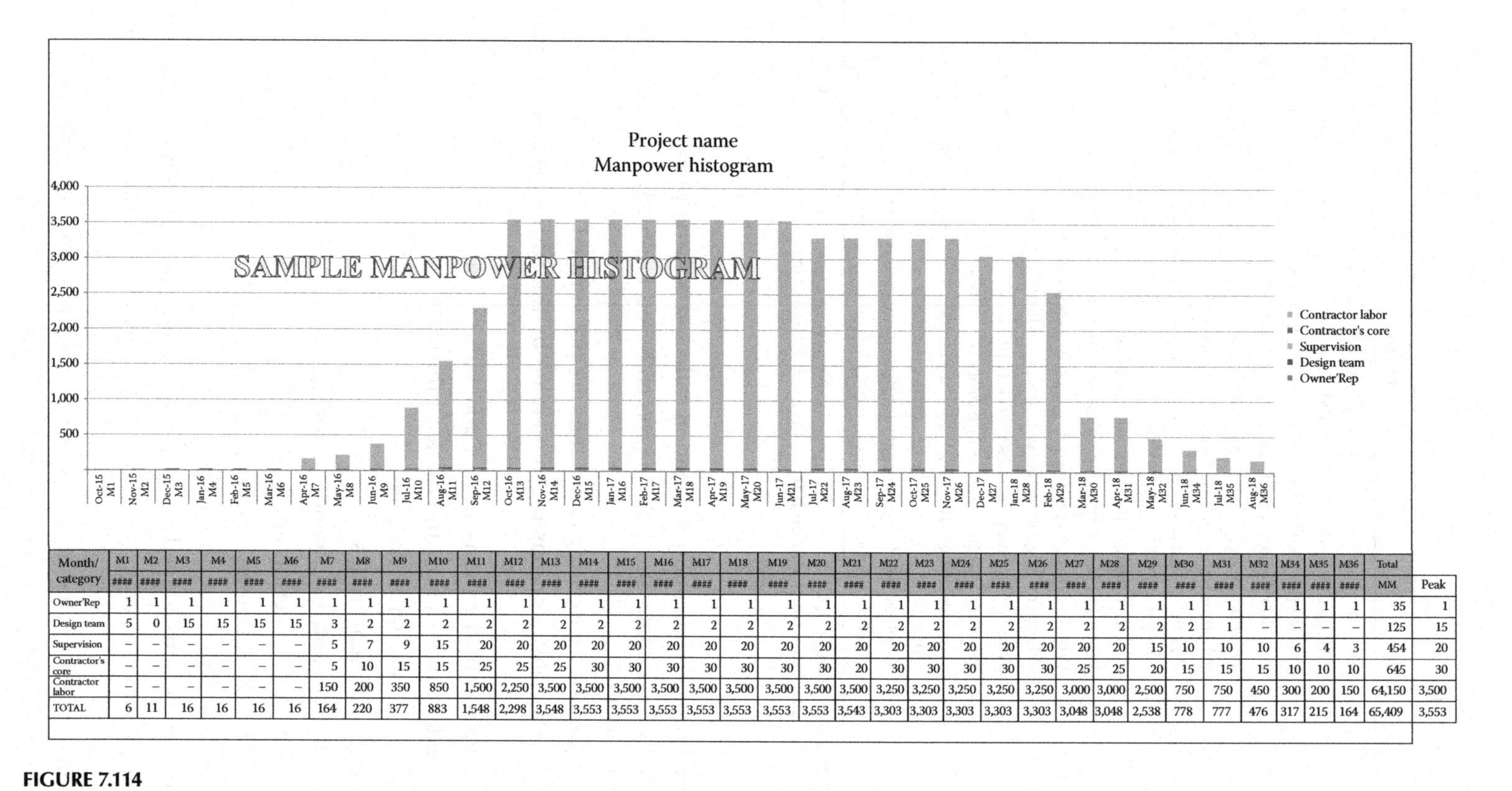

Month/ category	M1	M2	M3	M4	M5	M6	M7	M8	M9	M10	M11	M12	M13	M14	M15	M16	M17	M18
	####	####	####	####	####	####	####	####	####	####	####	####	####	####	####	####	####	####
Owner'Rep	1	1	1	1	1	1	1	1	1	1	1	1	1	1	1	1	1	1
Design team	5	0	15	15	15	15	3	2	2	2	2	2	2	2	2	2	2	2
Supervision	–	–	–	–	–	–	5	7	9	15	20	20	20	20	20	20	20	20
Contractor's core	–	–	–	–	–	–	5	10	15	15	25	25	25	30	30	30	30	30
Contractor labor	–	–	–	–	–	–	150	200	350	850	1,500	2,250	3,500	3,500	3,500	3,500	3,500	3,500
TOTAL	6	11	16	16	16	16	164	220	377	883	1,548	2,298	3,548	3,553	3,553	3,553	3,553	3,553

Month/ category	M19	M20	M21	M22	M23	M24	M25	M26	M27	M28	M29	M30	M31	M32	M34	M35	M36	Total	
	####	####	####	####	####	####	####	####	####	####	####	####	####	####	####	####	####	MM	Peak
Owner'Rep	1	1	1	1	1	1	1	1	1	1	1	1	1	1	1	1	1	35	1
Design team	2	2	2	2	2	2	2	2	2	2	2	2	1	–	–	–	–	125	15
Supervision	20	20	20	20	20	20	20	20	20	20	15	10	10	10	6	4	3	454	20
Contractor's core	30	30	20	30	30	30	30	30	25	25	20	15	15	15	10	10	10	645	30
Contractor labor	3,500	3,500	3,500	3,250	3,250	3,250	3,250	3,250	3,000	3,000	2,500	750	750	450	300	200	150	64,150	3,500
TOTAL	3,553	3,553	3,543	3,303	3,303	3,303	3,303	3,303	3,048	3,048	2,538	778	777	476	317	215	164	65,409	3,553

FIGURE 7.114
Manpower histogram.

- Project schedule is not updated as per detailed data and project assumptions.
- Errors in detail cost estimation.
- Number of drawings are not as per TOR requirements.

The designer has to take into account the above-mentioned risk factors while developing the detail design. Furthermore, the designer has to consider the following risk while planning the duration for completion of the design development phase:

- Impractical design development preparation schedule
- Duration to obtain authorities' approval

7.4.3.11 Monitor Work Progress

The designer has to plan a work schedule for preparing the detail design package in order to submit the same to the owner for review and approval. The detail design progress has to be monitored to meet the submission schedule. Regular meetings to monitor the progress and coordination among different trades and regulatory authorities are required to ensure timely submission of the detail design package.

7.4.3.12 Review Detail Design

The success of a project is highly correlated with the quality and depth of the engineering design prepared during this phase. Coordination and conflict resolution is an important factor during the development of design to avoid omissions and errors. The designer has to review the detail design for accuracy of drawings, interdisciplinary coordination, and documents before these are submitted to the owner/project manager for subsequent preparation of construction documents.

7.4.3.12.1 Review Drawings

Figure 7.115 illustrates design review steps for detail design drawings.

7.4.3.12.2 Perform Interdisciplinary Coordination

Table 7.70 illustrates the major points to perform interdisciplinary coordination.

7.4.3.12.3 Review Documents

The designer has to review contract documents and ensure all the requirements listed under TOR are taken care. The contract documents are prepared for the approved type of project delivery system and contract pricing method.

7.4.3.13 Finalize Detail Design

Final detail design is prepared incorporating the comments, if any, found during analysis, review, and interdisciplinary coordination of the drawings and documents for submission to the owner/client.

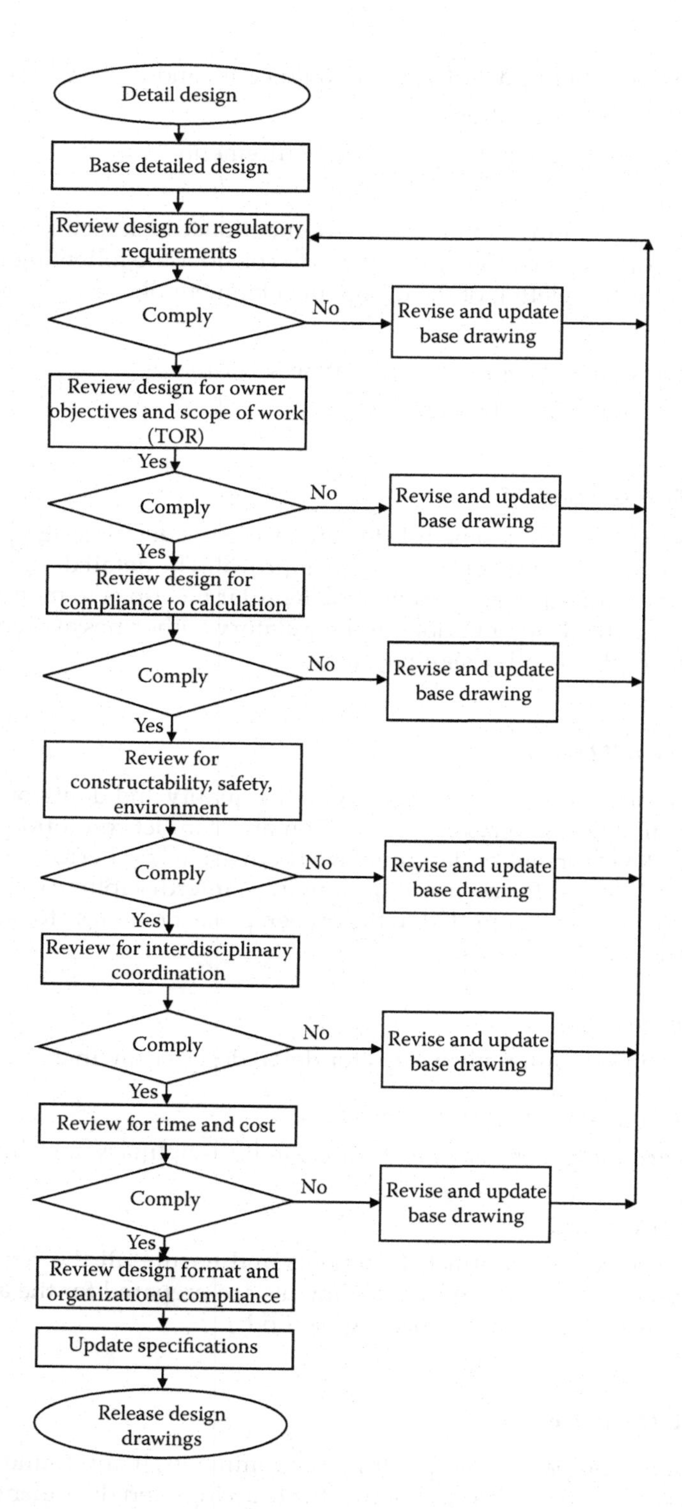

FIGURE 7.115
Design review steps. (Abdul Razzak Rumane (2013), *Quality Tools for Managing Construction Projects*, CRC Press, Boca Raton, FL. Reprinted with permission from Taylor & Francis Group.)

TABLE 7.70

Interdisciplinary Coordination

Sr. No.	Discipline	Architectural	Structural	Mechanical	HVAC	Electrical	External Works
		Discipline					
1	ARCHITECTURAL		1. Structural framing plans,	1. Pump room location and size of room	1. Plant room location and size	1. Location and size of substation and door sizing	1. Property limits
			2. Axis, grids, and levels	2. Void above false ceiling for piping	2. Void above false ceiling for HVAC equipment, duct and piping	2. Trenches for cables in substation, electrical room, and generator room	2. Location of outdoor equipment
			3. Location of columns and beams	3. Sprinkler in false ceiling	3. Access for maintenance of equipment	3. Location and size of electrical room and closets	3. Location of plants
			4. Modules to match with structural plan	4. Location of sanitary fixtures and accessories	4. K-value of thermal insulation, type of external glazing, and U-value	4. Location of electrical devices	4. Location of seating/relax area
			5. Location of stirs and fire exits	5. Location of fire hydrant, cabinet, and landing valves	5. Location of louvers, grills, and diffusers	5. Location of light fittings in the false ceiling	5. Location of maintenance room/area
			6. Expansion joints	6. Location of water tank	6. Location of thermosat and other devices	6. Void above false ceiling for cable tray and trunking	6. Location of manholes
			7. Building dimensions	7. Shaft for water supply, sanitary, and drainage pipes	7. Staircase pressurization system with respect to HVAC	7. Cable Tray, cable trunking route	7. Location of generator exhaust pipe
				8. Location of fuel filling point for fuel carrying tanker	8. Location of HVAC equipment on roof	8. Location and size of low-voltage rooms	
				9. Location of manholes	9. HVAC shaft requirement	9. Location and size of generator room	
						10. Ventilation of sub station and generator room	

(Continued)

TABLE 7.70 (*Continued*)

Interdisciplinary Coordination

Sr. No.	Discipline	Discipline: Architectural	Structural	Mechanical	HVAC	Electrical	External Works
2	STRUCTURAL	1. Structural framing plans		1. Opening for pipe crossing in the walls and slab	1. Shaft for piping and duct	1. Base for transformers	1. Manholes
		2. Axis, grids, and levels		2. Shaft for pipe risers (water supply, sanitary, and drainage)	2. Openings/sleeves for duct and piping	2. Base for generator	2. Foundation for light poles
		3. Location of columns and beams		3. Opening for roof drain	3. Operating weight of all HVAC equipment	3. Trenches for electrical cables	3. Manhole/ foundation for electrical panels
		4. Modules to match with structural plan		4. Openings/sleeves for piping	4. Floor height to accommodate equipment	4. Openings/ sleeves for cable tray, electrical bus duct	4. Manhole/ foundation for feeder pillars
		5. Location of stirs and fire exits		5. Opening for main circulation drain	5. Expansion joints requirements	5. Shaft for cable trays	5. Underground services tunnel
		6. Expansion joints		6. Water tank inlet location	6. Pump room equipment loads with HVAC equipment	6. Foundation for light poles	
		7. Building dimensions		7. Sanitary manholes		7. Manhole/ foundation for electrical panels	

(*Continued*)

TABLE 7.70 (*Continued*)

Interdisciplinary Coordination

		Discipline					
Sr. No.	**Discipline**	**Architectural**	**Structural**	**Mechanical**	**HVAC**	**Electrical**	**External Works**
3	MECHANICAL	1. Pump room location and size of room	1. Opening for pipe crossing in the walls and slab		1. Make up water requirements for HVAC	1. Power supply for pumps and other equipment	1. Irrigation system with external works
		2. Void above false ceiling for piping	2. Shaft for pipe risers (water supply, sanitary, and drainage		2. Connection of chilled water for plumbing works	2. Location of isolators for power supply	2. Area drain and road gully with external/ asphalt work
		3. Sprinkler in false ceiling	3. Opening for roof drain		3. Interface with building management system	3. Interface with fire alarm system	3. Storm water manholes with external works
		4. Location of sanitary fixtures and accessories	4. Openings/ sleeves for piping		4. HVAC/AHU drain with drainage system		4. External services to be hooked up with municipality route
		5. Location of fire hydrant, cabinet, and landing valves	5. Opening for main circulation drain				
		6. Location of water tank	6. Water tank inlet location				
		7. Shaft for water supply, sanitary, and drainage pipes	7. Sanitary manholes				
		8. Location of fuel filling point for fuel carrying tanker					
		9. Location of manholes					

(*Continued*)

TABLE 7.70 (*Continued*)

Interdisciplinary Coordination

Sr. No.	Discipline	Discipline: Architectural	Structural	Mechanical	HVAC	Electrical	External Works
4	HVAC	1. Plant room location and size	1. Shaft for ducts and piping	1. Make up water requirements for HVAC		1. Power supply for chillers, pumps, AHUs, and other equipment	1. Access for underground services
		2. Void above false ceiling for HVAC equipment, duct, and piping	2. Opening for ducts and piping in the wall and roof	2. Connection of chilled water for plumbing works		2. Location of isolators for power supply	2. Location of exhaust for underground ventilation system
		3. Access for maintenance of equipment		3. Interface with building management system		3. Heat dissipation from lighting and other electrical panels	
		4. K-value of thermal insulation, type of external glazing, and U-value		4. HVAC/AHU drain with drainage system		4. Three-phase/ single-phase power requirements	
		5. Location of louvers, grills, and diffusers				5. Power supply load during summer/ winter	
		6. Location of thermosat and other devices				6. Electrical power supply for equipment connected to generator	
		7. Staircase pressurization system with respect to HVAC				7. Interface with fire alarm system	
		8. Location of HVAC equipment on roof				8. Interface with building management system	
		9. HVAC shaft requirement					

(*Continued*)

TABLE 7.70 (*Continued*)

Interdisciplinary Coordination

Sr. No.	Discipline	Discipline: Architectural	Structural	Mechanical	HVAC	Electrical	External Works
5	ELECTRICAL	1. Location and size of substation and door sizing	1. Base for transformers	1. Power supply for pumps and other equipment	1. Power supply for chillers, pumps, AHUs, and other equipment		1. Location of lighting poles
		2. Trenches for cables in substation, electrical room, and generator room	2. Base for generator	2. Location of isolators for power supply	2. Location of isolators for power supply		2. Location of earth pits
		3. Location and size of electrical room and closets	3. Trenches for electrical cables	3. Interface with fire alarm system	3. Heat dissipation from lighting and other electrical panels		3. Location of electrical manholes, handholes
		4. Location of electrical devices	4. Openings/sleeves for cable tray and electrical bus duct		4. Three-phase/single-phase power requirements		4. Underground cable routes
		5. Location of light fittings and other devices in the false ceiling	5. Shaft for cable trays		5. Power supply load during summer/winter		5. Location of bollards
		6. Void above false ceiling for cable tray and trunking	6. Foundation for light poles		6. Electrical power supply for equipment connected to generator		6. Location of electrical panels, feeder pillars
		7. Cable tray, cable trunking route	7. Manhole/foundation for electrical panels		7. Interface with fire alarm system		
		8. Location and size of low-voltage rooms			8. Interface with building management system		
		9. Location and size of generator room					
		10. Ventilation of substation and generator room					

(*Continued*)

TABLE 7.70 (*Continued*)

Interdisciplinary Coordination

Sr. No.	Discipline	Discipline					
		Architectural	Structural	Mechanical	HVAC	Electrical	External Works
6	LANDSCAPE/EXTERNAL	1. Property limits	1. Manholes	1. Irrigation system with external works	1. Access for underground services	1. Location of lighting poles	
		2. Location of outdoor equipment	2. Foundation for light poles	2. Area drain, road gully with external/asphalt work	2. Location of exhaust for underground ventilation system	2. Location of earth pits	
		3. Location of plants	3. Manhole/foundation for electrical panels	3. Storm water manholes with external works		3. Location of electrical manholes and handholes	
		4. Location of seating/relax area	4. Manhole/foundation for feeder pillars	4. External services to be hooked up with municipality route		4. Underground cable routes	
		5. Location of maintenance room/area	5. Underground services tunnel			5. Location of bollards	
		6. Location of manholes				6. Location of electrical panels and feeder pillars	
		7. Location of generator exhaust pipe				7. Location of generator exhaust pipe	

Source: Abdul Razzak Rumane (2013), *Quality Tools for Managing Construction Projects*, CRC Press, Boca Raton, FL. Reprinted with permission from Taylor & Francis Group.

7.4.3.14 Submit Detail Design

Normally, the following items are submitted to the owner for their review and approval in order to proceed with development of construction documents for bidding and tendering purpose:

- Detail design drawings
- Project specifications
- BOQ
- Contract documents
- Project schedule
- Detailed cost estimate
- Detail design report
- Calculations
- Site investigations and survey reports
- Model

7.4.4 Construction Documents Phase

The construction documents phase is the fourth phase of construction project life cycle. During this phase, the drawings and specifications prepared during the design development phase are further developed into the working drawings. All the drawings, specifications, documents, and other related elements necessary for construction of the project are assembled and subsequently released for bidding and tendering. Figure 7.116 illustrates major activities relating to construction document phase developed based on the project management process groups methodology, and Figure 7.117 illustrates the logic flow process for the construction document phase.

7.4.4.1 Develop Construction Documents

The construction document phase provides a complete set of working drawings of all the disciplines, site plans, technical specifications, BOQ, schedule, except the standards specifications, documents for insertions normally added during bidding and tendering phase, and related graphic and written information to bid the project. It is necessary to take utmost care to develop and assemble all the documents and ensure accuracy and correctness to meet the owner's objectives.

7.4.4.1.1 Identify Construction Documents Requirements

In order to identify requirements to assemble contract documents, the designer has to gather the comments on the submitted detail design by the owner/project manager, collect TOR requirements, regulatory requirement, identify owner requirements, and all other related information to ensure that nothing is missed.

7.4.4.1.1.1 Gather Detail Design Comments The designer has to review the comments on detail design and coordinate with all the disciplines/trades and incorporate the same while preparing the construction documents.

Construction document phase

Management processes	Process management groups				
	Initiating process	Planning process	Execution process	Monitoring and Controlling process	Closing process
Integration management	Design development deliverables	Project management plan	Working drawings	Design calculations	Working drawings
	Comments on design development		Project specifications	Interdisciplinary coordination	
	Authorities' requirements		Bill of quantities	Compliance to TOR	
	Environmental		Tender documents		
	TOR requirements				
Stakeholder management		Project stakeholders			
		Stakeholders requirements		Design progress	
Scope management		Construction documents		Authorities approval	
		Construction documents deliverables		Compliance to owner's need	
		Tender documents		Stakeholders approval	
Schedule management		Project schedule		Project schedule	
		Construction document schedule deliverables		Construction document schedule	
Cost management		Bill of quantities price analysis		Project budget	
		Definitive estimate			
Quality management		Codes and standards	Design coordination with all disciplines	Design compliance to owner's goals and	
		Regulatory requirements	Assure design quality	Coordination with all disciplines	
		Design quality		Control design quality	
		Well defined specifications			
		Documents quality			
Resource management		Supervision team requirements	Manage team members from all discipline	Performance of team project members	Assign new project
		Contractor's core team			
		Contractor's manpower			
Communication management		Communication matrix	Liaison with all disciplines	Design status information	
			Coordination meetings		
Risk management		Identification of risk during bidding, construction, test-ing and commissioning		Design risk control	
Contract management		Bidding and tendering documents		Check for contracting system	Contract documents
					Tender documents
HSE management		Safety in design	Safety requirements		
		Environmental compatibility	Environmental requirements	HSE compliance in design	
Financial management		Designer/consultant payment		Payment to designer/consultant	
Claim management		Design change payment		Control changes	Settle designer's claim

FIGURE 7.116
Major activities relating to construction documents processes. *Note*: These activities may not be strictly sequential; however, the breakdown allows implementation of project management function more effective and easily manageable at different stages of the project phase.

7.4.4.1.1.2 Identify TOR Requirement Identify all the requirements listed under TOR. This includes the following:

- Final drawings to be prepared to the required scales, format with necessary logo, client name, location map, north orientation, project name, designer name, drawing title, drawing number, contract reference number, date of drawing, revision number, drawing scale, and duly signed by the designer.
- BOQ and schedule of rates

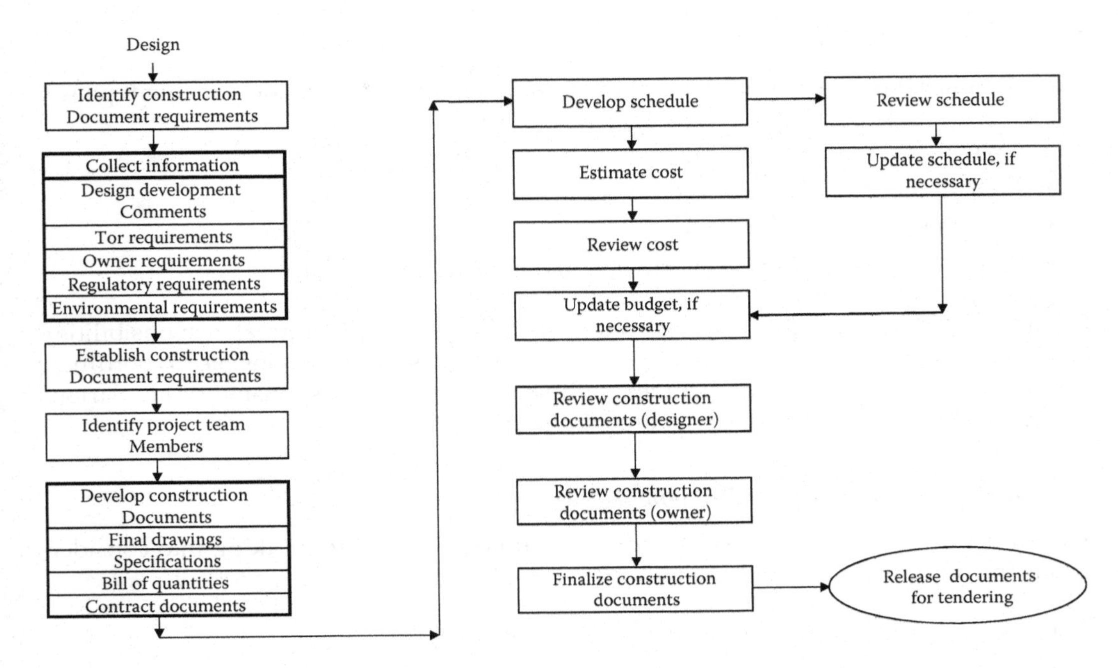

FIGURE 7.117
Logic flow process for construction document phase.

- Contract documents
- Project schedule
- Technical specifications
- Cost estimates
- Summary report

7.4.4.1.1.3 Collect Owner's Requirements The designer has to discuss with the owner to ascertain there are no changes to the earlier established owner's requirement and if there are any additional requirements or changes, then the scope has to be updated and incorporated in the final design documents before sending for tendering.

7.4.4.1.1.4 Identify Regulatory Requirements The designer has to ensure that there are no changes to existing regulatory requirements. If there are updates to regulatory requirements, then the designer has to incorporate the same as any changes during construction have adverse effect on the project.

7.4.4.1.1.5 Identify Environmental Requirements Environmental agencies always update their requirement to protect the environment. The designer has to verify that there are no changes to the requirements/assumptions considered during the detail design development phase.

7.4.4.1.2 Establish Construction Documents Requirements

The construction documents are developed taking into considerations all the items identified as per Section 7.4.4.1.1 and any other related information.

7.4.4.2 Identify Project Stakeholders

The following stakeholders have direct involvement in the construction document phase:

- Owner
- Consultant
- Designer
- Project manager/agency construction manager (as applicable)

During this phase, quantity surveyor/contract administrator has great responsibilities. His/her team under the leadership of project manager (design) is responsible to coordinate and assemble all the required documents. Table 7.71 illustrates responsibilities of various participants during construction document phase.

7.4.4.3 Develop Construction Documents Scope

The scope for development of construction documents is prepared taking into considerations the requirements established under Section 7.4.4.1.1, which are as follows:

- Approved design development phase documents
- TOR requirements
- Owner's preferred requirements
- Regulatory requirements
- Environmental requirements
- Approved project schedule
- Approved detail cost estimate
- Codes and standards

TABLE 7.71
Responsibilities of Various Participants (Design–Bid–Build Type of Contracts) during Construction Documents Phase

	Responsibilities		
Phase	**Owner**	**Designer**	**Regulatory Authorities**
Construction Document	• Approval of working drawings • Approval of tender documents • Approval of time schedule • Approval of budget	• Development of working drawings • Development of specifications • Development of contract documents • Project schedule • Project budget • BOQ • Development of tender documents • Review of construction documents	• Review and approval of project submittals

The purpose of construction documents is to provide sufficient information and detail to ensure that the bidders will be able to submit definitive cost for the project. There is no ambiguity in the drawings and specifications and the work to be performed by the contractor are properly identified and correctly addressed taking necessary measure to mitigate errors and omissions in the design. The scope of work during the construction document phase mainly comprise

- Preparation of working (final) drawings
- Technical specifications
- BOQ
- Project schedule
- Definitive cost estimate
- Authorities approvals
- Existing site conditions/site plans
- Site surveys
- Design calculations

7.4.4.3.1 Identify Construction Documents Deliverables

Table 7.72 lists construction document deliverables to be developed during this phase.

7.4.4.4 Develop Construction Documents

The following items are mainly developed during the construction document phase:

1. Working drawings
2. Technical specifications
3. Documents

7.4.4.4.1 Prepare Final (Working) Drawings

All the drawings prepared during the design development phase are reviewed to ensure all the related information and adjustments are carried out. The following is the list of major disciplines in building construction projects for which working drawings are developed:

1. Architectural design
2. Concrete structure
3. Elevator
4. Fire suppression
5. Plumbing
6. Drainage
7. HVAC works
8. Electrical system (light and power)
9. Fire alarm system
10. Information and communication system
11. Public address system

TABLE 7.72

Construction Documents Deliverables

Sr. No.	Deliverables
1	Document I
1.1	Tendering procedure i. Invitation to tender ii. Instructions to bidders iii. Forms for tender and appendix iv. List of equipment and machinery v. List of contractor's staff vi. Contractor's certificate of work statement vii. List of subcontractor(s) or specialist(s) viii. Initial bond ix. Final bond x. Forms of agreement
2	Document II
2.1	Conditions of contract II-1 General conditions II-2 Particular conditions II-3 Public tender laws
3	Document III III-1 General specifications III-2 Particular specifications III-3 Drawings III-4 Schedule of rates and bill of quantities III-5 Analysis of prices III-6 Addenda III-7 Tender requirements (if any) and any other instructions issued by the owner

12. Audio–visual system
13. Security system/CCTV
14. Security system/access control
15. Satellite/main antenna system
16. Integrated automation system
17. Landscape
18. External works (infrastructure and road)
19. Furnishings/furniture (loose)

Each of these trade drawings shall have the following:

- Detail drawings produced at different scales and format
- Plans
- Sections
- Elevations

- Schedule
- Drawing index

The following information will be included on all the drawings:

- Client name
- Client logo
- Location map
- North orientation
- Project name
- Drawing title
- Drawing number
- Date of drawing
- Revision number
- Drawing scale
- Contract reference number
- Signature block
- Signed by the designer for check and approval

7.4.4.4.2 Prepare Specifications

The designer has to prepare comprehensive technical specifications as per the division and section, taking into consideration related drawings. It is essential to have close coordination between working drawings and specifications. MasterFormat specification documents are used to prepare specifications for building project. These divisions and sections are divided into numbers of volumes for ease of references. BOQ for the project activities is prepared corresponding to these divisions and sections (see Chapter 6 for more information).

7.4.4.4.3 Prepare Tender Documents

The following documents are prepared during this phase:

1. Complete set of working (construction) drawings duly coordinated with other disciplines and technical specifications
2. Detailed BOQ
3. Technical specifications for all the activities shown on the drawings
4. Schedule
5. Cost estimate
6. Legal and contractual information
7. Contractor bidding requirements
8. Contract conditions
9. General specifications
10. Schedules
11. Reports

The above-listed documents are used as guidelines by the designer to prepare tender documents. These are as follows:

Document-I

1. Tendering procedure consists of the following:
   i. Invitation to tender
   ii. Instruction to bidders
   iii. Forms for tender and appendix
   iv. List of equipment and machinery
   v. List of contractor's staff
   vi. Contractor certificate of work statement
   vii. List of subcontractor (s) or specialist(s)
   viii. Initial bond
   ix. Final bond
   x. Form of agreement
   xi. List of tender documents

Document-II

1. II-1 General conditions
2. II-2 Particular conditions
3. II-3 Public tender laws

Document-III

1. III-1 General specifications
2. III-2 Particular specifications (Division 1-49)
3. III-3 Drawings
4. III-4 Schedule of rates and BOQ
5. III-5 Analysis of prices

7.4.4.5 Estimate Construction Schedule

The project schedule is developed using bottom-up planning details using key activities/events. It is also known as Class 1 schedule or Schedule Level 4 (see Tables 7.20 and 7.21).

Figure 7.118 illustrates typical construction schedule for a construction project. The example schedule is construction of a branch for a bank.

7.4.4.6 Estimate Construction Cost

The cost estimate during this phase is based on the detail costing methodology. During this phase, all the project activities are known and detail BOQ is available for costing purpose. It is also known as *detailed costing* (definitive estimate) (see Table 7.23).

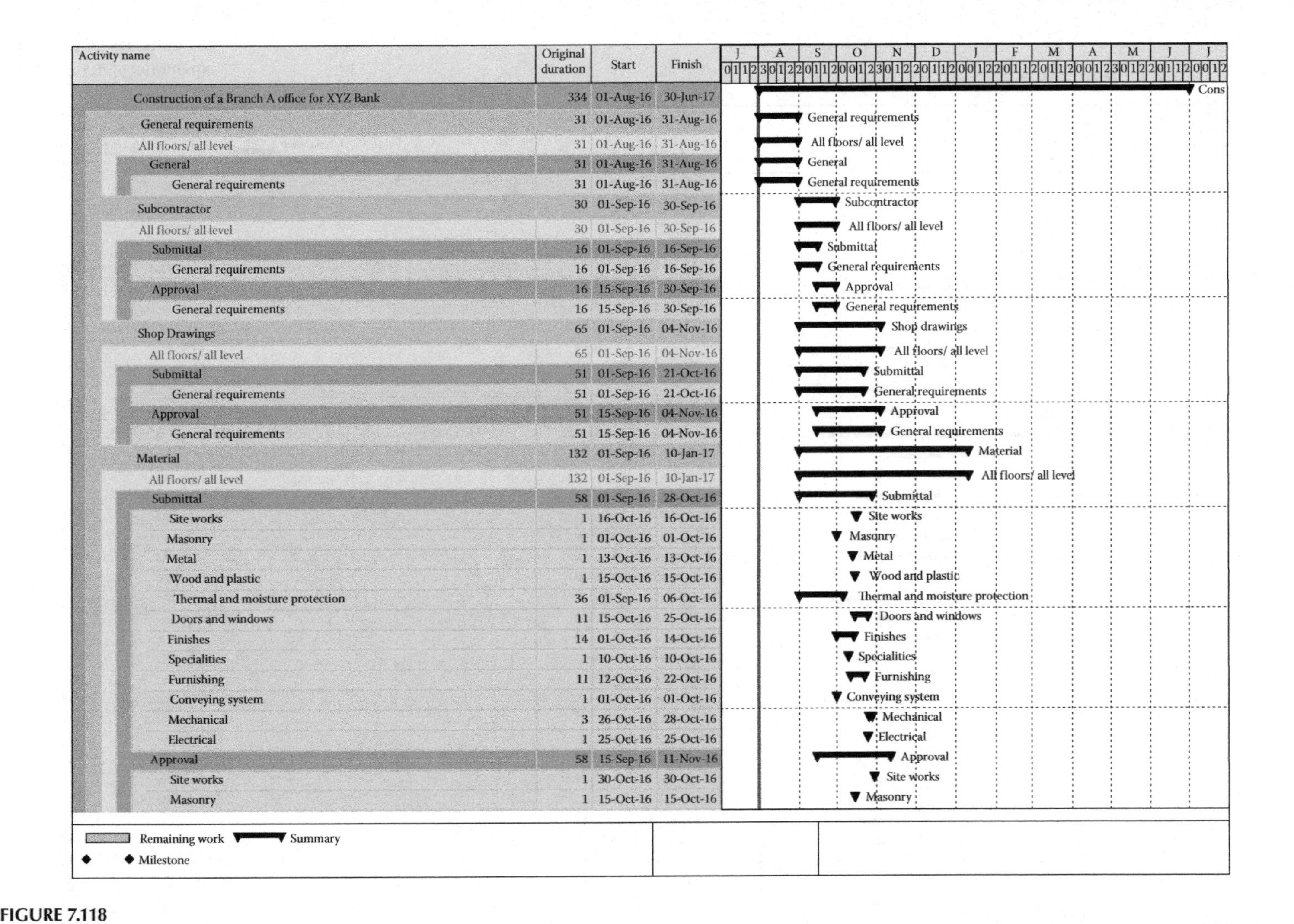

Activity name	Original duration	Start	Finish
Construction of a Branch A office for XYZ Bank	334	01-Aug-16	30-Jun-17
General requirements	31	01-Aug-16	31-Aug-16
All floors/ all level	31	01-Aug-16	31-Aug-16
General	31	01-Aug-16	31-Aug-16
General requirements	31	01-Aug-16	31-Aug-16
Subcontractor	30	01-Sep-16	30-Sep-16
All floors/ all level	30	01-Sep-16	30-Sep-16
Submittal	16	01-Sep-16	16-Sep-16
General requirements	16	01-Sep-16	16-Sep-16
Approval	16	15-Sep-16	30-Sep-16
General requirements	16	15-Sep-16	30-Sep-16
Shop Drawings	65	01-Sep-16	04-Nov-16
All floors/ all level	65	01-Sep-16	04-Nov-16
Submittal	51	01-Sep-16	21-Oct-16
General requirements	51	01-Sep-16	21-Oct-16
Approval	51	15-Sep-16	04-Nov-16
General requirements	51	15-Sep-16	04-Nov-16
Material	132	01-Sep-16	10-Jan-17
All floors/ all level	132	01-Sep-16	10-Jan-17
Submittal	58	01-Sep-16	28-Oct-16
Site works	1	16-Oct-16	16-Oct-16
Masonry	1	01-Oct-16	01-Oct-16
Metal	1	13-Oct-16	13-Oct-16
Wood and plastic	1	15-Oct-16	15-Oct-16
Thermal and moisture protection	36	01-Sep-16	06-Oct-16
Doors and windows	11	15-Oct-16	25-Oct-16
Finishes	14	01-Oct-16	14-Oct-16
Specialities	1	10-Oct-16	10-Oct-16
Furnishing	11	12-Oct-16	22-Oct-16
Conveying system	1	01-Oct-16	01-Oct-16
Mechanical	3	26-Oct-16	28-Oct-16
Electrical	1	25-Oct-16	25-Oct-16
Approval	58	15-Sep-16	11-Nov-16
Site works	1	30-Oct-16	30-Oct-16
Masonry	1	15-Oct-16	15-Oct-16

FIGURE 7.118
Construction schedule.

(Continued)

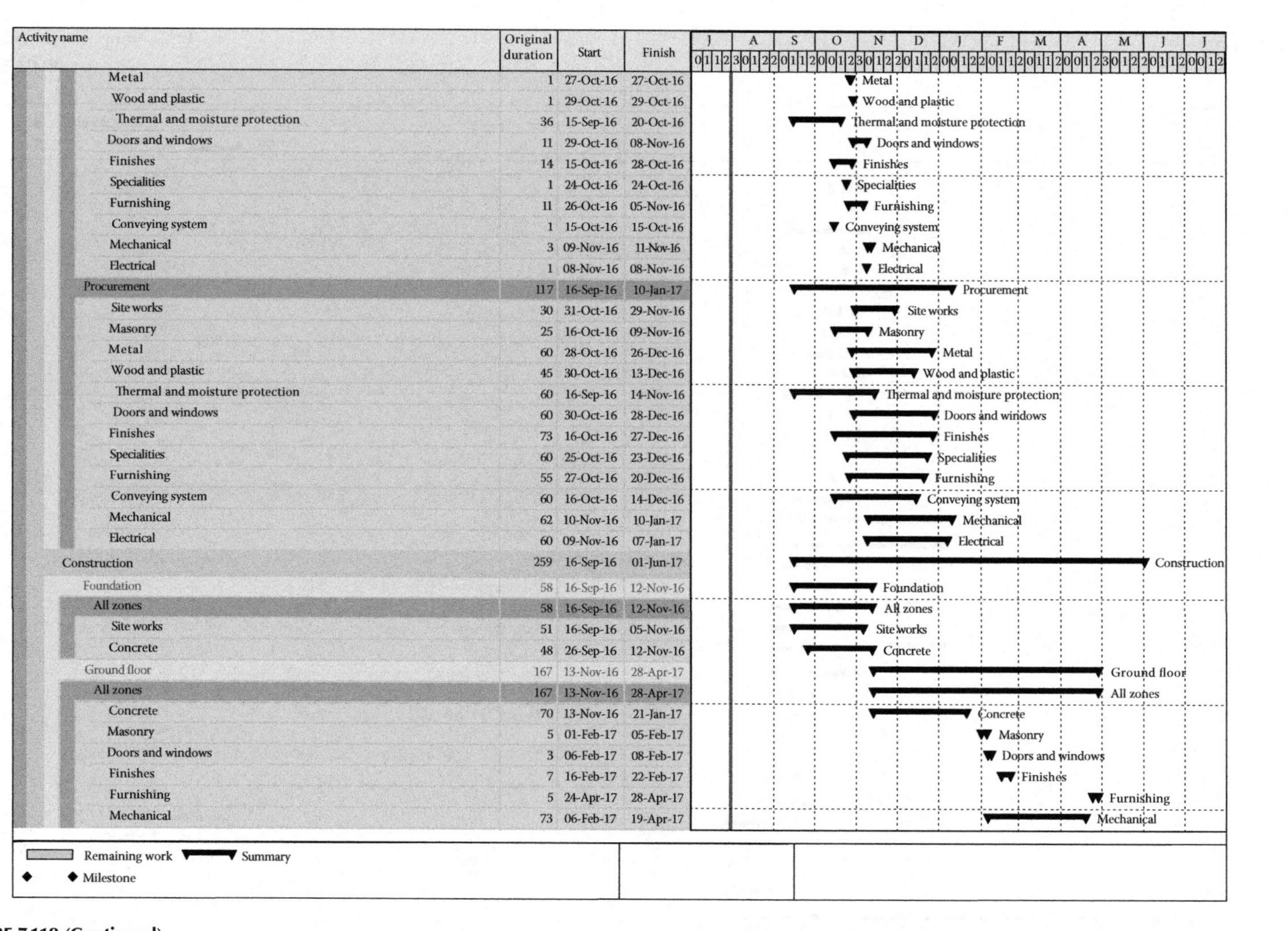

Activity name	Original duration	Start	Finish
Metal	1	27-Oct-16	27-Oct-16
Wood and plastic	1	29-Oct-16	29-Oct-16
Thermal and moisture protection	36	15-Sep-16	20-Oct-16
Doors and windows	11	29-Oct-16	08-Nov-16
Finishes	14	15-Oct-16	28-Oct-16
Specialities	1	24-Oct-16	24-Oct-16
Furnishing	11	26-Oct-16	05-Nov-16
Conveying system	1	15-Oct-16	15-Oct-16
Mechanical	3	09-Nov-16	11-Nov-16
Electrical	1	08-Nov-16	08-Nov-16
Procurement	117	16-Sep-16	10-Jan-17
Site works	30	31-Oct-16	29-Nov-16
Masonry	25	16-Oct-16	09-Nov-16
Metal	60	28-Oct-16	26-Dec-16
Wood and plastic	45	30-Oct-16	13-Dec-16
Thermal and moisture protection	60	16-Sep-16	14-Nov-16
Doors and windows	60	30-Oct-16	28-Dec-16
Finishes	73	16-Oct-16	27-Dec-16
Specialities	60	25-Oct-16	23-Dec-16
Furnishing	55	27-Oct-16	20-Dec-16
Conveying system	60	16-Oct-16	14-Dec-16
Mechanical	62	10-Nov-16	10-Jan-17
Electrical	60	09-Nov-16	07-Jan-17
Construction	259	16-Sep-16	01-Jun-17
Foundation	58	16-Sep-16	12-Nov-16
All zones	58	16-Sep-16	12-Nov-16
Site works	51	16-Sep-16	05-Nov-16
Concrete	48	26-Sep-16	12-Nov-16
Ground floor	167	13-Nov-16	28-Apr-17
All zones	167	13-Nov-16	28-Apr-17
Concrete	70	13-Nov-16	21-Jan-17
Masonry	5	01-Feb-17	05-Feb-17
Doors and windows	3	06-Feb-17	08-Feb-17
Finishes	7	16-Feb-17	22-Feb-17
Furnishing	5	24-Apr-17	28-Apr-17
Mechanical	73	06-Feb-17	19-Apr-17

FIGURE 7.118 (Continued)
Construction schedule.

(Continued)

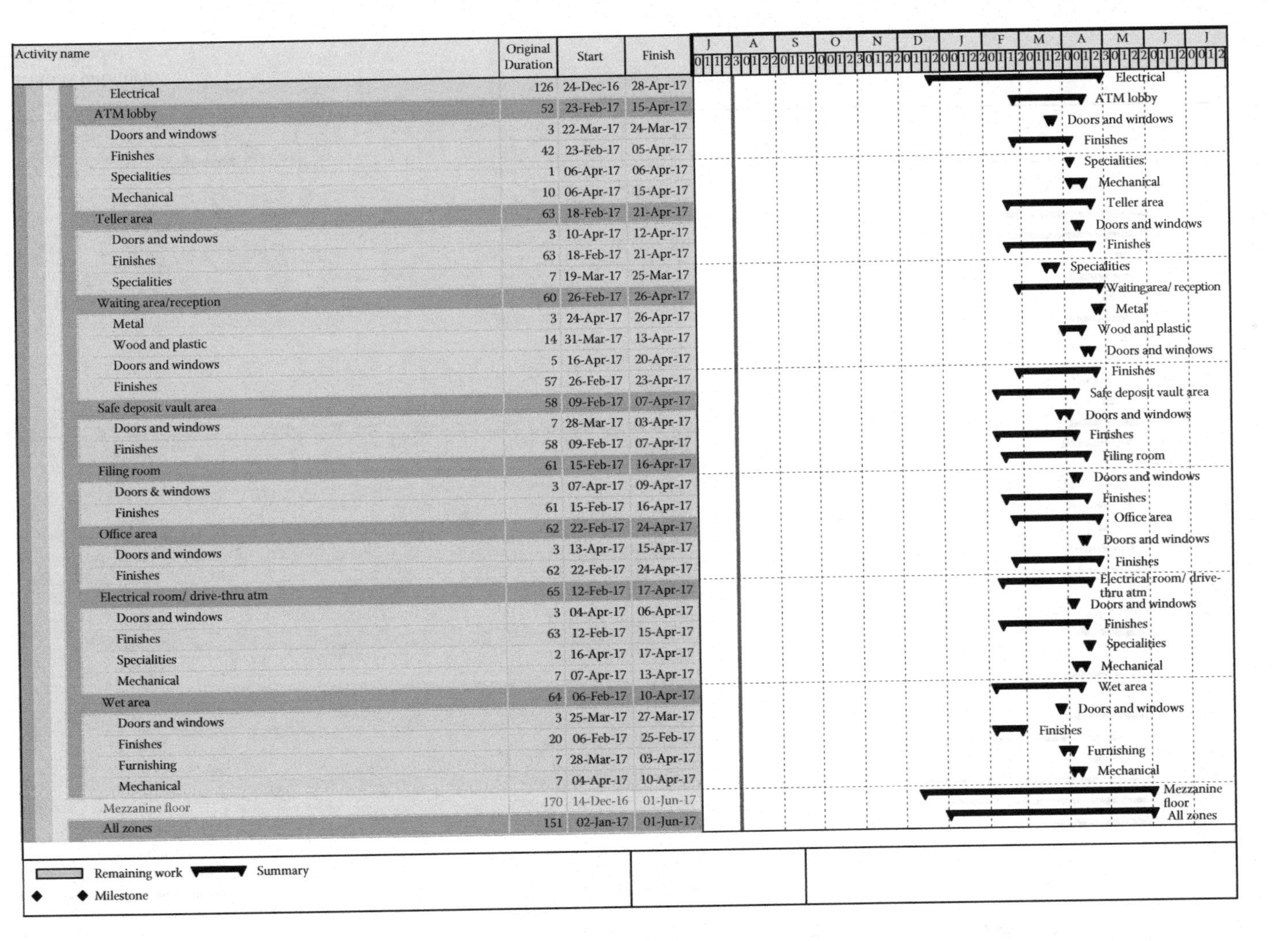

Activity name	Original Duration	Start	Finish
Electrical	126	24-Dec-16	28-Apr-17
ATM lobby	52	23-Feb-17	15-Apr-17
Doors and windows	3	22-Mar-17	24-Mar-17
Finishes	42	23-Feb-17	05-Apr-17
Specialities	1	06-Apr-17	06-Apr-17
Mechanical	10	06-Apr-17	15-Apr-17
Teller area	63	18-Feb-17	21-Apr-17
Doors and windows	3	10-Apr-17	12-Apr-17
Finishes	63	18-Feb-17	21-Apr-17
Specialities	7	19-Mar-17	25-Mar-17
Waiting area/reception	60	26-Feb-17	26-Apr-17
Metal	3	24-Apr-17	26-Apr-17
Wood and plastic	14	31-Mar-17	13-Apr-17
Doors and windows	5	16-Apr-17	20-Apr-17
Finishes	57	26-Feb-17	23-Apr-17
Safe deposit vault area	58	09-Feb-17	07-Apr-17
Doors and windows	7	28-Mar-17	03-Apr-17
Finishes	58	09-Feb-17	07-Apr-17
Filing room	61	15-Feb-17	16-Apr-17
Doors & windows	3	07-Apr-17	09-Apr-17
Finishes	61	15-Feb-17	16-Apr-17
Office area	62	22-Feb-17	24-Apr-17
Doors and windows	3	13-Apr-17	15-Apr-17
Finishes	62	22-Feb-17	24-Apr-17
Electrical room/ drive-thru atm	65	12-Feb-17	17-Apr-17
Doors and windows	3	04-Apr-17	06-Apr-17
Finishes	63	12-Feb-17	15-Apr-17
Specialities	2	16-Apr-17	17-Apr-17
Mechanical	7	07-Apr-17	13-Apr-17
Wet area	64	06-Feb-17	10-Apr-17
Doors and windows	3	25-Mar-17	27-Mar-17
Finishes	20	06-Feb-17	25-Feb-17
Furnishing	7	28-Mar-17	03-Apr-17
Mechanical	7	04-Apr-17	10-Apr-17
Mezzanine floor	170	14-Dec-16	01-Jun-17
All zones	151	02-Jan-17	01-Jun-17

FIGURE 7.118 (Continued)
Construction schedule.

(Continued)

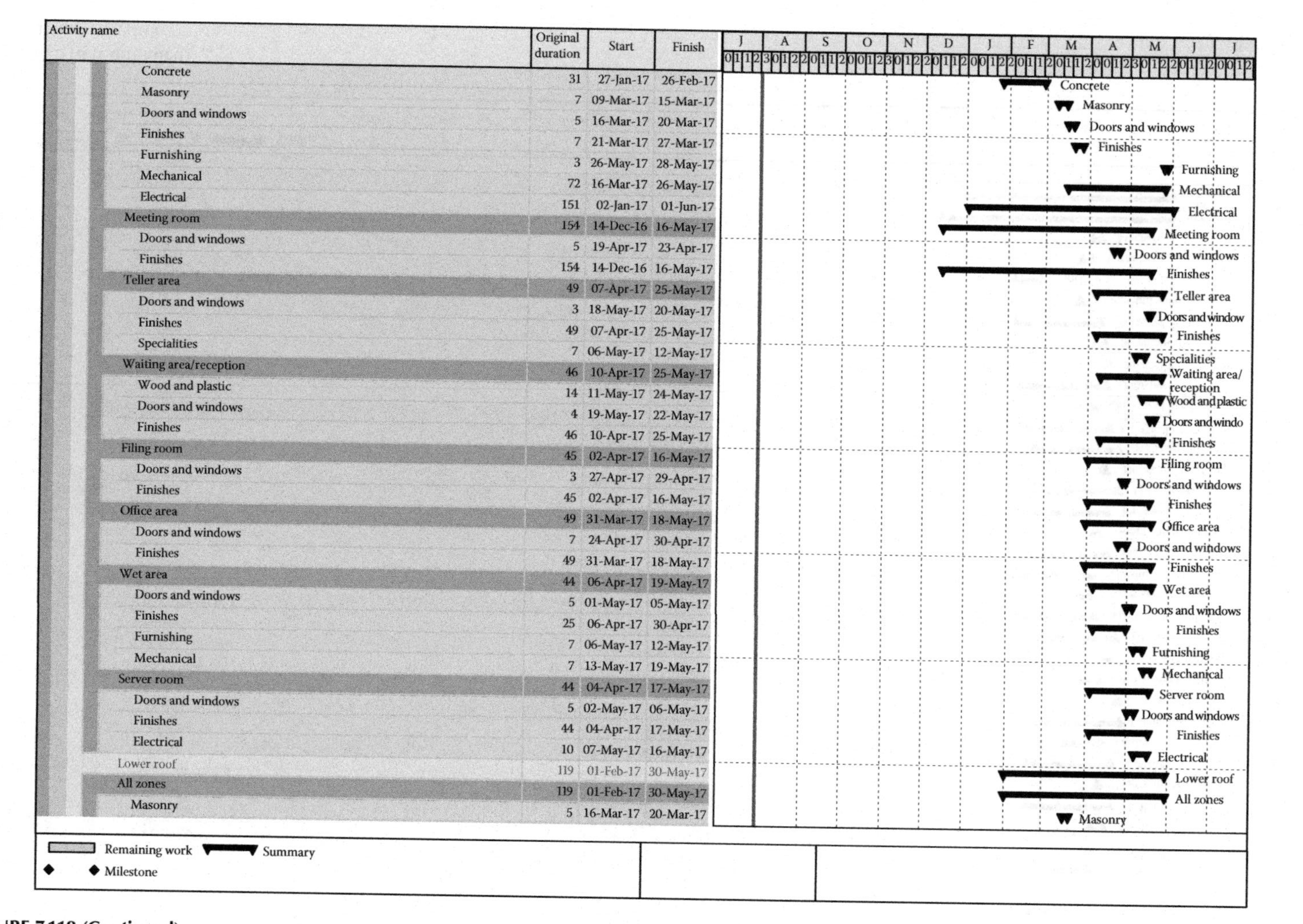

Activity name	Original duration	Start	Finish
Concrete	31	27-Jan-17	26-Feb-17
Masonry	7	09-Mar-17	15-Mar-17
Doors and windows	5	16-Mar-17	20-Mar-17
Finishes	7	21-Mar-17	27-Mar-17
Furnishing	3	26-May-17	28-May-17
Mechanical	72	16-Mar-17	26-May-17
Electrical	151	02-Jan-17	01-Jun-17
Meeting room	154	14-Dec-16	16-May-17
Doors and windows	5	19-Apr-17	23-Apr-17
Finishes	154	14-Dec-16	16-May-17
Teller area	49	07-Apr-17	25-May-17
Doors and windows	3	18-May-17	20-May-17
Finishes	49	07-Apr-17	25-May-17
Specialities	7	06-May-17	12-May-17
Waiting area/reception	46	10-Apr-17	25-May-17
Wood and plastic	14	11-May-17	24-May-17
Doors and windows	4	19-May-17	22-May-17
Finishes	46	10-Apr-17	25-May-17
Filing room	45	02-Apr-17	16-May-17
Doors and windows	3	27-Apr-17	29-Apr-17
Finishes	45	02-Apr-17	16-May-17
Office area	49	31-Mar-17	18-May-17
Doors and windows	7	24-Apr-17	30-Apr-17
Finishes	49	31-Mar-17	18-May-17
Wet area	44	06-Apr-17	19-May-17
Doors and windows	5	01-May-17	05-May-17
Finishes	25	06-Apr-17	30-Apr-17
Furnishing	7	06-May-17	12-May-17
Mechanical	7	13-May-17	19-May-17
Server room	44	04-Apr-17	17-May-17
Doors and windows	5	02-May-17	06-May-17
Finishes	44	04-Apr-17	17-May-17
Electrical	10	07-May-17	16-May-17
Lower roof	119	01-Feb-17	30-May-17
All zones	119	01-Feb-17	30-May-17
Masonry	5	16-Mar-17	20-Mar-17

FIGURE 7.118 (Continued)
Construction schedule.

(Continued)

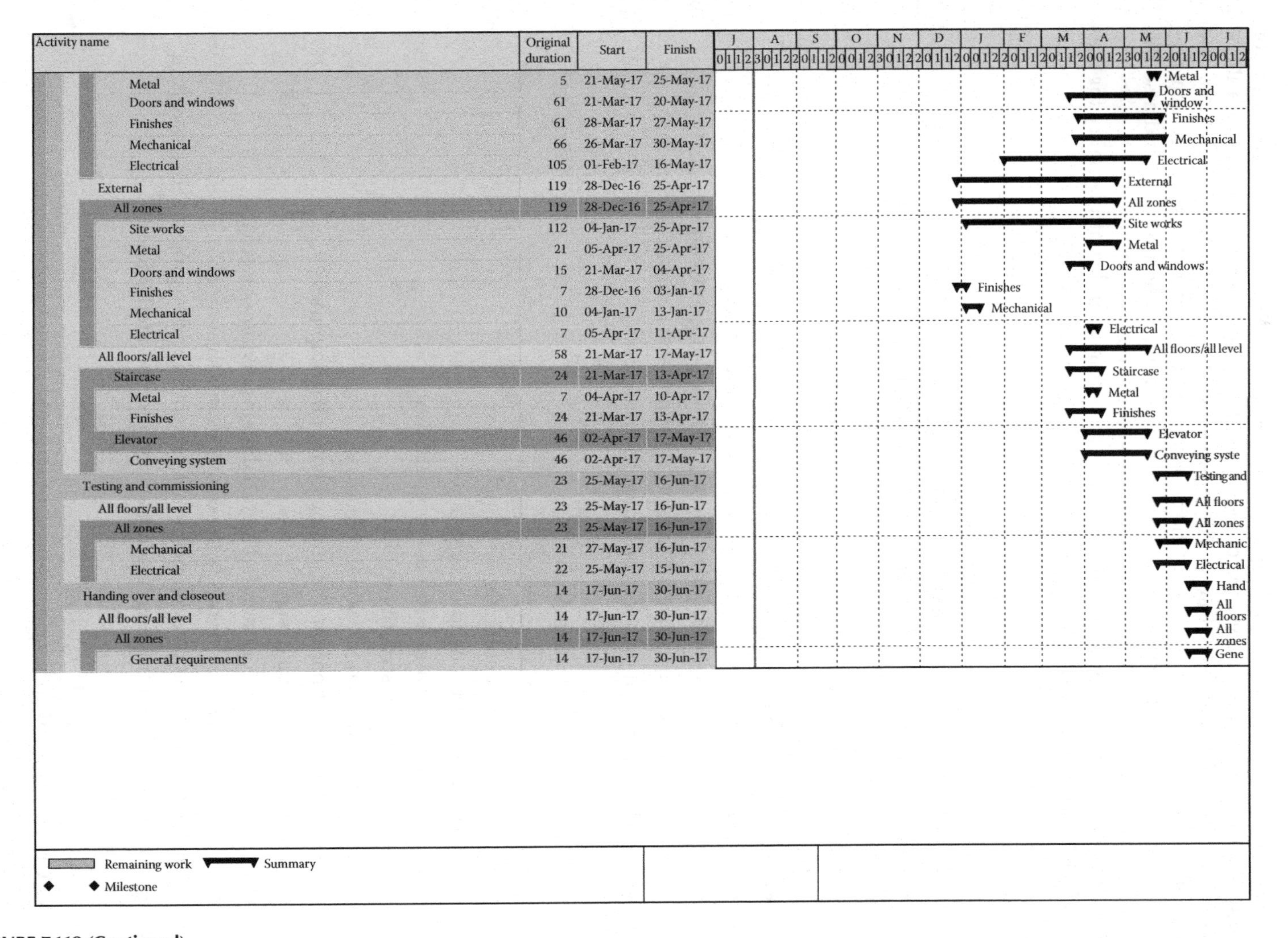

Activity name	Original duration	Start	Finish
Metal	5	21-May-17	25-May-17
Doors and windows	61	21-Mar-17	20-May-17
Finishes	61	28-Mar-17	27-May-17
Mechanical	66	26-Mar-17	30-May-17
Electrical	105	01-Feb-17	16-May-17
External	119	28-Dec-16	25-Apr-17
All zones	119	28-Dec-16	25-Apr-17
Site works	112	04-Jan-17	25-Apr-17
Metal	21	05-Apr-17	25-Apr-17
Doors and windows	15	21-Mar-17	04-Apr-17
Finishes	7	28-Dec-16	03-Jan-17
Mechanical	10	04-Jan-17	13-Jan-17
Electrical	7	05-Apr-17	11-Apr-17
All floors/all level	58	21-Mar-17	17-May-17
Staircase	24	21-Mar-17	13-Apr-17
Metal	7	04-Apr-17	10-Apr-17
Finishes	24	21-Mar-17	13-Apr-17
Elevator	46	02-Apr-17	17-May-17
Conveying system	46	02-Apr-17	17-May-17
Testing and commissioning	23	25-May-17	16-Jun-17
All floors/all level	23	25-May-17	16-Jun-17
All zones	23	25-May-17	16-Jun-17
Mechanical	21	27-May-17	16-Jun-17
Electrical	22	25-May-17	15-Jun-17
Handing over and closeout	14	17-Jun-17	30-Jun-17
All floors/all level	14	17-Jun-17	30-Jun-17
All zones	14	17-Jun-17	30-Jun-17
General requirements	14	17-Jun-17	30-Jun-17

FIGURE 7.118 (Continued)
Construction schedule.

7.4.4.7 Manage Documents Quality

In order to reduce errors and omissions, it is necessary to review and check the design for quality assurance by the quality control personnel from the project team through itemized review checklists to ensure that working drawings are suitable for construction. The designer has to ensure that the installation/execution specification details are comprehensively and correctly described and coordinated with working drawings and also the installation quality requirements for systems are specified in details.

The designer has to plan quality, perform quality assurance, and control quality for preparing contract documents. This will mainly consist of the following:

1. *Plan quality*:
    - Review comments on design development package
    - Determine the number of drawings to be produced
    - Establish scope of work for preparation of construction documents
    - Identify requirements listed under TOR
    - Identify quality standards and codes to be complied
    - Identify regulatory requirements
    - Identify environmental requirements
    - Establish quality organization with responsibility matrix
    - Develop review procedure for the produced working drawings
    - Develop review procedure for the specifications and contract documents
    - Establish submittal plan for construction documents
2. *Quality assurance*:
    - Prepare working drawings
    - Prepare detailed specifications
    - Prepare contract documents
    - Prepare BOQ and schedule of rates
    - Ensure functional and technical compatibility
    - Ensure design constructibility
    - Ensure meeting of operational objectives
    - Ensure full coordination of drawings with all disciplines
    - Ensure cost-effective design
    - Ensure selected/recommended material meets owner objectives
    - Ensure that design fully meets the owner's objectives/goals
    - Ensure that construction documents matches with approve project delivery system
    - Ensure type of contracting/pricing as per adopted methodology
3. *Control quality*:
    - Check quality of design drawings
    - Check accuracy and correctness of design
    - Verify BOQ for correctness as per working drawings

- Check complete specifications are prepared and coordinated to match working drawings and BOQ
- Check contract documents as per project delivery system
- Check for regulatory compliance
- Check project schedule
- Check project cost
- Check calculations
- Review studies and reports
- Check accuracy of design
- Check interdisciplinary requirements
- Check required number of drawings prepared drawing

The BIM tool can be used to control quality of the project (see Chapter 5 for more information).

Before the drawings are released for bidding and tendering, it is necessary to check the drawings for formatting, annotation, and interpretation. Table 7.73 lists the items to be checked for quality check (correctness) of design drawings.

7.4.4.8 Estimate Resources

At this stage, the designer can estimate resources having accuracy as more details are available to estimate exact resources.

7.4.4.9 Manage Risks

The following are typical risks that normally occur during the construction document phase:

- Design development deliverables and review comments are not taken into consideration while preparing the construction documents.
- Scope of work to produce construction documents is not properly established and is incomplete.
- Documents do not match as per project delivery system.
- Documents not as per type of contract/pricing methodology.
- Regulatory authorities' requirements are not taken into consideration.
- Latest environmental considerations are not considered.
- Conflict with different trades.
- Conflict between working drawings and specifications.
- Prediction of possible changes in design during the construction phase.
- Inadequate and ambiguous specifications.
- Project schedule not updated as per detailed data and project assumptions.
- Errors in definitive cost estimation.

TABLE 7.73
Quality Check for Design Drawings

Sr. No.	Points to Be Checked	Yes/No
1	Check for use of approved version of AutoCAD	
2	Check drawing for • Title frame • Attribute • North orientation • Key plan • Issues and revision number	
3	Client name and logo	
4	Designer (consultant name)	
5	Drawing title	
6	Drawing number	
7	Contract reference number	
8	Date of drawing	
9	Drawing scale	
10	Annotation • Text size • Dimension style • Fonts • Section and elevation marks	
11	Layer standards, including line weights	
12	Line weights, line type (continuous, dash, dot, etc.)	
13	Drawing continuation reference and match line	
14	Plot styles (CTB—color dependent plot style tables)	
15	Electronic CAD file name and project location	
16	XREF (X-reference) attachments (if any)	
17	Image reference (if any)	
18	Section references	
19	Symbols	
20	Legends	
21	Abbreviations	
22	General notes	
23	Drawing size as per contract requirements	
24	List of drawings	

Source: Abdul Razzak Rumane (2013), *Quality Tools for Managing Construction Projects*, CRC Press, Boca Raton, FL. Reprinted with permission from Taylor & Francis Group.

- Number of drawings not as per TOR requirements.
- It is likely that owner-supplied items, if any, are not included in the documents.

The designer has to take into account the above-mentioned risk factors while designing construction documents.

Furthermore, the designer has to consider the following risk while planning the duration for completion of the construction document phase:

- Impractical construction document preparation schedule

7.4.4.10 Monitor Work Progress

The designer has to prepare work schedule to prepare the construction document package in order to submit the same to the owner for review and approval. The progress to produce construction documents has to be monitored to meet the submission schedule. Regular meetings to monitor the progress and coordination among different trades are required to ensure timely submission of the detail design package. Any major deviations to the approved drawings and specifications produced during design development should be identified and discussed with the owner/project manager for specific consideration and approval by the client and accordingly all the related documents to be updated.

7.4.4.11 Review Construction Documents

Table 7.74 illustrates items to be reviewed for constructability of design.

7.4.4.12 Finalize Construction Documents

The final construction documents package is prepared taking into consideration review comments and identified risk by the designer and comments from the owner/project manager.

7.4.4.13 Release for Bidding

Normally, the following items are submitted to the owner for their review and approval in order to proceed for bidding and tendering phase of the project:

1. Working drawings
2. Project specifications
3. BOQ
   a. Priced
   b. Unpriced
4. Project schedule
5. Definitive cost estimate
6. Project summary report
7. Tender documents comprising
   I. Document-I—tendering procedure
   II. Document-II—condition of contract
   III. Document-III consisting of
      III-1—general specifications
      III-2—particular specifications
      III-3—drawings
      III-4—BOQ and schedules of rates
      III-5—analysis of prices
8. Soft copy of construction documents (DVD)

TABLE 7.74
Constructability Review for Design Drawings

Sr. No.	Items to Be Reviewed	Yes/No
1	Are there construction elements that are impossible or impractical to build?	
2	Does the design follow industry standards and practices?	
3	Is structural design per site conditions, soil conditions and bearing capacity?	
4	Are the site conditions verified and suitable with respect to access, availability of utility services?	
5	Will all the specified material shall be available during construction phase?	
6	Is the specified material available from single source or multiple sources and brands?	
7	Is the design suitable for construction using the specified, material, and equipment?	
8	Is the design suitable for construction using recommended method statement?	
9	Are the available labor resources capable of building the facility as per contract drawings and contracted methods and practices?	
10	Is the design fully coordinated with technical specifications and CSI format divisions?	
11	Is the specifications cover the material considered in the design?	
12	Does the design fully meet regulatory requirements?	
13	Is the drawings coordinated with all the trades and cross references are indicated wherever applicable?	
14	Is the design coordinated with adjacent land and its accessibility?	
15	Are requirements of general public and persons of special needs considered?	
16	Are construction schedule and milestone practical to achieve?	
17	Can application of QA/QC requirements be complied with?	
18	Has environmental impact and its mitigation considered?	
19	Is there space for temporary office facilities and parking space for workforce vehicles?	
20	Has availability of storage space for construction material considered?	
21	Is the design sustainable?	

Source: Abdul Razzak Rumane (2013), *Quality Tools for Managing Construction Projects*, CRC Press, Boca Raton, FL. Reprinted with permission from Taylor & Francis Group.

7.4.5 Bidding and Tendering Phase

During this phase, tender documents are released for bid, and the contract is awarded to successful bidder. Figure 7.119 illustrates major activities relating to bidding and tendering phase developed based on project management process groups methodology, and Figure 7.120 illustrates the logic flow process for bidding and tendering phase. This process is to be read in conjunction with Section 7.3.10.3.

7.4.5.1 Organize Tendering Documents

The owner hands over the approved construction documents/tender documents to the tender committee for further action. The bid documents are prepared as per the procurement method and contract strategy adopted during early stage of the project. Tendering procedure documents submitted by the designer are updated and necessary owner-related information is inserted in the tender documents. The bid advertisement material is prepared and upon approval from the owner, the bid notification is announced through different media as per the organization's/agency's policy.

Biddingand tendering phase

Management processes	Process management groups				
	Initiating process	Planning process	Execution process	Monitoring and controlling process	Closing process
Integration management	Construction documents	Organize tender	Tendering documents		Award contract
	TOR requirements				
	Owner's requirements				
	Authorities requirements				
Stakeholder	Identify bidders		Contractor selection		
Scope management		Bidder selection procedure	Bid review	Addendum	
		Bid review procedure			
Schedule management		Bid period		Monitor bid duration	
		Bid review duration		Monitor review duration	
Cost management		Estimate bid price		Control bid value	
Quality management					
Resource management					
Communication management		Advertize tender	Conduct meetings		
Risk management			Manage risk	Control risk	
Contract management		Select bidders	Prepare construction contract		Signed contract
HSE management					
Financial management				Update project finances	
Claim management					

FIGURE 7.119
Major activities relating to bidding and tendering processes. *Note*: These activities may not be strictly sequential; however, the breakdown allows implementation of project management function more effective and easily manageable at different stages of the project phase.

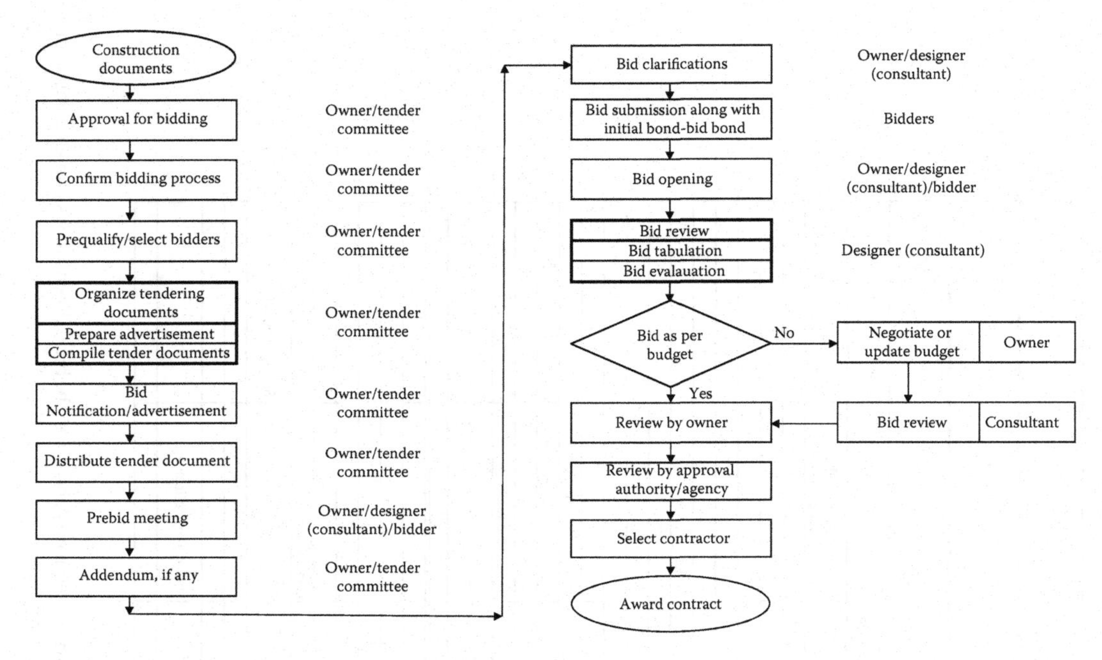

FIGURE 7.120
Logic flow process for bidding and tendering phase.

7.4.5.2 *Identify Stakeholders*

The following stakeholders have direct involvement in the bidding and tendering phase:

- Owner
- Tender committee
- Designer (consultant)
- Project manager/agency construction manager (as applicable)
- Bidders

Table 7.75 illustrates responsibilities of various participants during the construction document phase.

7.4.5.3 *Identify Tendering Procedure*

The owner has to identify the tendering procedure. Different types of tendering methods are discussed under Sections 2.4.1 and 7.3.10.3.2.

7.4.5.3.1 Define Bidder Selection Procedure

The owner has to select the bidder selection procedure such as low bid, quality-based. Refer to Sections 2.4.1, 7.3.10.3.2, and 7.3.10.3.3 for different types of contacting methods and contractor selection procedure.

TABLE 7.75
Responsibilities of Various Participants (Design–Bid–Build Type of Contracts) during Bidding and Tendering Phase

Phase	Responsibilities		
	Owner/Tender Committee	**Consultant (Designer)**	**Contractor**
Bidding and tendering	• Advertise bid • Distribute bid • Collect bid (proposal) • Negotiation • Approve contractor • Award contract	• Review/evaluate bid • Bid conference/meeting • Bid clarification • Recommend successful bidder • Prepare contract documents	• Collection of bid documents • Preparation of proposal • Submission of proposal

7.4.5.3.2 Establish Bid Review Procedure

The designer (consultant) establishes bid evaluation procedure in consultation with the owner.

7.4.5.4 Identify Bidders

Short listing of bidders is done with PQQs and their responses. This has been already discussed under Section 7.3.10.3.1.5.

7.4.5.4.1 Prequalify Bidders

Prequalification of contractor is already discussed under Section 7.3.10.3.1.6.

7.4.5.5 Manage Tendering Process

The tendering process involves the following activities:

- Bid notification
- Distribution of tender documents
- Prebid meeting(s)
- Issuing addendum, if any
- Bid submission

7.4.5.5.1 Advertise Tender

The tender is announced in different types of media such as newspaper, magazines, electronic media, as per the organization's/agency's policy.

7.4.5.5.2 Distribute Tender Documents

Normally, tender documents are distributed to eligible bidders against payment of fee announced in the bid notification, which is nonrefundable.

7.4.5.5.3 Conduct Meetings

The owner conducts prebid meeting to provide an opportunity for the contractors bidding the project to review and discuss the construction documents and to further discuss

- General scope of the project
- Any particular requirements of bidders that may have been difficult to specify
- Explain details of complex matters
- Engagement of subcontractor and specialist subcontractors
- Particular risks
- Any other matters that will contribute to the efficient delivery of project

The meeting is attended by the designer (consultant), bidders (contractors), project/construction manager, and tender committee member. Queries from the contractors pertaining to construction documents are noted and the designer (consultant) provides written response to these queries by clarifying all the points. The bidders have to consider the clarification points and incorporate the requirements while calculating the bid price. The responses recorded in the meeting become part of contract documents (part of addendum), which is signed by the owner and successful bidder. Figure 7.121 illustrates bid clarification form that becomes part of contract documents.

7.4.5.5.4 Submit/Receive Bids

Bids are received in accordance with the instructions to bidders section of the tender documents. The bid should be accompanied with initial bond in favor of owner/tender committee to be valid for a period mentioned in tendering procedures. All the bids received are documented and notified. The tender, which is submitted as sealed document, is opened as mentioned in tendering procedures.

Project name Project number Bid clarification form					
Sr. No.	Name of contractor	Item no. and clause reference	Queries	Owner/condultant's clarification	Remarks

FIGURE 7.121
Bid clarification.

7.4.5.6 Manage Risks

Please refer to Table 7.39 for typical risks affecting the contractor. It is essential that the contractor verify BOQ for correctness with respect to design drawings and specifications while estimating bid value. The following are typical risks that are likely to occur during this phase:

- Not all the qualified bidders taking part in bidding for the project
- Bidders noticing errors and omissions in construction documents resulting in delay in the submission of bids
- Amendment to construction documents
- Addendum
- Delay in submission of bids than the notified one
- Bid value exceeding the estimated definitive cost
- Successful bidder fails to submit performance bond

The owner/designer has to consider these risks and plan the phase duration accordingly.

7.4.5.7 Review Bid Documents

The designer (consultant) reviews the bids for compliance to tender requirements.

7.4.5.7.1 Evaluate Bids

The designer (consultant) evaluates the bid documents of each of the submitted bids. Please refer to Section 7.3.10.5.1 for bid evaluation process and Table 7.45 for bid review checklist.

7.4.5.7.2 Select Contractor

The contractor is selected based on the procurement strategy adopted by the owner.

7.4.5.8 Award Contract

Figure 7.122 illustrates contract award process.

7.4.6 Construction Phase

Construction is translating the owner's goals and objectives, by the contractor, to build the facility as stipulated in the contract documents, plans, specifications within budget and on schedule. Construction is the sixth phase of construction project life cycle and is an important phase in construction projects. A majority of total project budget and schedule is expended during construction. Similar to costs, the time required to construct the project is much higher than the time required for preceding phases. Construction usually requires a large number of work force and a variety of activities. Construction activities involve erection, installation, or construction of any part of the project. Construction activities are actually carried out by the contractor's own work force or by subcontractors. Construction therefore requires more detailed attention of its planning, organizations, monitoring, and controlling of project schedule, budget, quality, safety, and environment concerns. Figure 7.123 illustrates major activities relating to the construction phase developed based on project management process groups methodology, and Figure 7.124 illustrates the logic flow process for the construction phase.

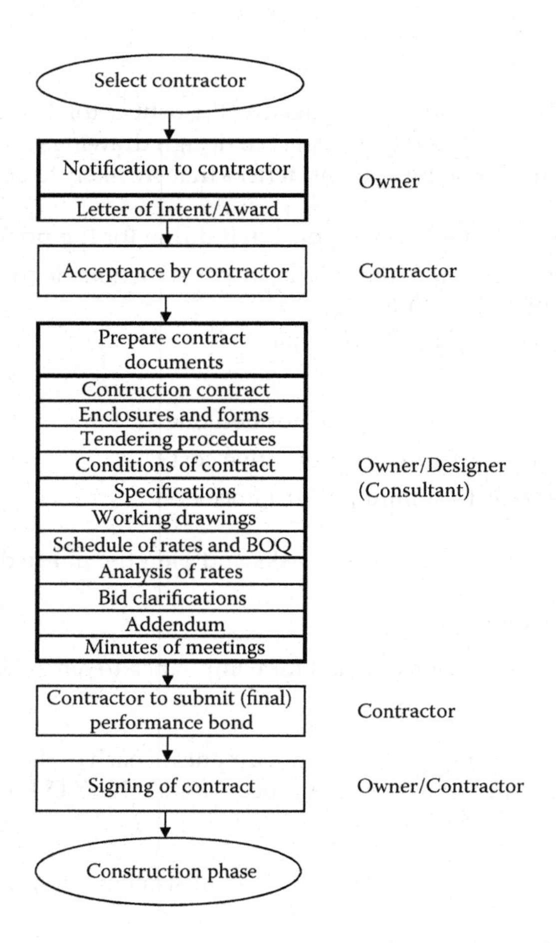

FIGURE 7.122
Contract award process.

7.4.6.1 Develop Project Execution Requirements

Once the contract is awarded to the successful bidder (contractor), then it is the responsibility of the contractor to respond to the needs of the client (owner) by building the facility as specified in the contract documents, drawings, and specifications within the budget and time.

7.4.6.1.1 Identify Contract Requirements

In order to develop contract requirements, the contractor has to review all the construction documents that are part of the contract that contractor has signed with the owner of the project. This includes the following:

- Working drawings
- Specification
- Contract documents
- Other related documents

Construction Phase

Management Processes	Project Management Process Groups				
	Initiating Process	**Planning Process**	**Execution Process**	**Monitoring & Controlling Process**	**Closing Process**
Integration Management	Contract Documents	Construction Management Plan	Mobilization	Compliance to Contract Documents	Executed Project
	Tender Documents		Submittals	Change Management	
	Notice to Proceed		Execution Process		
			Construction Work		
Stakeholder Management	Owner's Representative	Responsibility Matrix		Project Status/Performance	
	Supervision Team	Stakeholder Requirements		Payments	
	Contrator's Core Staff	Reports		Variation Orders	
	Subcontractor	Meetings		Conflict Resolution	
	Authorities				
Scope Management		Scope Change Management	Design Changes	Authorities Approval	
		Preventive and Corrective Actions		Stakeholders Approval	
				Scope Change Control	
				Alternate Material	
				Site Work Instruction	
				Variation Orders	
				Preventive and Corrective	
				Plan Updates	
Schedule Management		Contractor's Construction Schedule		Schedule Monitoring	
				Schedule Control	
				Work Progress Monitoring	
				Submittals Monitoring	
Cost Management		Contracted Value of Projec		Cost Control	
		Construction Budget		Cash Flow	
				Progress Payment	
				Variation Orders	
Quality Management		Contractor's Quality Control Plan	Quality Assurance	Quality Control	
			Shop Drawings	Quality Auditing	
			Builders Drawings	Material Inspection	
			Composite Drawings	Work Inspection/Testing	
			Material Approvals	Rework	
			Method Statement	Regulatory Compliance	
Resource Management		Resource Management Plan	Training of Project Team Members	Performance of Team Members	Demobilization of Workforce
		Project Manpower	Manage Project Team	Dispute Resolution	
		Sub contractor			
		Material and Equipment		Performance of Workforce	
Communication Management	Kick-off Meeting	Communication Plan	Site Administration Matrix	Meetings	
		Submittals			
		Documentation			
		Correspondance			
Risk Management		Risk Management Plan		Control Risk	
		Construction Risks Register		Risk Audit	
Contract Management		Contract Management	Contract Documents	Inspection	Finalize Work Performed
		Plan Purchase of Material/Equipment	Selection of Sub contractor(s)	Checklist	Finalize Material/Equipment Supplier's Contract
			Material, Systems, and Equipment		
HSE Management		Safety Management Plan	Site Safety	Accident Prevention Measures	
		Waste Management Plan	Temporary Fire Fighting	Loss Prevention during Construction	
Financial Management		Finance Management Plan	Contractor's Payments	Financial Control	Payment to Consultant
		Contractor Payment	Staff Payment		
		Material and Equipment Payments	Material and Equipment Payments		Payment to Contractor/ Sub contractor
			Progress (Interim Payment) Payment		Material and Equipment Payment
Claim Management		Claim Identification	Claim/Dispute Administration	Claim Prevention	Claim Payments
		Claim Quantification		Conflict Resolution	Settle Claims

FIGURE 7.123
Major activities relating to construction processes. *Note*: These activities may not be strictly sequential; however, the breakdown allows implementation of project management function more effective and easily manageable at different stages of the project phase.

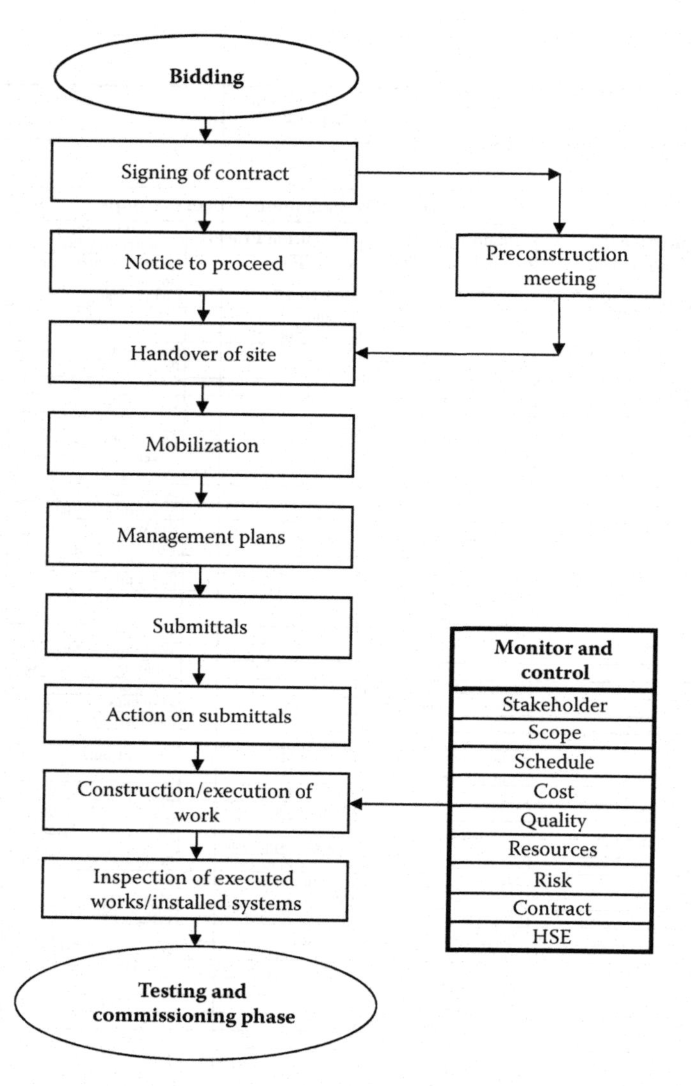

FIGURE 7.124
Logic flow process for construction phase.

7.4.6.1.1.1 Review Construction Drawings The contractor has to review the construction/working drawings to understand the project requirements. Each trade engineer has to review the drawing and understand the construction procedure to be followed to avoid omission and rework. The trade engineers have to

- Prepare a list of issue to be resolved and information needed from the designer (consultant).
- Prepare a list of conflicting items in the drawings.
- Compare the estimated bid activities and actual material for proper execution of project.
- Identify high risk items.

- Check constructability as per the specified method statement.
- Identify the execution process which can be simplified.
- Identify discrepancies between drawings and specifications.
- Identify any missing drawing required to execute the project.
- Identify value engineering change proposal.
- Identify complete scope of work.
- Prepare construction material takeoff.

7.4.6.1.1.2 Review Specifications The contractor has to review specifications and check for the following major points:

- Matching of construction drawings and specification requirements
- Any missing specifications
- Issues that need to be resolved
- Any discrepancies between specifications and drawings
- Codes and standards to be followed
- Submittal requirements
- Quality requirements
- Safety requirements
- Environmental requirements
- Installation procedure for owner supplied items

7.4.6.1.1.3 Review Contract Documents The contractor has to carefully study all the clauses of contract and identify

- High risk clauses
- Items having price difference between bid price and the actual price for the items to be installed for specified performance of work/system
- Items need to be resolved by raising RFI
- Clauses having conflict with regulatory requirements
- Schedule constraints
- Ambiguous clauses
- Clauses that favor owner
- Coordination among all the parties involved in the project
- Any hidden clause that will entitle the owner for compensation claim from the contractor
- Priorities among various construction documents
- Discrepancies and conflicting clauses
- Change order process
- Claim for extra work
- Payment procedure
- Damage and penalty for delay clauses
- Force majeure clause

7.4.6.1.2 Establish Construction Phase Requirements

Contract requirements are established taking into consideration items identified under Section 7.4.6.1.1. The following are the major requirements:

- Construction of the facility/project as per the specifications
- Complete the facility within the specified schedule
- Complete the facility within the contracted amount
- Complete the facility as per the quality requirements

7.4.6.2 Develop Construction Management Plan

Table 7.76 illustrates contents of contractor's construction management plan.

TABLE 7.76

Contents of Contractor's Construction Management Plan

Section	Topic
1	Introduction
	1.1 Project description
	1.2 Project scope
	1.3 Key milestone dates
2	Construction management organization
	2.1 Organizational chart
	2.2 Roles and responsibilities
	2.3 Coordination and liaison
3	Quality management plan
4	Risk management plan
5	Health safety and environmental management plan
	5.1 Site safety
	5.2 Hazard management
	5.3 Waste management
6	Construction schedule
7	Change management
8	Construction methodology
	8.1 Material submittals and approvals
	8.2 Shop drawing submittals and approvals
	8.3 Coordination and composite drawings
	8.4 Regulatory approvals
9	Subcontractor coordination
	9.1 Selection of subcontractor
	9.2 Coordination with subcontractor
10	Resources
	10.1 Workforce
	10.2 Equipment

(*Continued*)

TABLE 7.76 (*Continued*)

Contents of Contractor's Construction Management Plan

Section	Topic
11	Execution of work
	11.1 Method statement
	11.2 Mock up
	11.3 Work sequence
	11.4 Site activities (installation and coordination)
	11.5 Inspection of executed works
	11.6 Material storage and handling
12	Testing and commissioning
13	Site administration and finance management
14	Site access and security

7.4.6.3 Identify Stakeholders

Most stakeholders listed in Figure 7.23 have involvement in the project during construction phase. These are as follows:

1. Owner
   - Owner's representative/project manager
2. Construction supervisor
   - Construction manager (agency)
   - Consultant (designer)
   - Specialist contractor
3. Contractor
   - Main contractor
   - Subcontractor
   - Supplier
4. Regulatory authorities
5. End user

7.4.6.3.1 Identify Owner's Representative/Project Manager

The owner from his/her office deputes or hires from outside owner's representative (OR) to administer the overall project. The OR should have relevant experience and knowledge about the construction processes of similar nature of project. He/she should be able to manage the project with the help of supervision team members. In FIDIC terminology, this person is known as *engineer* who is appointed by the *employer* (owner) (see Chapter 6 for more information).

7.4.6.3.2 Identify Supervision Team

In a traditional type of contract, the client selects the same firm that has designed the project. The firm, known as *consultant*, is responsible for supervising the construction process and achieving project quality goals. The firm appoints a representative, who is acceptable and approved by the owner/client, to be on site and is often called as *resident engineer* (R.E.). The R.E. along with supervision team members is responsible to supervise, monitor and control, and implement the procedure specified in the contract documents and ensure completion of project within specified time, budget, and per defined scope of work.

In order to ensure smooth flow of supervision activities, R.E. has to follow the organization's supervision manual and contractual requirements. Depending on the type and size the project the supervision team usually consists of the following personnel:

1. R.E.
2. Contract administrator/quantity surveyor
3. Planning/scheduling engineer
4. Engineers from different trades such as architectural, structural, mechanical, HVAC, electrical, low-voltage system, landscape, and infrastructure
5. Inspectors from different trades
6. Interior designer
7. Document controller
8. Office secretary

The construction phase consists of various activities such as mobilization, execution of works, planning and scheduling, control and monitoring, management of resources/procurement, quality, and inspection. Table 7.77 illustrates major activities to be performed by the supervisor during the construction phase.

The owner may engage the construction manager (agency) to supervise complex and major projects. His or her role during construction stage is already discussed under Section 3.5.3.

7.4.6.3.3 Identify Contractor's Core Staff

Contract documents normally specify a list of minimum number of core staff to be available at site during the construction period. Absence of any of these staff may result in penalty to be imposed on the contractor by the owner.

The following is a typical list of contractor's minimum core staff needed during the construction period for execution of work of a major building construction project.

1. Project manager
2. Site senior engineer for civil works
3. Site senior engineer for architectural works
4. Site senior engineer for electrical works
5. Site senior engineer for mechanical works
6. Site senior engineer for HVAC works
7. Site senior engineer for infrastructure works
8. Planning engineer
9. Senior quantity surveyor/contract administrator

TABLE 7.77
Responsibilities of Supervision Consultant

Sr. No.	Description
1	Achieving the quality goal as specified
2	Review contract drawings and resolve technical discrepancies/errors in the contract documents
3	Review construction methodology
4	Approval of contractor's construction schedule
5	Regular inspection and checking of executed works
6	Review and approval of construction materials
7	Review and approval of shop drawings
8	Inspection of construction material
9	Monitoring and controlling construction expenditure
10	Monitoring and controlling construction time
11	Maintaining project record
12	Conduct progress and technical coordination meetings
13	Coordination of the owner's requirements and comments related to site activities
14	Project-related communication with contractor
15	Coordination with regulatory authorities
16	Processing of site work instruction for the owner's action
17	Evaluation and processing of variation order/change order
18	Recommendation of contractor's payment to owner
19	Evaluating and making decisions related to unforeseen conditions
20	Monitor safety at site
21	Supervise testing, commissioning, and handover of the project
22	Issue substantial completion certificate

Source: Abdul Razzak Rumane (2010), *Quality Management in Construction Projects*, CRC Press, Boca Raton, FL. Reprinted with permission from Taylor & Francis Group.

10. Civil works foreman
11. Architectural works foreman
12. Electrical works foreman
13. Mechanical works foreman
14. HVAC works foreman
15. Laboratory technician
16. Quality control engineer
17. Safety officer

7.4.6.3.4 Identify Regulatory Authorities

In certain countries, there is a regulation to submit electrical, mechanical, HVAC drawing for review and approval by the authorities. The contractor has to identify which drawings/documents are to be submitted to authorities during the construction phase. Necessary letter to the regulatory authorities/agencies is issued by the owner upon request for such letter from the contractor.

7.4.6.3.5 Identify Subcontractors

In most construction projects, the contractor engages special subcontractors to execute certain portion of contracted project works. Areas of subcontracting are generally listed in the particular conditions section of the contract document. Generally, the contractor has to submit subcontractors/specialist contractors to execute the following type of works:

1. Precast concrete works
2. Metal works
3. Space frame, roofing works
4. Wood works
5. Aluminum works
6. Internal finishes such as painting, false ceiling, tiling, and cladding
7. Furnishings
8. Waterproofing and insulation works
9. Mechanical works
10. HVAC works
11. Electrical works
12. Low-voltage systems/smart building system
13. Landscape
14. External works
15. Any other specialized works

The contractor has to submit their names for approval to the owner prior to their engagement to perform any work at the site. Table 7.78 is an example of subcontractor selection questionnaire.

7.4.6.3.5.1 Nominated Subcontractors In certain projects, the owner/client selects a contractor, known as *nominated subcontractor,* to carry out certain part of the work on behalf of the owner/client. Normally, the nominated subcontractor is imposed on the main contractor after the main contractor is appointed. Nominated subcontractor may be a specialist supplier, a specialist contractor nominated in accordance with the contract to be employed by the contractor for the supply of materials or services, or to execute the project work.

7.4.6.3.6 Establish Stakeholder's Requirements

A distribution matrix based on the interest and involvement of each stakeholder is prepared and the appropriate documents are sent for their action/information as per the agreed upon requirements. Table 7.79 shows an example matrix for site administration of a building construction project.

7.4.6.3.7 Develop Responsibility Matrix

Table 7.80 illustrates contribution of various participants during the construction phase.

TABLE 7.78

Subcontrator Prequalification Questionnaire

Instructions

Please type or write all your replies legibly. Attach additional sheets, if required.

Part I

I.1 Company Information

I.1.1 Name of organization:

I.1.2 Commercial registration no.:

I.1.3 Year of establishment:

I.1.4 Type of company:

I.1.5 Company address:

I.1.6 Affiliate company name(s) and address:

I.2 Subcontract Works (Please Tick-Mark All Interested)

Sr. No.	Work Description	Detail Design	Preparation of Shop Drawing	Construction	Inspection/ Auditing
1	Architectural work	☐			
2	Structural work				
3	Precast work				
4	Internal finishes				
5	External finishes				
6	Plumbing				
7	Drainage				
8	Firefighting				
9	HVAC				
10	Elevator				
11	Escalator				
12	Electrical work (power)				
13	Electrical work (low voltage)-specify				
14	Instrumentation				
15	Building management system				
16	Irrigation				
17	Landscape work				
18	Pavements				
19	Streets/roads				
20	Water proofing				

Part II

II.1 Financial Information

II.1.1 Provide copy of audited balance sheet:

II.1.2 Provide bonding capacity:

II.1.3 Provide insurance capacity:

II.1.4 Provide bank reference:

(Continued)

TABLE 7.78 (*Continued*)

Subcontrator Prequalification Questionnaire

PART III

III.1 Organization Details

III.1.1. Core business area:

III.1.2. Organizational chart:

III.1.3. ISO certification:

III.1.4. Years of experience:

III.2 Project Details

III.2.1. Project History for the Past 10 Years

Sr. No.	Name of Project	Type of Work	Value	Peak Workforce	Start Date	Finish Date
1						
2						
3						
4						

III.2.2 Current Projects

Sr. No.	Name of Project	Type of Work	Value	Peak Workforce	Start Date	Expected Finish Date
1						
2						
3						

PART IV

IV.1 Management Staff

IV.1.1 Provide list of project managers, project engineers, and engineers

IV.2 Workforce

Sr. No.	Work Description	Technicians	Foreman	Skilled	Unskilled
1	Architectural work				
2	Structural work				
3	Precast work				
4	Internal finishes				
5	External finishes				
6	Plumbing				
7	Drainage				
8	Firefighting				
9	HVAC				
10	Elevator				
11	Escalator				
12	Electrical Work (power)				
13	Electrical work (low voltage)—specify				
14	Instrumentation				

(*Continued*)

TABLE 7.78 (*Continued*)

Subcontrator Prequalification Questionnaire

Sr. No.	Work Description	Technicians	Foreman	Skilled	Unskilled
15	Building management system				
16	Irrigation				
17	Landscape work				
18	Pavements				
19	Streets/roads				
20	Water proofing				

Part V

V.1 Quality Management System

V.1.1 Provide copy of ISO certificate:

V.I.2 Person in charge of QA/QC activities:

V.I.3 Number of quality auditors:

Part VI

VI.1 HSE System

VI.1 Does the company have an HSE policy?

VI.1.2 Provide site accident records for the past 2 years

Declaration

We hereby declare that the information provided herein is true to our knowledge.

Note:

All relevant documents attached.

Signature of Authorized Person

Source: Abdul Razzak Rumane (2013), *Quality Tools for Managing Construction Projects*, CRC Press, Boca Raton, FL. Reprinted with permission from Taylor & Francis Group.

7.4.6.4 Develop Project Execution Scope

The following is the scope of works to be executed during the construction phase:

1. Execution of work
    - Site work such as cleaning and excavation of project site
    - Construction of foundations, including footings and grade beams
    - Construction of columns and beams
    - Forming, reinforcing, and placing the floor slab
    - Laying up masonry walls and partitions
    - Installation of roofing system
    - Finishes
    - Furnishings
    - Conveying system

TABLE 7.79

Matrix for Site Administration and Communication

Sr. No.	Description of Activities	Contractor	Consultant	Owner
1	**General**			
	1.1 Notice to proceed	–	–	P
	1.2 Bonds and guarantees	P	R	A
	1.3 Consultant staff approval	–	P	A
	1.4 Contractor's staff approval	P	R/B	A
	1.5 Payment guarantee	P	R	A
	1.6 Master schedule	P	R	A
	1.7 Stoppage of work	–	P	A
	1.8 Extension of time	–	P	A
	1.9 Deviation from contract documents	P	R	A
	a. Material			
	b. Cost			
	c. Time			
2	**Communication**			
	2.1 General correspondence	P	P	P
	2.2 Job site instruction	D	P	C
	2.3 Site works instruction	D	P/B	A
	2.4 Request for information	P	A	C
	2.5 Request for modification	P	B	A
3	**Submittals**			
	3.1 Subcontractor	P	B/R	A
	3.2 Materials	P	A	C
	3.3 Shop drawings	P	A	C
	3.4 Staff approval	P	B	A
	3.5 Premeeting submittals	P	D	C
4	**Plans and Programs**			
	4.1 Construction schedule	P	R	C
	4.2 Submittal logs	P	R	C
	4.3 Procurement logs	P	R	C
	4.4 Schedule update	P	R	C
5	**Monitor and Control**			
	5.1 Progress	D	P	C
	5.2 Time	D	P	C
	5.3 Payments	P	R/B	A
	5.4 Variations	P	R/B	A
	5.5 Claims	P	R/B	A
6	**Quality**			
	6.1 Quality control plan	P	R	C
	6.2 Checklists	P	D	C
	6.3 Method statements	P	A	C
	6.4 Mock up	P	A	B
	6.5 Samples	P	A	B
	6.6 Remedial notes	D	P	C
	6.7 Nonconformance report	D	P	C

(Continued)

TABLE 7.79 (*Continued*)

Matrix for Site Administration and Communication

Sr. No.	Description of Activities	Contractor	Consultant	Owner
	6.8 Inspections	P	D	C
	6.9 Testing	P	A	B
7	**Site Safety**			
	7.1 Safety program	P	A	C
	7.2 Accident report	P	R	C
8	**Meetings**			
	8.1 Progress	E	P	E
	8.2 Coordination	E	P	C
	8.3 Technical	E	P	C
	8.4 Quality	P	C	C
	8.5 Safety	P	C	C
	8.6 Closeout	–	P	
9	**Reports**			
	9.1 Daily report	P	R	C
	9.2 Monthly report	P	R	C
	9.3 Progress report	–	P	A
	9.4 Progress photographs	–	P	A
10	**Closeout**			
	10.1 Snag list	P	P	C
	10.2 Authorities approvals	P	C	C
	10.3 As-built drawings	P	D/A	C
	10.4 Spare parts	P	A	C
	10.5 Manuals and documents	P	R/B	A
	10.6 Warranties	P	R/B	A
	10.7 Training	P	C	A
	10.8 Handover	P	B	A
	10.9 Substantial completion certificate	P	B/P	A

P-Prepare/Initiate
B-Advise/Assist
R-Review/Comment
A-Approve
D-Action
E-Attend
C-Information

Source: Abdul Razzak Rumane (2010), *Quality Management in Construction Projects*, CRC Press, Boca Raton, FL. Reprinted with permission from Taylor & Francis Group.

- Installation of fire suppression system
- Installation of water supply, plumbing, and public health system
- Installation of heating, ventilating, and air-conditioning system
- Integrated automation system
- Installation of electrical lighting and power system
- Emergency power supply system
- Fire alarm system
- Communication system

TABLE 7.80

Responsibilities of Various Participants (Design–Bid–Build Type of Contracts) during Construction Phase

Phase	Responsibilities		
	Owner	**Supervisor**	**Contractor**
Construction	Approve subcontractor(s) Approve contractor's core staff Legal/regulatory clearance SWI V.O. Payments	Supervision Approve plan Monitor work progress Approve material Approve shop drawings Monitor schedule Control budget Recommend payment	Execution of work Contract management Selection of subcontractor(s) Planning Resources Procurement Quality Safety

- Electronic security and access control system
- Landscape works
- External works

2. Monitoring and control during construction
    - Stakeholder's requirements
    - Project schedule
    - Project cost
    - Project quality
    - Safety during construction
3. Inspection of executed works

7.4.6.4.1 Establish Construction Phase Deliverables

The construction phase deliverables are established based on project execution scope discussed under Section 7.4.6.4. These are:

1. Construction-installation/execution and inspection of following works;
    a. Structural work
    b. Architectural
    c. Internal finishes
    d. External finishes
    e. Conveying system
    f. MEP work
    g. Information and communication
    h. Low-voltage systems
    i. Landscape work
    j. External work
    k. Furnishings
2. Record drawings/updated working drawings (shop drawings)
3. Updated specifications

7.4.6.5 Develop Contractor's Construction Schedule

Prior to start of execution of project or immediately after the actual project starts, the contractor prepares the project construction plans based on the contracted time schedule of the project. Detailed planning is needed at the start of construction in order to decide the use of resources such as labors, plant, materials, finance, and subcontractors economically and safely to achieve the specified objectives. The plan shows the periods for all sections of the works and activities, indicating that everything can be completed by the date specified in the contract and ready for use or for installation of equipment by other contractors.

Project planning is a logical process to determine what work must be done to achieve project objectives and ensure that the work of project is carried out as follows:

- In an organized and structured manner
- By reducing uncertainties to minimum
- By reducing risk to minimum
- By establishing quality standards
- By achieving results within budget and scheduled time

Depending on the size of the project, the project is divided into multiple zones and relevant activities are considered for each zone to prepare the construction program. While preparing the program the relationships between project activities and their dependency and precedence is considered by the planner. These activities are connected to their predecessor and successor activity based on the way, the task is planned to be executed. There are four possible relationships that exist between various activities: finish-to-start relationship, start-to- start relationship, finish-to-finish relationship, and start-to-finish relationship.

Once all the activities are established by the planner and estimated, the duration of each activity has been assigned, the planner prepares detailed program fully coordinating all the construction activities.

The first step in the preparation of construction program is to establish the activities; the next step is to establish the estimated time duration of each activity. The deadline for each activity is fixed, but it is often possible to reschedule by changing the sequence in which the tasks are performed, while retaining the original estimated.

The activities to be performed during execution of project are grouped in a number of categories. Each of these categories has a number of activities. The following are the major categories of construction projects schedule:

1. **General activities**
   a. Mobilization
2. **Engineering**
   a. Subcontractor submittal and approval
   b. Materials submittal and approval
   c. Shop drawing submittal and approval
   d. Procurement

3. **Site activities**
   a. Site earth works
   b. Dewatering and shoring
   c. Excavation and backfilling
   d. Raft works
   e. Retaining wall works
   f. Concrete foundation and grade beams
   g. Water proofing
   h. Concrete columns and beams
   i. Casting of slabs
   j. Wall partitioning
   k. Interior finishes
   l. Furnishings
   m. External finishes
   n. Equipment
   o. Conveying systems works
   p. Fire suppression works
   q. Plumbing and public health works
   r. HVAC works
   s. Electrical works
   t. Fire alarm system works
   u. Information and communication system works
   v. Low-voltage systems works
   w. Landscape works
   x. External works
4. **Closeout**
   a. Testing and commissioning
   b. Completion and handover

The contractor also submits the following along with the construction schedule:

1. Resources (equipment and manpower) schedule
2. Cost loading (schedule of items' pricing based on BOQ)

Figure 7.125 illustrates the logic flow diagram for development of construction schedule (see also Figure 7.48) and Figure 7.126 illustrates an example of the contractor's construction schedule. The example schedule is construction of a branch for a bank.

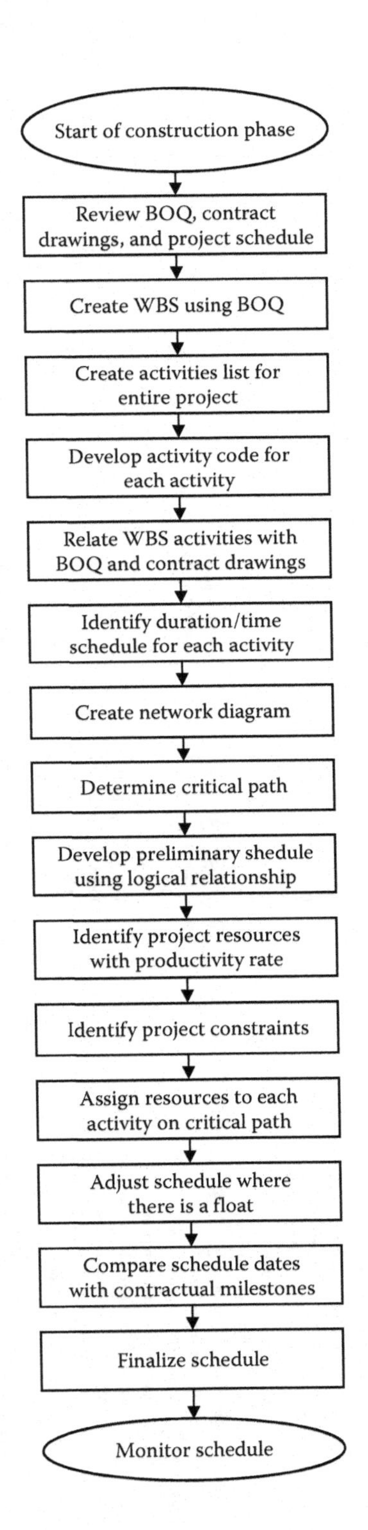

FIGURE 7.125
Logic flow diagram for development of construction schedule. (Abdul Razzak Rumane (2013), *Quality Tools for Managing Construction Projects*, CRC Press, Boca Raton, FL. Reprinted with permission from Taylor & Francis Group.)

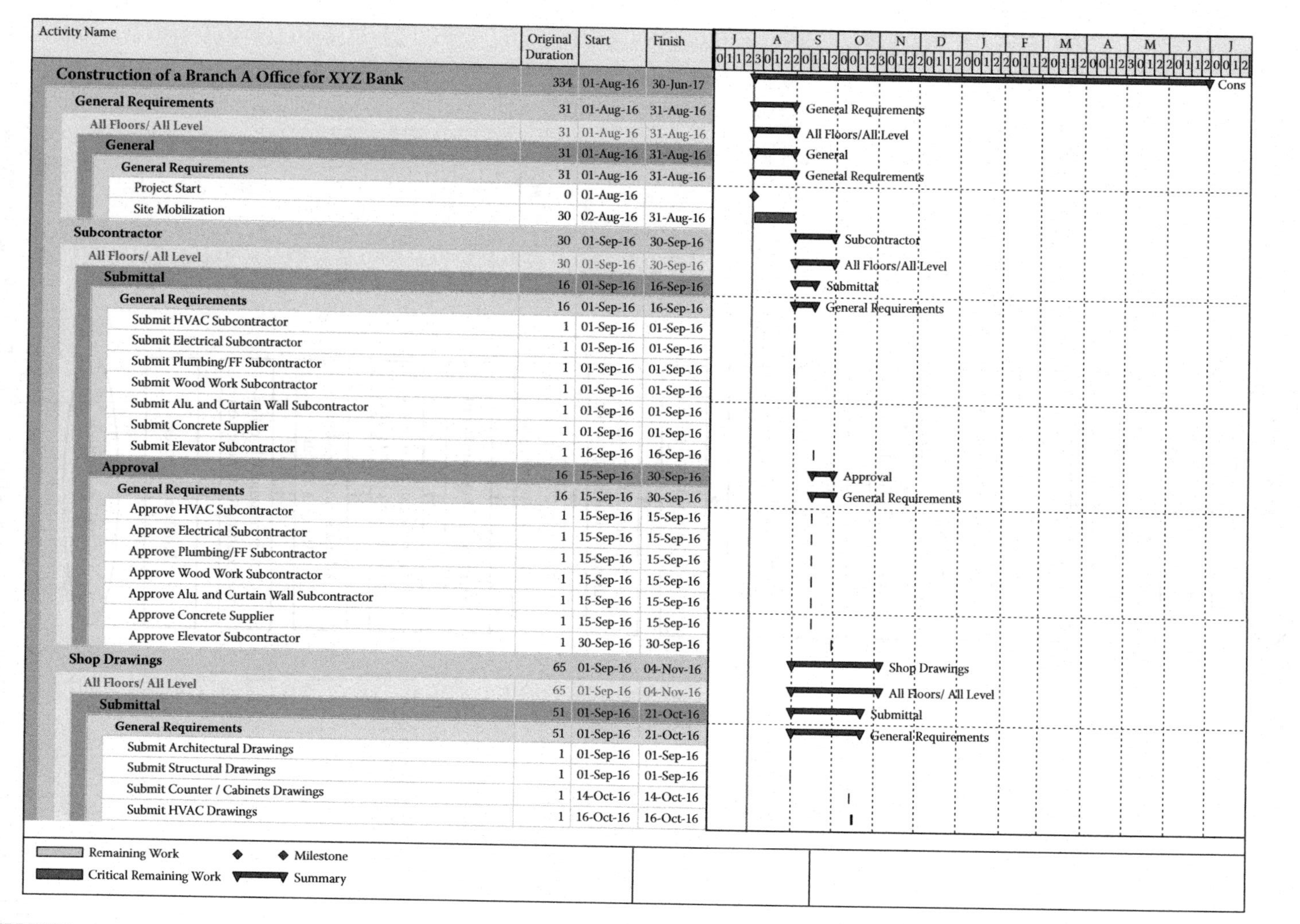

Activity Name	Original Duration	Start	Finish
Construction of a Branch A Office for XYZ Bank	334	01-Aug-16	30-Jun-17
General Requirements	31	01-Aug-16	31-Aug-16
All Floors/ All Level	31	01-Aug-16	31-Aug-16
General	31	01-Aug-16	31-Aug-16
General Requirements	31	01-Aug-16	31-Aug-16
Project Start	0	01-Aug-16	
Site Mobilization	30	02-Aug-16	31-Aug-16
Subcontractor	30	01-Sep-16	30-Sep-16
All Floors/ All Level	30	01-Sep-16	30-Sep-16
Submittal	16	01-Sep-16	16-Sep-16
General Requirements	16	01-Sep-16	16-Sep-16
Submit HVAC Subcontractor	1	01-Sep-16	01-Sep-16
Submit Electrical Subcontractor	1	01-Sep-16	01-Sep-16
Submit Plumbing/FF Subcontractor	1	01-Sep-16	01-Sep-16
Submit Wood Work Subcontractor	1	01-Sep-16	01-Sep-16
Submit Alu. and Curtain Wall Subcontractor	1	01-Sep-16	01-Sep-16
Submit Concrete Supplier	1	01-Sep-16	01-Sep-16
Submit Elevator Subcontractor	1	16-Sep-16	16-Sep-16
Approval	16	15-Sep-16	30-Sep-16
General Requirements	16	15-Sep-16	30-Sep-16
Approve HVAC Subcontractor	1	15-Sep-16	15-Sep-16
Approve Electrical Subcontractor	1	15-Sep-16	15-Sep-16
Approve Plumbing/FF Subcontractor	1	15-Sep-16	15-Sep-16
Approve Wood Work Subcontractor	1	15-Sep-16	15-Sep-16
Approve Alu. and Curtain Wall Subcontractor	1	15-Sep-16	15-Sep-16
Approve Concrete Supplier	1	15-Sep-16	15-Sep-16
Approve Elevator Subcontractor	1	30-Sep-16	30-Sep-16
Shop Drawings	65	01-Sep-16	04-Nov-16
All Floors/ All Level	65	01-Sep-16	04-Nov-16
Submittal	51	01-Sep-16	21-Oct-16
General Requirements	51	01-Sep-16	21-Oct-16
Submit Architectural Drawings	1	01-Sep-16	01-Sep-16
Submit Structural Drawings	1	01-Sep-16	01-Sep-16
Submit Counter / Cabinets Drawings	1	14-Oct-16	14-Oct-16
Submit HVAC Drawings	1	16-Oct-16	16-Oct-16

FIGURE 7.126
Contractor's construction schedule.

(Continued)

Activity Name	Original Duration	Start	Finish
Submit Plumbing/FF Drawings	1	16-Oct-16	16-Oct-16
Submit Electrical Drawings	1	16-Oct-16	16-Oct-16
Submit Low Voltage Drawings	1	21-Oct-16	21-Oct-16
Approval	51	15-Sep-16	04-Nov-16
General Requirements	51	15-Sep-16	04-Nov-16
Approve Architectural Drawings	1	15-Sep-16	15-Sep-16
Approve Structural Drawings	1	15-Sep-16	15-Sep-16
Approve Counter / Cabinets Drawings	1	28-Oct-16	28-Oct-16
Approve HVAC Drawings	1	30-Oct-16	30-Oct-16
Approve Plumbing/FF Drawings	1	30-Oct-16	30-Oct-16
Approve Electrical Drawings	1	30-Oct-16	30-Oct-16
Approve Low Voltage Drawings	1	04-Nov-16	04-Nov-16
Material	132	01-Sep-16	10-Jan-17
All Floors/ All Level	132	01-Sep-16	10-Jan-17
Submittal	58	01-Sep-16	28-Oct-16
Site Works	1	16-Oct-16	16-Oct-16
Submit Pavers, Kerbstone	1	16-Oct-16	16-Oct-16
Submit Plants	1	16-Oct-16	16-Oct-16
Masonry	1	01-Oct-16	01-Oct-16
Submit Block	1	01-Oct-16	01-Oct-16
Metal	1	13-Oct-16	13-Oct-16
Submit Steel Covers	1	13-Oct-16	13-Oct-16
Submit Handrail and Balustrade	1	13-Oct-16	13-Oct-16
Submit Wooden Cladding Material	1	13-Oct-16	13-Oct-16
SubmitExternal Door Matt (Metal)	1	13-Oct-16	13-Oct-16
Submit Canopy Materials (Roof and Glass)	1	13-Oct-16	13-Oct-16
Wood and Plastic	1	15-Oct-16	15-Oct-16
Submit Counter Wood materials / Accs.	1	15-Oct-16	15-Oct-16
Submit Wooden Material (Wall/Ceiling/Skirting)	1	15-Oct-16	15-Oct-16
Thermal and MoistureProtection	36	01-Sep-16	06-Oct-16
Submit Water Proofing Material	1	01-Sep-16	06-Oct-16
Submit Thermal Insulation b/h Stone Cladding	1	06-Oct-16	06-Oct-16
Doors and Windows	11	15-Oct-16	25-Oct-16
Submit Vault Door	1	15-Oct-16	15-Oct-16
Submit Door (Steel / Wooden / Sliding)	1	25-Oct-16	25-Oct-16
Submit Hardware (Doors)	1	25-Oct-16	25-Oct-16

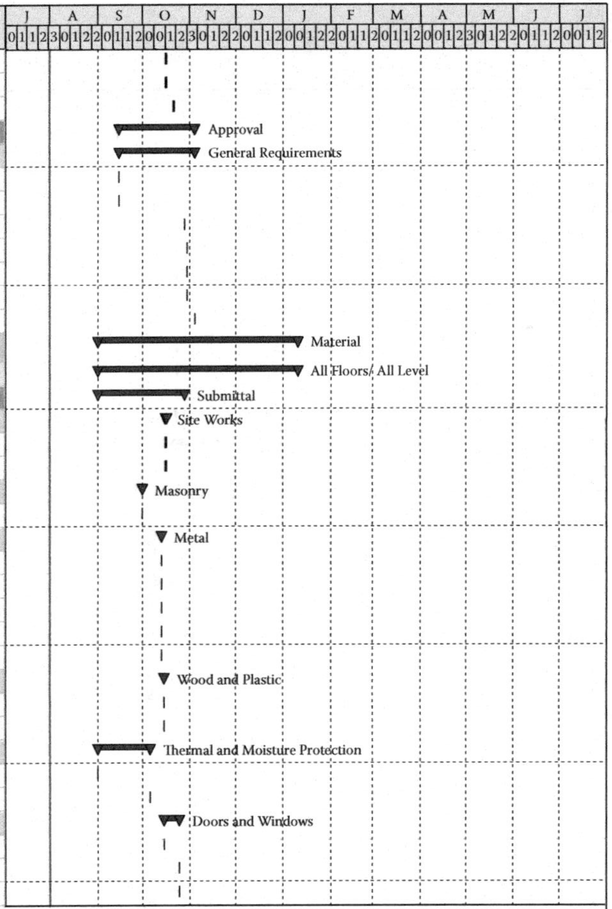

Remaining Work ◆ ◆ Milestone
Critical Remaining Work Summary

FIGURE 7.126 (Continued)
Contractor's construction schedule.

(Continued)

Activity Name	Original Duration	Start	Finish
Submit Curtain Wall	1	25-Oct-16	25-Oct-16
Submit Aluminum Composite Panel	1	25-Oct-16	25-Oct-16
Finishes	14	01-Oct-16	14-Oct-16
Submit Granite/Marble Flooring	1	01-Oct-16	01-Oct-16
Submit Terrazo Tiles	1	11-Oct-16	11-Oct-16
Submit Gypsum Board Partition	1	14-Oct-16	14-Oct-16
Submit Gypsum Ceiling and Accs	1	14-Oct-16	14-Oct-16
Submit Gypsum Tiles (60 × 60) and Accs	1	14-Oct-16	14-Oct-16
Submit Ceramic Tiles (Wall and Floor)	1	14-Oct-16	14-Oct-16
Submit Vinyl Tiles and Skirting	1	14-Oct-16	14-Oct-16
Submit Sandstone Cladding -Ext Walls	1	14-Oct-16	14-Oct-16
Submit Carpet	1	14-Oct-16	14-Oct-16
Submit Paints	1	14-Oct-16	14-Oct-16
Specialities	1	10-Oct-16	10-Oct-16
Submit Raised Floor System (Teller/Server)	1	10-Oct-16	10-Oct-16
Submit Toilet Accs	1	10-Oct-16	10-Oct-16
Submit Trash Box (ATM Area)	1	10-Oct-16	10-Oct-16
Submit Rubber Wall Guard (Below ATM)	1	10-Oct-16	10-Oct-16
Furnishing	11	12-Oct-16	22-Oct-16
Submit Curtains	1	12-Oct-16	12-Oct-16
Submit Fixed Furnitures	1	12-Oct-16	12-Oct-16
Submit Kitchen Cabinet	1	22-Oct-16	22-Oct-16
Conveying System	1	01-Oct-16	01-Oct-16
Submit Elevator	1	01-Oct-16	01-Oct-16
Mechanical	3	26-Oct-16	28-Oct-16
Submit Ducting Material and Insulation	1	26-Oct-16	26-Oct-16
Submit Roof Top Package Units	1	26-Oct-16	26-Oct-16
Submit Room Split Units	1	26-Oct-16	26-Oct-16
Submit VCD	1	26-Oct-16	26-Oct-16
Submit Diffusers, Registers, Grilles	1	26-Oct-16	26-Oct-16
Submit Electric Duct Heaters for Package Units	1	26-Oct-16	26-Oct-16
Submit Fans (TEF, KEF, EF)	1	26-Oct-16	26-Oct-16
Submit Condensate Drain Pipes	1	26-Oct-16	26-Oct-16
Submit FP Pipes and Accs	1	26-Oct-16	26-Oct-16
Submit FP Pumps and Accs	1	26-Oct-16	26-Oct-16
Submit Fire Cabinets	1	26-Oct-16	26-Oct-16

Remaining Work ◆ ◆ Milestone
Critical Remaining Work Summary

FIGURE 7.126 (Continued)
Contractor's construction schedule.

(Continued)

Activity Name	Original Duration	Start	Finish
Submit Fire Extinguishers	1	26-Oct-16	26-Oct-16
Submit Sprinkler and Spray System	1	26-Oct-16	26-Oct-16
Submit Plumbing Pipes and Accs	1	26-Oct-16	26-Oct-16
Submit SS access cover for Meter Box	1	26-Oct-16	26-Oct-16
Submit Soil and Waste Water Pipes	1	26-Oct-16	26-Oct-16
Submit Sanitary Fixtures	1	26-Oct-16	26-Oct-16
Submit Water Booster Pumps	1	26-Oct-16	26-Oct-16
Submit Electric Water Heater	1	26-Oct-16	26-Oct-16
Submit GRP Water Tank	1	26-Oct-16	26-Oct-16
Submit Air Curtain	1	28-Oct-16	28-Oct-16
Electrical	1	25-Oct-16	25-Oct-16
Submit MSB, SMSB, ATS	1	25-Oct-16	25-Oct-16
Submit Distribution Boards	1	25-Oct-16	25-Oct-16
Submit Cables, Trays, Conduits	1	25-Oct-16	25-Oct-16
Submit AVA Alarm Panel	1	25-Oct-16	25-Oct-16
Submit 50 KVAR Capacitor	1	25-Oct-16	25-Oct-16
Submit Socket, Switches, Jn Box, DP	1	25-Oct-16	25-Oct-16
Submit Trunking	1	25-Oct-16	25-Oct-16
Submit Light Fixtures and Accs	1	25-Oct-16	25-Oct-16
Submit UPS	1	25-Oct-16	25-Oct-16
Submit Data Commun. System	1	25-Oct-16	25-Oct-16
Submit Server Cabinet	1	25-Oct-16	25-Oct-16
Submit Standby Diesel Generator and Fuel Storage Tank	1	25-Oct-16	25-Oct-16
Submit Earthing System	1	25-Oct-16	25-Oct-16
Submit Fire Alarm System	1	25-Oct-16	25-Oct-16
Submit CCTV System	1	25-Oct-16	25-Oct-16
Submit Access Control System	1	25-Oct-16	25-Oct-16
Submit Intrusion Detection System	1	25-Oct-16	25-Oct-16
Submit Queing System	1	25-Oct-16	25-Oct-16
Approval	58	15-Oct-16	11-Oct-16
Site Works	1	30-Oct-16	30-Oct-16
Approve Pavers, Kerbstone	1	30-Oct-16	30-Oct-16
Approve Plants	1	30-Oct-16	30-Oct-16
Masonry	1	15-Oct-16	15-Oct-16
Approve Block	1	15-Oct-16	15-Oct-16
Metal	1	27-Oct-16	27-Oct-16

FIGURE 7.126 (Continued)
Contractor's construction schedule.

(Continued)

Activity Name	Original Duration	Start	Finish
Approve Steel Covers	1	27-Oct-16	27-Oct-16
Approve Handrail and Balustrade	1	27-Oct-16	27-Oct-16
Approve Wooden Cladding Material	1	27-Oct-16	27-Oct-16
Approve External Door Matt (Metal)	1	27-Oct-16	27-Oct-16
Approve Canopy Materials (Roof and Glass)	1	27-Oct-16	27-Oct-16
Wood and Plastic	1	29-Oct-16	29-Oct-16
Approve Counter Wood materials / Accs.	1	29-Oct-16	29-Oct-16
Approve Wooden Material (Wall/Ceiling/Skirting)	1	29-Oct-16	29-Oct-16
Thermal and Moisture Protection	36	15-Sep-16	20-Oct-16
Approve Water Proofing Material	1	15-Sep-16	15-Sep-16
Approve Thermal Insulation b/h Stone Cladding	1	20-Oct-16	20-Oct-16
Doors and Windows	11	29-Oct-16	08-Nov-16
Approve Vault Door	1	29-Oct-16	29-Oct-16
Approve Door (Steel / Wooden / Sliding)	1	08-Nov-16	08-Nov-16
Approve Hardware (Doors)	1	08-Nov-16	08-Nov-16
Approve Curtain Wall	1	08-Nov-16	08-Nov-16
Approve Aluminum Composite Panel	1	08-Nov-16	08-Nov-16
Finishes	14	15-Oct-16	28-Oct-16
Approve Granite/Marble Flooring	1	15-Oct-16	15-Oct-16
Approve Terrazo Tiles	1	25-Oct-16	25-Oct-16
Approve Gypsum Board Partition	1	28-Oct-16	28-Oct-16
Approve Gypsum Ceiling and Accs	1	28-Oct-16	28-Oct-16
Approve Gypsum Tiles (60x60) and Accs	1	28-Oct-16	28-Oct-16
Approve Ceramic Tiles (Wall and Floor)	1	28-Oct-16	28-Oct-16
Approve Vinyl Tiles and Skirting	1	28-Oct-16	28-Oct-16
Approve Sandstone Cladding -Ext Walls	1	28-Oct-16	28-Oct-16
Approve Carpet	1	28-Oct-16	28-Oct-16
Approve Paints	1	28-Oct-16	28-Oct-16
Specialities	1	24-Oct-16	24-Oct-16
Approve Raised Floor System (Teller/Server)	1	24-Oct-16	24-Oct-16
Approve Toilet Accs	1	24-Oct-16	24-Oct-16
Approve Trash Box (ATM Area)	1	24-Oct-16	24-Oct-16
Approve Rubber Wall Guard (Below ATM)	1	24-Oct-16	24-Oct-16
Furnishing	11	26-Oct-16	05-Nov-16
Approve Curtains	1	26-Oct-16	26-Oct-16
Approve Fixed Furnitures	1	26-Oct-16	26-Oct-16

FIGURE 7.126 (Continued)
Contractor's construction schedule.

(Continued)

Activity Name	Original Duration	Start	Finish
Approve Kitchen Cabinet	1	05-Nov-16	05-Nov-16
Conveying System	1	15-Oct-16	15-Oct-16
Approve Elevator	1	15-Oct-16	15-Oct-16
Mechanical	3	09-Nov-16	09-Nov-16
Approve Ducting Material and Insulation	1	09-Nov-16	09-Nov-16
Approve Roof Top Package Units	1	09-Nov-16	09-Nov-16
Approve Room Split Units	1	09-Nov-16	09-Nov-16
Approve VCD	1	09-Nov-16	09-Nov-16
Approve Diffusers, Registers, Grilles	1	09-Nov-16	09-Nov-16
Approve Electric Duct Heaters for Package Units	1	09-Nov-16	09-Nov-16
Approve Fans (TEF, KEF, EF)	1	09-Nov-16	09-Nov-16
Approve Condensate Drain Pipes	1	09-Nov-16	09-Nov-16
Approve FP Pipes and Accs	1	09-Nov-16	09-Nov-16
Approve FP Pumps and Accs	1	09-Nov-16	09-Nov-16
Approve Fire Cabinets	1	09-Nov-16	09-Nov-16
Approve Fire Extinguishers	1	09-Nov-16	09-Nov-16
Approve Sprinkler and Spray System	1	09-Nov-16	09-Nov-16
Approve Plumbing Pipes and Accs	1	09-Nov-16	09-Nov-16
Approve SS access cover for Meter Box	1	09-Nov-16	09-Nov-16
Approve Soil and Waste Water Pipes	1	09-Nov-16	09-Nov-16
Approve Sanitary Fixtures	1	09-Nov-16	09-Nov-16
Approve Water Booster Pumps	1	09-Nov-16	09-Nov-16
Approve Electric Water Heater	1	09-Nov-16	09-Nov-16
Approve GRP Water Tank	1	09-Nov-16	09-Nov-16
Approve Air Curtain	1	11-Nov-16	11-Nov-16
Electrical	1	08-Nov-16	08-Nov-16
Approve MSB, SMSB, ATS	1	08-Nov-16	08-Nov-16
Approve Distribution Boards	1	08-Nov-16	08-Nov-16
Approve Cables, Trays, Conduits	1	08-Nov-16	08-Nov-16
Approve AVA Alarm Panel	1	08-Nov-16	08-Nov-16
Approve 50 KVAR Capacitor	1	08-Nov-16	08-Nov-16
Approve Socket, Switches, Jn Box, DP	1	08-Nov-16	08-Nov-16
Approve Trunking	1	08-Nov-16	08-Nov-16
Approve Light Fixtures and Accs	1	08-Nov-16	08-Nov-16
Approve UPS	1	08-Nov-16	08-Nov-16
Approve Data Commun. System	1	08-Nov-16	08-Nov-16

FIGURE 7.126 (Continued)
Contractor's construction schedule.

(Continued)

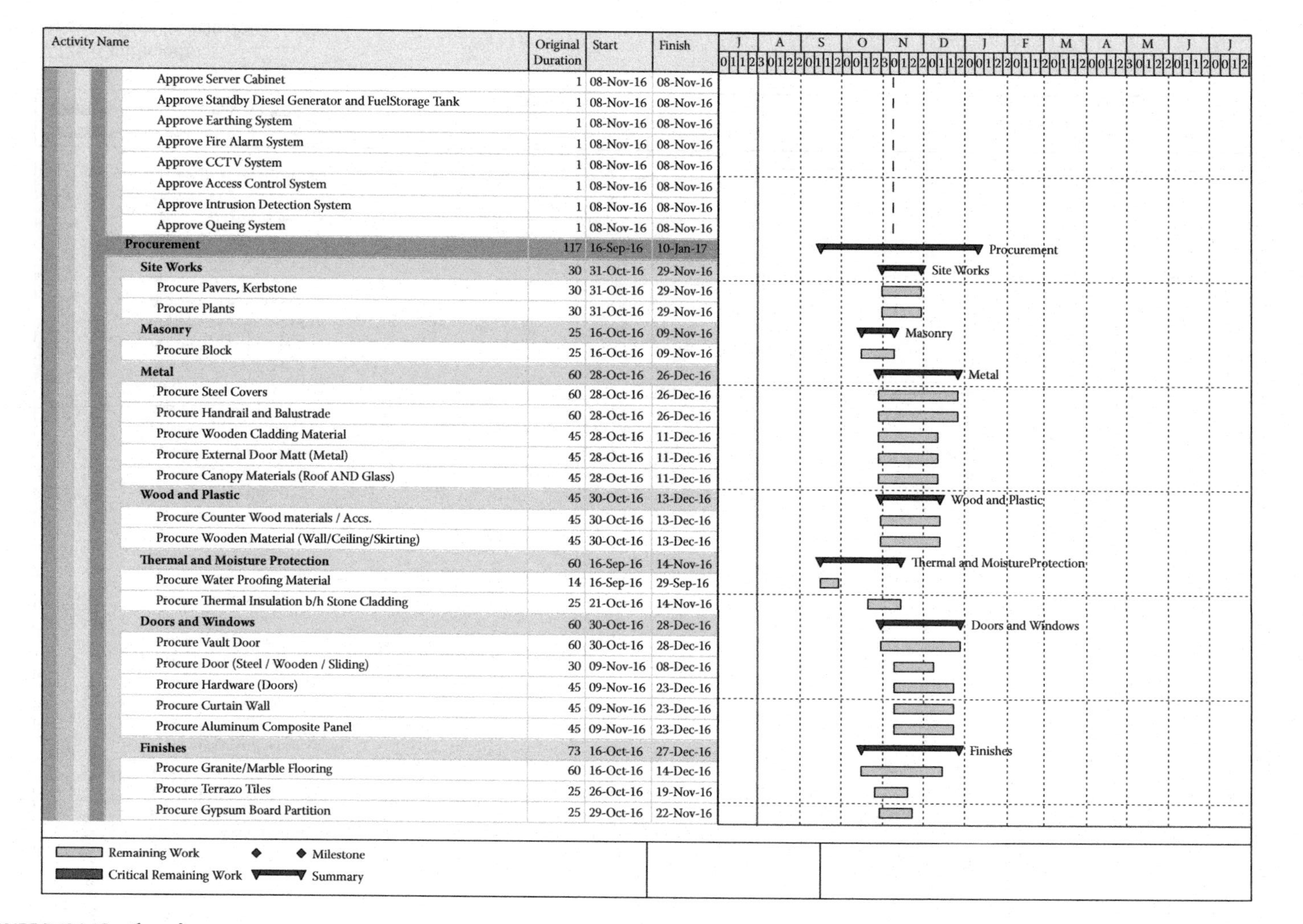

Activity Name	Original Duration	Start	Finish
Approve Server Cabinet	1	08-Nov-16	08-Nov-16
Approve Standby Diesel Generator and FuelStorage Tank	1	08-Nov-16	08-Nov-16
Approve Earthing System	1	08-Nov-16	08-Nov-16
Approve Fire Alarm System	1	08-Nov-16	08-Nov-16
Approve CCTV System	1	08-Nov-16	08-Nov-16
Approve Access Control System	1	08-Nov-16	08-Nov-16
Approve Intrusion Detection System	1	08-Nov-16	08-Nov-16
Approve Queing System	1	08-Nov-16	08-Nov-16
Procurement	117	16-Sep-16	10-Jan-17
Site Works	30	31-Oct-16	29-Nov-16
Procure Pavers, Kerbstone	30	31-Oct-16	29-Nov-16
Procure Plants	30	31-Oct-16	29-Nov-16
Masonry	25	16-Oct-16	09-Nov-16
Procure Block	25	16-Oct-16	09-Nov-16
Metal	60	28-Oct-16	26-Dec-16
Procure Steel Covers	60	28-Oct-16	26-Dec-16
Procure Handrail and Balustrade	60	28-Oct-16	26-Dec-16
Procure Wooden Cladding Material	45	28-Oct-16	11-Dec-16
Procure External Door Matt (Metal)	45	28-Oct-16	11-Dec-16
Procure Canopy Materials (Roof AND Glass)	45	28-Oct-16	11-Dec-16
Wood and Plastic	45	30-Oct-16	13-Dec-16
Procure Counter Wood materials / Accs.	45	30-Oct-16	13-Dec-16
Procure Wooden Material (Wall/Ceiling/Skirting)	45	30-Oct-16	13-Dec-16
Thermal and Moisture Protection	60	16-Sep-16	14-Nov-16
Procure Water Proofing Material	14	16-Sep-16	29-Sep-16
Procure Thermal Insulation b/h Stone Cladding	25	21-Oct-16	14-Nov-16
Doors and Windows	60	30-Oct-16	28-Dec-16
Procure Vault Door	60	30-Oct-16	28-Dec-16
Procure Door (Steel / Wooden / Sliding)	30	09-Nov-16	08-Dec-16
Procure Hardware (Doors)	45	09-Nov-16	23-Dec-16
Procure Curtain Wall	45	09-Nov-16	23-Dec-16
Procure Aluminum Composite Panel	45	09-Nov-16	23-Dec-16
Finishes	73	16-Oct-16	27-Dec-16
Procure Granite/Marble Flooring	60	16-Oct-16	14-Dec-16
Procure Terrazo Tiles	25	26-Oct-16	19-Nov-16
Procure Gypsum Board Partition	25	29-Oct-16	22-Nov-16

FIGURE 7.126 (Continued)
Contractor's construction schedule.

(Continued)

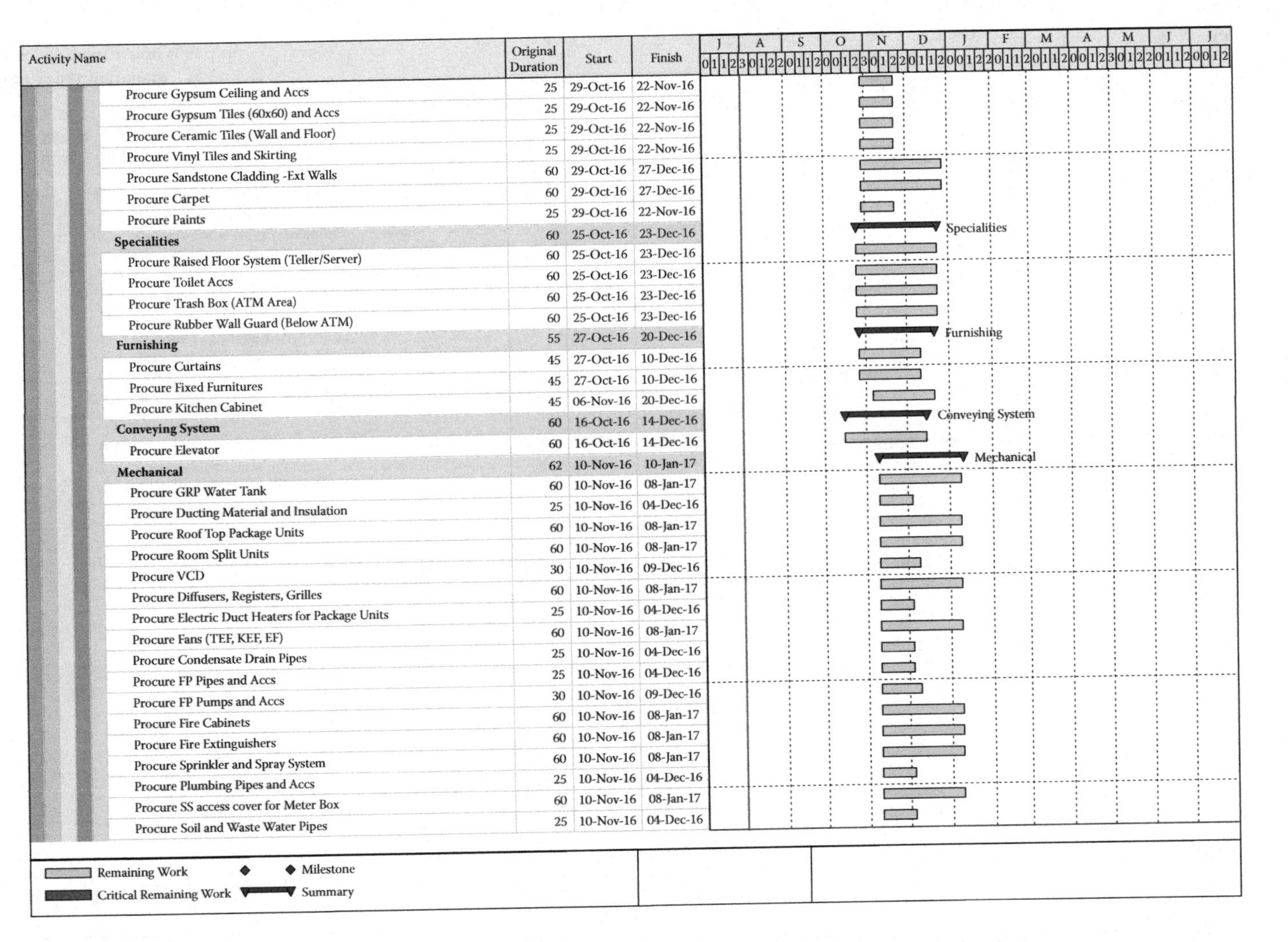

Activity Name	Original Duration	Start	Finish
Procure Gypsum Ceiling and Accs	25	29-Oct-16	22-Nov-16
Procure Gypsum Tiles (60x60) and Accs	25	29-Oct-16	22-Nov-16
Procure Ceramic Tiles (Wall and Floor)	25	29-Oct-16	22-Nov-16
Procure Vinyl Tiles and Skirting	25	29-Oct-16	22-Nov-16
Procure Sandstone Cladding -Ext Walls	60	29-Oct-16	27-Dec-16
Procure Carpet	60	29-Oct-16	27-Dec-16
Procure Paints	25	29-Oct-16	22-Nov-16
Specialities	60	25-Oct-16	23-Dec-16
Procure Raised Floor System (Teller/Server)	60	25-Oct-16	23-Dec-16
Procure Toilet Accs	60	25-Oct-16	23-Dec-16
Procure Trash Box (ATM Area)	60	25-Oct-16	23-Dec-16
Procure Rubber Wall Guard (Below ATM)	60	25-Oct-16	23-Dec-16
Furnishing	55	27-Oct-16	20-Dec-16
Procure Curtains	45	27-Oct-16	10-Dec-16
Procure Fixed Furnitures	45	27-Oct-16	10-Dec-16
Procure Kitchen Cabinet	45	06-Nov-16	20-Dec-16
Conveying System	60	16-Oct-16	14-Dec-16
Procure Elevator	60	16-Oct-16	14-Dec-16
Mechanical	62	10-Nov-16	10-Jan-17
Procure GRP Water Tank	60	10-Nov-16	08-Jan-17
Procure Ducting Material and Insulation	25	10-Nov-16	04-Dec-16
Procure Roof Top Package Units	60	10-Nov-16	08-Jan-17
Procure Room Split Units	60	10-Nov-16	08-Jan-17
Procure VCD	30	10-Nov-16	09-Dec-16
Procure Diffusers, Registers, Grilles	60	10-Nov-16	08-Jan-17
Procure Electric Duct Heaters for Package Units	25	10-Nov-16	04-Dec-16
Procure Fans (TEF, KEF, EF)	60	10-Nov-16	08-Jan-17
Procure Condensate Drain Pipes	25	10-Nov-16	04-Dec-16
Procure FP Pipes and Accs	25	10-Nov-16	04-Dec-16
Procure FP Pumps and Accs	30	10-Nov-16	09-Dec-16
Procure Fire Cabinets	60	10-Nov-16	08-Jan-17
Procure Fire Extinguishers	60	10-Nov-16	08-Jan-17
Procure Sprinkler and Spray System	60	10-Nov-16	08-Jan-17
Procure Plumbing Pipes and Accs	25	10-Nov-16	04-Dec-16
Procure SS access cover for Meter Box	60	10-Nov-16	08-Jan-17
Procure Soil and Waste Water Pipes	25	10-Nov-16	04-Dec-16

FIGURE 7.126 (Continued)
Contractor's construction schedule.

(Continued)

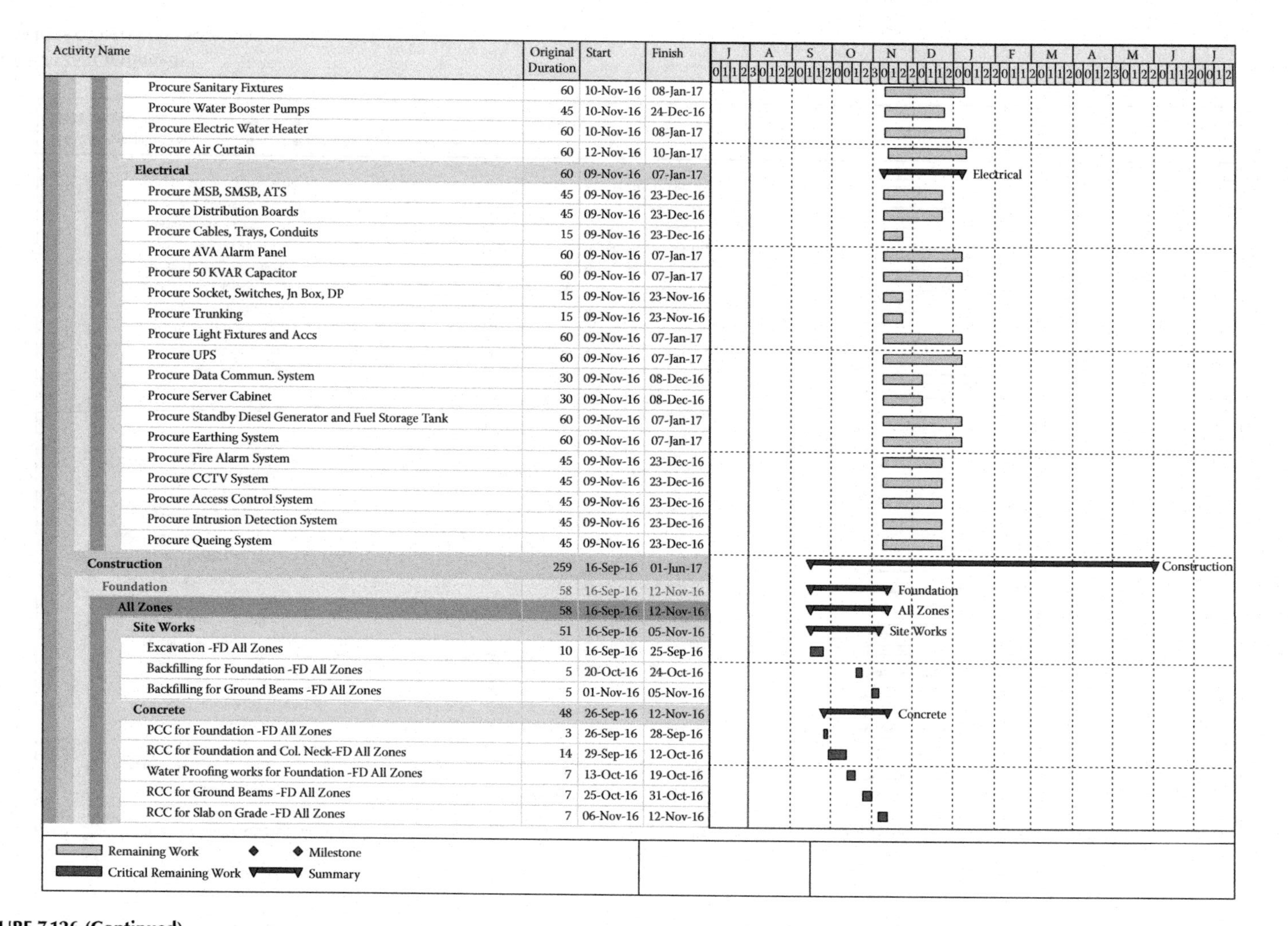

Activity Name	Original Duration	Start	Finish
Procure Sanitary Fixtures	60	10-Nov-16	08-Jan-17
Procure Water Booster Pumps	45	10-Nov-16	24-Dec-16
Procure Electric Water Heater	60	10-Nov-16	08-Jan-17
Procure Air Curtain	60	12-Nov-16	10-Jan-17
Electrical	60	09-Nov-16	07-Jan-17
Procure MSB, SMSB, ATS	45	09-Nov-16	23-Dec-16
Procure Distribution Boards	45	09-Nov-16	23-Dec-16
Procure Cables, Trays, Conduits	15	09-Nov-16	23-Dec-16
Procure AVA Alarm Panel	60	09-Nov-16	07-Jan-17
Procure 50 KVAR Capacitor	60	09-Nov-16	07-Jan-17
Procure Socket, Switches, Jn Box, DP	15	09-Nov-16	23-Nov-16
Procure Trunking	15	09-Nov-16	23-Nov-16
Procure Light Fixtures and Accs	60	09-Nov-16	07-Jan-17
Procure UPS	60	09-Nov-16	07-Jan-17
Procure Data Commun. System	30	09-Nov-16	08-Dec-16
Procure Server Cabinet	30	09-Nov-16	08-Dec-16
Procure Standby Diesel Generator and Fuel Storage Tank	60	09-Nov-16	07-Jan-17
Procure Earthing System	60	09-Nov-16	07-Jan-17
Procure Fire Alarm System	45	09-Nov-16	23-Dec-16
Procure CCTV System	45	09-Nov-16	23-Dec-16
Procure Access Control System	45	09-Nov-16	23-Dec-16
Procure Intrusion Detection System	45	09-Nov-16	23-Dec-16
Procure Queing System	45	09-Nov-16	23-Dec-16
Construction	259	16-Sep-16	01-Jun-17
Foundation	58	16-Sep-16	12-Nov-16
All Zones	58	16-Sep-16	12-Nov-16
Site Works	51	16-Sep-16	05-Nov-16
Excavation -FD All Zones	10	16-Sep-16	25-Sep-16
Backfilling for Foundation -FD All Zones	5	20-Oct-16	24-Oct-16
Backfilling for Ground Beams -FD All Zones	5	01-Nov-16	05-Nov-16
Concrete	48	26-Sep-16	12-Nov-16
PCC for Foundation -FD All Zones	3	26-Sep-16	28-Sep-16
RCC for Foundation and Col. Neck-FD All Zones	14	29-Sep-16	12-Oct-16
Water Proofing works for Foundation -FD All Zones	7	13-Oct-16	19-Oct-16
RCC for Ground Beams -FD All Zones	7	25-Oct-16	31-Oct-16
RCC for Slab on Grade -FD All Zones	7	06-Nov-16	12-Nov-16

FIGURE 7.126 (Continued)
Contractor's construction schedule.

(Continued)

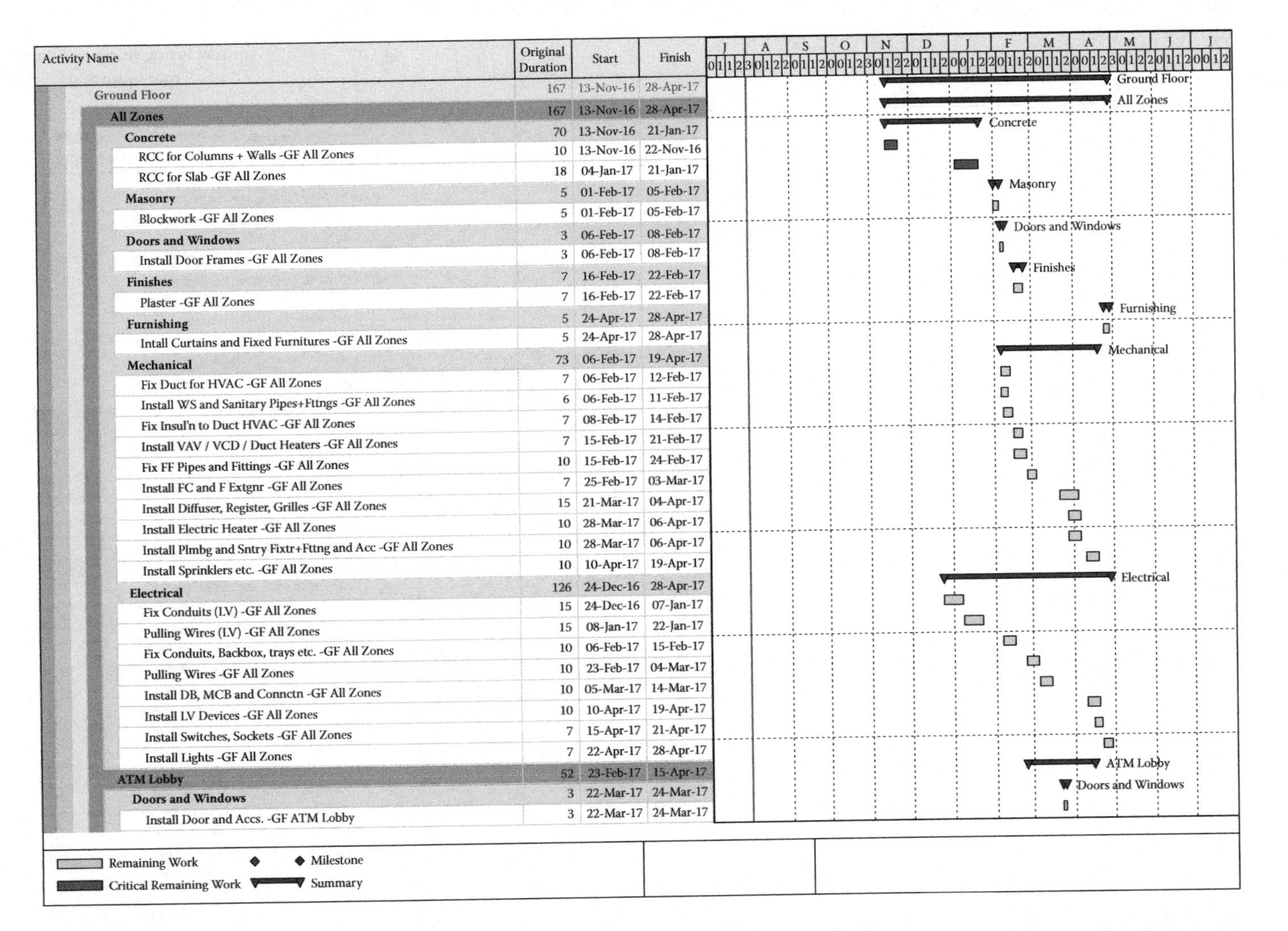

Activity Name	Original Duration	Start	Finish
Ground Floor	167	13-Nov-16	28-Apr-17
All Zones	167	13-Nov-16	28-Apr-17
Concrete	70	13-Nov-16	21-Jan-17
RCC for Columns + Walls -GF All Zones	10	13-Nov-16	22-Nov-16
RCC for Slab -GF All Zones	18	04-Jan-17	21-Jan-17
Masonry	5	01-Feb-17	05-Feb-17
Blockwork -GF All Zones	5	01-Feb-17	05-Feb-17
Doors and Windows	3	06-Feb-17	08-Feb-17
Install Door Frames -GF All Zones	3	06-Feb-17	08-Feb-17
Finishes	7	16-Feb-17	22-Feb-17
Plaster -GF All Zones	7	16-Feb-17	22-Feb-17
Furnishing	5	24-Apr-17	28-Apr-17
Intall Curtains and Fixed Furnitures -GF All Zones	5	24-Apr-17	28-Apr-17
Mechanical	73	06-Feb-17	19-Apr-17
Fix Duct for HVAC -GF All Zones	7	06-Feb-17	12-Feb-17
Install WS and Sanitary Pipes+Fttngs -GF All Zones	6	06-Feb-17	11-Feb-17
Fix Insul'n to Duct HVAC -GF All Zones	7	08-Feb-17	14-Feb-17
Install VAV / VCD / Duct Heaters -GF All Zones	7	15-Feb-17	21-Feb-17
Fix FF Pipes and Fittings -GF All Zones	10	15-Feb-17	24-Feb-17
Install FC and F Extgnr -GF All Zones	7	25-Feb-17	03-Mar-17
Install Diffuser, Register, Grilles -GF All Zones	15	21-Mar-17	04-Apr-17
Install Electric Heater -GF All Zones	10	28-Mar-17	06-Apr-17
Install Plmbg and Sntry Fixtr+Fttng and Acc -GF All Zones	10	28-Mar-17	06-Apr-17
Install Sprinklers etc. -GF All Zones	10	10-Apr-17	19-Apr-17
Electrical	126	24-Dec-16	28-Apr-17
Fix Conduits (LV) -GF All Zones	15	24-Dec-16	07-Jan-17
Pulling Wires (LV) -GF All Zones	15	08-Jan-17	22-Jan-17
Fix Conduits, Backbox, trays etc. -GF All Zones	10	06-Feb-17	15-Feb-17
Pulling Wires -GF All Zones	10	23-Feb-17	04-Mar-17
Install DB, MCB and Connctn -GF All Zones	10	05-Mar-17	14-Mar-17
Install LV Devices -GF All Zones	10	10-Apr-17	19-Apr-17
Install Switches, Sockets -GF All Zones	7	15-Apr-17	21-Apr-17
Install Lights -GF All Zones	7	22-Apr-17	28-Apr-17
ATM Lobby	52	23-Feb-17	15-Apr-17
Doors and Windows	3	22-Mar-17	24-Mar-17
Install Door and Accs. -GF ATM Lobby	3	22-Mar-17	24-Mar-17

FIGURE 7.126 (Continued)
Contractor's construction schedule.

(Continued)

Activity Name	Original Duration	Start	Finish
Finishes	42	23-Feb-17	05-Apr-17
Screed on Floor-GF ATM Lobby	2	23-Feb-17	24-Feb-17
Install Suspension for Flase Ceiling -GF ATM Lobby	5	25-Feb-17	01-Mar-17
Install False Gypsum Ceiling -GF ATM Lobby	5	02-Mar-17	06-Mar-17
Install Terrazo+ Vinyl Tiles + Skirting -GF ATM Lobby	5	07-Mar-17	11-Mar-17
Ceiling Paint (2 Coats) -GF ATM Lobby	5	12-Mar-17	16-Mar-17
Wall Paint (2 Coats) -GF ATM Lobby	5	17-Mar-17	21-Mar-17
FinalPainting -GF ATM Lobby	1	05-Apr-17	05-Apr-17
Specialities	1	06-Apr-17	06-Apr-17
Install Trash Box -GF ATM Lobby	1	06-Apr-17	06-Apr-17
Mechanical	10	06-Apr-17	15-Apr-17
Install Air Curtain -GF ATM Lobby	10	06-Apr-17	15-Apr-17
Teller Area	63	18-Feb-17	21-Apr-17
Doors and Windows	3	10-Apr-17	12-Apr-17
Install Door and Accs. -GF Teller Area	3	10-Apr-17	12-Apr-17
Finishes	63	18-Feb-17	21-Apr-17
Screed on Floor -GF Teller Area	4	18-Feb-17	21-Feb-17
Install Wooden Wall Cladding + Painting -GF Teller Area	7	22-Feb-17	28-Feb-17
Install Suspension for Flase Ceiling -GF Teller Area	5	01-Mar-17	05-Mar-17
Install Gypsum Ceiling -GF Teller Area	5	06-Mar-17	10-Mar-17
Install Terrazo Flooring + Skirting -GF Teller Area	5	14-Mar-17	18-Mar-17
Ceiling Paint (2 Coats) -GF Teller Area	5	26-Mar-17	30-Mar-17
Wall Paint (2 Coats) -GF Teller Area	5	31-Mar-17	04-Apr-17
Final Painting -GF Teller Area	1	17-Apr-17	17-Apr-17
Install Carpet -GF Teller Area	4	18-Apr-17	21-Apr-17
Specialities	7	19-Mar-17	25-Mar-17
Install Raised Flooring System -GF Teller Area	7	19-Mar-17	25-Mar-17
Waiting Area/Reception	60	26-Feb-17	26-Apr-17
Metal	3	24-Apr-17	26-Apr-17
Install External Door Floor Metal Matt -GF Waiting Area/Reception	3	24-Apr-17	26-Apr-17
Wood and Plastic	14	31-Mar-17	13-Apr-17
Install Wooden Counter -GF Waiting Area/Reception	14	31-Mar-17	13-Apr-17
Doors and Windows	5	16-Apr-17	20-Apr-17
Install Door and Accs. -GF Waiting Area/Reception	5	16-Apr-17	20-Apr-17
Finishes	57	26-Feb-17	23-Apr-17
Screed on Floor -GF Waiting Area/Reception	4	26-Feb-17	01-Mar-17

FIGURE 7.126 (Continued)
Contractor's construction schedule.

(*Continued*)

Activity Name	Original Duration	Start	Finish
Install Suspension for Flase Ceiling -GF Waiting Area/Reception	5	02-Mar-17	06-Mar-17
Install Gypsum Ceiling -GF Waiting Area/Reception	5	16-Mar-17	20-Mar-17
Install MarbleFlooring -GF Waiting Area/Reception	7	24-Mar-17	30-Mar-17
Ceiling Paint (2 Coats) -GF Waiting Area/Reception	5	05-Apr-17	09-Apr-17
Wall Paint (2 Coats) -GF Waiting Area/Reception	5	10-Apr-17	14-Apr-17
FinalPainting -GF Waiting Area/Reception	3	21-Apr-17	23-Apr-17
Safe Deposit Vault Area	58	09-Feb-17	07-Apr-17
Doors and Windows	7	28-Mar-17	03-Apr-17
Install Door and Accs. -GF Safe Deposit/Vault Area	7	28-Mar-17	03-Apr-17
Finishes	58	09-Feb-17	07-Apr-17
Screed on Floor -GF Safe Deposit/Vault Area	3	09-Feb-17	11-Feb-17
Install Suspension for Flase Ceiling -GF Safe Deposit/Vault Area	5	12-Feb-17	16-Feb-17
Install Gypsum Ceiling -GF Safe Deposit/Vault Area	5	22-Feb-17	26-Feb-17
Install Floor Marble/Terrazo/Vinyl + Skirting -GF Safe Deposit/Vault A	5	27-Feb-17	03-Mar-17
Ceiling Paint (2 Coats) -GF Safe Deposit/Vault Area	5	17-Mar-17	21-Mar-17
Wall Paint (2 Coats) -GF Safe Deposit/Vault Area	5	22-Mar-17	26-Mar-17
Final Painting -GF Safe Deposit/Vault Area	2	06-Apr-17	07-Apr-17
Filing Room	61	15-Feb-17	16-Apr-17
Doors and Windows	3	07-Apr-17	09-Apr-17
Install Door and Accs. -GF Filing Room	3	07-Apr-17	09-Apr-17
Finishes	61	15-Feb-17	16-Apr-17
Screed on Floor -GF Filing Room	3	15-Feb-17	17-Feb-17
Install Suspension for Flase Ceiling -GF Filing Room	5	20-Feb-17	24-Feb-17
Install 60x60 Ceiling -GF Filing Room	2	04-Mar-17	05-Mar-17
Install Terrazo Flooring + Skirting -GF Filing Room	5	09-Mar-17	13-Mar-17
Wall Paint (2 Coats)-GF Filing Room	4	14-Mar-17	17-Mar-17
Final Painting -GF Filing Room	1	16-Apr-17	16-Apr-17
Office Area	62	22-Feb-17	24-Apr-17
Doors and Windows	3	13-Apr-17	15-Apr-17
Install Door and Accs. -GF Office Area	3	13-Apr-17	15-Apr-17
Finishes	62	22-Feb-17	24-Apr-17
Screed on Floor -GF Office Area	4	22-Feb-17	25-Feb-17
Install Glass Wall Partition -GF Office Area	5	01-Mar-17	05-Mar-17
Install Suspension for Flase Ceiling -GF Office Area	5	06-Mar-17	10-Mar-17
Install Gypsum Ceiling -GF Office Area	5	11-Mar-17	15-Mar-17
Install Terrazo Flooring + Skirting -GF Office Area	5	19-Mar-17	23-Mar-17

FIGURE 7.126 (Continued)
Contractor's construction schedule.

(Continued)

Activity Name	Original Duration	Start	Finish
Ceiling Paint (2 Coats) -GF OfficeArea	5	31-Mar-17	04-Apr-17
Final Painting -GF OfficeArea	2	16-Apr-17	17-Apr-17
Install Carpet -GF OfficeArea	3	22-Apr-17	24-Apr-17
Electrical Room/Drive thru ATM	65	12-Feb-17	17-Apr-17
Doors and Windows	3	04-Apr-17	06-Apr-17
Install Door and Accs. -GF Electrical Room/Drive thruATM	3	04-Apr-17	06-Apr-17
Finishes	63	12-Feb-17	15-Apr-17
Screed on Floor -GF Electrical Room/Drive thru ATM	3	12-Feb-17	14-Feb-17
Install Suspension for Flase Ceiling -GF Electrical Room/Drive thru	5	15-Feb-17	19-Feb-17
Install 60x60 Ceiling -GF Electrical Room/Drive thru ATM	5	27-Feb-17	03-Mar-17
Install Terrazo Flooring + Skirting -GF Electrical Room/Drive thru AT	5	04-Mar-17	08-Mar-17
Wall Paint (2 Coats) -GF Electrical Room/Drive thru ATM	5	09-Mar-17	13-Mar-17
Final Painting -GF Electrical Room/Drive thru ATM	2	14-Apr-17	15-Apr-17
Specialities	2	16-Apr-17	17-Apr-17
Install Rubber Wall Guard (External) -GF Electrical Room/Drive thr	2	16-Apr-17	17-Apr-17
Mechanical	7	07-Apr-17	13-Apr-17
Install Mini Split Unit -GF Electrical Room/Drive thru ATM	7	07-Apr-17	13-Apr-17
Wet Area	64	06-Feb-17	10-Apr-17
Doors and Windows	3	25-Mar-17	27-Mar-17
Install Door and Accs. -GF Wet Area	3	25-Mar-17	27-Mar-17
Finishes	20	06-Feb-17	25-Feb-17
Screed on Floor -GF Wet Area	3	06-Feb-17	08-Feb-17
Install Ceramic Wall Tiles -GF Wet Area	5	09-Feb-17	13-Feb-17
Install Suspension for Flase Ceiling -GF Wet Area	4	14-Feb-17	17-Feb-17
Install 60x60 Ceiling -GF Wet Area	4	18-Feb-17	21-Feb-17
Install Ceramic Floor Tiles -GF Wet Area	4	22-Feb-17	25-Feb-17
Furnishing	7	28-Mar-17	03-Apr-17
Install Kitchen Cabinet -GF Wet Area	7	28-Mar-17	03-Apr-17
Mechanical	7	04-Apr-17	10-Apr-17
Install Fans (KEF,TEF) -GF Wet Area	7	04-Apr-17	10-Apr-17
Mezzanine Floor	170	14-Dec-16	01-Jun-17
All Zones	151	02-Jan-17	01-Jun-17
Concrete	31	27-Jan-17	26-Feb-17
RCC for Columns + Walls -MZ All Zones	10	27-Jan-17	05-Feb-17
RCC for Slab -MZ All Zones	21	06-Feb-17	26-Feb-17
Masonry	7	09-Mar-17	15-Mar-17

FIGURE 7.126 (Continued)
Contractor's construction schedule.

(*Continued*)

Activity Name	Original Duration	Start	Finish
Blockwork -MZ All Zones	7	09-Mar-17	15-Mar-17
Doors and Windows	5	16-Mar-17	20-Mar-17
Install Door Frames -MZ All Zones	5	16-Mar-17	20-Mar-17
Finishes	7	21-Mar-17	27-Mar-17
Plaster -MZ All Zones	7	21-Mar-17	27-Mar-17
Furnishing	3	26-May-17	28-May-17
Intall Curtains and Fixed Furnitures -MZ All Zones	3	26-May-17	28-May-17
Mechanical	72	16-Mar-17	26-May-17
Fix Duct for HVAC-MZ All Zones	10	16-Mar-17	25-Mar-17
Install WS and Sanitary Pipes+Fttngs -MZ All Zones	10	16-Mar-17	25-Mar-17
Fix Insul'n to Duct HVAC-MZ All Zones	7	21-Mar-17	27-Mar-17
Install VAV / VCD / Duct Heaters -MZ All Zones	5	28-Mar-17	01-Apr-17
Fix FF Pipes and Fittings -MZ All Zones	7	28-Mar-17	03-Apr-17
Install Diffuser, Register, Grilles -MZ All Zones	7	06-May-17	12-May-17
Install Electric Heater -MZ All Zones	5	06-May-17	10-May-17
Install Plmbg and Sntry Fixtr+Fttng and Acc -MZ All Zones	5	06-May-17	10-May-17
Install Sprinklers etc. -MZ All Zones	4	15-May-17	18-May-17
Install FC and F Extgnr -MZ All Zones	4	23-May-17	26-May-17
Electrical	151	02-Jan-17	01-Jun-17
Fix Conduits (LV)-MZ All Zones	15	02-Jan-17	16-Jan-17
Pulling Wires (LV)-MZ All Zones	15	17-Jan-17	31-Jan-17
Fix Conduits, Backbox, trays etc. -MZ All Zones	7	16-Mar-17	22-Mar-17
Pulling Wires -MZ All Zones	7	28-Mar-17	03-Apr-17
Install DB, MCB and Connctn -MZ All Zones	7	04-Apr-17	10-Apr-17
Install LV Devices -MZ All Zones	10	15-May-17	24-May-17
Install Switches, Sockets -MZ All Zones	7	19-May-17	25-May-17
Install Lights -MZ All Zones	7	26-May-17	01-Jun-17
Meeting Room	154	14-Dec-16	16-May-17
Doors and Windows	5	19-Apr-17	23-Apr-17
Install Door and Accs. -MZ Meeting Room	5	19-Apr-17	23-Apr-17
Finishes	154	14-Dec-16	16-May-17
Install Terrazo Flooring + Skirting -MZ Meeting Room	7	14-Dec-16	20-Dec-16
Install Glass Wall Partition -MZ Meeting Room	7	21-Dec-16	27-Dec-16
Install Suspension for Flase Ceiling -MZ Meeting Room	5	28-Dec-16	01-Jan-17
Screed on Floor -MZ Meeting Room	3	28-Mar-17	30-Mar-17
Install Gypsum Ceiling -MZ Meeting Room	5	04-Apr-17	08-Apr-17

Remaining Work ◆ Milestone
Critical Remaining Work Summary

FIGURE 7.126 (Continued)
Contractor's construction schedule.

(Continued)

Activity Name	Original Duration	Start	Finish
Ceiling Paint (2 Coats) -MZ Meeting Room	5	09-Apr-17	13-Apr-17
Wall Paint (2 Coats) -MZ Meeting Room	5	14-Apr-17	18-Apr-17
Final Painting -MZ Meeting Room	1	13-May-17	13-May-17
Install Carpet -MZ Meeting Room	3	14-May-17	16-May-17
Teller Area	49	07-Apr-17	25-May-17
Doors and Windows	3	18-May-17	20-May-17
Install Door and Accs. -MZ Teller Area	3	18-May-17	20-May-17
Finishes	49	07-Apr-17	25-May-17
Install Screed -MZ Teller Area	3	07-Apr-17	09-Apr-17
Install Suspension for Flase Ceiling -MZ Teller Area	4	10-Apr-17	13-Apr-17
Install Gypsum Ceiling -MZ Teller Area	4	27-Apr-17	30-Apr-17
Install Terrazo Flooring + Skirting -MZ Teller Area	5	01-May-17	05-May-17
Wall Paint (2 Coats)-MZ Teller Area	5	13-May-17	17-May-17
Final Painting -MZ Teller Area	1	21-May-17	21-May-17
Install Carpet -MZ Teller Area	4	22-May-17	25-May-17
Specialities	7	06-May-17	12-May-17
Install Raised Flooring System -MZ Teller Area	7	06-May-17	12-May-17
Waiting Area/Reception	46	10-Apr-17	25-May-17
Wood and Plastic	14	11-May-17	24-May-17
Install Wooden Counter -MZ Waiting Area/Reception	14	11-May-17	24-May-17
Doors and Windows	4	19-May-17	22-May-17
Install Door and Accs. -MZ Waiting Area/Reception	4	19-May-17	22-May-17
Finishes	46	10-Apr-17	25-May-17
Screed on Floor -MZ Waiting Area/Reception	3	10-Apr-17	12-Apr-17
Install Suspension for Flase Ceiling -MZ Waiting Area/Reception	5	13-Apr-17	17-Apr-17
Install Gypsum Ceiling -MZ Waiting Area/Reception	5	01-May-17	05-May-17
Install Marble Flooring -MZ Waiting Area/Reception	5	06-May-17	10-May-17
Ceiling Paint (2 Coats) -MZ Waiting Area/Reception	4	11-May-17	14-May-17
Wall Paint (2 Coats) -MZ Waiting Area/Reception	4	15-May-17	18-May-17
Final Painting -MZ Waiting Area/Reception	1	25-May-17	25-May-17
Filing Room	45	02-Apr-17	16-May-17
Doors and Windows	3	27-Apr-17	29-Apr-17
Install Door and Accs. -MZ Filing Room	3	27-Apr-17	29-Apr-17
Finishes	45	02-Apr-17	16-May-17
Screed on Floor -MZ Filing Room	2	02-Apr-17	03-Apr-17
Install Suspension for Flase Ceiling -MZ Filing Room	4	04-Apr-17	07-Apr-17

FIGURE 7.126 (Continued)
Contractor's construction schedule.

(Continued)

Activity Name	Original Duration	Start	Finish
Install 60x60 Ceiling -MZ Filing Room	4	15-Apr-17	18-Apr-17
Install TerrazoFlooring + Skirting -MZ Filing Room	4	19-Apr-17	22-Apr-17
Wall Paint (2 Coats)-MZ Filing Room	4	23-Apr-17	26-Apr-17
FinalPainting -MZ Filing Room	1	16-May-17	16-May-17
Office Area	49	31-Mar-17	18-May-17
Doors and Windows	7	24-Apr-17	30-Apr-17
Install Door and Accs.-MZ OfficeArea	7	24-Apr-17	30-Apr-17
Finishes	49	31-Mar-17	18-May-17
Screed on Floor-MZ OfficeArea	2	31-Mar-17	01-Apr-17
Install Glass Wall Partition -MZ OfficeArea	5	02-Apr-17	06-Apr-17
Install Suspension for Flase Ceiling -MZ OfficeArea	4	07-Apr-17	10-Apr-17
Install Gyps um Ceilng -MZ Office Area	4	11-Apr-17	14-Apr-17
Install TerrazoFlooring + Skirting -MZ OfficeArea	5	15-Apr-17	19-Apr-17
Ceiling Paint (2 Coats)-MZ OfficeArea	4	20-Apr-17	23-Apr-17
FinalPainting -MZ OfficeArea	2	14-May-17	15-May-17
Install Carpet -MZ OfficeArea	3	16-May-17	18-May-17
Wet Area	44	06-Apr-17	19-May-17
Doors and Windows	5	01-May-17	05-May-17
Install Door and Accs.-MZ Wet Area	5	01-May-17	05-May-17
Finishes	25	06-Apr-17	30-Apr-17
Screed on Floor-MZ Wet Area	2	06-Apr-17	07-Apr-17
Install Ceramic Wall Tiles -MZ Wet Area	4	08-Apr-17	11-Apr-17
Install Suspension for Flase Ceiling -MZ Wet Area	4	12-Apr-17	15-Apr-17
Install 60x60 Ceiling -MZ Wet Area	4	23-Apr-17	26-Apr-17
Install Ceramic Floor Tiles -MZ Wet Area	4	27-Apr-17	30-Apr-17
Furnishing	7	06-May-17	12-May-17
Install Kitchen Cabinet -MZ Wet Area	7	06-May-17	12-May-17
Mechanical	7	13-May-17	19-May-17
Install Fans (KEF,TEF) -MZ Wet Area	7	13-May-17	19-May-17
Server Room	44	04-Apr-17	17-May-17
Doors and Windows	5	02-May-17	06-May-17
Install Door and Accs.-MZ Server Room	5	02-May-17	06-May-17
Finishes	44	04-Apr-17	17-May-17
Screed on Floor-MZ Server Room	2	04-Apr-17	05-Apr-17
Install Suspension for Flase Ceiling -MZ Server Room	4	06-Apr-17	09-Apr-17
Install 60x60 Ceiling -MZ Server Room	4	19-Apr-17	22-Apr-17

FIGURE 7.126 (Continued)
Contractor's construction schedule.

(Continued)

Activity Name	Original Duration	Start	Finish
Install Terrazo Flooring + Skirting -MZ Server Room	5	23-Apr-17	27-Apr-17
Wall Paint (2 Coats) -MZ Server Room	4	28-Apr-17	01-May-17
Final Painting -MZ Server Room	1	17-May-17	17-May-17
Electrical	10	07-May-17	16-May-17
Install Server Cabinet -MZ Server Room	10	07-May-17	16-May-17
Lower Roof	119	01-Feb-17	30-May-17
All Zones	119	01-Feb-17	30-May-17
Masonry	5	16-Mar-17	20-Mar-17
Blockwork -LR All Zones	5	16-Mar-17	20-Mar-17
Metal	5	21-May-17	25-May-17
Install Steel Covers to Shaft Openings -LR All Zones	5	21-May-17	25-May-17
Doors and Windows	61	21-Mar-17	20-May-17
Install Door Frames -LR All Zones	4	21-Mar-17	24-Mar-17
Install Door and Accs. -LR All Zones	2	19-May-17	20-May-17
Finishes	61	28-Mar-17	27-May-17
Plaster -LR All Zones	5	28-Mar-17	01-Apr-17
Screed on Floor -LR All Zones	3	02-Apr-17	04-Apr-17
Install Suspension for Flase Ceiling -LR All Zones	4	18-Apr-17	21-Apr-17
Install 60x60 Ceiling -LR All Zones	4	06-May-17	09-May-17
Install Terrazo Flooring + Skirting -LR All Zones	4	11-May-17	14-May-17
Wall Paint (2 Coats) -LR All Zones	4	15-May-17	18-May-17
Final Painting -LR All Zones	2	26-May-17	27-May-17
Mechanical	66	26-Mar-17	30-May-17
Install Roof Top Package Units -LR All Zones	15	26-Mar-17	09-Apr-17
Install Water Booster Pump -LR All Zones	15	26-Mar-17	09-Apr-17
Install FP Pumps -LR All Zones	15	04-Apr-17	18-Apr-17
Install GRP Tank -LR All Zones	15	10-Apr-17	24-Apr-17
Install Mini Split Unit -LR All Zones	10	21-May-17	30-May-17
Electrical	105	01-Feb-17	16-May-17
Install AVA Alarm Panel -LR All Zones	7	01-Feb-17	07-Feb-17
Install 50KVR Capacitor -LR All Zones	7	01-Feb-17	07-Feb-17
Fix Conduits,Backbox,trays etc. -LR All Zones	7	23-Mar-17	29-Mar-17
Pulling Wires -LR All Zones	7	30-Mar-17	05-Apr-17
Install DB, MCB and Connctn -LR All Zones	9	06-Apr-17	14-Apr-17
Install Switches, Sockets -LR All Zones	7	15-Apr-17	21-Apr-17
Install Lights -LR All Zones	7	22-Apr-17	28-Apr-17

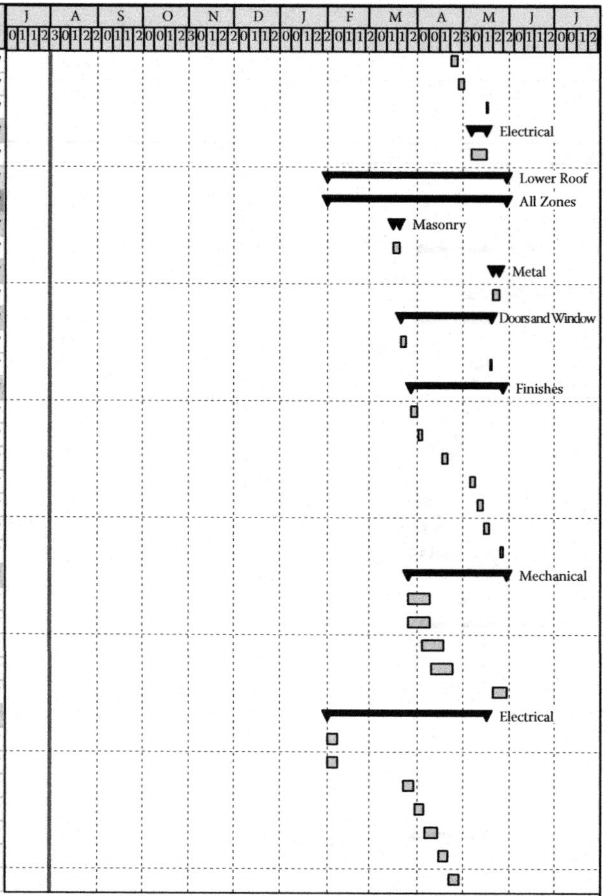

Remaining Work ◆ Milestone
Critical Remaining Work Summary

FIGURE 7.126 (Continued)
Contractor's construction schedule.

(Continued)

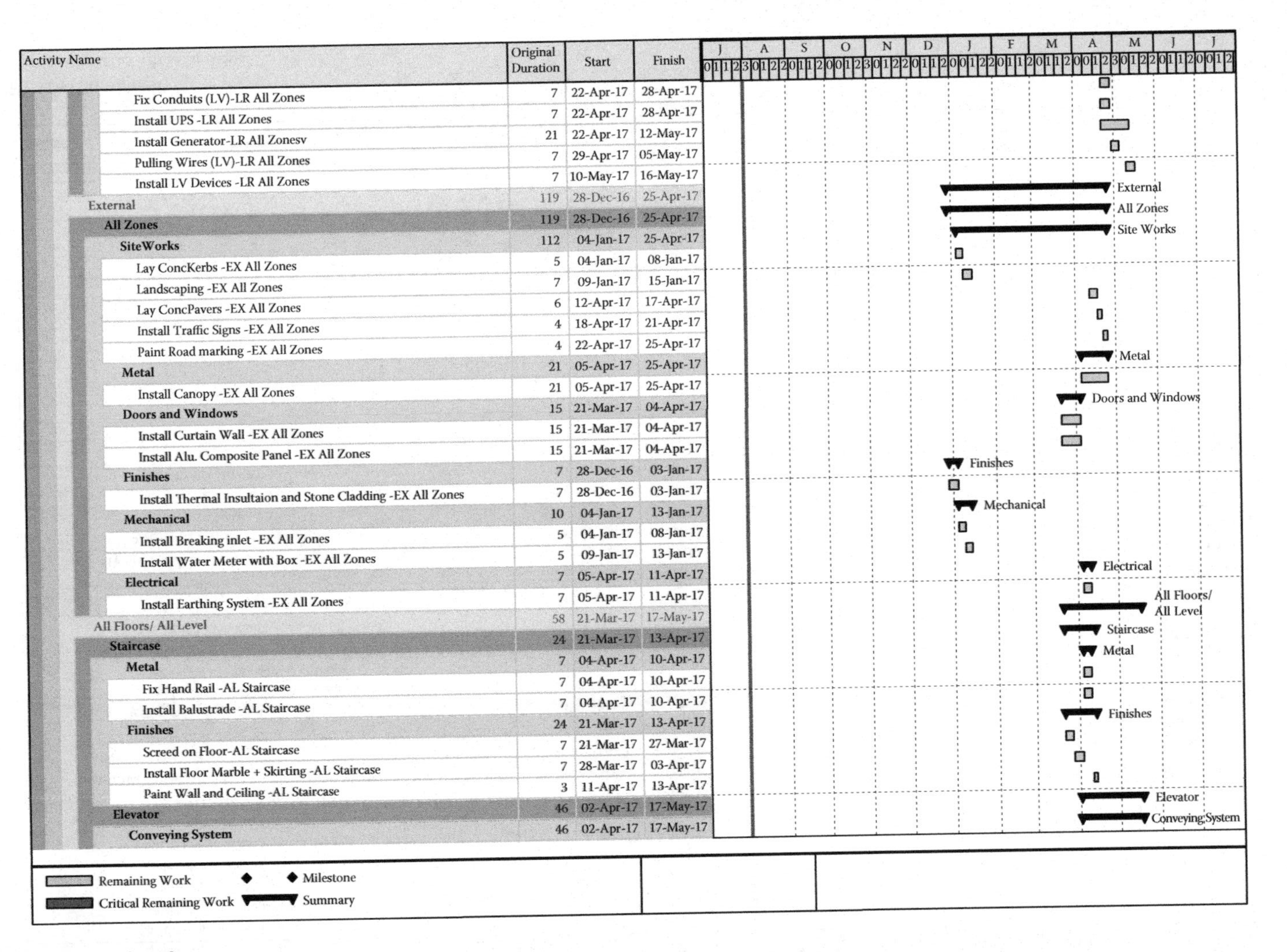

Activity Name	Original Duration	Start	Finish
Fix Conduits (LV)-LR All Zones	7	22-Apr-17	28-Apr-17
Install UPS -LR All Zones	7	22-Apr-17	28-Apr-17
Install Generator-LR All Zonesv	21	22-Apr-17	12-May-17
Pulling Wires (LV)-LR All Zones	7	29-Apr-17	05-May-17
Install LV Devices -LR All Zones	7	10-May-17	16-May-17
External	119	28-Dec-16	25-Apr-17
All Zones	119	28-Dec-16	25-Apr-17
SiteWorks	112	04-Jan-17	25-Apr-17
Lay ConcKerbs -EX All Zones	5	04-Jan-17	08-Jan-17
Landscaping -EX All Zones	7	09-Jan-17	15-Jan-17
Lay ConcPavers -EX All Zones	6	12-Apr-17	17-Apr-17
Install Traffic Signs -EX All Zones	4	18-Apr-17	21-Apr-17
Paint Road marking -EX All Zones	4	22-Apr-17	25-Apr-17
Metal	21	05-Apr-17	25-Apr-17
Install Canopy -EX All Zones	21	05-Apr-17	25-Apr-17
Doors and Windows	15	21-Mar-17	04-Apr-17
Install Curtain Wall -EX All Zones	15	21-Mar-17	04-Apr-17
Install Alu. Composite Panel -EX All Zones	15	21-Mar-17	04-Apr-17
Finishes	7	28-Dec-16	03-Jan-17
Install Thermal Insultaion and Stone Cladding -EX All Zones	7	28-Dec-16	03-Jan-17
Mechanical	10	04-Jan-17	13-Jan-17
Install Breaking inlet -EX All Zones	5	04-Jan-17	08-Jan-17
Install Water Meter with Box -EX All Zones	5	09-Jan-17	13-Jan-17
Electrical	7	05-Apr-17	11-Apr-17
Install Earthing System -EX All Zones	7	05-Apr-17	11-Apr-17
All Floors/ All Level	58	21-Mar-17	17-May-17
Staircase	24	21-Mar-17	13-Apr-17
Metal	7	04-Apr-17	10-Apr-17
Fix Hand Rail -AL Staircase	7	04-Apr-17	10-Apr-17
Install Balustrade -AL Staircase	7	04-Apr-17	10-Apr-17
Finishes	24	21-Mar-17	13-Apr-17
Screed on Floor-AL Staircase	7	21-Mar-17	27-Mar-17
Install Floor Marble + Skirting -AL Staircase	7	28-Mar-17	03-Apr-17
Paint Wall and Ceiling -AL Staircase	3	11-Apr-17	13-Apr-17
Elevator	46	02-Apr-17	17-May-17
Conveying System	46	02-Apr-17	17-May-17

FIGURE 7.126 (Continued)
Contractor's construction schedule.

(Continued)

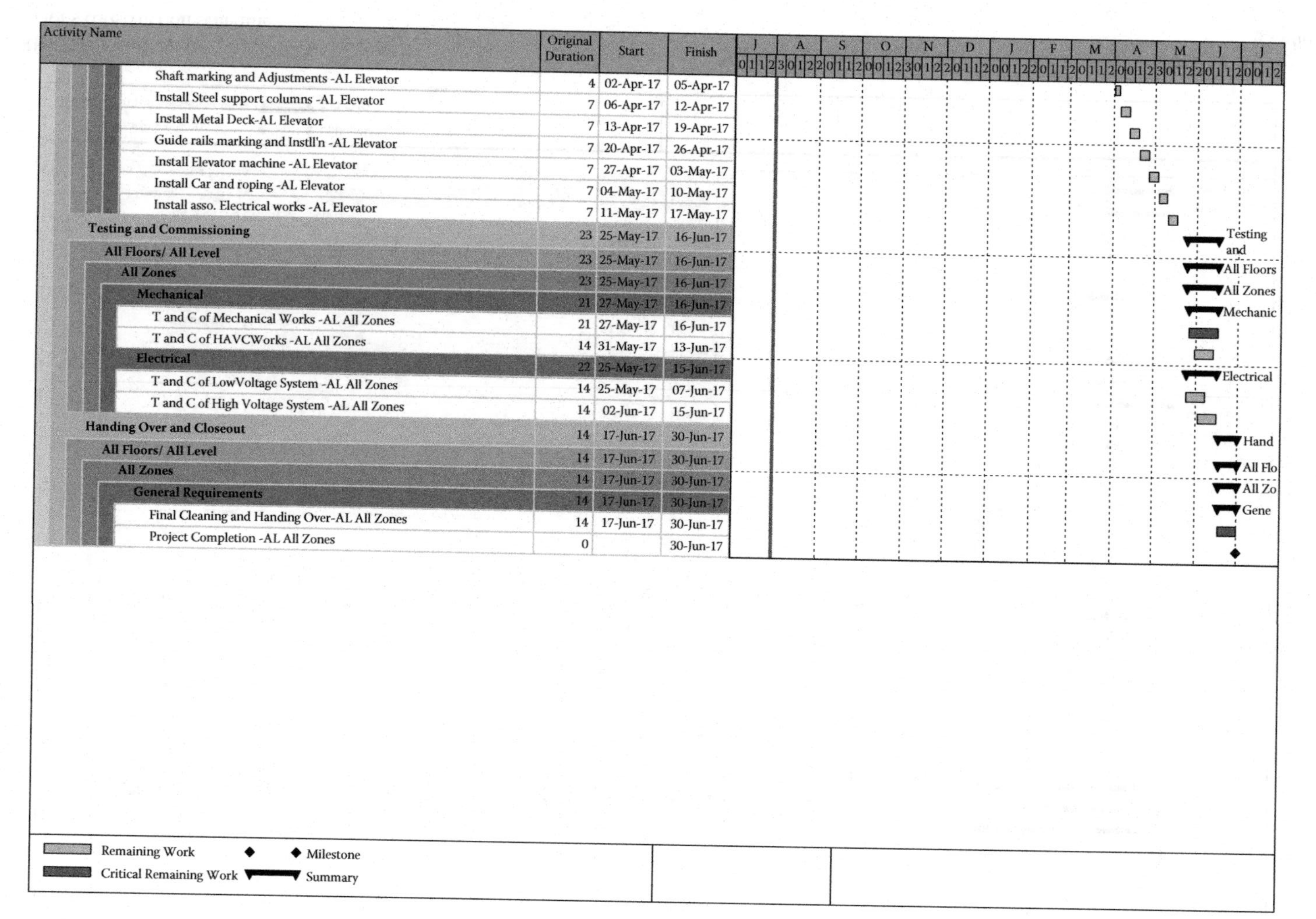

Activity Name	Original Duration	Start	Finish
Shaft marking and Adjustments -AL Elevator	4	02-Apr-17	05-Apr-17
Install Steel support columns -AL Elevator	7	06-Apr-17	12-Apr-17
Install Metal Deck-AL Elevator	7	13-Apr-17	19-Apr-17
Guide rails marking and Instll'n -AL Elevator	7	20-Apr-17	26-Apr-17
Install Elevator machine -AL Elevator	7	27-Apr-17	03-May-17
Install Car and roping -AL Elevator	7	04-May-17	10-May-17
Install asso. Electrical works -AL Elevator	7	11-May-17	17-May-17
Testing and Commissioning	23	25-May-17	16-Jun-17
All Floors/ All Level	23	25-May-17	16-Jun-17
All Zones	23	25-May-17	16-Jun-17
Mechanical	21	27-May-17	16-Jun-17
T and C of Mechanical Works -AL All Zones	21	27-May-17	16-Jun-17
T and C of HAVCWorks -AL All Zones	14	31-May-17	13-Jun-17
Electrical	22	25-May-17	15-Jun-17
T and C of LowVoltage System -AL All Zones	14	25-May-17	07-Jun-17
T and C of High Voltage System -AL All Zones	14	02-Jun-17	15-Jun-17
Handing Over and Closeout	14	17-Jun-17	30-Jun-17
All Floors/ All Level	14	17-Jun-17	30-Jun-17
All Zones	14	17-Jun-17	30-Jun-17
General Requirements	14	17-Jun-17	30-Jun-17
Final Cleaning and Handing Over-AL All Zones	14	17-Jun-17	30-Jun-17
Project Completion -AL All Zones	0		30-Jun-17

FIGURE 7.126 (Continued)
Contractor's construction schedule.

7.4.6.6 Develop Project S-Curve

The S-curve is a graphical display of cumulative cost, resources, or other quantities plotted against time. During the construction phase, the contractor uses the S-curve for forecasting cash flow that is based on the work (activities) the contractor is expected to complete and the quantity of amount (payment) he or she will receive. Conceptually, cash flow is a simple comparison of when revenue will be received and when the financial obligations must be paid. It is also an indication of the progress of work to be completed in a project. This is obtained by loading each activity in the approved schedule with the budgeted cost in the BOQ. The process of inputting schedule of values is known as *cost loading.* The graphical representation of the above is obtained as a curve and is known as S-curve. This also represents the planned progress of a project. Figure 7.127 illustrates the planned project S-curve prepared by the contractor, which is based on the construction schedule.

7.4.6.6.1 Cost Loaded Curve

Figure 7.128 illustrates cost loaded S-curve prepared by the contractor.

7.4.6.7 Develop CQCP

Quality management in construction is a management function. In general, quality assurance and control programs are used to monitor design and construction conformance to established requirements as determined by the contract specifications. Instituting quality management programs reduces costs while producing the specified facility.

The CQCP is the contractor's everyday tool to insure meeting the performance standards specified in the contract documents. The adequacy and efficient management of CQCP by contractor's personnel have great impact on both the performance of contract and owner's quality assurance surveillance of the contractor's performance.

CQCP is the documentation of contractor's process for delivering the level of construction quality required by the contract. It is a framework for the contractor's process for achieving quality construction. CQCP does not endeavor to repeat or summarize contract requirements. It describes the process that contractor will use to assure compliance with the contract requirements. The quality plan is virtually manual tailor-made for the project and is based on contract requirements.

In the quality plan, the generic documented procedures are integrated with any necessary additional procedures peculiar to the project in order to attain specified quality objectives. Application of various quality tools, methods, and principles at different stages of construction projects is necessary to make the project qualitative, economical, and meet the owner needs/specification requirements.

Based on contract requirements, the contractor prepares his or her quality control plan and submits the same to consultant for their approval. Figure 7.66 illustrates the logic flow for CQCP. This plan is followed by the contractor to maintain the project quality.

The CQCP is prepared based on project-specific requirement as specified in the contract documents. The plan outlines the procedures to be followed during the construction period to attain the specified quality objectives of the project fully complying with the contractual and regulatory requirements.

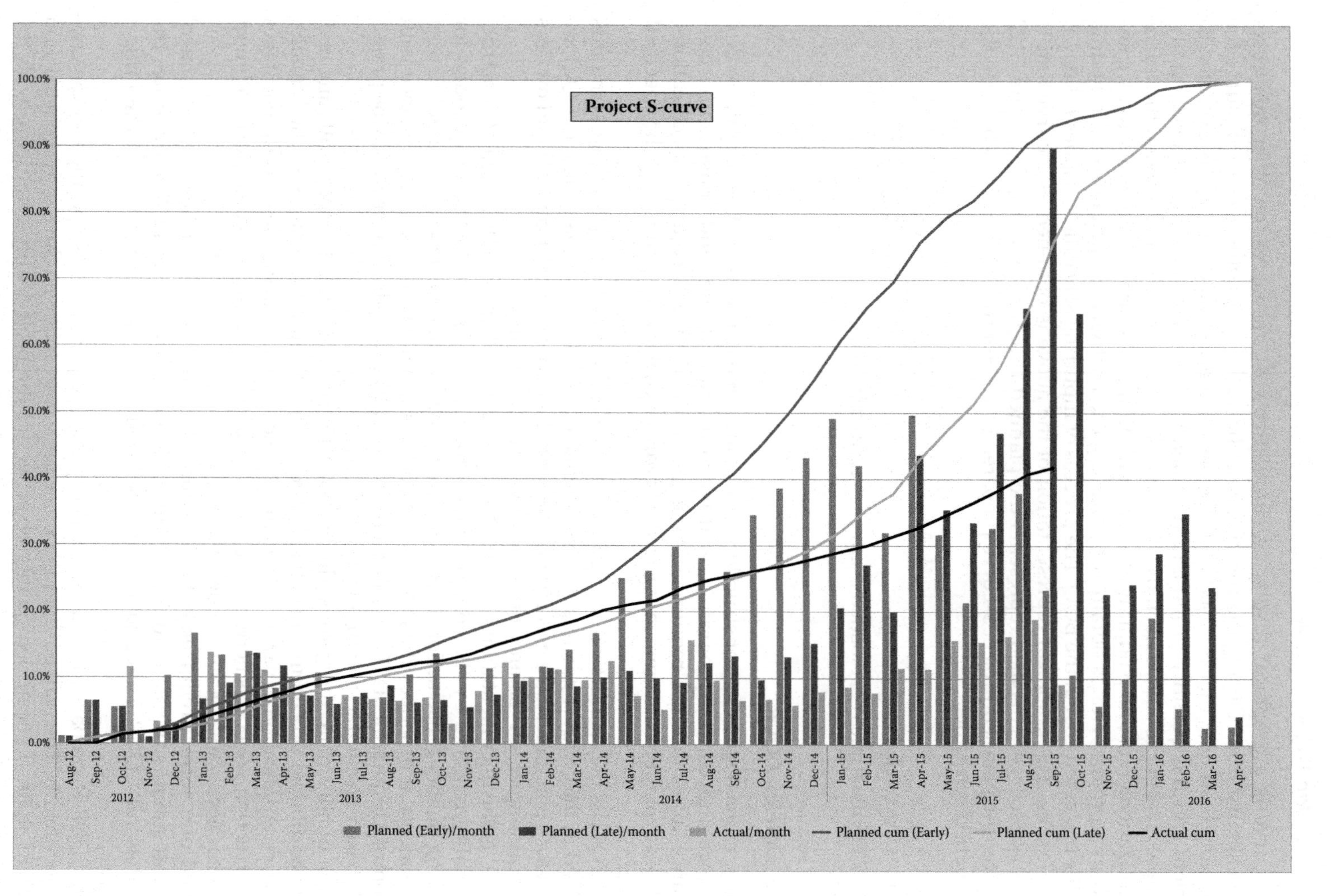

FIGURE 7.127
Project S-curve.

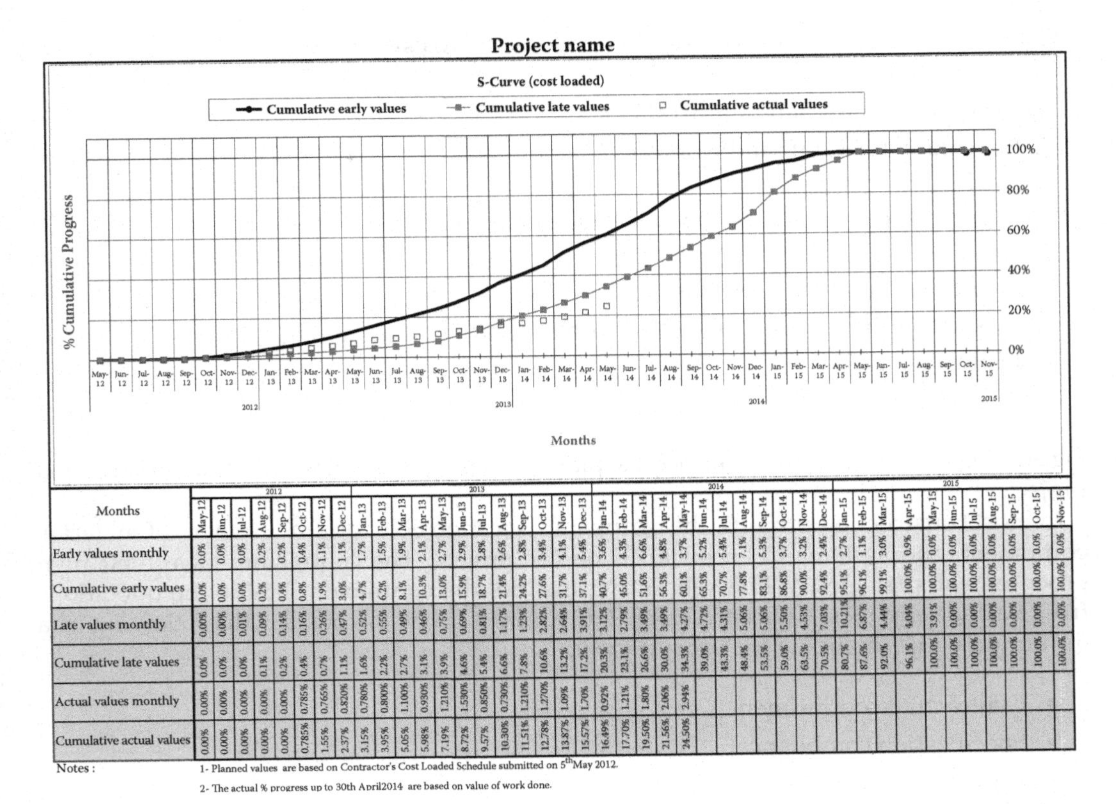

Months (2012)	May-12	Jun-12	Jul-12	Aug-12	Sep-12	Oct-12	Nov-12	Dec-12
Early values monthly	0.0%	0.0%	0.0%	0.2%	0.2%	0.4%	1.1%	1.1%
Cumulative early values	0.0%	0.0%	0.0%	0.2%	0.4%	0.8%	1.9%	3.0%
Late values monthly	0.00%	0.00%	0.01%	0.09%	0.14%	0.16%	0.26%	0.47%
Cumulative late values	0.0%	0.0%	0.0%	0.1%	0.2%	0.4%	0.7%	1.1%
Actual values monthly	0.00%	0.00%	0.00%	0.00%	0.00%	0.785%	0.765%	0.820%
Cumulative actual values	0.00%	0.00%	0.00%	0.00%	0.00%	0.785%	1.55%	2.37%

Months (2013)	Jan-13	Feb-13	Mar-13	Apr-13	May-13	Jun-13	Jul-13	Aug-13	Sep-13	Oct-13	Nov-13	Dec-13
Early values monthly	1.7%	1.5%	1.9%	2.1%	2.7%	2.9%	2.8%	2.6%	2.8%	3.4%	4.1%	5.4%
Cumulative early values	4.7%	6.2%	8.1%	10.3%	13.0%	15.9%	18.7%	21.4%	24.2%	27.6%	31.7%	37.1%
Late values monthly	0.52%	0.55%	0.49%	0.46%	0.75%	0.69%	0.81%	1.17%	1.23%	2.82%	2.64%	3.91%
Cumulative late values	1.6%	2.2%	2.7%	3.1%	3.9%	4.6%	5.4%	6.6%	7.8%	10.6%	13.2%	17.2%
Actual values monthly	0.780%	0.800%	1.100%	0.930%	1.210%	1.530%	0.850%	0.730%	1.210%	1.270%	1.09%	1.70%
Cumulative actual values	3.15%	3.95%	5.05%	5.98%	7.19%	8.72%	9.57%	10.30%	11.51%	12.78%	13.87%	15.57%

Months (2014)	Jan-14	Feb-14	Mar-14	Apr-14	May-14	Jun-14	Jul-14	Aug-14	Sep-14	Oct-14	Nov-14	Dec-14
Early values monthly	3.6%	4.3%	6.6%	4.8%	3.7%	5.2%	5.4%	7.1%	5.3%	3.7%	3.2%	2.4%
Cumulative early values	40.7%	45.0%	51.6%	56.3%	60.1%	65.3%	70.7%	77.8%	83.1%	86.8%	90.0%	92.4%
Late values monthly	3.12%	2.79%	3.49%	3.49%	4.27%	4.72%	4.31%	5.06%	5.06%	5.50%	4.53%	7.03%
Cumulative late values	20.3%	23.1%	26.6%	30.0%	34.3%	39.0%	43.3%	48.4%	53.5%	59.0%	63.5%	70.5%
Actual values monthly	0.92%	1.21%	1.80%	2.06%	2.94%							
Cumulative actual values	16.49%	17.70%	19.50%	21.56%	24.50%							

Months (2015)	Jan-15	Feb-15	Mar-15	Apr-15	May-15	Jun-15	Jul-15	Aug-15	Sep-15	Oct-15	Nov-15
Early values monthly	2.7%	1.1%	3.0%	0.9%	0.0%	0.0%	0.0%	0.0%	0.0%	0.0%	0.0%
Cumulative early values	95.1%	96.1%	99.1%	100.0%	100.0%	100.0%	100.0%	100.0%	100.0%	100.0%	100.0%
Late values monthly	10.21%	6.87%	4.44%	4.04%	3.91%	0.00%	0.00%	0.00%	0.00%	0.00%	0.00%
Cumulative late values	80.7%	87.6%	92.0%	96.1%	100.0%	100.0%	100.0%	100.0%	100.0%	100.0%	100.0%
Actual values monthly											
Cumulative actual values											

Notes :

1- Planned values are based on Contractor's Cost Loaded Schedule submitted on 5th May 2012.

2- The actual % progress up to 30th April2014 are based on value of work done.

FIGURE 7.128
S-curve (cost loaded).

Appendix I illustrates an outline of CQCP for a major building construction project. However, the contractor has to take into consideration requirements listed under contract documents depending on the nature and complexity of the project.

7.4.6.8 Develop Resource Management Plan

In most construction projects, the contractor is responsible to engage subcontractors, specialist installers, suppliers, arrange for materials, equipment, construction tools, and all type of human resources to complete the project as per contract documents and to the satisfaction of owner/owner's appointed supervision team. Workmanship is one of the most important factors to achieve the quality in construction; therefore, it is required that the construction workforce is fully trained and have full knowledge of all the related activities to be performed during the construction process.

Once the contract is awarded, the contractor prepares a detailed plan for all the resources he or she needs to complete the project. The contractor also prepares a procurement log based on the project completion schedule.

Contract documents normally specify a list of minimum number of core staff to be available at site during the construction period. Absence of any of these staff may result in penalty to be imposed on the contractor by the owner.

Contractor's human resources mainly consists of two categories:

1. Contractor's own staff and workers
2. Subcontractor's staff and workers

The main contractor has to manage all these personnel by

1. Assigning the daily activities
2. Observing their performance and work output
3. Daily attendance
4. Safety during construction process

Figure 7.129 illustrates the contractor's planned manpower chart for the construction project.

In most construction projects, the contractor is responsible for procurement of material, equipment, and systems to be installed on the project. The contractors have their own procurement strategies. While submitting the bid, the contractor obtains the quotations from various suppliers/subcontractors. The contractor has to consider the following as a minimum while finalizing the procurement:

- Contractual commitment
- Specification compliance
- Statutory obligations
- Time
- Cost
- Performance

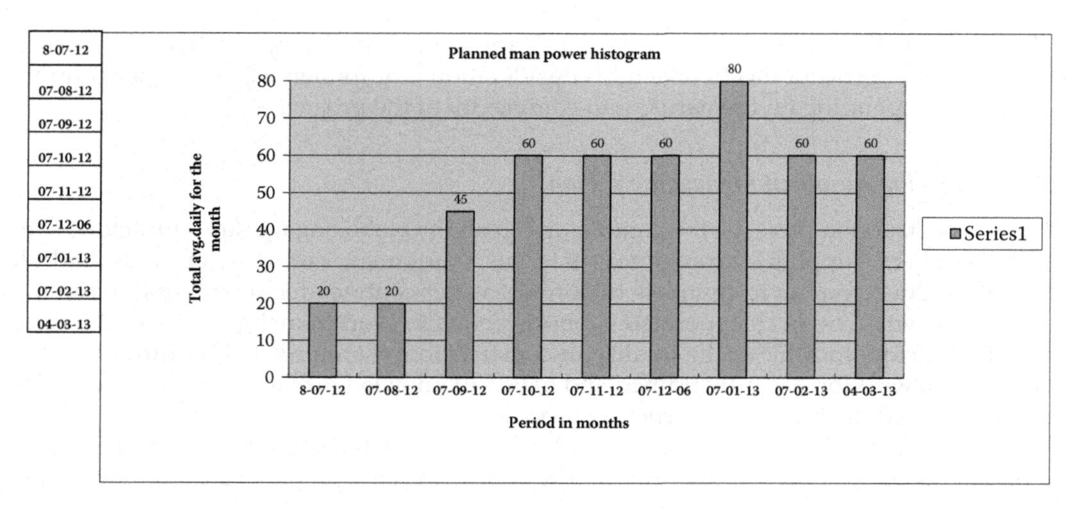

FIGURE 7.129
Manpower plan.

Please refer to Figure 7.70, which illustrates material management process for the construction project.

Likewise the contract documents specify that minimum number of equipment are to be available at site during the construction process to ensure smooth operation of all the construction activities. Figure 7.130 illustrates planned equipment schedule listing equipment the contractor has to make available at major building construction project.

7.4.6.9 Develop Communication Plan

For smooth flow of construction process activities during the construction phase, proper communication and submittal procedure need to be established between all the concerned parties at the beginning of the construction activities. The communication during the construction phase is mainly between the following three parties:

1. Owner
2. Construction supervisor (consultant)
3. Contractor

In certain cases, it may be the agency construction manager who is involved in the project on behalf of project owner. Each of these participants has to develop their own communication plan. Table 7.81 illustrates contents of communication plan.

Communication management process is already discussed in Section 7.3.8. The communication between the contractor and supervisor takes place through the transmittal form. Table 7.30 lists various types of transmittal forms used during construction project. Number of copies, distribution, and communication method is specified in the contract documents. Table 7.79 illustrates an example matrix for site administration of a building construction project. Proper adherence to these duties helps smooth implementation of project. Correspondence between the consultant and contractor is normally though letters or job site instructions. Figure 7.131 is a job site instruction form used by the consultant to communicate with the contractor.

7.4.6.10 Develop Risk Management Plan

Probability of occurrence of risk during the construction phase is very high when compared to design phases. During the construction phase, uncertainty comes from various sources, as this phase has involvement of various participants. Risk in construction projects has already been discussed in Section 7.3.9.3. Since the duration of the construction phase is longer than earlier phases, the contractor has to also consider occurrence of financial, economical, commercial, political, and natural risks. The contractor has to develop risk management plan as discussed under Section 7.3.9. Figure 7.86 illustrates risk management cycle (process) and Table 7.33 lists the probability of occurrence of various categories of risks during the construction project.

Table 7.82 lists major risks during the construction phase and mitigation actions. The contractor has to monitor the occurrence of these risks and take necessary measures to mitigate the effects on the project.

OWNER NAME
PROJECT NAME
EQUIPMENT SCHEDULE

CONTRACTOR NAME

	1	2	3	4	5	6	7	8	9	10	11	12	13	14	15	16	17	18	19	20	21	22	23	24	25	26	27	28	29
EQUIPMENT DESCRIPTION	Oct-15	Nov-15	Dec-15	Jan-15	Feb-15	Mar-15	Apr-15	May-15	Jun-15	Jul-15	Aug-15	Sep-15	Oct-15	Nov-15	Dec-15	Jan-17	Feb-15	Mar-15	Apr-15	May-15	Jun-15	Jul-15	Aug-15	Sep-15	Oct-15	Nov-15	Dec-15	Jan-17	Feb-15
BATCHING PLANT																					1-Jun								
AIR COMPRESSOR																													
BAR BENDING MACHINE																													
BAR CUTTING MACHINE																													
BULLDOZER																													
BUSES																													
FUEL TANKER																													
GENERATOR																													
JEEP																													
HALF LOORY																													
MOBILE CRANE																													
PICK UP																													
FLAT COMPACTOR																													
ROLLER COMPACTOR																													
BOB CAT																													
TIPPER TRAILER																													
TOWER LIGHT																													
ASPHALT FINISHER																													
BITUMEN SPRAYER																													
WHEEL LOADER																													
WELDING MACHINE																													
WATER TRUCK																													
SILO CEMENT																													
TRANSIT MIXER (CONCRETE)																													
FOAM PUMP																													
MARBLE CUTTER																													

FIGURE 7.130
Equipment schedule.

TABLE 7.81
Contents of Contractor's Communication Management Plan

Section	Topic
1	Introduction
	1.1 Project description
2	Stakeholders
	2.1 Project team directory 2.2 Project organizational chart 2.3 Roles and responsibilities 2.4 Stakeholders requirement
3	Communication methods
4	Communication management constraints
5	Communication matrix
6	Distribution of communication documents
	6.1 General correspondence 6.2 Submittals 6.3 Status reports 6.4 Meetings 6.5 Management plans 6.6 Change orders 6.7 Payments
7	Regulatory requirements
8	Communication plan update

7.4.6.11 Develop HSE Plan

In construction projects, the requirements to prepare a safety management plan by the contractor are specified under contract documents. The contractor has to submit the plan for review and approval by supervisor/consultant during mobilization stage of the construction phase. The following are outlines normally specified in the contract documents that are to be considered by the contractor to establish safety management system for construction project sites in order to raise the level of safety and health in the construction sites and prevent accidents:

1. Project scope detailing description of project and safety requirements
2. Safety policy statement documenting the contractor's/subcontractor's commitment and emphasis on safety
3. Regulatory requirements about safety
4. Roles and responsibilities of all individuals involved
5. Site communication plan detailing how safety information will be shared
6. Emergency evacuation plan
7. Accident reporting system
8. Hazard identification, risk assessment, and control
9. Accident investigation to document root causes and determine corrective and preventive actions
10. Measures for emergency situations

11. Plant, equipment maintenance, and licensing
12. Routine inspections
13. Continuous monitoring and regular assessment
14. Health surveillance
15. System feedback and continuous improvements
16. Safety assurance measures
17. Evaluation of subcontractor's safety capabilities
18. Site neighborhood characteristics and constraints
19. Safety audit
20. Documentation
21. Records
22. System education and training
23. System update

Table 7.83 illustrates contents of an HSE plan.

7.4.6.12 Execute Project Works

The contractor is given a few weeks to start the construction works after signing of the contract. A letter from the client/owner is issued to the contractor to begin the project work subject to the conditions of contract. This letter is known as *notice to proceed* letter.

7.4.6.12.1 Notice to Proceed

The notice to proceed authorizes the contractor to proceed with work to construct the project/facility as per the agreement. Prior to issuance of the notice to proceed, the owner has to ensure that

- Necessary permits have been obtained from the relevant authorities/agencies to hand over the construction site to the contractor
- Supervision staff/project manager/construction manager is selected to supervise the work
- Relevant departments have been informed about signing of contract for availability of funds
- Supervision undertaking guarantee is signed by the supervision consultant
- Owner's representative is selected
- Notice to proceed date is mutually discussed and agreed as per the conditions of contract
- All the copies of construction documents are distributed to concerned stakeholders
- Authorization letter to the owner's representative and engineer's representative is already issued

Project Name
Consultant Name

JOB SITE INSTRUCTION (JSI)

CONTRACTOR : ______________________ JSI No. : ____________

CONTRACT No.: ______________________ DATE : ____________

The work shall be carried out in accordance with the Contract Documents without change in Contract Sum or Contract Time. Proceeding with the work in accordance with these instructions indicates your acknowledgement that there will be no change in the Contract Sum or Contract Time.

Subject:

SAMPLE FORM

ATTACHMENTS: (List attached documents that support description.)

Signed: ____________ Resident Engineer	Received by Contractor : ____________ Date:

Distribution: ☐ Owner ☐ A/E ☐ Contractor

FIGURE 7.131
Job site instruction.

TABLE 7.82

Major Risks during Construction Phase and Mitigation Action

Sr. No.	Description of Risk	Mitigation Action
1	Incompetent subcontractor	Contractor has to monitor the workmanship and work progress.
2	Failure of team members not performing as expected	Select competent candidate. Provide training.
3	Low bid project cost	Contractor to try competitive material, improve method statement and higher production rate from its manpower
4	Delay in transfer of site	Contractor to adjust the construction schedule
5	Delay in mobilization	Adjust construction schedule accordingly
6	Scope/design changes	Resolve change order issues in order not to delay the project
7	Different site conditions to the information provided	Contractor to investigate site conditions prior to starting the relevant activity
8	Inadequate site investigation data	Contractor to investigate site conditions prior to starting the relevant activity
9	Conflict in contract documents	Amicably resolve the issue
10	Incomplete design	Raise request for information (RFI)
11	Incomplete scope of work	Raise RFI
12	Design changes	Follow contract documents for change order
13	Design mistakes	Raise RFI
14	Errors and omissions in contract documents	Raise RFI
15	Incomplete specifications	Raise RFI
16	Inappropriate construction method	Raise RFI and correct the method statement
17	Conflict with different trades	Coordinate with all trades while preparing coordination and composite drawings
18	Change in laws and regulations	Contractor to inform owner/consultant and raise RFI
19	Statutory/regulatory delay	Regular follow up by the contractor, owner with the regulatory agency
20	Project schedule	Compress duration of activities
21	Inappropriate schedule/plan	Contractor to prepare schedule taking into consideration site conditions all the required parameters
22	Delay in changer order negotiations	Request owner/supervisor/project manager to expedite the negotiations and resolve the issue
23	Resource availability(material)	Contractor to make extensive search
24	Resource (labor) low productivity	Contractor to engage competent and skilled labors
25	Equipment/plant productivity	Contractor hire/purchase equipment to meet project productivity requirements
26	Insufficient skilled workforce	Contractor arrange workforce from alternate sources
27	Failure/delay of machinery and equipment	Contractor to plan procurement well in advance
28	Failure/delay of material delivery	Contractor to plan procurement well in advance
29	Delay in approval of submittals	Notify owner/project manager
30	Delays in payment	Contractor to have contingency plans
31	Quality of material	Locate suppliers having proven record of supplying quality product

(*Continued*)

TABLE 7.82 (*Continued*)

Major Risks during Construction Phase and Mitigation Action

Sr. No.	Description of Risk	Mitigation Action
32	Variation in construction material price	Contractor to negotiate with supplier/manufacturer for best price. Contractor to request for change order if applicable as per contract
33	Damage to equipment	Regularly maintain the equipment. Take immediate action to repair damage equipment
34	Damage to stored material	Contractor to follow proper storage system
35	Structure collapse	Contractor to ensure that formwork and scaffolding is properly installed
36	Access to worksite	Access road to be planned in coordination with adjacent area and local authority
37	Leakage of hazardous material	Contractor to take necessary protect to avoid leakage. Store in safe area
38	Theft at site	Contractor to monitor access to site. Record Entry/Exit to the site. Provide fencing around project site.
39	Fire at site	Contractor to install temporary fire fighting system. Inflammable material to be stored in safe and secured place with necessary safety measures
40	Injuries	Contractor to keep first aid provision at site. Take immediate action to provide medical aid.
41	New technology	Owner/contractor to mutually agree for changes in the contract for better performance of project

Figure 7.132 is a sample notice to proceed.

7.4.6.12.2 Kick-off Meeting

The *kick-off meeting* is the first meeting with the owner/client and project team members. It is also called *preconstruction meeting*. This meeting provides opportunity to all project team members to interact and know each other. Figure 7.133 is an example of kick-off meeting agenda.

7.4.6.12.3 Mobilization

The activities to be performed during the mobilization period are defined in the contract documents. During this period, the contractor is required to perform many of the activities before the beginning of actual construction work at the site. Necessary permits are obtained from the relevant authorities to start the construction works. Upon receipt of construction site from the owner, the contractor starts mobilization works, which consist of preparation of site offices/field offices for owner, supervision team (consultant), and for contractor himself. This includes all the necessary on-site facilities and services necessary to carry out specific work and task. Mobilization activities usually occur at the beginning of a project but can occur anytime during a project when specific on-site facilities are required. During this time, the project site is handed over to the contractor. The contractor performs site survey and testing of soil to facilitate start of construction work.

TABLE 7.83

Contents of Contractor's HSE Plan

Section	Topic
1	Introduction
	1.1 Project description
2	HSE policy
3	Regulatory requirements
	3.1 Safety
	3.2 Environmental
	3.3 Health surveillance
4	HSE organization
	4.1 HSE organizational chart
	4.2 Roles and responsibilities
	4.3 Communication plan
5	Safety management plan
	5.1 Design phases
	5.2 Construction phase
	5.3 Project startup
6	Emergency evacuation plan
	6.1 Accident reporting
	6.2 Action plan
	6.3 Measures for emergency situations
	6.4 Disaster control
7	Training
	7.1 Awareness
	7.2 Meetings
	7.3 Drills
8	Monitoring
	8.1 Routing inspection
	8.2 Protective equipment
	8.3 Plant and equipment
	8.4 Safety measures
	8.5 Hygiene
9	Hazards management
	9.1 Safety hazards
10	Environmental protection measures
11	Waste management
12	Risk identification
13	Preventive action
14	Documentation
15	Record
16	System update

In anticipation of the award of contract, the contractor begins following activities much in advance, but these are part of contract documents and contractor's action is required immediately after signing of contract in order to start construction.

1. Mobilization of construction equipment and tools
2. Manpower to execute the project

LETTERHEAD

Ref: -------------------

Date: -----------------------

NOTICE TO PROCEED

To,

Contractor Name : SAMPLE LETTER

Address :

Subject : Contract Number------------

Attention : -----------------

Sir/Madam

You are hereby authorized to proceed with Project No. ------------ in accordance with construction contract dated -----------.This contract calls all the contracted works to be completed within --------- calendar days. The date of enterprise shall be -------------.

Sincerely,

Enclosures :

CC :

FIGURE 7.132
Notice to proceed.

7.4.6.12.3.1 Bonds, Permits and Insurance As per contract documents, the contractor has to

1. Submit advance payment guarantee
2. Permit from local authority (municipality)
3. Insurance policies covering the following areas:
   a. Contractor's all risks and third-party insurance policy
   b. Contractor's plant and equipment insurance policy

PROJECT NAME

Contract Number :			
Type of Meeting :		Date of Meeting :	
Place of Meeting :		Time of Meeting :	
Owner :			
Project/Construction Manager			
A/E (Consultant)			

AGENDA SAMPLE AGENDA
1.0 Points to be Discussed 1.1 Intoduction 1.2 Project goals and objectives 1.3 Scope of work 1.4 Permit, Bonds, Insurance 1.5 Site handover procedure 1.6 Mobilization 1.7 Contractor's organization chart 1.8 Construction Schedule 1.9 Communication and Correspondance 1.10 Transmittals and Submittal procedure 1.11 Meetings (Progress, Coordination) 1.12 Quality Management 1.13 Risk Management 1.14 Site safety 1.14 Payment 1.15 Nominated subcontractors 2.0 Any other business

Signed by :	Position : ....
Date :	

FIGURE 7.133
Kick-off meeting agenda.

c. Workmen's compensation insurance policy
d. Site storage insurance policy
4. Normally, the submitted originals are retained by the owner and the copies are kept with the R.E. or project/construction manager.

7.4.6.12.3.2 Temporary Facilities The requirements to set up temporary facilities are specified in the contract. The contractor has to submit layout plans, dimensions, and other pertinent details for temporary facilities to be constructed. These include the following:

1. Site offices for owner and supervisor (consultant, construction manager, and project manager)
2. Storage facilities
3. Toilets and wash rooms
4. Sanitary and drainage system
5. Drinking water facility
6. Site electrification
7. Temporary firefighting system
8. Site fence
9. Site access road
10. Signage
11. Fuel storage area
12. Guard room
13. Testing laboratory

Upon approval of plans, the contractor proceeds with construction of temporary facilities and necessary utilities. The contractor has to designate authority-approved dumping area for waste material.

7.4.6.12.3.3 Contractor's Core Staff The contractor's core staff requirements and their qualifications are listed under tendering procedures. A typical list of the contractor's core staff has already been discussed in Section 7.4.6.3.3. The absence of these staff from the project site without prior permission attracts penalty to the contractor. Normally, the penalty amount is specified in the contract documents. Upon signing of the contract, the contractor has to submit the names of the staff for the positions described in the contact documents for approval from the owner/consultant to work on the project. The contractor has to select an appropriate candidate to propose for the specified position (see Section 7.2.3 and Figures 7.18 and 7.19). Figure 7.134 is the staff approval request form used by the contractor to propose the staff to work for these positions and is submitted along with the qualification.

The contractor has to submit an organizational chart based on the typical minimum core staff list and approved staff.

7.4.6.12.3.4 Approval of Subcontractor As per the tendering procedures, the contractor has to declare names of subcontractors, specialist to perform the designated works. The subcontractors for various subprojects have to be submitted for approval in a timely and

Project Name
Consultant Name

REQUEST FOR SITE STAFF APPROVAL

CONTRACT NO :
CONTRACTOR :
NO :
DATE:

SAMPLE FORM

To : Owner

1. Name :
2. Profession :
3. Position No. in Document-I :
4. No. of years of Experience :
5. Membership of Professional Body : Valid ☐ Not Valid ☐
6. Requested Date of Commencement :
7. Remarks :

Signature Contractor's Project Manager

OWNER COMMENT APPROVED ☐ NOT APPROVED ☐

Owner Rep. Signature

Date

Distribution OWNER A/E CONTRACTOR

FIGURE 7.134
Request for staff approval.

orderly manner, as planned in the approved work program, so that work progresses in a smooth and efficient manner. The subcontractor submission log helps the project manager to remind the main contractor to submit the subcontractors on time.

The request includes all the related information to prove the subcontractor's capability of providing the services to meet the project quality, and have enough available resources to meet the specified schedule, past performance, and if any quality system is implemented. Figure 7.135 is a request for subcontractor approval to be submitted by the contractor for getting approval of any subcontractor proposed to work on the construction project. Sometimes, the owner/consultant nominates subcontractor(s) to execute portion of a contract and is known as nominated subcontractor.

7.4.6.12.3.5 Management Plans Upon signing of the contract, the contractor has to submit mainly the following plans to the supervisor (consultant) for their review and approval:

1. Contractor's construction schedule
2. CQCP
3. Site communication plan
4. HSE plan

Among these plans, the construction schedule is most important. This is the first and foremost program that the contractor has to submit for approval. The contractor cannot proceed with construction unless the preliminary construction schedule is approved. In certain cases, the progress payment is having relation with approval of the contractor's construction schedule. The contractor is not paid unless contractor's construction schedule is approved.

For smooth implementation of project, a proper communication system is established clearly identifying the submission process for correspondence and transmittals. Correspondence between consultant and contractor is normally done through job site instructions, whereas correspondence between owner, consultant, and contractor is normally done through letters. Figure 7.131 discussed earlier is a sample job site instruction form used by the consultant to communicate with the contractor, and Table 7.79 discussed earlier is an example matrix for site administration.

7.4.6.12.4 Manage Execution of Works

The contractor is responsible to execute the contracted works in accordance with contract drawings and specifications as specified in the contract documents. The contractor has to arrange necessary resources to complete the project within the schedule and contracted amount. The contractor has to maintain the executed works until handing over the project to the owner/end user and maintain for additional period if contracted to do so. During the construction period, the contractor has to protect executed/installed works to ensure that the works are not damaged. The contractor has to use new and approved material to construct the project/facility.

Construction activities mainly consist of the following:

- Site work such as cleaning and excavation of project site
- Construction of foundations, including footings and grade beams
- Construction of columns and beams

Project name

Contract No. :
Contractor :

CONTRACTOR REQUEST
FOR SUB-CONTRACTOR APPROVAL

Serial no : ....

Kindly approve the following as a SUB-CONTRACTOR in the above mentioned project

Sub-Contract Works ____________________
Sub Contractor ____________________
Address (Head Office) ____________________

Reference Letter ____________________

Attachments :

Commercial Register	☐	Foundation Contract	☐
Experience	☐	Resources	☐
Current Work	☐	Financial Status	☐
Others (List, if any)	☐		

....
....
....

Performance Bond (Yes/No) ____________ No. ____________

Signed by
Contractor's Representative .... Date: ....

OWNER'S NAME

Received by : .... Signature : .... Date : ....

SITE SUPERVISION CONSULTANT

Received by : .... Signature : .... Date : ....

Consultant's recommendation : ....
....
....
....
....
....

Signed by
Resident Engineer .... Date : ....

Distribution : OWNER (Original) Supervision Consultant (Copy) Contractor (Copy)

Note : Contractor shall submit original to OWNER with copy to Site Supervision Consultant.

FIGURE 7.135
Request for subcontractor approval.

- Forming, reinforcing, and placing the floor slab
- Laying up masonry walls and partitions
- Installation of roofing system
- Finishes
- Furnishings

- Conveying system
- Installation of fire suppression system
- Installation of water supply, plumbing, and public health system
- Installation of heating, ventilating, and air-conditioning system
- Integrated automation system
- Installation of electrical lighting and power system
- Emergency power supply system
- Fire alarm system
- Information and communication technology system
- Electronic security and access control system
- Landscape works
- External works

7.4.6.12.4.1 Perform Quality Assurance Quality assurance during construction phase is the activity(ies) performed by the contractor to provide evidence and confidence by the contractor that the works, materials and systems executed, installed shall meet owner's objectives as specified in the contract documents and to ensure and guarantee that performance of the project/facility is fully functional to the satisfaction of owner/end user.

The contractor performs the quality assurance process by

1. Selecting the materials and systems fully complying with contract specifications and installing the approved material and systems only.
2. Preparing the shop drawings detailing all the requirements included in the working drawings and installing/executing the works as per approved shop drawings.
3. Installing the works, materials, and systems as per specified method statement and as per recommendations from the manufacturer of products.

The approved materials, systems, and shop drawings are used by the supervision team to inspect the executed works and control the quality of works. The detail procedure for submitting shop drawings, materials, and samples is specified under section *Submittal* of contract specifications. The contractor has to submit the same to owner/consultant for review and approval. The consultant reviews the submittal and returns the transmittal to the contractor with an appropriate action (see Section 7.3.8.1.3).

Materials: The contractor has to install the material specified under particular specifications (document III-2) and contract drawings (III-3). The contractor has to submit the material specified in the contract documents prior to execution/installation of work.

The contractor has to submit following, as minimum, to the owner/consultant to get their review and approval of materials, products, equipment, and systems. The contractor cannot use these items unless they are approved for use in the project.

Product data: The contractor has to submit the following details:

- Manufacturer's technical specifications related to the proposed product
- Installation methods recommend by the manufacturer

- Relevant sheets of manufacturer's catalog(s)
- Confirmation of compliance with recognized international quality standards
- Mill reports (if applicable)
- Performance characteristic and curves (if applicable)
- Manufacturer's standard schematic drawings and diagrams to supplement standard information related to project requirements and configuration of the same to indicate product application for the specified works (if applicable)
- Compatibility certificate (if applicable)
- Single source liability (this is normally required for systems approval where different manufacturer's items are used)

Compliance statement: The contractor has to submit a specification comparison statement along with the material transmittal. The compliance statement form is normally included as part of contract documents. The information provided in the compliance statement helps the consultant to review and verify product compliance to the contracted specifications. In case of any deviation to that of specified item, the contractor has to submit a schedule of such deviation(s), listing all the points not conforming to the specification. In certain projects owner is involved in approval of materials.

Samples: The contractor has to submit (if required) the sample(s) from the approved material to be used for the work. The samples are mainly required to;

- Verify color, texture, and pattern
- Verify that the product is physically identical to the proposed and approved material
- To be used for comparison with products and materials used in the works

At times, it may be specified to install the sample(s) in a manner to facilitate review of qualities indicated in the specifications. Figure 7.136 illustrates procedure for selection of material/product, whereas Figure 7.137 illustrates material approval procedure.

Shop drawings: The contract drawings and specifications prepared by the design professionals are indicative and are generally meant for determining the tender pricing and for planning the construction project. In many cases, they are not sufficient for installation or execution of works at various stages. More details are required during the construction phase to ensure specified quality. These details are provided by the contractor on the shop drawings. Shop drawings are used by the contractor as reference documents to execute/install the works. Detailed shop drawing help contractor to achieve zero defect in installation at the first stage itself, thus avoiding any rejection/rework.

The number of shop drawings to be produced is mutually agreed between the contractor and consultant depending on the complexity of the work to be installed and to ensure that adequate details are available to execute the work.

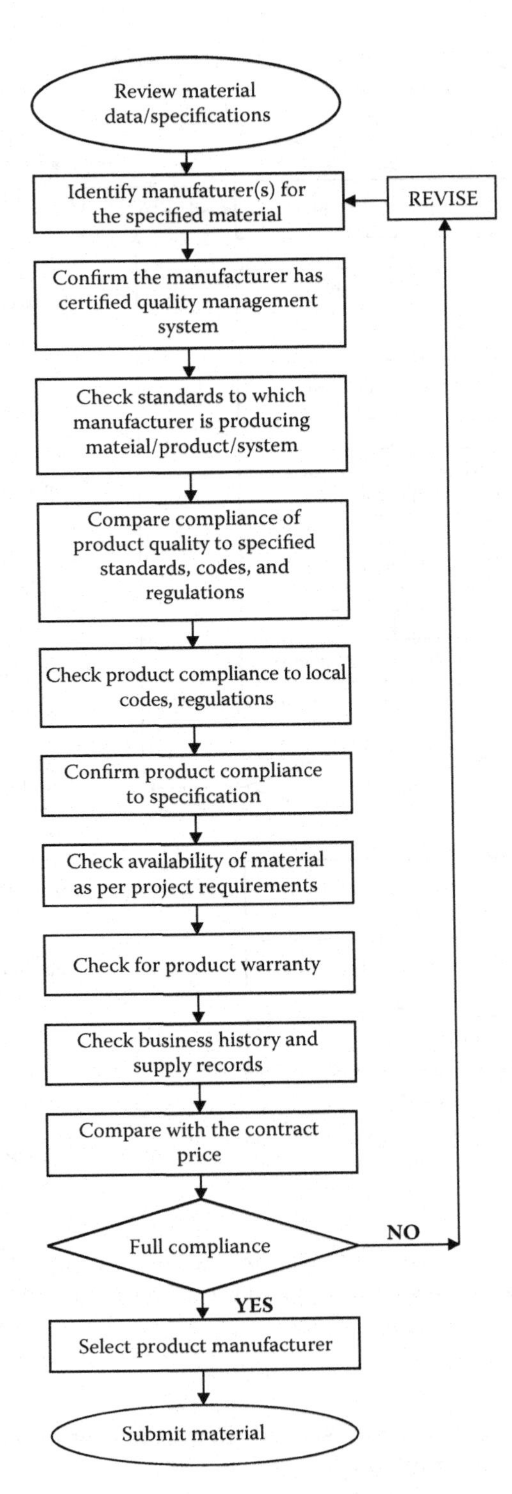

FIGURE 7.136
Material/product manufacturer selection procedure. (Abdul Razzak Rumane (2013), *Quality Tools for Managing Construction Projects*, CRC Press, Boca Raton, FL. Reprinted with permission from Taylor & Francis Group.)

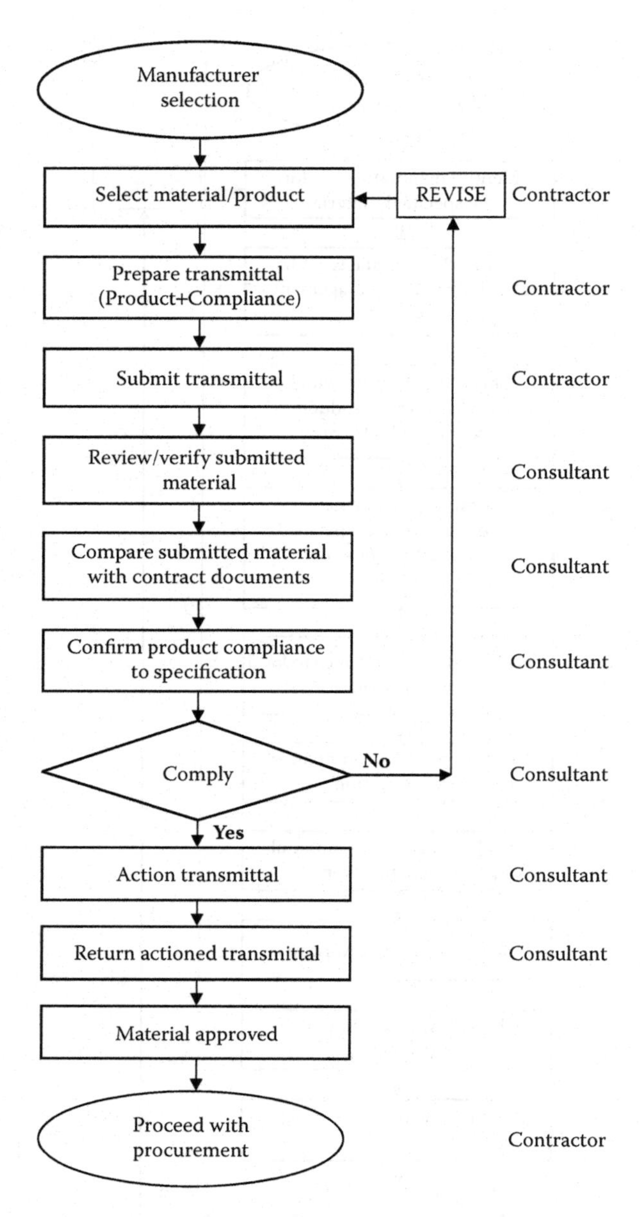

FIGURE 7.137
Material/product/system approval procedure. (Abdul Razzak Rumane (2013), *Quality Tools for Managing Construction Projects*, CRC Press, Boca Raton, FL. Reprinted with permission from Taylor & Francis Group.)

The contractor is required to prepare shop drawings taking into consideration the following as a minimum but not limited to

1. Reference to contract drawings. This helps A/E (consultant) to compare and review the shop drawing to that of contract drawing.

2. Detail plans and information based on the contract drawings.
3. Notes of changes or alterations from the contract documents.
4. Detail information about fabrication or installation of works.
5. All dimensions needed to verify at the jobsite.
6. Identification of product.
7. Installation information about the materials to be used.
8. Type of finishes, color, and textures.
9. Installation details relating to the axis or grid of the project.
10. Roughing in and setting diagram.
11. Coordination certification from all other related trades (subcontractors).

The shop drawings are to be drawn accurately to the scale and shall have project-specific information in it. The shop drawings shall not be reproduction of contract drawings.

Immediately after approval of individual trade shop drawings, the contractor has to submit builder's workshop drawings, composite/coordinated shop drawings by taking into consideration following as a minimum. Figure 7.138 illustrates shop drawing preparation and approval procedure (also see Figure 4.30).

Based on contract drawings, contractor prepares shop drawings and submits the same to consultant for approval. All the works are executed as per approved shop drawings. The contractor has to consider following point as the minimum, while developing shop drawings of different trades, to meet the design intents.

Architectural works: The architectural works shop drawings mainly covers masonry, doors and windows, cladding, partitioning, reflected ceiling, stone flooring, toilet details, stairs details, and roofing. A brief requirement to prepare shop drawings for these sections is as follows:

a. *Masonry*: The shop drawings for masonry works shall include minimum
   - Area layout
   - Guidelines
   - Height of masonry
   - Type and thickness of blocks used for masonry
   - Reinforcement details
   - Fixation details
   - Openings in the block works for other services
   - Sills
   - Lintels
   - Plastering details, if applicable

b. *Doors and windows*: The shop drawings for doors and windows shall include
   - Size of doors and windows
   - Type of material
   - Thickness of frames
   - Details of door leaves

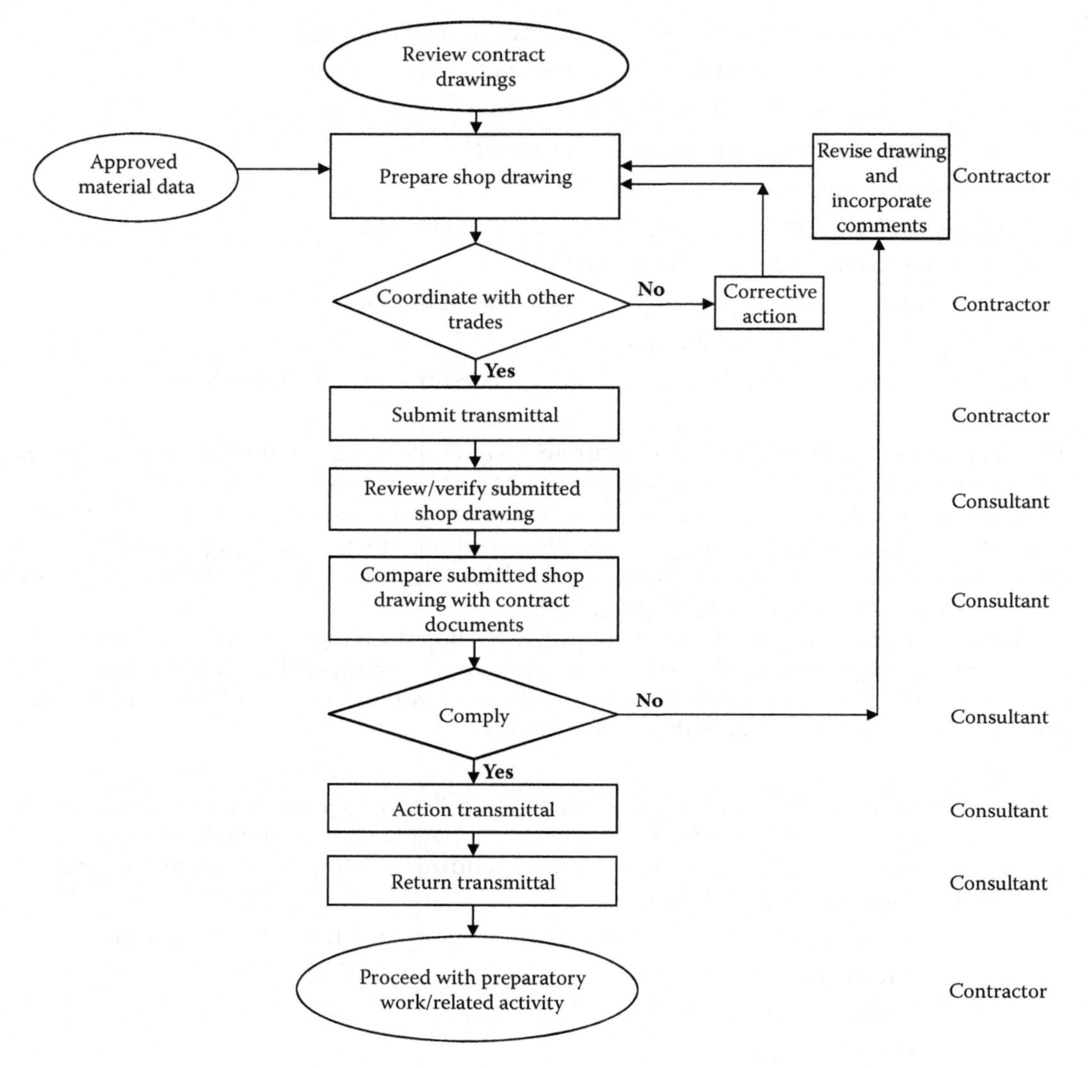

FIGURE 7.138
Shop drawing preparation and approval procedure. (Abdul Razzak Rumane (2013), *Quality Tools for Managing Construction Projects*, CRC Press, Boca Raton, FL. Reprinted with permission from Taylor & Francis Group.)

- Details of glazing
- Fixing details
- Schedule of doors and windows

c. *Cladding*: The shop drawings for cladding shall include
   - Type of cladding material
   - Size of panels/tiles and thickness
   - Elevations
   - Fixation method, anchorage, and supports
   - Openings for other services

d. *Partitioning*: The shop drawings for partitions shall include
   - Type of partitioning material
   - Size of panels
   - Frame size and its installation details
   - Partition support system
   - Fixation details of panels
e. *Reflected ceiling*: The shop drawings for reflected ceiling shall include
   - Type of ceiling material
   - Size of ceiling panels and thickness of material
   - Ceiling level
   - Suspension system and framing
   - Layout showing jointing layout, faceting, and boundaries between materials
   - Location of light fittings, detectors, sprinklers, and grills
   - Location of access panel
f. *Stone flooring*: The shop drawings for flooring shall include
   - Type of flooring material
   - Size of flooring tiles
   - Layout
   - Screeding details
   - Jointing details
   - Flooring pattern
   - Cut out for other services
   - Control joints and jointing method
   - Antislip inserts
   - Separators
   - Thresholds
g. *Toilet details*: The shop drawings for toilet details shall include
   - Layout
   - Plan
   - Sections
   - Installation details of sanitary wares and fixtures
   - Installation details of toilet accessories
   - Jointing method
   - Material finishes and surface textures
   - Installation details of toilet mirror
h. *Stairs details*: The shop drawings for stairs details shall include
   - Steps details, riser, treads, width, and height
   - Finishing details
   - Handrail details

- Sections
- Plan

i. *Roofing-water proofing*: The shop drawings for roofing details shall include
   - Insulation details
   - Installation details of insulation
   - Installation of water proofing material
   - Control joints/expansion joints details
   - Tiling details
   - Flashings

Structural works: The structural works shop drawings to be prepared according to ACI 315 and shall mainly cover reinforced concrete, formwork, precast concrete, and structural steel fabrication. A brief requirement to prepare these shop drawings is as follows:

a. *Reinforced concrete*: The reinforced concrete works shop drawings to be prepared according to ACI 315 and shall have the following details:
   - The size of reinforcement material
   - Bar schedule
   - Stirrup spacing
   - Bar bent diagram
   - Arrangement and support of concrete reinforcement
   - Dimensional details
   - Special reinforcement required for openings through concrete structure
   - Type of cement and its strength

b. *Formwork*: Formwork has great importance from the safety point of view. The shop drawing of formwork shall include
   - Details of individual panels
   - Position, size, and spacing of adjustable props
   - Position, size, and spacing of joints, soldiers, ties, and the like
   - Details of formwork for columns, beams, parapet, slabs, and kickers
   - Details of construction joints and expansion joints
   - Details of retaining walls, core walls, and deep beams showing the position and size of ties, joints, soldiers, and sheeting, together with detailed information on erection and casting sequences and construction joints
   - General assembly details, including propping, prop bearings, and through propping
   - All penetrations through concrete
   - Full design calculations

c. *Precast concrete*: The shop drawings for precast concrete works have great importance as the casting is carried out at precast concrete factory. All the required details need to be shown on these drawings to ensure precast elements are produced without any defects. The shop drawing should provide details of fabrication and installation of precast structural concrete units. It shall indicate member location, plans, elevations, dimensions, shapes, cross sections, openings, and type of reinforcement, including special reinforcement, if any. The following information also has to be indicated in the shop drawings:

- Welded connections by American Welding Society's standards symbol
- Details of loose and cast-in hardware, insets, connections, joints, and all types of accessories
- Location and details of anchorage device to be embedded in other construction
- Comprehensive engineering analysis, including fire-resistance analysis

d. *Structural steel fabrication*: The shop drawings for fabrication of structural steel shall include

- Details of cuts, connections, splices, camber, holes, and other pertinent data
- Welding standards symbols
- Size and length of bolts, indicating details of material and strength of bolts
- Details of steel bars, plates, angles, beams, and channels
- Method of erection
- Details of anchorage details, templates, and installation details of bolts

Elevator works

- Overall elevation (vertical) for elevator area
- Floor levels
- Hoist way plan, sections, and anchoring details for installation of rails
- Machine room plan and installation details for drive and controller
- Equipment layout in machine room/hoist way
- Cuttings/openings and sleeves required in the concrete slab/walls
- Details and level of cab overhead
- Details and levels about the pit
- Location and level of hall button and hall position indicator
- Openings for landing door, entrance view, and finishes level
- Finishes of cab, door, and indicators
- Power supply devices and equipment grounding
- Interface with elevator management system, if any

Mechanical works (fire suppression, water supply, and plumbing) The shop drawings for mechanical works shall include but is not limited to

- Sprinkler layout
- Hose reel details
- Hose reel cabinet size, location, and installation details
- Fire pump location and installation details
- Location of flow switches
- Riser diagram for water supply system
- Size of piping
- Piping route

- Piping levels and slope
- Pipes and sleeves for utility services
- Size of isolating valves and their location
- Equipment plan layout
- Pump room details
- Storage tank details
- Details of toilet accessories and connection details
- Riser diagram for rainwater system
- Drainage system piping
- Location of vents and traps
- Location of cleanouts
- Riser diagram for storm water system
- Storm water system
- Electrical power connection details
- Interface with fire alarm and building management system

HVAC works: The shop drawing for HVAC works shall include following but is not limited to

- Location of equipment and their configuration
- Piping size
- Piping route and levels
- Ducting size
- Ducting route and levels
- Insulation details
- Suspension/hanger details
- Equipment layout and plan/plant room details
- Riser diagram for chilled water system
- Installation details of equipment
- Size of diffusers and grills
- Installation details of grills
- Exhaust and ventilation fans layout and details
- Riser diagram for exhaust air system
- Return air opening details
- Equipment schedule
- Electrical connection and power supply details
- Control details
- Sequence of operation
- Schematic diagram for HVAC system
- Schematic diagram for building management system by configuring all the equipment and components

Electrical works: The shop drawings for electrical works shall indicate but is not limited to

- Size and type of conduits, raceways, and exact routing of the same
- Size of cable trays, cable trunking, and their installation methods and exact route indicating the level
- Size of wires and cables
- Small power layout
- Large power layout
- Wiring accessories with circuit references
- Lighting layout with circuit references
- Installation details of light fixtures
- Emergency lighting system
- External lighting layout
- Feeder pillar location and installation details
- Bus duct installation details
- Field installation wiring details for light, power, controls, and signals
- Installation details of lighting control panel
- Installation details of distribution boards, switch boards, and panels
- Location of distribution boards
- Panels, switch boards layout in low tension room
- Installation details of bus duct
- Schematic diagram showing the configuration of all the equipment
- Load schedules as per actual connected loads
- High-voltage panel layout
- Substation layout
- Interface with solar power/alternate energy system
- Voltage drop calculations
- Short circuit calculations
- Grounding (earthing) system
- Lightning protection system
- Diesel generator installation details
- Automatic transfer switch installation details
- Interface with other systems

HVAC electrical works: The shop drawings for HVAC electrical works shall include

- Wiring diagram of individual components and accessories
- MCC and starter panels
- Installation details of MCC panels and starter panels
- Schematic diagram including power and control wiring
- Type and size of conduit and number and size of wires in the conduit

- Type and size of cable
- Details of protection and interlocks
- Details of instrumentation
- Description of sequence of operation of equipment

Low-voltage systems: The shop drawings for low-voltage systems such as BMS, fire alarm system, communication system, public address system, audio–visual system, CCTV/security system, and access control system shall consist of

- Size of conduits and raceways, their routing and levels
- Location of components and wiring accessories
- Installation details of components and wiring accessories
- Wiring and cabling details
- Installation details of racks
- Installation details of equipment
- Location and installation details of panels
- Schematic/riser diagram configuring all the components and equipment used in the system
- Interface with other system(s)

Landscape works: The shop drawings for landscape shall include

- Boundary limit
- Excavation area
- Excavation level
- Type of soil in different areas
- Location of plants, trees, and shrubs
- Foundation details for plants, trees, and shrubs
- Location of sidewalk
- Location of driveways
- Foundation for paving
- Grass areas
- Location of services and ducts for utilities
- Location of light poles/bollards
- Foundation for light poles/bollard
- Details of special features
- Details of irrigation system

External works (infrastructure): The shop drawing for external works (infrastructure) shall include

- Width of the road
- Grading details
- Thicknesses of asphalt layers
- Pavement details
- Location of manholes and levels

- Services/utilities pipe routes
- Location of light poles
- Cable routes
- Trench details
- Road marking
- Traffic sign and signals

Builders workshop drawings: Builders workshop drawings indicating the openings required in the civil or architectural work for services and other trades. These drawings indicate the size of openings, sleeves, level references with the help of detailed elevation and plans. Figure 7.139 illustrates builders' workshop drawing preparation and approval procedure.

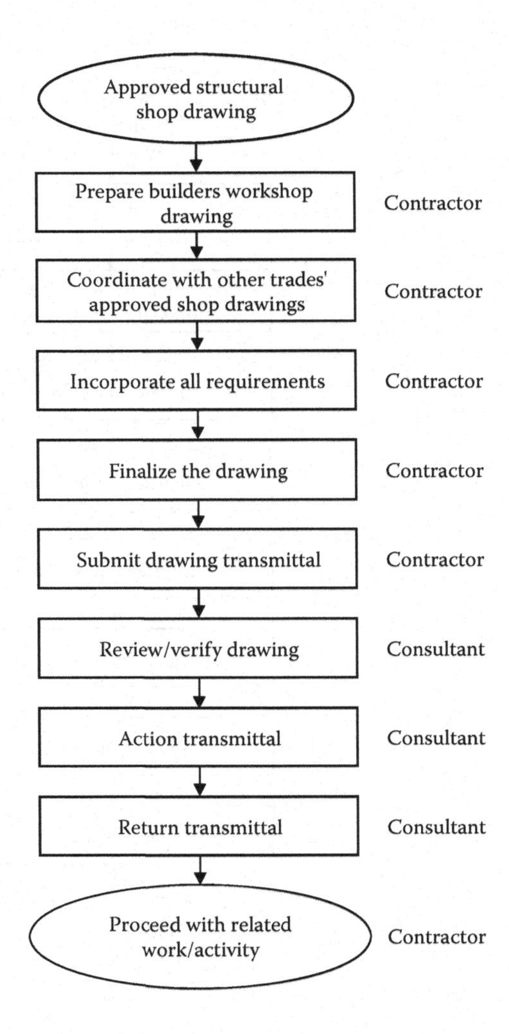

FIGURE 7.139
Builders workshop drawing preparation and approval procedure. (Abdul Razzak Rumane (2013), *Quality Tools for Managing Construction Projects*, CRC Press, Boca Raton, FL. Reprinted with permission from Taylor & Francis Group.)

Composite/coordination shop drawings: The composite drawings indicate relationship of components shown on the related shop drawings and indicate required installation sequence. Composite drawings shall show the interrelationship of all services with each other and with the surrounding civil and architectural work. Composite drawings shall also show the detailed coordinated cross sections, elevations, and reflected plans, resolving all conflicts in levels, alignment, access, and space. These drawings are to be prepared taking into consideration the actual physical dimensions required for installation within the available space. Figure 7.140 illustrates composite drawing preparation and approval procedure.

Method statement: The contractor has to execute the works as per the method statement specified in the contract. The contractor has to submit method statement to consultant for their approval as per the contract documents to ensure compliance with the contract requirements. The method statement shall describe the steps involved for execution/installation of work by ensuring safety at each stage. It shall have the following information:

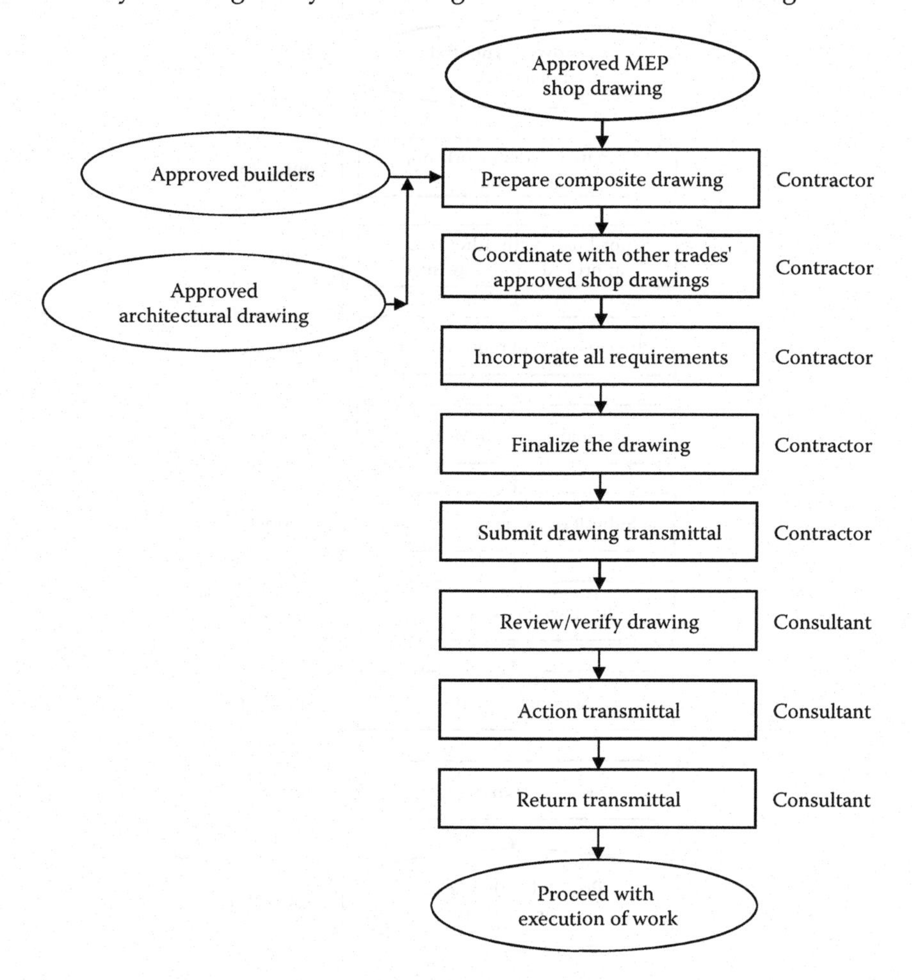

FIGURE 7.140
Composite drawing preparation and approval procedure. (Abdul Razzak Rumane (2013), *Quality Tools for Managing Construction Projects*, CRC Press, Boca Raton, FL. Reprinted with permission from Taylor & Francis Group.)

1. *Scope of work*: Brief description of work/activity
2. *Documentation*: Relevant technical documents to undertake this work/activity
3. Personnel involved
4. Safety arrangement
5. Equipment and plant required
6. Personal protective equipment
7. Permits/authorities approval to work
8. Possible hazards
9. *Description of the work/activity*: Detail method of sequence of each operation/key steps to complete the work/activity

Figures I.2 to I.9 are illustrative examples of work sequence of major activities, for different trades, performed during the construction process.

7.4.6.12.4.2 Manage changes It is common that during construction process, there will be some changes to the original contract. Even under most ideal circumstances, contract documents cannot provide complete information about every possible condition or circumstance that the construction team may encounter. The causes for changes have already been discussed in Section 7.3.3.6 (see Table 7.16).

These changes help build the facility to achieve project objective and are identified as the construction proceeds. Prompt identification of such requirements helps both the owner and contractor to avoid unnecessary disruption of work and its impact on cost and time. Refer Figure 7.39 RFI form which contractor submits to the consultant to clarify differences/errors observed in the contract documents, change in construction methodology, and change in the specified material.

These queries are normally resolved by the concerned supervision engineer. However, it is likely that the matter has to be referred to the designer, as RFI has many other considerations to be taken care that may be beyond the capacity of supervision team member to resolve. Such queries may result in variation to the contract documents. Figure 7.40 illustrates flow diagram for processing of RFI. It is in the interest of both the owner and contractor to resolve RFI expeditiously to avoid its effect on construction schedule. A site works instruction (SWI) is issued to the contractor, which gives instruction to the contractor to proceed with the change(s) (see Figure 7.43, which is a sample SWI form). All the necessary documents are sent along with the SWI to the contractor. The SWI is also used to instruct contractor for owner-initiated changes. Subsequently, the contractor submits variation order proposal form to the owner/consultant for approval of change(s) in the contract (see Figure 7.41).

Similarly, if the contractor requires any modification to the specified method, then the contractor submits a request for modification to the owner/consultant. Figure 7.44 illustrates the request for modification. Usually these modifications are carried out, by the contractor, without any extra cost and time obligation toward the contract.

It is the normal practice that, for the benefit of project, the engineer's representative assesses the cost and time related to SWI or request for change over and obtain preliminary approval from the owner and the contractor is asked to proceed with such changes. The cost and time implementation is negotiated and formalized simultaneously/later to issue the formal variation order. In all the circumstances, where a change in contract is necessary owner approval has to be obtained. Figure 7.46 illustrates the form used by the engineer's representative to obtain change order approval from owner.

Once cost and time implications are negotiated and finalized and both the owner and contractor approve the same, variation order is issued to the contractor and changes are adjusted with contract sum and schedule. Figure 7.47a illustrates variation order form issued to formalize the change order, and Figure 7.47b illustrates the attachment to change order.

Any change initiated by the owner is resolved as per the process discussed under Section 7.3.3.6 and illustrated by Figure 7.45.

Resolve conflict: It is essential that changes in the project are managed as quickly as possible in accordance with the conditions of contract. However, the disputes and conflicts in the construction projects are inevitable due to the fact that with all the precautionary steps, the discrepancies or errors do occur in the contract documents. The following methods are normally followed to resolve the conflict:

- *Negotiation*: The economical method to resolve the conflict is negotiation. Negotiation involves compromise. Both parties should discuss the issue by arranging the meetings of all the project team members involved and whose input to the issue will help resolve the conflict. The issue is to be analyzed with the help of related documents, substantiation for claim, and justification for claim. If agreement is not reached, then involve senior representatives from both the parties.
- *Mediation*: Mediation is a process in which a neutral third party group is involved to assist the parties to a dispute in reaching an amicable agreement that resolves the conflict.
- *Arbitration*: Arbitration is the voluntary submission of a dispute to one or more impartial persons for final and binding determination. There are certain agencies that certify arbitrators.
- *Litigation*: Litigation means to apply to the court to resolve the dispute.

7.4.6.12.4.3 Manage Construction Quality The construction project quality control process is a part of contract documents that provides details about specific quality practices, resources, and activities relevant to the project. The purpose of quality control during construction is to ensure that the work is accomplished in accordance with the requirements specified in the contract. Inspection of construction works is carried out throughout the construction period either by the construction supervision team (consultant) or appointed inspector agency. Quality is an important aspect of the construction project. The quality of construction project must meet the requirements specified in the contact documents. Normally, the contractor provides onsite inspection and testing facilities at the construction site. On a construction site, inspection and testing is carried out at three stages during the construction period to ensure quality compliance.

1. During construction process: This is carried with the checklist request submitted by the contractor for testing of ongoing works before proceeding to next step.
2. Receipt of material, equipment or services: This is performed by a material inspection request submitted by the contractor to the consultant upon receipt of material.
3. Before final delivery or commissioning and handover.

Quality management in construction is a management function. In general, quality assurance and control programs are used to monitor design and construction conformance to established requirements as determined by the contract specifications. Instituting quality management programs reduce costs, while producing the specified facility. CQCP discussed in the Appendix is followed throughout construction project. Table 7.84 illustrates the contractor's responsibilities to manage construction quality.

7.4.6.12.4.4 Manage Construction Resources The success of construction project depends largely on availability, performance, and utilization of resources. In the construction project, the resources are linked with duration of project and each activity is allocated a specific resource to be available at the specific time. The construction resource mainly consists of

1. Construction workforce
   i. Contractor's own staff and workers
   ii. Subcontractor's staff and workers
2. Construction equipment, machinery
3. Construction material, equipment, and systems to be installed on the project

In most construction projects, the contractor is responsible to engage all types of human resources to complete the project, subcontractors, specialist installers, suppliers, arrange equipment, construction tools, and materials as per contract documents and to the satisfaction of owner/owner's appointed supervision team. Workmanship is one of the most important factors to achieve the quality in construction; therefore, it is required that the construction workforce is fully trained and have full knowledge of all the related activities to be performed during the construction process.

Construction workforce: Once the contract is awarded, the contractor prepares a detailed plan for all the resources he or she needs to complete the project. Contract documents normally specify a list of minimum number of core staff to be available at site during construction period. Absence of any of these staff may result in penalty to be imposed on the contractor by the owner.

A typical list of the contractor's minimum core staff needed during the construction period for execution of work of a major building construction project is discussed in Section 7.4.6.3.3.

The contractor's human resources mainly consists of two categories:

1. Contractor's own staff and workers
2. Subcontractor's staff and workers

The human resources required to complete the projects are based on resource loading program. It is necessary that all the construction resources are coordinated and brought together at the right time in order to complete on time and within budget.

The main contractor has to manage all these personnel by

1. Assigning the daily activities
2. Observing their performance and work output (productivity)
3. Daily attendance
4. Safety during the construction process

TABLE 7.84
Contractor's Responsibilities to Manage Construction Quality

Sr. No.	Activity	Areas of Quality Control							
		Main Contractor		Subcontractors					
		Head Office/ Quality Manager	Project Site/ Project Manager	Structural	Interior	Mechanical (HVAC+PHFF)	Electrical	Landscape	External
1	Prepare quality control plan	□	■	□	□	□	□	□	□
2	Construction schedule	□	■	□	□	□	□	□	□
3	Mobilization	□	■	□	□	□	□	□	□
4	Staff approval	□	■	□	□	□	□	□	□
5	Prepare material submittal		■	■	■	■	■	■	■
6	Submit material transmittal		■	□	□	□	□	□	□
7	Prepare shop drawings		■	■	■	■	■	■	■
8	Submit shop drawing transmittal		■	□	□	□	□	□	□
9	Material sample	□	■	■	■	■	■	■	■
10	Receiving material inspection		■	■	■	■	■	■	■
11	Material testing		■	■	■	■	■	■	■
12	Mock up		■	■	■	■	■	■	■
13	Site work inspection		■	■	■	■	■	■	■
14	Quality of work		■	■	■	■	■	■	■
15	Prepare checklist		■	■	■	■	■	■	■
16	Submit checklist		■	□	□	□	□	□	□
17	Corrective/preventive action		■	■	■	■	■	■	■
18	Daily report		■	□	□	□	□	□	□
19	Monthly progress report		■	□	□	□	□	□	□
20	Progress payment	□	■	□	□	□	□	□	□
21	Site safety		■	■	■	■	■	■	■
22	Safety report		■	□	□	□	□	□	□
23	Waste disposal		■	■	■	■	■	■	■

(Continued)

TABLE 7.84 (*Continued*)

Contractor's Responsibilities to Manage Construction Quality

		Areas of Quality Control							
		Main Contractor		Subcontractors					
Sr. No.	Activity	Head Office/ Quality Manager	Project Site/ Project Manager	Structural	Interior	Mechanical (HVAC+PHFF)	Electrical	Landscape	External
24	Reply to job site instruction		■	□	□	□	□	□	□
25	Reply to nonconformance report		■	□	□	□	□	□	□
26	Documentation		■	□	□	□	□	□	□
27	Testing and commissioning		■	■	■	■	■	■	■
28	Project closeout documents	□	■	□	□	□	□	□	□
29	Punch list		■	■	■	■	■	■	■
30	Request for issuance of substantial completion letter	□	■	□	□	□	□	□	□

Source: Abdul Razzak Rumane (2013), *Quality Tools for Managing Construction Projects*, CRC Press, Boca Raton, FL. Reprinted with permission from Taylor & Francis Group.

■ Primary responsibility
□ Advise/assist

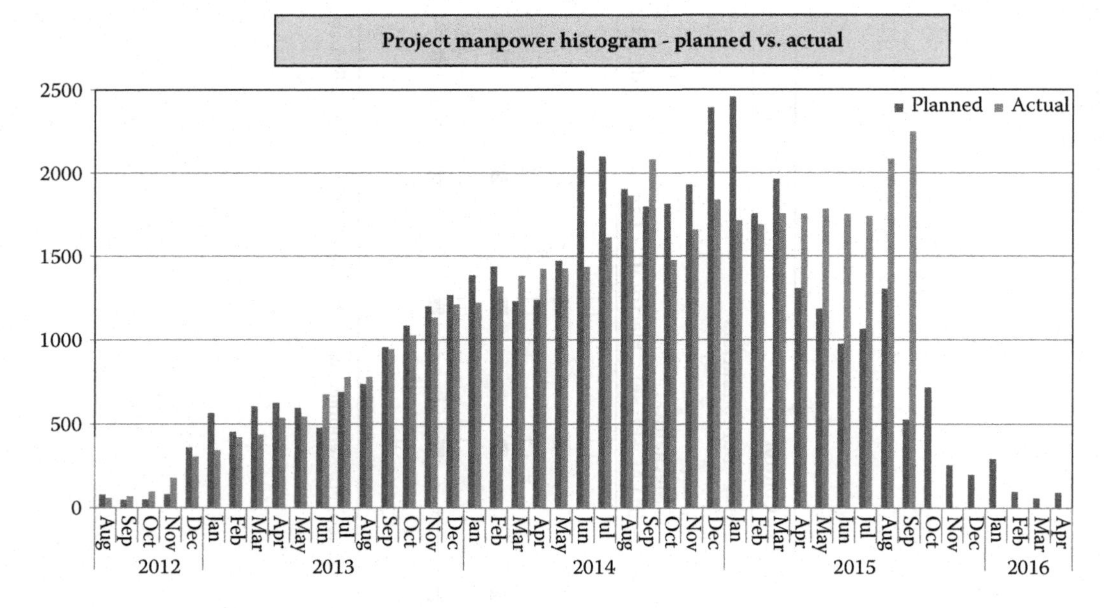

FIGURE 7.141
Contractor's manpower chart.

Figure 7.141 illustrates the contractor's actual manpower chart of a construction project.

Subcontractor: In most construction projects, main contractor engages subcontractors, specialist contractors, to execute certain portion of the contracted project work. The main contractor has to monitor the performance of the subcontractor throughout the project to ensure project success. In order to achieve project objectives, it is essential that main contractor and subcontractor maintain partnering relationship. In order for smooth execution of project, both parties should have cooperative and collaborative, with joint problem-solving attitude. Their aim should be to achieve successful project and maintain long-term business relationship.

The main contractor–subcontractor relationship starts the moment subcontractor is selected and a contract/agreement is signed to execute the project. It is important that necessary precautions and care are taken to prequalify and select the subcontractor. Table 7.78 lists questionnaires to prequalify the subcontractor.

In certain cases, the subcontractor is involved with the main contractor from the tendering stage. The main contractor takes into consideration the prices quoted by the subcontractor while submitting the proposal. The contractor price between main contractor and subcontractor can be

- Back to back on the main contractor prices keeping agreed upon margin for the main contractor.
- Negotiated prices with subcontractor in which case the main contractor's contract awarded prices are not known to the subcontractor.

In order for successful execution of the project, it is necessary to have subcontractor management plan in place to ensure that each of the subcontractors execute the project as specified without affecting the project quality and schedule.

The following is the typical contents of a subcontractor management plan:

a. Introduction
b. Organization
    i. Organizational chart
    ii. Roles
    iii. Responsibilities
c. Scope of work
d. Quality management plan
e. Project coordination method
f. Submittals
    i. Material
    ii. Shop drawings
g. Resource management
    i. Training
h. Change management
i. Communication
    i. Meetings
j. Risk management
k. HSE management
l. Documentation
m. Invoicing, payments
n. Closeout contract

Construction equipment: Likewise, the contract documents specify that minimum no. of equipment are to be available at site during construction process to ensure smooth operation of all the construction activities. Figure 7.142 illustrates equipment list and utilization schedule for a major building construction project.

Construction material: The contractor also prepares a procurement log based on the project completion schedule. In most construction projects, the contractor is responsible for procurement of material, equipment, and systems to be installed on the project. The contractors have their own procurement strategies. While submitting the bid, the contractor obtains the quotations from various suppliers/subcontractors. Figure 7.143 illustrates material procurement procedure to be followed by the contractor.

The contractor has to ensure that construction material is available at project site on time to avoid any delay. Delivery of long lead items have to be initiated at an early stage of the project and monitored closely. Late order placement for materials, results in delayed delivery of material, which in turn affects the timely completion of the project, is a common scenario in the construction projects. Hence, these logs have to be updated regularly and prompt actions have to be taken, to avoid delays. The contractor is required to provide twice a month or at any time requested by the owner/consultant, full and complete details of all products/systems procurement data relating to all the approved products, systems that have ordered and/or procured by the contractor for using in the construction project. The contractor maintains the contractor's procurement log E-2 (see Figure 7.71) to

Sr. No.	Item Description	Unit	2012 Aug	Sep	Oct	Nov	Dec	2013 Jan	Feb	Mar	Apr	May	Jun	Jul	Aug	Sep	Oct	Nov	Dec	2014 Jan	Feb	Mar	Apr	May	Jun	Jul	Aug	Sep	Oct	Nov	Dec	2015 Jan	Feb	Mar	Apr	May	Jun	Jul	Aug	Sep	Oct	Nov	Dec
1	Tower crane	Nos.	–	–	4	9	9	9	9	9	9	9	9	9	9	9	9	9	9	9	9	9	9	8	8	8	8	8	8	8	4	4	–	–	–	–	–	–	–	–	–	–	–
2	Construction hoist	Nos.	–	–	–	–	–	–	–	–	–	–	–	–	–	–	–	–	–	4	9	9	9	9	9	9	9	9	9	9	9	9	9	9	9	9	9	9	9	9	9	9	9
3	Mobile crane	Nos.	–	–	1	1	2	2	2	4	4	4	4	4	4	4	4	4	4	4	4	4	4	4	4	4	4	4	4	4	4	4	4	4	4	4	4	4	4	4	4	4	4
4	Bulldozer	Nos.	–	3	4	4	2	2	2	–	–	–	–	–	–	–	–	–	–	–	–	–	–	–	–	–	–	–	–	–	–	–	–	–	–	–	–	–	–	–	–	–	–
5	Dump truck	Nos.	–	4	4	4	4	4	4	–	–	–	–	–	–	–	–	–	–	4	4	4	–	–	–	–	–	–	–	–	–	–	–	–	–	–	–	–	–	–	–	–	–
6	Excavator	Nos.	–	2	2	2	2	2	2	2	2	–	–	–	–	2	–	–	–	–	–	–	–	–	–	–	–	–	–	–	–	–	–	–	–	–	–	–	–	–	–	–	–
7	JCB	Nos.	–	1	4	4	4	2	2	–	–	–	–	–	–	–	–	–	–	2	2	2	–	–	–	–	–	–	–	–	–	–	–	–	–	–	–	–	–	–	–	–	–
8	Concrete pump	Nos.	–	–	2	6	6	6	6	4	4	4	4	4	4	4	4	4	4	4	4	4	4	4	4	4	4	4	4	4	4	4	–	–	–	–	–	–	–	–	–	–	–
9	Transit mixer	Nos.	–	–	10	30	30	30	30	30	15	15	15	15	15	15	15	15	15	15	15	15	15	15	15	15	15	15	15	15	15	15	–	–	–	–	–	–	–	–	–	–	–
10	Fork lift	Nos.	–	–	2	2	2	2	2	2	2	2	2	2	2	2	2	2	2	2	2	2	2	2	2	2	2	2	2	2	2	2	2	2	2	2	2	2	2	2	2	2	2
11	Trailer	Nos.	–	–	2	4	4	4	4	4	2	2	2	2	2	2	2	2	2	2	2	2	2	2	2	2	2	2	2	2	2	2	2	2	2	2	2	2	2	2	2	2	2
12	Total station	Nos.	2	2	2	2	2	2	2	2	2	2	2	2	2	2	2	2	2	2	2	2	2	2	2	2	2	2	2	2	2	2	2	2	2	2	2	2	2	2	2	2	2
13	Theodolite	Nos.	2	2	2	2	2	2	2	2	2	2	2	2	2	2	2	2	2	2	2	2	2	2	2	2	2	2	2	2	2	2	2	2	2	2	2	2	2	2	2	2	2
14	Level instrument	Nos.	4	4	4	4	4	4	4	4	4	4	4	4	4	4	4	4	4	4	4	4	4	4	4	4	4	4	4	4	4	4	4	4	4	4	4	4	4	4	4	4	4
15	Precision level	Nos.	2	2	2	2	2	2	2	2	2	2	2	2	2	2	2	2	2	2	2	2	2	2	2	2	2	2	2	2	2	2	2	2	2	2	2	2	2	2	2	2	2
16	Bar bending machine	Nos.	–	–	4	6	6	6	6	6	6	6	6	6	6	6	6	6	6	6	6	6	6	6	6	6	6	6	6	6	6	6	–	–	–	–	–	–	–	–	–	–	–
17	Bar cutting machine	Nos.	–	–	4	6	6	6	6	6	6	6	6	6	6	6	6	6	6	6	6	6	6	6	6	6	6	6	6	6	6	6	–	–	–	–	–	–	–	–	–	–	–
18	Bar screw—tread	Nos.	–	–	2	2	2	2	2	2	2	2	2	2	2	2	2	2	2	2	2	2	2	2	2	2	2	2	2	2	2	2	–	–	–	–	–	–	–	–	–	–	–
19	Diesel generator	Nos.	–	–	4	4	4	4	4	4	4	4	4	4	4	4	4	4	4	4	4	4	4	4	4	4	4	4	4	4	4	4	4	4	4	4	4	4	4	4	4	4	4
20	Compactor	Nos.	–	–	2	2	2	2	2	2	2	2	2	2	2	2	2	2	2	4	4	4	4	2	2	2	2	2	2	2	2	2	–	–	–	–	–	–	–	–	–	–	–
21	Crawler compactor	Nos.	–	–	2	2	2	2	2	2	2	2	2	2	2	2	2	2	2	2	2	2	2	2	2	2	2	2	2	2	2	2	–	–	–	–	–	–	–	–	–	–	–
22	High pressure pump	Nos.	–	–	2	2	2	2	2	2	2	2	2	2	2	2	2	2	2	2	2	2	2	2	2	2	2	2	2	2	2	2	2	2	2	2	2	2	2	2	2	2	2
23	Manual hoist	Nos.	–	–	5	5	5	5	5	5	5	5	5	5	5	5	5	5	5	5	5	5	5	5	5	5	5	5	5	5	5	5	5	5	5	5	5	5	5	5	5	5	5
24	Concrete vibrator	Nos.	–	–	25	25	25	25	25	25	25	25	25	25	25	25	25	25	25	25	25	25	25	25	25	25	25	25	25	25	25	25	–	–	–	–	–	–	–	–	–	–	–
25	Mortar mixer	Nos.	–	–	–	–	–	–	–	–	–	–	–	–	6	6	6	6	6	6	6	6	6	6	6	6	6	6	6	6	6	6	6	6	6	6	6	6	6	6	6	6	6
26	Welding machine	Nos.	–	–	–	4	4	4	4	4	4	50	50	50	50	50	50	50	50	50	20	20	20	20	50	50	50	50	50	50	50	20	20	20	20	20	20	20	20	20	20	20	20

FIGURE 7.142
Equipment list and utilization schedule for major buildings project.

(Continued)

Sr. No.	Item Description	Unit	Numbers per Month																																								
			2012					2013												2014												2015											
			Aug	Sep	Oct	Nov	Dec	Jan	Feb	Mar	Apr	May	Jun	Jul	Aug	Sep	Oct	Nov	Dec	Jan	Feb	Mar	Apr	May	Jun	Jul	Aug	Sep	Oct	Nov	Dec	Jan	Feb	Mar	Apr	May	Jun	Jul	Aug	Sep	Oct	Nov	Dec
27	Circular saw	Nos.	–	–	6	6	6	6	6	6	6	6	6	6	6	6	6	6	6	6	6	6	6	6	6	6	6	6	6	6	6	6	6	6	6	6	6	6	6	6	6	6	6
28	Pipe painting machine	Nos.	–	–	–	–	–	–	–	–	–	–	2	2	2	2	2	2	2	2	2	2	2	2	2	2	2	2	2	2	2	2	2	2	2	2	2	2	–	–	–	–	–
29	Main distribution panel	Nos.	–	–	2	2	2	2	2	2	2	2	2	2	2	2	2	2	2	2	2	2	2	2	2	2	2	2	2	2	2	2	2	2	2	2	2	2	2	2	2	2	2
30	Duct fabrication machine	Nos.	–	–	–	–	–	–	–	–	–	–	–	–	–	2	2	2	2	2	2	2	2	2	2	2	2	2	2	2	2	2	2	2	2	2	2	2	–	–	–	–	–
31	Electric hammer	Nos.	–	–	4	4	4	4	4	4	4	4	4	4	4	12	12	12	12	12	12	12	12	12	12	12	12	12	12	12	12	12	12	12	12	12	12	12	12	12	12	12	12
32	Power float	Nos.	–	–	16	16	16	16	16	16	16	16	16	16	16	16	16	16	16	16	16	16	16	16	16	16	16	16	16	16	16	16	16	16	16	16	16	16	–	–	–	–	–
33	Angle grinder	Nos.	–	–	–	–	4	4	4	4	4	4	4	4	4	4	4	4	4	4	4	4	4	4	4	4	4	4	4	4	4	4	4	4	4	4	4	4	4	4	4	4	4
34	Electric hand drill	Nos.	–	–	12	12	12	12	12	12	12	12	12	12	12	12	12	12	12	12	12	12	12	12	12	12	12	12	12	12	12	12	12	12	12	12	12	12	12	12	12	12	12
35	Lighting equipment	Nos.	–	–	5	5	10	10	10	10	10	10	10	10	10	10	10	10	10	10	10	10	10	10	10	10	20	20	20	20	20	20	20	20	20	20	20	20	20	20	20	20	20
36	Air jack	Nos.	–	–	–	–	–	–	2	2	2	2	2	2	2	6	6	6	6	6	6	6	6	6	6	6	6	6	6	6	6	6	6	6	6	6	6	6	6	6	6	6	6
37	Angle saw	Nos.	–	–	–	6	6	6	6	6	6	6	6	6	6	6	6	6	6	6	6	6	6	6	6	6	6	6	6	6	6	6	6	6	6	6	6	6	6	6	6	6	6

FIGURE 7.142 (Continued)
Equipment list and utilization schedule for major buildings project.

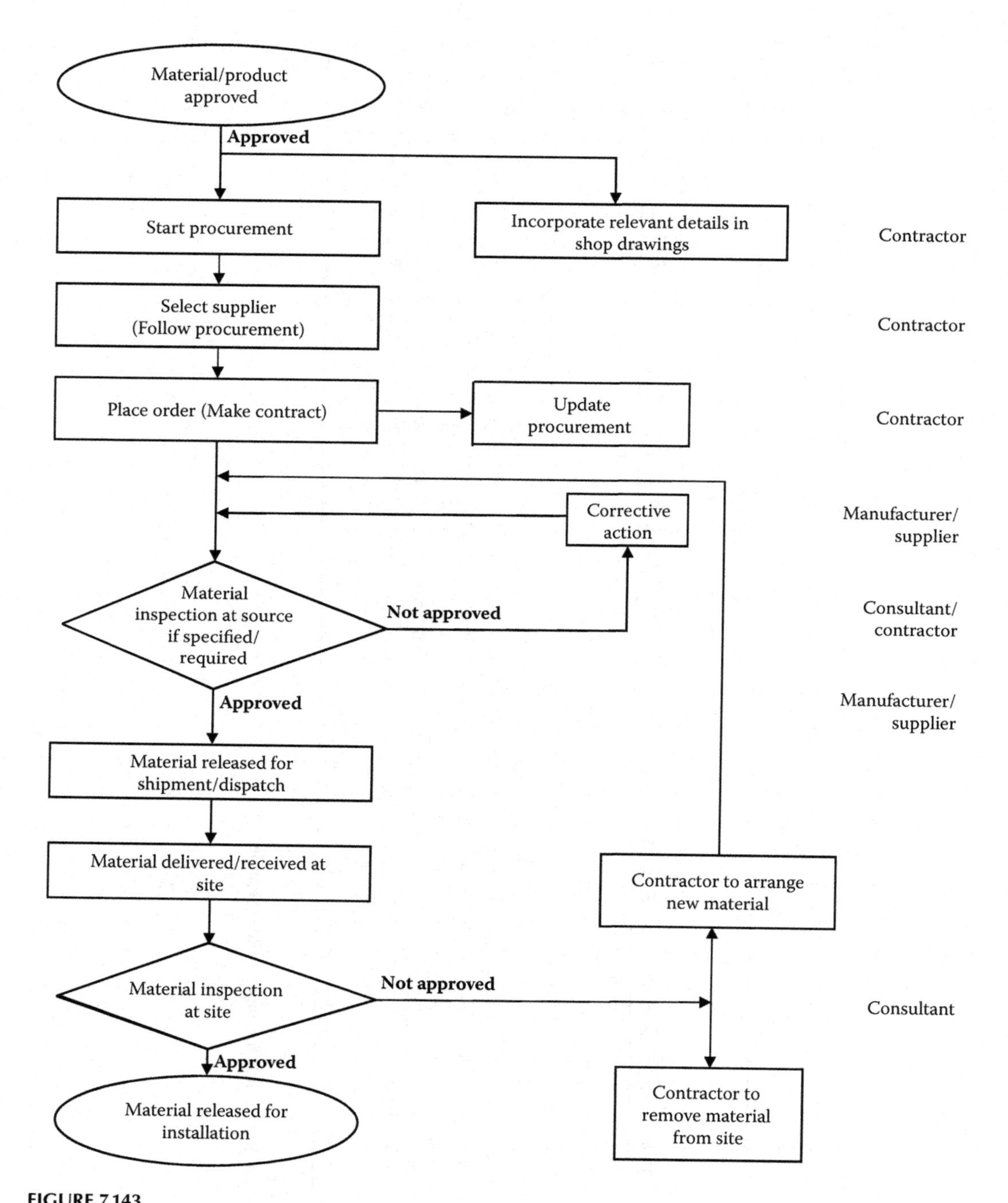

FIGURE 7.143
Material procurement procedure. (Abdul Razzak Rumane (2013), *Quality Tools for Managing Construction Projects*, CRC Press, Boca Raton, FL. Reprinted with permission from Taylor & Francis Group.)

keep track of material status. The log E-2 is normally submitted by the contractor along with monthly progress report.

Supply chain management: Supply chain management in construction project is managing and optimizing the flow of construction materials, systems, equipment, and resources to ensure timely availability of all the construction resources without affecting the progress of works at the site. Figure 7.144 illustrates supply chain management process in construction projects.

In construction projects, the supply chain management starts from the inception of project. The designer has to consider following while specifying the products (materials, systems, and equipment) for use/installation in the project:

- Quality management system followed by the manufacturer/supplier
- Quality of product
- Reliability of product
- Reliability of manufacturer/supplier
- Durability of product
- Availability of product for entire project requirement
- Price economy/cost efficient
- Sustainability
- Conformance to applicable codes and standards
- Manufacturing time
- Location of the manufacturer/supplier from the project site
- Interchangeability
- Avoid monopolistic product

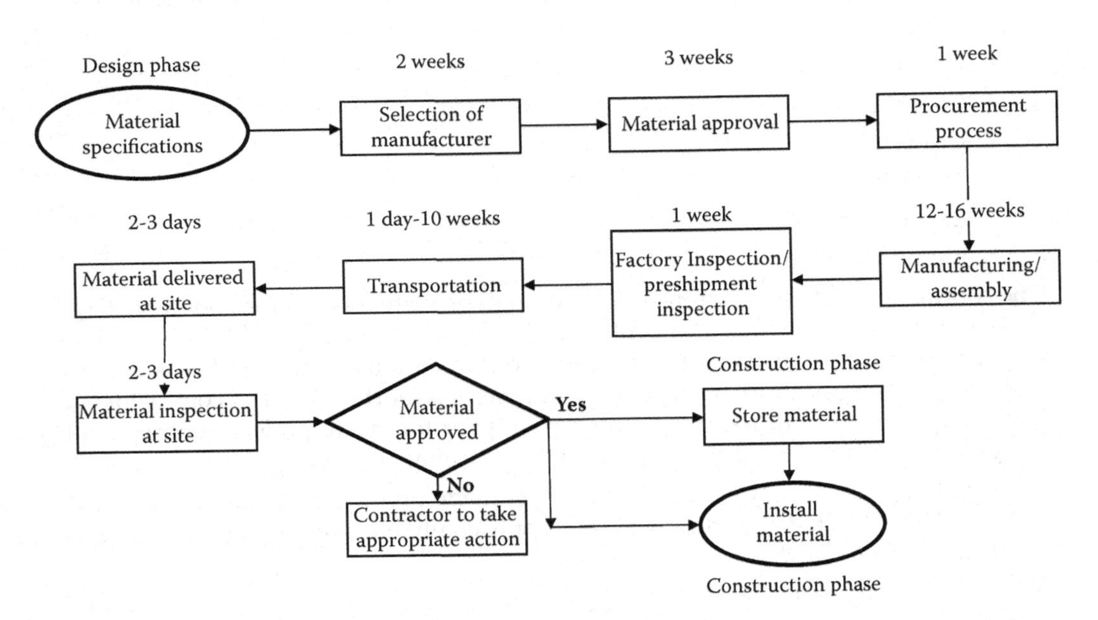

FIGURE 7.144
Supply chain process in construction project.

Product specifications are documented in the construction documents (particular specifications).In certain projects, the documents lists the names of recommended manufacturers/suppliers. However, in order of continuous and uninterrupted supply of specified product, the contractor has to consider the following:

- Quality management system followed by the manufacturer/supplier
- Historical rejection/acceptance record
- Reliability of the manufacturer/supplier
- Product certification
- Financial stability
- Proximity to the project site
- Manufacturing/lead time
- Availability of product as per the activity installation/execution schedule
- Manufacturing capacity
- Availability of quantity to meet the project requirements
- Timeliness of delivery
- Location of manufacturer/supplier
- Product cost
- Transportation cost
- Product certification
- Risks in delivery of product
- Responsiveness
- Cooperative and collaborative nature to resolve problem

In major construction projects, the following items are required in bulk quantities:

- Concrete
- Concrete block
- Conduit
- Utility pipes
- Light fixtures
- Electrical devices
- Plumbing fixtures

The contractor can follow just-in-time method to procure these items and avoid large inventory at project site. The contractor has to select reliable manufacturer/supplier for these products and to ensure that the manufacturer/supplier is capable of maintaining continuous flow of these items at a short notice. The contractor can sign agreement with the manufacturer/supplier for entire project quantity with agreed upon delivery schedule as per the requirement at site.

In order to ensure supply chain payments are made promptly, the cash flow system should be projected accordingly as the supply chain may affect due to interruption in payments toward supply of products.

Manage communication: For smooth implementation of project, proper communication system is established by clearly identifying the submission process for correspondence, submittals, minutes of meeting, and reports.

Correspondence: Normally all the correspondence between contractor and consultant is through submittal transmittal form (see Figure 7.79). Correspondence between consultant and contractor is normally done job site instructions whereas correspondence between owner, consultant, and contractor is normally done through letters. Figure 7.131 is a sample job site instruction form used by the consultant to communicate with the contractor.

Submittals: Contract documents specify number of copies to be submitted to various stakeholders (see Figure 7.78 for submittal process). Table 7.30 lists the different types of forms used in construction projects.

Meetings: There are various type of meetings conducted during the execution of project. These meetings are conducted at an agreed upon frequency. The meetings during construction phase are held for specific reasons. For example

a. Kick-off meeting is held to acquaint project team members and discuss project objectives, procedures, and other contract information.
b. Progress meetings to review work progress.
c. Coordination meetings to coordinate among different disciplines and resolve the issues.
d. Quality meetings to discuss onsite quality issues and improvements to the construction process.
e. Safety meetings to discuss site safety and environmental issues.

Table 7.85 lists points to be discussed during safety meeting.

TABLE 7.85

Points to Be Reviewed during Monthly Safety Meeting

Sr. No.	Points to Be Reviewed	Yes/No
1	Whether meetings are held as planned and attended by all the concerned persons	
2	Whether all the items from the previous meetings addressed and appropriate actions taken	
3	Whether all the accidents/incidents/near misses recorded and reviewed to identify common issues and learning points	
4	Whether training needs identified	
5	Whether records of training programs maintained	
6	Whether actions/feedback from weekly meetings reviewed and appropriate action taken	
7	Whether suggestions/comments from safety tours are implemented	
8	Whether safety awareness programs are regularly held	
9	Whether safety warning signs are displayed and operative	
10	Whether sirens and alarm bells are functioning properly	
11	Whether temporary firefighting system in action and all the equipment are updated	
12	Whether regular check and audits are performed	
13	Any latest regulation is introduced by authority and informed to all concerned	
14	Was there any visit by competent authority and their observations are taken care	
15	Whether escape route and assembly points are displayed	
16	All the workforce have personal protective equipment	
17	Whether first aid box has all the required medicines	

Apart from the above-discussed meeting, the contractor, consultant, and owner can call for a meeting to discuss any project relevant issue or information. Prior to conducting any meeting, an agenda is circulated to stakeholders and team members who attend the meeting. Figure 7.82 illustrates a sample agenda format for meeting. The proceedings of meeting are recorded and minutes of meeting is circulated among all attendees and others per the approved form for minutes of meeting. Figure 7.145 illustrates minutes of meeting format.

PROJECT NAME
Consultant Name

MINUTES OF MEETING

Contract No. : ________________ **Date** : ________

Contractor : ________________________________

____

Meeting Type		MOM No.	
		Date	
Meeting Location		Time	

ITEM	DESCRIPTION OF DISCUSSION	STATUS	PRIORITY	ACTION			
				By	Due	Started on	Closed on

Prepared by:

Distribution Owner Contractor A/E

FIGURE 7.145
Minutes of meeting.

Reports: During the construction phase, the following reports are submitted by the contractor:

- Progress reports such as
  - Daily report
  - Weekly report
  - Monthly progress report
- Safety report
- Risk report

7.4.6.12.4.6 Manage Risks Typical categories of risks in construction projects are listed in Table 7.35. During the construction project, the assigned project team members have to identify the risk, estimate likely occurrence of risk, and develop response plan to mitigate the risk. Table 7.86 lists probable risks that occur during the construction phase and its effects on scope, schedule, and cost.

TABLE 7.86

Potential Risks on Scope, Schedule, and Cost, during Construction Phase and Its Effects and Mitigation Action

Sr. No.	Potential Risk	Probable Effects	Control Measures/Mitigation Action
1.0 Scope			
1.1	Scope/design changes	• Project schedule • Project cost • Claim	• Compress schedule • Resolve change order issues in order not to delay the project
1.2	Different site conditions to the information provided	• Change in scope of work • Delay in project	• Contractor to investigate site conditions prior to starting the relevant activity
1.3	Inadequate site investigation data	• Additional work • Scope change	• Contractor to investigate site conditions prior to starting the relevant activity
1.4	Conflict in contract documents	• Project delay	• Amicably resolve the issue
1.5	Incomplete design	• Project scope • Project schedule • Project cost	• Raise RFI • Resolve issue in accordance with contract documents
1.6	Incomplete scope of work	• Project scope • Project schedule • Project cost	• Raise RFI • Resolve issue in accordance with contract documents
1.7	Design changes	• Project scope • Project schedule • Project cost	• Follow contract documents for change order
1.8	Design mistakes	• Project scope • Project schedule • Project cost	• Raise RFI • Resolve issue in accordance with contract documents
1.9	Errors and omissions in contract documents	• Project scope • Project schedule • Project cost	• Raise RFI • Resolve issue in accordance with contract documents
1.10	Incomplete specifications	• Project scope • Project schedule • Project cost	• Raise RFI • Resolve issue in accordance with contract documents

(*Continued*)

TABLE 7.86 (*Continued*)

Potential Risks on Scope, Schedule, and Cost during Construction Phase and Its Effects and Mitigation Action

Sr. No.	Potential Risk	Probable Effects	Control Measures/Mitigation Action
1.11	Conflict with different trades	• Project delay	• Coordinate with all trades while preparing coordination and composite drawings
1.12	Inappropriate construction method	• Project delay • Claim	• Raise RFI and correct the method statement
1.13	Quality of material	• Project delay	• Locate suppliers having proven record of supplying quality product
2.0 Schedule			
2.1	Incompetent subcontractor	• Project delay • Project quality	• Contractor has to monitor the workmanship and work progress.
2.2	Delay in transfer of site	• Project delay	• Contractor to adjust the construction schedule
2.3	Delay in mobilization	• Project delay	• Adjust construction schedule accordingly
2.4	Project schedule	• Project completion	• Compress duration of activities
2.5	Inappropriate schedule/plan	• Project delay	• Contractor to prepare schedule taking into consideration site conditions all the required parameters
2.6	Delay in changer order negotiations	• Project schedule	• Request owner/supervisor/project manager to expedite the negotiations and resolve the issue
2.7	Resource availability (material)	• Project delay	• Contractor to make extensive search
2.8	Resource (labor) low productivity	• Project quality • Project delay	• Contractor to engage competent and skilled labors
2.9	Equipment/plant productivity	• Project delay	• Contractor hire/purchase equipment to meet project productivity requirements
2.10	Insufficient skilled workforce	• Project duration	• Contractor arrange workforce from alternate sources
2.11	Failure/delay of machinery and equipment	• Project delay	• Contractor to plan procurement well in advance
2.12	Failure/delay of material delivery	• Project delay	• Contractor to plan procurement well in advance
2.13	Delay in approval of submittals	• Project delay	• Notify owner/project manager
2.14	Delays in payment	• Project delay • Claim	• Contractor to have contingency plans • Owner to pay as per contract
2.15	Statutory/regulatory delay	• Project delay	• Regular follow up by the contractor and owner with the regulatory agency
3.0 Cost			
3.1	Low bid project cost	• Project quality	• Contractor to try competitive material, improve method statement and higher production rate from its manpower

(*Continued*)

TABLE 7.86 (*Continued*)

Potential Risks on Scope, Schedule, and Cost during Construction Phase and Its Effects and Mitigation Action

Sr. No.	Potential Risk	Probable Effects	Control Measures/Mitigation Action
3.2	Variation in construction material price	• Project quality • Project cost	• Contractor to negotiate with supplier/manufacturer for best price. Contractor to request for change order if applicable as per contract
3.3	Damage to equipment	• Schedule	• Regularly maintain the equipment. Take immediate action to repair damage equipment
3.4	Damage to stored material	• Project delay • Material quality	• Contractor to follow proper storage system
3.5	Structure collapse	• Injuries • Project delays	• Contractor to ensure that formwork and scaffolding is properly installed
3.6	Leakage of hazardous material	• Safety hazards	• Contractor to take necessary protect to avoid leakage. Store in safe area
4.0 General			
4.1	Failure of team members not performing as expected	• Project quality • Project delay	• Select competent candidate. Provide training
4.2	Change in laws and regulations	• Scope/specification changes • Variation order	• Contractor to inform owner/consultant and raise RFI
4.3	Access to worksite	• Extra/additional time to access site	• Access road to be planned in coordination with adjacent area and local authority
4.4	Theft at site	• Project delay	• Contractor to monitor access to site. Record entry/exit to the site.
4.5	Fire at site	• Project delay	• Contractor to install temporary fire fighting system. Inflammable material to be stored in safe and secured place with necessary safety measures
4.6	Injuries	• Project delay	• Contractor to keep first aid provision at site. Take immediate action to provide medical aid
4.7	New technology	• Scope change • Schedule • Cost	• Owner/contractor to mutually agree for changes in the contract for better performance of project

7.4.6.12.4.7 Manage Contracts Contract management during the construction phase is an organizational method, process, and procedure to manage all contract agreements involved between the owner, contractor, subcontractor, manufacturers, and suppliers. During the construction phase, contracts are managed mainly by the following parties who are directly involved for the execution of project:

- Consultant/construction (project) manager
- Contractor

Apart from these two parties, subcontractor and vendors also have their contract management system.

The contract management process starts once the contract is signed. The consultant is responsible to manage the contract on behalf of the owner. The consultant monitors the scope, schedule, cost, and quality of the construction to ensure that contract conditions are met. The contractor is responsible to ensure that all project works are executed within the agreed upon time and cost in accordance with the contract conditions and specification.

For successful contract management, the contractor as well as the consultant/CM/PM has to consider the following points while executing the project:

1. Use of RFI to get clarification some aspects of the project. There are two parts in RFI. These are as follows:
   i. *Question* by the contractor
   ii. *Answer* by the owner (consultant)
2. Executing project works using specified and approved materials, equipment, and systems.
3. Developing project execution plan considering realistic duration for each activity.
4. Executing contracted works in a timely manner in accordance with agreed upon schedule.
5. Dealing variations to the specified product, method, work in accordance with related specification, contract clauses and by providing substantiation and justifications that has resulted proposing alternative or substitute material.
6. Managing errors, omissions, and additions strictly in accordance with contract terms and avoiding any delays to the project.
7. Conducting meetings to monitor progress and clarify prevailing project issues.
8. Cooperating with all team members to fulfill their contractual obligations.
9. Resolving disputes in an amicable way by adopting cooperative approach.
10. Providing resources to ensure timely availability of competent workforce as per resource schedule.
11. Taking action on all the transmittals within agreed upon period.
12. Timely reply to all correspondences and queries.
13. Communicating issues and problems well in advance.
14. Not to ignore problems/issues with the hope that they might go away.
15. Arranging payment of monthly progress payment as per contractual entitlement within stipulated time.
16. Maintaining list of claims on monthly basis.
17. Settling claims in an accordance with contract terms.
18. Maintaining proper logs and records.

7.4.6.12.4.8 Manage Site Safety and Protect Environment The construction industry has been considered as dangerous for long time. The nature of works at site alwa ys presents some dangers and hazards. There is relatively high number of injuries and accidents at construction sites. Safety represents important aspect of construction projects. Every project manager tries to ensure that the project is completed without major accident on the site.

The construction site should be a safe place for those who are working at sites. Necessary measures are always required to ensure safety of all those working at the construction site. Effective risk control strategies are necessary to reduce and prevent accidents.

Contract documents normally stipulate that the contractor, upon signing of contract, has to submit safety and accident prevention program. It emphasizes that all the personnel have to put efforts to prevent injuries and accidents. In the program, the contractor has to incorporate requirements of safety and health requirements of local authorities, manuals of accident prevention in construction, and all other local codes and regulations. The contractor has to also prepare emergency evacuation plan (EEP). The EEP is required to protect personnel and to reduce the number of fatalities in case of major accidents at site. The evacuation routes have to be displayed at various locations in a manner. Transfer points and gathering points have to be designated, and sign boards have to be displayed all the time. Evacuation sirens to be sounded on regular basis in order to ensure smooth functioning of evacuation plan.

A safety violation notice is issued to the contractor/employee if the contractor or any of his employees are not complying with safety requirements. Figure 7.146 illustrates safety violation notice, which is to be actioned by the contractor.

Penalties are also imposed on the contractor for noncompliance with the site safety program. Figure 7.147 illustrates sample disciplinary notice form for breach of safety rules. Different colors of card may be issued along with the notice. Table 7.87 illustrates concepts of issuance of different colors of card.

Penalties are also imposed on contractor for noncompliance with the site safety program. The safety program shall embody the prevention of accidents, injury, occupational illness and property damage. The contract specifies that a safety officer is engaged by the contractor to follow safety measures. The safety officer is normally responsible for

1. Conducting safety meetings
2. Monitor on-the-job safety
3. Inspect the works and identify hazardous area
4. Initiate safety awareness program
5. To ensure availability of first aid and emergency medical services as per local code and regulations
6. To insure that the personnel are using protective equipment such as hard hat, safety shoes, protective clothing, life belt, and protective eye coverings
7. To ensure that temporary fire fighting system is working
8. To ensure that work areas and access are free from trash and hazardous material
9. Housekeeping

Construction sites have many hazards that can cause serious injuries and accidents. The contractor has to identify these areas and ensure that all site personnel and subcontractor employees working at site are aware of unsafe act, potential, and actual hazards and the immediate corrective action to be taken and adhere to safety plan

Project Name
Consultant Name

Contract No.:
Contractor:

SVN No.
Date :
Time:

SAFETY VIOLATION REPORT

SAFETY RELATED ITEMS

Sr.No.	Description	Sr.No.	Description
1	Access Facilities	14	Hygieninc
2	Barricade/Railing	15	Poor lighting
3	Construction Equipment	16	Protective Equipment
4	Crane	17	Lifting Gears
5	Earthwork/Excavation	18	Poor lighting
6	Electrical	19	Protective Equipment
7	Fire Fighting/Protection	20	Safety Gears
8	First Aid	21	Scaffolding
9	Formwork	22	Site Fencing
10	Hand and Power Tools	23	Storage Facilities
11	Hazars/Imflamable Material	24	Vehicles
12	Hoist	25	Welding/Hot Work
13	House Keeping	26	Others

VIOLATION DESCRIPTION **Action code:** ☐

Item No.	Location	Description

ORIGINATOR: **RESIDENT ENGINEER:**

CONTRACTOR'S ACTION

Item No.	Location	Action	Date	Time

SAFETY OFFICER: **CONTRACTOR'S PROJECT MANAGER:**

Action Code: A For immediate action /( )hours B Within ( ) days

FIGURE 7.146
Safety violation notice.

PROJECT NAME	
CONSULTANT NAME	
Safety Disciplinary Notice	
Notice No.:	Date:
Name of Employee: Contractor Name: Area/Floor: Date & Time of Observance:	SAMPLE FORM
Type of Notice	
☐	**First/Verbal Warning** (White Card)
☐	**Second/Written Warning** (Yellow Card)
☐	**Suspension from Site** (Red Card)
Reason for Issuance of Notice:	
Action Required by Recipient:	
Date by Which Action is Required:	
Issued by: Signature:	Date:
Reveived by:	Date:

CC: Owner ☐ Resident Engineer ☐ Project Manager ☐

FIGURE 7.147
Safety disciplinary notice.

and procedures. The following is the list of some of the common areas that can cause injuries/accidents at site:

1. Unsafe access
2. Ladders
3. No barricades around excavated areas, trenches, openings, holes, and platforms
4. Nonbarricades/railings on the stairs
5. Scaffolding
6. Lifting gear
7. Crane
8. Hoist

TABLE 7.87
Concept of Safety Disciplinary Action

Sr. No.	Card Color	Type of Disciplinary Action	Warning Validity	Reasons for Disciplinary Action
1	White	Verbal followed by safety discipline notice	One month to three months	1. Failure to use personal protective equipment 2. Failure to used define access 3. Working on plant, crane, vehicle without license 4. Working with unsafe scaffolding 5. Working on unsafe platform 6. Using unsafe sling or ropes for lifting 7. Working on unsafe ladders
2	Yellow	Issuance of safety discipline notice and suspension from the work for rest of the day	Six months	1. Repetition of activities listed under *white card* within one month of issuance of first notice 2. Failure to observe HSE-related instructions 3. Failure to work as per instructed method of work, as per given task 4. Failure to follow storage principles about hazardous materials
3	Red	Issuance of safety disciplinary notice and suspension from site for one month	One year	1. Breach of safety rules where there is risk to life 2. Removal of safety devices, interlocks, guardrails, and barriers without any authority 3. Deliberately exposing public to danger by not complying with agreed safe methods of work 4. Disposal of hazardous material in unsafe area

9. Welding
10. Hand and power tools
11. Poor lighting
12. Fire

The following are general safety guidelines contractor may be followed to avoid accidents/injuries:

1. Safe access
2. Ensure ladders are in good condition and properly secured
3. Provide barricades around openings, holes, and platform
4. Choose right system of scaffolding for the job to be performed
5. Checking lifting gears for capacity, condition of wire rope, slings, hooks, eyebolts, shackles, proprietary lighting equipment, spreader beam

6. Ensure the crane is
    - On a firm, leveled ground and outriggers are fully extended
    - Check safe working load against the load to be lifted
    - Ensure that load swing is minimum
    - Select right type of chain
    - Keep the load clear of personnel
7. Hoist is certified by third party and the certificate is valid
8. Wear all protective equipment and garments necessary to be safe on the job
9. Wear and use eye protection coverings
10. Use safety shoes
11. Use hard hat
12. Use safety belts
13. Use respiratory mask whenever required
14. Use right size of hand tools
15. Check any plant or mechanized equipment is certified for safe operating conditions before using
16. Before using electric tools
    - Check it is properly earthed
    - Cable, plugs, or connectors are sound and properly wired up
17. Use proper guards while using power tools such as circular saws, portable grinders, and bench grinders
18. Use safe loading/unloading techniques
19. All material stored at site to be stacked properly to ensure that it is stable and secured against sliding or collapse
20. Work areas and means of access should be maintained in a safe and orderly condition
21. Mark access and escape route
22. Keep passageway and access way free from materials, supplies, and obstructions all the time
23. Mark all the hazardous areas
24. Prohibit storage of flammable and combustible material
25. No smoking sign to be displayed
26. Install sirens and alarm bells at site
27. Ensure temporary fire fighting system is working all the time
28. All formwork, shoring, and bracing should be designed, fabricated, erected, supported, braced, and maintained so that it will safely support all vertical and lateral loads that might be applied, until such loads can be supported by the structure
29. While placing concrete
    - Make use of protective clothing and equipment
    - Use appropriate gloves

- Use rubber boots
- Wear protective goggles
- Take necessary precautions while using concrete skips or concrete

30. Display safety sign such as *danger, caution,* and *warning* on all live electrical panels
31. Post safety, warning signs, and notices of weekly project safety record
32. Conduct *safety awareness* programs

The contractor is responsible for ongoing maintenance of accident/incident records on the construction site and their notification to the consultant. Figure 7.148 is accident reporting form, and Figure 7.149 is a summary procedure to be followed once an accident takes place at the site (see Figure 4.34 for site safety improvement).

7.4.6.12.4.9 Manage Project Finance In construction projects, maximum amount is expended during the construction phase. During this phase

1. Owner has to make payments to
   a. Main contractor
   b. Supervisor (consultant)
   c. Construction/project manager, if applicable
   d. Specialist consultant
   e. Specialist contractor
   f. Any other party such as direct appointments
   g. Owner supplied items, if any
2. Main contractor has to make payments to
   a. Subcontractors
   b. Suppliers (material procurement)
   c. Designer, if any design work is involved
   d. Workforce
   e. Rent (equipment rent and rental vehicles)
3. Subcontractor has to make payment to
   a. Suppliers
   b. Specialist

Owner's payment is mainly related to progress payment claimed and approved by the consultant, advance payments (if any as per contract) to the contractor, monthly fees to the consultant's construction/project manager.

Contractor and subcontractor's payments are linked to the approved executed works. The management of finance for project is done through project cash flow forecast.

Cash flow:

Cash flow = Cash in-Cash out

Forecasting of cash flow is important for the smooth functioning of contract to ensure that an appropriate level of funding is in place and suitable draw-down facilities are available. Cash flow prediction is normally made with the help of S-curve.

Project Name

ACCIDENT REPORT

CONTRACT : ______________________ CONTRACT NO. : ______________

CONTRACTOR : ______________________ REPORT # : ______________

SUBCONTRACTOR : ______________________ REPORT DATE : ______________

SAMPLE FORM

ACCIDENT DATE			
ACCIDENT TIME			
ACCIDENT LOCATION			
INJURED PERSON		I.D.#	
ADDRESS			AGE
ACCIDENT DETAILS : - CAUSE - PERSONAL INJURY - PROPERTY DAMAGE			
ACCIDENT REPORTED BY		I.D.#	
ACCIDENT REPORTED TO		I.D.#	
WITNESSED BY		I.D.#	
INVESTIGATED BY		I.D.#	
ACTION TAKEN : - MEDICAL AID - FOLLOW UP - LEGAL ACTION			

NOTE : IDENTIFY ENTRY PASS NUMBERS FOR ALL INDIVIDUALS INVOLVED

____
CONTRACTOR REP. SIGNATURE.

FIGURE 7.148
Accident report.

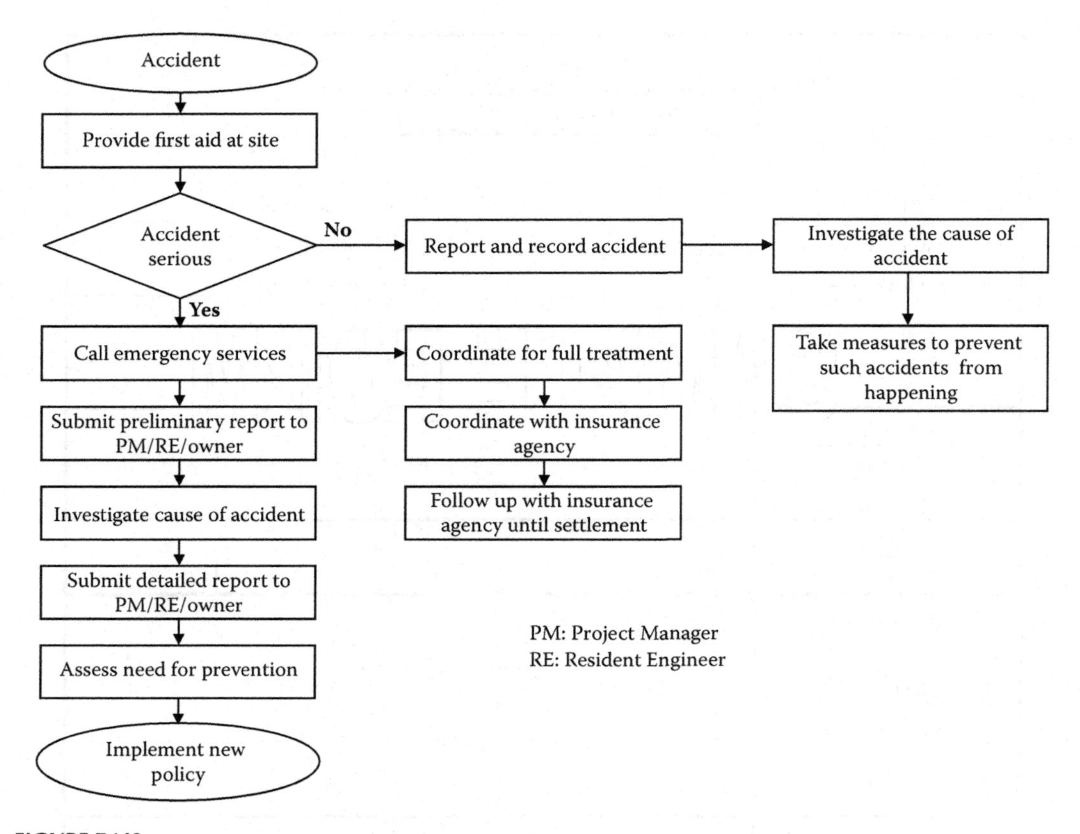

FIGURE 7.149
Summary procedure for actions after accident.

The S-curve stands for *standard* curve but the name has also taken into account the shape of letter *S*. The contractor prepares *S* based on expected work progress on monthly basis.

Forecasting of cash flow is important to

- Ensure that sufficient fund are available to meet the demand during construction to fulfill the commitment/obligations
- Sufficient funds are available when it is required (timing)
- Ensure maximum utilization of funds

In order to prepare the S-curve, accurate information of the following elements is required:

- Total expenses
- Total income
- Timing of payment

The resource loaded construction schedule can be used to calculate cash flow forecast. The S-curve for the owner is different than that of contractor. The owner has to take into consideration the following points while preparing cash flow:

- Errors and omissions
- Effects of change order or variation

- Effect of inflation
- Effect of international exchange rate
- Change in the sequence of work schedule to expedite or to mitigate delay
- Payment toward material at site or off site
- Provisional sum in the contract

Similarly, while preparing the cash flow, the contractor has to consider the following points:

- Delay in execution of work
- Productivity less than estimated
- Delay in receipt of payment
- Advance payment, if any, for material
- Different site conditions
- Stoppage of work due bad weather
- Resequencing of work
- Delay in material delivery at site resulting payment cannot be claimed
- Higher material price than estimated
- Fluctuation in international exchange rate

Figure 7.150 is illustrative S-curve for contractor.

Payments: The contractor's payment is related to the approved executed works. The contractor submits the payment on monthly basis. Figure 7.151 illustrates progress payment submission format generally used by the contractor, Figure 7.152 illustrates process for progress payment approval, and Figure 7.153 illustrates payment certificate format normally prepared by R.E./CM to recommend payment to the contractor.

7.4.6.12.4.10 Manage Claims Most claims in construction projects are due to

- Errors and omissions in the contract documents
- Incomplete design
- Design changes
- Delay in payment by the owner
- Delay in transfer of site
- Delay in approval of submittals

Table 7.50 lists major causes of claim.

Identify claims: The contractor during the construction phase has to identify the errors or omissions in the contract, if any, and follow the contractual procedure to resolve the issues. Table 7.51 lists the claims in construction projects and their effects on the construction project.

Prevent claims: Table 7.52 has listed actions to taken to prevent/mitigate claims.

Resolve claims: Any claim submitted by the contractor has to be resolved by the project team members amicably as per the condition of contract.

Figure 7.154 illustrates claim resolution process.

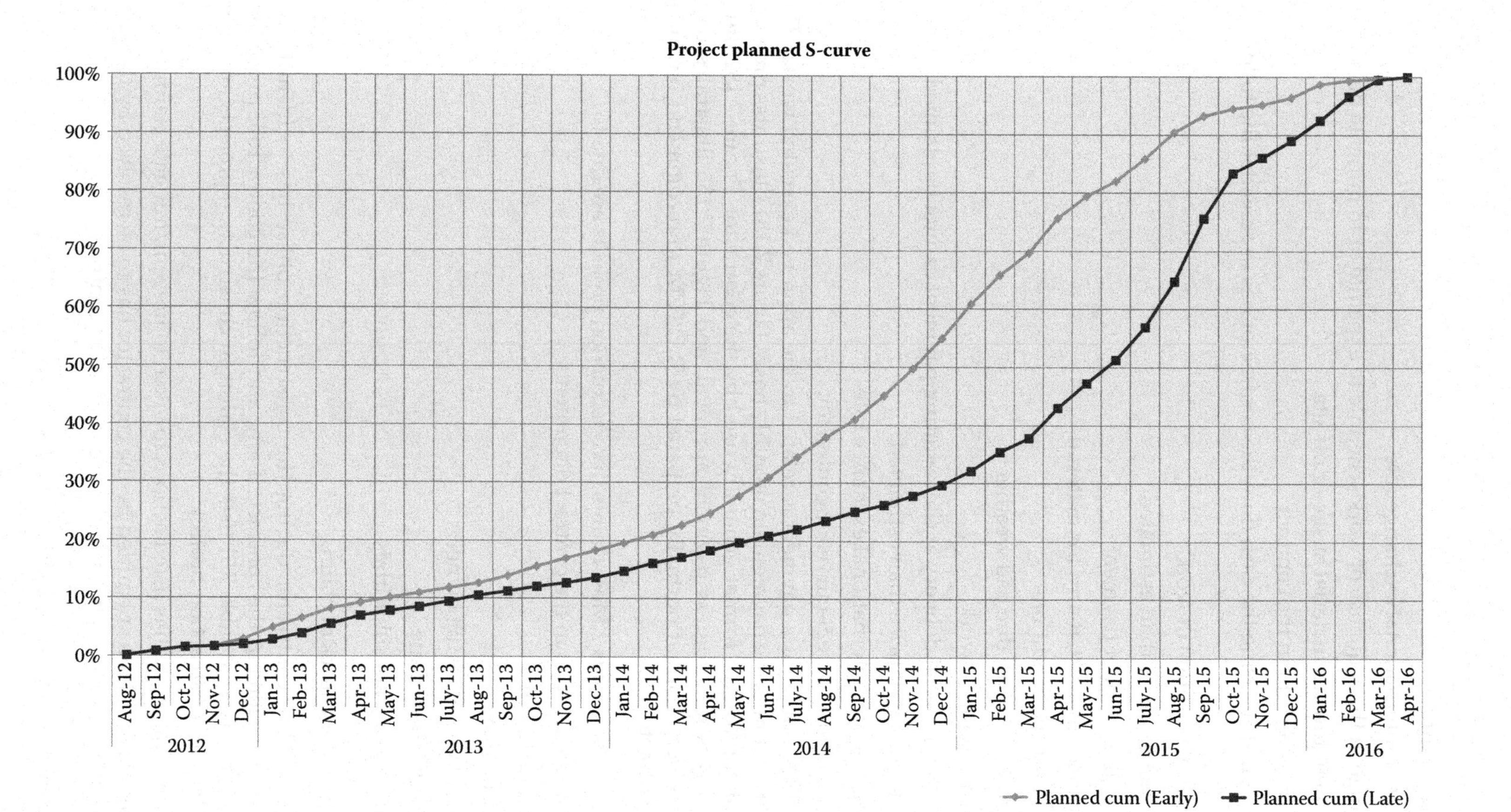

FIGURE 7.150
Contractor's planned S-curve.

Project Name

Contractor Name :
Project Number :

Sr. No.	BOQ Reference	Description	Original Reference				Work Completed (Quantity/Percentage)							Cummulative Work			Approved Amount
							Previous			Current				Previous+Current			
			Unit	BOQ Quantity	Unit Price	Total Amount	Quantity	Percentage (%)	Total Amount	Quantity	Percentage (%)	Checklist/ Approval Refrence	Total Amount	Quantity	Percentage (%)	Total Amount	

FIGURE 7.151
Progress payment submission format.

7.4.6.13 Monitor and Control Project Works

Monitoring and controlling of project works depends on the type of project delivery system. In major projects, the owner engages construction/project management firm that is responsible to supervise, monitor and control construction activities performed by the contractor. However in most cases supervision of the project during the construction phase is carried out by the consultant who is involved in designing the project. The R.E. along with supervision team members is responsible to supervise, monitor and control, implement the procedure specified in the contract documents and ensure completion of project within specified time, budget, and per defined scope of work. In order to ensure smooth flow of supervision activities, the R.E. has to follow the organization's supervision manual and contractual requirements. Table 7.88 illustrates an example checklist of items to be verified by the R.E. to ensure availability of all the necessary documents, information to facilitate smooth flow of supervision work.

Monitoring and controlling the project is an ongoing process. It starts from the inception of the project and continue till handover of the project. Monitoring and control of construction project is operative during the execution of the project and its aim is to recognize any obstacles encountered during execution and to apply measures to mitigate these difficulties and to ensure that the goals and objectives on the project are being met.

Monitoring is collecting, recording, and reporting information concerning any and all aspects of project performance that the project manager or others in the organization need to know. Monitoring of construction project is normally done collecting and recording the status of various activities and compiling them in the form of progress reports. These are prepared by the consultant or contractor and distributed to the concerned members of project team.

Monitoring involves not only tracking time but also budget, quality, resources, and risk. Monitoring in construction projects is normally done by compiling status of various activities in the form of progress reports. These are prepared by contractor, supervision

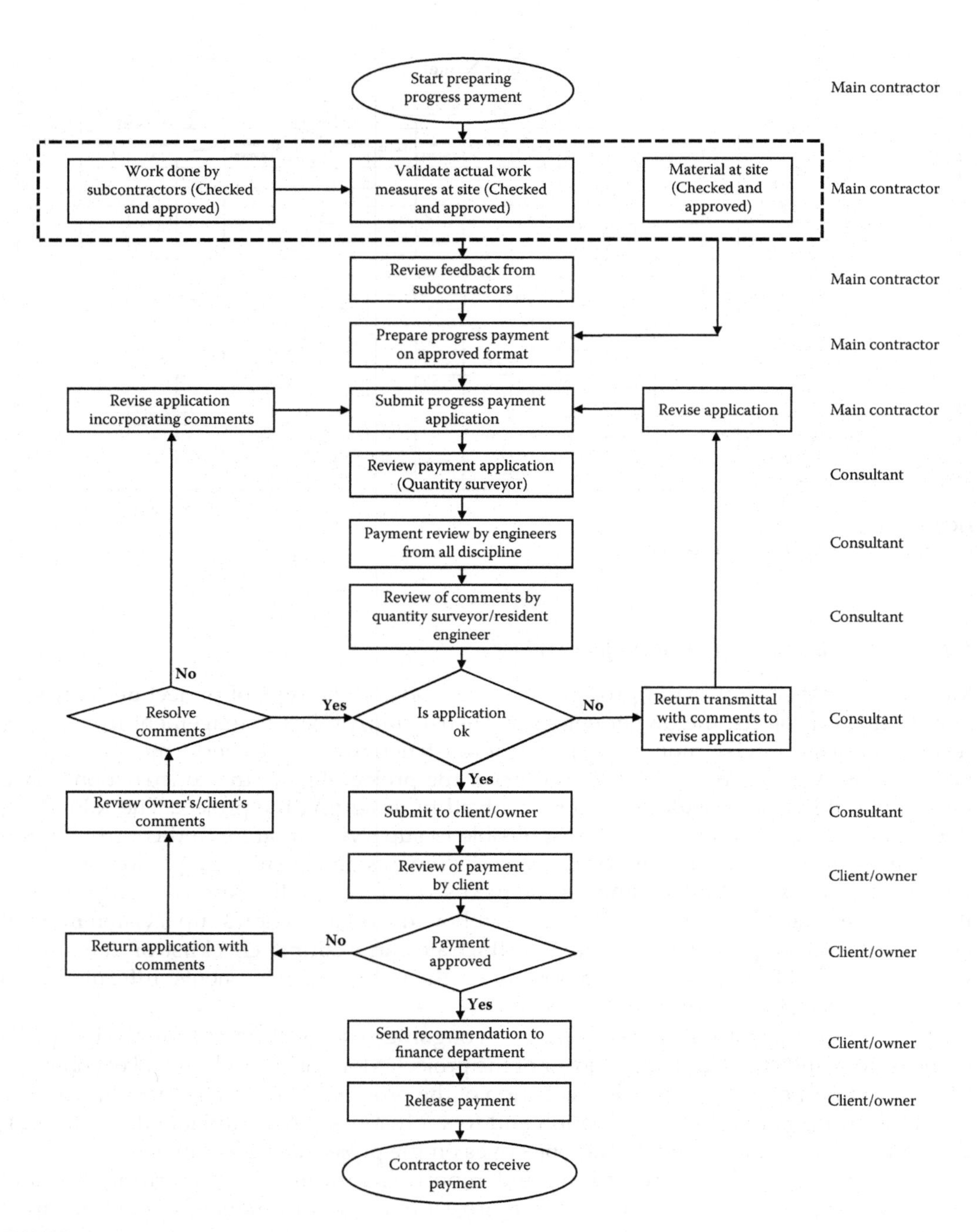

FIGURE 7.152
Progress payment approval process.

PROJECT NAME

PAYMENT CERTIFICATE

CONTRACT NO. :
CONTRACTOR :
PAYMENT CERTIFICATE NO. : ____________
DATE : ____________

SAMPLE FORM

Original Contract Value		Commencement Date:	
Orovisional Items	:	Contract Period (days):	
Variation Orders Additon	:	Extension of Time (days):	
Variation Order (Omission)	: ____________	Date of Completion:	
Current Contract Value	:		

			Amount
Work Executed up to:			
Work Executed of Original Contract			
Work Executed of Variation Orders			
Work Executed of Provisional Sum			
Total Material on Site	X 80%		____________
Total Amount Due			
Additions/Deductions:			
Amount of Retention	X 10%	Deduct	____________
Advance Payment (1st Installment)	% of Contract Value		
Advance Payment (2nd Installment)	% of Contract Value		
Total			
Repayment ____________	% of Total Amount Due		
	Balance (Add)	Add	____________
Release of Retention		Add	____________
Less Omissions			
Deduct Previous Payments (including advance payment up to PC # 1)		Deduct	____________
	NET AMOUNT DUE TO CONTRACTOR		

Amount in Word:

(Note: The amount to be paid to the Contractor's Account No. xxxxxxxxxxx)

RECOMMENDED BY RESIDENT ENGINEER ____________ DATE

____________ CONTRACTOR'S REPRESENTATIVE ____________ DATE

FIGURE 7.153
Payment certificate.

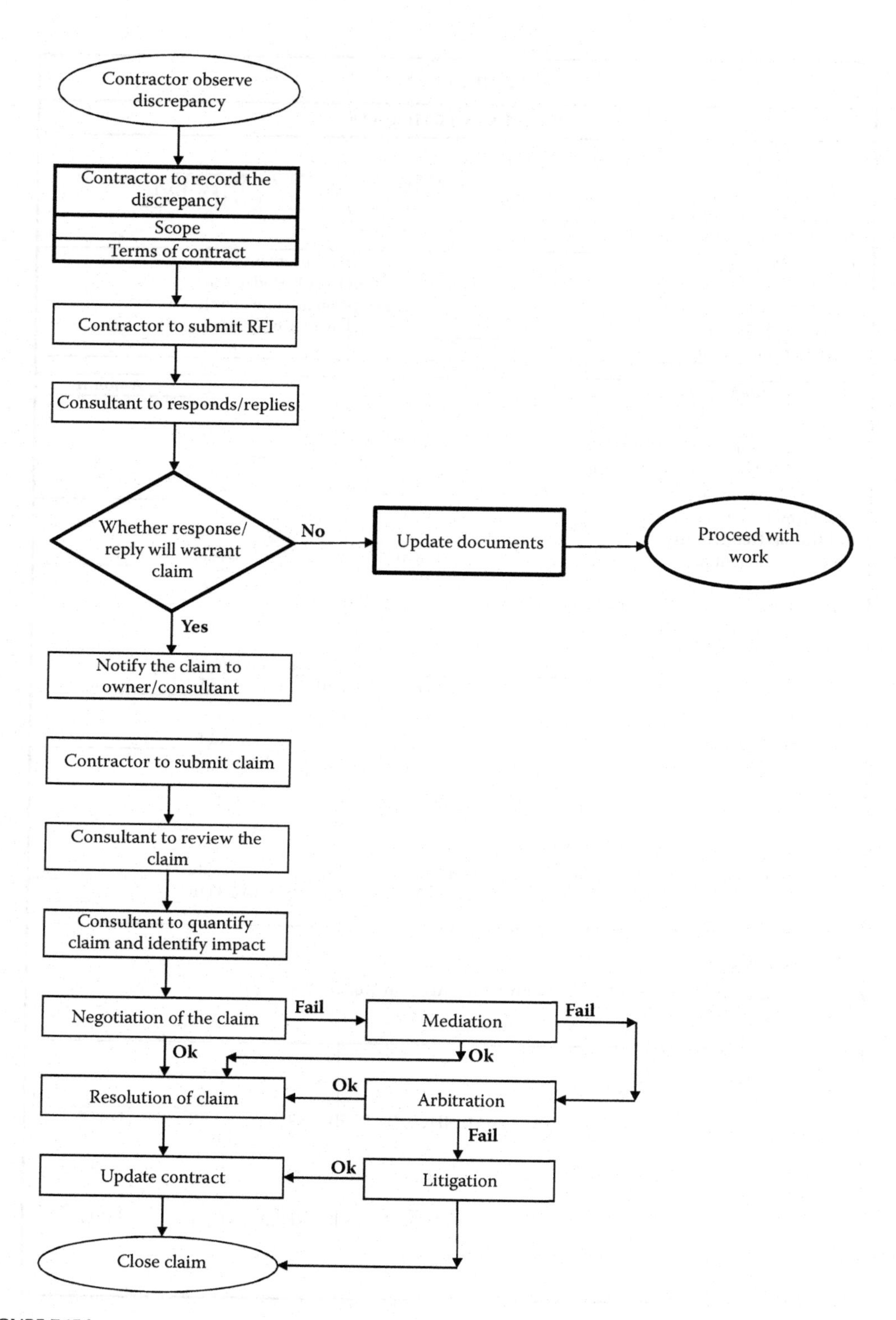

FIGURE 7.154
Claim resolution process.

TABLE 7.88

Consultant's Checklist for Smooth Functioning of Project

Sr. No.	Items to Be Checked/Verified
I	**Project Details**
	I.1 Scope of work
	I.2 Project objectives
	I.3 Project deliverables
II	**Project Organization**
	II.1 Organizational chart and roles and responsibilities of defined supervision staff
	II.2 Supervision staff deployment matching with project requirements
	II.3 Contractor's staff deployment plan approved as per contract requirements
	II.4 Responsibility matrix prepared and approved by the client and distributed among all project parties
	II-5 Project directory
III	**Mobilization**
	III.1 Site permit from authorities available
	III.2 Project plot boundaries are marked as per the permit
	III.3 Project commencement order issued
	III.4 Copy of permit issued to the contractor
	III.5 Temporary site offices drawings approved
	III.6 Temporary firefighting plan approved by respective authority
	III.7 Copies of the contractor's performance bond, guarantees, insurance policies and licenses available at site
	III.8 Copies of the consultant's performance bond, guarantees, insurance policies and licenses available at site
	III.9 Preconstruction meeting conducted and submittal and approval procedures discussed and agreed
IV	**Project Administration**
	IV-1 Contract Documents
	IV-1.1 Signed copy of the contract between the owner and contractor available at site
	IV-1.2 Copies of contract documents available at site
	IV-1.3 Contracted bill of quantity (BOQ) is available
	IV-1.4 All volumes of particular specifications available
	IV-1.5 Contracted drawings are available
	IV-1.6 Authority approved drawings, duly stamped, available
	IV-1.7 Addendum, if any, to the contract available
	IV-1.8 Replies to tender queries available
	IV-1.9 Copy of signed contract documents and drawings handed over to contractor and has acknowledged the same
	IV-1.10 Log for codes and standards available
	IV-2 Document Management
	IV-2.1 Document control system is in place
	IV-2.2 Filing index is available
	IV-2.3 Material submittal log is available
	IV-2.4 Shop drawing submittal log is available
	IV-2.5 Logs for correspondence between various parties available
	IV-2.6 Log for checklist (request for inspection) available
	IV-2.7 Log for job site instruction available
	IV-2.8 Log for site work instruction available
	IV-2.9 Log for request for information available
	IV-2.10 Log for variation order available
	IV-2.11 Log for nonconformance report available
	IV-2.12 Material sample log and place identified

(*Continued*)

TABLE 7.88 (*Continued*)

Consultant's Checklist for Smooth Functioning of Project

Sr. No.	Items to Be Checked/Verified
	IV-2.13 Log for equipment test certificate available
	IV-2.14 Log for visitor's at site
	IV-2.15 Contractor's staff approval log in place
	IV-2.16 Subcontractor's approval log in place
	IV-2.17 Consultants staff approval in place
	IV-2.18 Overtime request log available
V	**Communication**
	V-1 Communication matrix established and agreed by all the parties
	V-2 Distribution system for transmittals/submittals agreed
VI	**Project Monitoring and Control**
	VI-1 Daily report log in place
	VI-2 Weekly report log in place
	VI-3 Monthly report log in place
	VI-4 Progress meetings log in place
	VI-5 Minutes of meetings log in place
	VI-6 Progress payment log in place
	VI-7 Construction schedule log in place
VII	**Construction**
	VII-1 Quality control plan log in place
	VII-2 Safety management plan log in place
	VII-3 Risk management plan log in place
	VII-4 Method statement submittal log in place
	VII-5 Accident and fire report
	VII-6 Off-site inspection visits
	VII-7 Location of gathering point established
VIII	**General**
	VIII-1 Correspondence between site and head office
	VIII-2 Staff-related matters
	VIII-3 Copy of supervision manual available
	VIII-4 Emergency contact telephones and contact details displayed at site

Source: Abdul Razzak Rumane (2013), *Quality Tools for Managing Construction Projects,* CRC Press, Boca Raton, FL. Reprinted with permission from Taylor & Francis Group.

team (consultant), and construction/project management team. The objectives of project monitoring and control are as follows:

1. To report the necessary information in details and in appropriate form that can be interpreted by management and other concerned personnel to provide with the information about how the resources are being used to achieve project objectives
2. To provide an organized and efficient means of measuring, collecting, verifying, and quantifying data reflecting the progress and status of execution of project activities, with respect to schedule, cost, resources, procurement, and quality
3. To provide an organized, efficient, and accurate means of converting the data from the execution process into information
4. To identify and isolate the most important and critical information about the project activities to enable decision making personnel to take corrective action for the benefit of the project
5. To forecast and predicting about future progress of activities to be performed

Figure 7.155 illustrates the logic flow diagram for monitoring and control process and Table 7.89 illustrates monitoring and control references for construction projects.

Construction project control is exercised through knowing where to put the main efforts at a given time and maintaining good communication. There are mainly three areas where project control is required:

1. Quality (scope)
2. Schedule
3. Budget

All of these areas are to be kept in balance to achieve project objectives. In order to accomplish the project objectives in construction projects, monitoring and control is done through various tools and methods. There are mainly three elements that need monitoring and control:

1. Quality (scope)
2. Schedule (work progress)
3. Budget (cost control)

7.4.6.13.1 Control Scope

The project requirements set out in the contract documents determine the project scope, which is described in terms of project deliverables in the form of specifications and drawings. During the construction phase, the contractor has to execute/install the works as defined in the project specifications and contract drawings. The contractor has to make certain that the project changes do not result in compromising the intended project deliverables in terms of required quality and performance as defined in the scope baseline.

7.4.6.13.2 Monitor Progress

Approved contractor's construction schedule is the performance baseline for construction projects, which is achieved by collecting information through different methods. Work progress is monitored through various types of logs, S-curves, reports, and meetings.

7.4.6.13.2.1 Logs There are various types of logs used in construction projects to monitoring and control construction activities. The main logs used in a construction project are as follows:

1. Subcontractors submittal and approval log
2. Submittal status log
3. Shop drawings and materials logs—E1
4. Procurement log—E2
5. Equipment log
6. Manpower logs

These logs provide necessary information about the status of subcontractors, materials, shop drawings, procurement, and availability of contractor's resources and help determine its effects on project schedule and project completion.

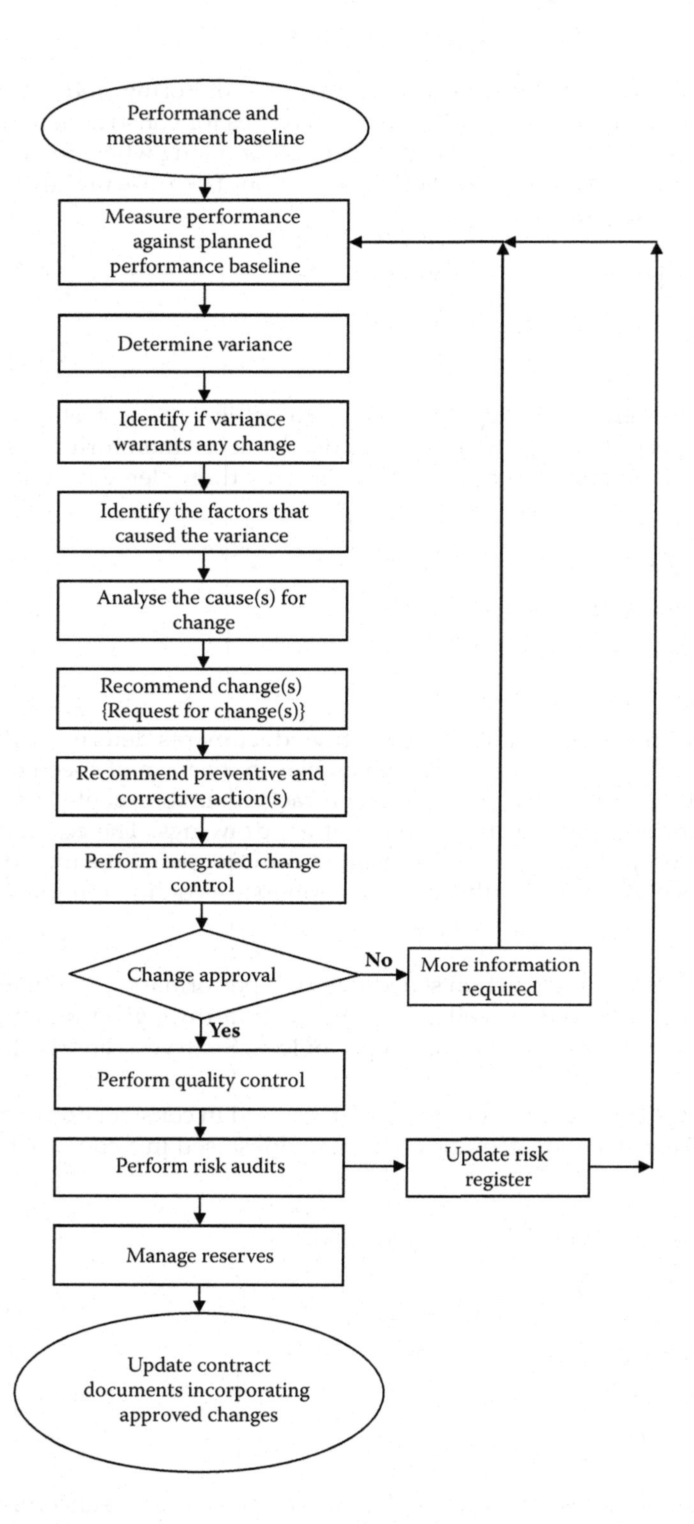

FIGURE 7.155
Logic flow diagram for monitoring and control process. (Abdul Razzak Rumane (2013), *Quality Tools for Managing Construction Projects*, CRC Press, Boca Raton, FL. Reprinted with permission from Taylor & Francis Group.)

TABLE 7.89

Monitoring and control plan references for construction projects

Sr. No.	Elements	Contract Reference	Contractor Reference
1	Performance Baseline	Schedule1	1. Contractor's Construction Schedule
		Specifications, Drawings	1. Approved Materials 2. Approved Shop Drawings 3. Approved Composite Drawings
		Cost	1. Approved S-Curve
2	Data Collection Methods	Reports	1. Daily Report 2. Weekly Report 3. Monthly Report 4. SafetyReport 5. Risk Report 6. Accident Report 7. Checklist 8. Risk Report
3	Frequency of Data Collection	1. Daily 2. Weekly 3. Monthly	1. Daily Report 2. Weekly Report 3. Monthly Report
4	Status Information Collection	1. Logs 2. Reports 3. Meetings 4. Checklists	1. Logs 2. Reports 3. Meetings 4. Checklists
5	Comparison between Planned and Actual (Variance)	1. S-Curves 2. Milestones	1. Progress Reports 2. Progress Payment 3. Milestones
6	Analysis1	1. Price Analysis	1. Construction Schedule Attachment 2. Progress Payment
7	Corrective Action	1. Comments by Consultant	1. Incorporate Comments
8	Change Order	1. Variation Order1	1. Request for Information 2. Request for Modification 3. Request for Variation
9	Document Updates	1. On regular basis	1. Reports

Source: Abdul Razzak Rumane (2013), *Quality Tools for Managing Construction Projects*, CRC Press, Boca Raton, FL. Reprinted with permission from Taylor & Francis Group.

7.4.6.13.2.2 S-Curves Figure 7.156 illustrates S-curve for actual work versus planned work.

7.4.6.13.2.3 Reports

Progress reports: Apart from different types of logs and submittals, progress curves, time control charts, contractor's progress is monitored through various types of reports and meetings. These are as follows:

a. Daily report
b. Weekly report
c. Monthly repot

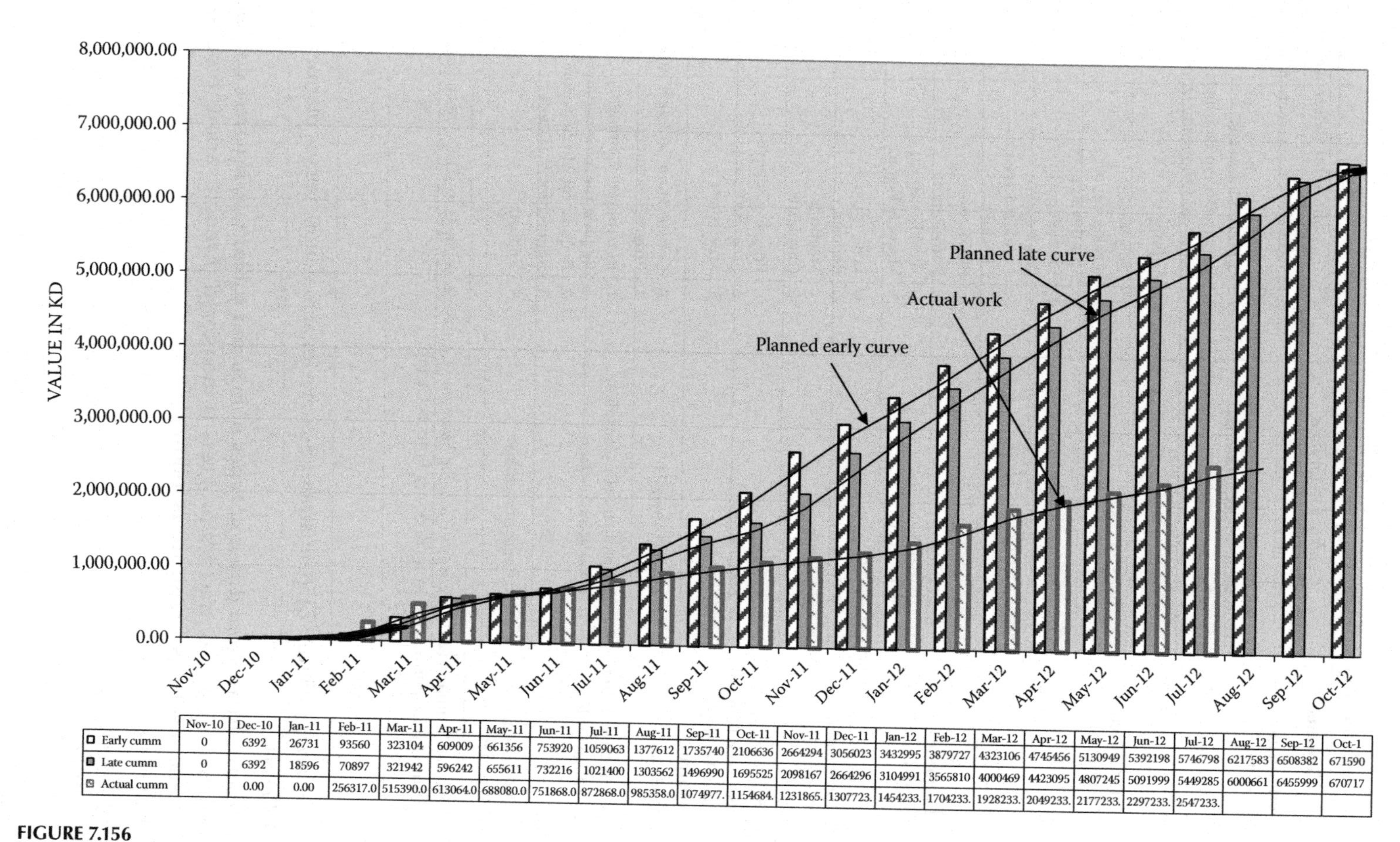

	Nov-10	Dec-10	Jan-11	Feb-11	Mar-11	Apr-11	May-11	Jun-11	Jul-11	Aug-11	Sep-11	Oct-11	Nov-11	Dec-11	Jan-12	Feb-12	Mar-12	Apr-12	May-12	Jun-12	Jul-12	Aug-12	Sep-12	Oct-1
Early cumm	0	6392	26731	93560	323104	609009	661356	753920	1059063	1377612	1735740	2106636	2664294	3056023	3432995	3879727	4323106	4745456	5130949	5392198	5746798	6217583	6508382	671590
Late cumm	0	6392	18596	70897	321942	596242	655611	732216	1021400	1303562	1496990	1695525	2098167	2664296	3104991	3565810	4000469	4423095	4807245	5091999	5449285	6000661	6455999	670717
Actual cumm		0.00	0.00	256317.0	515390.0	613064.0	688080.0	751868.0	872868.0	985358.0	1074977.	1154684.	1231865.	1307723.	1454233.	1704233.	1928233.	2049233.	2177233.	2297233.	2547233.			

FIGURE 7.156
Planned versus actual.

The contractor's daily progress is monitored through daily progress report submitted by the contractor on the morning of the working day following the day which the report relates. It gives the status of all the resources available at site for that particular day. It shows the details of contractor's staff and manpower, contractor's plant and equipment, and material received at site. Details of subcontractor's work and resources are also included in the report. Figure 7.157 illustrates daily progress report of a building construction project.

Along with the daily report contractor submit work in progress report. Figure 7.158 illustrates the same.

Figure 7.159 illustrates checklist status report that is also to be submitted along with the daily report. This will help contractor as well as supervision consultant to monitor quality of works on daily basis by knowing how many checklists are approved and how much of the works are not conforming to the specified requirements.

Monthly report giving details of all the site activities along with the photographs is submitted by the contractor to the consultant/owner for their information to know the progress of work during the month. Table 7.90 illustrates contents of contractor's monthly progress report and Table 7.91 illustrates contents of consultant's monthly progress report.

Safety report: Contractor submits safety report every month listing important activities with photographs.

Risk report: Identify occurrence of new risk and report the same if this will have any effect on project progress and performance.

7.4.6.13.2.4 Meetings

Progress meetings: Progress meetings are conducted at an agreed upon interval to review the progress of works and discuss about the problems, if any, for smooth progress of construction activities. Contractor submits premeeting submittal to project manager/consultant normally two days in advance of the scheduled meeting date. The submittal consists of

a. List of completed activities
b. List of current activities
c. Two weeks look ahead
d. Critical activities
e. Materials submittal log
f. Shop drawings submittal log
g. Procurement log

Apart from the issues related to progress of works and programs, site safety and quality-related matters are also discussed in these meetings. These meetings are normally attended by the owner's representative, designer/consultant staff, contractor's representative, and subcontractor's responsible personnel.

Coordination meetings: Coordination meetings are held from time to time to resolve coordination matters among various trades.

Quality meetings: Quality meetings are conducted to discuss quality issues at site and how to improve the construction process to avoid/reduce rejection and rework.

Project Name

Consultant Name

CONTRACTOR'S DAILY PROGRESS REPORT

Contract No.:- **Contract Day No.**
Contractor :- **Date**

SAMPLE FORM

This daily report to be completed on both sides and submitted to the Resident Engineer following the report date.

Contractor's Staff and Manpower Required			Contractor's Staff and Manpower Required		
Job Description	No.	Actual	Job Description	Skilled	Un-skilled
Contractor's Representative	1		Secretary		
Project Manager	1		Store keeper		
Deputy Project Manager	1		Carpenter		
Planning Manager	1		Steel Bender		
Deputy Planning Manager	1		Concrete workers		
Quality Control Manager	1		Mason		
Quality Control Engineer	1		Plasterer		
Quantity Surveyor	1		Tiler		
Assistant Quantity Surveyor	1		Marble		
Site Engineer (Architect)	2		Ceramic		
Site Civil Engineer	2		Stone		
Site Engr. (Water and Sewerage)	2		Precast		
HVAC Engineer	2		Safety officer		
Mechanical/Fire Fighting Engr.	1		Painter		
Electrical Engineer	2		Plumber		
Communications Engineer	1		HVAC		
Site Engineer (Marine)	1		Fire system		
Site Engr. (Roads and Services)	1		Seaman (Diver)		
Material Engineer	1		Mechanical supervisor		
Assistant Material Engineer	1		Driver		
Landscape Gardner	1		Operator		
Safety Engineer	1		Welder		
Coordinator	1		Electrician		
Surveying Engineer	1		Mech. and Elec. Workshop labour		
Surveyor	1		Labour		
Supervisor	4		Others		
Computer Programmer	1				
Computer Draftsman	2				
Draftsman	1				
Laboratory Engineer	1				
Laboratory Technician	1				
	40				

Distribution
Original: Resident Engineer
CC: Owner

Contractor...

FIGURE 7.157
Daily progress report.

(Continued)

Project Name

Consultant Name

CONTRACTOR'S DAILY PROGRESS REPORT

Contract No.:- **Contract Day No.**

Contractor :- **Date**

SAMPLE FORM

This daily report to be completed on both sides and submitted to the Resident Engineer following the report date.

Contractor's Plant and Equipment Required			Contractor's Plant and Equipment Additional	
Description of Item	No.	Actual	Description of Item	No.
Tower Crane	3		Loader	
Crane	4		Rock body truck	
Tipper Truck	12		Boat with crew and radio	
Excavator	4		Radio communication system	
Grader	2		Bob cat	
Well point system with WP.	4		Fork-lift	
Water Tanker	4		Crane	
Compactor (Plate)	8		Transit mixer	
Vibrator	8		Flat bed truck	
Conc. Testing Equipment	1		Floating crane 120 tonne	
Soil Testing Equipment	1		Pile driving machine	
Compressor	5		Side crane	
Transit Mixer	6		Tug	
Water Pump	2		Tractor	
Vibrator Compact Roller	4		Truck with crane	
Automatic Batching Plant	1		Gantry crane	
Concrete Pump	2		Bulldozer	
Asphalt Roller	4		Pick-up	
Welding Machine	4		Car	
Generator	4		Bus	
Bulldozer	2		Mini bus	
Barge	1		Tug boat	
Split Barges	1		Motor grader	
Crane Pontoon	1			
Grab	2			
Diving Equipment	4			
Automatic Tide Gauge	1			

These items are provided by supplier

Distribution

Original: Resident Engineer

CC: Owner

Contractor...

FIGURE 7.157 (Continued)
Daily progress report.

(Continued)

Project Name

Consultant Name

CONTRACTOR'S DAILY PROGRESS REPORT

Contract No.:- **Contract Day No.**
Contractor :- **Date**

This daily report to be completed on both sides and submitted to the Resident Engineer following the report date.

Material Delivered to the Site		
Description of Material	Quantity	Unit

Distribution
Original: Resident Engineer
CC: Owner

Contractor..

FIGURE 7.157 (Continued)
Daily progress report.

Project Name
Consultant Name

WORK IN PROGRESS REPORT

Contract No.: **Contract Day No.:**
Contractor: **Date:**

On Site Activities									
During the Day					Expected Next Day				
No.	Description	Area	Unit	Qty.	No.	Description	Area	Unit	Qty.
Reasons for Delay, if any.									

Off Site Work/Activities							
During the Day				Expected Next Day			
No.	Subcontractor Name	Work Description	Qty.	No.	Subcontractor Name	Work Description	Qty.
Reasons for Delay, if any.							

FIGURE 7.158
Work in progress.

Safety meetings: Safety meetings are also held to discuss related health, site safety, and environmental matters.

Frequency of conducting meetings is agreed between all the parties at the beginning of the construction phase. Normally, the construction manager/R.E. prepares the agenda for the meeting and circulates to all the participants. The contractor informs the R.E. in advance about the points the contractor would like to discuss, which are included in the agenda. The minutes of meetings are recorded and circulated among all the attendees and other per the approved responsibility/site communication matrix.

Project Name
Consultant Name

DAILY CHECKLIST STATUS

Contract No.: **Contract Day No.:**
Contractor: **Date:**

Sr. No.	Checklist No.	Description	Activity	Area/Location	Action	Remedial Action*	Remark

* For Unapproved Checklist

FIGURE 7.159
Daily checklist status.

7.4.6.13.2.5 Digitized Monitoring of Work Progress The work progress is normally monitored through daily and monthly progress reports. Monthly progress report consists of progress photographs to document physical progress of work. These photographs are used to compare compliance with the planned activities and actual performance. Figure 7.160 illustrates the traditional monitoring system.

With the advent of technology, it is possible to monitor and evaluate construction activities using cameras and related software technologies. In this process, digital images are captured through use of cameras. These photographs are processed using photo modeler software and developing 3D model view of the digital picture captured from the site. The captured as-built data is compared with the planned activities by interfacing through integrated information modeling system. The use of the system

- Improves the accuracy of information
- Avoids delays in getting the information
- Improves communication among all parties
- Improves effective control of the project
- Improves document recording
- Helps reduce claims

Figure 7.161 illustrates the schematic for digitized monitoring system.

TABLE 7.90

EXAMPLE CONTENTS

Monthly Progress Report

Contents of Monthly Report			
Sr. No.		**Contents**	**Description**
1	**Executive Summary. - Tabular**		
	1.1	Summary Status Report	Brief descriptionof the Project Status up to date i.e.Manpower,cash, activities
	1.2	NOC's Report	No Objection Certificate Report
	1.3	Project Manager Narratives	Narratives description of Project Status up to date
2	**Progress Layouts**		
	2.1	Updated Milestone Table	Comparison of Planned Vs. Actual for Contractual Milestones per Construction Unit \ Design, You track the delays for major trades through the color theme
	2.2	Updated Major of Events Table	Comparison of Planned Vs. Actual for Major trades per Construction Unit \ Design, Track the delays for major trades through the color theme
	2.3	Updated Layouts	Same as (2.1) above but presentation per milestone phase (Drawing)
3	**Updated Execution Program.**		
	3.1	Updated Milestone Schedule - Roll up "Update versus latest Target"	To indicate the status of the Control & Key Milestones comparing the current status with the baseline
	3.2	Updated Detailed Schedule	All activities in details
	3.3	One month look Ahead Program	Same as (3.2), but includes only the detailed activities for the coming month BUT ON EXCEL FORMAT
4	**Submittal Status Report (E1 Log).**		Updated status of submittals
	4.1	Submittal Status Report	Briefly describe the Project Submittal Status up to date
	4.2	Detailed E1 Log	Detailed describe the Project Submittal Status up to date
5	**Procurement Status Report (E2 Log).**		Updated Procurement Status
6	**Status of Information Requested.**		
	6.1	RFI Status report	Request for Information (RFI)
	6.2	NCR Report	Non Conformance Request (NCR)
	6.3	PCO & NOV Log	Potential Change Order & Notice of Variation Summary
7	**Updated Cost Loaded Program.**		
	7.1	T-7 Updated Status Report	Money Progress = Physical * Budgeted Cost for each running or completed activity
	7.2	Updated Cost Loaded Schedule	Same like T-7 but money values not percentages
	7.3	Updated Work In Place (%) Report	Histogram & Cumulative Curve
	7.4	Udated Cash Flow Report	Histogram & Cumulative Curve
8	**Updated Manpower Histogram.**		Histogram & Cumulative Curve
9	**Updated Schedule of Construction Equipment & Vehicles**		Tabular Report
10	**Updated Critical Indicators.**		
	10.1	Shop Drawings Status Report	Histogram & Cumulative Curve
	10.2	Material Status Report	Histogram & Cumulative Curve
	10.3	Construction. Leading Indicators	Each trade alone
	10.4	Line of Balance Diagram	It indicates all progress of the major trades as line cumulative chart
11	**Updated Progress Photographs.**		-
12	**Updated Safety Inspection Checklist.**		Tabular Report
13	**Contractor Information**		Organization Chart, Tabular Report

Source: Abdul Razzak Rumane (2013), Quality Tools for Managing Construction Projects, CRC Press, Boca Raton, FL. Reprinted with permission of Taylor & Francis Group.

TABLE 7.91

Contents of Progress Report

1.0	Contract Particulars
	1.1 Project description
	1.2 Project data
2.0	Construction Schedule
3.0	Progress of Works
	3.1 Temporary facilities and mobilization
	3.2 Summary of construction progress
	—3.2.1 Status
	—3.2.2 On-shore progress
	—3.2.3 Off-shore progress
4.0	Time Control
	4.1 CPM schedule—level one (target vs. current)—summary by building/marine
	4.2 CPM schedule—level two (target vs. current)—summary by building/division
	4.3 30 days look-ahead schedule
	4.4 Time control conclusion
5.0	Cost Control
	5.1 Financial progress
	5.2 Cash flow curve and histogram
	5.3 Work-in-place S-curve and histogram
	5.4 Cost control conclusion
6.0	Status of Contractor's Submittals
	6.1 Material status
	6.2 Shop drawing status
7.0	Subcontractors
8.0	Consultant's Staff
9.0	Quality Control
10.0	Meetings
11.0	Site Work Instructions
12.0	Variation Orders
13.0	Construction Photographs
14.0	Contractors Resources
15.0	Others Matters
	15.1 Safety
	15.2 Weather conditions
	15.3 Important developments/proposals/submissions

7.4.6.13.2.6 Monitoring of Submittals It is required that the contractor's submittals are processed and response is sent within the specified period mentioned in the contract. The consultant is required to maintain the log for material submittals, shop drawing submittals, and also all the correspondence between the contractor and client. Figure 7.162 illustrates an example log for monitoring material/shop drawing submittals.

7.4.6.13.3 Control Schedule

Completion of construction project within defined schedule is most important. Time control status is prepared in different formats to monitor the project completion time. Figure 7.163

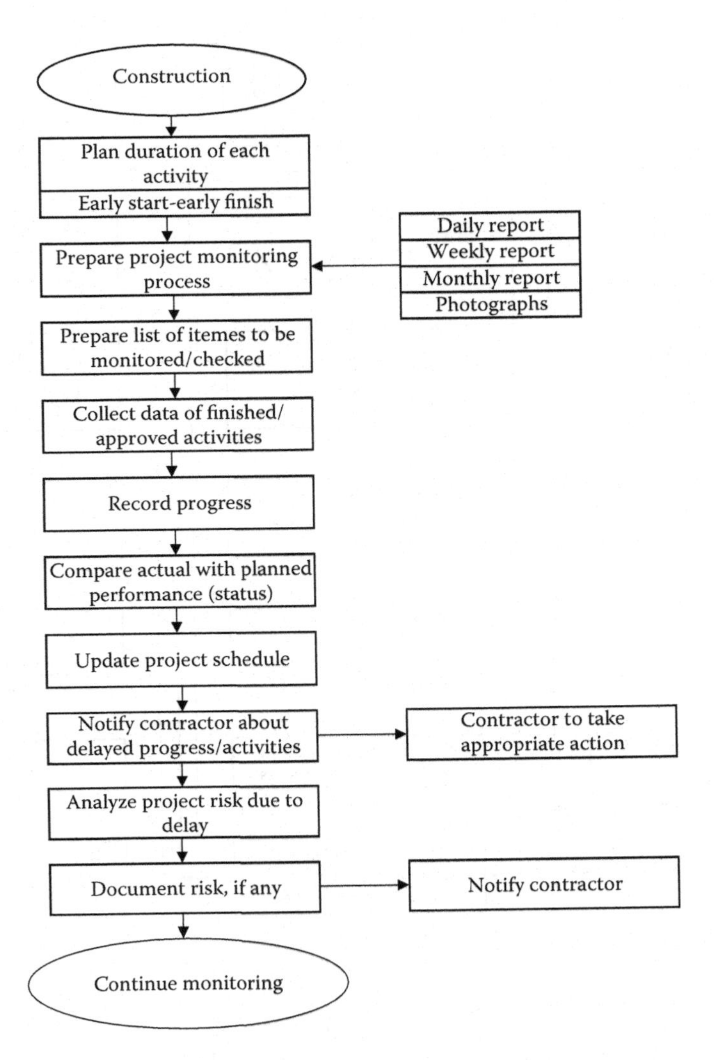

FIGURE 7.160
Traditional monitoring system.

illustrates the project progress status of a building construction project. This chart presents the overall picture of the elapsed period of the project and remaining period of the scheduled project duration, and actual progress versus planned progress.

7.4.6.13.4 Control Cost

Monitoring and control of project payment is essential with the budgeted amount. This is done through monitoring cash flow with the help of S-curves and progress curves, which gives exact status of payment and also identifies if it is exceeding the budget. Uninterrupted cash flow is one of the most important elements in the overall success of the project. Figure 7.164 illustrates the planned and actual cost S-curve.

7.4.6.13.4.1 Project Payment/Progress Curve (S-Curve) Cash flow is a simple comparison of when revenue will be received and when the financial obligations must be paid. It is also an indication of the progress of work to be completed in a project. This is obtained by

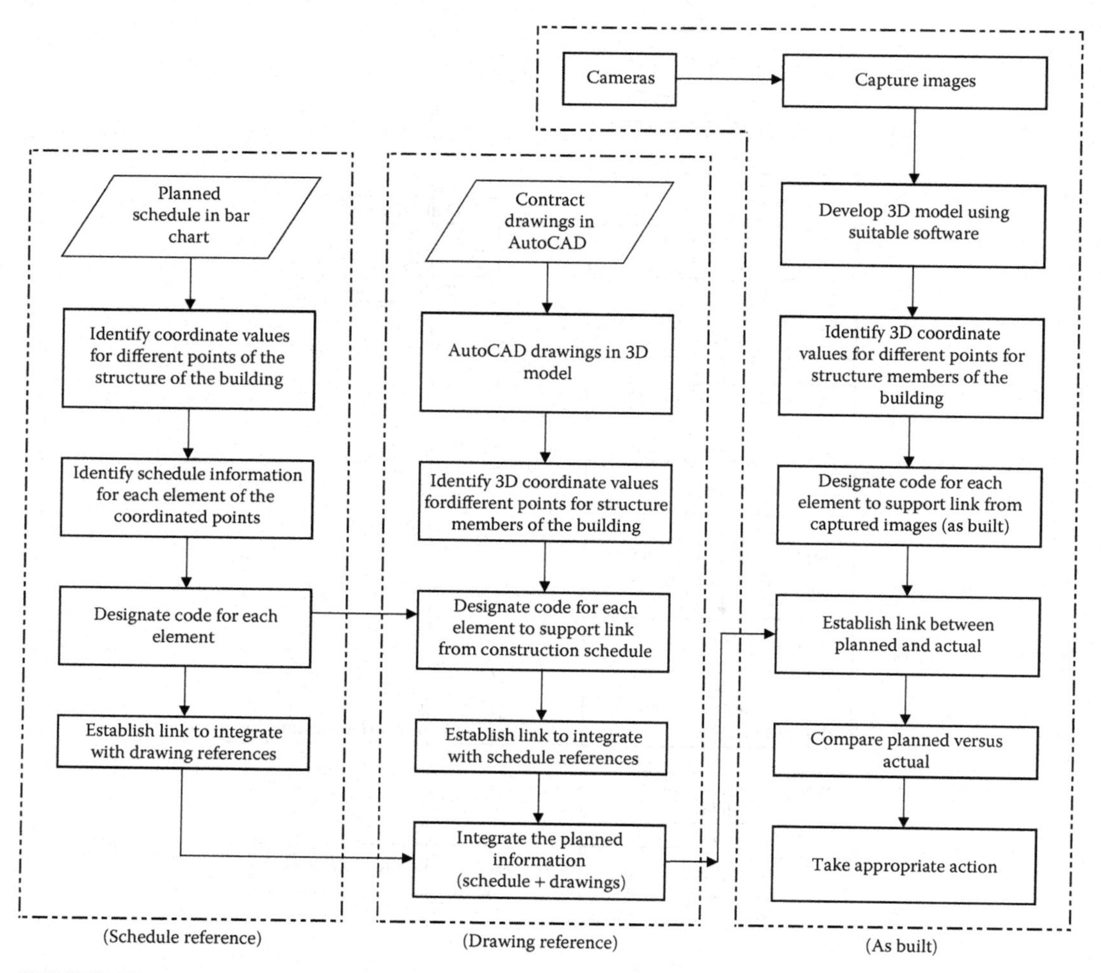

FIGURE 7.161
Digitized progress monitoring. (Abdul Razzak Rumane (2013), *Quality Tools for Managing Construction Projects*, CRC Press, Boca Raton, FL. Reprinted with permission from Taylor & Francis Group.)

loading each activity in the approved schedule with the budgeted cost in the BOQ. The process of inputting schedule of values is known as *cost loading*. The graphical representation of the above is obtained as a curve and is known as S-curve.

7.4.6.13.4.2 Earned Value Management EVM is a methodology used to measure and evaluate project performance against cost, schedule, and scope baseline. It compares the amount of planned work with what is actually accomplished to determine whether the project is progressing as planned. EV analysis is used to

- Measure progress of the project budget, schedule, scope to know how much percentage of
    - Budget is spent
    - Time has elapsed
    - Work is done

Project Name
Consultant Name

Contractor Name:
Week Start Date:
Contract No:

Transmittal No.	Rev. No.	Transmittal Date	Date Received	Description	Document Reference	Responsible Engineer	Action Code	Reply Date	Action Pending	Remark/ Reason for Pending

FIGURE 7.162
Submittal monitoring form.

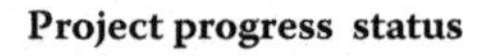

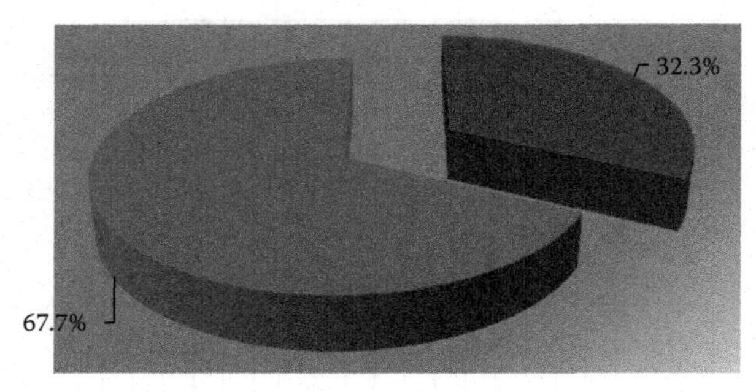

Time elapsed-days	413	32.30%
Time remaining-days	867	67.70%

Current project status			
	June 15, 2015		
Completion period (days)	1280		
Duration elapsed (days)	413		
% elapsed duration	32.30%		
Remaining duration (days)	867		
% remaining duration	67.70%		
Total contract amount (KD)	142,061,073.44		
Earned value in KD upto date	Early	Late	Actual
	31,494,939.98	25,613,611.54	19,422,904
Project completion % till date	22.17%	18.03%	13.67%

FIGURE 7.163
Project progress status.

- Forecast its completion date and cost
- Provide budget and schedule variances

The following are the basic terminologies used in EVM.

1. BCWS → budgeted cost of work scheduled or planned value (PV)
   It is the planned cost of the total amount of work scheduled to be performed by the milestone date.
2. BCWP → budgeted cost of work performed or EV
   It is the actual cost incurred to accomplish the work that has been done to date.
3. ACWP → actual cost of work performed or actual cost (AC)
   It is the planned cost to complete the work that has been done.

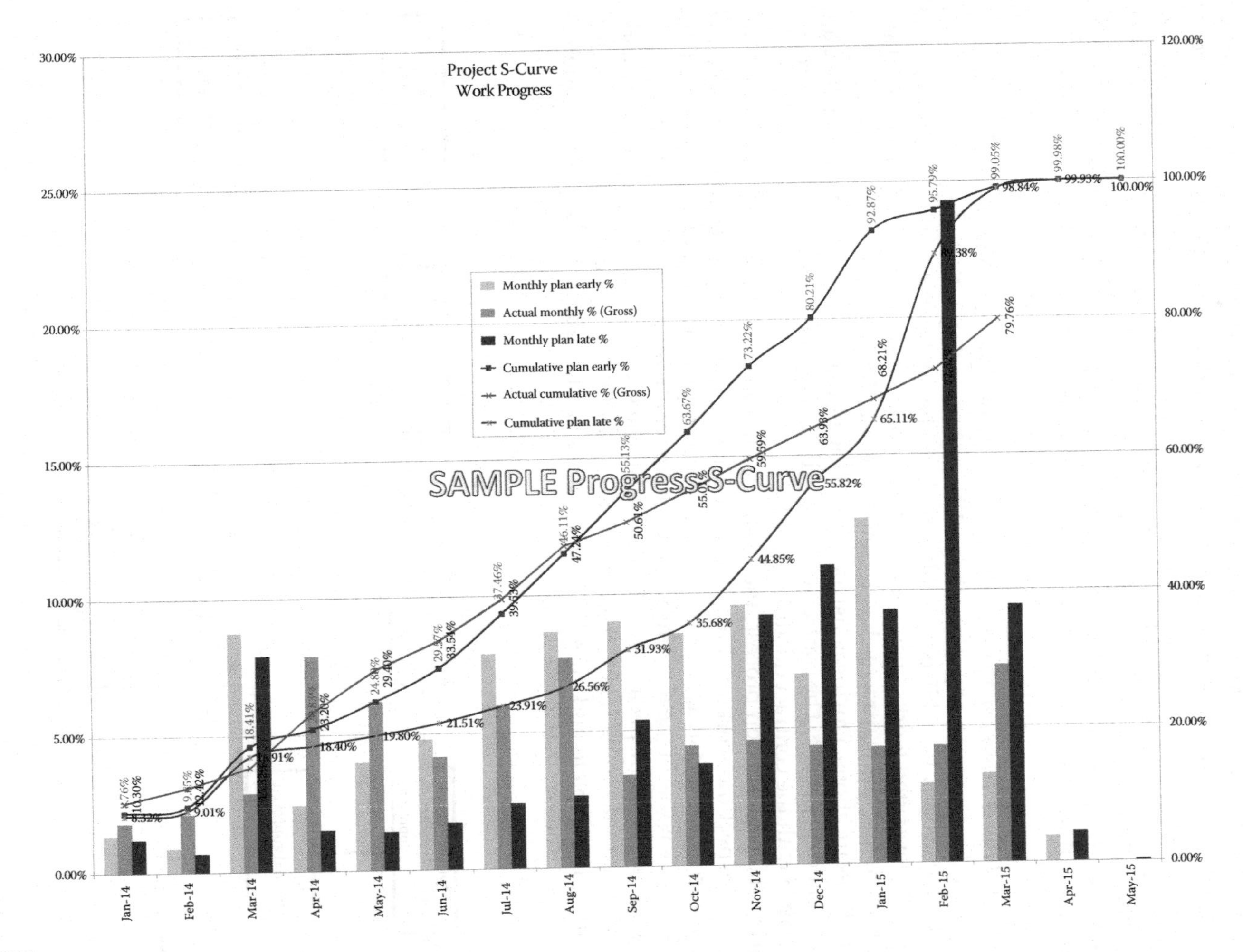

FIGURE 7.164
S-curve (work progress).

7.4.6.13.5 *Control Quality*

In order to achieve quality in construction projects, all the works have to be executed as per approved shop drawings using approved material and fully coordinating with different trades. Proper sequencing and method statement for installation of work must be followed by the contractor to avoid rejection/rework. Figure 7.165 illustrates the sequence of execution of work.

Following this sequence will help the contractor to avoid the rejection of works. Rejection of checklist will result in rework, which will need time to redo the works and cost implication to the contractor. Frequent rejection of works may delay the project ultimately affecting the overall completion schedule.

7.4.6.13.5.1 Inspect Executed Works The inspection of construction works is performed throughout the execution of project. Inspection is an ongoing activity to physically check the installed works. Checklists are submitted by the contractor to the consultant, who inspects the executed works/installations. If the work is not carried out as specified, then it is rejected, and the contractor has to rework or rectify the same to ensure compliance with the specifications. During construction, all the physical and mechanical activities are accomplished on the site. The contractor carries final inspection of the works to ensure full compliance with the contract documents.

During the construction process, the contractor has to submit the checklists to the consultant to inspect the works. Submission of check or request for inspection is an ongoing activity during the construction process to ensure proper quality control of construction. Concrete work is one of the most important components of building construction. The concrete work has to be inspected and checked at all the stages to avoid rejection or rework. Necessary care has to be taken right from the control of design mix of the concrete till the casting is complete and cured. The contractor has to submit a checklist at different

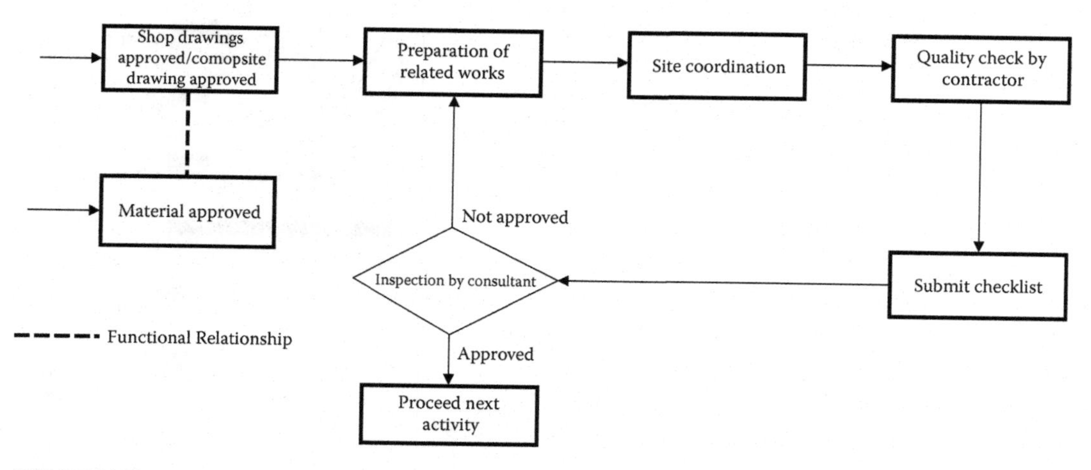

FIGURE 7.165
Sequence of execution of works. (Abdul Razzak Rumane (2010), *Quality Management in Construction Projects*, CRC Press, Boca Raton, FL. Reprinted with permission from Taylor & Francis Group.)

stages of the concrete work and has to make certain tests, specified in the contract, during casting of concrete. In order to ensure that structure concrete works are executed without any defects or rejection and achieve concrete strength as specified, proper sequencing of works is important. Figure 7.166 illustrates work sequence for formwork and Figure 7.167 illustrates process for concrete casting.

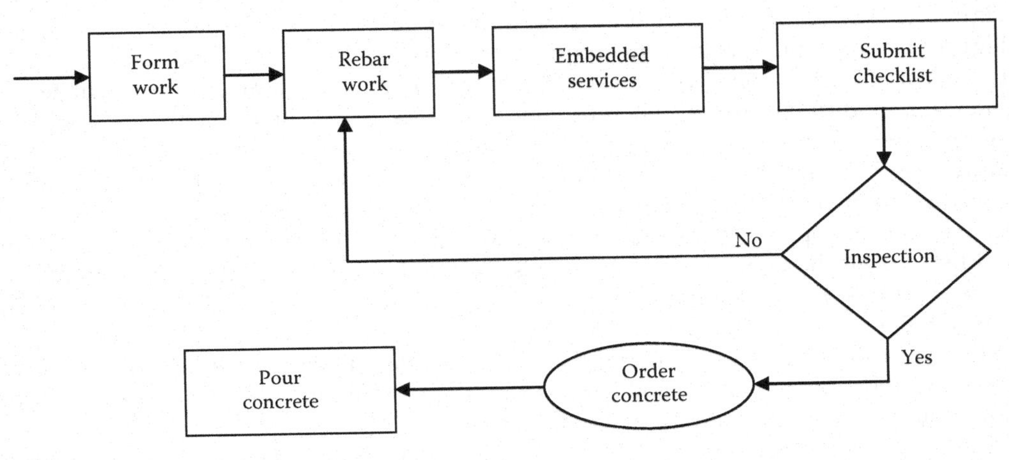

FIGURE 7.166
Flowchart for concrete casting. (Abdul Razzak Rumane (2010), *Quality Management in Construction Projects*, CRC Press, Boca Raton, FL. Reprinted with permission from Taylor & Francis Group.)

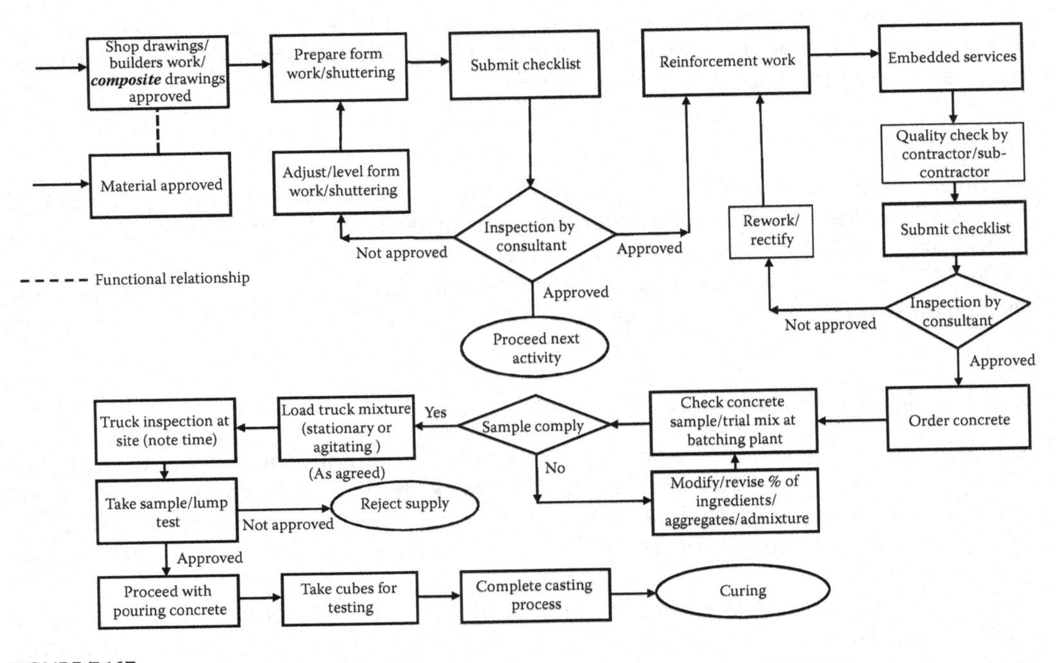

FIGURE 7.167
Process for structural concrete work. (Abdul Razzak Rumane (2013), *Quality Tools for Managing Construction Projects*, CRC Press, Boca Raton, FL. Reprinted with permission from Taylor & Francis Group.)

The following are the checklists that a contractor submits at different stages of concreting process. Figure 7.168 illustrates a checklist for quality control of formwork.

Figure 7.169 illustrates a notice for daily concrete casting.

Figure 7.170 illustrates a checklist for concrete casting.

Figure 7.171 illustrates a checklist for quality control of concreting.

Figure 7.172 illustrates a report on concrete casting.

Figure 7.173 illustrates a notice for testing at lab.

Figure 7.174 illustrates a concrete quality control form.

In certain projects, the owner asks the contractor to involve an independent testing agency for quality control of concrete. The agency is responsible for quality control of materials, perform tests, and submit test reports to the owner.

Figure 7.175 illustrates a cause-and-effect diagram to analyze causes for rejection of concrete work not meeting the specified requirements.

In order to get approval of executed/installed work/system, the contractor has to submit checklists for all the installations and works executed as per the contract documents and the approved shop drawings and materials. Figure 7.176 illustrates a checklist for general works to be inspected by the consultant.

If the consultant finds that an item has been executed at site not as per contract documents, specification, or general code of practice, then a remedial note is issued by the consultant. Figure 7.177 illustrates a remedial note.

The contractor is required to reply on the same form after taking necessary action. Upon finalization of the issue, the withdrawal notice is issued by the consultant/supervision staff.

The consultant's supervision staff always makes a routine inspection during the construction process. A nonconformance report is prepared and sent to the contractor to make corrective/preventive action toward this activity. Figure 7.178 illustrates a nonconformance report used for a building construction project.

7.4.6.13.6 Control Risk

The contractor has to prepare risk plan and the response to control, to mitigate the risk. Table 7.92 illustrates a risk plan for material handling at site, and Table 7.93 illustrates a risk plan for material delivery using a mobile crane.

Table 7.82 discussed earlier has listed probable risks during the construction phase, and Table 7.86 illustrates risk affecting scope, schedule, and cost and its effect and mitigation plan.

7.4.6.14 Validate Executed Works

The inspection of construction works is performed throughout the execution of project. Inspection is an ongoing activity to physically check the installed works. Checklists are submitted by the contractor to the consultant to inspect the executed works/installations. Table 7.94 lists the items to be checked by the consultant and probable reasons for rejection of executed works.

If the work is not carried out as specified, then it is rejected and the contractor has to rework or rectify the same to ensure compliance with the specifications. During construction all the physical and mechanical activities are accomplished on the site. The contractor carries final inspection of the works to ensure full compliance with the contract documents.

Project Name

Consultant Name

QUALITY CONTROL OF FORMWORK / FALSEWORK / DIMENSIONS / LEVELS

CONTRACTOR: DATE: / /

CONTRACT NO:

Site Engineer : ______________________

Inspected Element : ______________________

(N.B.: This form is to be prepared by the Site Engineer and submitted to the R.E.)

(A=Acceptable, N=Needs Adjustment, U=Unsatisfactory)

1) Form Dimensions & Levels:

		A	N	U
1.1	Setting Out			
1.2	Top of Concrete Level Ready for Casting			
1.3	Dimensions			
1.4	Heights & Levels			
1.5	Chamfers			

2) Falsework:

		A	N	U
2.1	Supports			
2.2	Rigidity			
2.3	Bracing			
2.4	Screw Jacks			
2.5	Timber Straightness			
2.6	Splices of Vertical Members			

3) Formwork:

		A	N	U
3.1	Rigidity			
3.2	Water Tightness			
3.3	Steel Bolts / Rods / Ties			
3.4	Openings & Inserts			
3.5	Cleanliness			
3.6	Oiling			
3.7	Working Plat forms and Walkways			

R.E's Comments:

____

____

____

____

Signature of Resident Engineer ______________________ Date: ______________

FIGURE 7.168
Checklist for form work.

Project Name
Consultant Name

NOTICE OF DAILY CONCRETE CASTING

CONTRACTOR:
CONTRACT NO:

DATE:

The following is the tentative schedule of our concrete casting for elements already inspected and accepted for casting this day:

NO.	LOCATION	ELEMENT	GRADE Mpa	QTY m^3	TENTATIVE CASTING TIME	
					START	FINISH
TOTAL						

Your action for supervision of the above would be very much appreciated.

SITE ENGINEER	CONTRACTOR MANAGER
______________	______________

CONSULTANT RECEIVED: ______________ Date: TIME:____

R. E's COMMENTS:-

Resident Engineer Date

FIGURE 7.169
Notice for daily concrete casting.

Project Name
Consultant Name

CHECKLIST FOR CONCRETE CASTING

CONTRACTOR: ______________________

CONTRACT NO: ______________________ **Date:** / /

Building/Structure	:	
Element	:	

No.	Description	Status	
		Availability	Detail
	PLANT & TOOLS		
1	Concrete Pumps		
2	Standby Concrete Pumps		
3	Cranes		
4	Truck Mixers		
5	Vibrators		
6	Trowlers		
7	Lighting		
8	Access Means		
9	Communications		
	QUALITY CONTROL		
1	Cubes		
2	Slump Apparatus		
3	Thermometer		
	STAFF & LABOUR		
1	Engineer		
2	Foreman		
3	Carpenter		
4	Steel Fixer		
5	Electrician		
6	Mechanic		
7	Vibrating Labour		
8	Trowling Labour		
9	Ordinary Labour		

SITE ENGINEER **CONTRACTOR MANAGER**

CONSULTANT RECEIVED: ______________________ DATE: / / TIME: ________

COMMENTS:

Signature of Resident Engineer: ______________________ **Date:** ______________

FIGURE 7.170
Checklist for concrete casting.

Project Name
Consultant Name

QUALITY CONTROL OF CONCRETING

CONTRACTOR: **DATE:** / /
CONTRACT NO:

Site Engineer :

Inspected Element :

(N.B.: This form is to be prepared by the Site Engineer and submitted to the Consultant)

Starting Time
Maximum Recorded Temperature

Finishing Time
Total Concrete Quantity
Average Slump
Cubes IDs

Number of Vibrators Available through Process Compaction (S=Satisfactory, U=Unsatisfactory)	S	U
Preparation of Previous Construction Joints	Yes	No
Stopped at Preplanned Locations	Yes	No
Preparation of New Construction Joints	Yes	No
Measurements of Falsework / Formwork Deformation after casting (S=Satisfactory, U=Unsatisfactory)	S	U

Curing Starting Time

R.E's Comments:

Resident Engineer: **Date:**

FIGURE 7.171
Quality control of concreting.

Owner Name
Consultant Name

REPORT ON CONCRETE CASTING AT SITE (CC)

CONTRACTOR: CC No.:___________

CONTRACT NO: Date:___________

Time:___________

Location:	Zone:	Approved Checklist Ref: No. Date :	
Type of Concrete	(Required Strength)		
Starting Time			
Finishing Time			
Approximate Quantity	m3	No. of Trucks	
Test Cubes / Cylinders	Total No.	Sr. Nos.	
	Cylinder Nos.	Slump	Average
SLUMP IN CENTIMETERS			
Engineer on Duty	Name:		Time:

SAMPLE FORM

Resident Engineer's Remarks:

Machinery: ___________

Manpower: ___________

Workmanship: ___________

Rate of Casting: ___________

Others:

Distribution: ☐ Engineer ☐ Resident Engineer ☐ Contractor

FIGURE 7.172
Report on concrete casting.

Project Name
Consultant Name

NOTICE FOR TESTING AT SITE LAB.

CONTRACTOR:
CONTRACT NO:

Date: / /

The following is the tentative schedule for testing activities in the Site Laboratory.

I. **SOIL TESTING**

NO.	LOCATION	SAMPLE	TYPE OF TESTING

II. **CONCRETE CUBES TESTING**

NO.	LOCATION	AGE OF CONCRETE	NO. OF CUBES

We request your to witness the testing at..........A.M./P.M.

SITE ENGINEER	CONTRACTOR MANAGER
________________	________________

CONSULTANT RECEIVED:____________________ Date: / / Time:____

R.E's Comments: ____________________

Signature of Resident Engineer Date:

FIGURE 7.173
Notice for testing at lab.

Project Name

Consultant Name

CONCRETE QUALITY CONTROL FORM

CONTRACTOR: Serial No.: ____________

CONTRACT NO:

Building	____________	Drawing Refer.	____________	Concrete Temp.	________
Particular Area	____________	Type of Concrete	____________	Slump (mm)	________
Date of Cube	____________	Special Material	____________	Drawing Refer.	________
Conc. Supplier	____________	Truck #	____________		

Date of Test ___/___/______ Age of Cubes ________ Days

Cube No.	Width cm	Length cm	Height cm	Area cm2	Weight gram	Volume cm3	Density g/cm3	Failing Load kN	Compressive N/mm2 Strength	Type of Failure
1										
2										
3										

Required 7 Day Strength ________ Average Comp. Strength ________

Laboratory Technician Consultant Contractor

Date of Test ____________ Age of Cubes ________ Days

Cube No.	Width cm	Length cm	Height cm	Area cm2	Weight gram	Volume cm3	Density g/cm3	Failing Load kN	Compressive N/mm2 Strength	Type of Failure
1										
2										

Required 14 Day Strength ________ Average Comp. Strength ________

Laboratory Technician Consultant Contractor

Date of Test ____________ Age of Cubes ____________

Cube No.	Width cm	Length cm	Height cm	Area cm2	Weight gram	Volume cm3	Density g/cm3	Failing Load kN	Compressive N/mm2 Strength	Type of Failure
1										
2										

Required 28 Day Strength ________ Average Comp. Strength ________

Laboratory Technician Consultant Contractor

Distribution: ☐ Owner ☐ A/E ☐ Contractor

FIGURE 7.174
Concrete quality control form.

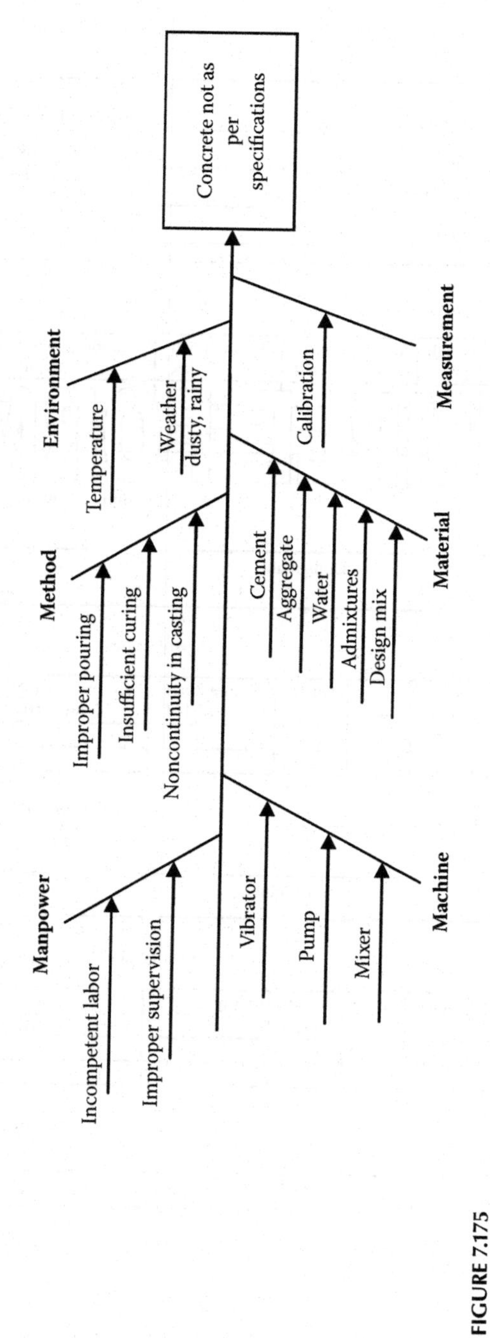

FIGURE 7.175
Cause-and-effect diagram for concrete.

Project Name
Project Name
CHECKLIST

CONTRACTOR: ____________________ CHECKLIST No.: [ ]

CONTRACT No.: ____________________ PREVIOUS C.L. No.: [ ]

TO : **Resident Engineer**

CCS ACTIVITY NO: __________ SPECIFICATION DIVISION: ______ SECTION: ______

AREA: [ ] Building Works [ ] Electrical Works [ ] Mechanical Works

[ ] HVAC Works [ ] Finishes Works [ ]

Please inspect the following :-

Location:

Work :

SAMPLE FORM

Sketch(es) attached { } No.

The work to be inspected has been coordinated with all related subcontractors.

Estimated Quantity of Work: __________ Date & Time Inspection Required: __________

Contractor Signature: __________ Date & Time __________

Received By: __________ Date & Time __________

C.C.: Owner Rep: __________ Date & Time __________

(All request must be submitted at least 24 hours prior to the required inspection)

Reply: The above is Approved/Not approved for the following:-

Inspected by __________ Date & Time ______ Resident Engineer ______ Date & Time ______

Received by Contractor __________ Date & Time __________

C.C.: Owner Rep: __________ Date & Time __________

FIGURE 7.176
Checklist.

Project Name
Consultant Name

REMEDIAL NOTE (RN)

Contractor: ______________________ R.N. No.: __________

Contract No.: ______________________ DATE: ____________

Your attention is drawn to the following works which have not been carried out in accordance with the Contract and are therefore not acceptable. Failure to carry out remedial works within a reasonable period of time may result either in additional work at your expense, or the Employer may elect to invoke Clause — of the General Conditions of Contract.

LOCATION:

DEFECTS:

SAMPLE FORM

Signed: ______________ Received by: ______________
Resident Engineer Contractor/Date

Distribution: ☐ Owner ☐ A/E ☐ Contractor ☐

FIGURE 7.177
Remedial note.

Project Name
Consultant Name

NON-CONFORMANCE REPORT
NO. [ ]

CONTRACTOR: DATE: / /

CONTRACT NO:

Location of Non-Conformance: ______

Drawing / Specification: ______

SAMPLE FORM

Description of Non-Conformance: ______

Resident Engineer: ______

Corrective/Preventive Action: (Proposed by Contractor)

Quality Control Engineer: ______

Contractor Manager: ______ Date: ______

Comments:

Resident Engineer: ______ Date: ______

FIGURE 7.178
Nonconformance report.

TABLE 7.92

Risk Plan for Material Handling

Risk Assessment Owner Name Project Name										
Activity = Material Handling										
Sr. No.	Hazard	Potential Accident or Identified Risk	Initial Risk Rating: Probability	× Severity	Risk Factor	Control Measures in Place/Mitigation Actions	Final Risk Rating: Probability	× Severity	Residual Risk Factor	Risk Acceptable
1	Fall of materials	Serious injury or permanent disability	3	3	9	(a) Secure materials stored in tiers by stacking, racking, blocking, or interlocking to prevent them from falling. (b) Stack the materials properly in marked location. (c) When lifting objects, lift with your legs, keep your back straight, do not twist, and use handling aids. (d) Loads with sharp or rough edges, wear gloves or other hand and forearm protection. (e) Dock boards must have handholds, or other effective means for safe handling.	2	1	2	Low
2	Fall of personnel	Injury to the personnel	3	3	9	(a) Store materials safely to avoid struck by/crushed by hazards. (b) Do not stand in corners where emergency movement is restricted. (c) When lifting objects, lift with your legs, keep your back straight, do not twist, and use handling aids. (d) Wear PPE. (e) Use taglines to control the load movement. (f) Avoid lifting above the shoulder level.	3	1	3	Low

RPN—Risk Priority Number

8–16	High risk
4–6	Medium risk
1–3	Low risk

Source: Abdul Razzak Rumane (2013), *Quality Tools for Managing Construction Projects*, CRC Press, Boca Raton, FL. Reprinted with permission from Taylor & Francis Group.

TABLE 7.93

Risk Plan for Material Delivery using Mobile Crane

Risk Assessment Owner Name Project Name								
Activity = Material Delivery using Mobile Crane								
SL.NO.	**Hazard**	**Potential Accident or Identified Risk**	**Initial Risk Rating** **Probability × Severity**	**Risk Factor**	**Control Measures in Place/Mitigation Actions**	**Final Risk Rating** **Probability × Severity**	**Residual Risk Factor**	**Risk Acceptable**
	(a) Fall of object from the height during lifting. (b) Failure of slings or lifting gears.	May cause permanent disability or death to the person. Damage to the property	4 4	16	(a) Plan the load and lift accordingly. (b) Check all the lifting equipments and materials before commencing the work. (c) Adopt proper signaling method and ensure proper communication and coordination. (d) Ensure proper guiding of the load to avoid jerks and swaying of the load. (e) Ensure that the crane and lifting gears have been tested and examined by the approved third-party agencies. (f) Ensure that the rigger or signal man are certified competent enough for lifting operations. (g) Wear appropriate PPE (h) No one should stand under the suspended load. (i) Area to be cordoned off by tiger or warning tape. (j) Alert the other employees by displaying the warning signs/posters during hoisting the truss.	2 1	2	**Low**

(*Continued*)

TABLE 7.93 (*Continued*)

Risk Plan for Material Delivery using Mobile Crane

Risk Assessment Owner Name Project Name								
Activity = Material Delivery using Mobile Crane								
			Initial Risk Rating			Final Risk Rating		
Sr. No.	Hazard	Potential Accident or Identified Risk	Probability × Severity	Risk Factor	Control Measures in Place/Mitigation Actions	Probability × Severity	Residual Risk Factor	Risk Acceptable
					(k) Ensure that the SWL is not exceeded. (l) Never leave the load suspended without an operator at the controls. (m) Ensure the slings used are suitable for the purpose intended. (n) Ensure that the movement of the load is clear of all obstructions and overhead hazards.			

RPN—Risk Priority Number

8–16	High risk
4–6	Medium risk
1–3	Low risk

TABLE 7.94

Reasons for Rejection of Executed Works

Sr. No.	Description of Works	Probable Reasons for Rejection
Shoring		
1	Shoring	No adequate support for shoring. Vertical. plumb level not proper. No proper bracing. Not enough depth. Anchor test not approved
2	Dewatering	Water level not under control.
Earth Works		
1	Excavation	Not as per specified level. Surface is not even. Excavated material not removed from the site.
2	Backfilling	Compaction is not proper. Backfilling thickness is not as specified. Soil is rubbish and loose. Soil test (strength) failed.
Concrete Substructure		
1	Blinding concrete	Thickness not as specified. Concrete strength not as specified.
2	Termite control	Uneven spray.
3	Reinforcement of steel	Reinforcement arrangement not as specified. Support not proper. Water stops are not provided. Construction joints are not provided.
4	Concrete casting	Concrete strength not as specified.
5	Shuttering for beams and columns	Shuttering dimensions not as specified. Shuttering is not strong enough. Shutters are not vertical. Shuttering height not proper.
Concrete Super Structure		
1	Reinforcement of steel for beams and columns	Reinforcement arrangement not as specified. Support not proper.
2	Shuttering for columns and beams	Shuttering dimensions not as specified. Shuttering is not strong enough. Shutters are not vertical. Shuttering height not proper.
3	Concrete casting of beams and columns	Concrete strength not as specified.
4	Formwork for slab	Props spacing not correct. Props are not sturdy. Formwork surface is not clean.
5	Reinforcement for slab	Reinforcement arrangement not as specified. Reinforcement is not properly placed and secured. Spacers are not provided. Minimum concrete cover not provided. Construction joints not provided.
6	Concrete casting of slab	Concrete strength not as specified. Casting level in not proper.
7	Precast panels	Panels are not fixed properly. Load test on panel not performed before erection.
Masonry		
1	Block work	Block alignment is not proper. Joints are not aligned. Guidelines are missing. Anchor beads are not provided. Reinforcement mesh is not provided. Mortar is not as specified.
2	Concrete unit masonry	Block alignment is not proper. Joints are not aligned. Reinforcement mesh is not provided. Mortar is not as specified.
Partitioning		
1	Installation of frames	Stud spacing not correct. Fixation method is not as specified. Insulation is not provided.
2	Installation of panels	Alignment not proper. Joints are not proper. Panels are not painted.

(Continued)

TABLE 7.94 (*Continued*)

Reasons for Rejection of Executed Works

Sr. No.	Description of Works	Probable Reasons for Rejection
Metal Work		
1	Structural steel work	Anchorage and fixing not proper. Method of erection not as specified. Fire protection is not applied. Finishing is different than specified.
2	Installation of cat ladders	Fixation is not proper. Alignment is not done. Finishing is not as specified.
3	Installation of balustrade	Fixation is not proper. Alignment is not done. Finishing is not as specified.
4	Installation of space frame	Fixation is not proper. Alignment is not done. Finishing is not as specified.
5	Installation of handrails and railings	Fixation is not proper. Alignment is not done. Finishing is not as specified.
Thermal and Moisture Protection		
1	Applying waterproofing membrane	Number of layers not as specified. There is no overlap between the rolls. Application not done properly. Number of coats as per specs not applied. There is no skirting. Leakage test failed.
2	Installation of insulation	Fixation is not properly. Insulation thickness is not as specified.
Doors and Windows		
1	Aluminum windows and doors	Dimensions are not as specified. Accessories and hardware are different than specified. Windows are not air/water tight and not sealed properly. Doors are not opening properly.
2	Glazing	Color and type is different than specified. Fixing method is not proper. Cracks are observed in some of the units.
3	Steel windows and doors	Alignment not proper. Hardware and accessories not as specified. Finishing is not proper.
4	Wooden frames	Dimensions are different. Finishing is not as specified.
5	Wooden doors	Alignment not proper. Finishing is not as specified. Fixation hardware is not as specified.
6	Curtain wall	Pattern is not as specified. Glazing is different than specified. Drainage system is not provided. Expansion joints not provided. Wind pressure and water test failed.
7	Hatch doors	Door is not aligned. Finishing is not matching as per architectural requirements. Fixing is not proper.
8	Louvers	Alignment and levels not proper. Accessories not as specified.
Internal Finishes		
1	Plastering	Cracks in the plaster. Voids are observed. Specified accessories not used. Hollow sound observed. Curing is not enough.
2	Painting	Number of layers not as specified. Color and texture is not proper.
3	Cladding—ceramic	Alignment, angles and joints are not proper. Grouting is not done properly. Color and texture pattern is not matching.
4	Cladding—marble	Fixation is not proper. Alignment, angles, and joints are not proper. Lines are not matching.
5	Ceramic tiling	Alignment, angles, and joints are not proper. Grouting is not done properly. Color and texture pattern is not matching.
6	Stone flooring	Fixation is not proper. Alignment, angles, and joints are not proper.

(Continued)

TABLE 7.94 (*Continued*)

Reasons for Rejection of Executed Works

Sr. No.	Description of Works	Probable Reasons for Rejection
7	Acoustic ceiling	Suspension system is not as specified. Alignment, levels, and joints are not proper. Services openings are not provided. Ceiling height not matching with approved level.
8	Demountable partitions	Fixation is not proper. Alignment is not done properly.
Furnishing		
1	Carpeting	Joints are not smooth. Carpet fixed without adhesive. Edges are not trim properly. Color and texture not matching.
2	Blinds	Stacking/rolling not proper. Fixing is not secured.
3	Furniture	Location is not as per the approved layout. Dimensions are not correct. Finishing is not proper. Fixation is not secured properly. Accessories are not matching with the furniture.
External Finishes		
1	Painting	Number of layers not as specified. Color and texture is not proper.
2	Brickwork	Brick alignment is not proper. Joints are not aligned. Grouting is not proper.
3	Stone	Fixation is not proper. Alignment, angles, and joints are not proper. Color and texture are not matching.
4	Cladding—aluminum	Alignment is not proper. Cladding is not leveled and aligned. Fixation is not secured. Joints are without sealant.
5	Cladding—granite	Fixation is not proper. Alignment, angles, and joints are not proper. Color and texture are not matching.
6	Curtain wall	Accessories are not properly fixed. Leveling and alignment not done. Drainage system is not provided. Wind pressure and water tests failed.
7	Glazing	Glass type is different than specified and approved. Fixation is not secured. Leveling and alignment not done properly. Joints have gap. Frame finishing is not as per approved sample.
Equipment		
1	Installation of maintenance equipment	Brackets are not fixed at specified location.
2	Installation of kitchen equipment	Equipment are not fixed as per approved layout.
3	Installation of parking control	Barrier is not fixed at specified location. Control panel is not terminated.
Roof		
1	Parapet wall	Parapet level is not correct. Finishing is not as specified.
2	Thermal insulation	Type of insulation and thickness is different than specified.
3	Waterproofing	Number of layers are less than specified. Application is not smooth. There is no overlap between the sheets. Skirting is not provided. Water test failed.
4	Roof tiles	Thickness of tiles is less than specified. Tiles are not aligned and leveled. Joints are not proper.
5	Installation of drains	Drains are not located as per approved shop drawings. Fixation is not proper.
6	Installation of gutters	Type of gutters is different than specified. Gutters are not aligned and leveled.

(*Continued*)

TABLE 7.94 (*Continued*)

Reasons for Rejection of Executed Works

Sr. No.	Description of Works	Probable Reasons for Rejection
Elevator System		
1	Installation of rails	Fixation is not proper.
2	Installation of door frames	Frame is not aligned properly.
3	Installation of cabin	Cabin is not leveled.
4	Cabin finishes	Cabin finishes are incomplete.
5	Installation of wire rope	Rope joints are weak.
6	Installation of drive machine	Drive is leveled properly. Equipment earthing is not done.
7	Installation of controller	Location of controller is not proper. Controller is not terminated. Equipment earthing is not provided. Identification label is not provided.
HVAC Works		
1	Installation of piping	Hanger supports are not properly fixed. Invert level is not correct. Piping is rusty, not clean and not painted.
2	Installation of ducting	Duct level not correct. Duct metal not clean. Suspension system not properly fixed. No proper sealant around duct joints.
3	Installation of insulation	Insulation has cuts. Straps not fixed at specified interval.
4	Installation of dampers, grills, and diffusers	Material is squeezed. Color is faded. Material not matching with architectural requirements.
5	Installation of cladding	Cladding surface is not even. Cladding is not overlapped and sealed properly.
6	Installation of fans	Fans fixing is not secured.
7	Installation of fan coil units (FCU)	FCU level not correct. FCU fixed without spring isolators. No slope for drain pipe.
8	Installation of air handling units (AHU)	No flexible connector between chilled water piping and AHU. AHU body is damaged. Body paint is peeled off.
9	Installation of pumps	Pumps are not leveled. Cable termination not complete. Equipment earthing not done.
10	Installation of chillers	Rubber isolators not installed. Anchorage not fixed properly.
11	installation of cooling towers	Vibration springs not installed. Damage in cooling tower surface. Spray nozzle damaged.
12	Installation of thermostat and controls	Location not as specified. Thermostat body damaged.
13	Installation of starters	Location to be near the equipment. Installation height not as specified.
14	Building management system (BMS)	All the components are not installed. Termination not complete. Wiring not dressed and bunched properly.
Mechanical Works		
A. Water Supply		
1	Installing of piping system	Pipe supports required to keep the pipe in straight position.
2	Installation of pumps	Pumps are not leveled. Cable termination not complete. Equipment earthing not done.
3	Filter units	Filter installed without bypass line and tab point for pressure gauge.
4	Toilet accessories	Towell rod not fixed.
5	Installation of insulation	Insulation has cuts. Insulation is not covered with canvas.
6	Water heaters	Water heater drain not installed.
7	Hand dryers	Installation height is more than specified. Termination is not proper.
8	Water tank	Float switch for pump not installed.

(*Continued*)

TABLE 7.94 (*Continued*)

Reasons for Rejection of Executed Works

Sr. No.	Description of Works	Probable Reasons for Rejection
B. Drainage System		
1	Installation of pipes below grade	Drainage pipe to be tightened properly. Slope not proper.
2	Installation of pipes above grade	Slope not proper. More support is required.
3	Installation of manholes	Manhole level is not correct.
4	Installation of clean out, floor drains	Floor drains and clean out are not flush with the floor finish.
5	Installation of sump pumps	Height of float switch need adjustment.
6	Installation of gratings	Grating frame not flush with floor finish level.
C. Irrigation System		
1	Installation of piping system	Pipe crossing under road need sleeves.
2	Installation of pumps	Pumps installed without suction strainer.
3	Installation of controls	Controllers are not weatherproof. Installation height is not proper. Fixing method is not proper.
D. Fire Fighting System		
1	Installation of piping system	Pipes are dirty and need cleaning and painting.
2	Installation of sprinklers	Sprinkler spacing not as per authorities approved shop drawing.
3	Installation of foam system	Foam agent concentration level not proper.
4	Installation of pumps	Eccentric reducer need to fixed at the suction side of the pump. Pumps are not leveled. Cable termination not complete. Equipment earthing not done.
5	Installation of hose reels/cabinet	Cabinet door not closing properly. Color of cabinet to be as per authorities requirement.
6	Installation of fire hydrant	Fire hydrant to be fixed properly at specified location.
7	External fire hose cabinet	External fire hose cabinet need hose rack.
Electrical Works		
1	Conduiting—raceways	Method of installation not proper. Conduit run is not parallel. Minimum clearance from other services less than specified. Supports are not secured, straps, or clamp not provided. Spacing between embedded conduit and concrete aggregate is not as specified. Fastening of conduit with reinforcement steel is not secured. Location of sleeves not as per approved shop drawings. Location of boxes is not as specified and is not coordinated.
2	Cable tray—Trunking	Cable trays are not aligned properly. Cable tray run is not parallel and no proper bends. Supports are not as per approved shop drawings. Minimum spacing between any cable tray and other services not maintained. Supports are not secured.
3	Floor Boxes	Location is not coordinated. Level of floor boxes is not proper.

(*Continued*)

TABLE 7.94 (*Continued*)

Reasons for Rejection of Executed Works

Sr. No.	Description of Works	Probable Reasons for Rejection
4	Wiring	Number of wires in the conduit are not as per approved shop drawing. Circuit wiring for switches not properly done. Termination method is not proper.
5	Cabling	Distance between two cables on cable tray is less than twice the diameter of larges cable diameter. Cable tie is not provided. Circuit identification is missing.
6	Installation of bus duct	Supports are not fixed properly. connection between the ducts is not proper. Level of installation is not as specified. Spacing between parallel is less than specified. Supports are not secured.
7	Installation of wiring devices/accessories	Installation height is not as specified. Coordination with architectural requirements not done to match with the finishes.
8	Installation of light fittings	Installation methods is not correct. Proper size of hangers and support not provided. Location and levels are not matching with architectural requirements.
9	Grounding	Welding method for joining tapes is not thermo-weld type. Connection methods between tape and clamp to be as specified. Provide clamp at specified distance.
10	Distribution switch boards	Installation height of boards not as specified. Termination of wires/cables not done properly. Shrouding for cable not provided. Wires and cables do not have continuity. Wires in distribution board is not properly dressed and bunched.
Fire Alarm System		
4	Installation of detectors, bells, pull stations, interface modules	Location is not as specified. Height of installation not matching with shop drawing and architectural requirements.
5	Installation of repeater panel	Installation height not as specified and method of installation not proper. Termination of cables is not done.
6	Installation of mimic panel	Installation height not as specified. Method of installation not proper. Termination of cables is not done.
7	Installation of fire alarm panel	Installation height not as specified. Method of installation not proper. Termination of cables is not done.
Telephone/Communication System		
1	Installation of racks	Method of installation not as per approved shop drawing. Termination of cables not done properly. Racks are not leveled. Cables are not labeled.
2	Installation of switches	Switches are not properly installed. Switched are not installed in sequence. Cabling is not complete. Identification label not provided for cables.
Public Address System		
5	Installation of speakers	Fixing of speakers is not secured. Height of installation is not as specified. Speakers installed in false ceiling do not have proper support.
6	Installation of racks	Method of installation not as per approved shop drawing. Termination of cables not done properly. Racks are not leveled. Cables are not labeled.
7	Installation of equipment	Equipment are not properly installed. Cabling is not proper.

(*Continued*)

TABLE 7.94 (*Continued*)

Reasons for Rejection of Executed Works

Sr. No.	Description of Works	Probable Reasons for Rejection
Audio–Visual System		
5	Installation of speakers	Fixing of speakers is not secured. Height of installation is not as specified. Speakers installed in false ceiling do not have proper support.
6	Installation of monitors/screens	Method of fixing nor proper. Installation height is not as per approved shop drawings.
7	Installation of racks	Method of installation not as per approved shop drawing. Termination of cables not done properly. Racks are not leveled. Cables are not labeled.
8	Installation of equipment	Equipment are not properly installed. Cabling is not proper.
CCTV/Security System		
3	Installation of cameras	Fixing of cameras not proper. Installation height is not as specified. Termination of cameras not proper.
5	Installation of panels	Panels not properly fixed. Cabling not complete.
7	Installation monitors/screens	Method of fixing nor proper. Installation height is not as per approved shop drawings.
Access Control		
3	Installation of RFID proximity readers and fingerprint readers	Readers not installed properly.
4	Installation of magnetic locks, release buttons, and door contacts	Magnetic locks not fixed properly. Release buttons not installed at specified height and location.
5	Installation of panel	Panels not properly fixed. Cabling not complete.
6	Installation of server	Cable termination not proper. Cables to be properly dressed. Identification labels not provided.
Systems Integration		
1	Installation of switches	Switches not stacked as per sequence. Identification labels not provided.
2	Installation of servers	Cable termination not proper. Cables to be properly dressed. Identification labels not provided.
External Works		
1	Site works	Compaction failure. Grading level not proper. Slope is not provided.
2	Asphalt work	Asphalt levels not proper. Asphalt material not as specified.
3	Pavement works	Pavement limits not as per approved layout. Curb stones not laid properly.
4	Piping works	Pipe not laid at specified depth. No protection on pipes. Pipe joints not done well. Leakage observed.
5	Electrical works	Cable not buries properly. Location of light poles not correct. Light poles are not vertically installed.
6	Manholes	Location of manholes not correct. Manhole level not correct.
7	Road marking	Marking color not as specified.

Source: Abdul Razzak Rumane (2010), *Quality Management in Construction Projects*, CRC Press, Boca Raton, FL. Reprinted with permission from Taylor & Francis Group.

7.4.7 Testing, Commissioning, and Handover Phase

Testing, commissioning, and handover is the final phase of the construction project life cycle. This phase involves testing of electromechanical systems, commissioning of the project, obtaining authorities' approval, training of user's personnel, handing over of technical manuals, documents, and as-built drawings to the owner/owner's representative, During this period, the project is transferred/handed over to the owner/end user for their use and substantial completion certificate is issued to the contractor. Figure 7.179 illustrates major activities relating to testing, commissioning, and handover phase developed based on project management process groups methodology.

7.4.7.1 *Identify Testing and Startup Requirements*

Testing and startup requirements are specified in the contract documents. It is essential to inspect and test all the installed/executed works prior to handing over the project to the owner/end user. Generally, all works are checked and inspected on regular basis while the construction is in progress; however, there are certain inspection and tests to be carried out by the contractor in the presence of the owner/consultant. These are especially for electromechanical systems, conveying systems, electrical works, low-voltage systems,

Testing, Commissioning, and Handover Phase

Management Processes	Project Management Process Groups				
	Initiating Process	Planning Process	Execution Process	Monitoring and Controlling Process	Closing Process
Integration Management	Executed Project/Facility	Testing, Commisioning, and Handover Plan	Testing and Commissioning	Punch List/Snag List	Handover of Project/ Facility
	Testing and Commissioning Program		Authorities Approval	Tesing and Commisioning Requirements	
	Contract Documents		Punch List/Snag List		
			As Built Drawings		
			Manuals		
			Spare Parts		
Stakeholder Management	Identify Stakeholders	Stakeholders Requirements	Move In Plan		Project Acceptance/ Takeover
Scope Management		Contract Documents		Authorities Approval	Lesson Learned
				Stakeholders Approval	
Schedule Management		Testing Schedule			
		Commissioning Schedule			
Cost Management					
Quality Management				Project Quality	
Resource Management		Demobilization Plan			Assign New Project/ Termination
Communication Management		Test Results			
Risk Management		Plan Start-up Risk		Control Risk	
Contract Management		Prepare Contract Closeout Documents	Finalize Closeout Documents	Check Documents for Compliance to Contract Requirements	Close Contracts
HSE Management					
Financial Management		Financial Administration and Records	Payment to All Contractors and Sub contractors		Payment to All Contractors and Sub contractors
					Payment toward all Purchases
Claim Management		Calim Resolution		Check for Claims	Settlement of Claims

FIGURE 7.179
Testing, commissioning, and handover. *Note*: These activities may not be strictly sequential; however, the breakdown allows implementation of project management function more effective and easily manageable at different stages of the project phase.

information-and-technology-related products, emergency power supply system, and electrically operated equipment, which are energized after connection of permanent power supply. Testing of all these systems start after completion of installation works. By this time, the facility is connected to permanent electrical power supply and all the equipment are energized.

7.4.7.2 Develop Testing, Commissioning, and Handover Plan

The testing is mainly carried out mainly electromechanical works/systems and electrically operated equipment/systems, which are energized after connection of permanent power supply to the facility. These include the following:

1. Conveying system
2. Water supply, plumbing, and public health system
3. Fire suppression system
4. HVAC system
5. Integrated automation system (building automation system)
6. Electrical lighting and power system
7. Grounding (earthing) and lightning protection system
8. Fire alarm system
9. Telephone system
10. Information and communication system
11. Electronic security and access control system
12. Public address system
13. Emergency power supply system
14. Electrically operated equipment

The testing of these works/systems is essential to ensure that each individual work/system is fully functional and operate as specified. The tests are normally coordinated and scheduled with specialist contractors, local inspection authorities, third-party inspection authorities and manufacturer's representatives. Sometimes, owner's representative may accompany the consultant to witness these tests.

Test procedures are submitted by the contractor along with the request for final inspection. Standard forms, charts, and checklists are used to record the testing results.

7.4.7.3 Manage Stakeholders

The following stakeholders are involved during testing, commissioning, and handover phase:

1. Owner
   - Owner's representative/project manager
2. Construction supervisor
   - Construction manager (agency)
   - Consultant (designer)
   - Specialist contractor

3. Contractor
    - Main contractor
    - Subcontractor
    - Testing and commissioning specialist
4. Regulatory authorities
5. End user
6. Third-party inspecting agency

7.4.7.3.1 Select Team Members

During the testing and commissioning phase, it is essential to select team members having experience in testing and commissioning of major projects. The team members can be from the same supervision team that was involved during execution of the project, if they have experience to carry out testing and commissioning. In most cases, the manufacturer's representative is involved to test the supplied systems/equipment. The owner may engage specialist firms to perform startup activities and commission the project.

7.4.7.3.2 Select TAB Subcontractor

The contract documents generally specify to engage specialist firm to perform testing, adjusting, and balancing of HVAC works. The firm is responsible to confirm that the HVAC installations are meeting the specified cooling and heating system requirements.

7.4.7.3.3 Develop Responsibility Matrix

Table 7.95 lists responsibilities of consultant during the closeout phase and Table 7.96 illustrates contribution of various participants during the testing, commissioning, and handover phase.

7.4.7.4 Develop Scope of Work

The contract documents specify the testing and commission works to be performed by contractor, subcontractor, and specialist supplier of equipment/systems. The following is the main scope of work to be carried out during this phase:

1. Testing of all systems
2. Commissioning of all systems
3. Obtaining authorities' approvals
4. Submission of as-built drawings
5. Submission of technical manuals and documents
6. Submission of warranties and guarantees
7. Training of owner's/user's personnel
8. Handover of spare parts
9. Handover of facility to owner/end user
10. Occupancy/operation of new facility
11. Preparation of punch list
12. Issuance of substantial certificate
13. Lessons learned

TABLE 7.95

Typical Responsibilities of Consultant during Project Closeout Phase

Sr. No.	Responsibilities
1	Ensure that occupancy permit from respective authorities is obtained
2	Ensure that all the systems are functioning and operative
3	Ensure that job site instruction (JSI) and nonconformance Report (NCR) are closed
4	Ensure that site is cleaned and all the temporary facilities and utilities are removed
5	Ensure that master keys handed over to the owner/end user
6	Ensure that guarantees, warrantees, bonds are handed over to the client
7	Ensure that operation and maintenance manuals handed over to the client
8	Ensure that test reports, test certificates, inspection reports handed over to the client
9	Ensure as-built drawings handed over to the client/end user
10	Ensure that spare parts are handed over to the client
11	Ensure that snag list prepared and handed over to the client
12	Ensure that training for client/end user personnel completed
13	Ensure that all the dues of suppliers, subcontractors, and contractor paid
14	Ensure that retention money is released
15	Ensure that substantial completion certificate issued and maintenance period commissioned
16	Ensure that supervision completion certificate from the owner is obtained
17	Lesson learned documented

Source: Abdul Razzak Rumane (2013), *Quality Tools for Managing Construction Projects*, CRC Press, Boca Raton, FL. Reprinted with permission from Taylor & Francis Group.

TABLE 7.96

Responsibilities of Various Participants (Design–Bid–Build Type of Contracts) during Testing, Commissioning, and Handover Phase

	Responsibilities		
Phase	**Owner**	**Designer**	**Contractor**
Testing, commissioning, and handover	• Acceptance of project • Takeover • Substantial completion certificate • Training • Payments	• Witness tests • Check closeout requirements • Recommend take over • Recommend issuance of substantial completion certificate	• Testing • Commissioning • Authorities' approvals • Documents • Training • Handover

Table 7.97 lists the items to be tested and commissioned prior to handing over the project.

7.4.7.5 Establish Testing and Commissioning Quality Procedure

Commissioning is the orderly sequence of testing, adjusting, and balancing the system, and bringing the systems and subsystems into operation; this starts when the construction and installation of works are complete. Commissioning is normally carried out by contractor or specialist in the presence of the consultant and owner/owner's representative and user's operation and maintenance personnel to ascertain proper functioning of the systems to the specified standards.

TABLE 7.97

Major Items for Testing and Commissioning of Equipment

Sr. No.	Discipline	Items
1	Elevator	1. Power supply 2. Speed 3. Capacity to carry design load 4. Number of stops 5. Emergency landing 6. Emergency call system 7. Elevator management system 8. Interface with fire alarm system
2	Mechanical	
2.1	Mechanical (Fire Suppression)	1. Sprinklers 2. Piping 3. Fire pumps 4. Power supply and Controls 5. Emergency power supply for fire pumps 6. Hydrants 7. Hose reels 8. Water storage facility 9. Gaseous protection system for communication rooms 10. Fire protection system for diesel generator room 11. Interface with fire alarm system 12. Interface with BMS
2.2	Mechanical (Public Health)	1. Piping 2. Pipe flushing and cleaning 3. Pumps 4. Boilers 5. Hot water system 6. Water supply and purity 7. Fixtures 8. Power supply for equipment 9. Controls 10. Drainage system 11. Irrigation system
3	HVAC	1. Pipe cleaning and flushing 2. Chemical treatment 3. Pumps 4. Duct work 5. Air handling Unit 6. Heat recovery Unit 7. Split units 8. Chillers 9. Cooling towers 10. Heating system (controls, piping, pumps) 11. Fans (ventilation, exhaust) 12. Humidifiers 13. Starters 14. Variable frequency drive 15. Motor control centers (MCC panels) 16. Chiller control panels 17. Building management system (BMS) 18. Interface with fire alarm system 19. Thermostat 20. Air balancing

(Continued)

TABLE 7.97 (*Continued*)

Major Items for Testing and Commissioning of Equipment

Sr. No.	Discipline	Items
4	Electrical	
4.1	Electrical (Power)	1. Lighting illumination levels 2. Working of photo cells and controls 3. Wiring devices (power sockets) 4. Lighting control panels 5. Electrical distribution boards 6. Electrical bus duct system 7. Main switch boards/sub main switch boards 8. Min low tension panels 9. Isolators 10. Emergency switch boards 11. Motor control centers (MCC panels) 12. Audio visual alarm panel 13. Diesel generator 14. Automatic transfer switch (ATS) 15. UPS (uninterrupted power supply) 16. Earthing (grounding) system 17. Lightning protection system 18. Surge protection system 19. Power supply to equipment (HVAC, mechanical, elevators, others) 20. IP rating of out door switches, isolators, switch boards 21. Interface with BMS 22. Emergency power system 23. Exhaust emission of generator
4.2	Electrical (Low Voltage)	1. Fire alarm system 2. Communication system 3. CCTV system 4. Access control system 5. Public address system 6. Audio visual system 7. Master satellite antenna system
5	External Works	1. Lighting poles 2. Boundary wall lighting 3. Lighting bollards 4. Irrigation system 5. Electrical distribution boards
6	General	1. Power supply for gate barriers 2. Automatic gates 3. Rolling shutters 4. Window cleaning system 5. Gas detection system 6. Water/fluid leak detection system

Source: Abdul Razzak Rumane (2013), *Quality Tools for Managing Construction Projects*, Reprinted with permission of Taylor & Francis Group.

The commissioning of the construction project is a complex and intricate series of startup operations, and may extend over many months. Testing, adjusting, and operation of the systems and equipment are essential in order to make the project operational and useful for the owner/end user. It requires extensive planning and preparatory work, which commences long before the construction is complete. All the contracting parties are involved in

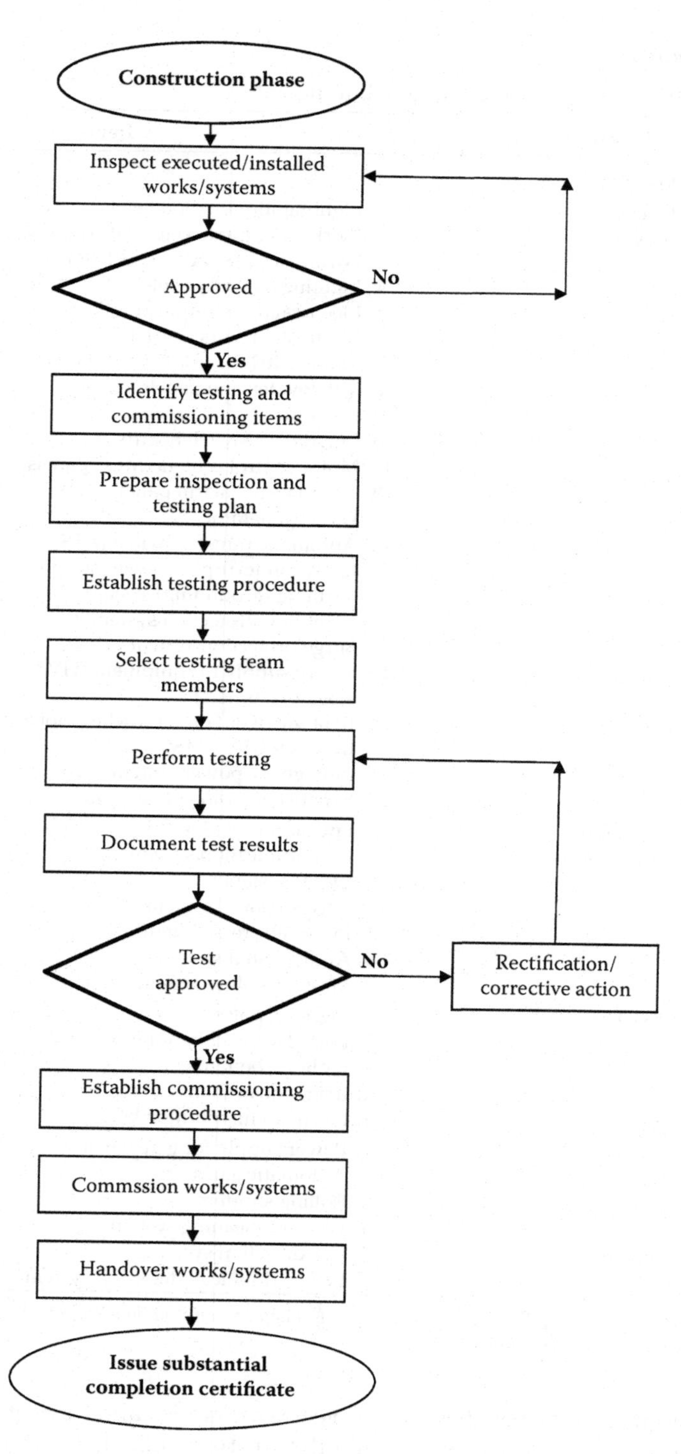

FIGURE 7.180
Logic flow process for testing, commissioning, and handover phase.

the startup and commissioning of the project as it is required for the project to be handed over to the owner/user. Figure 7.180 illustrates the testing and commissioning process.

7.4.7.6 Execute Testing and Commissioning Works

Figure 7.181 illustrates the flowchart for inspection and testing plan, and Figure 7.182 is a request form for final inspection of electromechanical works.

7.4.7.7 Develop Documents

Figure 7.183 illustrates project closeout documents to be submitted for project closeout.

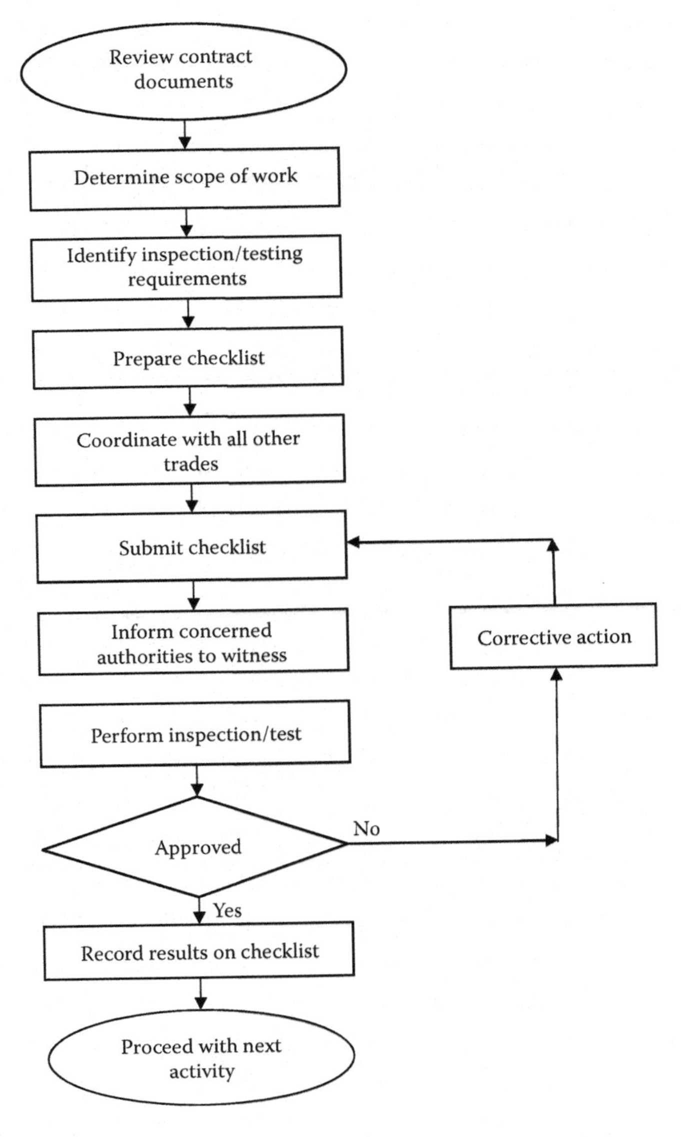

FIGURE 7.181
Development of inspection and test plan.

Project Name
Project Name
ON SITE TESTING OF ELECTRO MECHANICAL WORKS

CONTRACTOR: ______________________ CHECKLIST No.: [ ]

SUBCONTRACTORt: ______________________ DATE: [ ]

FOLLOWING WORKS ARE READY FOR INSPECTION ON:__________ TIME: __________

[ ] ELECTRICAL [ ] MECHANICAL

LOCATION: __________ DRAWING REF: __________

SPECIFICATION NO: __________ DIVISIONS:__________ ECTION: __________

Description of Work/System:

TEST PROCEDURE: YES [ ] NO [ ]

METHOD STAEMENT: YES [ ] NO [ ]

CONTRACTOR SIGNATURE: __________ DATE: __________

A/E REMARKS:

CONSULTANT ENGINEER: __________ RESIDENT ENGINEER: __________

DATE: __________ DATE: __________

Distribution [ ] OWNER [ ] R.E. [ ] CONTRACTOR

FIGURE 7.182
Checklist for testing of electromechanical works.

Description	Testing and Commissioning	As-Built Drawings	Operation and Maintenance	Guarantees	Warranties	Government Authorities	Record Documents	Test Certificate	Samples	Spare Parts	Punch Lists	Final Cleaning	Training	Taking Over Certificate	Remarks
ARCHITECTURAL WORKS															
CIVIL WORKS															
MECHANICAL WORKS															
HVAC WORKS															
ELECTRICAL WORKS (LIGHT and POWER)															
ELECTRICAL WORKS (LOW VOLTAGE)															
FINISHES															
EXTERNAL WORKS															

SAMPLE FORM

FIGURE 7.183
Project closeout report.

7.4.7.7.1 As-Built Drawings

Most contracts require the contractor to maintain a set of record drawings. These drawings are marked to indicate the actual installation where the installation varies appreciably from installation shown in the original contract. Revisions and changes to the original contract drawings are almost certain for any construction project. All such revisions and changes are required to be shown on the record drawings. As-built drawings are prepared by incorporating all the modifications, revisions, and changes made during the construction. These drawings are used by the user/operator after taking over the project for their reference purpose. It is the contractual requirements that the contractor handover as-built drawings along with record drawings, record specifications, and record product data to the owner/user before handing over of the project and issuance of substantial completion certificate. In certain projects, the contractor has to submit field records on excavation and underground utility services detailing their location and levels.

7.4.7.7.2 Technical Manuals and Documents

Technical manuals, design and performance specifications, test certificates, warranties and guarantees of the installed equipment are required to be handed over to the owner as part of contractual conditions.

Systems and equipment manuals submitted by the contractor to the owner/end user generally consist of

- Source information
- Operating procedures
- Manufacturer's maintenance documentation
- Maintenance procedure
- Maintenance and service schedules
- Spare parts list and source information
- Maintenance service contract
- Warranties and guarantees

Procedure for submission of all these documents is specified in the contract document.

7.4.7.7.3 Warranties and Guarantees

The contractor has to submit warranties and guarantees in accordance to the contract documents. Normally the guarantee for water proofing woks varies from 15 to 20 years. Similarly warranty for diesel generator is set at 5 years.

7.4.7.8 Monitor Work Progress

Schedule for the testing, commissioning, and handover phase has to be prepared and all the activities have to be performed as per agreed upon plan.

7.4.7.9 Train Owner's/End User's Personnel

Normally, training of user's personnel is part of contract terms. The owner's/user's commissioning, operating, and maintenance personnel are trained and briefed before commissioning starts in order to familiarize the owner's/user's personnel about the installation works and also to ensure that the project is put into operation rapidly, safely, and effectively without any interruption. Timings and details of training vary widely from project to project. Training must be completed well in advance of the requirement to make the operating teams fully competent to be deployed at the right time during commissioning. This needs to be planned from project inception, so that the roles and activities of the commissioning and operating staff are integrated into a coherent team to maximize their effectiveness.

7.4.7.10 Hand over the Project

Once the contractor considers that the construction and installation of works have been completed as per the scope of contract, and final tests have been performed and all the necessary obligations have been fulfilled, the contractor submits a written request to the owner/consultant for handing over of the project and for issuance of substantial completion certificate. This is done after testing and commissioning is carried out, and it is established that project can be put in operation or owner can occupy the same. In most construction projects, there is a provision for partial hand over of the project.

The contractor starts handing over of all completed works/systems, which are fully functional and the owner has agreed to take over the same. A handing over certificate is prepared and signed by all the concerned parties. Figure 7.184 illustrates a sample handing over certificate.

Project Name
Project Name
HANDING OVER CERTIFICATE

CONTRACTOR: ______________________ CERTIFICATE No. : [ ]

SUBCONTRACTOR: ______________________ DATE [ ]

SPECIFICATION NO : ____________ DIVISION : ____________ SECTION : ________

DRAWING No. BOQ REF: ____________

AREA : [ ] Building Works [ ] Electrical Works [ ] Mechanical Works

[ ] HVAC Works [ ] Finishes Works [ ]

Description of Work/System:

The work/system mentioned above is completed by the contractor as specified and has been inspected and tested as per contract documents. The work/system is fully functional to the satisfation of owner/end user. The contractor hand over the said work/system the owner/end user as on---------------. The guarentee/warranty of work/system shall start as of ----------- and shall be valid for a period of ------------------years(duration) from the date of issuance of substantial completion certificate. The contractor shall be liab contractually till the end of warranty/guarentee period.

SIGNED BY:

OWNER/END USER: ______________________ CONTRACTOR: ______________________

CONSULTANT: ______________________ SUBCONTRACTOR: ______________________

FIGURE 7.184
Handing over certificate. (Reprinted with permission from Rumane, A.R., *Quality Management in Construction Projects*, CRC Press, Boca Raton, FL, 2010.)

7.4.7.10.1 Obtain Authorities' Approval

Necessary regulatory approvals from the respective concerned authorities are obtained, so that the owner can occupy the facility and start using/operating the same. In certain countries, all such approvals are needed before electrical power supply is connected to the facility. It is also required that the building/facility is certified by the related fire department authority/agency that it is safe for occupancy.

7.4.7.10.2 Hand over Spare Parts

Most contract documents include the list of spare parts, tools, and extra materials to be delivered to the owner/end user during the closeout stage of the project. The contractor has to properly label these spare parts and tools clearly indicating the manufacturer's name and model number if applicable. Figure 7.185 illustrates spare parts handing over form used by the contractor.

7.4.7.10.3 Accept/Take over the Project

Normally, a final walk through inspection is carried out by the committee that consists of the owner's representative, design and supervision personnel, and the contractor to decide the acceptance of works and that the project is complete enough to be put in use and operational. If there are any minor items remains to be finished, then such list is attached with the certificate of substantial completion for conditional acceptance of the project. Issuance of substantial completion certifies acceptance of works. If the remaining works are of minor nature, then the contractor has to submit a written commitment that he shall complete said works within the agreed upon period. A memorandum of understanding is signed between the owner and the contractor that the remaining works will be completed within an agreed upon period. Table 7.47 illustrates a list of activities need to be considered for project closeout.

7.4.7.11 Close Contract

Table 7.47 lists all the activities to be completed to close the contract.

7.4.7.11.1 Prepare Punch List

The owner/consultant inspects the works and inform the contractor of unfulfilled contract requirements. A punch list (snag list) is prepared by the consultant listing all the items still requiring completion or correction. The list is handed over to the contractor for rework/correction of the works mentioned in the punch list. The contractor resubmits the inspection request after completing or correcting previously notified works. A final snag list is prepared if there are still some items that need corrective action/completion by the contractor; such remaining works are to be completed within the agreed period to the satisfaction of the owner/consultant. Table 7.98 is a sample form for preparation of punch list.

7.4.7.11.2 Prepare Lesson Learned

The construction/project manager, consultant, and contractor has to prepare lesson learned and document the same for future references to improve the processes and organizational performance. This includes the following:

- Reasons for delay
- Resigns for rejection/rework

Project Name
Project Name
HANDING OVER OF SPARE PARTS

CONTRACTOR: ______________________ CERTIFICATE No. : [ ]

SUBCONTRACTOR: ______________________ DATE [ ]

SPECIFICATION NO : ______________ DIVISION ________ SECTION : ________

DRAWING No. ______________ BOQ Ref. ________

AREA : [ ] Building Works [ ] Electrical Works [ ] Mechanical Works

[ ] HVAC Works [ ] Finishes Works [ ]

Following Spare Parts have been handed over to the owner/end user

Description of Spare Parts

Sr.No.	Description	BOQ Reference	Spec. Ref.	Manufacturer	Specified Qty	Delivered Qty

(Attach additional sheet,if required)

SIGNED BY:

OWNER/END USER: ______________ CONTRACTOR: ______________

CONSULTANT: ______________ SUBCONTRACTOR: ______________

FIGURE 7.185
Handing over of spare parts. (Reprinted with permission from Rumane, A.R., *Quality Management in Construction Projects*, CRC Press, Boca Raton, FL, 2010.)

TABLE 7.98
Punch List

Owner Name		
Project Name		
Punch List		
Punch List Number:		**Date:**
Building Name		**Floor**
Area/Zone		**Room Number**
Sr. No.	**Item**	**Remark**
1	Flooring	
2	Skirting	
3	Walls	
4	Ceiling	
5	Door	
6	Windows	
7	Door hardware	
8	Window hardware	
9	Electrical light fixture	
10	Wiring devices (sockets)	
11	HVAC grill	
12	HVAC diffuser	
13	Thermostat	
14	Sprinklers	
15	Fire alarm detectors	
16	Speakers	
17	Communication system devices	
18	TV outlets	
19	Any other item	

Source: Abdul Razzak Rumane (2013), *Quality Tools for Managing Construction Projects*, CRC Press, Boca Raton, FL. Reprinted with permission from Taylor & Francis Group.

- Preventive/corrective actions
- Cost overrun
- Causes for claims
- Reasons for conflict

7.4.7.11.3 Issue Substantial Completion Certificate

A substantial certificate is issued to the contractor once it is established that the contractor has completed works in accordance with the contract documents and to the satisfaction of the owner. The contractor has to submit all the required certificates and other documents to the owner before issuance of the certificate. Figure 7.186 illustrates procedure for issuance of substantial certificate normally used for building construction projects in Kuwait. Authorities' approval requirements mentioned in this figure may vary from country to country; however, other requirements need to be completed by the contractor for issuance of substantial completion certificate.

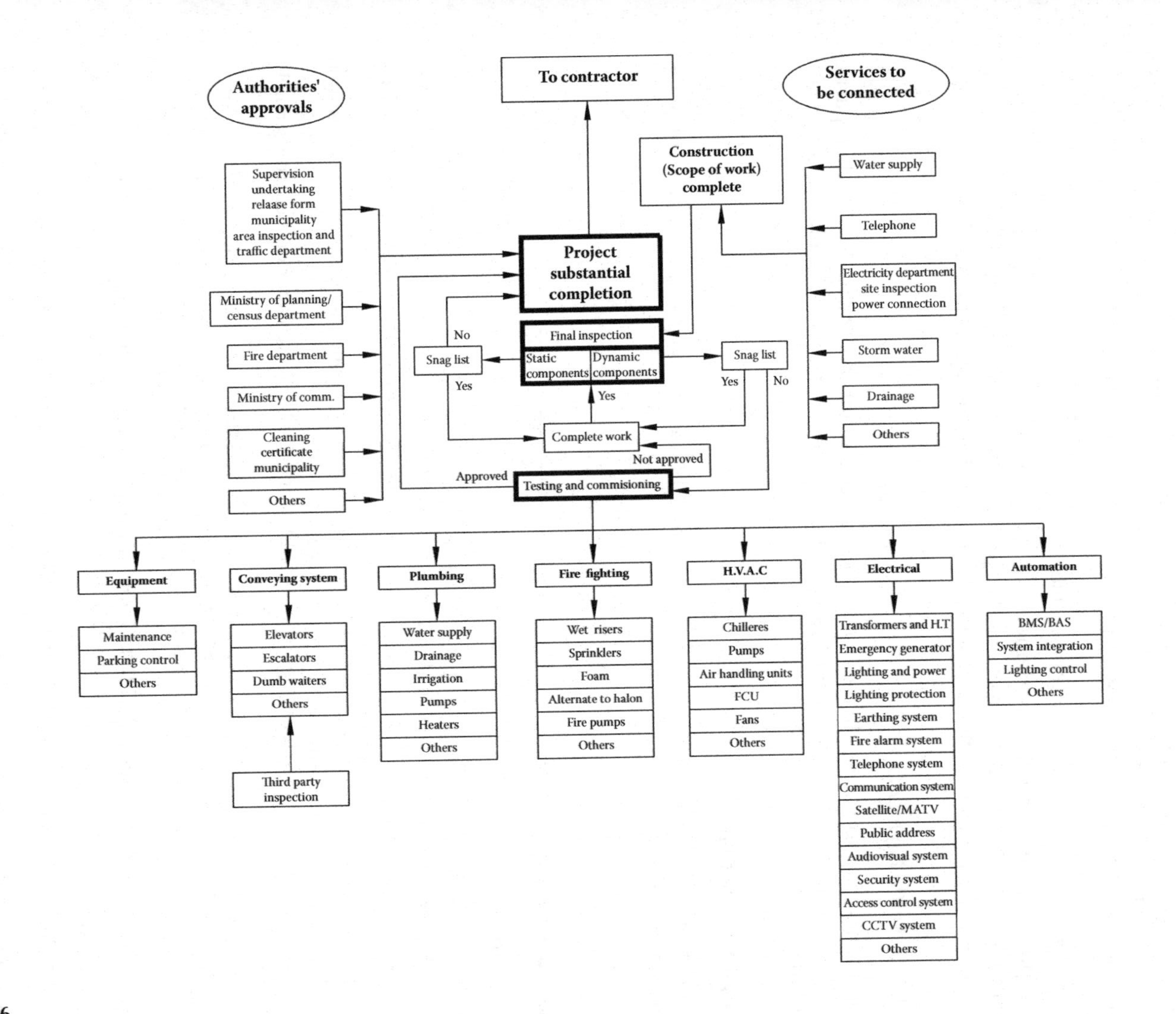

FIGURE 7.186
Project substantial completion procedure.

The certificate of substantial completion is issued to the contractor, and the facility is taken over by the owner/end user. By this stage, the owner/end user already takes possession of the facility and operation and maintenance of the facility commences. The project is declared complete and is considered as the end of the construction project life cycle. The defect liability period starts after issuance of substantial certificate.

During this period, the contractor has to complete the punch list items and also to rectify the defects identified in the project/facility.

7.4.7.11.4 Terminate/Demobilize Team Members

The project team members are demobilized from the project and their services are either terminated or shifted to other projects, if positions are available to accommodate. In certain cases, the project members join the member of functional team at the project office.

7.4.7.12 Settle Payments

The owner has to settle all the dues toward consultant, contractor, and other parties involved. Similarly, the contractor has to settle payments due to their subcontractors and suppliers.

7.4.7.13 Settle Claims

The entire project related claims have to be amicably settled as per contract conditions to close the project.

References

1. Stanleigh, M. (2008). Dealing with Conflict in Project Teams. *Optimize Organizational Performance.* http://www.bia.ca/articles/DealingwithConflictinProjectTeams.htm.
2. Public Citizen. (2012). *The Price of Inaction: A Comprehensive Look at the Costs of Injuries and Fatalities in Maryland's Construction Industry.*
3. OSHA's $afety Pays Program, https://www.osha.gov/dcsp/smallbusiness/safetypays/estimator.html (sprain or strain with 3% profit margin).
4. American Society of Safety Engineers (ASSE). (June 2002). White Paper on Return on Safety Investment.
5. NHTSA Traffic Safety Facts, Distracted Driving 2010, http://www.distraction.gov/download/researchpdf/. Distracted-Driving-2009.pdf (there were approximately 1000 annual fatalities from texting and thus an implied rate of 0.5 per 100,000 among 200 million licensed U.S. drivers).
6. CPWR, OSHA, and NIOSH. (2012). Graphic from Campaign to Prevent Falls in Construction.
7. OSHA Directorate of Training and Education. https://www.osha.gov/dte/outreach/outreach_growth.html.
8. United States Department of Labor, OSHA, 1999. https://www.osha.gov/oshstats/commonstats.html.
9. CPWR, OSHA, and NIOSH. (2012). Campaign to Prevent Falls in Construction.

8

Lean Construction

Zofia K. Rybkowski and Lincoln H. Forbes

CONTENTS

8.1 Lean Construction

Lean project delivery (LPD) or Lean construction—a project delivery philosophy that aims to reduce waste and add value using continuous improvement in a culture of respect—was developed to address many of the productivity problems currently plaguing the architecture, engineering, and construction (AEC) sector. The purpose of this chapter is to provide an overview of the challenges the sector is currently facing, a brief history of the emergence of lean construction as a countermeasure response to these challenges, and principles, tools, and processes typically implemented during LPD.

8.2 Current Challenges in the AEC Industry

8.2.1 Low Overall Productivity

The Bureau of Labor Statistics does not maintain an official productivity index for the construction industry—the only major industry that is not tracked consistently. Productivity in construction is usually expressed in qualitative terms, and cannot easily be analyzed empirically. A number of independent studies estimated that construction productivity increased at a rate of 33%—or 0.78% per year—between 1966 and 2003.

This productivity growth is less than one-half of that of the US nonagricultural productivity gains during the same period, which averaged 1.75% annually. An estimate by the National Institute of Science and Technology (NIST) points to a decline in the US construction productivity over a 40-year period to 2007 that was −0.6% per year, versus a positive nonfarm productivity growth of 1.8% per year. The increased productivity in nonfarm-related industries may be attributable to mechanization, automation, and prefabrication.

Similarly, most of the improvements in construction productivity have been the result of research and development work in the manufacturing industry related to construction machinery. Earthmoving equipment has become larger and faster and power saws have replaced handsaws. However, despite increased productivity of individual activities, evidence suggest that "overall" construction productivity has declined over the past 50 years. Again, this is in clear contrast to observable increases in productivity for nonfarm-related industries (Teicholz from Eastman et al. 2008). As shown in Figure 8.1, the decline

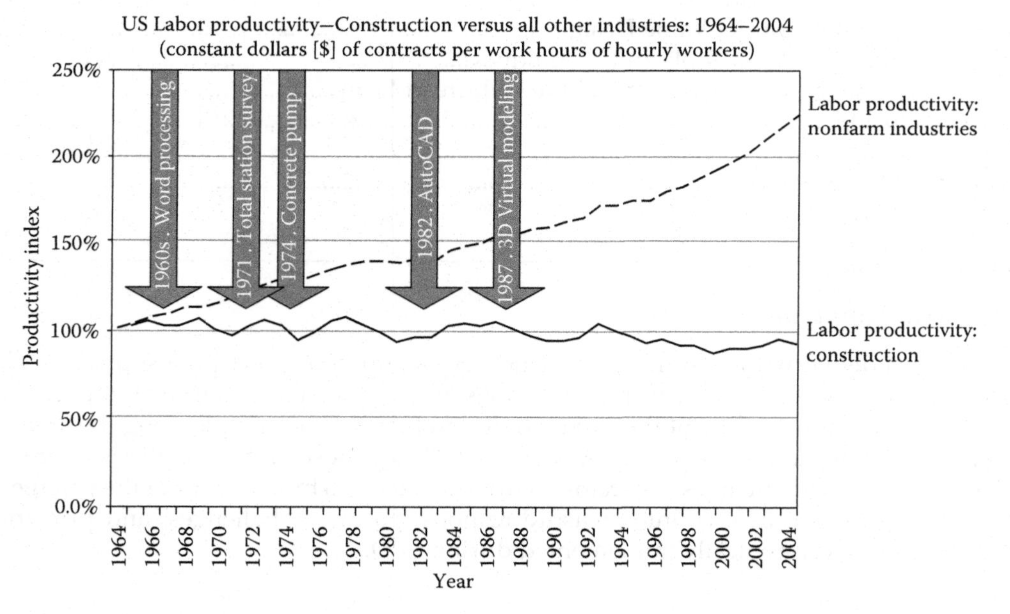

FIGURE 8.1
The entry year of several technological advances, such as word processing, total station surveying, concrete pump, AutoCAD and 3D modeling (ArchiCAD) juxtaposed onto the indexes of labor productivity for construction and nonfarm industries, 1964–2004. (Adapted from Teicholz, P., *J. Const. Eng. Manag.*, 127(5), 427–429, 2001 [as cited in Figure 1.3, p. 8 of Eastman, C. et al., *BIM Handbook: A Guide to Building Information Modeling for Owners, Managers, Designers, Engineers, and Contractors*, John Wiley and Sons, Hoboken, NJ, 2008; US Bureau of Labor Statistics, Department of Commerce]. Illustration by author [Rybkowski].)

continues despite the introduction of significant technological advances, suggesting that improving the situation may have more to do with healing an increasingly fragmented, noncollaborative industry than with only enhancing the productivity of its individual components.

8.2.2 Indications of Industry Failure

In order to better understand problems plaguing the industry, construction management research has focused on identifying sources of failure. For example, Josephson and Hammarlund (1999) observed seven building projects during a six-month period, analyzed nearly 3000 defects, and then identified the root causes of the defects. Perhaps surprisingly, the researchers discovered that the greatest numbers of defects were not induced by stress or risk, but instead by a lack of motivation and knowledge among four participant categories surveyed: designers, site managers, workers, and subcontractors. Researchers found a strong alignment of ranking among the top two causes within each of the four categories. Table 8.1 illustrates the same.

Other researchers have focused less on defects, and more on the adversarial nature of the construction industry. For example, Black et al. (2000) surveyed over 78 consultants, contractors, and clients. They found that respondents perceived traditional design–bid–build systems as failing in a number of ways, including exploitation is common, specifications are rigid, decisions are made with limited knowledge, and focus is placed on short-term (rather than long-term) success. Such failings appear to be pervasive and common to the design–bid–build–delivery method—regardless of geography or nationality (Proverbs et al. 2000).

Research has pointed to a significantly high level of wasted resources in the construction industry—both human and material. An article titled "Construction and the Internet" in *The Economist* of January 15, 2000, noted that up to 30% of construction costs is due to inefficiencies, mistakes, delays, and poor communications.

The Construction Industry Institute (CII) estimates that in the United States:

- 10% of project cost is spent on rework
- 25–50% of construction costs is lost to waste and inefficiencies in labor and materials control

TABLE 8.1

Causes of Defects

	Cause of Defects (% of Defect Cost per Category)			
	Workmanship	**Site Management**	**Subcontractors**	**Design**
Motivation	69	50	47	35
Knowledge	12	31	27	44
Information	2	8	13	18
Stress	1	6	3	2
Risk	16	5	10	1

Source: Reprinted from *Automation in Construction*, 8, Josephson, P.-E. and Hammarlund, Y., The causes and costs of defects in construction: A study of seven building projects, 681–687 (Table 6, p. 686), 1999, with permission from Elsevier.

- Losses are incurred in errors in information in translating designs to actual construction, and
- Inadequate interoperability accounts for losses of $17–$36 billion per year due to communication difficulties between software used by different actors in the design and construction supply chain (Ballard and Kim 2007; Gallaher et al. 2004)

The observations of a Construction Industry Cost Effectiveness (CICE) study in 1983 are still relevant today. The study teams noted that "more than half the time wasted in construction is attributable to poor management practices." They concluded that "if only the owners who pay the bills are willing to take the extra pains and pay the often small extra cost of more sensible methods they will reap the benefit of more construction for their dollars."

The construction industry is not unified on the measurement of productivity or overall performance. Many researchers define construction project success in terms of (a) cost—within budget; (b) schedule—on-time completion within schedule; (c) safety—high safety levels—few or no accidents; and (d) quality—conformance with specifications within a few defects. Some view quality as one point of a triangle with cost and schedule; quality is often the first to be sacrificed in favor of cost savings and schedule reductions. Many contractors believe that it is impossible to meet the desirable benchmarks in all four factors simultaneously and that there is a zero-sum relationship between them. In other words, an accelerated schedule is expected to result in an increased cost and lower levels of safety and quality.

In other words, the concept of construction performance does not emphasize productivity and quality initiatives. The work of many researchers has revealed an industry tendency to measure performance in terms of the following: completion on time, completion within budget, and meeting quality requirements (i.e., conformance to plans, specifications, and construction codes).

From an industry-wide perspective, little attention has been directed to owner satisfaction as a performance measure. The emerging adoption of Lean construction is beginning to change attitudes very gradually, since Lean practices emphasize providing customer value, but that process is likely to take many years. Although the Malcolm Baldrige National Quality Award was established in 1987, in the period of more than 20 years that have elapsed since its inception, the construction industry has been greatly underrepresented.

There is a general lack of productivity/quality awareness in the industry among all parties, including owners. Owners have come to accept industry pricing; prices have simply become higher on a per unit basis. By contrast, manufacturing activities have become cheaper over time on a per unit basis.

Traditional architect/engineer (A/E) contracts are said to be unclear with respect to professional standards of performance, often leading to unmet expectations. Construction owners feel that typical A/E contracts protect designers at the owner's expense. For example, prevailing contract language relieves designers of any role in the case of a lawsuit or arbitration between an owner and contractor. An outgrowth of this is the practice of "substantial completion" where a job is usable but has 5% of the remaining work in the form of a "punch list." An owner often has a very difficult time in persuading a contractor to finish that work. Other stakeholders, too, feel frustrated. For example, architects often feel undercompensated for the amount of work they are asked to do. Also, specialty contractors sometimes argue they are forced to carry an unfair share of risk and blame if something goes wrong.

8.3 The Lean Construction Response

8.3.1 Lean Construction Defined

Lean construction emerged as a response to such frustrations with construction productivity, as well as to concerns about commonly observed errors, delays, cost overruns, and safety (Forbes and Ahmed 2010). Founded on a belief that problems in the construction industry are a byproduct of deep-seated systemic inefficiencies in the way increasingly specialized stakeholders interact, Lean construction aims to address these interactions in a holistic way.

Lean construction is an innovative project delivery approach that addresses many of the shortcomings of traditional project management. It has several interpretations, including LPD, integrated project delivery (IPD), and collaborative project delivery (CPD). It is based on the "Lean" manufacturing principles that are a foundation of the Toyota Production System (TPS) (Howell 1999). Lean "production" pursues the ideal to "(1) do what the customer wants, (2) in no time, and (3) with nothing in stores" (Tommelein 2015). Lean construction practitioners seek to apply Lean production theories to both design and construction practices. Its three tenets are as follows: minimize waste and add value in all forms, continuously improve products and systems, and maintain respect for people. The Lean construction philosophy views a project as a "promise delivered by people working in a network of commitments" (Macomber and Howell 2005).

Lean construction is described as a way to design production systems to minimize waste of materials, time, and effort in order to generate the maximum possible amount of value (Koskela et al. 2002). In his tranformation flow value (TFV) theory of Lean construction, Koskela argued that all three aspects of construction must be managed simultaneously for a successful project delivery (Tommelein 2015). Abdelhamid (2013) defines Lean construction as "a holistic facility design and delivery philosophy with an overarching aim of maximizing value to all stakeholders through systematic, synergistic and continuous improvement in the contractual arrangements, the product design, the construction process design and methods selection, the supply chain, and the workflow reliability of site operations."

Current definitions of Lean construction are varied, in part, because the Lean construction thinking continually emerges from a grass roots, communal, and collaborative base. There is a general reluctance of the Lean construction pioneers to define a concept that might limit or exclude the contributions of the community Lean construction ideology that is intended to embrace and motivate. Nevertheless, some definitions have been advanced and are summarized in Table 8.2.

More recently, Lean construction has come to be defined by a triangle that addresses the way a system must be restructured in order to function as Lean (Figure 8.2).

The three sides of this triangle include the organization and commercial terms that are both underlain by a unique operating system. For example, while traditional construction is "profit"-driven, LPD is "value"-driven. In traditional construction, the operating system is "push"-driven, whereas in LPD, it is "pull"-driven; the organization suboptimizes each "part", operating by "command and control", whereas LPD optimizes the "whole", facilitating "collaboration"; and in traditional construction, the commercial terms are "transactional", whereas in LPD they are "relational" (Mossman 2014).

Another way to envision Lean construction is through the lens of a graphic definition proposed by Rybkowski et al. (2013). The authors challenged Lean industry practitioners

TABLE 8.2

Definitions of Lean Construction by Lean Construction Pioneers and Societies

1	Lean construction is a "way to design production systems to minimize waste of materials, time, and effort in order to generate the maximum possible amount of value."	Koskela et al. (2002)
2	"Lean Construction is a production management-based project delivery system emphasizing the reliable and speedy delivery of value. It challenges the generally-accepted belief that there are always trade-offs between time, cost and quality."	LCI (2016)
3	"Lean Construction is a set of ideas, practiced by individuals in the construction industry, based in the holistic pursuit of continuous improvements aimed at minimizing costs and maximizing value to clients in all dimensions of the built and natural environment: planning, design, construction, activation, operations, maintenance, salvaging, and recycling."	AGC (2013)
4	Lean construction is "A holistic facility design and delivery philosophy with an overarching aim of maximizing value to all stakeholders through systematic, synergistic, and continuous improvements in the contractual arrangements, the product design, the construction process design and methods selection, the supply chain, and the workflow reliability of site operations."	Abdelhamid (2013)
5	"While current project management approaches reduce total performance by attempting to optimize each activity, Lean construction succeeds by optimizing at the project level as opposed to the less effective current project management approaches, which reduce total performance by attempting to optimize each activity."	Forbes and Ahmed (2010, p. 19)
6	Lean construction is "the improving of the construction process to profitably deliver what the customer needs while having the right systems, resources, and measure to deliver things right the first time. It is focused on value from the customer's perspective, more than on cost, and seeks to remove all nonvalue-adding activities (wastes)."	Hayes (2014, p. 226)
7	"The ideas of Lean thinking comprise a complex cocktail of ideas including continuous improvement, flattened organization structures, teamwork, the elimination of waste, efficient use of resources and cooperative supply chain management. Within the U.K. construction industry, the language of Lean thinking has since become synonymous with best practice."	Green (2002)
8	"Lean stems from the recognition that the world is subject to all kinds of variation and that one can distinguish good variation from bad variation. Good variation is variation that customers intentionally want (e.g., people want buildings to not all look the same); all other variation is bad. To improve performance, Lean starts by relentlessly driving out bad variation and then buffering where needed to protect the system from the impact of remaining variation."	Tommelein (2015)

and academics to graphically illustrate their understanding of Lean thinking on the backs of cocktail napkins. A resulting composite diagram situates the P-D-C-A (plan–do–check–act) continuous improvement cycle as the engine for incremental transformations (Figure 8.3).

The black and white inverted triangles illustrate that, as waste is reduced, capital spent on waste can instead be diverted toward added value for the project. The OAEC acronym within each PDCA cycle signifies that all transformations from current to improved future states are accomplished collaboratively with OAEC team members (owner, architect, engineer,

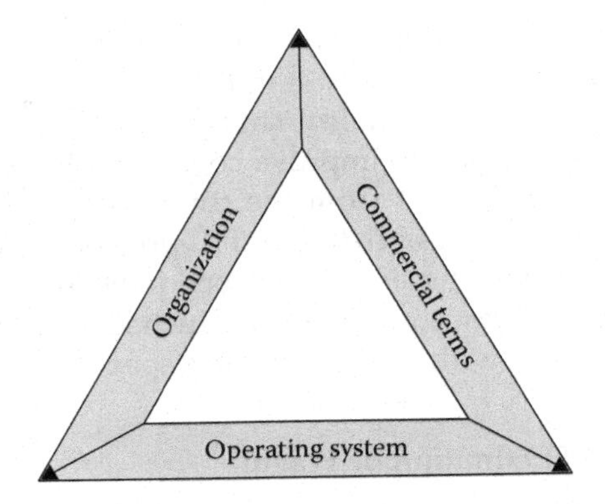

FIGURE 8.2
Three characteristics used to draw distinctions between Lean project delivery and traditional (design–bid–build) project delivery. (Adapted from Mossman, A., Traditional construction and Lean project delivery—A comparison, The Change Business, Stroud, 2014 [as cited on p. 1]. Reprinted with permission from Lean Construction Institute.)

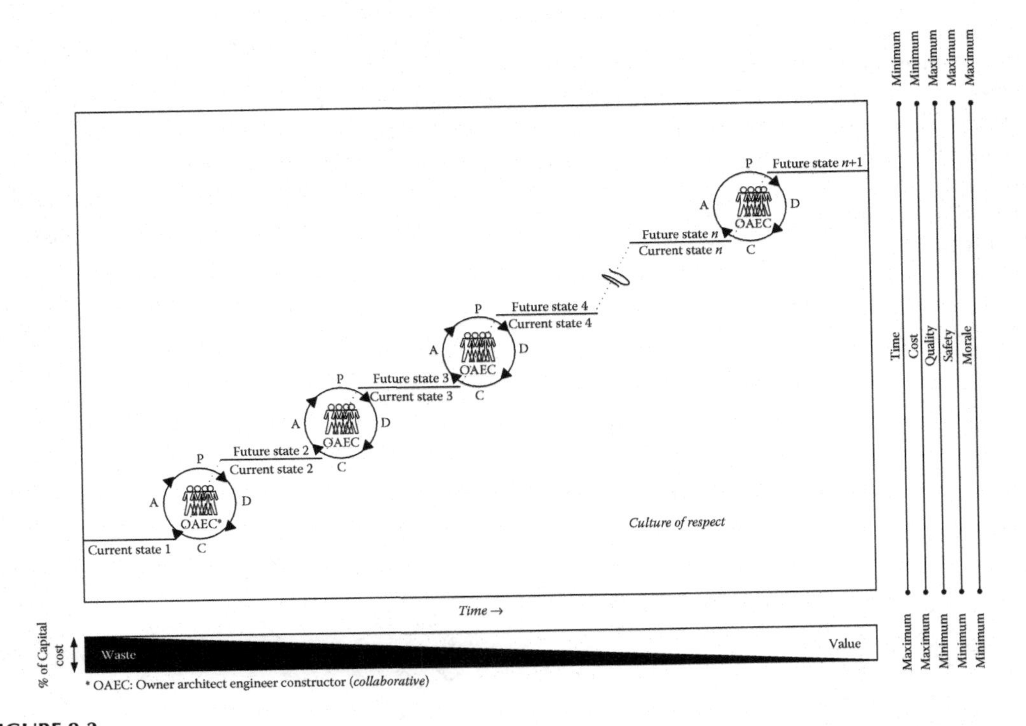

FIGURE 8.3
The "Kaizen stairway"—A chain of continuous improvement. (Adapted from Rybkowski, Z.K., Abdelhamid, T., and Forbes, L., On the back of a cocktail napkin: An exploration of graphic definitions of lean construction, *Proceedings of the 21st Annual Conference for the International Group for Lean Construction*, Fortaleza, Brazil, July 31–August 2, 2013 [as cited in Figure 4 and 5, p. 88]; Fernandez-Solis, J.L. and Rybkowski, Z.K., A theory of waste and value, *International Journal of Construction Project Management*, 4, 89–105, 2012 [as cited in Figure 1, p. 94]; Rybkowski, Z.K. and Kahler, D., Collective kaizen and standardization: The development and testing of a new lean simulation, *Proceedings of the 22nd Annual Conference for the International Group for Lean Construction*, Oslo, Norway, June 25–27, 2014 [as cited in Figure 1, p. 1260]. Reprinted with permission from Kahler.)

and constructors). All actions are preformed within a culture of respect. Horizontal lines between PDCA cycles represent the need to standardize improvements. If done according to Lean principles, overall time, cost, quality, safety, and morale will improve simultaneously. The ability for all five metrics to improve concurrently qualifies Lean construction as a paradigm shift, distinguishing it from the time–cost–trade–off phenomenon often observed in more traditional (i.e., design–bid–build) project delivery methods (Feng et al. 1997; Hegazy 1999; LCI 2016; Siemens 1971), described in Section 2.2. The diagram visually captures the final objective of Lean construction, which is to "reduce waste and add value using continuous improvement within a culture of respect".

8.3.2 Lean Goals and the Elimination of Waste

In the manufacturing world, the goal of Lean is to produce a product that satisfies the customer's requirements—while minimizing waste and maximizing value. At the risk of appearing overly simplistic, it may be useful to draw an analogy between Lean construction and a Lean animal. The word "Lean" might be associated with the lithe body of the cheetah which has evolved to build adequate muscle and minimize fat, enabling it to optimize speed while hunting. But an arctic seal insulated by a thick layer of blubber also conforms to the Lean ideal. The utlimate Lean goal is to create a product that is "fit for use" or to customer satisfaction, while minimizing waste and maximizing value.

In construction, waste is everywhere—and waiting to be eliminated. For example, adversarial relationships, claims and disputes, demotivated workers, and defects, as discussed in Section 2.2, can all be considered sources of waste because these actions do not add value to the final product. One of the benefits of waste reduction is that resources that would have been spent on waste can be reallocated to enhancement of value, as suggested by Figure 8.4.

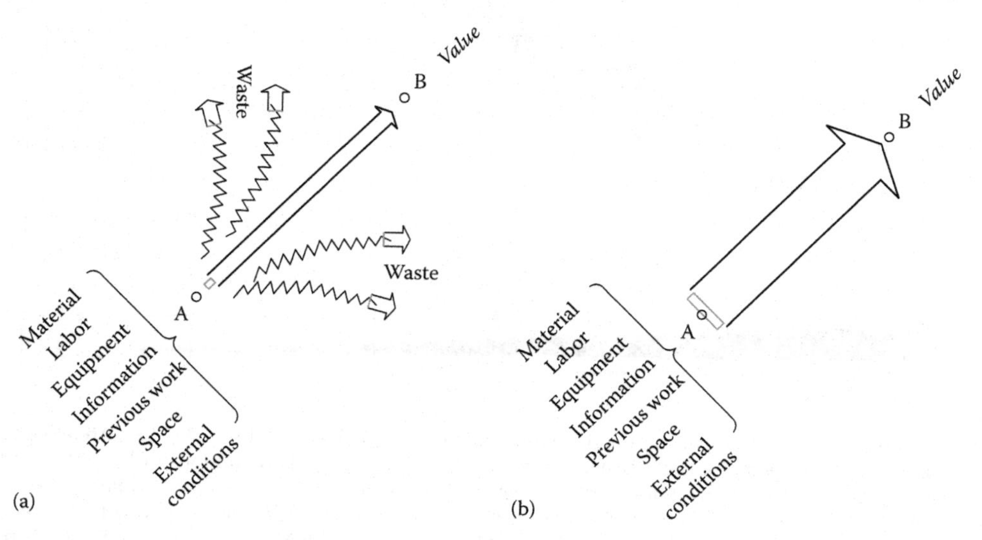

FIGURE 8.4
Recapturing waste as value. Traditional delivery (a) and Lean delivery (b), illustrate how recaptured waste can be transformed into value. (Adapted from Rybkowski, Z.K., The application of root cause analysis and target value design to evidence-based design in the capital planning of healthcare facilities, Doctoral dissertation, University of California, Berkeley, CA, 2009 [as cited in Figure 71, p. 216]. Illustration by author [Rybkowski].)

Lean construction maximizes value and reduces waste and applies specific techniques in an innovative project delivery approach including supply chain management and just-in-time (JIT) techniques as well as the open sharing of information and close collaboration between all the parties involved in the production process. Lean manufacturing is an outgrowth of the TPS that was developed by Taiichi Ohno in the 1950s (Womack et al., 1990). Ohno had observed mass production at Ford Motor Corporation's manufacturing facilities in the United States and recognized that there was much waste (muda) everywhere. Ohno and Shigeo Shingo (Ohno 1988), an industrial engineer at Toyota identified "seven wastes" in mass production systems:

1. Overproducing
2. Idle time waste—waiting time/queue time
3. Transporting/conveyance waste
4. Processing waste—waste in the work itself
5. Inventory waste—having unnecessary stock on hand
6. Wasted operator motion—using unnecessary motion
7. Producing defective goods—waste of rejected production

Ohno sought to develop a delivery process that met customers' needs with very little inventory because inventory is costly to store and maintain. It is also risky to hold because it may physically degrade with time and its functions may become obsolete.

To eliminate wasteful inventory, the TPS was based on the "just-in-time" philosophy. Its three tenets were minimizing waste in all forms, continuous improvement of processes and systems, and maintaining respect for all workers. Its benefits include reduced inventories (and space), higher human productivity, shorter cycle times, shorter lead times, fewer errors, and higher morale. In the mid-1970s, Toyota reduced the time needed to produce a car from 15 days to one day, using JIT. It is important to note that Ohno's improvement of Toyota's production process was not necessarily a new technology, but rather the result of involving all participants in a new philosophy of avoiding waste.

8.3.3 Brief History of Lean Construction

Lean construction became grafted onto a rich value tradition and knowledge base of antecedent thought. Although some authors argue the TPS planted the flag of Lean thinking, TPS itself emerged from pioneers of process productivity, such as Frederick Taylor, Frank and Lillian Gilbreth, Henry Ford, and W. Edwards Deming. In Japan, the auto manufacturer Toyota rose from the devastation of World War II by drawing heavily on the teachings of statistics gurus, Walter A. Shewhart and W. Edwards Deming, the latter who paid many visits to the recovering nation. Toyota heeded Deming's teachings, and perceptive writings of the Japanese businessman, Taiichi Ohno, reflect his indebtedness to Deming, Ford, and other productivity pioneers (Ohno 1988). MIT International Motor Vehicle Program coauthors James P. Womack, Daniel T. Jones, and Daniel Roos (1990) described successful theories and methods used by Toyota in their revolutionary book, *The Machine That Changed the World* (Womack et al. 1990). The term "Lean" was coined by John Krafcik in a 1988 article derived from his MIT Master's thesis. Toyota's successes in terms of improved revenue, speed of production, and refined product quality prompted manufacturers in other industries to investigate ways to duplicate their success. Jeffrey Liker assembled key elements of Toyota's practices into his book, *The Toyota Way*—now read and discussed during company brown bag lunches and sometimes described as "the bible of Lean". Other similarly themed productivity-enhancement theories also emerged, including Eliyahu Goldratt's fictitious work, *The Goal,* which introduced his theory of constraints to the manufacturing community, and which has since joined Womack, Jones and Roos,'s and Liker's books as necessary reading in the Lean canon.

In the now renowned CIFE Technical Report #72 entitled: *Application of the New Production Philosophy to Construction* (Koskela 1992), Lauri Koskela stood on the principles of a manufacturing movement which he termed "the new production philosophy" and applied its principles to the construction industry. Koskela spent a year at Stanford University as a visiting scholar and authored a study entitled "Applications of the New Production Philosophy to Construction." He drew parallels between both fields by characterizing construction as a form of production. Koskela looked to the manufacturing industry for a new direction for construction. Specifically, he modeled this new production philosophy after the Toyota's highly successful production system TPS. Researchers had already recommended solutions to the construction industry's underperformance including industrialization (prefabrication and modularization), automation and robotics, and information technology to reduce fragmentation. Koskela proposed a new approach that was not based on technology, but rather on the principles of a production philosophy. He noted its evolution through three stages:

1. Tools, such as kanban and quality circles
2. A manufacturing method
3. A management philosophy (Lean production, JIT/total quality control (JIT/TQC)

Koskela inferred from a number of productivity-based studies on the US and European manufacturing plants that the most successful methods were based on the JIT philosophy. Manufacturing studies by other researchers such as Schonberger (1986) and Harmon and Peterson (1990) reinforced this observation.

In a typical production process, material is converted (transformed) in a number of discrete stages. It is also inspected, moved from one operation to another, or made to wait. Inspection and waiting are considered flow activities. Conversion activities are considered to add value while flow activities do not add value.

Koskela visualized Lean construction as a flow process combined with conversion activities, and noted that only the conversion activities add value. This was the transformation-flow-value (TVF) theory of construction. Production improvements can be derived by eliminating or reducing flow activities while making conversion activities more efficient. He pointed to earlier studies that showed that only 3% to 20% of the steps in a typical process add value and that their share of the total cycle time is only 0.5% to 5% (Koskela 1992).

Although mentioned third in the TFV lineup, the creation of "value"—to design a product or building to customer satisfaction—is arguably the most critical of the three, since it only makes sense to design a building within budget and on time if it serves the function for which it was intended (Ballard 2009a).

Koskela attributed the predominance of nonvalue-adding activities to three root causes: design, ignorance, and the inherent nature of production. Design was due to the subdivision of tasks; each added subtask increased the incidence of inspecting, waiting, and moving. A natural tendency for processes to evolve over time without close analysis leads to ignorance of their inherent wastefulness. The inherent nature of production is that events such as defects and accidents add to nonvalue-added steps and time between different conversion activities.

Incorporating Koskela's concepts, Glenn Ballard and Greg Howell cemented their own observations of the need to enhance reliability of project planning and founded the Lean Construction Institute in August 1997 (Lean Construction Institute 2009), formalizing a collaboration that has shaped the discourse on Lean construction.

Howell and Ballard's experimental field work applied Lean theory to construction processes; they collected metrics to substantiate results obtained during the playing of live simulations, and eventually drew validations from computer simulations of UC Berkeley civil engineering academician, Iris D. Tommelein (Tommelein et al. 1999). Collecting data from live and computerized simulations confirmed the mathematical logic of Lean principles and gave assurance to practitioners that the successes enjoyed during case studies could be trusted as generalizable and repeatable. A substantial acceleration point came when attorney Will Lichtig joined the pioneers, and enshrined Lean principles into a legal document entitled "The Integrated Form of Agreement" (IFOA; Lichtig 2004). Unlike the "partnering" movement of the 1990s, which brought stakeholders together for a typical weekend retreat and encouraged participants to sign a legally unenforceable declaration of collaboration, the IFOA instead incentivized synergistic behavior with financial rewards and backed collaboration with legal consequence. In their book, *Modern Construction: Lean Project Delivery and Integrated Practices*, Forbes and Ahmed (2010) captured the range and breadth of developments in Lean construction that had taken place since its inception.

Ballard's research led to the development of the "Last Planner System of Production Control™", a method in which participants in the design and/or construction processes are empowered to engage in detailed planning and coordination needed to secure near-seamless interaction. The methodology includes procedures for removing constraints before work is initiated, as well as measurement systems to gauge performance, and feedback to introduce corrective actions at frequent intervals during a project when they can have the maximum effect on outcomes.

8.3.4 The Lean Project Delivery System (LPDS): A Closer Look

The "Lean Project Delivery System" (LPDS)™ provides a means of addressing industry shortcomings and improving the entire design and construction process (Ballard 2000b; Ballard and Howell 2003). It was developed by Glenn Ballard in 2000, and subsequently refined. While traditional industry practice separates the roles of designers and constructors, the LPDS sees the activities of these professionals as a continuum for project management to achieve three fundamental goals (Koskela 2000):

- Deliver the product
- Maximize value
- Minimize waste

8.3.4.1 The Structure of the LPDS

The LPDS model improves project delivery via the following characteristics:

- Downstream stakeholders are involved in front-end planning and design through cross-functional teams.
- Project control has the job of execution as opposed to reliance on after-the-fact variance detection.
- Pull techniques are used to govern the flow of materials and information through networks of collaborating specialists.
- Capacity and inventory buffers are minimized and only used to absorb variability that cannot be removed.
- Feedback loops are incorporated at every level, dedicated to rapid system adjustment, that is, learning (Ballard and Howell 2003).

The LPDS comprises a number of phases that are represented as inverted triangles. Figure 8.5 illustrates the same.

The system juxtaposes the phases in such a manner as to apply production system design principles to enhance the delivery of the entire project from predesign to completion and use.

The phases include

1. Project definition
2. Lean design

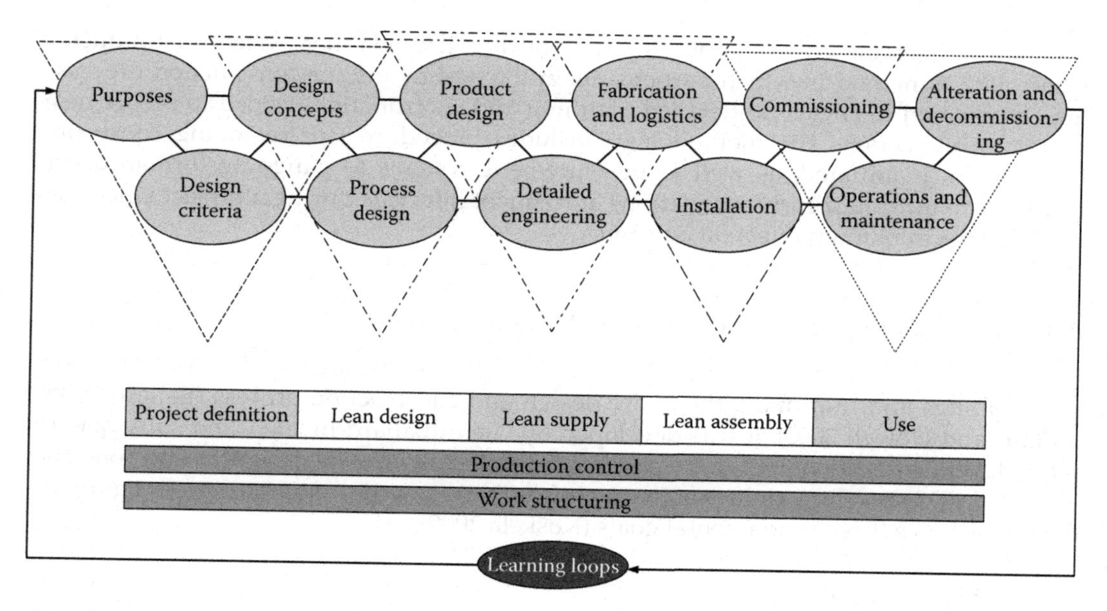

FIGURE 8.5
The Lean project delivery system. (Adapted from Ballard, G., *LCI White Paper-8: Lean Project Delivery System*, Lean Construction Institute, Ketchum, ID, 7pp, September 23, 2000b [as cited in Figure 3, p. 5]; Ballard, G., and Howell, G., Lean project management, *Building Research and Information*, 31, 119–133, 2003 [as cited in Figure 2, p. 3]. Reprinted with permission from Lean Construction Institute [LCI].)

3. Lean supply
4. Lean assembly
5. Use/completion

Each LPDS project phase is described in greater detail below.

The "project definition phase" typically involves developing project alternatives at a conceptual level, analyzing project risks and the economic payoff, and developing a financial plan. Effective project scope definition enables all involved parties to understand the owners needs, and to work toward meeting those needs.

The project definition phase comprises:

- Needs and values determination
- Design criteria
- Conceptual design

In the needs and values determination, design professionals assist the owner/client in clarifying a value proposition, that is, the purpose of the project and the needs to be served. A design criteria document describes specific needs to be met, such as size, space proximities/adjacencies, and energy efficiency requirements. Conceptual design uses the design criteria and value proposition to define an outline design that serves as a starting point for the design phase.

The "Lean design phase" comprises conceptual design, process design, and product design. It builds on the output from the project definition phase, but with a deviation from traditional design practice. The design team often starts by designing the design process; they use adhesive notes ("stickies") on a wall to ensure that design assignments have necessary prerequisite work completed and that no constraints will delay the process. Design is done with both the construction product and the process in mind. Constructability reviews and value engineering are not seen as tools to apply in a problem-solving mode, but instead are continually integrated with decision-making in the design process.

The third phase, the "Lean supply phase", comprises product design, detailed engineering, fabrication, and logistics (Ballard and Howell 2003); it requires up-front product and process design to define what is needed and when it should be delivered. This is especially important with engineered-to-order components as utilized in engineering procurement and construction (EPC) projects. Lean supply also includes reducing the lead time for project information requirements. Lean supply addresses these problems through three main approaches:

1. Improving workflow reliability
2. Using web-based project management software to increase transparency across value streams and
3. Linking production workflow with material supply

The fourth phase, the "Lean assembly phase", is practiced in the actual construction of a project, putting materials, systems and components in place to create a completed facility.

In the last planner system (LPS) of production control, work structuring culminates in the form of schedules that represent specific project goals. Schedules are created for each phase of the project, beginning at the design phase and ending at project completion.

The production control provided by the LPS deploys the activities necessary to accomplish those schedules. The horizontal bars in the diagram labeled "production control" and "work structuring" refer to the management of production throughout the project.

The commissioning process provides quality assurance by ensuring that all systems have been installed by the contractor as promised in keeping with the designer's plans and specifications. It improves the probability that the facility will meet the owner's project requirements and performs as expected, to provide user satisfaction.

The fifth and last phase, the "use phase", refers to a completed facility. Following successful commissioning, the facility should undergo a protracted operations and maintenance phase as it is used. The circle labelled "alteration and decommissioning" refers to a future activity when the facility may be repaired, renovated, or taken out of service.

"Learning loops" refer to the application of root cause analysis to the LPS on a weekly basis to review percent planned complete (PPC) values and commitment reliability.

Learning is also accomplished through the process of post occupancy evaluation (POE) in which a facility is surveyed after occupancy/acceptance to review the consequences of decisions made during the execution of the project. POE enables project participants to "learn from the past."

8.3.4.2 Facilitating Flow through Work Structuring and Production Control

Achieving "flow" is one of the hallmarks of Lean, and various tools have been developed by the Lean Construction Community to help facilitate it. Work structuring and production control are used throughout a project to manage production. The term "work structuring" was developed by the Lean Construction Institute (Ballard 2000b) to describe construction-related process design. It is a process of subdividing work so that pieces are different from one production unit to the next to promote flow and throughput, and to have work organized and executed to benefit the project as a whole. The LPS and target value design (TVD) are methods that have been developed by LCI to promote flow. But before these tools can be described, it is important to help the reader recognize situations of "flow" and "nonflow". It is possible to represent activity flows graphically so that managers can identify opportunities to enhance it.

8.3.4.2.1 Representing and Visualizing Flow

Every Lean project is initiated by clearly defining "value" through rigorous consultation with the owner. Once value has been defined, a Lean project management process is identifiable by the way it "flows" toward the owner's defined value. For the JIT processing, the right elements need to be brought together at the right place, at the right time (i.e., in the right sequence), and in the right quantities. Once this happens, a construction project process—like that of a product assembly line—can start to flow. By contrast, failure to bring together the right elements, at the right place, at the right time, and in the right quantities, introduces constraints that block the flow process. Unless constraints are removed, the construction process encounters a bottleneck, and the success metrics of Lean—namely time, cost, quality, safety, and morale (Liker 2004; Figure 3.3, p. 33)—start to suffer.

Production control typically relies on a bar chart, where a horizontal bar extends from the very inception of the project to its conclusion. One of the simplest ways to visualize project flow is via a "line-of-balance" (LOB) schedule (also called a location-based or velocity schedule). A LOB schedule is similar to a typical bar (Gantt) chart except that the vertical axis has been transformed into an abscissa (the y-axis) onto which each activity's expected location is recorded. As with a bar chart, activity durations are represented along the ordinate (the x-axis).

By way of example, Figure 8.6 shows a bar chart that was used to schedule the installation of drywall. When converted into an LOB schedule (Figure 8.7), activity rates and potential conflicts (Figure 8.8) become clear. Note that rate is represented by the slope of the line; as the slope increases (the line becomes more vertical), the rate increases.

Similarly, as the slope decreases (the line become more horizontal), the rate decreases. If there is a need to complete a project sooner, rates of individual activities may be increased by adding labor and equipment. Gaps between activities may be minimized when there is a reduction of risk. Also, if a project's completion time is not urgent, the project manager can more easily identify opportunities to reduce labor or equipment dedicated to specific activities, confident that doing so will not adversely affect overall project completion time. On a LOB schedule, flow starts when activity rates align. When flows become parallel, they approach the Lean ideal (Figures 8.9 and 8.10).

LOB schedules may be used for heavy civil (horizontal) construction, as well as building (vertical) projects. For the latter, the y-axis is labeled according to floor levels or areas of repetitive work. Here too, LOB scheduling makes it easier for managers to visualize constraints to flow. Once constraints have been identified and removed, productivity may be enhanced and flow facilitated (Figures 8.9 and 8.10).

Interestingly, achieving flow does not necessarily demand access to cutting-edge technology. For example, although the Empire State Building was constructed in the late 1920s in New York City without contemporary forms of technology, near parallel flows were achieved, and the skyscraper was completed in just 16 months (Figure 8.11).

In order to understand how flow can be reached by implementing Lean principles, it helps to begin with some of the insights of W. Edwards Deming that have contributed to contemporary Lean thought.

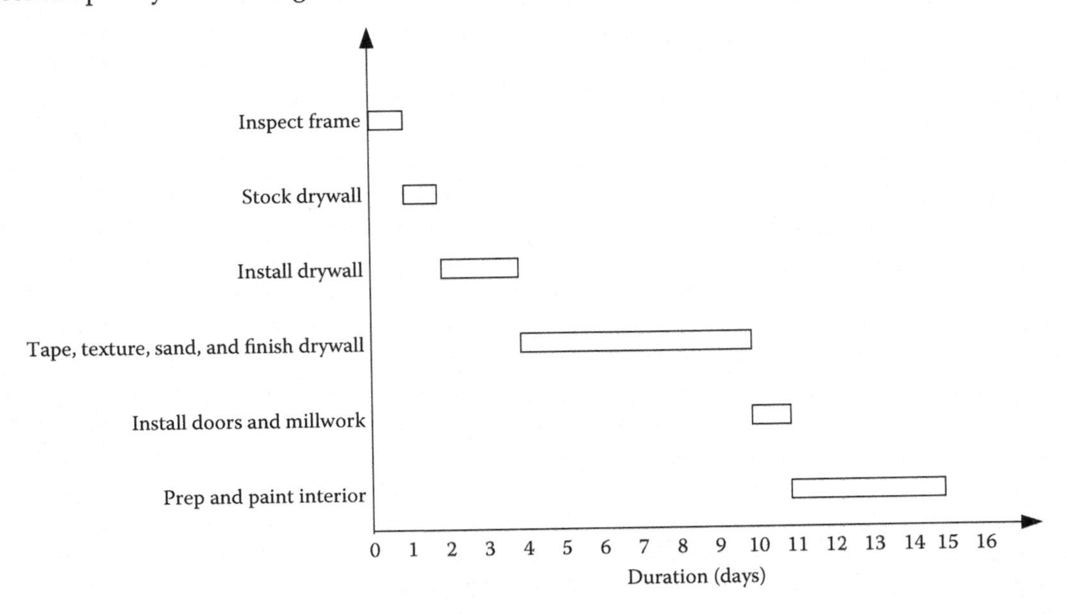

FIGURE 8.6
Example of bar chart used to sequence activities involved in the installation and finishing of interior walls. (Adapted from Bigelow, B.F., personal communication, approximate schedule for a single-story home, 1600–3000 SF, 2015. With permission.)

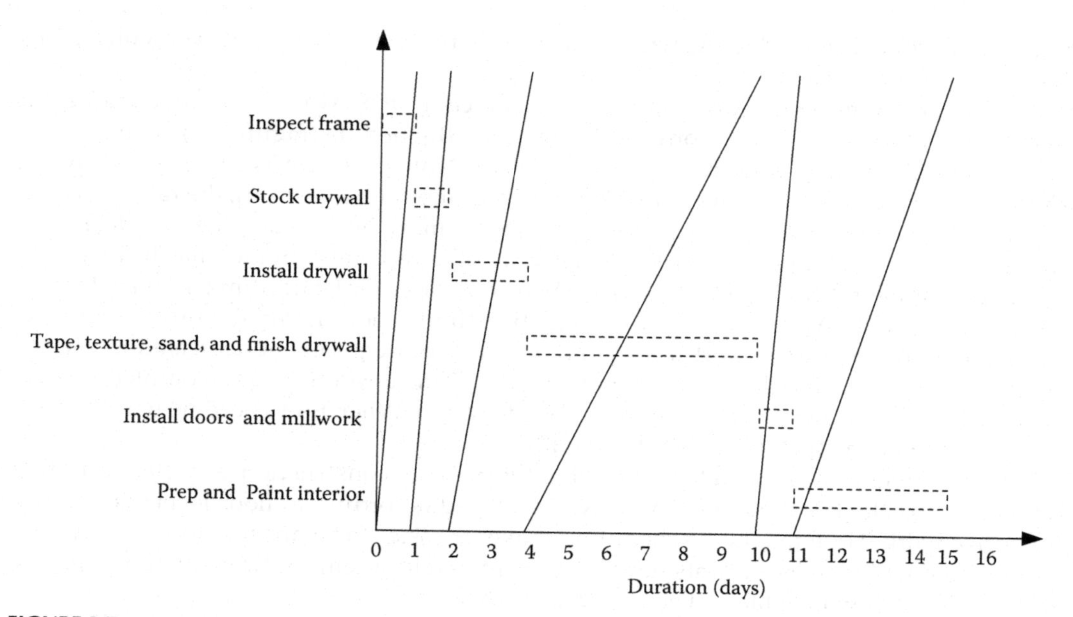

FIGURE 8.7
Example of a bar chart representing the installation and finishing of interior walls and its transformation into a line-of-balance (LOB) schedule. (Adapted from Bigelow, B.F., personal communication, approximate schedule for a single-story home, 1600–3000 SF, 2015. With permission.)

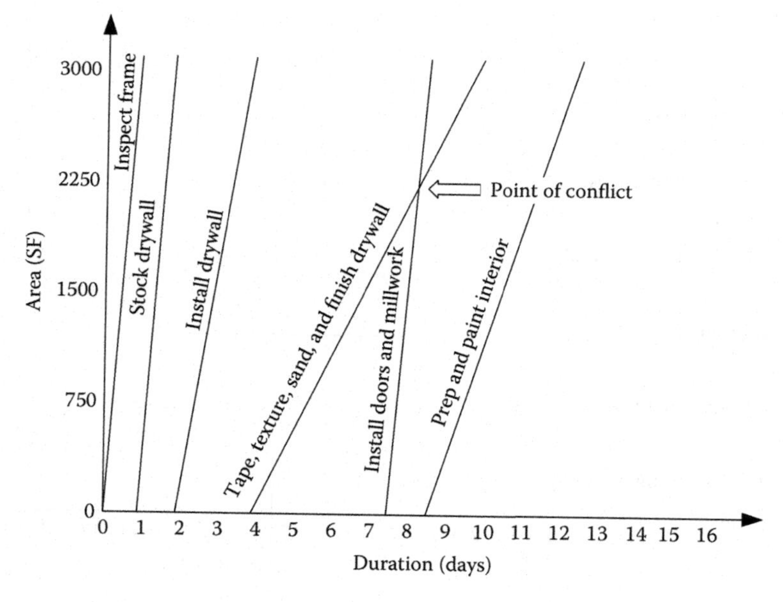

FIGURE 8.8
Line-of-balance reveals time and location of a potential scheduling conflict if start time of "Install Doors & Millwork" is begun sooner than shown in Figure 8.7. (Adapted from Bigelow, B.F., personal communication, approximate schedule for a single-story home, 1600–3000 SF, 2015. With permission.)

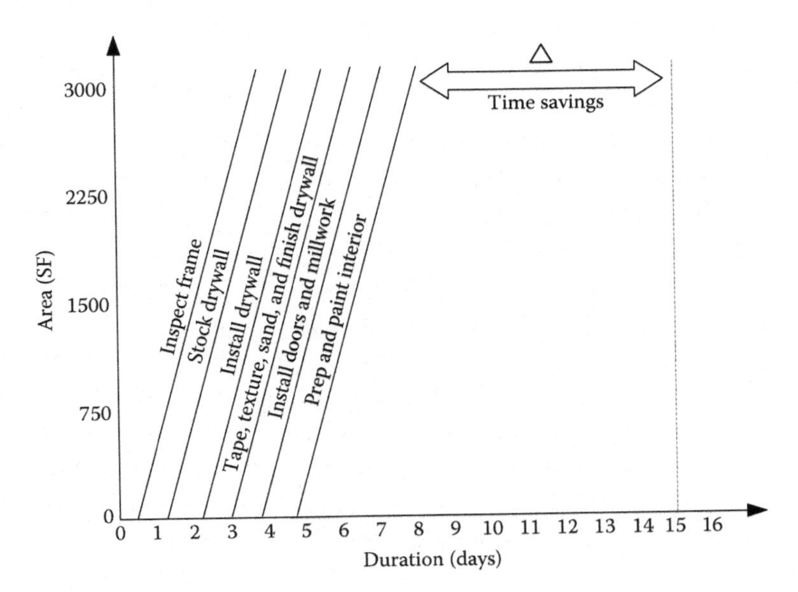

FIGURE 8.9
Parallel flows are revealed using LOB scheduling. Acceleration of individual activities can lead to significant time savings (Δ). (Adapted from Bigelow, B.F., personal communication, approximate schedule for a single-story home, 1600–3000 SF, 2015. With permission.)

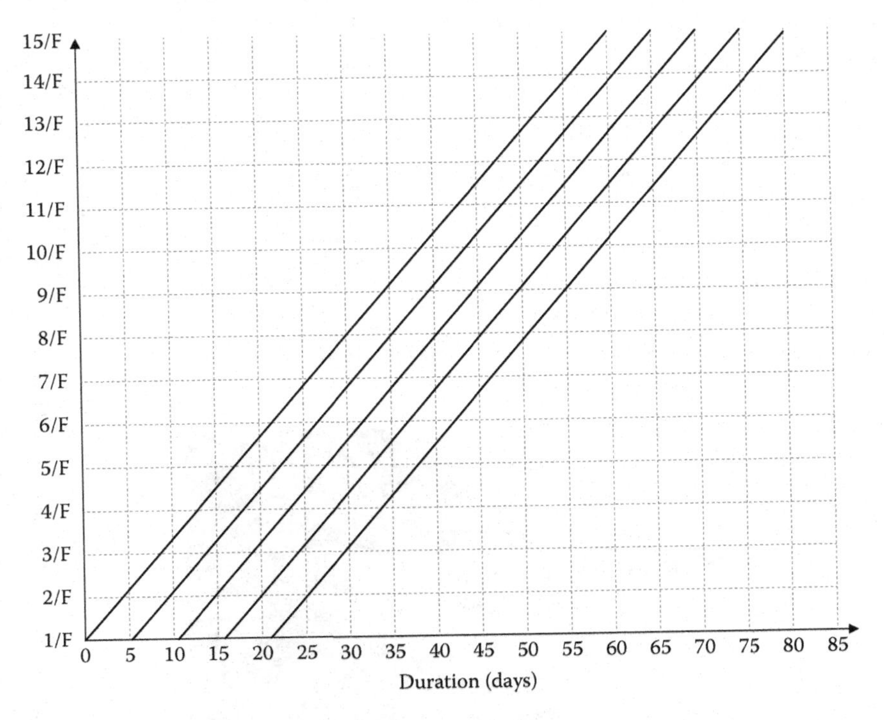

FIGURE 8.10
Example of LOB applied to a 15-storey high-rise building. (Adapted from Priven, V., Sacks, R., Seppanen, O., and Savosnick, J., *Lean Workflow Index for Construction Projects Survey [handout]*, Technion: Israel Institute of Technology, VC Lab: Seskin Virtual Construction Laboratory, Haifa, Israel, 2014 [as cited in Figure 5, p. 726]. Reprinted with permission from Vitaliy Priven.)

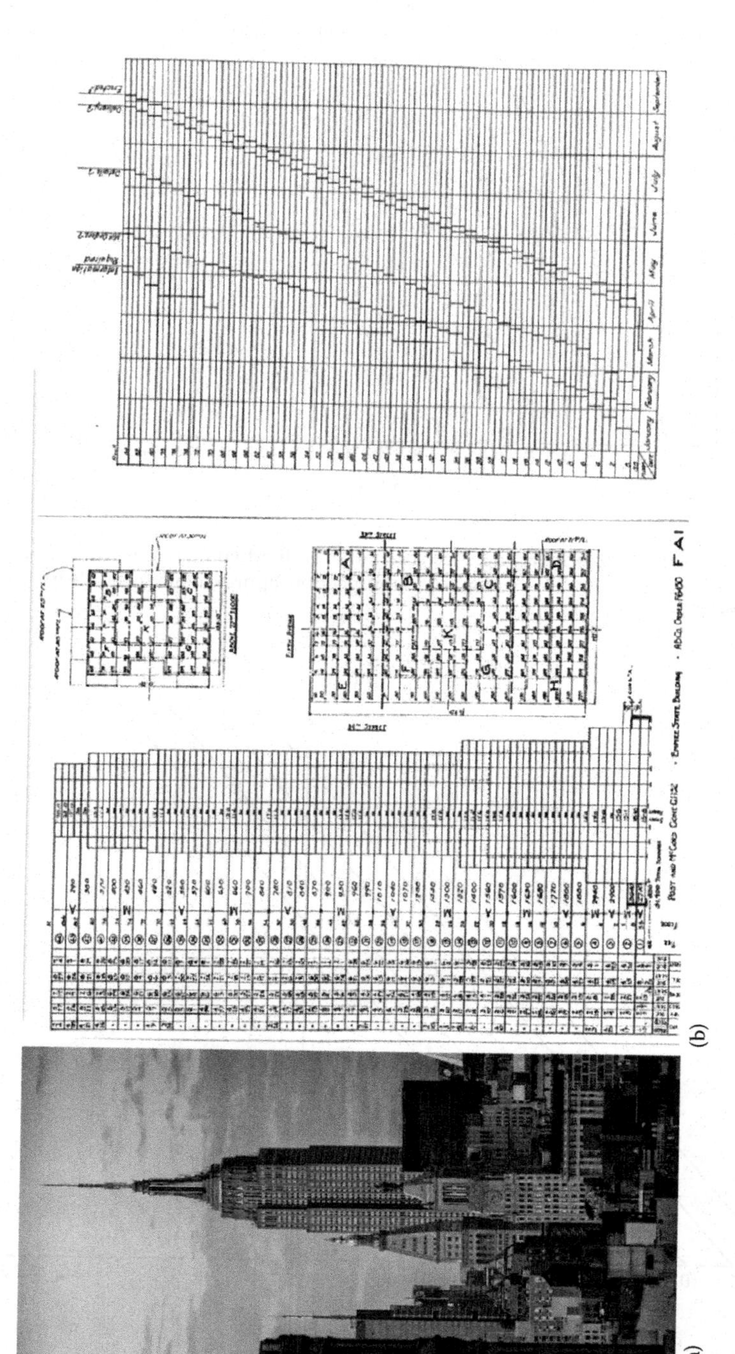

FIGURE 8.11
(a) Empire State Building overall view. (b) Building section and LOB schedule of the Empire State Building depicting near parallel flows (Left: location based measurement at structural steel design; Right: location based schedule for structural steel and control dates per level). (Adapted from Shreve, R.H., The empire state building: Organization, *Architectural Forum*, 52(6), 773, June 1930.)

Lean thinking is predicated on observations that blockages to flows often have less to do with worker incompetency, as is often believed, than to inherent inefficiencies of a system in which workers are asked to perform. This is because, unlike traditional project management that depends on the lone project manager to orchestrate construction processes, Lean construction focuses on respect for the judgment and experience of those who actually perform the work. This belief lies at the heart of the work of one of the pioneers of prelean thinking, mid-twentieth century statistician W. Edwards Deming. Deming exported his principles to post-World War II Japan, aiding the country's rapid economic recovery and subsequent quality revolution. Later in life, Deming facilitated workshops for business disciples, and recruited volunteers during these workshops to play his famous "red bead experiment". During the simulation, participants were asked to fill 50 depressions by dipping a dimpled paddle into a bin of red and white beads. Their task?—to minimize the drawing of red ("problem") beads. Deming played task-master; he reprimanded workers for drawing too many red beads, but also showered them with praise and promises of bonuses whenever a draw revealed a small number of red beads. Unaware that the ratio of red and white beads in the bin made success more a statistical fluke than a measure of actual accomplishment, participants struggled in vain to find ways to lower their red bead count. Deming used the simulation game as a metaphor to represent existing business practices in corporate America.

Like Deming, proponents of Lean construction argue that traditional construction delivery methods are rigged. Through the collection of data, they demonstrate that efforts to improve productivity by stakeholders—owners, architects, engineering, contractors, specialty contractors, suppliers, and attendant industries—are continually frustrated because traditional construction delivery methods are riddled with wasteful practices and constraints that systematically clog process flow. As a result, attempts to improve time, cost, quality, safety, and morale routinely fail.

8.3.4.2.2 The LPS of Production Control

Lean construction practitioners recognize that perfect flow toward value can be attained only if wasteful practices are eliminated, and variability and constraints systematically removed. To streamline processes, Lean practitioners first map existing practices in order to identify and eliminate "waste"—anything that does not add value for which the owner is willing to pay. Ohno (1988) included the following seven outcomes and practices as waste: defects, overproduction, waiting, transporting, movement, inappropriate processing, and inventory. Additional outcomes and practices have since also been identified as waste by members of the Lean construction community. But the message is clear: wasteful activities unnecessarily drive up cost, extend processing time, imperil safety, degrade quality, and damage morale.

LPD methods, such as the LPS, were developed so that critical project stakeholders can systematically identify waste and constraints far in advance and remove each before activity work is scheduled to begin (Ballard 2000a). By way of example, in order to assemble prefabricated, flat-packed kitchen cabinets in a house, the following prerequisites are required: wall and floor onto which the cabinetry will be hung and installed; sufficient laydown area; shelves, cabinet doors, hinges, etc.; skilled labor; appropriate tools and equipment; weather that permits transportation of materials and labor, freedom from labor strikes, permitting issues, sufficient time; and an assembly instruction sheet. In other words, the following resources must be made available for assembly to take place: "prerequisite work, space, materials, labor, tools and equipment, external conditionals, and information". Absence or delay of any of these elements represents a "constraint" that must first be removed for the process to flow.

Lean construction pioneers realized early on that one of the chief wastes in the industry is an inaccurate scheduling process. The reason traditional scheduling is so poorly informed is that knowledge is dispersed among those who actually complete the work. Scheduling centralized in the project manager is typically flawed because it has been developed by someone who does not actually complete the work. By contrast, LPS overcomes this limitation by engaging "boots-on-the-ground" personnel to collectively develop and inform the project schedule. Not to be confused with software solutions, LPS is often most effective as a "low-tech" application, engaging participants by asking them to arrange activities with colored notes ("stickies") on a wall. LPS uses "pull-scheduling" (i.e., critical completion milestones are first established, and activities are scheduled backward in time from these milestones). In addition to including stakeholders in the advance for identification and removal of constraints, flow is aided by engaging them in an accountability exercise known as PPC, or "percent-planned complete", where the percentage of promises-kept versus promises-made is regularly recorded. The simple act of recording and publicly posting daily PPC has been shown to reduce variability of scheduled work, and consequently enhance project flow (Ballard 2000a).

8.3.4.2.2.1 Reverse-Phase Scheduling LPS is a method for applying Lean techniques to construction. It emphasizes control by proactively assisting activities to conform to plan, as opposed to the construction tradition of monitoring progress against schedule and budget projections. A production planning and control system, it is the brainchild of Lean Construction Institute co-founder, Glenn Ballard (Ballard 2000a). The term "Last Planner" refers to the front line supervisor. Ballard initiated pull in construction by asking construction partners to engage in a process known as "reverse-phase scheduling" (RPS). Once a client's time constraint has been established, the deadline is fixed to a wall with a self-adhesive notecard. Team members then plan activities collaboratively and collectively—also on the wall using self-adhesive notecards—and backward from the posted deadline. The deadline establishes the basis for a type of "takt time", which is the rate at which individual activities need to be accomplished in order to meet the client's required deadline. It must be mentioned here that, unlike a manufacturing assembly line where the final design is known before manufacturing the product, the Last Planner is applied while design of a building is under development. This means precisely that the construction times of various phases of the building cannot be more than estimates and the term "takt time" must be applied loosely to Last Planner as a general rate at which a project must be "pulled" in order to meet the required time contraints. Nevertheless, the analogy is helpful for understanding how Lean construction principles intersect with those of Lean manufacturing.

There are four components of the LPS—master scheduling, phase scheduling, lookahead planning, and commitment/weekly work plan—as graphically depicted in Figure 8.12 (Hamzeh 2009). The last two phases, lookahead planning and weekly work plan, are of special interest to us here.

RPS or "pull planning" is used as a starting point for the LPS. It is a detailed work plan that specifies the handoffs between trades for each project phase (Ballard and Howell 2003). In practice, subcontractors participate extensively in RPS on a wall-mounted board to plan their schedules in concert with the schedule for each project phase. Traditional construction emphasizes pushing tasks to meet this optimistic schedule, with little thought to what CAN and WILL be done. In LPS, the master schedule is refined as a reverse-phase schedule by using a team to work backward from the expected completion date. This team comprises "last planners" from different trades/subcontractors. The master schedule and reverse-phase schedule indicate what work SHOULD be done in order to meet the schedule.

During "lookahead planning", constraints are systematically removed to ensure that which SHOULD be done CAN be done. The "weekly work planning" process then assures that which CAN be done actually WILL be done (Figure 8.12).

The percent plan complete (PPC) "public" declaration accounting system incentivizes those who commit to perform various tasks to follow through with their commitments. Root cause analysis of problems encountered on that which the team actually DID, helps generate countermeasures that will enshrine learning and advance the team toward a more advanced step of the Lean plan-do-check-act cycle—ensuring never-ending levels of continuous improvement.

It might be argued that a critical purpose of Last Planner is to serve as a series of conceptual kanbans, where metaphorical carts have been replaced by a scheduling directive called the "weekly work plan". Although last-planner-as-kanban is an imperfect metaphor, the two processes share some common traits. On the weekly work plan, the "last planner"—the individual responsible for organizing final work assignments for the overall project—divides work into defined (often day-long) batch sizes. The last planner then "fills the kanban carts"—assigning work to each day of the work week. Like the bus driver who must wait at a stop to conform to an overall transit plan if he arrives ahead of schedule, no member of a team may perform work either before or after his turn has been designated. Team members are, in effect, informed by the last planner facilitator about when to get on the bus—not a moment before and not a moment after the appropriate time. This is the essence of the JIT system, so integral to Lean thinking (Rybkowski 2010).

Most construction projects are unique to their site and function. Because of the one-off nature of most construction projects, no one individual—not even an experienced project manager—can know all that is required to fill the metaphorical kanban carts. Last Planner acknowledges this by engaging the "big room" concept of meetings common to Lean thinking. The term big room refers to the need to bring together all those who are critical to the design of a building so that their knowledge can inform that which needs to be done during a regular specified time period. The day or half-day of a weekly work plan in the last planner can be imagined as an empty kanban cart waiting to be filled with resources

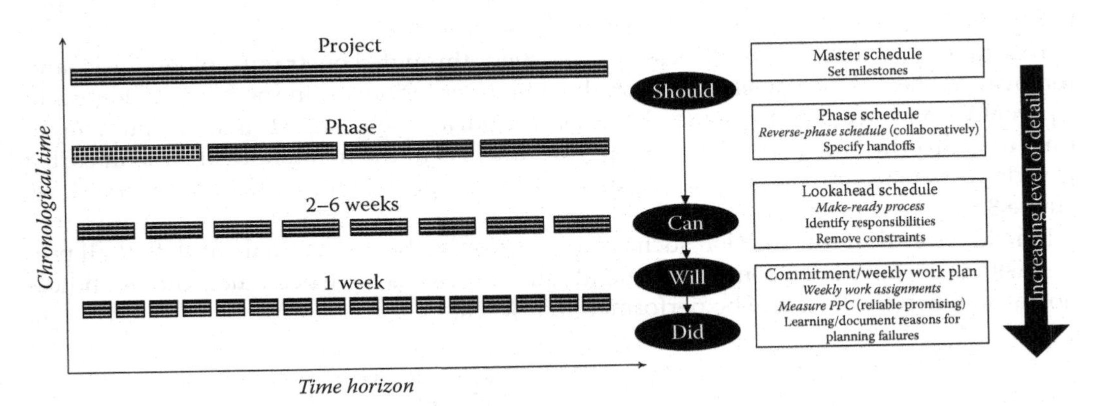

FIGURE 8.12
The Last Planner system of production control. (Adapted from Ballard, G., The last planner system of production control. PhD dissertation, University of Birmingham, Birmingham, 2000a [as cited in Figure 3.2, pp. 3–5]; Hamzeh, F.R., Improving construction workflow: The role of production planning and control, Doctoral dissertation, Civil and Environmental Engineering, University of California, Berkeley, CA, 273 pp, 2009 [as cited in Figure 5.2, p. 129]. Reprinted with permission from Farook Hamzeh.)

that will be transformed at designated stations. In Figure 8.13, the collective experience of team members in the Big Room is symbolized by a cloud of shared knowledge.

Naturally, one risk of a JIT delivery system is that it may place an unfair burden on those who must fill a cart. Anyone who has heard the words "I need it tomorrow" or, worse yet—"give it to me now"—knows how unreasonable such directives can be. Responding to this, Lean practitioners frequently use the phrase "last responsible moment" instead of JIT. The lookahead plan of the LPS focuses on constraints analysis and removal, making JIT achievable, as shown in Figure 8.14.

8.3.4.2.2.2 Removing Constraints By the time a task is committed during a weekly work plan meeting, the expectation is that it will be completed as scheduled to maintain a predictable flow of work through the network of specialists. Therefore, anything that might hinder completion of the task needs to be cleared before it is assigned. During lookahead planning (two to six weeks before weekly work plan assignments are made), tasks are "made ready". In a landmark paper on shielding, Ballard and Howell proposed five "quality criteria" against which a task must be checked before it is allowed into the weekly work plan (Ballard and Howell 1998). These are

1. *Definition:* Is the task specific? Will it be clear when it has been finished?
2. *Soundness:* Are all materials available, including completed prerequisite work, for the task to be performed?
3. *Sequence:* Is the task being performed in the correct order?
4. *Size:* Is the task sized to the capacity of the crew?
5. *Learning:* When assignments are not completed, are they tracked and reasons identified?

The facilitator of the meeting checks for these conditions in order to ensure that the "customer" of any task (the trade that immediately follows) is furnished with all that is necessary to complete the task successfully. Because all downstream work suffers when a task cannot be completed, it is crucial that the facilitator rigorously honor this checklist. Once a task has been made ready, it can safely be assigned to enter the flow.

The quality criterion "soundness" is satisfied through constraints of analysis and removal. In the public transit metaphor, the quality criterion "soundness" is analogous to a parent who wakes up, dresses, feeds a school child, and sends her/him to the bus stop in time to board the bus at its scheduled arrival time. As in the metaphorical kanban cart, a weekly work plan signals a request for a task to be completed for the customer that follows (Table 8.3).

It includes critical information such as: description of the task, a final check that all prerequisite tasks have been completed and all quality criteria have been met, and an indication as to when the task will be performed that week.

8.3.4.2.2.3 Percent Planned Complete Variability is undesirable when attempting to achieve flow. To test this principle, Tommelein (1997, 1998, 2000) developed two computer models to simulate variability within manufacturing processes. The researcher compared the total time required to complete a process when individual component tasks were assigned deterministic (set) times versus when they were assigned stochastic (distributed) completion times. The results demonstrated the detrimental impact of variability on flow. Figure 8.15 illustrates the same.

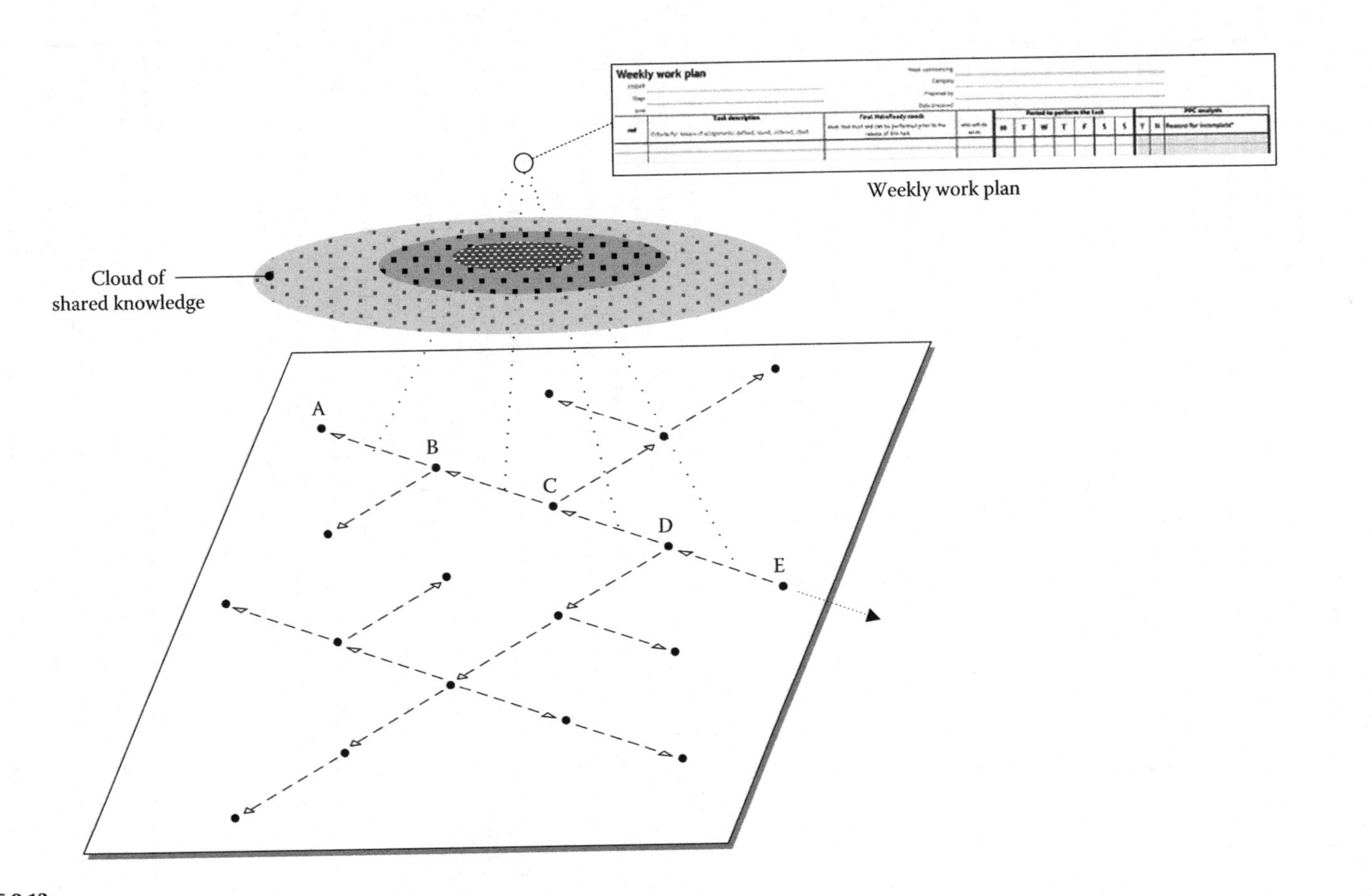

FIGURE 8.13
The Last Planner serves as a "kanban" that pulls activities, as informed by a cloud of shared knowledge. (Adapted from Rybkowski, Z. K., The application of root cause analysis and target value design to evidence-based design in the capital planning of healthcare facilities, Doctoral dissertation, University of California, Berkeley, CA, 2009 [as cited in Figure 80, p. 230]. Illustration by author [Rybkowski].)

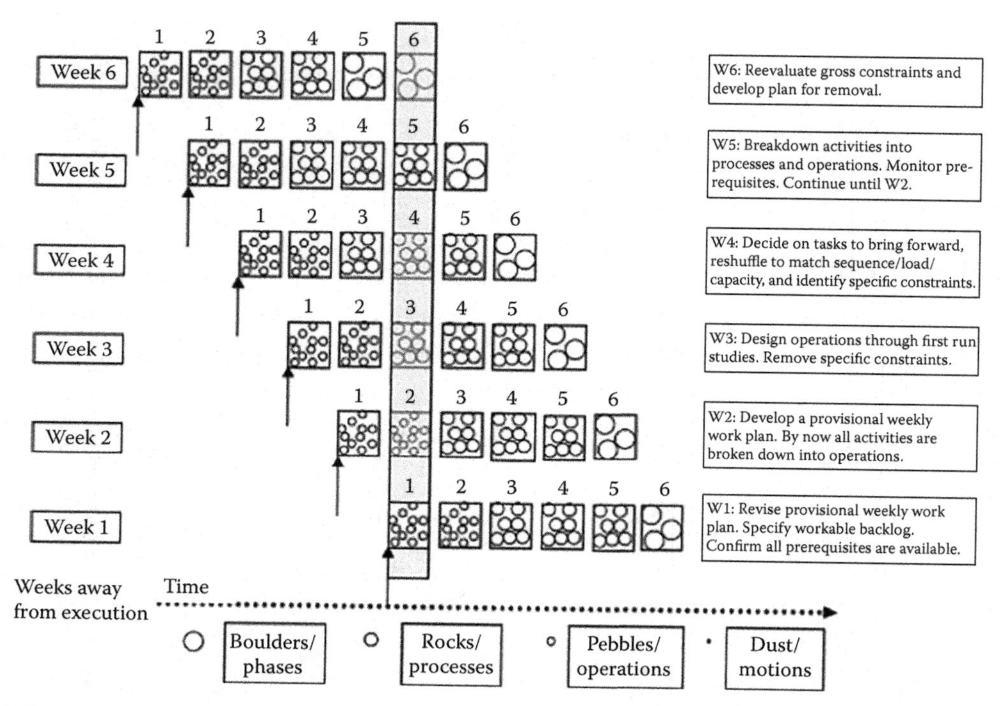

FIGURE 8.14
Six-week look-ahead planning process. (Adapted from Hamzeh, F.R., Improving construction workflow: The role of production planning and control, Doctoral dissertation, Civil and Environmental Engineering, University of California, Berkeley, CA, 273pp, 2009 [as cited in Figure 5.13, p. 148]; Hamzeh, F.R., Ballard, G., and Tommelein, I.D., Rethinking lookahead planning to optimize construction workflow, *Lean Construction Journal*, 15–34, 2012 [as cited in Figure 4, p. 27]. Reprinted with permission from Farook Hamzeh and Lean Construction Institute [LCI].)

The impact of variability is important because work cannot be infinitely buffered. The weekly work plan kanban "batch" of one day, for example, still requires a defined time limit that should not be exceeded if flow is to be maintained. The assumption is that some work will be accomplished more quickly and other work will take longer than planned. Using a public bus route as a metaphor, a scenario may be envisioned where traffic is so light that the driver arrives at stops ahead of schedule. Alternatively, a bus driver may be trapped behind an unforeseen traffic accident—making the vehicle arrive at stops later than any reasonable amount of buffering could have accommodated. But in addition to some events that may be beyond a driver's control, buses can also be

TABLE 8.3
Weekly Work Plan

Ref	Task description	Final make ready needs Work that must be performed prior to the release of this task	Who will do work	Period to perform the task							PPC analysis		
				M	T	W	T	F	S	S	Y	N	Reasons for incomplete

Source: Adapted from Lean Construction Institute, 2009. With permission of the Lean Construction Institute.

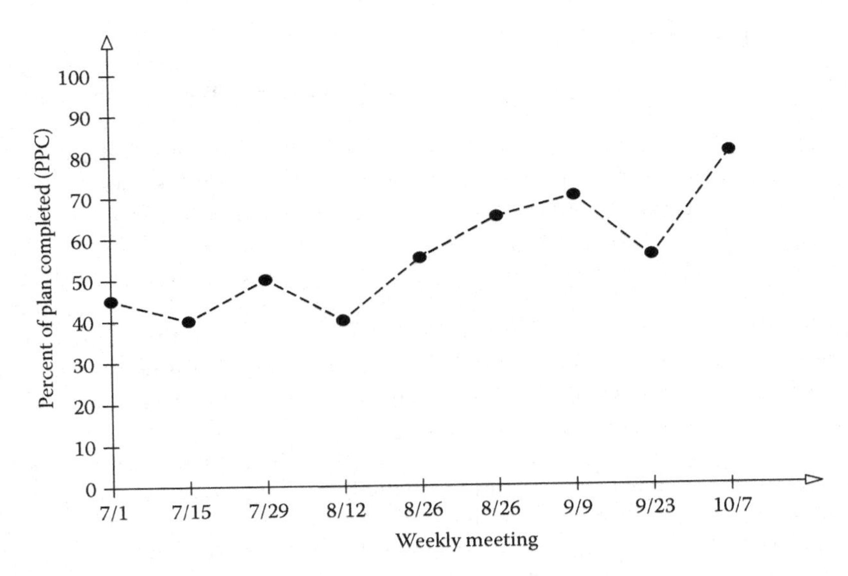

FIGURE 8.15
A percent plan complete (PPC) chart. (Adapted from Rybkowski, Z.K., The application of root cause analysis and target value design to evidence-based design in the capital planning of healthcare facilities, Doctoral dissertation, University of California, Berkeley, CA, 2009 [as cited in Figure 84, p. 236]. Illustration by author [Rybkowski].)

delayed by unmotivated or careless bus drivers. Accountability is important because it increases reliability and reduces variability. For public transportation networks in the United States to operate according to a reliable schedule, drivers operating ahead of schedule often pause longer at specified stops to realign their actual schedule with the publicly posted schedule.

The critical nature of worker accountability is recognized by the LPS. For example, a measure of workflow reliability called *PPC* is embedded in the weekly work plan process; PPC is used to increase the reliability of planning by reducing variability. The idea is that specific tasks designated to be completed before the next "big room" Last Planner meeting are listed. During Last Planner meetings, each list is checked for its level of completion. Research has demonstrated that when more disciplined screening of potential commitments is used in combination with urgent expectations and peer pressure to make reliable promises, PPC increases. This increase suggests that the reliability of planning increases (Ballard 1999; Ballard and Howell 1998).

Because the variability of completion times for dependent activities negatively affects project schedules, it is advantageous to reduce variability. A publicly posted PPC chart increases the reliability of deliverables by motivating workers to maximize their PPC ratings.

It is important to note that PPC should not be mistaken as an indicator of productivity. In fact, stakeholders sometimes underpromise simply to boost their PPC score. The role of PPC is instead to enhance the reliability of the work promised, making future planning more reliable.

When a task is not completed as planned, the weekly work plan includes a section to indicate the cause for the divergence under "reason for noncompletion" in order to incorporate learning into the process should a similar situation arise again.

8.3.4.3 Target Value Design

TVD involves designing to a specific estimate instead of estimating based on a detailed design. It seeks to address the problem that affects many projects, namely, designing and constructing a building that costs more than expected. Traditionally, various design disciplines work from a common schematic design to develop designs in their areas of expertise. Working in their respective offices, they are subject to "project creep." With little cross-functional collaboration, project designs often become overpriced, un-constructable, and behind schedule (Forbes and Ahmed 2010). Corrective action may include a misuse of value engineering to radically cut the scope of a project, or to suppress certain features that are desirable but unaffordable. Furthermore, the lack of collaboration often results in early design decisions that are later found to be suboptimal, but difficult to change. Ultimately, much time and effort are wasted, and the design cycle is longer than it should be. These wastes run counter to the Lean philosophy.

Since TVD applies target costing to building construction, it helps to first define target costing. According to Cooper and Slagmulder (1997), "Target Costing is a disciplined process for determining and realizing the total cost at which a proposed product with specified functionality "must" be produced to generate the desired profitability at its anticipated selling price in the future." It is perhaps simplest to illustrate target costing as it applies to product design and then highlight how target costing differs from traditional product costing. In traditional product costing, a manufacturer may add a profit markup to a product's production cost to establish its selling price. The problem with this method is there is no guarantee that buyers will be willing to pay the asking price. The process of target costing, by contrast, implements a reverse strategy; the market price is "first" established by determining how much buyers might be willing to pay, by using focus group research or by looking to similar products on the market, for example. A desired profit is then subtracted to give product designers the cost to which they must design the final product:

Target cost = Target price − Target margin (Clifton et al. 2004)

The concept of target costing, as applied to product design, can also be envisioned diagrammatically, as shown in Figure 8.16.

TVD builds on the concept of target costing but represents target costing applied to construction rather than product design. While it is helpful to understand the genesis of target costing in product development, the terms used by the TVD community differ somewhat in their meaning and include four distinct components, as defined by Glenn Ballard: "market cost, allowable cost, expected cost", and "target cost". They are defined as follows (Figure 8.17).

"Market cost" is a benchmark cost; it consists of the cost per square foot that would be expected for comparable construction project. "Allowable cost" represents the maximum cost that must not be exceeded; if the project team cannot design to the allowable cost, the project must be cancelled because it would, by definition, become financially unfeasible. "Expected cost" is the estimated cost of the project in its current state during the TVD process; the expected cost is continually recalculated with each new iteration of design. *Target cost* is the stretch goal for the project, meaning it is usually set below the allowable cost (Ballard 2009a; Rybkowski 2009).

TVD also proposes that designers should "do it right the first time" and build constructability into their designs, as opposed to designing first and then evaluating constructability later. TVD recommends concurrent design, with various disciplines in ongoing contact, as opposed to periodic reviews. Solution sets should be carried forward in the design process to ensure that good alternatives can be available later—a process known as "set-based design".

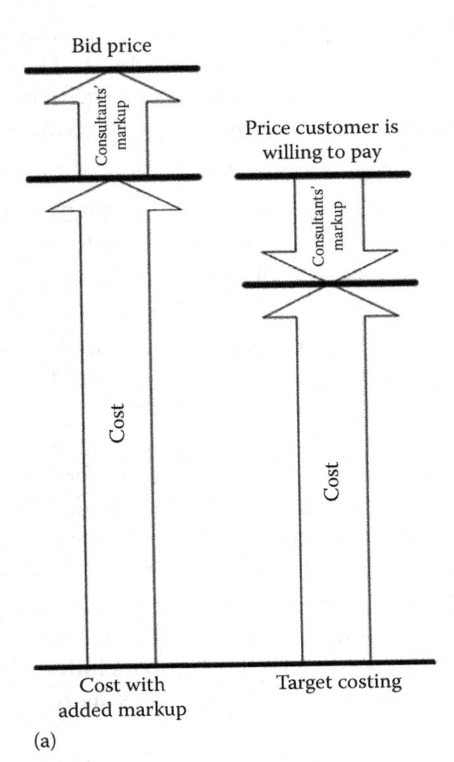

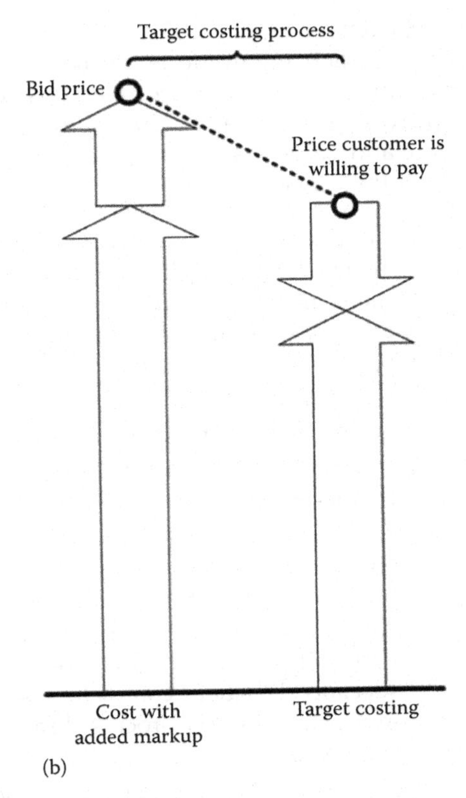

FIGURE 8.16
Cost with added markup (a) versus target costing (b). (Adapted from Rybkowski, Z.K., The application of root cause analysis and target value design to evidence-based design in the capital planning of healthcare facilities, Doctoral dissertation, University of California, Berkeley, CA, 2009 [as cited in Figure 46, p. 129]. Illustration by author [Rybkowski].)

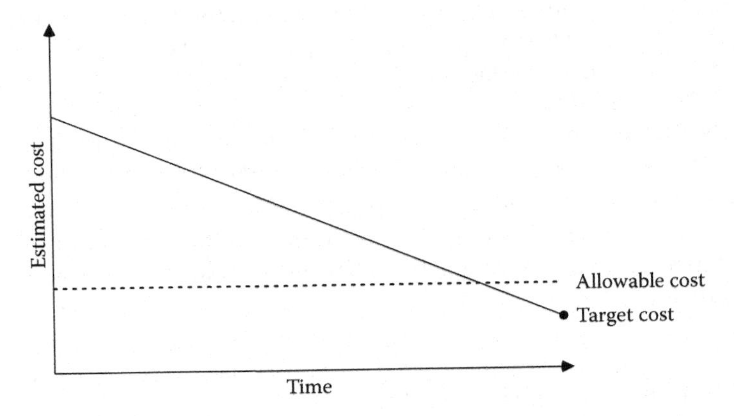

FIGURE 8.17
Diagram of the target value design process. (Adapted from Rybkowski, Z.K., The application of root cause analysis and target value design to evidence-based design in the capital planning of healthcare facilities, Doctoral dissertation, University of California, Berkeley, CA, 2009 [as cited in Figure 47, p. 131]. Illustration by author [Rybkowski].)

Macomber et al. (2005; 2008) proposed a number of foundational practices for TVD. These practices promote "design conversation" as they see design as a social activity that involves several professionals focusing on meeting the needs of the client. This approach is especially effective in light of the fact that the client's needs can change over time and value assessments need to be repeatedly made to ensure that design decisions meet these needs.

1. Conduct design activities in a big room ("obeya" in Japanese; see Section 4.3.2.3.2). Toyota has used the obeya concept successfully, especially in product development, to enhance effective and timely communication. The obeya is similar in concept to traditional "war rooms."
2. Work closely with the client to establish the "target value." Designers should guide clients to establish what represents value, and how that value is produced. They should ensure that clients are active participants in the process, not passive customers.
3. Once the "target value" is established, use it to work with a detailed estimate. Have the design team develop a method for estimating the cost of design alternatives as they are developed. Deviations should not continue unchecked; if a particular design feature exceeds the budget allocated for it, that design should be adjusted promptly in order not to abort further design work that cannot be accepted.
4. Apply concurrent design principles to design both the product and the process that will produce it. This work should be done as a collaboration between A/Es, specialty designers, contractors/subcontractors, and the owner. Be flexible to include innovation in this process. Practice reviewing and approving design work as it progresses.
5. Practice set-based design where multiple potential design solutions are carried forward, ensuring time and resources are not wasted if scope changes or previously unknown constraints are discovered later in the design process.
6. Work in small, diverse groups. Groups of eight or fewer people facilitate better group dynamics—it is easier to create a spirit of collegiality and trust that lead in turn to more innovation and learning.
7. Design with the "customer" in mind. Focus on designing in the sequence of the discipline that will use it. Use the "pull" approach with each design assignment to serve the next discipline. Lean is obtained by meeting downstream needs as opposed to producing what is convenient. Overproduction increases the possibility that the work so produced may not be what is needed for the next discipline to maintain its schedule, and it may lack the collaboration necessary for constructability.
8. Collaboratively plan and replan the project. Planning should involve all stakeholders to continually maintain an actionable schedule. Joint planning will refine practices of coordinating action. This will avoid delay, rework, and out-of-sequence design.
9. Lead the design effort for learning and innovation. Expect the team to learn and produce something surprising. Also expect surprise events to upset the current plan and require more replanning.
10. Learn by carrying out conversations on the results of each design cycle. Include all project participants in order to capture knowledge on success factors. Use this information as a part of the PDCA cycle, and use formal measurement systems, if possible.

The steps involved in TVD are shown in Figure 8.18. Note that each step of the TVD process requires a reevaluation of whether or not the project team should proceed based on a realistic assessment of whether or not the stakeholder group can meet the agreed allowable cost that had been validated during the project definition phase.

8.3.4.3.1 Integrated Project Delivery and TVD

Waste in design–bid–build construction occurs mostly in the interaction between trades. Traditional contracts are written with the intent of transferring risk to one party at the expense of the others. The result is multiple parties working for their own self-interest rather than for the good of the project. By contrast, IPD contracts are written to provide

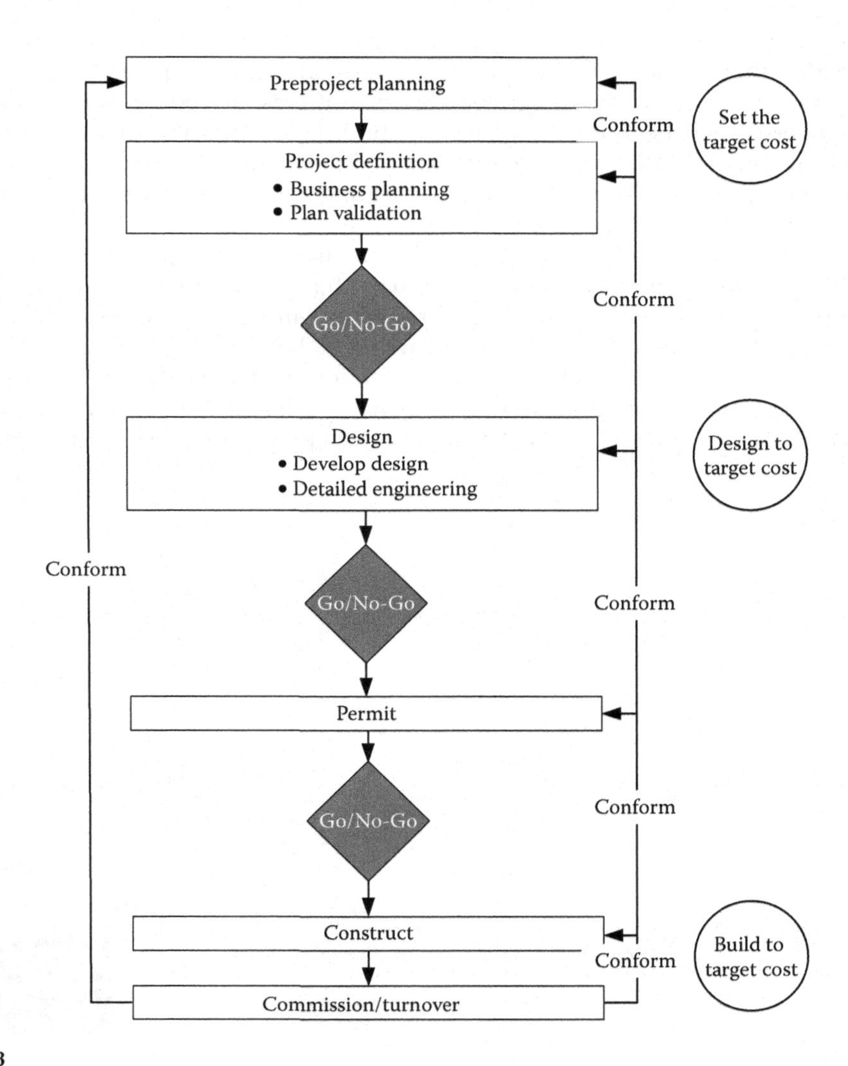

FIGURE 8.18
The target value design process. (Adapted from Ballard, G., The lean project delivery system: An update, *Lean Construction Journal*, 1–19, 2008 [as cited in Figure 5, p. 8]; Rybkowski, Z.K., The application of root cause analysis and target value design to evidence-based design in the capital planning of healthcare facilities, Doctoral dissertation, University of California, Berkeley, CA, 2009 [as cited in Figure 80, p. 230]. Reprinted with permission from Lean Construction Institute [LCI].)

financial incentives for behavior that is good for the overall project, harnessing natural self-interest as a beneficial project delivery driver. IPD involves a contractual combination of LPD and an integrated team that is expected to improve project performance in a number of dimensions. If the process is successfully executed, owners benefit from reduced time and cost as well as improved quality and safety. Designers and contractors derive increased profits, improved owner satisfaction as well as greater employee satisfaction.

With traditional design–bid–build–delivery, stakeholders are often brought into the process too late—after the ability to affect cost and function has already passed. By contrast, with IPD, key stakeholders are involved early enough in the process to influence critical design decisions. This relationship is famously represented by the MacLeamy curve (Figure 8.19).

IPD is designed to support the values and principles of LPD that is, to reduce waste and provide value for the owner. While traditional contracts anticipate adverse events and focus on transferring risk, the Integrated Agreement seeks to reduce risk by empowering team members to use Lean thinking and collaborative approaches. In fact, the collaborative problem-solving skills of the project leadership team, that is, the owner, designer, and contractor enable it to transcend unexpected challenges and function as an agile, high-performing team that delivers projects successfully.

IPD was developed to improve innovation by moving money across boundaries (Alarcon et al. 2013). IPD is defined as a delivery system that seeks to align interests, objectives and practices, even in a single business, through a team-based approach (LCI 2015). The primary team members include the owner, architect, key technical consultants, a general contractor, and key subcontractors. An organization is created that is able to apply the principles and practices of the Lean Project Delivery System™. The American Institute of Architects (AIA) has recognized IPD as an effective project delivery method and defines

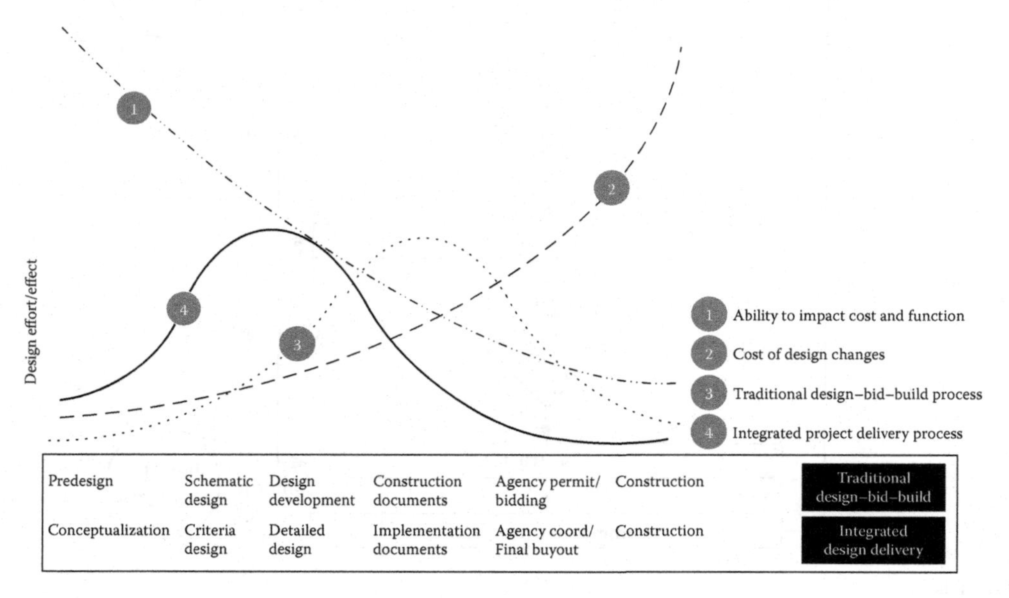

FIGURE 8.19
The MacLeamy curve. (Adapted from MacLeamy, P., Collaboration, integrated information, and the project lifecycle in building design and construction and operation, *Introduced at the Construction User's Roundtable*, WP-1202, August 2004 [as cited on p. 4]; MSA., MacLeamy curve, <http://msa-ipd.com/MacleamyCurve.pdf>, Accessed on February 21, 2015, 2004. Reprinted with permission from HOK.)

it as "a project delivery approach that integrates people, systems, business structures and practices into a process that collaboratively harnesses the talents and insights of all participants to optimize project results, increase value, to the owner, reduce waste, and maximize efficiency through all phases, design, fabrication, and construction." The first IPD application involved a central chilled water plant built by Westbrook Air Conditioning and Plumbing of Orlando, Florida (Forbes and Ahmed 2010; Matthews and Howell 2005). Westbrook held a design–build contract with the owner, but developed a separate business entity with design professionals and construction subcontractors, termed "Primary Team Members" (PTMs). Through teamwork, innovation, and optimization, work and costs moved between different companies to improve performance and reduce overall costs. The team saved 10% of the contract cost (Matthews and Howell 2005).

Some versions of IPD involve an Integrated Form of Agreement (IFOA) as a single contract between the owner, the architect, and the construction manager or general contractor. The IFOA involves a core group of representatives of the owner, architect, and GC/CM to administer the project. As shown in Figure 8.20, the core group selects the remaining members of the IPD team including engineers, technical consultants, key subcontractors, and suppliers.

Sutter Health is a not-for-profit community-based health care and hospital system headquartered in California. Faced with a state mandate (and penalties) to upgrade inadequate facilities, they embarked on a very tightly budgeted $6.5 billion design and construction project starting in 2004. Sutter Health embraced the Lean philosophy (as a leap of faith) and engaged Lean Project Consulting Inc. to guide the adoption of IPD. McDonough Holland & Allen PC, Attorneys at Law, provided legal services especially with developing an early version of the IFOA. In order to derive the greatest synergy from team integration, the CM/GC entity was engaged immediately after architect selection. Major subcontractors were also engaged as early as the schematic design phase. While this approach incurred expenditure for professional services earlier than traditional construction practice, the Sutter organization felt that the resulting collaboration led to improved constructability and savings that far exceeded the added cost.

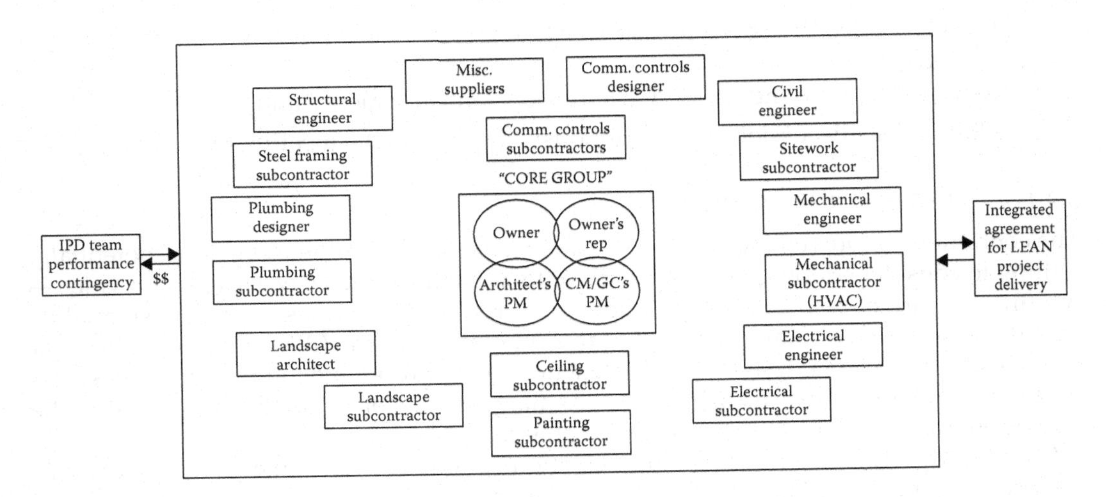

FIGURE 8.20
The Sutter Health integrated project delivery relationship. (Adapted from Forbes, L.H., and Ahmed, S.M., *Modern Construction: Lean Project Delivery and Integrated Practices*, CRC Press, Boca Raton, FL, 2010. Reprinted with permission from Taylor & Francis Group.)

The core group served in an administrative oversight role, scheduling meetings and problem solving sessions. The group maintained open information flow, and ensured that all parties worked together harmoniously, using 3D BIM, TVD, and the LPS. They promoted the network of commitments needed to achieve reliable workflow. In IPD, this reliability is critical to maintaining a smooth handoff of work from one crew to another. Cost savings and owner satisfaction attest to Sutter Health's Lean project successes (Forbes 2009, 2014).

8.3.4.3.2 The Role of the "Big Room" in TVD

"Big room" (obeya) is a term that describes a space where project team members can work collaboratively, and in close proximity as opposed to being located remotely. While it may not be feasible for all teams to colocate continuously, the big room environment can be created for project meetings through the use of site trailers that are appropriately equipped with adjacent meeting rooms for "break out" sessions. The equipment may consist of wall-mounted schedule boards, A3 charts, flip charts mounted on easels, dry-erase white boards, projection equipment such as a Smart Board, and audiovisual equipment for long distance meetings with remote locations.

The big room houses project leaders and key staff in close but comfortable proximity to shorten the communication path and promote an effective PDCA cycle. Spontaneity comes easily as specialists can collaborate readily in key design or construction decisions.

Big room activities include

- Revolving schedules with agendas for meetings of the complete team including designers, suppliers, subcontractors, the management team, Last Planners, and others
- Agendas which vary with the needs of each project and ensure the right people are included
- Status reviews, presentations, and pull-planning sessions
- Learning opportunities, such as orientation, onboarding, and ongoing Lean training
- "Gemba Walks" to the site to "Go and See"—a Lean practice and
- "Plus/Delta" discussions for continuously improving the meeting process

8.3.4.3.3 The Role of BIM in TVD

The continuous improvement culture of Lean means that a Lean project team is potentially willing to consider adopting any approach or tool that may reduce waste and add value—within a culture of respect—and so improve performance in terms of time, cost, quality, safety, and morale. For example, building information modeling (BIM) supports initiatives that are critical to Lean design and construction. As the fundamental principle of Lean is to reduce or eliminate waste, BIM addresses many aspects of waste that occur first in the design phase, and later in the construction phase.

Lean design promotes the active participation of construction stakeholders as early as the project definition phase. As the design concept is developed, designers, owners, and constructors work interactively to make decisions that influence the overall project—concurrently and in real time. Traditional design reviews are treated as sequential events, long after significant design decisions have been made. At that point, changes can quickly become time-consuming and expensive. Furthermore, as many disciplines are involved in the design of a project, changes in some elements of the design may not be fully represented by all the disciplines. In traditional projects, that oversight often manifests itself as errors and omissions—classic examples of waste.

A Lean construction process known as TVD is enhanced with BIM. Cost impacts of design are quickly determined in a concurrent manner, instead of relying on the traditional estimating approach. The "big room" or "obeya" concept is adopted in Lean construction from the TPS, and brings together cross-functional teams under one roof to explore problems. During TVD, team members generate synergy by collaborating on not just the design, but on the construction process required to bring it to fruition. BIM provides a critical platform for big room meetings. "What-if" games can be played with various design approaches and the results can be evaluated immediately.

Clash detection is easily accomplished with BIM—design errors often include having different building systems compete for the same limited space in ceilings and building penetrations from floor to floor. Air conditioning ducts and plumbing/fire protection piping typically compete for that space. When these clashes are detected during field installation, corrective action can have significant consequences. Bends in ducts and pipes that were not part of the original design increase their equivalent length, and may restrict the flow of air or water below the design levels defined by the mechanical engineer. That in turn leads to suboptimal building performance.

BIM has features that promote prefabrication—an important component of Lean construction. As the BIM data are machine-readable, CNC machines can be readily provided with instructions for making sections of ductwork, piping, or other building elements under controlled conditions in fabrication shops. Such offsite work is more cost-effective and more accurate than on-site work, yet the accuracy provided with BIM allows such prefabricated components to literally be unloaded from a truck and mounted in place with fewer work hours than otherwise possible with traditional methods.

Lean Design with BIM was successfully used on the El Camino Medical Group's campus in Mountain View in California. A 250,000 SF medical office building was included in the project, as well as a 420,000 SF parking structure, at a cost of $94.5 million. The project schedule was highly accelerated for early occupancy, hence it was necessary to utilize concurrent engineering approaches to start construction activity while the design was still in progress. The general contractor and a number of mechanical subcontractors were engaged during the design phases at approximately the same time as the architect and the structural engineers. The contractors and designers collaborated on the design to derive maximum constructability, lower cost, and an aggressive construction schedule. The team collaborated with a virtual model of the project. Design focused not only on the product but also on the construction schedule, including material supply and prefabrication activities.

Results at the end of the project were as follows:

- The project was completed six months earlier than would have been achieved using the traditional design–bid–build project delivery method without BIM and Lean techniques;
- It was completed under budget
- Labor productivity was 15% to 30% better than the industry standard
- There were no change orders related to field conflict issues and
- There were no field conflicts between the systems coordinated using BIM

As was experienced in this project, 3D modeling is one of the tools used to improve both the design and construction processes. It not only helps designers to visualize and avoid potential design conflicts between different trades, it simultaneously generates bills of materials. 3D modeling can also be used to simulate facility design and also the construction and fabrication process.

8.3.4.4 Lean Tools and Techniques

The fundamental engine of Lean thought—continuous improvement (kaizen)—requires a project manager to think clearly about the meaning and nature of improvement and standardization. The promise of Lean thinking is that if improvement is done collaboratively with key team members in a culture of respect, with an aim to reduce waste and add value, then all project metrics—that is, time, cost, quality, safety, and morale—should improve simultaneously. If only four of the five metrics improve, the recommended improvement is "not" Lean and needs to be rethought. This is why Lean construction may be considered to be a paradigm shift. For example, although both Frederick Taylor and Henry Ford developed early theories of production flow on which Lean theory is partially based, Taiichi Ohno sharply criticized Ford for his disrespect of the worker. As Norman Bodek wrote in his foreword to Ohno's book: "Manufacturing must be both efficient and also have respect for the person running the machine" (Ohno 1988, p. x: publisher's "Foreword").

8.3.4.4.1 Plan-Do-Check-Act Cycle: The Continuous Improvement Engine

The design engine of Lean construction—the PDCA cycle—operates within a culture of continuous improvement, as suggested by Figure 8.21, and played a central role in the work of Shewhart and Deming (1939). For this reason the PDCA cycle is also variously called the "Deming" or "Shewhart Cycle."

The PDCA circle really represents the scientific process of developing a hunch or hypothesis of how a challenge may be met (plan), testing the hypothesis through experimentation (do), checking to see if the results of the experiment validate the hypothesis (check), and then modifying the hypothesis to better explain the results obtained (act). Since there is ever more to know, the circle is perceived as continuous and never-ending.

The PDCA engine (Figure 8.22) may be represented as part of an overall chain of continuous improvement and standardization, as shown previously in Figure 8.3.

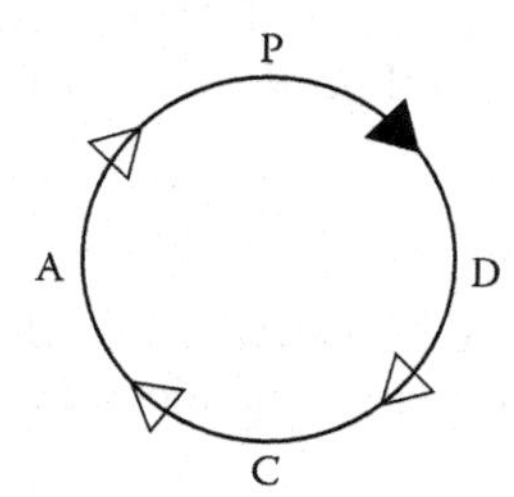

FIGURE 8.21
The PDCA or Deming cycle. (Adapted from Rybkowski, Z.K., Abdelhamid, T., and Forbes, L., On the back of a cocktail napkin: An exploration of graphic definitions of lean construction, *Proceedings of the 21st Annual Conference for the International Group for Lean Construction*, Fortaleza, Brazil, July 31–August 2, 2013 [as cited in Figure 3, p. 87]. Illustration by author [Rybkowski].)

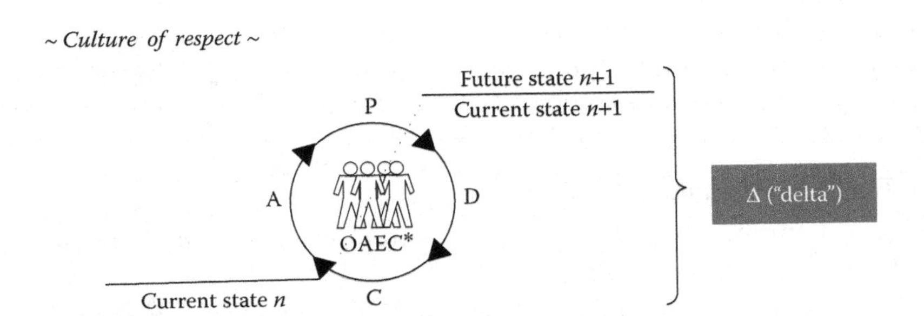

FIGURE 8.22
Graphic definition of the Lean construction PDCA engine. (Adapted from Rybkowski, Z.K., Abdelhamid, T., and Forbes, L., On the back of a cocktail napkin: An exploration of graphic definitions of lean construction, *Proceedings of the 21st Annual Conference for the International Group for Lean Construction*, Fortaleza, Brazil, July 31–August 2, 2013 [as cited in Figure 3, p. 87]. Illustration by author [Rybkowski].)

In Lean construction, the process of continuous improvement, also known as "kaizen", is conducted collaboratively and often in a big room, so that key stakeholders representing owner, architect, engineers, and constructors participate in the process.

8.3.4.4.2 Plus-Delta Chart (+/Δ)

To feed the PDCA cycle and recognition of areas that can be improved, a number of tools have been developed.

Before improvement can take place, the current state situation needs to be documented, and a desired future state proposed. Several tools are used by Lean construction stakeholders to facilitate this. One of the most popular tools is called a "Plus-Delta chart" (+/Δ; Figure 8.23).

Most Lean construction meetings end with a +/Δ debriefing exercise. Although seemingly simple, a +/Δ exercise is quite effective. During a +/Δ session, a facilitator invites all meeting participants to openly offer what they feel worked effectively during a meeting as well as that which they feel can be improved. Several rules must be obeyed: the facilitator must record all comments proffered (i.e., she may paraphrase but not edit). This is important because doing so motivates participants to speak up; some of the best ideas emerge when an environment is perceived as safe and nonconfrontational. In the plus (+) column, the facilitator records those items which participants feel worked well and which should be repeated. Note that the column headings are written as +/Δ rather than +/−. The distinction, though seemingly subtle, is actually significant. While minus (−) implies fault-finding, Lean principles are designed to reinforce a culture of collaboration and to motivate teams to continually improve toward a higher future

+	Δ

FIGURE 8.23
Plus-delta chart (+/Δ) chart used to facilitate continuous improvement. (Illustration by author [Rybkowski].)

state—a process which is antithetical to fault-finding. Unlike (–), (Δ) involves visioning an improved future situation. The distinction is exemplified by the difference between saying "the meeting food was terrible" (–), and instead saying "it would be nice to serve more vegetables at future meetings" (Δ).

8.3.4.4.3 The Ishikawa (Cause-and-Effect) Diagram

The fishbone diagram or cause-and-effect diagram was developed by Kaoru Ishikawa. The diagram also bears his name. It serves as a tool to find the sources of quality problems, or in fact, almost any type of problem (Forbes and Ahmed 2010). It focuses attention on causes, not just symptoms. The diagram looks like the skeleton of a fish, with the problem represented by the head of the fish. The ribs represent possible causes, and the smaller bones are subcauses. The process works best with a group of people who are familiar with the problem, or who may be involved with it. A facilitator works with the group, helping them to work backward from the observed effect caused by the problem, asking "Why?" The diagram serves as an important communications tool as users examine target processes in great detail in order to find a solution to the problem that is associated with it.

Typical "ribs" might include: machines, materials, methods, and people, as many problems can be traced to the influence of one or more of those factors.

Guidelines for problem-solving with the diagram are:

1. State the problem clearly in the head of the fish
2. Draw the backbone and the ribs
3. Working with the facilitator, work backward from the head of the fish, and write along each rib a potential cause of the problem

A generic diagram is shown in Figure 8.24.

8.3.4.4.4 Pareto Analysis

Not all causes are equal in their impact. After a fishbone diagram is constructed, it is important to identify the causes that have the greatest impact and to focus on developing countermeasures to address the most significant causes. This can be done by creating a Pareto chart.

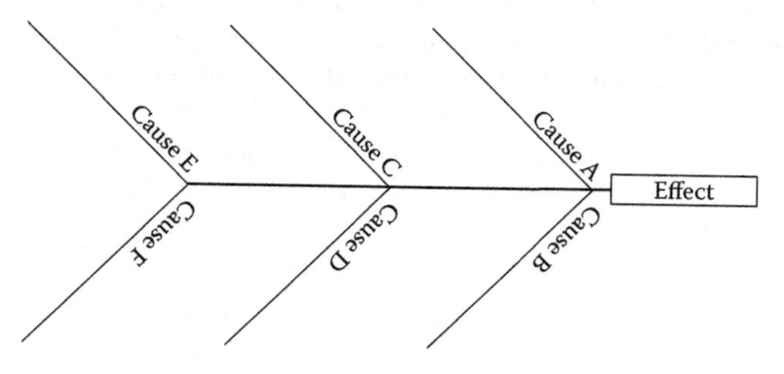

FIGURE 8.24
An Ishikawa fishbone or cause-and-effect diagram. (Illustration by author [Rybkowski].)

The Pareto chart is named after an Italian economist, Vilfredo Pareto (1848–1923), who studied the distribution of wealth in Europe. He proposed an economic theory based on the observation that a few people were very wealthy, while large numbers of people had relatively little means. Essentially, 20% of the people in Italy held 80% of the wealth, hence the "80–20" rule. Joseph Juran applied this concept to the field of quality, distinguishing between the "vital few" and the "useful many." The Pareto chart helps users to identify those 20% attributes that represent 80% of the benefit. Faced with a variety of problems, a user can find the 20% that, when addressed, provide 80% of the benefit.

The example shown in Table 8.4 represents a window installation operation (Forbes and Ahmed 2010). The data provided a list of the number of problems relating to window installation and their unit cost. For example, when a window is installed in a manner that violates the code, it costs $1100. Four occurrences are priced at $4,400. In performing Pareto analysis, one has to decide whether to use the number of occurrences or the costs of these occurrences. In this case, the analysis was based on total costs.

The procedure for creating a table and Pareto chart is as follows:

1. Set up a table to record the values of the factors that are being studied.
2. List the values of the factors in descending order of magnitude (i.e., the largest cost first, and the smallest last. The category "not finished on time" costs $19,625. By contrast, the category "crew very rude" costs $800 and is listed last).
3. Total the cost for all categories (i.e., $44,824.10).
4. For each category, divide its cost by this total to obtain a percentage.

For the Pareto chart, plot the table values onto a combined histogram and superimposed graph where the left-hand y-axis is used to scale histograms for each factor and the right hand y-axis is to scale a superimposed cumulative cost graph, where total cumulative cost reaches 100% (Table 8.4 and Figure 8.25).

TABLE 8.4

Example of Table Used to Develop a Pareto Chart

Item*	Category	Cost ($)	Percentage	Cumulative Percentage
c	Not finished on time	19,625.00	44%	44%
g	Wrong window installed	7,000.00	16%	59%
f	Job not done properly	5,000.00	11%	71%
a	Code infraction	4,400.00	10%	80%
h	Damage during installation	3,500.00	8%	88%
e	Not starting on time	2,250.00	5%	93%
b	Leave job site dirty	2,249.00	5%	98%
d	Crew very rude	800.00	2%	100%
		44,824.00	100%	

Source: Adapted from Forbes, L.H., and Ahmed, S.M., *Modern Construction: Lean Project Delivery and Integrated Practices*, CRC Press, Boca Raton, FL, 2010 [as cited in Figure 9.13, p. 270]. Reprinted with permission from Taylor & Francis Group.

**Note:* Items have been re-sorted in the order of decreasing magnitude to help the project manager focus on items responsible for the largest magnitude of impact.

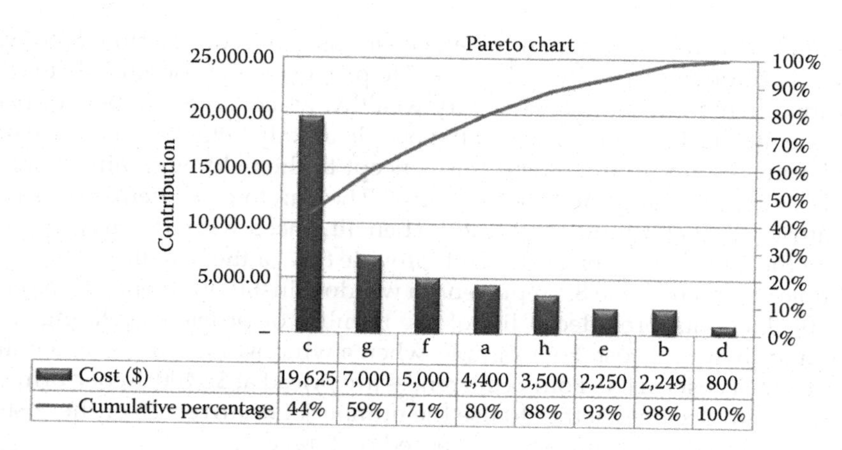

FIGURE 8.25
Pareto chart based on calculations in Table 9.3. (Adapted from Forbes, L.H., and Ahmed, S.M., *Modern Construction: Lean Project Delivery and Integrated Practices*, CRC Press, Boca Raton, FL, 2010 [as cited in Figure 9.13, p. 270]. Reprinted with permission from Taylor & Francis Group.)

The completed chart shows that the first four factors represent 80% of the cost of complaints.

In essence, the Pareto chart allows the team to focus on the largest ribs of the fish—thus mitigating 80% of the problems (Figure 8.26).

8.3.4.4.5 Five Whys/Root Cause Analysis

Once the team agrees on the most significant possible causes, the next step is to identify strategies and set countermeasures that address the largest causes. This can be done using the root cause analysis, also known in the Lean community as the "five whys."

The intent of root cause analysis is to "drill down" to the root of a problem. The assumption is that, by eliminating the root cause of a problem, the problem itself can be resolved. Lean construction borrows heavily from Lean manufacturing, described by Jeffrey Liker in *The Toyota Way* (Liker 2004). Liker offers an example of root cause analysis in the form of a "Five-Whys" chart (Liker 2004), after presenting a challenge: "There is a puddle of oil on the shop floor" (Table 8.5).

If we ask, "Why is this so?," the answer may be: "because the machine is leaking oil." If we are to again ask, "Why is 'this:' so?," the response may be "because the gasket has

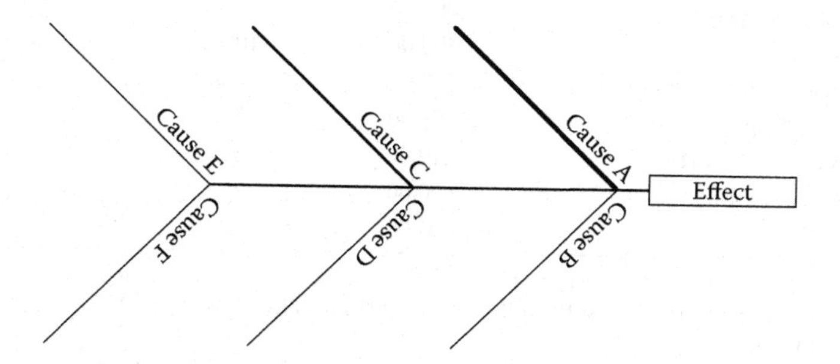

FIGURE 8.26
Causes transferred from the Pareto chart to the Ishikawa fishbone diagram. (Illustration by the author [Rybkowski].)

TABLE 8.5
Example of Root Cause Analysis Using 5 Whys/Root Cause Analysis

Level of Problem	Corresponding Level of Countermeasure	Result If Action Taken at This Point
There is a puddle of oil on the shop floor	Clean up the oil	Short-term solution
Because the machine is leaking oil	Fix the machine	"
Because the gasket has deteriorated	Replace the gasket	Mid-term solution
Because we bought gaskets made of inferior material	Change gaskets specifications	"
Because we got a good deal (price) on those gaskets	Change purchasing policies	"
Because the purchasing agent gets evaluated on short-term cost savings	Change the evaluation policy for purchasing agents	Long-term solution

Source: Liker, J.K., *The Toyota Way*, McGraw-Hill, New York, 2004 (as cited in Figure 20.1, p. 253). Reprinted with permission from McGraw-Hill.

deteriorated." Each time we reach a new level of causal understanding, we decide whether or not to take action at that point or to continue with our line of inquiry. For example, a "knee-jerk" response following the discovery of leaky oil might be to clean up the oil. Or, upon realizing that the gasket has deteriorated, we may elect to replace the gasket. Each level of causal analysis brings with it a new potential solution. However, note that upper-level solutions are often temporary. Cleaning up the oil will not arrest the leak; the oil will likely need to be cleaned up again. Although replacing the gasket will stop the leak from reoccurring for a while longer, a poor quality gasket replaced by another poor quality gasket only forestalls another leak. In other words, each successive level of inquiry brings with it a longer term solution. Not until we reach the lowest level of the Liker figure do we arrive at a solution of some permanence.

The logic behind the Five-Whys technique is that short-term solutions require that fixes must be repeated multiple times over a given period, while a long-term solution demands a singular fix. Despite its sometimes larger first cost, the Five-Whys solution is often less expensive than the short-term one in the long run—and should therefore be preferred (Figure 8.27).

The question then may be, at what stage in the cascade of questioning does one stop a root cause analysis? The five-whys technique is not intended to literally suggest specifically stopping after asking "why" five times, but rather after reaching "an actionable cause."

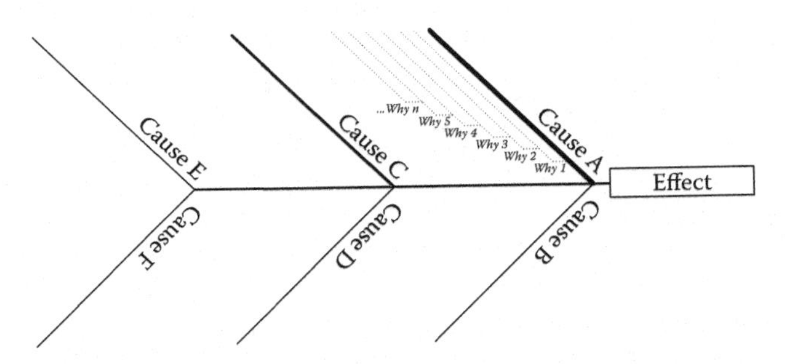

FIGURE 8.27
Conducting five whys/root cause analysis on the "largest bone" of the cause–effect diagram. (Illustration by author [Rybkowski].)

Ideally, one should take action at the moment when the number of repeated fixes matches the needs of the situation at hand. For example, in the case of the oil leak, the short-term solution might be most appropriate if the machine needs to be fixed only long enough to use it for two hours (as opposed to two years).

In Lean construction, +/Δ charts, Pareto charts, and "five whys" root cause analysis are tools used following PPC exercises on the LPS of production control to determine the causes of failure when PPC objectives have not been met so that adjustments may be made when encountering a similar situation in the future. Incorporating learning after each activity facilitates an organization's improvement. Once a process has been bettered, it needs to be standardized. The cycle is then repeated. This is how the continuous improvement stairway is climbed.

8.3.4.4.6 Swimlane Diagrams

Lean practitioners are aided by tools such as "value-stream maps" and "swimlane diagrams" to document and analyze "current states" of processes and to propose desired "future states" that eliminate constraints and wasteful practices. For example, a swimlane diagram documenting the request for information (RFI) process required by traditional, design–bid–build contracts reveals unnecessary causes for delay (Figure 8.28).

In this circumstance, a subcontractor working on site, finding an error, cannot query an owner directly, but must communicate in writing along the document chain required by a legal contract. By contrast, if critical stakeholders instead meet collectively in a "big room" meeting, communication is liberated and issues can be resolved much more quickly. Lean tools such as swimlane diagrams help Lean project managers visualize and quantify before-and-after impacts of their interventions. In the RFI example just discussed, project managers can collect data from time stamps on documents to benchmark their system's current state—and then map a recommended and improved future state. In this way, the swimlane diagram can help a manager quantify actual savings.

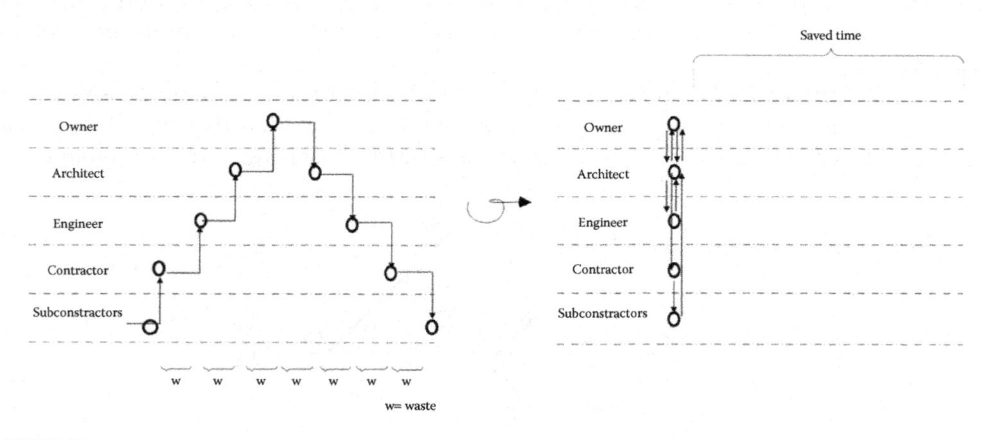

FIGURE 8.28
Swimlane diagrams comparing the RFI communication process during traditional design–bid–build projects (left) versus a typical lean project delivery "big room" meeting (right). The horizontal axis represents time. (Adapted from Rybkowski, Z.K., *Lean Construction Training Modules*, Department of Construction Science, Texas A&M University Construction Industry Advisory Council [DVD], 2012. Illustration by author [Rybkowski].)

8.3.4.4.7 The Five-Step Plan (5S)

The Five-Step Plan is a Japanese approach to improving operations. It helps to create an organized environment that promotes the application of Lean thinking. It involves a sequential process based on the five Ss.

Step 1: Seiri ("sort"): This involves getting rid of unnecessary items in order to be more organized. This step is applied to such areas as tools, work in process, products, and documents.

Step 2: Seiton ("set in order"): Neatness is obtained by rearrangement of the work area and identification of proper locations. The object of the exercise is not just to maintain a neat appearance. Workers can reliably find what they need to do in a job without wasting time by looking. They should put tools and equipment in a manner that will improve the flow of work. Everything should be kept in its designated location. Equipment should be returned to its proper place at the end of each job.

Step 3: Seiso ("shine"): Cleaning and daily inspection avoid the confusion of an untidy work area. Unnecessary items should be disposed of properly in order not to create clutter that inhibits the effective execution of work tasks. After each job, tools and equipment should be cleaned and restored to their proper locations. Note that when items such as tools and equipment are kept clean, malfunctions, such as leaks, are easier to detect. The root causes of waste and uncleanliness should be investigated to ensure that they do not reemerge.

Step 4: Seiketsu ("standardize"): This refers to standardizing locations for tools, files, equipment, and materials. Color coding and labeling can help to standardize locations. Employees' creativity is tapped by having them actively involved in developing standardized systems.

Step 5: Shitsuke ("sustain"): This "S" provides motivation for employees to sustain the other four Ss. Management can use recognition programs to provide motivation.

Note that the order in which these tasks are accomplished generally "does" matter. A bit of thought will make this apparent. For example, it is usually wasteful to clean items before they are sorted and set in order.

8.3.4.5 Collaboration and the Culture of Lean

As discussed in Section 3.4.2.4.1, the Lean model of continuous improvement is based on Shewhart and Deming's PDCA Circle. "Kaizen" is also the Japanese concept of continual incremental improvement. Literally, "kai" means "change," and "zen" means "good"; hence, kaizen stands for ongoing changes for beneficial reasons on a never-ending basis. It involves all members of an organization actively participating in making improvements on an ongoing basis. Masaaki Imai of Japan credits kaizen as the "single most important concept in Japanese management, and Japan's competitive success." Kaizen is rooted in Japanese philosophy which regards improvement as being closely related with the concept of change. According to Masaaki Imai (1986), in the Japanese way of thinking, change is a basic condition of life and should be incremental in order to be healthy, whereas sudden change is seen as unnatural.

The kaizen approach encourages employees/stakeholders to develop and implement ongoing improvements to the systems that they are most involved with, thereby improving their job performance. It seeks to standardize processes and eliminate or reduce waste. It starts with recognizing a problem, and subsequently a need for improvement. It has been said that complacency is the arch-enemy of kaizen.

One important consideration for the adaptation of kaizen to construction-related activities is that improvements should be followed by a period of time for new methods or processes to be standardized.

8.3.4.5.1 Role of People in the Kaizen Process

Kaizen practitioners study a process firsthand by visiting the involved work area ("gemba") and observing the activities involved. This approach clarifies the difference between value-added and nonvalue-added process steps (Forbes and Ahmed 2010). Kaizen teams reduce nonvalue-added steps through a prescribed approach: to combine, simplify, or eliminate in order to reduce waste. Documenting current processes using swimlane diagrams and value-stream maps is helpful to identify areas of waste.

There are two primary approaches to kaizen improvement activities—flow kaizen and process kaizen. Flow kaizen emphasizes value stream improvement, while process kaizen focuses on the elimination of waste. While kaizen involves ongoing, long-term dedication to improvement of the organization and its people, the process periodically conducts short-term activities termed "kaizen events or kaizen breakthrough" methodology. These events are a cross-functional, team-based process for rapid improvement, and are often a week-long in duration. They involve a bias for action that harnesses creativity to obtain results. The areas selected for kaizen events have direct impact on organizational performance.

Full engagement of employees' hearts and minds to the kaizen philosophy is critical for the needed organizational transformation. Kaizen events are selected for maximum visibility and impact. They have several advantages. Benefits of kaizen events are obtained very rapidly and include, for example:

- Quick implementation sends a positive message to all stakeholders
- The events provide hands-on training for participants and
- The events allow management to gauge workers' resistance to change

To be effective, multiple layers of management and staff should be engaged in the kaizen process as follows:

"Senior executives and managers" must first embrace kaizen as an organizational strategy and communicate this policy to all stakeholders. As with other performance improvement initiatives, top management's commitment, leadership, and active support are critical to the success of kaizen.

"Middle managers" need to deploy the kaizen policies established by senior executives. They ensure that staff members have the appropriate training and preparation. They establish implementation milestones for supervisory staff.

"Supervisors" should work closely with frontline workers to implement kaizen at the functional level. They need to be actively involved in collecting kaizen suggestions for improvement from employees and coaching respective teams to promote success.

"Employees" should be expected to actively pursue self-development through education and training in order to support kaizen team activities.

8.3.4.5.2 *Typical Structure of the Kaizen Process*

The following steps are recommended when engaging participants in a kaizen event. They include

1. Select area/work unit for kaizen event based on business requirements and potential for improvement
2. Establish project objectives: document baseline data and set metrics for improvement (i.e., numerical goals)
3. Appoint kaizen team members: a team leader and subteam leader
4. Publicize the event: meet with the involved work unit team, explain roles and expectations, engage hearts and minds for the event (Forbes and Ahmed 2010, p. 122)

A typical kaizen event lasts one week and brings everyone together in a focused series of activities, starting with orientation and training on the first day. It often culminates on day five with a presentation of recommendations and results to management, followed by a celebration of kaizen event accomplishments. Kaizen events are short term in duration, but have proven to be very effective. They focus everyone's effort for a few days, but the benefits in improved performance far outweigh the cost of lost production. They are also one of the primary reasons that adoption of Lean thinking in an organization improves morale as well as efficiency. The adage is: "With every pair of hands comes a free brain."

8.3.4.5.3 *The Role of Leadership*

Leadership plays an indispensable role in the successful implementation of the Lean philosophy to construction projects. Consultants and coaches can provide knowledge and expertise, but success depends heavily on the hearts and minds of the project team being both invested and engaged. Therefore, the leaders of the stakeholder organizations—owners, designers, and constructors—must lay the groundwork with their respective teams. Leaders need to address the top three barriers to successful Lean implementation. As observed in a UK study (Sarhan and Fox 2012), three key issues influence the readiness of organizations for undertaking a Lean journey. They are: lack of adequate Lean awareness and training; lack of top management commitment; and culture and human attitudinal issues.

Leaders must ensure that their respective organizations are ready for the Lean journey, and that they meet the following attributes (Forbes 2009):

- *A willingness to change.* Lean methods are a departure from conventional methods and their adoption requires changing the behavior of people. Cultural change is the most compelling quest along with the physical transformation of an organization. One cannot force change on people; they have to be engaged so that the intrinsic satisfaction of outstanding performance will motivate them.
- *A commitment to training and learning.* Stakeholders at all levels need to be trained in Lean techniques in order to become successful participants in Lean projects. Lean implementation also requires that completed assignments have to be continually examined as a source of learning for future improvements instead of being a search for sources of blame.
- *A quality-oriented culture is needed for successful application of Lean techniques.* JIT, in particular, demands discipline, as there is no room for unreliable suppliers. JIT does not work in an atmosphere of suspicion, distrust, and internal competition.

- *A "shared vision" is essential to have all stakeholders on the same page.* Hal Macomber and Greg Howell (2005) promote the importance of a "shared vision" in which a workforce aligns itself with the direction set by a leader. This alignment is far different from carrying out orders. It is based on a sharing of beliefs and a common view of a future state that benefits everyone and makes them receptive to the changes necessary to reach that future state.
- A commitment to reducing or eliminating waste is a fundamental principle of Lean construction (Polat and Ballard 2004). A commitment to improving safety is critical to Lean implementation as construction accidents are rivalled only by the mining industry.
- *A commitment to cost and performance measurement.* These measurements are important indicators of the impacts of Lean in construction projects. Measurements of PPC are essential as a foundation for continuous improvement.
- *A willingness to implement Lean during the design stages.* The LPD system (LPDS) links designers with constructors, beyond the norms of traditional practice to include elements of construction process design in their scope of work. Lean designers also engage in TVD to meet the client's value proposition within a defined budget.
- *Collaborative relationships.* Lean requires close collaboration between the parties whereas the standard forms of contract are adversarial in nature. Lean projects apportion responsibilities and benefits of the contract fairly and transparently, based on trust and partnership between the parties. Working relationships are improved, and the improved efficiency and reduction in conflict generally lead to improved financial returns for all stakeholders.
- Leaders must develop clear and explicit conditions of satisfaction (CoS) in collaboration with key stakeholders. IPD is based on a network of promises, commitments, or agreements among the project team. A promise can only be meaningful when it has clear, mutually agreed CoS. They are measurable statements that explain what tests must be passed to create success. They represent the customer's value proposition in measureable terms, and should be posted for all to share in understanding.

8.4 Conclusion

The definition of Lean construction is not static. Similar to a Wikipedia page, its definition grows and evolves through the communal input of interested parties.

When Lauri Koskela published the now seminal *Technical Report Number 72*, he wrote of a "new production philosophy" being applied to manufacturing that might solve the safety and productivity problems facing construction (Koskela 1992, 2000). Koskela's report and theory of TFV came to the attention of like-minded thinkers Greg Howell and Glenn Ballard, who also saw the need to revamp a highly unproductive and litigious construction industry. They developed LPS—a planning method that stabilizes projects and is almost routinely the first step companies now adopt when deciding to implement Lean. Early Lean construction pioneers drew on lessons learned from Lean manufacturing, to be sure, but also from productivity improvement advances made by Frederick Taylor and Henry Ford, the time-motion studies of Frank Gilbreth, an underlying respect for the individual advocated by pioneering industrial psychologist, Lilian Gilbreth, and the continuous improvement control cycles (PDCA) put forth by statisticians Shewhart and Deming.

In 1993, the International Group of Lean Construction (IGLC) was founded and held its first conference in Espoo, Finland, with a handful of participants. Proceedings from the first three IGLC conferences were published in a book entitled *Lean Construction* (Alarcón 2013). In 2014—21 years after the first conference—125 Lean construction research papers were presented and published, and hundreds of academic, as well as industry-related attendees, participated in presentations, at the IGLC conference in Oslo, Norway. The Lean Construction Institute was founded in 1997 (LCI 2016), and today hosts regular trainings and meetings; these sessions openly welcome multiple stakeholders, including building owners, architects, engineers, attorneys, constructors, trade partners, and suppliers. In 2004, *The Lean Construction Journal* was founded and has become a publishing venue for dissemination of research results from Lean experimentation (LCJ 2014).

Contributors to Lean construction theory are both academicians and practitioners and, unlike conventional academic conference arenas, it is common for papers from the Lean construction community to be coauthored collaboratively by academics and practitioners. There is a fundamental acknowledgment that building project delivery is an applied science and that it is only when ideas are tested on an actual design and construction project that they are imbued with useful meaning. It is also a foundational belief that the inclusion of multiple stakeholders and the bodies of knowledge they bring to a problem-solving table contribute to Lean construction's success and worldwide adoption. As of this writing, at least 14 countries now have active, formally affiliated Lean construction networks, and at least 21 additional countries are practicing Lean in some fashion (Mossman 2015). This number continues to grow.

The story of Lean construction is multifaceted; as in any grassroots movement, hundreds have paved the way, but their names and specific contributions are too numerous to mention here. Like a lithe cheetah that leaves evidence of its astounding speed by grasses that were parted during its run, one likely sign that "Lean construction was here" is the orchestrated choreography of "flow" that can be wondrous to behold.

References

Abdelhamid. (March 21, 2013). Lean construction, <https://www.msu.edu/user/tariq/Learn_Lean.html>.

Alarcón, L., ed. (2013). *Lean Construction*, A. A. Balkema, Rotterdam, the Netherlands.

Ballard, G. (1999). Improving work flow reliability, *Proceedings of the 7th Annual Conference of the International Group for Lean Construction*, Berkeley, CA, 275–286.

Ballard, G. (2000a). The last planner system of production control. PhD dissertation, University of Birmingham, Birmingham.

Ballard, G. (September 23, 2000b). *LCI White Paper-8: Lean Project Delivery System*, Lean Construction Institute, Ketchum, ID, 7 pp.

Ballard, G. (2008). The lean project delivery system: An update. *Lean Construction Journal*, 1–19.

Ballard, G. and Howell, G. (1998). Shielding production: Essential step in production control, *Journal of Construction Engineering and Management*, 124(1), 11–17.

Ballard, G. and Howell, G. (2003). Lean project management, *Building Research and Information*, 31, 119–133.

Ballard, G. and Kim, Y. W. (2007). *Roadmap for Lean Implementation at the Project Level*. Research Report 234-11, Construction Industry Institute, Austin, TX, 426.

Black, C., Akintoye, A., and Fitzgerald, E. (2000). An analysis of success factors and benefits of partnering in construction, *International Journal of Project Management*, 18, 423–434.

Cooper, R. and Slagmulder, R. (1997). *Target Costing and Value Engineering*, Productivity Press, Portland.

Eastman, C., Teicholz, P., Sacks, R., and Liston, K. (2008). *BIM Handbook: A Guide to Building Information Modeling for Owners, Managers, Designers, Engineers, and Contractors*, John Wiley & Sons, Hoboken, NJ.

Feng, C., Liu, L., and Burns, S. (1997). Using genetic algorithms to solve construction time-cost trade-off problems, *ASCE Journal of Computing in Civil Engineering*, 11(3), 184–189.

Fernandez-Solis, J. L. and Rybkowski, Z. K. (2012). A theory of waste and value, *International Journal of Construction Project Management*, 4(2), 89–105.

Forbes, L. H. (2009). "Lean construction principles, prerequisites, and strategies for implementation, *Proceedings of the 5th International Conference on Construction in the 21st Century*, Istanbul, Turkey.

Forbes, L. H. (April 25, 2014). Harnessing lean and integrated project delivery to optimize design and construction performance, *Hong Kong Institute of Project Management Conference, Novel Project Delivery Systems: Current Status and The Way Forward*, Hong Kong, China.

Forbes, L. H. and Ahmed, S. M. (2010). *Modern Construction: Lean Project Delivery and Integrated Practices*, CRC Press, Boca Raton, FL.

Gallaher, M. P., O'Connor, A. C., Dettbarn Jr., J. L., and Gilday, L. T. (2004). Cost analysis of inadequate interoperability in the U.S. capital facilities industry, National Institute of Standards and Technology, Gaithersburg, MD, 210 pp.

Hamzeh, F. R. (2009). Improving construction workflow: The role of production planning and control, Doctoral dissertation, Civil and Environmental Engineering, University of California, Berkeley, CA, 273 pp.

Hamzeh, F. R., Ballard, G., and Tommelein, I. D. (2012). Rethinking lookahead planning to optimize construction workflow, *Lean Construction Journal*, 15–34.

Harmon, R. D. and Peterson, L. D. (1990). *Reinventing the Factory: Productivity Breakthroughs in Manufacturing Today*, The Free Press, New York.

Hayes, S. (2014). *The Simple Lean Pocket Guide for Construction: Tools for Elimination of Waste in the Design-Bid-Build Construction Project Cycle!* MCS Media Chelsea, MI.

Hegazy, T. (1999). Optimization of construction time-cost trade-off analysis using genetic algorithms, *Canadian Journal of Civil Engineering*, 26(6), 685–697.

Imai, M. (1986). *Kaizen*, McGraw-Hill, New York.

Josephson, P.-E., and Hammarlund, Y. (1999). The causes and costs of defects in construction: A study of seven building projects, *Automation in Construction*, 8, 681–687.

Koskela, L. (1992). Application of the new production philosophy to construction. Technical Report No. 72, Center for Integrated Facility Engineering (CIFE), Stanford University, Stanford, CA.

Koskela, L. (April 3, 2000). An exploration towards a production theory and its application to construction, D. Tech. thesis, Helsinki University of Technology, Espoo, Finland, <http://www.vtt.fi/inf/pdf/publications/2000/P408.pdf>.

Lean Construction Institute. (2016). A Conceptual History of LCI-Plus thoughts on the rationale and history of 'Lean Construction'. <http://leanconstruction.org/about-us/history/>.

Lean Construction Institute. (2015). Glossary. <http://leanconstruction.org/training/glossary/#i>.

Lean Construction Journal. (2014). Lean Construction Journal: Aims.<http://leanconstruction.org/training/lcj>.

Lichtig, W. A. (2004). The integrated agreement for lean project delivery, *Construction Lawyer*, 26(3), 1–8.

Liker, J. K. (2004). *The Toyota Way*, McGraw-Hill, New York.

MacLeamy, P. (August 2004). Collaboration, integrated information, and the project lifecycle in building design and construction and operation, *Introduced at the Construction User's Roundtable*, WP-1202.

Macomber, H. and Howell, G. (2005). *Using Study Action Teams to Propel Lean Implementations*, Lean Project Consulting, Louisville, CO.
Macomber, H., Howell, G., and Barberio, J. (2005). Target value design: Seven foundational practices for delivering surprising client value. *Lean Project Consulting*, Louisville, CO, 1–2.
Macomber, H., Howell, G., and Barberio, J. (2008). Target value design: Nine foundational practices for delivering surprising client value. *Lean Project Consulting*, Louisville, CO, 1–2.
Matthews, O. and Howell, G. A. (2005). Integrated project delivery: A example of relational contracting, *Lean Construction Journal*, 2(1), 46–61.
Mossman, A. (2014). *Traditional Construction and Lean Project Delivery—A Comparison*, The Change Business, Stroud.
Mossman, A. (February 7, 2015). The global lean construction community on the web. <http://db.tt/0xozjL5> and <http://bit.ly/LCweb-global>.
MSA. (February 21, 2004). MacLeamy Curve, <http://msa-ipd.com/MacleamyCurve.pdf>.
Ohno, T. (1988). *Toyota Production System: Beyond Large Scale Production*, Productivity Press, Cambridge, MA.
Polat, G. and Ballard, G. (2004). Waste in Turkish construction: Need for Lean Construction techniques, *Proceedings of the 12th Annual Conference of the International Group for Lean Construction*, Copenhagen, Denmark.
Priven, V., Sacks, R., Seppanen, O., and Savosnick, J. (2014). *Lean Workflow Index for Construction Projects Survey* (handout), Technion: Israel Institute of Technology, VC Lab: Seskin Virtual Construction Laboratory, Haifa, Israel.
Proverbs, D. G., Holt, G. D., and Cheok, H. Y. (2000). Construction industry problems: The views of UK construction directors. *In*: Akintoye, A. (Ed.), *16th Annual ARCOM Conference*, 6-8 September 2000, Glasgow Caledonian University. Association of Researchers in Construction Management, Vol. 1, 73–81.
Rybkowski, Z. K. (2009). The application of root cause analysis and target value design to evidence-based design in the capital planning of healthcare facilities, Doctoral dissertation, University of California, Berkeley, CA.
Rybkowski, Z. K. (July 14–16, 2010). Last planner and its role as conceptual kanban, *Proceedings of the 18th Annual Conference of the International Group for Lean Construction*, Haifa, Israel, 10 pp.
Rybkowski, Z. K., Abdelhamid, T., and Forbes, L. (July 31–August 2, 2013). On the back of a cocktail napkin: An exploration of graphic definitions of lean construction, *Proceedings of the 21st Annual Conference for the International Group for Lean Construction*, Fortaleza, Brazil.
Rybkowski, Z. K. and Kahler, D. (June 25–27, 2014). Collective kaizen and standardization: The development and testing of a new lean simulation, *Proceedings of the 22nd Annual Conference for the International Group for Lean Construction*, Oslo, Norway.
Sarhan, S. and Fox, A. (2012). Trends and challenges to the development of a lean culture among UK construction organisations," *Proceedings of the 20th Annual Conference of the International Group for Lean Construction*, San Diego, CA, July 18–20, 2012.
Schonberger, R. J. (1986). *World Class Manufacturing: The Lessons of Simplicity Applied*. Free Press, New York.
Shewhart, W. A. and Deming, W. E. (1939). *Statistical Method from the Viewpoint of Quality Control*, The Graduate School, The Department of Agriculture, Washington, DC.
Siemens, N. (1971). A simple CPM time-cost tradeoff algorithm, *Management Science*, 17(6), B-354–363.
Teicholz, P. (September/October, 2001). U.S. construction labor productivity trends, 1970–1998, *Journal of Construction Engineering and Management*, 127(5), 427–429.
The Economist. (2000). Construction and the Internet, *The Economist*, January 15.
Tommelein, I. D. (1997). Discrete-event simulation of a pull-driven materials-handling process that requires resource-matching: Example of a pipe-spool installation, *Technical Report 97-2*, Construction Engineering and Management Program, Civil and Environmental Engineering, University of California, Berkeley, CA.

Tommelein, I. D. (1998). Pull-driven scheduling for pipe-spool installation: Simulation of lean construction technique, *Journal of Construction Engineering and Management*, 124(4), 279–288.
Tommelein, I. D. (2000). Impact of variability and uncertainty on product and process development, *Proceedings, Construction Congress VI*, ASCE, Orlando, FL, 969–976.
Tommelein, I. D. (2015). Journey toward Lean Construction: Pursuing a paradigm shift in the AEC industry, *Journal of Construction Engineering and Management*, 141(6), 12 pp.
Tommelein, I. D., Riley, D., and Howell, G. (1999). Parade game: Impact of work flow variability on trade performance. *Journal of Construction Engineering Management*, 125(5), 304–310.
Womack, J. P., Jones, D. T., and Roos, D. (1990). The machine that changed the world, Rawson Associates, New York.

Glossary of Lean Terms

A3 report: The A3 report is a disciplined and rigorous system for implementing PDCA management. As used by Toyota, the report fits on a single sheet of A3 paper, measuring 11 inches by 17 inches approximately, and documents problems and the available information about them in a manner that helps users to understand their ramifications quickly. It enables users to focus on improving processes and solving problems.

autonomation: Lean construction applies the principle of autonomation that is an important ingredient of the Toyota Production System (TPS); the people closest to the work are empowered to stop production if they determine that the upstream production is defective. In using the LPS, the last planners and their crews fulfill this role.

basis of design (BOD): The owner's basis of design (BOD) reflects the owner's needs and wants that must be satisfied by the design of a project. This may include the use of spaces, their sizes, finishes, and activities to be performed within them.

benchmarking: Involves comparing actual or planned project practices to those of other projects to generate ideas for improvement and to provide a standard by which to measure performance.

buffer: The means (capacity, inventory, time) used to cushion against the shock of variation in a process. Two types of buffers may be used to shield downstream construction processes from flow variation. Plan buffers are inventories of workable assignments. Schedule buffers are materials, tools, equipment, manpower, and time.

building information modeling (BIM): BIM is the process of generating and managing building data during its life cycle. It is a model-based technology linked with a database of project information. A BIM carries all information related to a facility, including its physical and functional characteristics and project life cycle information, in a series of "smart objects."

commissioning: A systematic process of assuring by verification and documentation, from the design phase to a minimum of one year after construction, that all facility systems perform efficiently and in accordance with the design documentation and intent, and meet the owner's operational needs.

commissioning agent (CxA): The commissioning agent (or authority) is a professional that serves as an objective advocate for the owner and is responsible for (a) directing the commissioning team and process; (b) coordinating, overseeing, and/or performing the commissioning testing; and (c) reviewing the results of the system performance verification.

commitment planning: Commitment planning is a method for defining criteria that lead to the selection of quality assignments. It results in commitments to deliver that other actors in the production system can rely on, by following a prerequisite that only "sound" assignments should be made or accepted.

commitment reliability: Commitment reliability is a measure applied to the commitment made by producers (contractors and designers) in the design/construction supply chain. Downstream producers depend on hand-offs from upstream producers, and secure their commitment to dates certain in order to meet schedules. Commitment reliability is calculated by comparing completed work (DID) with planned work (WILL).

constraint: A constraint is an obstacle that inhibits the execution of a task that is required in the look-ahead plan. It should also be beyond the control of the Last Planner.

constraint analysis: Foremen, project managers, schedulers, and other staff communicate to identify constraints, and the reasons for their occurrence. This enables them to guide the scheduling process and remove constraints so that "sound" work tasks can be assigned.

consensus Docs 300: A standard form of contract for the construction industry that is based on an Integrated Form of Agreement (IFOA). Twenty-two leading construction associations united to publish a consensus set of contract documents in 2007, which they felt was fair to all parties.

continuous flow: Producing and moving one item at a time (or a small and consistent batch of items) through a series of processing steps as continuously as possible, with each step making just what is requested by the next step. Also called *one-piece flow* or *single-piece flow*.

continuous improvement: Small improvements in products or processes to reduce costs and ensure consistency of performance of products and services.

cycle time: The time it takes for a product or unit of built work to be completed from beginning to end.

design–bid–build: A traditional contracting method in which the architect and contractor secure separate contracts with the owner to provide specified services (Construction Management Association of America, 2009).

design–build: A contracting method in which an architect and contractor provide design and construction services under a single responsibility contract to an owner (Construction Management Association of America, 2009).

design criteria: Design criteria for buildings define the owner's requirements, including those that may exceed applicable codes. They primarily serve as a performance-based guide to designers to ensure that ensuing designs meet the functional needs and preferences of the owner. A criteria document typically describes the building systems and materials to be used. It may also define project-specific requirements such as building orientation, space descriptions, and adjacencies to meet owners' operating needs.

first run studies: These studies involve trying out new construction ideas on a pilot basis and applying the PDCA methodology to determine their success. First run studies also identify the best means, methods, and sequencing for a specific activity.

fishbone diagram: Also called the Ishikawa diagram or cause-and-effect diagram. The diagram illustrates how various causes and subcauses relate to creating potential problems. It may be used to determine the root causes of observed problems.

five whys: A method of root cause analysis whereby the investigator repeatedly probes operators and other employees with the simple question "why?" until the root cause of a problem is uncovered.

five S (5S): One of the guiding principles of lean meant to achieve standardization and stability in the workplace through visual management.

flowcharting: A flowchart is any diagram that shows how various elements of a system relate (cause-and-effect diagrams, system, or process flowcharts).

gemba: A Japanese term that means the place where things are actually happening. Gembashugi: shop floor-oriented (Shugi = orientation or philosophy).

heijunka: Creating basic stability in processes. Leveling short-term customer demand at a pacemaker point to produce outputs at a constant and predictable rate.

IPD team performance contingency: A provision in integrated project delivery—for the sharing of costs and benefits based on the performance of the IPD team.

Integrated Form of Agreement (IFOA): The IFOA is a legal agreement that seeks to align the commercial relationships of a construction project's design and construction participants that are assembled as a temporary production system. It also requires collaborating throughout design and construction, planning, and managing the project as a network of commitments. As a result it optimizes the project as a whole, rather than any particular piece.

integrated project delivery (IPD): IPD is a relational contracting approach that aligns project objectives with the interests of key participants. It creates an organization able to apply the principles and practices of the lean project delivery system. The fundamental principle of IPD is the close collaboration of a team that is focused on optimizing the entire project as opposed to seeking the self-interest of their respective organizations.

Ishikawa (cause-and-effect) diagram: This line and text diagram, developed by Kaoru Ishikawa, evokes the image of a fish and its ribs. It is used to facilitate itemizing—through brainstorming—all potential causes of an effect. In the diagram, the effect is written at the location of the fish "head," and all potential underlying causes of that effect are written along the "ribs."

just-in-time (JIT) production: A system of production that makes and delivers just what is needed, just when it is needed, and just in the amount needed. JIT and jidoka are the two pillars of the Toyota Production System.

kaizen: Continuous improvement of an entire value stream or an individual process to create more value with less waste. There are two levels of kaizen: (1) system or flow kaizen focuses on the overall value stream and (2) process kaizen focuses on individual processes.

kanban: A signaling device that gives authorization and instructions for the production or withdrawal (conveyance) of items in a pull system. The term is Japanese for sign or signboard.

last planner: The person or group that makes assignments that direct workers—"squad boss" and "discipline lead" are common names in design processes. Superintendent (for small jobs) or foreman are common names for last planners in construction processes.

Last Planner System of Production Control™ (LPS): The LPS is an important subset of the lean project delivery system and is critical to its effective deployment. The LPS accommodates project variability and smoothens workflow so that labor and material resources can be maximally productive. It uses lean methods to provide improved project control.

lean construction: Extends to the construction industry the lean production revolution started in manufacturing. This approach maximizes value delivered to the customer while minimizing waste (Lean Construction Institute 2009).

lean production: A business system for organizing and managing product development, operations, suppliers, and customer relations that requires less human effort, less space, less capital, and less time to make products with fewer defects to precise customer desires, compared with the previous system of mass production. It was pioneered by Toyota after the World War II.

lean project delivery system: The LPDS (Ballard, 2000) is a model that integrates a number of phases to facilitate the design and delivery of construction project, using lean techniques. These phases are: (1) project definition, (2) lean design, (3) lean supply, (4) lean assembly, and (5) use. The LPDS is based on a close collaboration between the members of the project delivery team. They are bound by codes of conduct—both written and unwritten—that focus on the success of the overall project, rather than their individual success.

muda: Seven classes of waste in processing: overproduction, delays, transportation, over processing, excess inventory, wasted motion, and defective parts.

mura (variation): mura internal to the enterprise does include day-to-day variation in customer demand as long as it is not part of a long-term trend. It also includes gyrations in orders and operational performance progressing up a value stream—through production facilities and suppliers—that are caused by the internal dynamics of the process. These variations are the norm in modern production systems, leading to firefighting and muri (overburden of employees and technologies).

muri: Overburden of employees and technologies.

obeya: A Japanese word that translates to "big room." At Toyota it has become a major project management tool, used especially in product development, to enhance effective and timely communication. Similar in concept to traditional "war rooms," an obeya will contain highly visual charts and graphs depicting program timing, milestones, and progress to date and countermeasures to existing timing or technical problems.

owner's project requirements (OPR): (See design criteria.)

pacemaker process: Any process along a value stream that sets the pace for the entire stream. (The pacemaker process should not be confused with a bottleneck process which necessarily constrains downstream processes due to a lack of capacity.)

Pareto chart: A diagram that is used to display the relative importance of each of a number of factors. In evaluating problems, it shows the types of problems and the frequency of their occurrence or cost impact based on the so-called 80/20 rule. The chart enables users to readily prioritize needed corrective action.

percent planned complete (PPC): In lean construction, work accomplishment is recorded as a graphical plot of PPC; it shows the percentage of the assigned plans, that is, commitments that were completed (fractional completion is not considered). Some lean construction practitioners refer to PPC as commitment reliability.

plan–do–check–act (PDCA): An improvement cycle based on the scientific method of proposing a change in a process, implementing the change, measuring the results, and taking appropriate action. The PDCA cycle has four stages: (1) plan, (2) do, (3) check, and (4) act.

poka yoke: Japanese term which means "mistake" (or "error") proofing. A poka yoke device is one that prevents incorrect parts from being made or assembled, or easily identifies a flaw or error. Error-proofing is a manufacturing technique of preventing errors by designing the manufacturing process, equipment, and tools so that an operation literally cannot be performed incorrectly.

prevention over inspection: A well-known mantra, emphasized by proponents of quality. Prior to the emergence of the quality movement, the main focus of quality was on inspection. Both research and experience pointed to the fact that the net cost of inspecting is so high that it is better to spend money on preventing problems. "Quality must be planned in, not inspected in."

productivity: Productivity is the measure of how well resources are brought together and utilized for accomplishing a set of goals. Productivity is measured as the ratio of outputs to inputs. In the construction environment, it may be represented as the constant-in-place value divided by inputs such as the dollar value of material and labor.

quality assurance: Planned and systematic actions to help assure that project components are being designed and constructed in accordance with applicable standards and contract documents.

quality control: The review of project services, construction work, management, and documentation for compliance with contractual and regulatory obligations and accepted industry practices.

reasons for noncompletion (RNC): In conjunction with lean-based construction projects, incomplete work plans are studied each week to determine the root causes. At each weekly meeting, time is devoted to learning why certain tasks were not completed in order to improve the effectiveness of future work plans.

relational contracting: Relational contracting is a transaction or contracting mechanism that apportions responsibilities and benefits of the contract fairly and transparently, based on trust and partnership between the parties. It provides a more efficient and effective system for construction delivery in projects that require close collaboration for execution. The relationship between the parties transcends the exchange of goods and services and displays the attributes of a community with shared values and trust-based interaction.

reliability: The extent to which a commitment is fulfilled.

reverse-phase scheduling: A strategy used in lean construction to develop the schedule of a project by first anchoring the desired delivery date and scheduling activities backward toward the start of the project.. It is a detailed work plan that specifies the hand-offs between trades for each project phase, that is, what SHOULD be done at each phase.

set-based design: Set-based design is a design management approach that defers design decisions to the "last responsible moment" to allow for the evaluation of alternatives that improve constructability. Toyota has used this technique to design new models in a much shorter time than the industry standard. They have learned that the best solutions are hybrids of original design options.

supply chain: The term "supply chain" encompasses all the activities that lead to having an end user provided with a product or service—the chain is comparable to a network that provides a conduit for flows in both directions, such as materials, information, funds, paper, and people.

sustainability: The ability of a society to operate indefinitely into the future without depleting its resources. Sustainability includes concepts of green building design and construction, reuse and recycling of materials, reduced use of material and energy resources for building construction and operation, water conservation, and responsible stewardship of the surrounding environment.

takt time: In a manufacturing assembly line, it is the maximum time allowed per unit to meet demand; it sets the pace of the assembly line.

target value design: Target value design (TVD) involves designing to a specific estimate instead of estimating based on a detailed design. It seeks to address the problem that affects many projects—various design disciplines work from a common schematic design to do design development in their areas of expertise.

team dynamics: Problem solving, communications skills, conflict resolution.

throughput time (also cycle time): The time required for a product to move all the way through a process from start to finish. At the plant level, this is often termed "door-to-door time". The concept can also be applied to the time required for a design to progress from start to finish in product development or for a product to proceed from raw materials all the way to the customer.

total quality management (TQM): A philosophy that encourages companies and their employees to focus on finding ways to continuously improve the quality of their business practices and products.

Toyota production system (TPS): The production system developed by Toyota Motor Corporation to provide best quality, lowest cost, and shortest lead time through the elimination of waste. TPS is comprised of two pillars, Just-In-Time production and jidoka. TPS is maintained and improved through iterations of standardized work and kaizen, following the scientific method of the PDCA cycle.

trend analysis: Involves using mathematical techniques to forecast future outcomes based on results. It is often used to monitor technical and cost/schedule performance.

value-added and nonvalue-added work activities: Value-added activities provide desirable outcomes for a customer, that is, they meet the customer's value proposition. Conversely, nonvalue-added activities do not contribute to meeting a customer's expectations. For example, searching for a tool to perform a task does not get the task done, and does not provide value.

value stream: All of the actions, both value-creating and nonvalue-creating, required to bring a product from concept to launch and from order to delivery. These include actions to process information from the customer and actions to transform the product on its way to the customer.

value stream mapping (VSM): A simple diagram of every step involved in the material and information flows needed to bring a product from order to delivery. A current-state map follows a product's path from order to delivery to determine the current conditions. A future-state map shows the opportunities for improvement identified in the current-state map to achieve a higher level of performance at some future point.

variation: The divergence of a process from an intended plan. The lean philosophy seeks to reduce process variation in order to obtain reliable flow through a process.

visual management: Using visual displays for better communication. It works in conjunction with empowerment so that anyone, not just a supervisor, can detect process anomalies and take prompt corrective action.

waste: Any activity that consumes resources but creates no value for the customer. There are seven wastes as identified in lean: Overproduction, Excessive Inventory, Unnecessary Conveyance, Overprocessing, Excessive Motion, Waiting, and Corrections (of defects).

workable backlog: Workable backlog is used to describe assignments that have met all quality criteria, but that are not assigned to be performed during the active week in the weekly work plan (WWP). Workable backlog enables crews to continue working productively if there is a constraint that prevents the completion of an item on the WWP.

work package: A compilation of information regarding the availability of material, equipment, resources, and other pertinent information that is necessary to perform a specific scope of work. The work package is coordinated with other interdependent work packages and requirements.

work structuring: The process of defining and organizing activities into logical, practical groups of work. It is a process of subdividing work so that the pieces are different from one production unit to the next to promote flow and throughput, and to have work organized and executed to benefit the project as a whole.

9

ISO Certification in the Construction Industry

Shirine L. Mafi, Marsha Huber, and Mustafa Shraim

CONTENTS

9.1 Introduction

In the construction sector today, quality workmanship provides a competitive advantage in terms of customer satisfaction and loyalty. Quality organizations provide a safe working environment for their crews and a safe living environment for future residents. Although construction site accidents are not as common as in developed countries, even one accident is one too many. The loss of human life, property, credibility, and resources are often the result of poor quality control.

Quality in the construction management has a lengthy history comparable to manufacturing, but with its own unique characteristics (Arditi and Gunaydin, 1997):

- Almost all construction projects are different from each other and constructed in varying conditions
- The life cycle of a construction project is longer than the life cycle of most manufactured products
- There is no uniform standard used to evaluate overall construction quality as in manufacturing; therefore, evaluations of construction projects tend to be more subjective

- Various participants in construction—project-owner, designer, general contractor, subcontractor, and materials supplier—have different ideas of what quality workmanship is

Because of these distinguishing characteristics, the construction industry has not been able to implement quality procedures as effectively as the manufacturing sector. Defect prevention, however, is desirable in both sectors. The rule of quality cost called *1–10–100*, meaning it only costs one unit to detect a quality issue in the design phase. It will cost 10 times more if the defect is identified during the installation stage and 100 times more if the defect is discovered after the product is installed or purchased by the customer (Sowards, 2013).

Another problem is that the costs of poor construction are not really known because defects are often not recorded. In order to reduce defects, construction managers need to implement a system to better manage their processes. One way to do this is to adopt a quality framework, such as the International Organization for Standardization (ISO) framework for planning and design, load computations, operations, and maintenance.

9.2 Importance of Standards

Quality measures can be embedded in various philosophies, management systems, tools, and techniques. ISO certification can help construction companies with customer or regulatory compliance, gain a competitive advantage, improve operational efficiency, and send a credible signal of quality to current and potential clients (Kale and Arditi, 2006). Although ISO does not define product quality, it ensures that customer feedback is captured and measured to improve future processes (Ollila, 2011).

Large construction companies have taken advantage of ISO certification, whereas small and medium-sized firms have not because of fewer resources such as staff, money, and time. "We have a lot of skilled engineers, engineers … [who] like to focus on engineering and are somewhat resistant to documentation and procedures," stated a vice president of a manufacturing/design firm (Libby, 2007, p. 102). Therefore, many cost overruns, reworks, delays, and disputes can be traced back to inferior design, poor contract administration, and lax supervision by the client (Tan and Lu, 1995; Chini and Valdez, 2003).

9.3 Standards Organizations

Many organizations want to systematize the implementation of business management systems in terms of different functions and operations such as quality improvement (ISO 9000, TS 16949, QS 9000, EAQF, VDA, etc.), environmental impact (ISO 14000 and EMAS), and corporate social responsibility (SA 8000, AA 1000, and ISO 26000). When comparing individual countries' standards to ISO, ISO appears to be the common denominator for most standardization attempts. For instance, the ASN of Senegal, SAC of China, ELOT of Greece, BSI of the United Kingdom, BIS of India, ABNT of Brazil, SCC of Canada, and ANSI of the United States have ISO in common.

9.4 International Standards Organization

To enhance efficiency, competitiveness, and customer satisfaction, an increasing number of organizations are adopting the philosophy of total quality management (TQM). In many instances, this philosophy is operationalized through the implementation of a quality management system (QMS) that uses ISO 9000 series of standards.

The ISO, founded in Europe in 1947, is headquartered in Geneva, Switzerland. ISO's purpose is to aid worldwide commerce in implementing international quality standards for products and services (Stevenson and Barnes, 2001). The underlying premise of ISO 9000 certification is the creation of products and services that utilize a system of inputs and outputs that adds and measures value throughout the manufacturing process.

Unlike TQM that originated in Japan, ISO certification began in Europe in 1987 and quickly spread to North America, Japan, and the rest of the world. ISO certification is a quality management program that encourages companies to shift their mindsets from the traditional reactive approach to a proactive management approach (Kale and Arditi, 2006). The ISO's certification process provides worldwide standards to support external quality assurance processes and to aid in global commerce. Having met the certification requirement, partners, buyers, and customers gain assurance that a company's products meet certain standards of quality.

The ISO series consists of two basic types of standards: (1) those addressing quality assurance and (2) those addressing quality management (Arditi and Gunaydin, 1997). The quality assurance standards designed for contractual and assessment purposes are ISO 9001, ISO 9002, and ISO 9003. The quality management standard is ISO 9004, which provides guidance for companies developing and implementing quality systems.

To receive ISO certification, an accredited third party (an approved external auditor) evaluates the company's processes and documentation and determines whether the organization is systematically following the policies and procedures necessary to produce high quality products or not. Companies using QMS will generally experience improvement of the quality of products and services, a reduction in quality-related costs, more efficient internal business processes, and the creation of corporate memory through proper documentation and retention (Vrassidas and Chatzistelios, 2014).

According to Renuka and Venkateshwara (2006), the benefits of ISO certification are as follows:

- Higher productivity and sales
- Better control of business, reduced costs, and fewer customer complaints
- Improved quality of work life
- Increased customer preference
- Improved company image, customer relationships and competitiveness in the marketplace
- Streamlined procedures and documentation
- Increased consciousness for preventive and corrective actions

On the other hand, critics of ISO argue that it does more harm than good. Seddon (1998) criticizes ISO as an inspection-oriented tool, not as a development-oriented process. Since quality is defined by external auditors, critics believe ISO can make matters worse for customers by not delivering promised results. Others believe certification is too costly, representing a pursuit of certification rather than quality (Brown, 1994). Some view ISO as a paper-driven, overly bureaucratic process (Curkovic and Pagell, 1999; Quazi et al., 2002; Boiral and Amara, 2009; Boiral, 2011).

Critics argue that inconsistencies are inherent in the evaluation process (Stevenson and Barnes, 2001). For instance, since regulations and implementation of the standards are left up to each participating country, no single set of guidelines exists, and the amount of required work for ISO varies. Furthermore, the cost of ISO implementation is significant, particularly for a small organization with limited employee participation (Martinez-Costa and Martinez-Lorente, 2007a, b).

Since ISO 9000 is not industry-specific, critics feel that it fails to address the unique problems inherent in some industries (Stevenson and Barnes, 2001). The transportation, chemical, oil, computer, and electronics industries have traditionally had strong ties to ISO 9000. On the other hand, some industries like tool manufacturers, auto manufacturing, and steel believe that the ISO 9000 certification does not address their specific needs and have developed their own sets of standards.

Symonds (1998) offers counterarguments in favor of ISO. He believes people who criticize ISO are usually those not directly involved with its implementation and should not blame the tool, but rather poor execution of the tool. Harrington (1997) studied 60 organizations to analyze best practices. He concluded that failure or success depends on management—if management understands how to implement ISO, then it will succeed. Otherwise, regardless of the quality tools the management uses, improvements will not succeed.

Proponents of ISO 9000 further argue that proper implementation leads the organization to better performance by increasing sales volume, lowering customer complaints, reducing the variance in production process, and increasing competitive advantage.

Despite the discussion for and against certification, the number of certifications around the world is increasing. As of December 2015, ISO have members from 161 countries. The number of ISO 9000 QMS certifications in developing countries is also increasing. In fact, the number of certifications in developed countries has stabilized or even decreased, whereas certifications in developing countries such as China, India, and some countries in Africa have increased. The top 10 countries with the highest number of ISO certification are listed in Table 9.1.

TABLE 9.1

Top Ten Countries with ISO Certification as of Fall 2014

Sr. No.	Country	Number of Certification
1	China	334,032
2	Italy	137,390
3	Spain	59,418
4	Germany	51,809
5	Japan	50,339
6	United Kingdom	44,670
7	France	31,631
8	India	29,402
9	United States of America	26,177
10	Brazil	25,791

9.5 ISO Implementation Process

According to Kale and Arditi (2006), the ISO implementation process follows three stages:

1. Mapping or documenting routines that underlie an organization's internal and external relationships
2. Improving internal and external relationships by making changes and setting up essential business processes in order to bring customers and suppliers together, to achieve continuous flow of accurate and reliable information from all parties to reduce guesswork and to increase speed and flexibility
3. Adhering to improved internal and external organizational routines

Unlike the manufacturing sector, the construction industry has its unique challenges. Construction companies provide services in ever-changing locations, deal with different stakeholders, and lack fringe benefits to employees, making it more difficult to incorporate standardization. Thus, the very nature of this industry makes systemization, standardization, and process automation difficult. When implementing a QMS, the industry struggles with the following (Cachadinha, 2009):

1. Systematization and structuring
2. Document control
3. Defining and maintaining procedures
4. Client satisfaction including evaluation, analysis, and action
5. Interaction between quality and production departments
6. Cost and assignment of human resources

During the first half of the twentieth century, engineers and architects controlled the design phase of construction (Ardili and Gunaydin, 1997). They ensured that clients (owners) received quality construction. In the 1950s and 1960s, clients became increasingly concerned with costs, but still with emphasis on quality. At about the same time, however, the practice of sealed competitive bidding forced many general contractors to cut costs in order to maintain profitability. As mechanical and electrical systems became more complex, the general contractor turned responsibility for such work over to subcontractors, giving up general oversight regarding the quality of workmanship.

Owing to increased complexity and stricter regulations in the 1980s and 1990s, construction management firms began to perform managerial functions on behalf of the client from inception to the completion of the construction project. Consequently, inspections once performed by architects and engineers are now being done by the construction management firms.

The following case is about a construction management firm that developed and applied a QMS that was certified to ISO 9001:2008 in 2013.

9.6 Case Study—ISO Implementation at Resource International Inc.

This case study is based on direct participation of the authors as ISO consultants for Resource International Inc. (Rii). Incorporated in 1973, Farah Majidzadeh, CEO and chairperson, founded Rii in the basement of her home in Columbus, Ohio. Rii is a family-run business. Dr. K. Majidzadeh, a retired Ohio State University professor of civil and structural engineering, provides engineering oversight, whereas his wife, Farah, oversees the business aspects of Rii. All four children of the founder work in top management positions.

The company based in Columbus, Ohio, has additional offices in Cleveland and Cincinnati, and has 150 employees. It is characterized by a strong, centralized hands-on leadership style. Rii offers a full-service consulting, engineering, and construction management firm, assisting public and private clients in the following areas:

- Civil engineering
- Construction administration, engineering, and inspection services
- Construction management
- Environmental services
- Geotechnical and material engineering services
- Ground-penetrating radar
- Information technology (IT)
- Asset management and pavement engineering
- Survey, mapping, and subsurface utility engineering (SUE)

Rii took the ISO path for three reasons:

1. To enhance its professional image and assurance of quality
2. To gain a competitive edge by improving its internal quality systems
3. To improve its capability to bid on larger projects, both public and private, especially because many major public works projects are awarded to companies with a certified QMS (Cachadinha, 2009)

9.6.1 Methodology

Mafi and Shraim followed the model (Figure 9.1) that was used by the consulting company Rii hired to help design and implement a QMS. Rii management also wanted ISO 9001 system that minimized the long-term burdens of sustaining it.

9.6.1.1 Design and Implementation of the QMS

Mafi and Shraim met with Rii leadership to devise a plan and timeline for this project. The authors asked the leadership to identify an individual to serve as a change agent who had the authority and credibility to spearhead this initiative, and a project coordinator who could oversee the details of the project and serve as the point person for all the questions and documents related to the certification process.

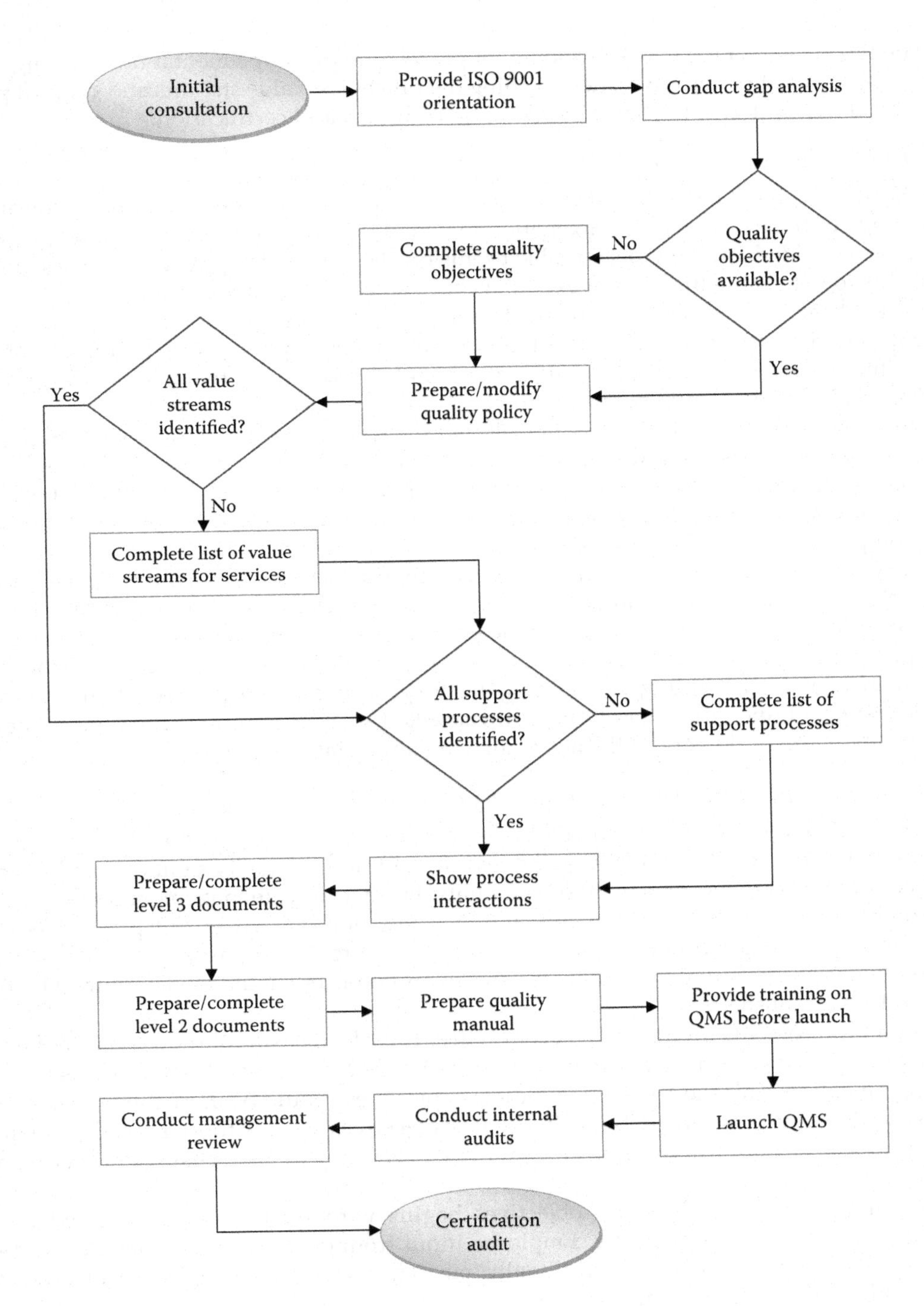

FIGURE 9.1
ISO 9001 implementation model.

The five phases of the QMS development process are pre-implementation, defining and establishing quality objectives, determining the processes (value streams and support processes), documentation (all levels), and the QMS launch and certification.

9.6.1.1.1 Pre-Implementation

At the initial consultation, the authors gathered general information about Rii to determine the scope of the project. Such information included size, locations, number of employees, services provided, and any permissible exclusion. This helped the authors prepare a gap analysis to determine the current extent of compliance to the standards.

At that time, the company was trying to decide whether to apply for certification or not. Skeptics in the company did not think the company was ready for such a project. Others thought the process would add significant amounts of administrative work to the organization, and therefore could not be justified.

Interestingly enough, Rii leadership selected the CFO of the company, a CPA, to be the change agent for this initiative. When asked about this decision, Dr. Majidzadeh justified it stating, "If we can get the most skeptical person to buy into this process, then it is easy to convince everyone else." The social media/marketing staff served as the coordinators of the company-wide ISO process.

The CFO was the most critical person pertaining to the success of the initiative. Once he met with the consultants and shared his concerns, they asked him to give the process a chance to succeed. At the beginning, he estimated the chance of success to be 10%. As the process progressed, his expectations rose to 100%. He began to meet directors of each department at lunch meetings, explaining the purpose of the ISO certification and answering questions. Lunch time was strategically selected so as to minimize the time spent on certification and build goodwill since employees appreciate a free lunch.

9.6.1.1.2 Defining and Establishing Quality Objectives and Policy

In defining the quality policy and objectives, it is important to review the specific requirements stated in the ISO 9001:2008 standards. In addition, a quality policy must be written, communicated, and reviewed for suitability. Furthermore, management must commit to comply with the ISO requirements and its continuous improvement process, which includes developing a framework for establishing and reviewing quality objectives. This requirement suggests that the quality objectives and policies should be determined before the QMS is established.

As for requirements for quality objectives, they must be measurable and consistent with the policy. These objectives are to be established at relevant functions and levels within the organization. This means that higher level objectives should be implemented at lower levels with measurable goals. Typically, quality objectives include those related to quality of services as well as timely delivery. Other objectives may include customer service and responsiveness.

Defining a quality policy and objectives begins with listing input generated from employees as well as from clients. Employee input from different operations can lead to better policies and measures. Additionally, customer surveys can be given in order to understand what is important to clients.

9.6.1.1.3 Determining Processes

Utilizing the process approach is a critical element when it comes to effective management of operations. As described under clause 0.2 of ISO 9001:2008, the "adoption of a process approach when developing, implementing and improving the effectiveness of a quality management system, ... enhance[s] customer satisfaction by meeting customer requirements."

Additionally, clause 4.1 requires the organization to determine processes needed for effective operations as well as the sequence and interactions among such processes. It should be noted that processes can be either value stream or support-oriented. A value stream process includes all actions required to bring about the product or service through the main flows (Rother and Shook, 2003). On the other hand, support processes ensure that value streams are effective and meet or exceed customer requirements.

Separating processes into value stream and support categories are beneficial when applying Lean management techniques. For example, value stream maps can show both current and future states so that improvement can be easily implemented and documented.

9.6.1.1.4 Value Stream Processes

In general, the approach Mafi and Shraim used for designing and implementing a QMS starts with identifying the services being provided. For example, what revenue-generating services does Rii provide to its client base?

To address this question, all department heads should be brought together for a session to report all of the sources of generating revenue for the company. After a list is created, these services are grouped based on commonalities. For example, an IT company might provide services such as website design and development, search engine optimization, network system design and integration, barcode systems, warehouse automation, online support, and staffing of IT professionals. This list might then be grouped into families as follows:

- *Website services*: Website design, web hosting, and search engine optimization
- *Warehouse services*: Warehouse automation and barcode systems
- *Network services*: Network system design and integration
- *Expert services*: Online support and staffing of IT professionals

Value stream services were grouped into the following families at Rii:

- Design engineering
- Environmental services
- Construction management
- Surveying services
- Geotechnical services
- Construction, inspection, and testing
- IT services

Each of these families represented services that are managed in a similar manner. For example, the construction, inspection, and testing family includes specific tests conducted in the lab and on-site. What is common among these tests is that they are managed in the same way. Although each family has separate *how to* documents, the overall management process is the same.

9.6.1.1.5 Support Processes

Support processes are divided into two groups:

- Processes/procedures required by the standard
- Processes not required by the standard but important for sustaining the value streams as well as the quality policy and objectives

The processes required by the procedures include document control, internal audits, control of nonconformance, and corrective and preventive actions. Additional processes were developed based on importance of achieving results influenced by regulations, company policies, and value stream operations. Such processes include the following:

- Management review
- Feasibility review
- Purchasing and supplier control
- Training
- Client feedback management

After determining all value stream and support processes needed for the QMS, the authors had to determine the relationships and interactions among the processes. Table 9.2 illustrates how process relationships, interactions, and performance measurements are documented. Performance of all processes must be linked (directly or indirectly) to quality objectives and ultimately to the performance of Rii.

9.6.1.2 Documentation

In general, QMSs begin with the writing of a quality manual. The quality manual serves as a roadmap for the QMS. It is more practical to build the *road* before preparing the *map*. The manual is generic at first and as the QMS develops, the manual is updated.

TABLE 9.2

An Example of a Process Inputs, Activities, Outputs, and Measurements

Process Name	Input	Process Activities and Reference	Primary Output	Output Measurement and Records
Document and Record Control	>ISO 9001 requirements >Other standards >Process requirements >Legal requirements	(QSP-1)	Controlled Documents	>Doc Control Database Reports >Internal and External Audits
Management Review	>Review input	(QSP-2)	>Decisions and Actions >Improvement Plans	>Management Review Reports >Action Database Status >Internal and External Audits
Purchasing and Supplier Control	>Quotes >Subconsultant proposals >Subconsultant performance	(QSP-3)	>PO issued >Signed Contracts >Received Items	>Internal/External Audits >Supplier Evaluation

A team including the manager and director responsible for each value stream process meet to document the existing value stream processes. The team is provided with a template to document value stream processes in the following format:

- Service provided
- Responsibility
- Outsourced or subcontracted
- Input
- Activities
- Output/record
- Measurement of performance

After completing this first step, the plans are reviewed to ensure that everything is accounted for, including external references, forms, and records. Additionally, process performance measures were verified and modified. If other documents exist that are a part of the value stream process, they are referenced at this time. At Rii, all existing forms and records were utilized with minor adjustments.

There are three levels of documents in the QM manual.

1. *Level 3 documents*: This level includes internal work instructions, external testing procedures, and other *how to* standards.

   Some of the internal documents already existed, but did not include controls. Since all external testing procedures, standards, and manuals were already available, the authors were able to

   a. Take full inventory of all of available documents (hardcopies and electronic).
   b. Remove redundant documents.
   c. Assign responsibilities for creating new documents and adjusting existing ones.
   d. Document controls.
2. *Level 2 documents*: This level includes documents that describe value stream and support processes. If internal work instructions or other external level 3 documents are available for any of these processes, they are referenced at this level. In summary, each of the procedures at this level (value stream or quality system procedures) is considered the parent document for relevant level 3 documents.
3. *Level 1 documents*: After the level 2 and 3 documents are prepared, the quality manual is prepared in level 1. The quality manual for Rii addresses the following:
   a. Requirements in the ISO standard
   b. Legal and company requirements
   c. Reference to procedures required by the standard
   d. A table describing the interactions between QMS processes

9.6.1.2.1 QMS Manual

A quality system has to cover all the activities leading to turn out final product or service. The quality system depends entirely on the scope of operation of the organization and particular circumstances such as number of employees, type of organization, and physical size of the premises of the organization. Adoption of quality system is a strategic decision of the organization. The quality manual is the document that identifies and describes the QMS. The QMS is based on the guidelines for performance improvement as per ISO 9004:2000 and the quality management requirements. ISO 9000:2000 standard outlines the necessary steps to implement the QMS. These are as follows:

1. Identify the process (activities and necessary elements) needed for QMS.
2. Determine the sequence and interaction of these processes and how they fit together to accomplish quality goals.
3. Determine how these processes are effectively operated and controlled.
4. Measure, monitor, and analyze these processes and implement action necessary to correct the process and achieve continual requirement.
5. Ensure that all information is available to support the operation and monitoring of the process.
6. Display the most options, thus helping make the right management system.

These documents are regularly updated and revised to meet the changing industry requirements due to

- Globalization
- Higher levels of expectations by customers
- Higher performance requirements
- Complex nature of environment and works
- Specialized services requirements

9.6.1.3 QMS Launch and Certification

Before the QMS is launched, all documents are reviewed for corrections. Applicable copies of the final QMS are made available and accessible at all locations.

Although everyone is aware of their responsibilities, QMS training is provided for all associates. In addition, associates receive additional training on their specific value streams. At this point, the QMS is launched as scheduled. In addition, a date for the certification audit is established with the selected registrar.

Prior to certification, an internal audit is required for the entire QMS. In this case, minor issues were found and fixed. Results of the internal audits, as well as other inputs, were evaluated at the management review session. Now, Rii is ready for the certification audit.

When the audit occurred, Rii associates were well prepared with everyone assuming ownership of the process. The audit team was very impressed with the QMS and its results. The team reported no nonconformities and recommended a few opportunities for improvement. Rii received its certification in 2013, the same year the company celebrated its 40th anniversary.

9.6.1.4 Lessons Learned

Even a quarter of a century after the debut of the ISO 9000 standards, the standards have had a worldwide impact on trade and quality. Despite the costs and complexity of the certification process, research shows that certification process and its maintenance is a worthwhile investment that pays off many times over, if done correctly. In the case of Rii, the long-term impact of ISO certification is not known yet, but management has been pleased with the results so far.

During an interview, the CFO commented that ISO process did not bring drastic changes to the Rii culture, but has had a significant positive impact on documentation control and maintenance bringing consistency to their documentation processes. Furthermore, ISO helped Rii catch lapses in QMS that would have gone unnoticed in the past (i.e., customer complaints). Now all complaints are tracked and recorded so that improvements can be made.

Most employees especially the engineers were reluctant to add yet another *job* to their daily routine. As they experienced how ISO improved their work, their resistance diminished. Rii leadership hopes that the ISO processes will become a second nature for employees. ISO certification has also helped improve communications within the organization and external parties thereby improving client relationships. Standardization, consistency, and accountability across the entire organization now exist at Rii.

The authors learned the following valuable lessons on how to better implement a QMS.

- Management buy-in and commitment serve as the cornerstone for completing an ISO project on time. If top management leads the implementation efforts, the entire organization is more likely to participate in the process
- Unlike manufacturing, value streams in service organizations are more difficult to identify because they are not related to a tangible product. Value stream identification was difficult because Rii provides many different services
- Managers found brainstorming sessions helpful in outlining their value stream processes. Each brainstorming session addressed services offered, inputs and outputs, and performance measures
- Rii subcontracted its internal audits to the local chapter of the American Society for Quality (ASQ) to ensure the audits were effective in uncovering issues. Using the local ASQ chapter is less expensive than starting and maintaining an internal auditing program. The local ASQ chapter provides a list of members who volunteer their services for continuing education credits
- In the beginning, ISO efforts might meet resistance. Providing basic training on requirements and answering questions at the onset helps to alleviate anxieties and improves participation

9.6.1.5 Conclusion

Clearly a primary target in construction safety is zero on-the-job accidents and zero quality defects. Most construction managers, even if they agree on these standards, still accept defects in construction as commonplace, thinking that it is impossible to deliver a defect-free product. The best target, however, is no defects, no reworks, and no punch lists (Sowards, 2013). This case chapter demonstrates the journey Rii took to implement ISO standards.

Findings of quality management practices are as applicable today as they were back when they were first created nearly 30 years ago (Ardili and Gunaydin, 1997). Top management commitment to quality in each phase of the construction process, education of basics of quality management, statistical methods, and problem solving tools in monitoring construction progress to architects and engineers lead to TQM. Facilitation of teamwork among manufacturers, subcontractors, main contractors, vendors, designers, project manager, and clients (owners) are also part of the TQM process.

For success, the requirements of the client (owner) must be clearly defined at the beginning of the project. The more time spent in planning, the more smoothly the project will run. Similarly, drawings and specifications received from designers should be clear, concise, and uniform because they are the only documents a constructor sees regarding the design concept, size, and scope of the job.

No system can replace human knowledge and know-how, but a QMS can help organizations by documenting human knowledge in a systematic way. Going through the journey of ISO certification is not an easy one, but well worth it. Some organizations, after the initial success, however, fail to sustain their momentum and fall back into the old habits. Thus, bi-annual management review helps companies maintain and improve their TQM programs.

9.7 ISO 14001:2004

ISO 14001:2004 gives the requirements for environmental management systems by identifying measurable performance goals that verify improvement. This standard is especially important for the construction sector because this sector builds the infrastructures that house human life and can adversely affect the environment with land deterioration, resource depletion, waste generation, air pollution, noise pollution, and water pollution (Turk, 2009). For example, the annual solid waste generated from construction activities in China is around 30% to 40% of the total municipal waste (Zeng et al., 2003), and the amount of annual construction waste per capita in the EU countries is estimated to be 0.5 to 1 ton (Ekanayake and Ofori, 2004). It is also estimated that construction activities consume approximately 40% of the world's material and energy flow (Lenssen and Roodman, 1995). In addition, many of the materials are irreplaceable raw materials. These and similar issues attest to the importance of the construction sector compliance with ISO 14000:2004.

A study by Turk (2009) found construction companies in Turkey seek certification for the following reasons: easier access to the international markets, per the requests from clients, to beat other competitors, and in anticipation that ISO 14000 standards will be mandatory in the near future for reasons similar to those described for ISO 9001.

9.8 ISO 27001:2013

ISO 27001:2013 gives the requirements for information security management systems (ISMS). As with ISO 9001 and 14001, if the process is correct, then outcome will be satisfactory (Gillies, 2011). Organizations can develop and implement a framework for managing the security of their information assets, including financial information, intellectual property, and employee details, or information entrusted to them by clients or third parties. These standards prepare companies for an independent assessment of ISMS as applied to the protection of information.

TABLE 9.3

Number of Certifications by Categories

Standard	Number of Certifications in 2012	Number of Certifications in 2013	Number of Certifications in 2014
ISO 9001	1,096,987	1,126,460	1,138,155
ISO 14001	284,654	301,622	324,148
ISO 27001	19,620	22,349	23,972

Source: International Organization for Standardization (2014). The ISO Survey of Management System Standard Certification.

Studies suggest that ISO 27001 adoption has been slower compared to other ISO standards (Gillies, 2011). Costs appear to be a major barrier for most small companies. However, with further globalization of businesses, there is more legitimacy for ISO 27001 certification either by government regulations (i.e., Japan), supplier/buyer demands, or the necessity for outsourcing and offshoring (Gillies, 2011).

Another important driver for this certification is to provide evidence to partners that the company has identified and measured their security risks, implementing a series of procedures to minimize those risks. A further incentive is lower insurance premiums for ISO 27001 certified companies (von Solms and von Solms, 2005).

The more recent requirements have now replaced the earlier 2005 version. Information security has always been a national and international issue, and this new version of the standard reflects eight years of improvements in the understanding of effective information security management. It also takes account of the evolution in the cyber threat landscape over that period, and allows for a new range of best practices. Information security is now a management issue and governance responsibility. Table 9.3 lists the number of certifications for the major categories as of Fall 2014.

9.9 ISO 9001:2015

The ISO 9001:2015 was published in September 2015. It has several changes to earlier ISO 9001:2008.

The ISO 9001:2015 is the new guidelines for all management system standards (MSS). The modifications to previous guidelines are as follows:

The first modification is to promote a consistent look, language, and feel to all the QMSs. This is an improvement over the 2008 standards because this allows organizations that have or seek multiple certifications to use a common language and structure for all their certification processes. Additionally, the word *leadership* is no longer avoided as in previous standards. The previous clause *management responsibility* or *top management* has now become *leadership*, where the emphasis is clearly on the demonstration of leadership.

The second change is to incorporate risk-based thinking into the management system by considering the context of the organization. In other words, all processes are not equal for all organizations with some being more critical than others, resulting in different levels of risk. Therefore, it is recommended that organizations must fully explore and understand the internal and external conditions under which they function.

The complexity of today's business practices along with advance in technology makes the speed of change faster. ISO 9001: 2015 requires the organization to "… determine external and internal issues that are relevant to its purpose and its strategic direction and that affect its ability to achieve the intended result(s) of its quality management systems" (ISO 9001: 2015). In order to do so, organizations must have a clear understanding of their strategic direction. Often, organizations' strategies are uncoordinated and misaligned without clear direction. This lack of direction has a direct impact on its QMS, thus aligning strategies is a great input to any QMS (West and Cianfrani, 2014). Earlier versions of Draft International Standard (DIS) did not emphasize this alignment. Although the standards mandate such an inquiry, it is the responsibility of the organization to decide on the extent to evaluate the internal and external conditions pertaining to its performance. This is a challenge, although the concept is fairly simple.

A third addition is the use of the terms *interested parties* and *needs and expectations.* Previously, only the customers' voices were addressed. The change ensures that voices other than customers are relevant. For instance, interested parties may include regulatory bodies, suppliers, the community, and the society at large. This modification does not negate the voice of the customers, but rather acknowledges the voices of other stakeholders. It is up to the organization to decide who the stakeholders are and what issues they have that could potentially impact the organization's performance.

The last major modification proposed is the system thinking and process approach. Although the use of process approach in developing and maintaining QMS is not new, the ISO 9001:2015 requires organizations to identify their processes, interactions, and resources needed to operate. These processes need to be controlled, measured, and improved. ISO 9001:2015 also reiterates that people are key to improving processes and system thinking serves as the catalyst to make this happen. It is critical to understand how the entire system works rather than search only for direct cause–effect relationships.

Changes in ISO 9001:2015 are an opportunity to revisit organizational areas that yet need to be improved. An awareness of the upcoming changes in ISO 9001:2015 will enable quality professionals to better prepare for the future. ISO 9001:2008 was revised by various committees, societies, and institutes, and ISO 9001:2015 was published in September 2015. It has the following clauses:

1. Context of organization/QMS
2. Leadership
3. Planning for QMS
4. Support
5. Operation
6. Performance evaluation
7. Improvement

Table 9.4 lists the clauses of ISO 9001:2015 and ISO 9001:2008 and their correlations. Table 9.5 lists example contents of quality management manual for engineering consultant (design and supervision) and Table 9.6 lists example contents of QMS manual for building construction organization (contractor).

TABLE 9.4

Correlation between ISO 9001:2015 and ISO 9001:2008

Clause	ISO 9001:2015	Correlation between ISO 9001:2015 to ISO 9001:2008	Clause	ISO 9001:2008
1.0	**Scope**		1.0	**Scope**
	1.1 General			1.1 General
	1.2 All exclusions from ISO 9001:2008 Clause 1.2 removed			1.2 Application
2.0	**Not valid**		2.0	**Normative References**
3.0	**Included into Primary Standard, ISO 9000**		3.0	**Terms and Definitions**
4.0	**Context of the Organization/Quality Management System**	**1.0**	4.0	**Quality Management System**
	4.1 Understanding the organization and its context	**1.1**		4.1 General requirements
	4.2 Understanding the needs and expectations of both the parties	**1.1**		4.2 Documentation requirements
				4.2.1 General
				4.2.2 Quality manual
				4.2.3 Control of documents
				4.2.4 Control of records
	4.3 Determine the scope of quality management system	**1.2,4.2.2**		
	4.4 Quality management system and its processes	**4.0,4.1,4.2.2**		
5.0	**Leadership**	**5.0,5.5**	5.0	**Management Responsibility**
	5.1 Leadership and commitment	**5.1**		5.1 Management commitment
	5.1.1 Leadership and commitment to quality management system	**5.1**		
	5.1.2 Customer focus	**5.2**		
	5.2 Quality policy	**5.3**		5.2 Customer focus
	5.2.1 Requirements of top management in respect of quality policy			
	5.2.2 Specific requirements in respect of organization's policy			

(Continued)

TABLE 9.4 (*Continued*)

Correlation between ISO 9001:2015 and ISO 9001:2008

Clause	ISO 9001:2015	Correlation between ISO 9001:2015 to ISO 9001:2008	Clause	ISO 9001:2008
5.3	Organizational roles, responsibilities, and authorities	**5.5.1,5.5.2**	5.3	Quality policy
			5.4	Planning
			5.4.1	Quality objective
			5.4.2	Quality management system planning
			5.5	Responsibility, authority, and communication
			5.5.1	Responsibility and authority
			5.5.2	Management perspective
			5.5.3	Internal communication
			5.6	Management review
			5.6.1	General
			5.6.2	Review input
			5.6.3	Review output
6.0	**Planning for Quality Management System**	**5.4,5.4.2**	**6.0**	**Resource Management**
6.1	Actions to address risks and opportunities	**5.4.2,8.5.3**	6.1	Provision of resources
6.1.1	Organization's context when planning for their QMS			
6.1.2	How to address risks and opportunities			
6.2	Quality objectives and planning to achieve them	**5.4.1**	6.2	Human resources
6.2.1	Enhancement and extension of ISO 9001:2008 requirements		6.2.1	General
6.2.2	Enhancement of ISO 9001:2008 sub-clause 5.4.2		6.2.2	Competence, training, and awareness
6.3	Planning of changes	**5.4.2**	6.3	Infrastructure
		6.0	6.4	Work environment

(*Continued*)

TABLE 9.4 (*Continued*)

Correlation between ISO 9001:2015 and ISO 9001:2008

Clause	ISO 9001:2015	Correlation between ISO 9001:2015 to ISO 9001:2008	Clause	ISO 9001:2008
7.0	**Support**	6.0	7.0	**Product Realization**
7.1	Resources	6.0	7.1	Planning of product realization
7.1.1	General	6.1		
7.1.2	People	6.1		
7.1.3	Infrastructure	6.3		
7.1.4	Environment for the operation of processes	6.4		
7.1.5	Monitoring and measuring resources	7.6		
7.1.6	Organizational knowledge	New		
7.2	Competence	5.2,6.2.1,6.2.2	7.2	Customer-related processes
			7.2.1	Determination of requirements related to the product
			7.2.2	Review of requirements related to the product
			7.2.3	Customer communication
7.3	Awareness	6.2.2	7.3	Design and development
			7.3.1	Design and development planning
			7.3.2	Design and development input
			7.3.3	Design and development output
			7.3.4	Design and development review
			7.3.5	Design and development verification
			7.3.6	Design and development validation
			7.3.7	Control of design and development changes
7.4	Communication	5.5.3	7.4	Purchasing
			7.4.1	Purchasing process
			7.4.2	Purchasing information
			7.4.3	Production and service provision

(*Continued*)

TABLE 9.4 (*Continued*)

Correlation between ISO 9001:2015 and ISO 9001:2008

Clause	ISO 9001:2015	Correlation between ISO 9001:2015 to ISO 9001:2008	Clause	ISO 9001:2008
	7.5 Documented information	4.2		7.5 Production and service provision
	7.5.1 General	4.2.1,4.2.2		7.5.1 Control of production and service provision
	7.5.2 Creation and updating	4.2.3,4.2.4		
	7.5.3 Control and documented information	4.2.3,4.2.4		7.5.2 Validation of process for production and service provision
	7.5.3.1 Availability of document when needed			7.5.3 Identification and traceability
	7.5.3.2 Distribution, access, and retrieve			7.5.4 Customer property
				7.5.5 Preservation of product
				7.6 Control of monitoring and measuring equipment
8.0	**Operation**	**7.0**	**8.0**	**Measurement, Analysis, and Improvement**
	8.1 Operational planning and control	7.1		8.1 General
	8.2 Determination of requirements for products and services	7.2		8.2 Monitoring and measurement
	8.2.1 Customer communication	7.2.3		8.2.1 Customer satisfaction
	8.2.2 Determination of requirements related to products and services	7.2.1		8.2.2 Internal audit
	8.2.3 Review of requirements related to products and services	7.2.2		8.2.3 Monitoring and measurement of process
				8.2.4 Monitoring and measurement of product
	8.3 Design and development of products and services	7.3,7.3.1		8.3 Control of nonconforming product
	8.3.1 General	7.3.1		
	8.3.2 Design and development planning	7.3.1		
	8.3.3 Design and development inputs	7.3.2		
	8.3.4 Design and development controls	7.3.4,7.3.5,7.3.6		
	8.3.5 Design and development outputs	7.3.3		
	8.3.6 Design and development changes	7.3.7		

(Continued)

TABLE 9.4 (*Continued*)

Correlation between ISO 9001:2015 and ISO 9001:2008

Clause	ISO 9001:2015	Correlation between ISO 9001:2015 to ISO 9001:2008	Clause	ISO 9001:2008
8.4	Control of externally provided products and services	**7.4,7.4.1**	8.4	Analysis data
8.4.1	General	**7.4.1**		
8.4.2	Type and extent of control of external provision	**7.4.1,7.4.3**		
8.4.3	Information for external providers	**7.4.2**		
8.5	Production and service provision	**7.3,7.5**	8.5	Improvement
8.5.1	Control of production and service provision	**7.5.1**	8.5.1	Continual improvement
8.5.2	Identification and traceability	**7.5.3**	8.5.2	Corrective action
8.5.3	Property belonging to customers or external providers	**7.5.4**	8.5.3	Prevention action
8.5.4	Preservation	**7.5.5**		
8.5.5	Post-delivery activities	**7.5.1**		
8.5.6	Control of changes	**7.3.7**		
8.6	Release of products and services	**8.2.4,7.4.3**		
8.7	Control of nonconforming process outputs, products, and services	**8.3**		

(*Continued*)

TABLE 9.4 (*Continued*)

Correlation between ISO 9001:2015 and ISO 9001:2008

Clause	ISO 9001:2015	Correlation between ISO 9001:2015 to ISO 9001:2008	Clause	ISO 9001:2008
9.0	**Performance Evaluation**	**New**		
	9.1 Monitoring, measurement, analysis, and evaluation	**8.0**		
	9.1.1 General	**8.1,8.2.3**		
	9.1.2 Customer satisfaction	**8.2.1**		
	9.1.3 Analysis and evaluation	**8.4**		
	9.2 Internal audit	**8.2.2**		
	9.2.1 Organization's requirement to carry out internal audit			
	9.2.2 Requirements to how audit programs must be structured			
	9.3 Management review	**5.6**		
	9.3.1 Review of QMS by top management	**5.6.2**		
	9.3.2 Specific requirements in respect of management review	**5.6.3**		
10	**Improvement**	**8.5**		
	10.1 General	**8.5.1**		
	10.2 Nonconformity and corrective action	**8.3,8.5.2**		
	10.2.1 How the organization to act when nonconformity is identified			
	10.2.2 Documented information relating to nonconformity	**8.3,8.5.2**		
	10.3 Continual improvement	**8.5.1**		

TABLE 9.5

List of Quality Manual Documents–Consultant (Design and Supervision)

Document No.	Title Quality Document	Relevant Clause in 9001:2015	Correlated Clause in ISO 9001:2008	Version/Revision Date
QC-0	Circulation list			
QC-00	Records of revision			
QC-1.1	Understanding the organization and its context	4.1	1.1	
QC-1.2	Monitoring and review of internal and external issues	4.1	1.1	
QC-2.1	Relevant requirements of stakeholders	4.2	1.1	
QC-2.2	Monitoring and review of stakeholder's information	4.2	1.1	
QC-3	Scope of quality management system	4.4	4.1	
QC-4	Project quality management system	4.4	4.1	
QC-5	Management responsibilities	5.1	5.5.1	
QC-6	Customer focus	5.1.2	5.2	
QC-7.1	Quality policy (organization)	5.2	5.3	
QC-7.2	Quality policy (project)	5.2.2	5.3	
QC-8	Organizational roles, responsibilities, and authorities (organization chart)	5.3	5.5.1/2	
QC-9	Preparation and control of project quality plan	6.0	5.4.2	
QC-10.1	Project risk (during design)	6.1	5.4.2,8.5.3	
QC-10.2	Project risk (during construction)	6.1	5.4.2,8.5.3	
QC-11	Project quality objective	6.2	5.4.1	
QC-12.1	Change management (during design phase)	6.3	5.4.2	
QC-12.2	Change management (during construction phase)	6.3	5.4.2	
QC-13	Office resources (human resources, office equipment, and design software)	7.1	6.1	
QC-14	Infrastructure	7.1.3	6.3	
QC-15	Work environment	7.1.4	6.4	
QC-16	Human resources (design team, supervision team)	7.2	6.2.1	
QC-17.1	Training in quality system	7.2	6.2.2	
QC-17.2	Training in quality auditing	7.2	6.2.2	
QC-17.3	Training in operational/technical skills	7.2	6.2.2	
QC-18	Communication internal and external	7.4	5.5.3	

(Continued)

TABLE 9.5 (*Continued*)

List of Quality Manual Documents–Consultant (Design and Supervision)

Document No.	Title Quality Document	Relevant Clause in 9001:2015	Correlated Clause in ISO 9001:2008	Version/Revision Date
QC-19.1	Control of documents for general application	7.5.2/3	4.2.3	
QC-19.2	Control of documents for specific projects	7.5.2/3	4.2.3	
QC-20	Records updates	7.5.2	4.2.3/4	
QC-21	Control of quality records	7.5.3	4.2.4	
QC-22.1	Planning of engineering design and quality plan	8.1/8.3/8.3.1/2/3	7.1/7.3.1/7.3.7	
QC-22.2	Design development (design–bid–build)	8.1/8.3/8.3.1/2/3	7.1/7.3.1/7.3.7	
QC-22.3	Design development (design–build)	8.1/8.3/8.3.1/2/3	7.1/7.3.1/7.3.7	
QC-23	Evaluation of subconsultant and selection	8.4	7.4.1	
QC-23	Communication with subconsultant	8.4.3	7.4.1/2/3	
QC-24	Engineering design procedure	8.5	7.5.1/2	
QC-25.1	Construction supervision procedure	8.5	7.5.1/2	
QC-25.2	Project management procedure	8.5	7.5.1/2	
QC-25.3	Construction management procedure	8.5	7.5.1/2	
QC-26.1	Control of nonconforming work (design errors)	8.7	8.3	
QC-27	Project review (management and control)	9.1	8.1	
QC-28	Internal quality audits	9.2	8.2.2	
QC-29	Management review	9.3	5.6	
QC-30.1	Corrective action	10.2.2	8.5.2	
QC-30.2	Preventive action	10.3	8.5.3	
QC-32	Control of client complaints	10.2.2	8.3	

TABLE 9.6

List of Quality Manual Documents–Contractor

Document No.	Title Quality Document	Relevant Clause in 9001:2015	Correlated Clause in ISO 9001:2008	Version/ Revision Date
QC-0	Circulation list			
QC-00	Records of revision			
QC-1.1	Understanding the organization and its context	4.1	1.1	
QC-1.2	Monitoring and review of internal and external issues	4.1	1.1	
QC-2.1	Relevant requirements of stakeholders	4.2	1.1	
QC-2.2	Monitoring and review of stakeholder's information	4.2	1.1	
QC-3	Scope of quality management system	4.3/4	4.0,4.1	
QC-4	Project quality management system	4.4	4.1	
QC-5	Management responsibilities	5.1	5.5.1	
QC-6	Customer focus	5.1.2	5.2	
QC-7.1	Quality policy (organization)	5.2	5.3	
QC-7.2	Quality policy (project)	5.2.2	5.3	
QC-8	Organizational roles, responsibilities, and authorities (organization chart)	5.3	5.5.1/2	
QC-9	Preparation and control of project quality plan	6.0	5.4.2	
QC-10	Project risk management	6.1	5.4.2,8.5.3	
QC-11	Project quality objectives	6.2	5.4.1	
QC-12.1	Change management (scope)	6.3	5.4.2	
QC-12.2	Change management (variation orders, site work instructions)	6.3	5.4.2	
QC-13	Construction resources (human resources, equipment, and machinery)	7.1	6.1	
QC-14	Infrastructure	7.1.3	6.3	
QC-15	Work environment	7.1.4	6.4	
QC-16	Control of construction, material, measuring, and test equipment	7.1.5	7.6	
QC-17	Control of human resources	7.2	6.2.1	
QC-18.1	Training and development in quality system	7.2	6.2.2	
QC-18.2	Training in quality auditing	7.2	6.2.2	
QC-18.3	Training in operational/technical skills	7.2	6.2.2	
QC-19	Communication internal and external	7.4	5.5.3	

(Continued)

TABLE 9.6 (*Continued*)

List of Quality Manual Documents–Contractor

Document No.	Title Quality Document	Relevant Clause in 9001:2015	Correlated Clause in ISO 9001:2008	Version/ Revision Date
QC-20.1	Control of documents for general application	7.5.2/3	4..2.3	
QC-20.2	Control of documents for specific projects	7.5.2/3	4..2.3	
QC-21	Records updates	7.5.2	4.2.3/4	
QC -22	Control of quality records	7.5.3	4.2.4	
QC-24	Documents control (logs)	7.5.3.2	4.2.3/4	
QC-25	Project planning and control	8.1	7.1	
QC-26	Project-specific requirements	8.2	7.2	
QC-27	Project-specific quality control plan	8.2	7.2	
QC-28-1	Tender documents	8.2	7.2	
QC-28.2	Tender review	8.2	7.2.1/2/3	
QC-28.3	Contract review	8.2.1	7.2.1/2/3	
QC-29	Variation review	8.2.3	7.2.2	
QC-30	Construction processes	8.2.2	7.2.1	
QC-31	Engineering and shop drawings	8.3	7.3.1	
QC-32	Design developments for design–build projects	8.3		
QC-33	Evaluation of subcontractors and selection	8.4	7.4.1	
QC-34	Communication with subcontractors, material suppliers, and vendors	8.4.3	7.4.1/2/3	
QC-35.1	Inspection of subcontracted work	8.4.2	7.4.3	
QC-35.2	Incoming material inspection and testing	8.4.3	7.4.3	
QC-36	Installation procedures	8.5	7.5.1/2	
QC-37	Product identification and traceability	8.5.2	7.5.3	
QC-38	Identification of inspection and test status	8.5.2	7.5.3	
QC-39	Control of owner-supplied items	8.5.3	7.5.4	
QC-40	Handling and storage	8.5.4	7.5.5	
QC-41	Construction inspection, testing, and commissioning	8.6	8.2.4	
QC-42.1	Control of nonconforming work	8.7	8.3	
QC-42.2	Control of nonconforming work	8.7	8.3	

(*Continued*)

TABLE 9.6 (*Continued*)

List of Quality Manual Documents–Contractor

Document No.	Title Quality Document	Relevant Clause in 9001:2015	Correlated Clause in ISO 9001:2008	Version/ Revision Date
QC-42.1	Project performance review	9.1.1	8.1	
QC-42.2	Project quality assessment and measurement	9.1.2	8.1	
QC-43	Internal quality audits	9.2	8.2.2	
QC-44	Management review	9.3	5.6	
QC-45	New technology in construction	10.1	8.5.1	
QC-46.1	Corrective action	10.2.2	8.5.2	
QC-46.2	Preventive action	10.3	8.5.3	
QC-47	Control of client complaints	10.2.2	8.3	

References

ANSI/ISO/ASQ Q9001:2008 Quality Management Systems—Requirements, Quality Press, Milwaukee, WI.

Arditi, D. and Gunaydin, H. M. (1997). Total quality management in the construction process. *International Journal of Project Management,* 15(4), 235–243.

Boiral, O. (2011). Managing with ISO systems: Lessons from practice. *Long Range Planning,* 44(3), 197–220.

Boiral, O. and Amara, N. (2009). Paradoxes of ISO 9000 performance: A configurational approach. *The Quality Management Journal,* 16(3), 36–60.

Brown, R. (1994). Does America need ISO 9000? *Machine Design,* June 6, 70–74.

Cachadinha, N. M. (2009). Implementing quality management systems in small and medium construction companies: A contribution to a road map for success. *Leadership and Management in Engineering,* 9(1), 32–39.

Chini, A. R. and Valdez, H. E. (2003). ISO 9000 and the U.S. Construction Industry. *Journal of Management in Engineering,* 19(2), 69–77.

Curkovic, S. and Pagell, M. (1999). A critical examination of the ability of ISO 9000 certification to lead to a competitive advantage. *Journal of Quality Management,* 4(1), 51–67.

Draft International Standard (DIS) ISO 9001: 2015.

Ekanayake, L. L. and Ofori, G. (2004). Building waste assessment score: Design-based tool. *Building and Environment,* 39(7), 851–861.

Gillies, A. (2011). Improving the quality of information security management systems with ISO27000. *The TQM Journal,* 23(4), 367–376.

Harrington, H. J. (1997). The fallacy of universal best practices. *The TQM Magazine,* 9(1), 61–75.

International Organization for Standardization (2014). The ISO Survey of Management System Standard Certification.

International Organization for Standardization (2015). ISO 9001:2015, Draft International Standards-Quality Management System.

Kale, S. and Arditi, D. (2006). Diffusion of ISO 9000 certification in the precast concrete industry. *Construction Management and Economics,* 24(5), 485–495.

Lenssen, N. and Roodman, D. M. (1995). Making better buildings. In L. R. Brown, C. F. Flavin, H. French, L. Starke, D. Denniston, H. Kane et al. (Eds), *State of the World, A Worldwatch Institute Report on Progress Toward Sustainable Society* (pp. 95–112). New York: W.W. Norton.

Libby, J. (2007). Adding credibility. *Manufacturing Today* (November/December), 102–103.

Martinez-Costa, M. and Martinez-Lorente, A. R. (2007a). A triple analysis of ISO 9000 effects on company performance. *Journal of Productivity and Performance Management,* 56(5–6), 484–499.

Martinez-Costa, M. and Martinez-Lorente, A. R. (2007b). ISO 9000:2000: The key to quality? An exploratory study. *The Quality Management Journal,* 14(1), 7–18.

Ollila, A. (2011). The role of quality system in performance improvement—Is ISO 9001 up-to-date? *Global Conference on Business and Finance Proceedings,* 6(2), 304–314.

Quazi, H. A., Hong, C. W., and Meng, C. T. (2002). Impact of ISO 9000 certification on quality management practices: A comparative study. *Total Quality Management,* 13(1), 53–67.

Renuka, S. D. and Venkateshwara, B. A. (2006). A comparative study of human resource management practices and technology adoptions of SMEs with and without ISO certification. *Singapore Management Review,* 28(1), 41–61.

Rother, M. and Shook, J. (2003). *Learning to See. The Lean Enterprise Institute,* Brookline, MA.

Seddon, J. (1998). The case against ISO 9000. *ISO 9000 News,* (July/August), 12–15.

Sowards, D. (2013). Quality is key to lean's success. *Contractor Magazine,* (May), 44, 52.

Stevenson, T. H. and Barnes, F. C. (2001). Fourteen years of ISO 9000: Impact, criticisms, costs, and benefits. *Business Horizons,* 44, 45–51.

Symonds, J. (1998). ISO 9000- case sensitive. *ISO News*, 6, 12–15.
Tan, R. R. and Lu, Y. G. (1995). On the quality of construction engineering design projects: Criteria and impacting factors. *International Journal of Quality & Reliability Management*, 16(6), 562–574.
Turk, A. M. (2009). ISO 14000 environmental management system in construction: An examination of its application in Turkey. *Total Quality Management*, 20(7), 713–733.
von Solms, B. and von Solms, R. (2005). From information security to … business security. *Computers & Security*, 24(4), 271–273.
Vrassidas, L. and Chatzistelios, G. (2014). Quality management systems development based on product systems taxonomy. *The TQM Journal*, 26(2), 215–229.
West, J. E. and Cianfrani, C. A. (2014). Managing the system revision introduces focus on organizational operating conditions. *Quality Progress* (August), 53–56.
Zeng, S., Tan, C., Deng, Z., and Tan, V. (2003). ISO 14000 and the construction industry: Survey in China. *Journal of Management Engineering*, 19(3), 107–115.

Appendix I: Contractor's Quality Control Plan

The contractor's quality control plan (QCP) is prepared based on project-specific requirements as specified in the contract documents.

The plan outlines the procedures to be followed during the construction period to attain the specified quality objectives of the project and to fully comply with the contractual and regulatory requirements.

Following is an outline of the contractor's QCP based on the contract documents.

I.1 Introduction

The contractor's QCP is developed to meet the contractor's QC requirements of (project name) as specified under clause (___) and section (___) of the contract documents.

The plan provides the mechanism to achieve the specified quality by identifying the procedures, control, instructions, and tests required during the construction process to meet the owner's objectives. This QCP does not endeavor to repeat or summarize contract requirements. It describes the process M/S ABC (contractor name) will use to assure compliance with those requirements.

I.2 Description of Project

(Owner name) has contracted with M/S ABC to construct (name of facility) located at (site plan). The facility consists of approximately (___) m^2 gross building area and approx (___) m^2 of two basements for car parking and other services.

The facility is to be used as office premises to accommodate (___) personnel. The architecture at the building is of very high quality with spacious atrium between two towers duly interconnected with a bridge inside the building, well-designed internal landscape area with plants and sky garden to provide pleasant view at the building. The building has glazed wall and curtain walling for the atrium area.

Vertical transportation is through panoramic elevators and a designated elevator for VIP. Additionally, there is a goods elevator and firefighters elevator to meet the emergency situation.

The structure of the building is by reinforced concrete and the atrium is of structural steel. External cladding includes spandrel curtain wall to the block structure and special curtain walling to the atrium. Interior finishes area of painted plastered wall to marble and stone finish. Internal partitions have flexibility to adjust the office area. Entire building has raised floor system, and space under raised floor is used for distribution of electromechanical services.

HVAC system is through water-cooled chillers, and electric duct heaters are provided for heating during winter. BMS takes care for full control of HVAC. Building is provided with all the safety measures against fire. Fire protection system, smoke exhaust fans, and fire alarm system with voice evacuation system are provided to meet the emergency situation. Fire protection system also includes automatic sprinklers, hose reels, extinguishers, and foam system.

Building is equipped with all the amenities to take care for public health system. Plumbing system is consistent with the requirement of the building and includes cold and hot water, drainage, rainwater, and irrigation system. Water fountains are provided at various locations for drinking purpose.

Electrical systems consist of energy saving lighting system having centralized control system. Electrical distribution has all the safety features in it. Diesel generator system and uninterrupted power supply system is provided to meet the emergency due to power failure from main electricity company. The building to be constructed shall be equipped with latest technological systems. It shall have IP-based communication system and all the low voltage systems shall be fully integrated and authorized persons shall have access from anywhere either from within the building or any other place.

The building has fully integrated low voltage systems.

Apart from training and conference rooms, a fully functional auditorium having a capacity for (___) people for conferencing is available in the building. It has sophisticated conferencing system with rear projection screen.

The contract documents consist of total (___) contract drawings. These are as follows:

Architectural	(___) Nos
Architectural interior	(___) Nos
Structural	(___) Nos
Electrical	(___) Nos
Mechanical (HVAC)	(___) Nos
Public health	(___) Nos
Firefighting	(___) Nos
Smart building	(___) Nos
Traffic signage	(___) Nos
Landscape	(___) Nos

I.3 Quality Control Organization

A quality control (QC) organization is independent of those persons actually performing the work. They will be responsible to implement the quality plan for the entire contract/project-related activities by scheduling the inspection, testing, sampling, and preparation of mockup and will ensure that the work is performed as per the approved shop drawings and contract documents.

An organization chart showing the line of authority and functional relationship is shown in Figure I.1.

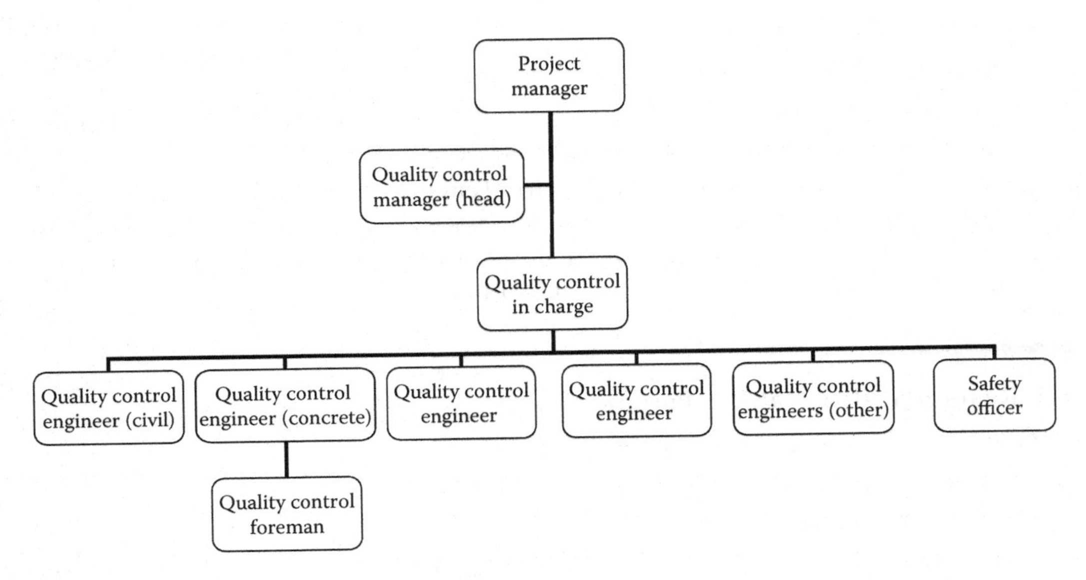

FIGURE I.1
Site quality control organization.

The QC in charge will be responsible to ensure implementation of project quality. He will be supported by QC engineers as follows:

1. QC engineer (Civil works)
2. QC engineer (Concrete works)
3. QC engineer (Mechanical works)
4. QC engineer (Electrical works)
5. Foreman (Concrete works)

These engineers will be responsible to implement three phases of quality system at the site. The QC incharge will coordinate with the company's head office for all the support and necessary actions. Respective QC engineers will be responsible to implement the quality program in their respective fields.

I.4 Qualification of QC Staff

All the QC personnel have adequate experience in their respective field. Following is the qualification of each of the QC staff. Additional staff will be provided if required.

QC in charge: He shall be a qualified civil engineer having minimum 10 years of experience as QC engineer in major construction projects

QC engineer (civil works): Graduate civil engineer with minimum three years' experience as quality engineers on similar projects

QC engineer (concrete works): Graduate civil engineer with minimum three years' experience as quality engineers on similar projects

QC engineer (mechanical works): Graduate mechanical engineer with minimum three years' experience as quality engineers on similar projects

QC engineer (electrical works): Graduate electrical engineer with minimum three years' experience as quality engineers on similar projects

Foreman (concrete works): Diploma in civil engineering with minimum five years' experience as quality supervisor on similar projects

I.5 Responsibilities at QC Personnel

Project manager: Overall project responsibilities

QC in charge: He/she will be responsible for the following:

- Preparation of QCP
- Responsible for overall QC/QA responsibilities
- Responsible for monitoring and evaluation QCP
- Implementation of QC plan
- Responsible for implementing QC procedure
- Maintain QC records and documents
- Inspection of works
- Responsible for off-site inspection
- Inspection of incoming material
- Responsible for subcontractor's QC plan
- Coordinating with safety officer to implement safety plan
- Calibration of measuring instruments
- Monitoring equipment certification

QC engineers: Each trade shall have responsible QC engineers to ensure that the work is carried out in accordance with the contract documents. Following QC engineers shall be available at the site. A foreman will assist the QC engineers to maintain quality of concrete works.

1. QC engineer (civil works)
2. QC engineer (concrete works)
3. QC engineer (mechanical works)
4. QC engineer (electrical works)
5. QC engineer (other subcontractors)
6. Foreman (concrete works)

The above personnel will be responsible for following:

- Overall quality of the respective trade
- Preparation of method statement

- Preparation of mock up
- Monitor and inspect quality of site works
- Inspection of incoming material
- Coordinate with quality incharge for preparation of QCP
- QC records and documents

I.6 Procedure for Submittals

M/S ABC shall be responsible for timely submission of subcontractors, shop drawings. and materials to be used in the project.

I.6.1 Submittals of Subcontractor(s)

Prior to submission of subcontractor's name to the owner/consultant, M/S ABC shall review the capabilities of the subcontractor to perform the contracted works properly as per the specified quality and within the time allowed. M/S ABC shall also consider the contractor's performance in the previously executed project. Companies implementing quality management system plan and having ISO certification shall be given preference while selecting the subcontractor(s) to work on this project.

I.6.2 Submittals of Shop Drawings

Before the start of any activity, M/S ABC shall submit shop drawings based on contract drawings and other related documents. The number of shop drawings shall depend on the requisite details to be prepared against each contract drawings. The shop drawings shall be prepared on the agreed format and shall have coordination certification from all the trades. All the shop drawings shall be submitted through site transmittal for shop drawings. Shop drawing approval transmittals shall be numbered serially for each section of the particular specification. M/S ABC shall be responsible for shop drawings related to subcontracted works.

I.6.3 Submittals of Materials

M/S ABC shall submit materials, products, and systems for owner/consultant's approval from the recommended/nominated manufacturers specified in the contract documents. If material is proposed from a manufacturer who is not nominated, then all the documents to justify that proposed product is approved and equal to the specified shall be submitted along with the site transmitted for materials approval.

I.6.4 Modification Request

If during construction a contractor observes that some work requires modification, then such modification request shall be submitted to the owner/consultant for their review and approval. Financial obligations shall be clearly specified in the request form.

I.6.5 Construction Program

M/S ABC shall prepare and submit construction program based on the contracted time schedule and submit the same to the consultant for approval. Progress of the work shall be monitored on regular basis and the construction program shall be updated as and when required to overcome the delay (if any). Following reports shall be submitted to the consultant as per the schedule given below:

1. Daily report
2. Weekly report
3. Monthly report
4. Monthly progress photographs

I.7 QC Procedure

I.7.1 Procurement

Prior to placing a purchase order or supply contract with the supplier/vendor, M/S ABC will ensure that the material has been approved and that it complies with the requirements of the contract. A procurement log shall be maintained for all the items for which the action has been taken. The procurement log shall be submitted to the owner/consultant every two weeks for their review and information.

I.7.2 Inspection of Site Activities (Checklist)

All the work at the site shall be performed as per the approved shop drawings by using the approved material. At least three phases of control shall be conducted by the QC incharge for each activity prior to submission of the checklist to the consultant.

These will be

Preparatory phase: The concerned engineer shall review the applicable specifications, references and standards, approved shop drawings and materials, and other submittals (method of statements) to ensure compliance with the contract.

Startup phase: The concerned engineer shall discuss with the foreman responsible to perform the work and shall establish the standards of workmanship.

Execution phase: The work shall be executed by continuous inspection during this phase.

All the execution/installation works shall be carried out as per the approved shop drawings utilizing approved material. All the works will be checked for quality before submitting the checklist to the consultant to inspect the work.

The following main categories of site work shall be performed by conducting the QC as per above-mentioned phases.

I.7.2.1 Definable Feature of Work

The procedure for scheduling, reviewing, certifying, and managing QC for definable feature of work shall be agreed upon during the coordination meetings. The definable features of work for concrete shall include formwork, reinforcement and embedded items, design mix, placement of concrete, and concrete finishes.

Likewise, a detailed list for other trades shall be prepared during the coordination meetings. This will consist of embedded conducting work, embedded sleeves, ducts, and underground piping.

I.7.2.2 Earth Works and Site Works

This category shall consist of the following subgroups:

1. Excavation and backfilling
2. Compaction
3. Dewatering

The checklist shall be submitted to the consultant at every stage of the work, that is, after backfilling and compaction. Samples shall be taken to make compaction test. If the compaction test fails the required specifications, then remedial action shall be taken to obtain specified results.

I.7.2.3 Concrete

M/S (approved subcontractor to supply ready-mix concrete) will provide ready-mix concrete for the entire concrete structured works of the project. M/S (ready-mix subcontractor) have their own QC system to maintain the mix design.

Reinforcement steel shall be inspected upon receipt of the material at the site for proper size, type, and the factory test certificates received with the supply. Small pieces of sample shall be taken from each lot and size for laboratory testing. The testing shall be performed by an approved testing laboratory.

Formwork and reinforcement steel process shall be performed under a qualified civil engineer and a foreman. Both will continuously inspect the work for each operation and shall submit their report to the consultant for their approval.

Placement of concrete shall be supervised by the civil engineer along with the foreman and other team members. The respective consultant will witness concrete casting and take concrete samples during casting for laboratory testing and crushing test. Slump test shall be performed on the concrete to verify the strength. Curing shall be supervised by the respective foreman.

In case the test results do not comply with specification limit, the result shall be discussed with the consultant engineer. Failed test results will be followed by appropriate corrective (reworking) efforts, and retesting and remedial measures.

M/S ABC shall maintain all the records at the laboratory test and results regarding the concrete works and submit the same to the consultant for their information.

I.7.2.4 Masonry

The concrete block bricks used on the project shall be procured from the approved source. Material for the mortar aggregate and other accessories, such as inserts, reinforcement material, and wire mesh shall be submitted to the consultant for approval.

Prior to the start of the masonry work, all the related and coordinated drawings shall be reviewed. Marking shall be made for the masonry layout. Necessary mockup shall be submitted for each type of unit of the masonry work. A checklist shall be submitted for visual and other types of inspection.

I.7.2.5 Metal Works

Metal fabrication and installation work shall be carried out by certified welders, as per manufacturer's recommendations, and as per the approved shop drawings. Samples of fittings, brackets, fasteners, anchors, and different types of metal members including plates, bars, pipes, tubes, and any other type of material to be used in the project shall be submitted to the consultant prior to the start of fabrication work at the site. Finishing material such as primer and paint shall be submitted for approval to the consultant/owner.

In case the fabrication is carried out at the subcontractor(s)'s workshop, then M/S ABC shall take full responsibility to control the quality.

A checklist shall be submitted at different stages of the work to ensure compliance with the specifications and to avoid any rework at the end of the completion of fabrication and installation.

I.7.2.6 Wood Plastics and Composite

Prior to the start of any wood work, the material shall be physically and visually inspected for type of the timber quality to confirm its compliance with the type of the approved material and also to ensure that the wood is free from decay, insects, and that necessary treatment has already been done on the wood to be used for the project. Product certificate signed by the woodwork manufacturer, certifying that products comply with the specified requirements shall be submitted along with the material. All types of fasteners and other hardware to be used shall be submitted for approval as per applicable standards and checklists shall be submitted at various stages of work prior to applying the paint or any other finishes.

I.7.2.7 Doors and Windows

This category shall consist of the following subcategories:

1. Wooden doors and windows
2. Steel doors
3. Aluminum windows
4. Glazing

All doors and windows shall be fabricated from a specified type of material and as per the approved shop drawings. Fire-rated type of doors shall comply with relevant standards and local regulations. Samples of materials used in fabrication of doors and windows shall be submitted to the consultant/owner for their approval.

Wooden doors and windows shall be fabricated at the approved subcontractor's workshop for carpentry works and aluminum doors and windows shall be fabricated at the approved subcontractor's workshop for aluminum works. Finishes shall be as approved by the owner/consultant. Samples of hardware coating material and finishing material shall be submitted for approval. Special care shall be taken to maintain the acoustic nature of the door and its sound transmission properties.

Glass and glazing material shall be submitted for approval and shall take care of heat transfer properties to maintain internal air temperature of the building. The glass used in the project shall comply with the specified strength, safety, and impact performance requirements. Inspection at different stages shall be arranged by field inspection at the factory.

Entire fabrication shall be carried out under the supervision of M/S ABC to control the quality of finished products. Doors and windows fabrication shall be coordinated for security requirements. Mockup shall be prepared before the start of installation work.

I.7.2.8 Finishes

This category shall consist of the following subgroups:

1. Acoustic ceiling
2. Specialty ceiling
3. Masonry flooring
4. Tiling
5. Carpet
6. Wall cladding
7. Wall partitioning
8. Paints and coating

All the finishes shall be as specified and approved. Execution of finishing work shall be coordinated with all other trades. Samples and mockup shall be submitted for approval before the start of any finishing work. All the finishing work shall be performed by skilled workmen. Specialist subcontractor(s) shall be submitted material for approval to carryout finishing works. QC shall be conducted under the direct supervision of M/S ABC.

I.7.2.9 Landscape

Landscape work shall be executed by specialist-approved subcontractor as per the approved shop drawings under the direct supervision of M/S ABC. Samples of each and every type of plants, and trees shall be submitted for approval. Soil preparation shall be done as per the specified standards.

I.7.2.10 Furnishing

All the furnishing shall be from the specified recommended manufacturer. Special care shall be taken to protect the furnishings. Furnishing material sample(s) shall be submitted for approval. Fabrication of furniture shall be at the approved subcontractor's factory. M/S ABC shall be responsible to control the quality.

I.7.2.11 Equipment

This category shall consist of the following subgroups:

1. Maintenance equipment, such as windows cleaning and facade cleaning
2. Stage equipment
3. Parking control equipment
4. Kitchen equipment

All the equipment shall be obtained from specified manufacturers. If the items specified are commercial items, that is, the materials manufactured and sold to the public as against the materials made to specifications, necessary precaution shall be taken to verify that the materials shall be installed as per the manufacturer's recommended procedure under the supervision of a qualified engineer.

Items to be manufactured as per the contract documents shall be procured from the manufacturer producing products complying with recognized quality standards. The product and manufacturer's data shall be submitted for approval with all the technical details. Installation of the equipment shall be carried out by the skilled workmen under the supervision of the manufacturer's authorized representative and under the control of M/S ABC.

I.7.2.12 Conveying System

This category consists of the following subgroups:

1. Passenger elevators
2. Goods elevators
3. Kitchen elevators
4. Escalators

Prior to submission of the shop drawings, full technical details along with the catalogues and compliance statement shall be submitted for approval. The elevators shall be from one of the recommended manufacturers specified in the contract documents.

Fabrication of conveying systems shall comply with relevant codes and standards as applicable. Regulatory approval shall be obtained from local authorities for installation of the conveying systems to assure their compliance with local codes and regulations.

Finishing material for cab, landing door, car control station, and car position indicator shall be submitted for approval.

Coordinated factory shop drawings shall be submitted for approval. Size of shaft and door openings shall be coordinated during structural work. Necessary tests shall be performed by the manufacturer's authorized personnel.

Factory fabrication works shall be performed as per manufacturer's QC plan and installation at the site shall be carried out by personnel authorized by the manufacturer. Third-party inspection shall be arranged as per specification requirements.

I.7.2.13 Mechanical Works

This category shall consist of the following subgroups:

1. Fire suppression
2. Water supply, plumbing, and public health
3. HVAC

Mechanical works shall be executed by the approved mechanical subcontractor. All the work shall be carried out as per the approved shop drawings using approved material.

A mechanical engineer from the subcontractor shall be responsible for controlling the quality of work. QC engineer (mechanical works) shall coordinate all quality-related activities on behalf of M/S ABC. Fire suppression or firefighting works shall be carried out as per the approved drawings by the relevant authority. Materials shall be from the manufacturers approved by the relevant authority having jurisdiction over such materials. Fire pumps and firefighting materials shall comply with regulatory requirements.

Piping for water supply, plumbing, and public health shall be installed with approved material. Leakage and pressure test shall be performed as specified.

Checklists shall be submitted after installation of piping, accessories, and fixing of equipment. Drainage systems shall be executed as per the approved shop drawings.

HVAC works shall be supervised by a qualified mechanical engineer. Ductwork and duct accessories shall be fabricated at the subcontractor's workshop and shall be installed at the site as per the approved shop drawings. A mechanical engineer from the subcontractor shall be responsible to control the quality at the workshop. QC engineer (mechanical works) shall coordinate all quality-related activities on behalf of M/S ABC. Duct material shall be submitted for approval prior to the start of fabrication works. Chilled/hot water piping shall be from the approved manufacturer. Installation of piping shall be carried out by certified pipe fitters.

Selection of HVAC pumps shall be done with the help of performance curves and technical data from the pump manufacturer. Chillers shall be from the specified manufacturer. Installation of chillers shall be carried out as per the manufacturer's recommendation.

I.7.2.14 Automation System

Automation work shall be carried out under the supervision of a specialist engineer. The system shall be obtained from the specified manufacturer.

Shop drawings and schematic diagrams shall be submitted for approval, configuring approved components/items/equipment.

I.7.2.15 Electrical Works

This category shall consist of the following subgroups:

1. Electrical lighting and power
2. Fire alarm system
3. Communication system (telephone)
4. Public address system
5. Access control and security system

6. Audiovisual system
7. MATV system
8. Emergency generating system

All types of electrical works shall be executed by the approved electrical subcontractor. All the materials/products to be used shall be from the specified manufacturer and shall be installed by a skilled workman under the supervision of a qualified engineer.

An electrical engineer from the subcontractor shall be responsible to control the quality at the workshop. QC engineer (electrical works) shall coordinate all quality-related activities on behalf of M/S ABC. Work shall be carried out as per the approved shop drawings. Quality of the works shall be controlled at different stages such as after the installation of conducting/raceways, pulling of wires/cables, installation of accessories/equipment/panels. Work shall comply with the specifications and recognized codes and standards.

Specified tests shall be performed at every stage. Coordination will be done with other trades to ensure that the required power supply for their equipment will be available at the designated locations. Checklists shall be submitted after completion of the activities at each stage.

I.7.3 Inspection and Testing Procedure for Systems

The QC procedure for systems (low voltage/low current) and their equipment/components shall be performed in the following stages.

Submission of proposed manufacturer's complete data giving full details of the capability of the proposed manufacturer to substantiate the compliance with the specifications and to prove that the proposed manufacturer is following the quality management system and manufacturing the product as per specified quality standards. Components/Equipment data shall be submitted for all the specified items of the system.

Submission of a schematic diagram, configuring all the items for the system shall be submitted along with detailed technical specifications and confirmation from the manufacturer that it complies with the contract documents and is equal or better than the specified products.

Submission of shop drawing(s) fully co-coordinating with other trades, showing the system location and the raceways used for installation of the system. All the works at the site shall be submitted for inspection at different stages of execution of works.

Cable testing shall be performed prior to installation of the equipment. A checklist shall be submitted to the consultant to witness these tests. Installation of the equipment shall be done as per the manufacturer's recommendations and as per approved installation methods. A checklist along with the testing procedure shall be submitted for final inspection and testing.

I.7.4 Off-Site Manufacturing, Inspection, and Testing

QC procedure for off-site manufacturing/assembling shall be as per contract documents. Factory control inspection for items fabricated or assembled shall be carried out to comply with all the specified requirements and standards under the responsibility of QC personnel at the fabrication/assembly plant. A specified test shall be carried out as per the manufacturer's quality procedures and as per the specification requirements. Test reports for items fabricated/assembled off-site shall be submitted along with the delivery of material at site. Witnessing/observation of inspection and tests by the consultant/client, if it is required per specifications, shall be arranged. The procedure for such tests shall be submitted and inspection shall be carried out in advance of factory/plant visit.

QC for *Off-the-Shelf* items shall be standard QC practices followed by the manufacturer.

I.7.5 Procedure for Laboratory Testing of Materials

M/S ABC shall submit name(s) of testing laboratories for approval for performing tests specified in the contract documents.

I.7.6 Inspection of Material Received at Site

All the material received at the site shall be submitted for inspection by the consultant. Material on-site inspection request (MIR) shall be submitted to the consultant. Prior to dispatch of material from the manufacturer's premises, all the tests shall be carried out and test certificates shall be submitted along with MIR. Apart from this, all the relevant documents such as packing list, delivery note, certificate of compliance to recognized standards, country of origin, and any other documents shall be submitted along with the MIR.

Material received at the site shall be properly stored at designated locations and shall maintain original packing until its installation/use.

Inventory records shall be maintained for all the material stored at the site. In case of materials to be directly installed at the site upon its receipt, the inspection shall be carried out before its use or installation on the project.

In case of concrete ready mix, samples will be tested as per the contract documents before pouring the concrete for castings. Specified tests shall be performed after the delivery of material at the site.

I.7.7 Protection of Works

All the works shall be protected at the site as per the specified methods in the contract documents.

I.7.8 Material Storage and Handling

M/S ABC shall arrange for a storage facility within the project site to keep the material in safe condition. Material at the site shall be fully protected from damages and properly stored. Necessary care should be taken to store paints and liquid type of materials by maintaining appropriate temperature to prevent damage and deterioration of such products. Materials, products, and equipment shall be properly packed and protected to prevent damage during transportation and handling.

I.8 Method Statement for Various Installation Activities

A method statement shall be submitted to the consultant for their approval as per the contract documents. The method statement shall describe the steps involved for execution/installation of work by ensuring safety at each stage. It shall have the following information:

1. Scope of work: Brief description of work/activity
2. Documentation: Relevant technical documents to undertake this work/activity
3. Personnel involved
4. Safety arrangement
5. Equipment and plant required
6. Personal protective equipment
7. Permits/Authorities' approval to work
8. Possible hazards
9. Description of the work/activity: Detail method of sequence of each operation/key steps to complete the work/activity

Figures I.2 through I.9 are illustrative examples of the method of sequence for major activities of different trades performed during the construction process.

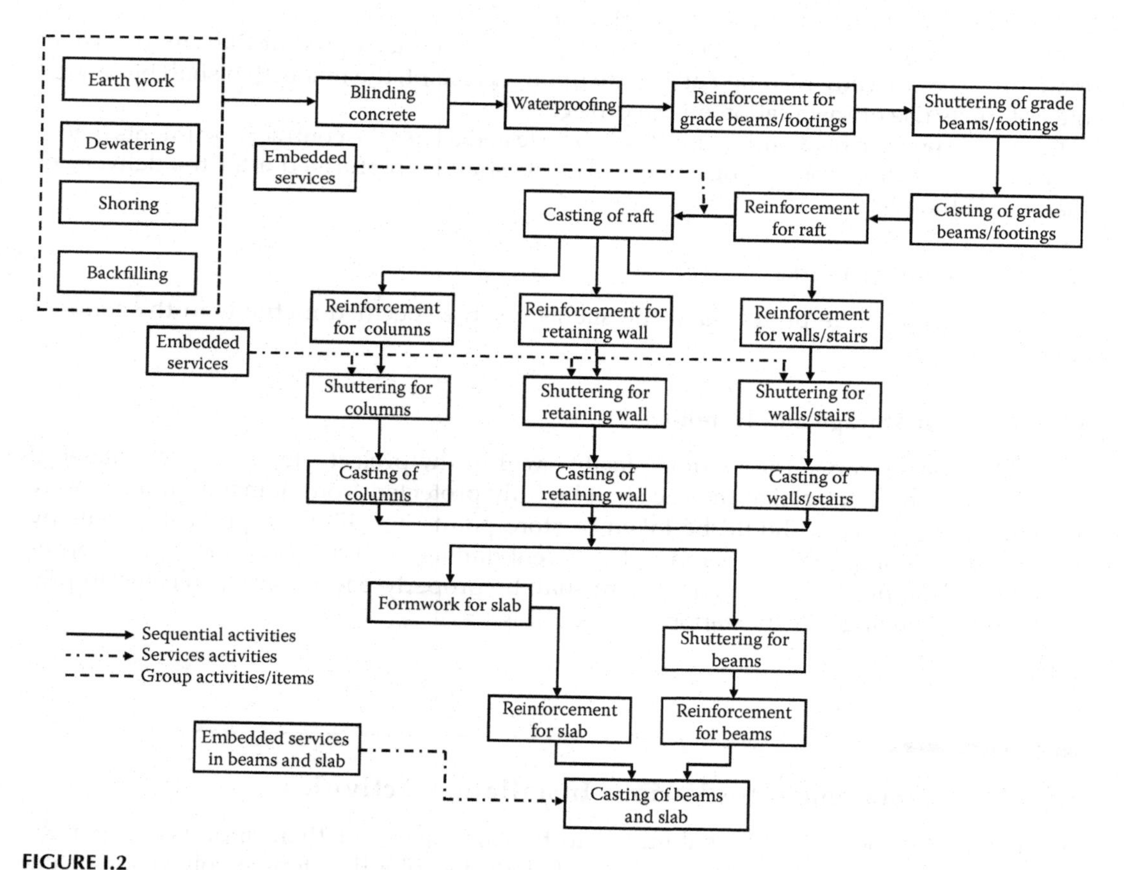

FIGURE I.2
Method of sequence for concrete structure work. (Abdul Razzak Rumane (2010), *Quality Management in Construction Projects*, CRC Press, Boca Raton, FL. Reprinted with permission from Taylor & Francis Group.)

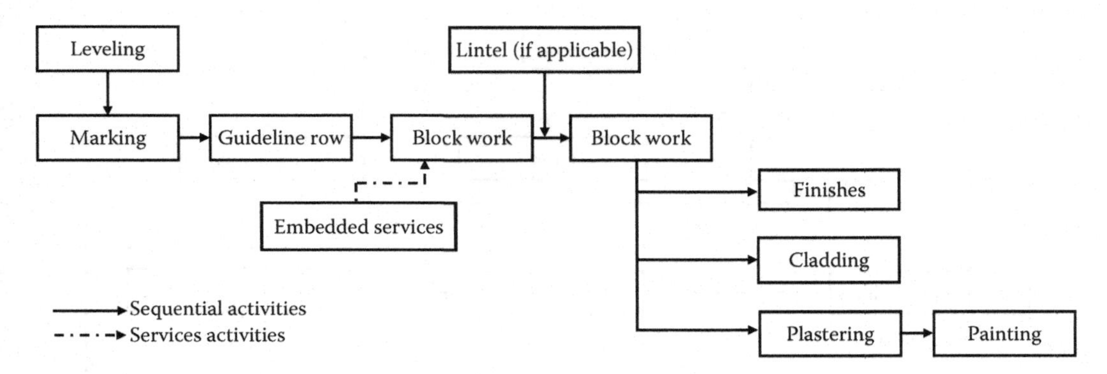

FIGURE I.3
Method of sequence for block masonry work. (Abdul Razzak Rumane (2010), *Quality Management in Construction Projects*, CRC Press, Boca Raton, FL. Reprinted with permission from Taylor & Francis Group.)

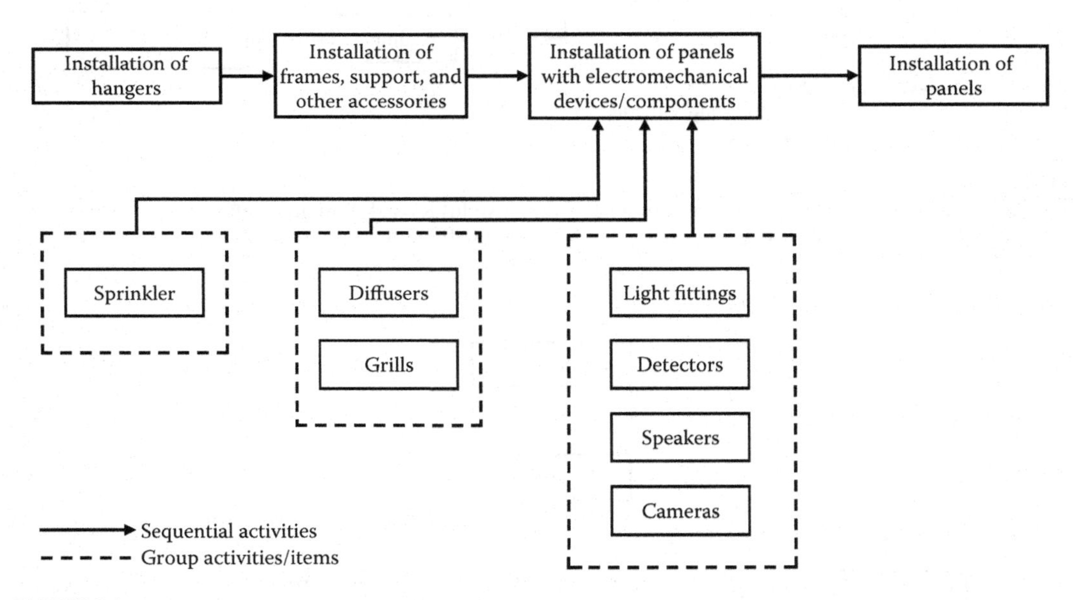

FIGURE I.4
Method of sequence for false ceiling work. (Abdul Razzak Rumane (2010), *Quality Management in Construction Projects*, CRC Press, Boca Raton, FL. Reprinted with permission from Taylor & Francis Group.)

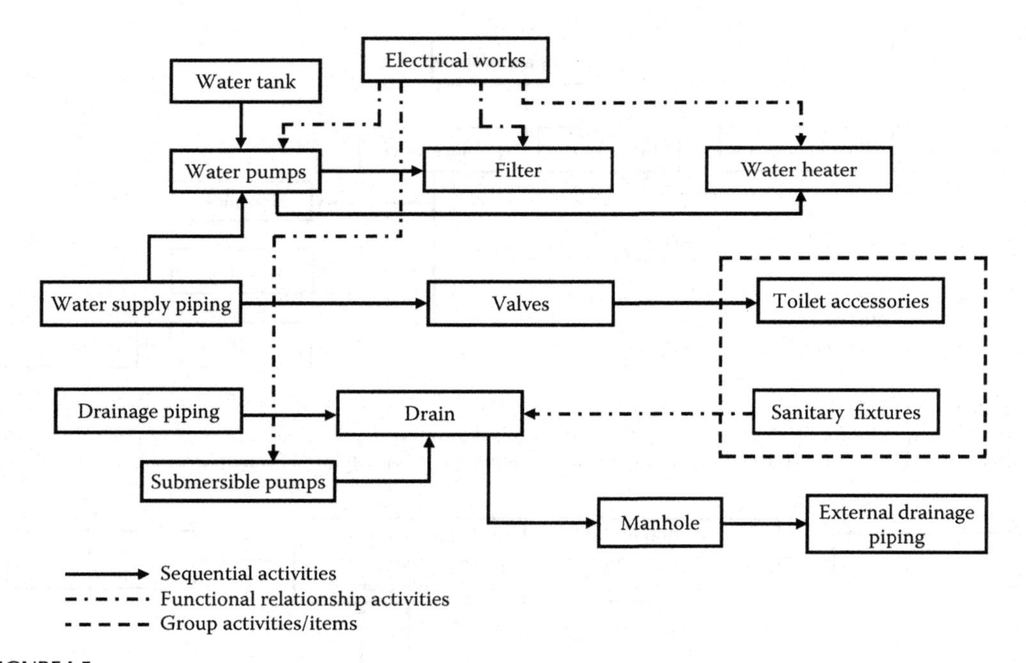

FIGURE I.5
Method of sequence for mechanical work (public health). (Abdul Razzak Rumane (2010), *Quality Management in Construction Projects*, CRC Press, Boca Raton, FL. Reprinted with permission from Taylor & Francis Group.)

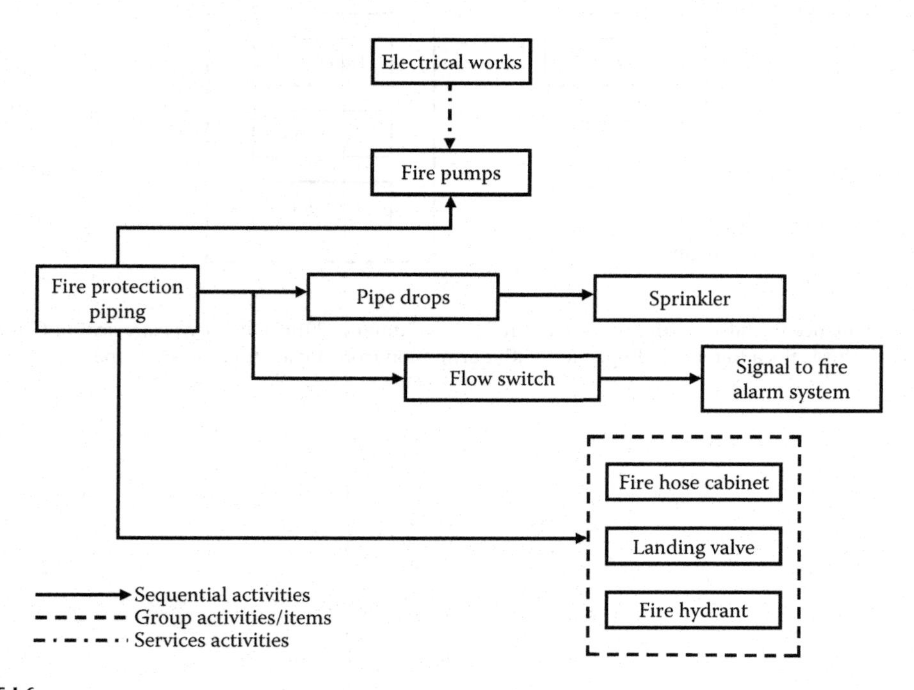

FIGURE I.6
Method of sequence for mechanical work (fire protection). (Abdul Razzak Rumane (2010), *Quality Management in Construction Projects*, CRC Press, Boca Raton, FL. Reprinted with permission from Taylor & Francis Group.)

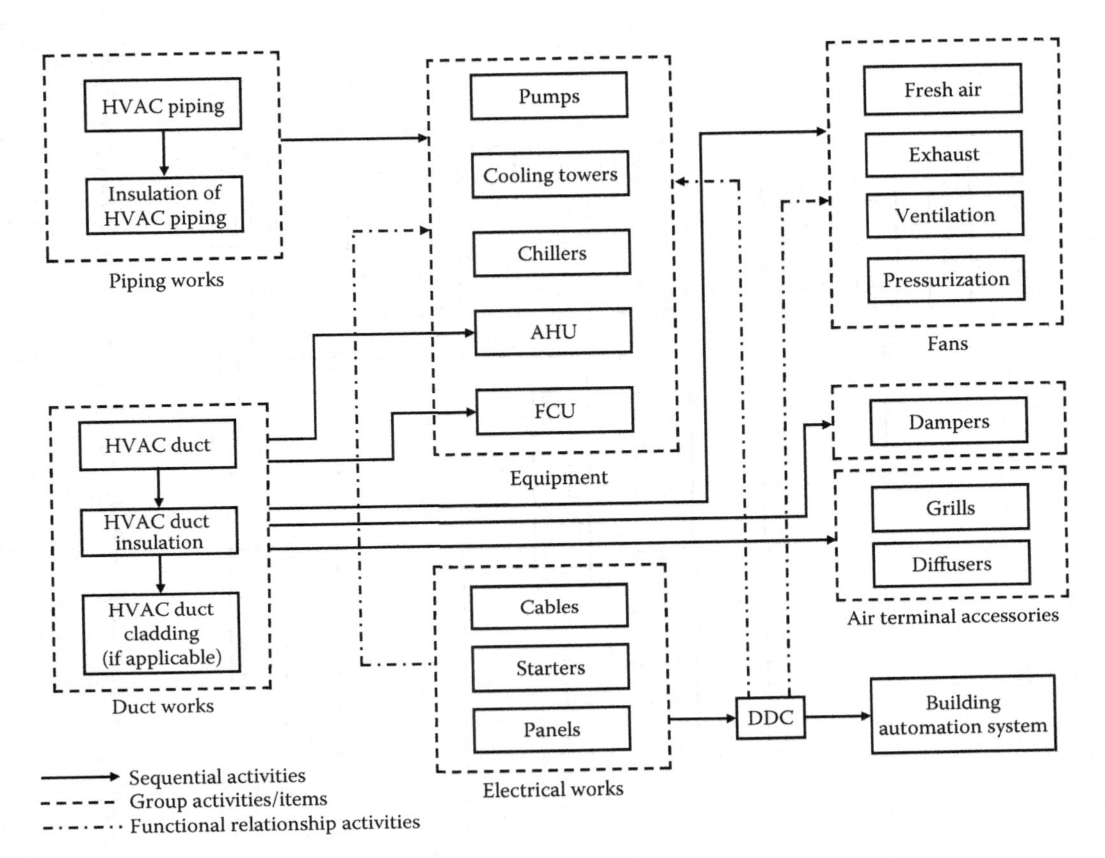

FIGURE I.7
Method of sequence for HVAC work. (Abdul Razzak Rumane (2010), *Quality Management in Construction Projects*, CRC Press, Boca Raton, FL. Reprinted with permission from Taylor & Francis Group.)

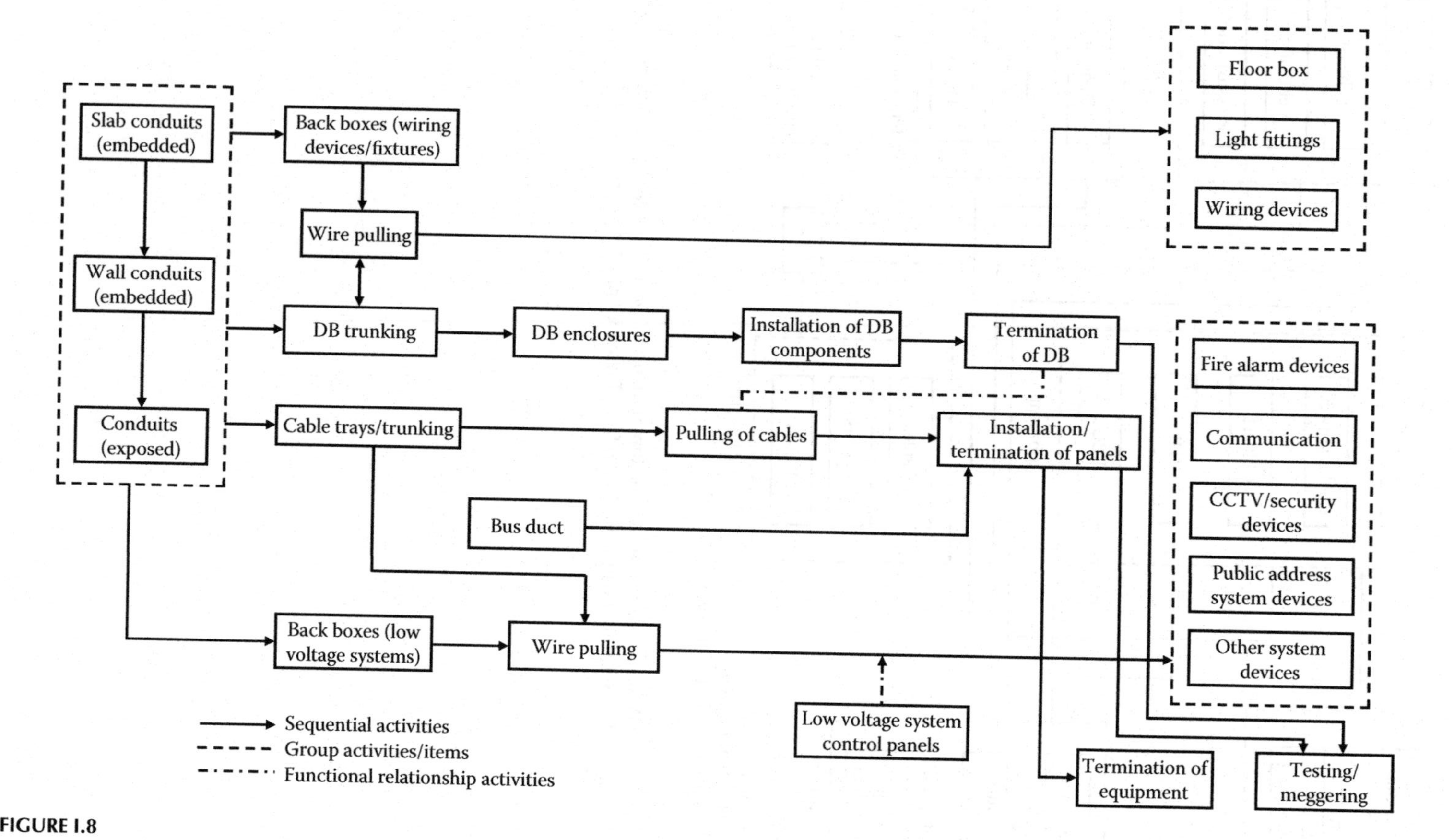

FIGURE I.8
Method of sequence for electrical work. (Abdul Razzak Rumane (2010), *Quality Management in Construction Projects*, CRC Press, Boca Raton, FL. Reprinted with permission from Taylor & Francis Group.)

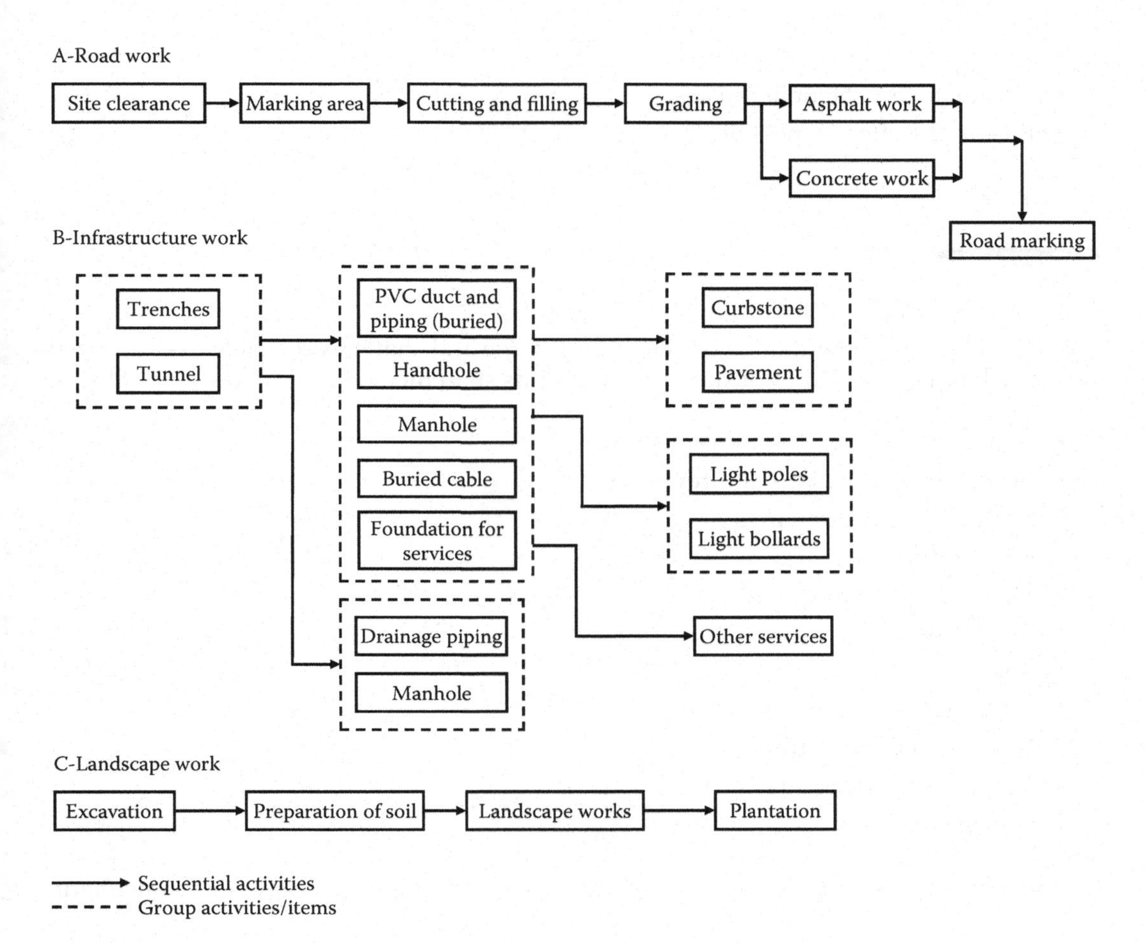

FIGURE I.9
Method of sequence for external works. (Abdul Razzak Rumane (2010), *Quality Management in Construction Projects*, CRC Press, Boca Raton, FL. Reprinted with permission from Taylor & Francis Group.)

I.9 Project-Specific Procedures for Site Work Installations Remedial Notice

QC for such installations shall be as specified in the contract documents or as mentioned in the instructions.

A remedial notice received shall be actioned immediately and replied to inspect the works having corrected/performed as per contract documents. Similarly nonconformance report shall be actioned and replied.

All repaired and reworked items shall be inspected by the concerned engineer and QC engineer before submitting *Request for Inspection* or *Checklist.*

Appropriate preventive actions shall be taken to avoid repetition of nonconformance work. Following steps shall be taken to deal with any problem requiring preventive action.

- Detect
- Identify the potential cause(s)

- Analyze the potential cause(s)
- Eliminate the potential cause(s)
- Ensure the effectiveness of preventive action

I.10 QC Records

QC records shall be maintained by the QC in charge and shall be accessible to authorized personnel for review and information. The records shall include:

- Contract drawings and revisions
- Contract specifications and revisions
- Approved shop drawings
- Record drawings
- Material approval record
- Checklists
- Test reports
- Material inspection reports
- Minutes of QC meetings
- Site works instructions
- Remedial notes

Other records specified in the contract documents or as requested by the owner/consultant.

I.11 Company's Quality Manual and Procedures

The following documents from the company's manual shall be part of the contractor's QCP:

1. Documents for specific projects
2. Subcontractor evaluation
3. Supplier evaluation
4. Client-supplied items
5. Receiving inspection and testing
6. Site inspection and testing
7. Final inspection, testing, and commissioning
8. Material storage and handling
9. Control of quality records
10. Internal audit

I.12 Periodical Testing of Construction Equipment

Hoist, crane, and other lifting equipment used at the site shall be tested periodically by third-party inspection authority, and test certificates shall be submitted to the consultant. Measuring instruments shall have calibration test performed by the approved testing agencies/government agencies at regular intervals and calibration certificate shall be submitted to the consultant. Other equipment shall be tested/calibrated as per contract requirements and as per the schedule for such tests.

I.13 Quality Updating Program

QC program shall be reviewed every six months and necessary updates shall be done, if required. All such information shall be given to the consultant and a record shall be maintained for such revision.

I.14 Quality Auditing Program

Internal auditing shall be performed by the company's internal auditor to ensure that the specified quality procedures are followed by the site quality personnel.

I.15 Testing, Commissioning, and Handover

The test shall be carried out as per contract documents. Necessary test procedures shall be submitted prior to the start of testing the installed systems/equipment.

Engineers from the respective trades will be responsible to arrange for an orderly performance of testing and commissioning of the systems/equipment.

I.16 Health, Safety, and Environment

A safety officer shall be responsible for implementing safety and accident preventive methods. All the personnel at the site shall be provided with safety gears. Regular meetings and awareness trainings shall be conducted at the site on a regular basis. M/S ABC shall comply with all the safety measures specified in the contract documents. M/S ABC will also follow up regularly with the local environmental protection agency and follow waste management plan.

Bibliography

AACE International Recommended Practice No. 27R-03 (2010).

AACE International Recommended Practice No. 37R-06 (2010).

ANSI/ISO/ASQ Q9001:2008 (2003). *Quality Management Systems—Requirements*, Quality Press, Milwaukee, WI.

Arditi, D. and Gunaydin, H. M. (1997). Total quality management in the construction process. *International Journal of Project Management*, 15(4), 235–243.

Besha, B., Kitaw, D., and Alemu, N. (2013). Significance of ISO 9000 quality management system for performance improvement in developing countries. *KCA Journal of Business Management*, 5(1), 44–51.

Boiral, O. (2011). Managing with ISO systems: Lessons from practice. *Long Range Planning*, 44(3), 197–220.

Boiral, O. and Amara, N. (2009). Paradoxes of ISO 9000 performance: A configurational approach. *The Quality Management Journal*, 16(3), 36–60.

Business Improvement Architects. http://www.bia.ca/article/DealingwithConflictinProjectTeam.htm.

Cachadinha, N. M. (2009). Implementing quality management systems in small and medium construction companies: A contribution to a road map for success. *Leadership and Management in Engineering*, 9(1), 32–39.

California Department of Transportation (2007). *Project Management Handbook*, 2nd Edition, California Department of Transportation, Sacramento, CA.

Chini, A. R. and Valdez, H. E. (2003). ISO 9000 and the U.S. Construction Industry. *Journal of Management in Engineering*, 19(2), 69–77.

Curkovic, S. and Pagell, M. (1999). A critical examination of the ability of ISO 9000 certification to lead to a competitive advantage. *Journal of Quality Management*, 4(1), 51–67.

Ekanayake, L. L. and Ofori, G. (2004). Building waste assessment score: Design-based tool. *Building and Environment*, 39(7), 851–861.

Federation of Internationale Des Ingenieurs-Conseils. http://www.fidic.org.

Gillies, A. (2011). Improving the quality of information security management systems with ISO27000. *The TQM Journal*, 23(4), 367–376.

Harrington, H. J. (1997). The fallacy of universal best practices. *The TQM Magazine* 9(1), 61–75.

International Organisation for Standardisation. http://www.iso.org.

Kale, S. and Arditi, D. (2006). Diffusion of ISO 9000 certification in the precast concrete industry. *Construction Management and Economics*, 24(5), 485–495.

Lenssen, N. and Roodman, D. M. (1995). Making better buildings. In L. R. Brown, C. F. Flavin, H. French, L. Starke, D. Denniston, H. Kane et al. (Eds), *State of the World*, A Worldwatch Institute report on progress toward sustainable society (pp. 95–112). W.W. Norton, New York.

Martinez-Costa, M. and Martinez-Lorente, A. R. (2007a). A triple analysis of ISO 9000 effects on company performance. *Journal of Productivity and Performance Management*, 56(5–6), 484–499.

Martinez-Costa, M. and Martinez-Lorente, A. R. (2007b). ISO 9000:2000: The key to quality? An exploratory study. *The Quality Management Journal*, 14(1), 7–18.

Master Format (2014). *The Construction Specifications Institute and Construction Specification*, Canada. http://www.csi.com.

New Engineering Contract. http://www.nec.org.

Ollila, A. (2011). The role of quality system in performance improvement—Is ISO 9001 up-to-date? *Global Conference on Business and Finance Proceedings*, 6(2), 304–314.

Project Management Institute (2007). *Construction-Extension-PMBOK® Guide*, 3rd Edition. Project Management Institute, Newtown Square, PA.

Project Management Institute (2013). *A Guide to the Project Management Body of Knowledge (PMBOK® Guide)*, 5th Edition. Project Management Institute, Newtown Square, PA.

Quazi, H. A., Hong, C. W., and Meng, C. T. (2002). Impact of ISO 9000 certification on quality management practices: A comparative study. *Total Quality Management*, 13(1), 53–67.

Renuka, S. D. and Venkateshwara, B. A. (2006). A comparative study of human resource management practices and technology adoptions of SMEs with and without ISO certification. *Singapore Management Review*, 28(1), 41–61.

Rother, M. and Shook, J. (2003). Learning to See. *The Lean Enterprise Institute*, Brookline, MA.

Abdul Razzak Rumane (2010). *Quality Management in Construction Projects*, CRC Press, Boca Raton, FL.

Abdul Razzak Rumane (2013). *Quality Tools for Managing Construction Projects*, CRC Press, Boca Raton, FL.

Seddon, J. (1998). The case against ISO 9000. *ISO 9000 News*, July/August, 12–15.

Sowards, D. (2013). Quality is key to lean's success. *Contractor Magazine*, May, 44, 52.

Stevenson, T. H. and Barnes, F. C. (2001). Fourteen years of ISO 9000: Impact, criticisms, costs, and benefits. *Business Horizons*, 44, 45–51.

Symonds, J. (1998). ISO 9000- case sensitive. *ISO News*, 6, 12–15.

Tan, R. R. and Lu, Y. G. (1995). On the quality of construction engineering design projects: Criteria and impacting factors, *International Journal of Quality & Reliability Management*, 16(6), 562–574.

The Engineers Joint Contract Documents Committee (2007). http://www.ejcdc.org.

Turk, A. M. (2009). ISO 14000 environmental management system in construction: An examination of its application in Turkey. *Total Quality Management*, 20(7), 713–733.

von Solms, B. and von Solms, R. (2005). From information security to … business security. *Computers & Security*, 24(4), 271–273.

Vrassidas, L. and Chatzistelios, G. (2014). Quality management systems development based on product systems taxonomy. *The TQM Journal*, 26(2), 215–229.

West, J. E. and Cianfrani, C. A. (2014). Managing the system revision introduces focus on organizational operating conditions. *Quality Progress*, August, 53–56.

Zeng, S., Deng, Z., and Tam, V. (2003). ISO 14000 and the construction industry: Survey in China. *Journal of Management Engineering*, 19(3), 107–115.

Author Index

Note: Page numbers followed by t refer to tables.

Subject Index

Note: Page numbers followed by f and t refer to figures and tables, respectively.

G

H

I

J

K

L

M

S

U

V

W

X

Y

Printed in the United States
by Baker & Taylor Publisher Services